LINEAR ALGEBRA FOR THE 21ST CENTURY

Linear Algebra for the 21st Century

A. J. ROBERTS

University of Adelaide
South Australia, 5005

OXFORD
UNIVERSITY PRESS

OXFORD
UNIVERSITY PRESS

Great Clarendon Street, Oxford, OX2 6DP,
United Kingdom

Oxford University Press is a department of the University of Oxford.
It furthers the University's objective of excellence in research, scholarship,
and education by publishing worldwide. Oxford is a registered trade mark of
Oxford University Press in the UK and in certain other countries

First Edition published in 2020

Impression: 1

Published in the United States of America by Oxford University Press
198 Madison Avenue, New York, NY 10016, United States of America

British Library Cataloguing in Publication Data

Data available

Library of Congress Control Number: 2019957932

ISBN 978–0–19–885639–9 (hbk.)
ISBN 978–0–19–885640–5 (pbk.)

Printed and bound by
CPI Group (UK) Ltd, Croydon, CR0 4YY

To Barbara for her patience, understanding, and support during this long project, and to our children and grand-children, with the aim that this approach to mathematics will serve them well.

Preface

Traditional courses in linear algebra make considerable use of the reduced row echelon form (RREF), but the RREF is an unreliable tool for computation in the face of inexact data and arithmetic. The [Singular Value Decomposition] SVD can be regarded as a modern, computationally powerful replacement for the RREF.[1]   *Cleve Moler, MathWorks (2006)*

The Singular Value Decomposition (SVD) is sometimes called the *jewel in the crown of linear algebra*. Traditionally the SVD is introduced and explored at the end of several linear algebra courses. Question: Why were students required to wait until the end of the course, if at all, to be introduced to the beauty and power of this jewel? Answer: the limitations of hand calculation.

This book establishes a new route through linear algebra, one that reaches the SVD jewel in linear algebra's crown very early, in Section 3.3. Thereafter its beautiful power both explores many modern applications, especially in data science, and also develops traditional linear algebra concepts, theory, and methods. No rigour is lost in this new route: indeed, this book demonstrates that most theories are better proved with an SVD rather than with the traditional RREF. This new route through linear algebra becomes a preferred approach both because of the ready availability of ubiquitous computing and because of the importance of data science in the 21st century.

As so many other disciplines use the SVD, it is not only important that mathematicians understand what it is, but also teach it thoroughly in linear algebra and matrix analysis courses.
                                                                                    Turner et al. (2015)

---

[1] http://au.mathworks.com/company/newsletters/articles/professor-svd.html [Oct 2019]

Acknowledgements

I acknowledge with thanks the work of many others who inspired much design and details here, including the stimulating innovations of calculus reform (e.g. Hughes-Hallett et al., 2013), the comprehensive efforts behind recent reviews of undergraduate mathematics and statistics teaching (e.g. Alpers et al., 2013; Bressoud et al., 2014; Turner et al., 2015; Horton et al., 2014; Schumacher et al., 2015; Bliss et al., 2016), the books of Anton and Rorres (1991); Davis and Uhl (1999); Holt (2013); Larson (2013); Lay (2012); Nakos and Joyner (1998); Poole (2015); Will (2004). I also thank the entire LaTeX team, especially Knuth, Lamport, Feuersänger, and the AMS. Lastly, I thank my son Ben for valuable proofreading and suggestions, and thank reviewers for their constructive comments.

Aims for students

How should mathematical sciences departments reshape their curricula to suit the needs of a well-educated workforce in the twenty-first century?

… The mathematical sciences themselves are changing as the needs of big data and the challenges of modeling complex systems reveal the limits of traditional curricula.

Bressoud et al. (2014)

Linear algebra is packed with compelling results for application in science, engineering and computing, and with answers for the twenty-first century needs of big data and complex systems. This book provides the conceptual understanding of the essential linear algebra of vectors and matrices for modern engineering and science. The traditional linear algebra course is reshaped herein to meet modern demands.

Crucial is to inculcate the terms and corresponding relationships met most often in professional life, often when using professional software. For example, the manual for the engineering software package *Fluent* most often invokes the following linear algebra concepts: diagonal, dot product, eigenvalue, least square, orthogonal, projection, principal axes, symmetric, and unit vector. Engineers need to understand these concepts. What such useful terms mean, their relationships, and their use in applications are central to the mathematical development in this book: you will see them introduced early and used often.

For those who study more mathematics, the development here also provides a great foundation of key concepts, relationships, and transformations necessary for higher mathematics.

Important for all is to develop facility in manipulating, interpreting, and transforming between visualizations, algebraic forms, and vector-matrix representations—of problems, working, solutions, and interpretation. In particular, one overarching aim of the book is to encourage your formulation, thinking, and operation at the crucial system-wide level of matrix/vector operations.

In view of ubiquitous computing, this book explicitly integrates computer support for developing the concepts and their relations. The central computational tools to understand are the operation A\ for solving straightforward linear equations; the function svd() for difficult linear equations, for approximation, and for data science; and the function eig() for probing structures. This provides a framework to understand key computational tools to effectively utilize the so-called third arm of science: namely, computation.

Throughout the book, examples (many graphical) introduce and illustrate the concepts and relationships between them. Working through these helps form the mathematical relationships essential for application. Interspersed throughout the text are questions labelled "Activity": these help form and test your developing understanding of the concepts.

Also included are many varied applications, described to varying levels of detail. These applications indicate how the mathematics will empower you to answer many practical challenges in engineering and science.

> The main contribution of mathematics to the natural sciences is not in formal computations …, but in the investigation of those non-formal questions where the exact setting of the question (what are we searching for and what specific models must be used) usually constitute half the matter.
>
> *Arnold (2014)*

Background for teachers

Depending upon the background of your students, your class should pick up the story somewhere in the first two or three chapters. Some students will have previously learnt some vector material in just 2D and 3D, in which case refresh the concepts in nD.

As a teacher you can use this book in several ways:

- as a reasonably rigorous mathematical development of concepts and interconnections by invoking its definitions, theorems, and proofs, all interleaved with examples.
- as the development of practical techniques and insight for application orientated science and engineering students via the motivating examples to appropriate definitions, theorems and applications.
- or any mix of these two.

One of the aims of this book is to organize the development of linear algebra so that if a student only studies part of the material, then s/he still obtains a powerful and useful body of knowledge for science or engineering.

The book typically introduces concepts in two or three dimensional cases, and subsequently develops the general theory of the concept. This is to help focus the learning, empowered by visualization, while also making a preformal connection to be strengthened subsequently. People are not one dimensional; knowledge is not linear. Cross-references make many connections, explicitly recalling earlier learning (although references are sometimes forward to material not yet 'covered').

> information that is recalled grows stronger with each retrieval ... spaced practice is preferable to massed practice.
> *Halpern and Hakel (2003)*

One characteristic of the development is that the concept of linear independence does not appear until relatively late, namely in Chapter 7. This is good for several reasons. First, orthogonality is much more commonly invoked in science and engineering than is linear independence. Second, it is well documented that students struggle with linear independence:

> there is ample talk in the math ed literature of classes hitting a 'brick wall', when linear (in)dependence is studied in the middle of such a course
> *Uhlig (2002)*

Consequently, here we learn the more specific orthogonality before the more abstract linear independence. Many modern applications, especially in data science, are opened up by the relatively early introduction of orthogonality.

In addition to many exercises at the end of each section, throughout the book are questions labelled "Activity". These are for the students to do to help form the concepts being introduced with a small amount of work. These activities may be used in class to foster active participation by students (perhaps utilizing clickers or web tools such as that provided by http://www.quizsocket.com). Such active learning has positive effects (Pashler et al., 2007).

On visualization

All Linear Algebra courses should stress visualization and geometric interpretation of theoretical ideas in 2- and 3-dimensional spaces. Doing so highlights "algebraic and geometric" as "contrasting but complementary points of view,"
Schumacher et al. (2015)

Throughout, this book also integrates visualization. This visualization reflects the fundamentally geometric nature of linear algebra. It also empowers learners to utilize different parts of their brain and integrate the knowledge together from the different perspectives. Visualisation also facilitates greater skills at interpretation and modelling—skills essential in applications. But as commented by Fara (2009) [p.249] "just like reading, deciphering graphs and maps only becomes automatic with practice." So, lastly, visual exercise questions develop understanding without a learner being able to defer the challenge to online tools, as yet.

Visual representations are effective because they tap into the capabilities of the powerful and highly parallel human visual system. We like receiving information in visual form and can process it very efficiently: around a quarter of our brains are devoted to vision, more than all our other senses combined ...researchers (especially those from mathematic backgrounds) see visual notations as being informal, and that serious analysis can only take place at the level of their semantics. However, this is a misconception: visual languages are no less formal than textual ones
Moody (2009)

On integrated computation

Cowen argued that because "no serious application of linear algebra happens without a computer," computation should be part of every beginning Linear Algebra course. ... While the increasing applicability of linear algebra does not require that we stop teaching theory, Cowen argues that "it should encourage us to see the role of the theory in the subject as it is applied."
Schumacher et al. (2015)

We need to empower students to use computers to improve their understanding, learning, and application of mathematics; not only integrated in their study but also in their later professional career.

One often expects that it should be easy to sprinkle a few computational tips and tools throughout a mathematics course. This is not so—extra computing is difficult. There are two reasons for the difficulty: first, the number of computer language details that have to be learned is surprisingly large; second, for students it is a genuine intellectual burden to learn and relate both the mathematics and the computations.

Consequently, this book chooses a computing language where it is as simple as reasonably possible to perform linear algebra operations: Matlab/Octave appears to answer this criterion.[1] Further, we are as ruthless as possible in invoking herein the smallest feasible set of commands and functions from Matlab/Octave so that students have the minimum to learn. Most teachers will find that many of their favourite commands are missing—this omission is all to the good in focussing upon useful mathematical development aided by only essential integrated computation.

This book does not aim to teach computer programming: there is no flow control, no looping, no recursion, nor even function definitions. The aim herein is to use short sequences of declarative assignment statements, coupled with the power of vector and matrix data structures, to learn core mathematical concepts, applications, and their relationships in linear algebra.

The internet is now ubiquitous and pervasive. So too is computing power: students can execute Matlab/Octave not only on laptops, but also on tablets and smart phones, perhaps using university or public servers, `octave-online.net`, Matlab-Online or Matlab-Mobile. We no longer need separate computer laboratories. Instead, expect students to access computational support simply by reaching into their pockets or bags.

> long after Riemann had passed away, historians discovered that he had developed advanced techniques for calculating the Riemann zeta function and that his formulation of the Riemann hypothesis—often depicted as a triumph of pure thought—was actually based on painstaking numerical work.
>
> *Donoho and Stodden (2015)*

---

[1] To compare popular packages, just look at the length of expressions students have to type in order to achieve core computations: Matlab/Octave is almost always the shortest (Nakos and Joyner, 1998, e.g.). (Of course, be wary of this metric: e.g., APL would surely be too concise!)

Linear algebra for statisticians

This book forms an ideal companion to modern statistics courses, especially data science. The recently published Curriculum Guidelines for Undergraduate Programs in Statistical Science, by Horton et al. (2014), emphasizes that linear algebra courses must provide "matrix manipulations, linear transformations, projections in Euclidean space, eigenvalues/eigenvectors, and matrix decompositions". These are all core topics in this book, especially the statistically important SVD factorization (Chapters 3 and 5). Furthermore, this book explicitly makes the recommended "connections between concepts in these mathematical foundations courses and their applications in statistics" (Horton et al., 2014, p.12).

Moreover, with the aid of some indicative statistical applications along with the sustained invocation of "visualization" and "basic programming concepts", this book helps to underpin the requirement to "Encourage synthesis of theory, methods, computation, and applications" (Horton et al., 2014, p.13).

Contents

1 Vectors

This chapter provides a relatively concise introduction to vectors, their properties, and a little computation with MATLAB/Octave. Skim or study as needed.

Mathematics started with counting. The natural numbers $1, 2, 3, \ldots$ quantify how many objects have been counted. Historically, there were many existential arguments over many centuries about whether negative numbers and zero are meaningful. Nonetheless, eventually negative numbers and the zero were included to form the **integers** $\ldots, -2, -1, 0, 1, 2, \ldots$. In the meantime people needed to quantify fractions such as two and a half bags, or a third of a cup, which led to the rational numbers such as $\frac{1}{3}$ or $2\frac{1}{2} = \frac{5}{2}$. Now **rational numbers** are defined as all numbers writeable in the form $\frac{p}{q}$ for integers p and q (q nonzero). Over two thousand years

Linear Algebra for the 21st Century. A. J. Roberts, Oxford University Press (2020). © A. J. Roberts.
DOI: 10.1093/oso/9780198856399.003.0001

ago, Pythagoras[1] was forced to recognize that for many triangles the length of a side could not be rational, and hence there must be more numbers in the world about us than rationals could provide. To cope with non-rational numbers such as $\sqrt{2} = 1.41421\cdots$ and (pi) $\pi = 3.14159\cdots$, mathematicians define the **real numbers** to be all numbers which in principle can be written as a decimal expansion such as $\sqrt{2}, \pi$,

$$\tfrac{9}{7} = 1.285714285714\cdots \quad \text{or} \quad e = 2.718281828459\cdots .$$

Such decimal expansions may terminate or repeat or may need to continue on indefinitely (as denoted by the three dots, called an ellipsis). The frequently invoked symbol $\mathbb{R}$ denotes the *set* of all possible real numbers.

In the sixteenth century, Gerolamo Cardano[2] developed a procedure to solve cubic polynomial equations. But the procedure involved manipulating $\sqrt{-1}$ which seemed a crazy figment of imagination. Nonetheless the procedure worked. Subsequently, many practical uses were found for $\sqrt{-1}$, now denoted by i (or j in some disciplines). Consequently, many areas of modern science and engineering use **complex numbers** which are those of the form $a + bi$ for real numbers a and b. The symbol $\mathbb{C}$ denotes the set of all possible complex numbers. This book mostly uses integers and real numbers, but eventually (Chapter 7) we need the marvellous complex numbers.

This book uses the term **scalar** to denote a number that could be integer, real, or complex. In this book, and before Chapter 7, a scalar is almost always real valued. The term 'scalar' arises because such numbers are often used to scale the length of a 'vector'.

1.1 Vectors have magnitude and direction

There are more things in heaven and earth, Horatio, than are dreamt of in your philosophy.
(Hamlet I.5:159–167)

In the eighteenth century, astronomers needed to describe both the position and the velocity of the planets. Such a description required quantities which have both a magnitude and a direction. Step outside, a wind blowing at 8 m/s (metres per second) from the south-west has both a magnitude and a direction. Quantities that have the properties of both a magnitude and a direction are called **vectors** (from the Latin for *carrier*).

---

[1] Pythagoras was born on Samos, Greece, in 569 BC, studied in Egypt, imprisoned in Babylon, then settled in Crotona to found a philosophical and religious school. Pythagoras and his school established Pythagoras' Theorem, the sum of the angles of a triangle are two right-angles, properties of polygons and polyhedra, and solved equations geometrically.

[2] Considered one of the great mathematicians of the Renaissance, Cardano was one of the key figures in the foundation of probability, and the earliest introducer of the binomial coefficients and the binomial theorem in the western world. . . . He made the first systematic use of negative numbers, published with attribution the solutions of other mathematicians for the cubic and quartic equations, and acknowledged the existence of imaginary numbers. (*Wikipedia 2015, Westfall (2012)*)

Example 1.1.1 *(displacement vector)* An important class of vectors are the so-called **displacement vectors**. Given two points in space, say A and B, the displacement vector $\overrightarrow{AB}$ is the directed line segment from the point A to the point B—as illustrated by the two displacement vectors $\overrightarrow{AB}$ and $\overrightarrow{CD}$ to the right. For example, if your home is at position A and your school at position B, then travelling from home to school is to move by the amount of the displacement vector $\overrightarrow{AB}$.

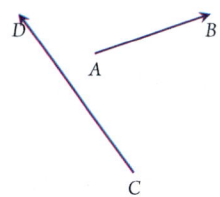

To be able to manipulate vectors we describe them with numbers. For such numbers to have meaning they must be set in the context of a coordinate system. So choose an origin for the coordinate system, usually denoted O, and draw coordinate axes in the plane (or space), as illustrated for the above two displacement vectors. Here the displacement vector $\overrightarrow{AB}$ goes three units to the right and one unit up, so we denote it by the ordered pair of numbers $\overrightarrow{AB} = (3,1)$. Whereas the displacement vector $\overrightarrow{CD}$ goes three units to the left and four units up, so we denote it by the ordered pair of numbers $\overrightarrow{CD} = (-3,4)$. Our choice of the origin O does not affect the number representation of these vectors. □

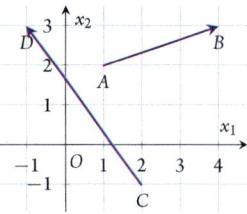

Example 1.1.2 *(position vector)* The next important class of vectors are the **position vectors**. Given some chosen fixed origin in space, usually denoted O, then $\overrightarrow{OA}$ is the position vector of the point A. This picture illustrates the position vectors of four points in the plane (A,B,C,D), from the given origin O.

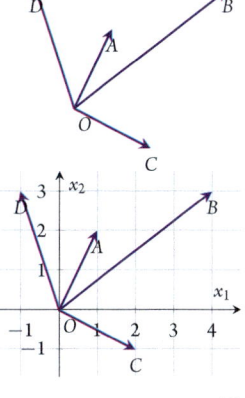

Again, to be able to manipulate such vectors we describe them with numbers, and such numbers have meaning via a coordinate system. So draw coordinate axes in the plane (or space), as illustrated for the above four position vectors. Here the position vector $\overrightarrow{OA}$ goes one unit to the right and two units up so we denote it by $\overrightarrow{OA} = (1,2)$. Similarly, the position vectors $\overrightarrow{OB} = (4,3)$, $\overrightarrow{OC} = (2,-1)$, and $\overrightarrow{OD} = (-1,3)$. Recognize that the ordered pairs of numbers in the position vectors are exactly the coordinates of each of the specified end-points. □

Example 1.1.3 *(velocity vector)* Consider an airplane in level flight at 900 km/hr (kilometres per hour) to the east-north-east. Choosing coordinate axes oriented to the east and the north, the direction of the airplane is at an angle $22.5°$ from the east, as illustrated on the right. Trigonometry then tells us that the eastward part of the speed of the airplane is $900\cos(22.5°) = 831.5$ km/hr, whereas the northward part of the speed is $900\sin(22.5°) = 344.4$ km/hr (as indicated quantitatively). Further, the airplane is in level flight, not going up or down, so in the third direction of space (vertically) its speed component is zero. Putting these together forms the velocity vector $(831.5, 344.4, 0)$ in km/hr in space.

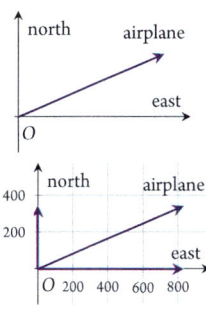

Another airplane takes off from an airport at 360 km/hr to the northwest and climbs at 2 m/s. The direction northwest is 45° to the east-west lines and 45° to the north-south lines. Trigonometry then tells us that the westward speed of the airplane is $360\cos(45°) = 360\cos(\frac{\pi}{4}) = 254.6$ km/hr, whereas the northward speed is $360\sin(45°) = 360\sin(\frac{\pi}{4}) = 254.6$ km/hr as illustrated.

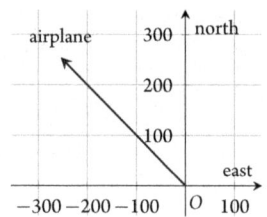

But west is the opposite direction to east, so if the coordinate system treats east as positive, then west must be negative. Consequently, together with the climb in the vertical, the velocity vector is $(-254.6 \text{ km/hr}, 254.6 \text{ km/hr}, 2 \text{ m/s})$. But we should avoid mixing units within a vector, so here convert all speeds to m/s: here 360 km/hr upon dividing by 3600 secs/hr and multiplying by 1000 m/km gives 360 km/hr = 100 m/s. Then the north and west speeds are both $100\cos(\frac{\pi}{4}) = 70.71$ m/s. Consequently, the velocity vector of the climbing airplane should be described as $(-70.71, 70.71, 2)$ in m/s. □

In applications, as these examples illustrate, the 'physical' vector exists before the coordinate system. It is only when we choose a specific coordinate system that a 'physical' vector gets expressed by numbers. Throughout, unless otherwise specified, this book assumes that vectors are expressed in what is called a **standard coordinate system**.

- In the two dimensions of the plane the standard coordinate system has two coordinate axes, one horizontal and one vertical, at right-angles to each other, often labelled x_1 and x_2 respectively (as illustrated), although labels x and y are also common.

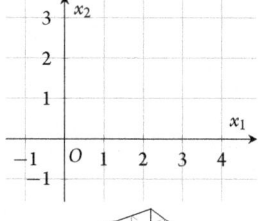

- In the three dimensions of space the standard coordinate system has three coordinate axes, two horizontal and one vertical, all at right-angles to each other, often labelled x_1, x_2, and x_3 respectively (as illustrated), although labels x, y, and z are also common.

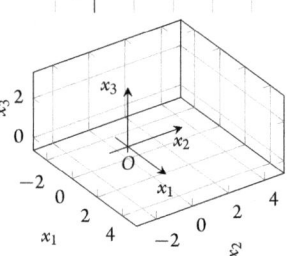

- Correspondingly, in so-called 'n dimensions' the standard coordinate system has n coordinate axes, all at right-angles to each other, and often labelled x_1, x_2, ..., x_n, respectively.

Definition 1.1.4 *Given a standard coordinate system with n coordinate axes, all at right-angles to each other, a **vector** is an ordered n-tuple of real numbers x_1, x_2, ..., x_n equivalently written either as a row in parentheses or as a column in brackets,*

$$(x_1, x_2, \ldots, x_n) = \begin{bmatrix} x_1 \\ x_2 \\ \vdots \\ x_n \end{bmatrix}$$

(they mean the same; it is just more convenient to usually use a row in parentheses in text, and a column in brackets in displayed mathematics). The real numbers x_1, x_2, ..., x_n are called the **components** *of the vector, and the number of components is termed its* **size** *(here n). The components are determined such that letting X be the point with coordinates $(x_1, x_2, ..., x_n)$ then the position vector $\overrightarrow{OX}$ has the same magnitude and direction as the vector denoted $(x_1, x_2, ..., x_n)$.*

Two vectors of the same size are **equal**, $=$, *if all their corresponding components are equal (vectors with different sizes are never equal).*

Robert Recorde invented the equal sign circa 1557 "bicause noe two thynges can be moare equalle". He also invented the term "sine" and the method of extracting the square-root by hand.

Examples 1.1.1 and 1.1.2 introduced some vectors and wrote them as a row in parentheses, such as $\overrightarrow{AB} = (3,1)$. In this book, exactly the same thing is meant by the columns in brackets: for example,

$$\overrightarrow{AB} = (3,1) = \begin{bmatrix} 3 \\ 1 \end{bmatrix}, \quad \overrightarrow{CD} = (-3,4) = \begin{bmatrix} -3 \\ 4 \end{bmatrix},$$

$$\overrightarrow{OC} = (2,-1) = \begin{bmatrix} 2 \\ -1 \end{bmatrix}, \quad (-70.71, 70.71, 2) = \begin{bmatrix} -70.71 \\ 70.71 \\ 2 \end{bmatrix}.$$

However, as defined subsequently, a row of numbers within brackets is quite different: for two examples, $(3,1) \neq \begin{bmatrix} 3 & 1 \end{bmatrix}$, and $(831,344,0) \neq \begin{bmatrix} 831 & 344 & 0 \end{bmatrix}$.

The *ordering* of the components is very important. For example, as illustrated to the right, the vector $(3,1)$ is very different from the vector $(1,3)$; similarly, the vector $(2,-1)$ is very different from the vector $(-1,2)$.

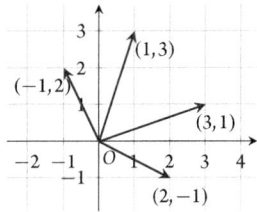

Definition 1.1.5 *The set of all vectors with n components is denoted $\mathbb{R}^n$. The vector with all components zero, $(0,0,...,0)$, is called the* **zero vector** *and denoted by* **0**.

Example 1.1.6

- All the vectors we can draw and imagine in the two-dimensional plane form $\mathbb{R}^2$. Sometimes we write that $\mathbb{R}^2$ is the plane because of this very close connection.

- All the vectors we can draw and imagine in three-dimensional space form $\mathbb{R}^3$. Again, sometimes we write that $\mathbb{R}^3$ is three-dimensional space because of the close connection.

- The set $\mathbb{R}^1$ is the set of all vectors with one component, and that one component is measured along one axis. Hence $\mathbb{R}^1$ is effectively the same as the set of real numbers labelling that axis. □

As just introduced for the zero vector **0**, this book generally denotes vectors by a bold letter (except for displacement vectors). The other common notation you may see elsewhere is to denote vectors by a small over-arrow such as in the "zero vector $\vec{0}$". Less commonly, some books and articles use an over- or under-tilde ($\sim$) to denote vectors. Be aware of this different notation in reading other books.

Question: why do we need vectors with n components, in $\mathbb{R}^n$, when the world around us is only three-dimensional? Answer: because vectors can encode much more than just spatial structure. The next example illustrates another use of vectors.

Example 1.1.7 *(linguistic vectors)* Consider the following four sentences.

(a) The dog sat on the mat.

(b) The cat scratched the dog.

(c) The cat and dog sat on the mat.

(d) The dog scratched.

These four sentences involve up to three objects, cat, dog, and mat, and two actions, sat and scratched. Some characteristic of each of the sentences is captured simply by counting the number of times each of these three objects and two actions appears in each sentence, and then forming a vector from the counts. Let's use vectors $w = (N_{cat}, N_{dog}, N_{mat}, N_{sat}, N_{scratched})$ where the various N are the counts of each word (w for words). The previous statement implicitly specifies that we use five coordinate axes, perhaps labelled "cat", "dog", "mat", "sat", and "scratched", and that distance along each axis represents the number of times the corresponding word is used. These word vectors are in $\mathbb{R}^5$ as there are five components in each vector w. Then

(a) "The dog sat on the mat" is summarized by the vector $w = (0,1,1,1,0)$.

(b) 'The cat scratched the dog" is summarized by the vector $w = (1,1,0,0,1)$.

(c) "The cat and dog sat on the mat" is summarized by the vector $w = (1,1,1,1,0)$.

(d) "The dog scratched" is summarized by the vector $w = (0,1,0,0,1)$.

(e) An empty sentence is the zero vector $w = (0,0,0,0,0)$.

(f) Together, the two sentences "The dog sat on the mat. The cat scratched the dog." are summarized by the vector $w = (1,2,1,1,1)$.

Using such crude summary representations of some text, even of entire documents, empowers us to use powerful mathematical techniques to relate documents together, compare and contrast, express similarities, look for type clusters, and so on. In application we would not just count words for objects (nouns) and actions (verbs), but also qualifications (adjectives and adverbs).[3]

People generally know and use thousands of words. Consequently, in practice, such word vectors typically have thousands of components corresponding to coordinate axes of thousands of distinct words. To cope with such vectors of many components, modern linear algebra has been developed to powerfully handle problems involving vectors with thousands, millions, or even an 'infinite number' of components. □

---

[3] Look up Latent Semantic Analysis, such as at https://en.wikipedia.org/wiki/Latent_semantic_analysis [July 2019]

King – man + woman = queen

Computational linguistics has dramatically changed the way researchers study and understand language. The ability to number-crunch huge amounts of words for the first time has led to entirely new ways of thinking about words and their relationship to one another.

This number-crunching shows exactly how often a word appears close to other words, an important factor in how they are used. So the word Olympics might appear close to words like running, jumping, and throwing but less often next to words like electron or stegosaurus. This set of relationships can be thought of as a multidimensional vector that describes how the word Olympics is used within a language, which itself can be thought of as a vector space.

And therein lies this massive change. This new approach allows languages to be treated like vector spaces with precise mathematical properties. Now the study of language is becoming a problem of vector space mathematics.[a]   *Technology Review, 2015*

---

[a] http://www.technologyreview.com/view/541356 [Oct 2019]

Activity 1.1.8 Given word vectors $w = (N_{\text{cat}}, N_{\text{dog}}, N_{\text{mat}}, N_{\text{sat}}, N_{\text{scratched}})$ as in Example 1.1.7, which of the following has word vector $w = (2,2,0,2,1)$?

(a) "Which cat sat by the dog on the mat, and then scratched the dog."

(b) "The dog scratched the cat on the mat."

(c) "A dog and cat both sat on the mat which the dog had scratched."

(d) "A dog sat. A cat scratched the dog. The cat sat."

Definition 1.1.9 *(Pythagoras)* *For every vector $v = (v_1, v_2, \ldots, v_n)$ in $\mathbb{R}^n$, define the **length**, or **magnitude**, of a vector v to be the real number (≥ 0)*

$$|v| := \sqrt{v_1^2 + v_2^2 + \cdots + v_n^2}.$$

*A vector of length one is called a **unit vector**. (Many people and books denote the length of a vector with a pair of double lines, as in $\|v\|$. Either notation is good.)*

Remember that the term *size* denotes the number of components in a vector (Definition 1.1.4) and so "size" should not be used for the length/magnitude of a vector.

Example 1.1.10 Find the lengths of the following vectors: $a = (-3,4)$; $b = (3,3)$; $c = (1,-2,3)$; $d = (1,-1,-1,1)$.

Solution:

- $|a| = \sqrt{(-3)^2 + 4^2} = \sqrt{25} = 5.$
- $|b| = \sqrt{3^2 + 3^2} = \sqrt{18} = 3\sqrt{2}.$
- $|c| = \sqrt{1^2 + (-2)^2 + 3^2} = \sqrt{14}.$
- $|d| = \sqrt{1^2 + (-1)^2 + (-1)^2 + 1^2} = \sqrt{4} = 2.$ □

Example 1.1.11 Write down three different vectors, all three with the same number of components, for each of the following cases: (a) of length 5, (b) of length 3, and (c) of length -2.

Solution:

(a) Humans knew of the $3 : 4 : 5$ right-angled triangle thousands of years ago, so perhaps one answer could be $(3,4)$, $(-4,3)$, and $(5,0)$.

(b) One answer might be $(3,0,0)$, $(0,3,0)$, and $(0,0,3)$. A more interesting answer might arise from knowing $1^2 + 2^2 + 2^2 = 3^2$ leading to an answer of perhaps $(1,2,2)$, $(2,-1,2)$, and $(-2,2,1)$.

(c) Since the length of a vector is $\sqrt{\cdots}$, which is always positive or zero, the length cannot be negative, so there is no possible answer to this last case. □

Activity 1.1.12 What is the length of the vector $(2,-3,6)$?

(a) 5   (b) 11   (c) $\sqrt{11}$ (d) 7

Theorem 1.1.13 *The zero vector is the only vector of length zero:* $|v| = 0$ *if and only if* $v = 0$.

Proof. First establish the zero vector has length zero. From Definition 1.1.9, in $\mathbb{R}^n$,

$$|0| = \sqrt{0^2 + 0^2 + \cdots + 0^2} = \sqrt{0} = 0.$$

Second establish that if a vector has length zero then it must be the zero vector. Let vector $v = (v_1, v_2, \ldots, v_n)$ in $\mathbb{R}^n$ have zero length. By squaring both sides of the Definition 1.1.9 for length we then know that

$$\underbrace{v_1^2}_{\geq 0} + \underbrace{v_2^2}_{\geq 0} + \cdots + \underbrace{v_n^2}_{\geq 0} = 0.$$

Being squares, all terms on the left are non-negative, so the only way they can all add to zero is if they are all zero. That is, $v_1 = v_2 = \cdots = v_n = 0$. Hence, the vector v must be the zero vector 0.

1.1.1 Exercises

Exercise 1.1.1 For each case: on the plot, draw the displacement vectors $\overrightarrow{AB}$ and $\overrightarrow{CD}$, and the position vectors of the points A and D.

(c)

(d)

Exercise 1.1.2 For each case: roughly estimate (to say ± 0.2) each of the two components of the four position vectors of the points A, B, C, and D.

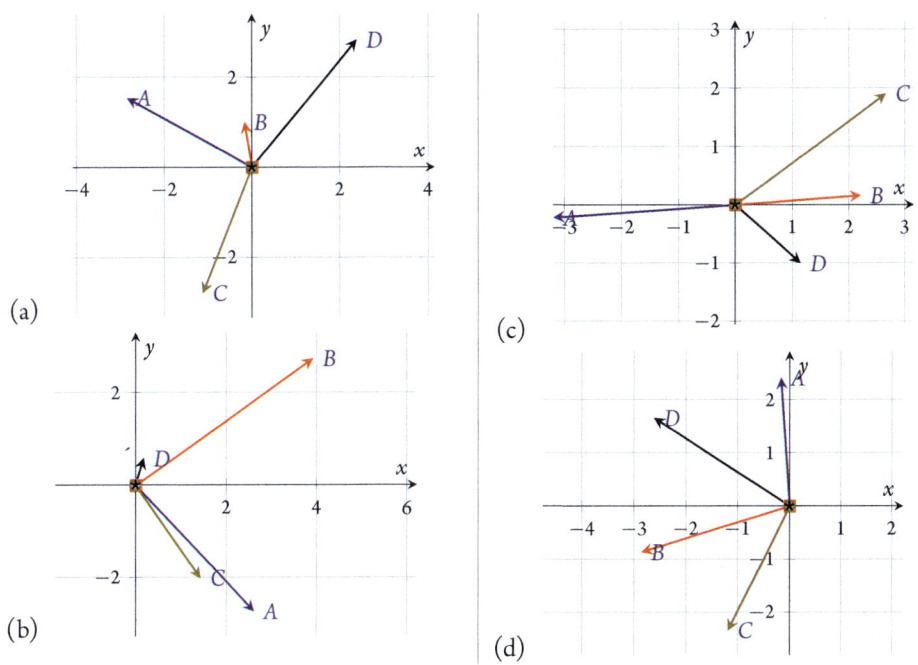

(a)

(b)

(c)

(d)

Exercise 1.1.3 For each case plotted in Exercise 1.1.2: from your estimated components of each of the four position vectors, calculate the length (or magnitude) of the four vectors. Also use a ruler (or otherwise) to directly measure an estimate of the length of each vector. Confirm that your calculated lengths reasonably approximate your measured lengths.

Exercise 1.1.4 Below are the titles of eight books that The Society of Industrial and Applied Mathematics (SIAM) reviewed recently.

(a) Introduction to Finite and Spectral Element Methods using MATLAB
(b) Derivative Securities and Difference Methods
(c) Iterative Methods for Linear Systems: Theory and Applications
(d) Singular Perturbations: Introduction to System Order Reduction Methods with Applications
(e) Risk and Portfolio Analysis: Principles and Methods
(f) Differential Equations: Theory, Technique, and Practice
(g) Contract Theory in Continuous-Time Models
(h) Stochastic Chemical Kinetics: Theory and Mostly Systems Biology Applications

Make a list of the five significant words that appear more than once in this list (not including the common nontechnical words such as "and" and "for", and not distinguishing between words with a common root). Being consistent about the order of words, represent each of the eight titles by a word vector in $\mathbb{R}^5$.

Exercise 1.1.5 In a few sentences, answer/discuss each of the following.

(a) Why is a coordinate system important for a vector?

(b) Describe the distinction between a displacement vector and a position vector.

(c) Why do two vectors have to be the same size in order to be equal?

(d) What is the connection between the length of a vector and Pythagoras' theorem for triangles?

(e) Describe a problem that would occur if the ordering of the components in a vector was not significant?

(f) Recall that a vector has both a magnitude and a direction. Comment on why the zero vector is the only vector with zero magnitude.

(g) In what other courses have you seen vectors? What was the same and what was different?

1.2 Adding and stretching vectors

We want to be able to make sense of statements such as "king – man + women = queen". To do so, we need to define operations on vectors. Useful operations on vectors are those that are physically meaningful. Then our algebraic manipulations derive powerful results in applications. The first two vector operations are addition and scalar multiplication.

1.2.1 Basic operations

Example 1.2.1 Vectors of the same size are added component-wise. Equivalently, obtain the same result by geometrically joining the two vectors 'head-to-tail' and drawing the vector from the start to the finish.

(a) Let's add the two vectors shown below-left: $(1,3) + (2,-1) = (1+2, 3+(-1)) = (3,2)$ as illustrated below-middle, where the vector $(2,-1)$ is drawn from the end of $(1,3)$, and the end-point of the result determines the vector addition $(3,2)$.

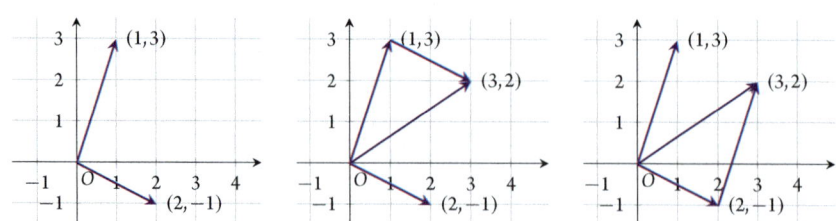

This result $(3,2)$ is the same if the vector $(1,3)$ is drawn from the end of $(2,-1)$ as shown above-right. That is, $(2,-1)+(1,3)=(1,3)+(2,-1)$. That the order of addition is immaterial is the commutative law of vector addition. Theorem 1.2.19(a) establishes this law in general.

(b) $(3,2,0)+(-1,3,2)=(3+(-1),2+3,0+$
$2)=(2,5,2)$ as illustrated below where (given the two vectors as plotted to the right) the vector $(-1,3,2)$ is drawn from the end of $(3,2,0)$, and the end-point of the result determines the vector addition $(2,5,2)$. As below, find the same result by drawing the vector $(3,2,0)$ from the end of $(-1,3,2)$.

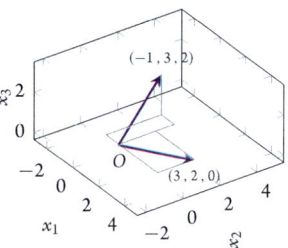

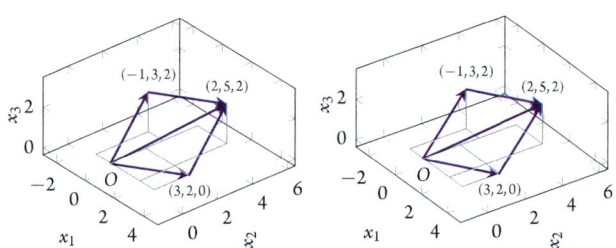

As drawn above, many of the three-D plots in this book are **stereo pairs**, drawing the plot from two slightly different viewpoints:[4] cross your eyes to merge two of the images, and then focus on the pair of plots to see the 3D effect. With practice viewing such 3D stereo pairs becomes less difficult![5]

(c) The addition $(1,3)+(3,2,0)$ is not defined and cannot be done because the two vectors have a different number of components; they have different sizes.

Example 1.2.2 To multiply a vector by a scalar, a number, multiply each component by the scalar. Equivalently, visualize the result through stretching the vector by a factor of the scalar.

(a) Let the vector $u=(3,2)$ then, as illustrated to the right,

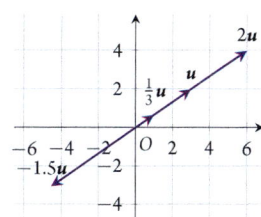

$$2u=2(3,2)=(2\cdot3,2\cdot2)=(6,4),$$
$$\tfrac{1}{3}u=\tfrac{1}{3}(3,2)=(\tfrac{1}{3}\cdot3,\tfrac{1}{3}\cdot2)=(1,\tfrac{2}{3}),$$
$$(-1.5)u=(-1.5\cdot3,-1.5\cdot2)=(-4.5,-3).$$

---

[4] I implement such cross-eyed stereo so that these stereo images are useful both printed and when projected on a screen.
[5] To help keeping your eyes crossed while you try to focus on the plots, hold a finger or pen in front of your nose so that your right-eye sight across the top of the finger/pen aligns with the centre of the left-picture, and simultaneously your left-eye sight across the top of the finger/pen aligns with the centre of the right-picture.

(b) Let the vector $v = (2,3,1)$ then, as illustrated below in cross-eyed stereo,

$$2v = 2\begin{bmatrix} 2 \\ 3 \\ 1 \end{bmatrix} = \begin{bmatrix} 2 \cdot 2 \\ 2 \cdot 3 \\ 2 \cdot 1 \end{bmatrix} = \begin{bmatrix} 4 \\ 6 \\ 2 \end{bmatrix},$$

$$\left(-\tfrac{1}{2}\right)v = -\tfrac{1}{2}\begin{bmatrix} 2 \\ 3 \\ 1 \end{bmatrix} = \begin{bmatrix} -\tfrac{1}{2} \cdot 2 \\ -\tfrac{1}{2} \cdot 3 \\ -\tfrac{1}{2} \cdot 1 \end{bmatrix} = \begin{bmatrix} -1 \\ -\tfrac{3}{2} \\ -\tfrac{1}{2} \end{bmatrix}.$$

Activity 1.2.3 Combining multiplication and addition, what is $u + 2v$ for vectors $u = (4,1)$ and $v = (-1,-3)$?

(a) $(3,-2)$ (b) $(2,-5)$ (c) $(5,-8)$ (d) $(1,-8)$

Definition 1.2.4 *Let two vectors in $\mathbb{R}^n$ be $u = (u_1 , u_2 , \ldots , u_n)$ and $v = (v_1 , v_2 , \ldots , v_n)$, and let c be a scalar. Then the **sum** or **addition** of u and v, denoted $u + v$, is the vector obtained by joining v to u 'head-to-tail', and is computed as*

$$u + v := (u_1 + v_1, u_2 + v_2, \ldots, u_n + v_n).$$

*The **scalar multiplication** of u by c, denoted cu, is the vector of length $|c| \, |u|$ in the direction of u when $c > 0$ but in the opposite direction when $c < 0$, and is computed as*

$$cu := (cu_1, cu_2, \ldots, cu_n).$$

*The **negative** of u denoted $-u$, is defined as the scalar multiple $-u := (-1)u$, and is a vector of the same length as u but in exactly the opposite direction. The **difference** $u - v$ is defined as the sum $u + (-v)$ and is equivalently the vector drawn from the end of v to the end of u.*

Example 1.2.5 For the vectors u and v shown to the right, draw the vectors $u + v, v + u, u - v, v - u, \tfrac{1}{2}u$, and $-v$.

Solution: Drawn below.

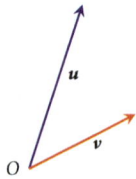

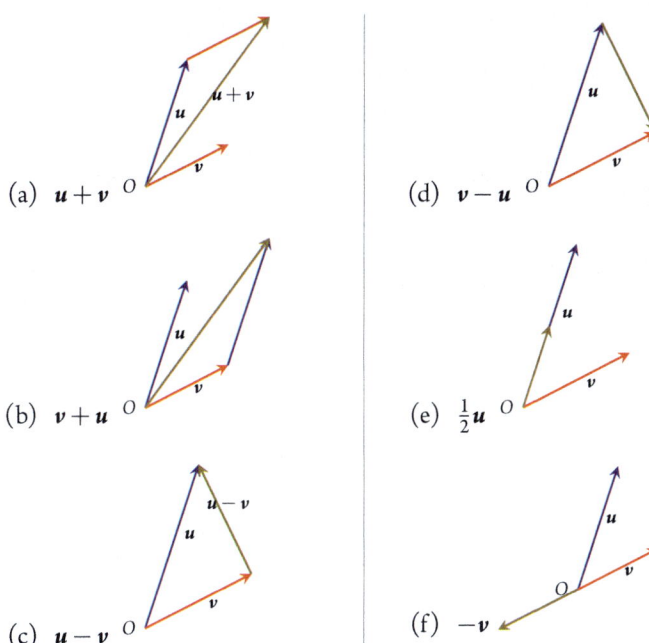

(a) $u+v$

(b) $v+u$

(c) $u-v$

(d) $v-u$

(e) $\frac{1}{2}u$

(f) $-v$

Activity 1.2.6 For the vectors u and v shown to the right, what is the result vector, which is also shown?

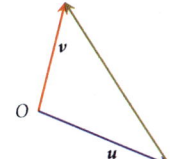

(a) $v-u$ (b) $u-v$ (c) $v+u$ (d) $u+v$

Using vector addition and scalar multiplication, we often write vectors in terms of so-called standard unit vectors. In the plane, as drawn right, are the two unit vectors i and j defined to be of length one and in the direction of the two coordinate axes, respectively. Hence $i=(1,0)$ and $j=(0,1)$, as shown. Then, for example,

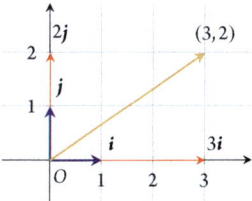

$$(3,2) = (3,0)+(0,2) \quad \text{(by addition)}$$
$$= 3(1,0)+2(0,1) \quad \text{(by scalar mult)}$$
$$= 3i+2j \quad \text{(by definition of } i \text{ and } j).$$

Similarly, in three-dimensional space, we often write vectors in terms of the three vectors i, j, and k, defined to be each of length one, aligned along the three coordinate axes. Hence $i=(1,0,0), j=(0,1,0)$, and $k=(0,0,1)$. For example,

$$(2,3,-1) = (2,0,0) + (0,3,0) + (0,0,-1) \quad \text{(by addition)}$$
$$= 2(1,0,0) + 3(0,1,0) - (0,0,1) \quad \text{(by scalar mult)}$$
$$= 2\boldsymbol{i} + 3\boldsymbol{j} - \boldsymbol{k} \quad \text{(by definition of } \boldsymbol{i}, \boldsymbol{j}, \text{ and } \boldsymbol{k}).$$

The next definition generalizes these standard unit vectors to vectors in $\mathbb{R}^n$ for every size n.

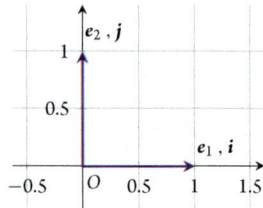

Definition 1.2.7 *Given a standard coordinate system with n coordinate axes, all at right-angles to each other, the* ***standard unit vectors*** *e_1, e_2, ..., e_n are the vectors of length one in the direction of the corresponding coordinate axis (as illustrated to the right for $\mathbb{R}^2$ and below for $\mathbb{R}^3$). That is,*

$$e_1 = \begin{bmatrix} 1 \\ 0 \\ \vdots \\ 0 \end{bmatrix}, \quad e_2 = \begin{bmatrix} 0 \\ 1 \\ \vdots \\ 0 \end{bmatrix}, \quad \dots, \quad e_n = \begin{bmatrix} 0 \\ 0 \\ \vdots \\ 1 \end{bmatrix}.$$

In $\mathbb{R}^2$ and $\mathbb{R}^3$, the symbols $\boldsymbol{i}$, $\boldsymbol{j}$, and $\boldsymbol{k}$ are often used as synonyms for e_1, e_2, and e_3, respectively (as illustrated below).

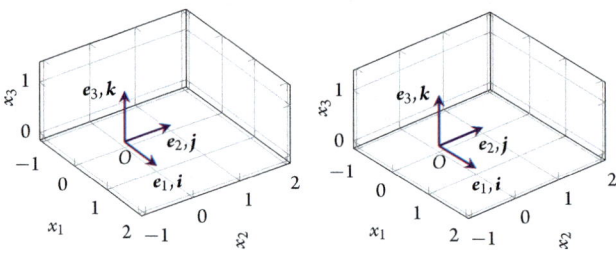

That is, for three examples, the following are equivalent ways of writing the same vector:

- $(3,2) = \begin{bmatrix} 3 \\ 2 \end{bmatrix} = 3\boldsymbol{i} + 2\boldsymbol{j} = 3e_1 + 2e_2;$

- $(2,3,-1) = \begin{bmatrix} 2 \\ 3 \\ -1 \end{bmatrix} = 2\boldsymbol{i} + 3\boldsymbol{j} - \boldsymbol{k} = 2e_1 + 3e_2 - e_3;$

- $(0,-3.7,0,0.1,-3.9) = \begin{bmatrix} 0 \\ -3.7 \\ 0 \\ 0.1 \\ -3.9 \end{bmatrix} = -3.7e_2 + 0.1e_4 - 3.9e_5.$

Activity 1.2.8 Which of the following is the same as the vector $3e_2 + e_5$?

(a) $(0,3,0,0,1)$ (b) $(5,0,2)$ (c) $(3,1)$ (d) $(0,3,0,1)$

1.2.2 Distance

Defining a 'distance' between vectors empowers us to concisely compare vectors.

Example 1.2.9 We would like to say that $(1.2, 3.4) \approx (1.5, 3)$ to an error 0.5 (as illustrated to the right). Why is the error 0.5? Because the difference between the vectors $(1.5, 3) - (1.2, 3.4) = (0.3, -0.4)$ has length $\sqrt{0.3^2 + (-0.4)^2} = 0.5$.

Conversely, we would like to recognize that vectors $(1.2, 3.4)$ and $(3.4, 1.2)$ are very different (as also illustrated)—there is a large 'distance' between them. Why is there a large 'distance'? Because the difference between the vectors $(1.2, 3.4) - (3.4, 1.2) = (-2.2, 2.2)$ has length $\sqrt{(-2.2)^2 + 2.2^2} = 2.2\sqrt{2} = 3.1113$, which is relatively large.

This concept of distance between two vectors $\boldsymbol{u}$ and $\boldsymbol{v}$, directly corresponding to the distance between two points, is the length $|\boldsymbol{u} - \boldsymbol{v}|$.

Definition 1.2.10 *The **distance** between vectors $\boldsymbol{u}$ and $\boldsymbol{v}$ in $\mathbb{R}^n$ is defined to be the length of their difference, $|\boldsymbol{u} - \boldsymbol{v}|$.*

Example 1.2.11 Given three vectors $\boldsymbol{a} = 3\boldsymbol{i} + 2\boldsymbol{j} - 2\boldsymbol{k}$, $\boldsymbol{b} = 5\boldsymbol{i} + 5\boldsymbol{j} + 4\boldsymbol{k}$, and $\boldsymbol{c} = 7\boldsymbol{i} - 2\boldsymbol{j} + 5\boldsymbol{k}$ (shown right in stereo) use the concept of distance between vectors to answer the following: which pair are the closest to each other? And which pair are furthest from each other?

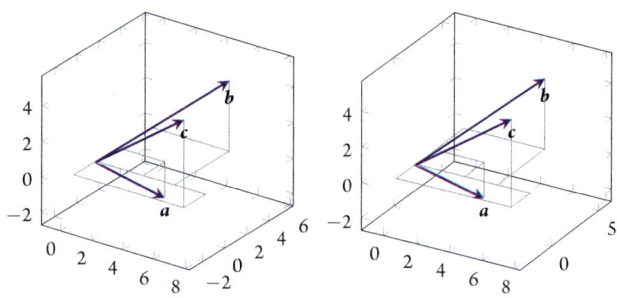

Solution: Compute the distances between each pair.

- $|\boldsymbol{b} - \boldsymbol{a}| = |2\boldsymbol{i} + 3\boldsymbol{j} + 6\boldsymbol{k}| = \sqrt{2^2 + 3^2 + 6^2} = \sqrt{49} = 7$.
- $|\boldsymbol{c} - \boldsymbol{a}| = |4\boldsymbol{i} - 4\boldsymbol{j} + 7\boldsymbol{k}| = \sqrt{4^2 + (-4)^2 + 7^2} = \sqrt{81} = 9$.
- $|\boldsymbol{c} - \boldsymbol{b}| = |2\boldsymbol{i} - 7\boldsymbol{j} - \boldsymbol{k}| = \sqrt{2^2 + (-7)^2 + (-1)^2} = \sqrt{54} = 7.3485$.

The smallest distance of 7 is between $\boldsymbol{a}$ and $\boldsymbol{b}$ so these two are the closest pair of vectors. The largest distance of 9 is between $\boldsymbol{a}$ and $\boldsymbol{c}$ so these two are the furthest pair of vectors. □

Activity 1.2.12 Which pair of the following vectors are closest—have the smallest distance between them? $a = (7,3)$, $b = (4,-1)$, $c = (2,4)$

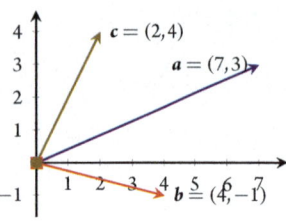

(a) b, c

(c) two of the pairs

(b) a, b

(d) a, c

1.2.3 Parametric equation of a line

We are familiar with lines in the plane, and equations that describe them. Let's now consider such equations from a vector view. The insights empower us to generalize the descriptions to lines in space, and then to lines in any number of dimensions.

Example 1.2.13 Consider the line drawn to the right in some chosen coordinate system. Recall that one way to find an equation of the line is to find the intercepts with the axes, here at $x = 4$ and at $y = 2$, then write down $\frac{x}{4} + \frac{y}{2} = 1$ as an equation of the line. Algebraic rearrangement gives various other forms, such as $x + 2y = 4$ or $y = 2 - x/2$.

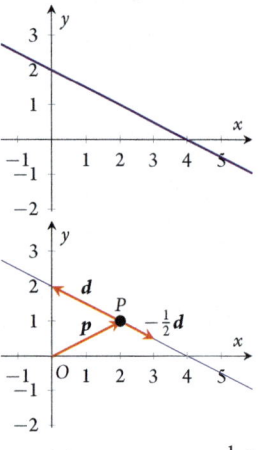

The alternative is to describe the line with vectors. Choose any point P on the line, such as $(2,1)$ as drawn to the right. Then view every other point on the line as having position vector that is the vector sum of $\overrightarrow{OP}$ and a vector aligned along the line. Denote $\overrightarrow{OP}$ by p as drawn. Then, for example, the point $(0,2)$ on the line has position vector $p + d$ for vector $d = (-2,1)$ because $p + d = (2,1) + (-2,1) = (0,2)$. Other points on the line are also given using the same vectors, p and d: for example, the point $(3,\frac{1}{2})$ has position vector $p - \frac{1}{2}d$ (as drawn) because $p - \frac{1}{2}d = (2,1) - \frac{1}{2}(-2,1) = (3,\frac{1}{2})$; and the point $(-2,3)$ has position vector $p + 2d = (2,1) + 2(-2,1)$. In general, every point on the line may be expressed as $p + td$ for some scalar t.

For every given line, there are many possible choices of p and d in such a vector representation. A different looking, but equally valid, form is obtained from any pair of points on the line. For example, one could choose point P to be $(0,2)$ and point Q to be $(3,\frac{1}{2})$, as drawn to the right. Let position vector $p = \overrightarrow{OP} = (0,2)$ and the vector $d = \overrightarrow{PQ} = (3,-\frac{3}{2})$, then every point on the line has position vector $p + td$ for some scalar t:

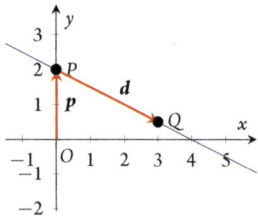

- $(2,1) = (0,2) + (2,-1) = (0,2) + \frac{2}{3}(3,-\frac{3}{2}) = p + \frac{2}{3}d$;
- $(6,-1) = (0,2) + (6,-3) = (0,2) + 2(3,-\frac{3}{2}) = p + 2d$;
- $(-1,\frac{5}{2}) = (0,2) + (-1,\frac{1}{2}) = (0,2) - \frac{1}{3}(3,-\frac{3}{2}) = p - \frac{1}{3}d$.

Other choices of points P and Q give other valid vector equations for a given line.

Activity 1.2.14 Which one of the following is *not* a valid vector equation for the line plotted to the right?

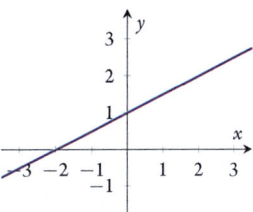

(a) $(-2,0) + (-4,-2)t$

(c) $(-1,1/2) + (2,-1)t$

(b) $(0,1) + (2,1)t$

(d) $(2,2) + (1,1/2)t$

Definition 1.2.15 *A **parametric equation** of a line is of the form*
*$x = p + td$ where p is the position vector of some point on the line, the so-called **direction vector** d is parallel to the line ($d \neq 0$), and the scalar **parameter** t varies over all real values, to give all position vectors x on the line.*

Beautifully, this definition applies for lines in any number of dimensions by using vectors with the corresponding number of components.

Example 1.2.16 Given that the line drawn to the right in space goes through points $(-4,-3,3)$ and $(3,2,1)$, find a parametric equation of the line.

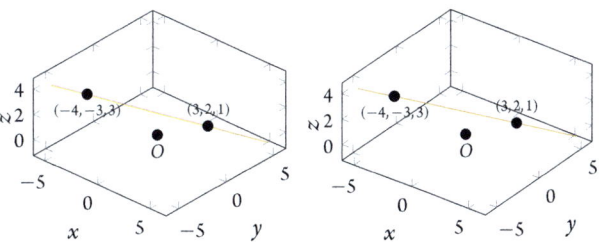

Solution: Let's call the points $(-4,-3,3)$ and $(3,2,1)$ as P and Q, respectively, and as shown below. First, choose a point on the line, say P, and set its position vector $p = \overrightarrow{OP} = (-4,-3,3) = -4i - 3j + 3k$, as drawn. Second, choose a direction vector to be, say, $d = \overrightarrow{PQ} = (3,2,1) - (-4,-3,3) = 7i + 5j - 2k$, also drawn. A parametric equation of the line is then $x = p + td$, specifically

$$x = (-4i - 3j + 3k) + t(7i + 5j - 2k)$$
$$= (-4 + 7t)i + (-3 + 5t)j + (3 - 2t)k.$$

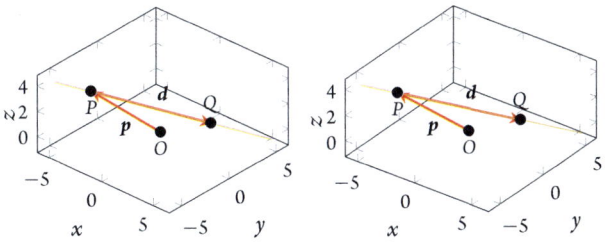

$\square$

Example 1.2.17 Given the parametric equation of a line in space is $x = (-4 + 2t, 3 - t, -1 - 4t)$, find the value of the parameter t that gives each of the following points on the line: $(-1.6, 1.8, -5.8)$, $(-3, 2.5, -3)$, and $(-6, 4, 4)$.

Solution:

- For the point $(-1.6, 1.8, -5.8)$ we need to find the parameter value t such that $-4 + 2t = -1.6$, $3 - t = 1.8$ and $-1 - 4t = -5.8$. The first of these requires $t = (-1.6 + 4)/2 = 1.2$, the second requires $t = 3 - 1.8 = 1.2$, and the third requires $t = (-1 + 5.8)/4 = 1.2$. All three agree that choosing parameter $t = 1.2$ gives the required point.

- For the point $(-3, 2.5, -3)$ we need to find the parameter value t such that $-4 + 2t = -3$, $3 - t = 2.5$ and $-1 - 4t = -3$. The first of these requires $t = (-3 + 4)/2 = 0.5$, the second requires $t = 3 - 2.5 = 0.5$, and the third requires $t = (-1 + 3)/4 = 0.5$. All three agree that choosing parameter $t = 0.5$ gives the required point.

- For the point $(-6, 4, 4)$ we need to find the parameter value t such that $-4 + 2t = -6$, $3 - t = 4$ and $-1 - 4t = 4$. The first of these requires $t = (-6 + 4)/2 = -1$, the second requires $t = 3 - 4 = -1$, and the third requires $t = (-1 - 4)/4 = -1.25$. Since these three require different values of t, namely -1 and -1.25, it means that there is no single value of the parameter t that gives the required point. That is, the point $(-6, 4, 4)$ cannot be on the line. Consequently the task is impossible.[6]

□

1.2.4 Manipulation requires algebraic properties

It seems to be nothing other than that art which they call by the barbarous name of 'algebra', if only it could be disentangled from the multiple numbers and inexplicable figures that overwhelm it …

Descartes

To unleash the power of algebra on vectors, we need to know the properties of vector operations. Many of the following properties are familiar, as they directly correspond to familiar properties of arithmetic operations on scalars. Moreover, the proofs show that the vector properties follow directly from the familiar properties of arithmetic operations on scalars.

Example 1.2.18 Let vectors $u = (1, 2)$, $v = (3, 1)$, and $w = (-2, 3)$, and let scalars $a = -\frac{1}{2}$ and $b = \frac{5}{2}$. Verify the following properties hold:

(a) $u + v = v + u$ (commutative law);

Solution: $u + v = (1, 2) + (3, 1) = (1 + 3, 2 + 1) = (4, 3)$, whereas $v + u = (3, 1) + (1, 2) = (3 + 1, 1 + 2) = (4, 3)$ is the same. □

(b) $(u + v) + w = u + (v + w)$ (associative law);

Solution: $(u + v) + w = (4, 3) + (-2, 3) = (2, 6)$, whereas $u + (v + w) = u + ((3, 1) + (-2, 3)) = (1, 2) + (1, 4) = (2, 6)$ is the same. □

---

[6] Section 3.5 develops how to treat such inconsistent information in order to 'best solve' such impossible tasks.

(c) $u + 0 = u$;

    Solution: $u + 0 = (1,2) + (0,0) = (1 + 0, 2 + 0) = (1,2) = u.$ □

(d) $u + (-u) = 0$;

    Solution: Recall $-u = (-1)u = (-1)(1,2) = (-1,-2)$, and so $u + (-u) = (1,2) + (-1,-2) = (1 - 1, 2 - 2) = (0,0) = 0.$ □

(e) $a(u + v) = au + av$ (a distributive law);

    Solution: $a(u + v) = -\frac{1}{2}(4,3) = (-\frac{1}{2} \cdot 4, -\frac{1}{2} \cdot 3) = (-2, -\frac{3}{2})$, whereas $au + av = -\frac{1}{2}(1,2) + (-\frac{1}{2})(3,1) = (-\frac{1}{2}, -1) + (-\frac{3}{2}, -\frac{1}{2}) = (-\frac{1}{2} - \frac{3}{2}, -1 - \frac{1}{2}) = (-2, -\frac{3}{2})$ which is the same.   □

(f) $(a + b)u = au + bu$ (a distributive law);

    Solution: $(a + b)u = (-\frac{1}{2} + \frac{5}{2})(1,2) = 2(1,2) = (2 \cdot 1, 2 \cdot 2) = (2,4)$, whereas $au + bu = (-\frac{1}{2})(1,2) + \frac{5}{2}(1,2) = (-\frac{1}{2}, -1) + (\frac{5}{2}, 5) = (-\frac{1}{2} + \frac{5}{2}, -1 + 5) = (-2, 4)$ which is the same.   □

(g) $(ab)u = a(bu)$;

    Solution: $(ab)u = (-\frac{1}{2} \cdot 52)(1,2) = (-\frac{5}{4})(1,2) = (-\frac{5}{4}, -\frac{5}{2})$, whereas $a(bu) = a(\frac{5}{2}(1,2)) = (-\frac{1}{2})(\frac{5}{2}, 5) = (-\frac{5}{4}, -\frac{5}{2})$ which is the same.   □

(h) $1u = u$;

    Solution: $1u = 1(1,2) = (1 \cdot 1, 1 \cdot 2) = (1,2) = u.$ □

(i) $0u = 0$;

    Solution: $0u = 0(1,2) = (0 \cdot 1, 0 \cdot 2) = (0,0) = 0.$ □

(j) $|au| = |a| \cdot |u|$.

    Solution: Now $|a| = |-\frac{1}{2}| = \frac{1}{2}$, and the length $|u| = \sqrt{1^2 + 2^2} = \sqrt{5}$ (Definition 1.1.9). Consequently, $|au| = |(-\frac{1}{2})(1,2)| = |(-\frac{1}{2}, -1)| = \sqrt{(-\frac{1}{2})^2 + (-1)^2} = \sqrt{\frac{1}{4} + 1} = \sqrt{\frac{5}{4}} = \frac{1}{2}\sqrt{5} = |a| \cdot |u|$ as required.   □

Now let's state and prove these properties in general.

Theorem 1.2.19 *For all vectors* u, v, *and* w *with* n *components (that is, in* $\mathbb{R}^n$*), and for all scalars* a *and* b, *the following properties hold:*

(a) $u + v = v + u$ *(commutative law);*

(b) $(u + v) + w = u + (v + w)$ *(associative law);*

(c) $u + 0 = 0 + u = u$;

(d) $u + (-u) = (-u) + u = 0$;

(e) $a(u + v) = au + av$ *(a distributive law);*

(f) $(a + b)u = au + bu$ *(a distributive law);*

(g) $(ab)u = a(bu)$;

(h) $1u = u$;

(i) $0u = 0$;

(j) $|au| = |a| \cdot |u|$.

Proof. We prove 1.2.19(a), and leave the proof of other properties as exercises. The approach is to establish the properties of vector operations using the known properties of scalar operations.

Property 1.2.19(a) is the commutativity of vector addition. Example 1.2.1(a) shows graphically how the equality $u + v = v + u$ in just one case, and the diagram here shows another case. In general, let vectors $u = (u_1, u_2, \ldots, u_n)$ and $v = (v_1, v_2, \ldots, v_n)$ then

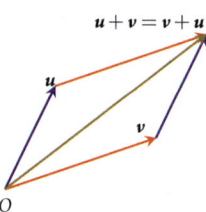

$u + v = v + u$

$u + v$

$= (u_1, u_2, \ldots, u_n) + (v_1, v_2, \ldots, v_n)$

$= (u_1 + v_1, u_2 + v_2, \ldots, u_n + v_n)$ (by Definition 1.2.4)

$= (v_1 + u_1, v_2 + u_2, \ldots, v_n + u_n)$ (commutative scalar add)

$= (v_1, v_2, \ldots, v_n) + (u_1, u_2, \ldots, u_n)$ (by Definition 1.2.4)

$= v + u$.

Example 1.2.20 Which of these two diagrams best illustrates the associative law 1.2.19(b)? Give reasons.

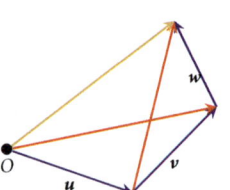
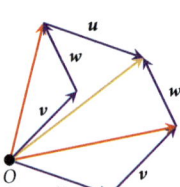

Solution: The left diagram.

- In the left diagram, the two red vectors represent $u + v$ (left) and $v + w$ (right). Thus the left-red followed by the blue w represents $(u + v) + w$, whereas the u followed by the right-red represents $u + (v + w)$. The brown vector shows that they are equal: $(u + v) + w = u + (v + w)$.

- The right-hand of the two diagrams invokes the commutative law as well. The top-left part of the diagram shows $(v + w) + u$, whereas the bottom-right part shows $(u + v) + w$. That these are equal, the brown vector, requires both the commutative and associative laws.

☐

We frequently use the algebraic properties of Theorem 1.2.19 in rearranging and solving vector equations.

Example 1.2.21 Find the vector x such that $3x - 2u = 6v$.

Solution: Using Theorem 1.2.19, all the following equations are equivalent:

$$
\begin{aligned}
3x - 2u &= 6v; \\
(3x - 2u) + 2u &= 6v + 2u \quad \text{(add } 2u \text{ to both sides)}; \\
3x + (-2u + 2u) &= 6v + 2u \quad \text{(by 1.2.19(b), associativity)}; \\
3x + 0 &= 6v + 2u \quad \text{(by 1.2.19(d))}; \\
3x &= 6v + 2u \quad \text{(by 1.2.19(c))}; \\
\tfrac{1}{3}(3x) &= \tfrac{1}{3}(6v + 2u) \quad \text{(multiply both sides by } \tfrac{1}{3}); \\
\tfrac{1}{3}(3x) &= \tfrac{1}{3}(6v) + \tfrac{1}{3}(2u) \quad \text{(by 1.2.19(e), distributivity)}; \\
(\tfrac{1}{3} \cdot 3)x &= (\tfrac{1}{3} \cdot 6)v + (\tfrac{1}{3} \cdot 2)u \quad \text{(by 1.2.19(g))}; \\
1x &= 2v + \tfrac{2}{3}u \quad \text{(by scalar operations)}; \\
x &= 2v + \tfrac{2}{3}u \quad \text{(by 1.2.19(h))}.
\end{aligned}
$$

Generally we do not write down all such details. Generally the following shorter derivation is acceptable. The following are equivalent:

$$
\begin{aligned}
3x - 2u &= 6v; \\
3x &= 6v + 2u \quad \text{(adding } 2u \text{ to both sides)}; \\
x &= 2v + \tfrac{2}{3}u \quad \text{(dividing both sides by 3)}.
\end{aligned}
$$

But exercises and examples in this section sometimes explicitly require full details and justification.

☐

1.2.5 Exercises

Exercise 1.2.1 For each of the pairs of vectors u and v shown below, draw the vectors $u + v$, $v + u$, $u - v$, $v - u$, $\tfrac{1}{2}u$, and $-v$.

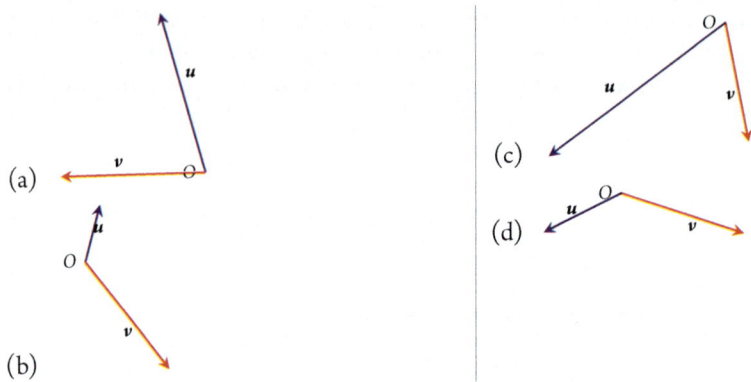

(a)

(b)

(c)

(d)

Exercise 1.2.2 For each of the following pairs of vectors shown below, use a ruler (or other measuring stick) to directly measure the distance between the pair of vectors.

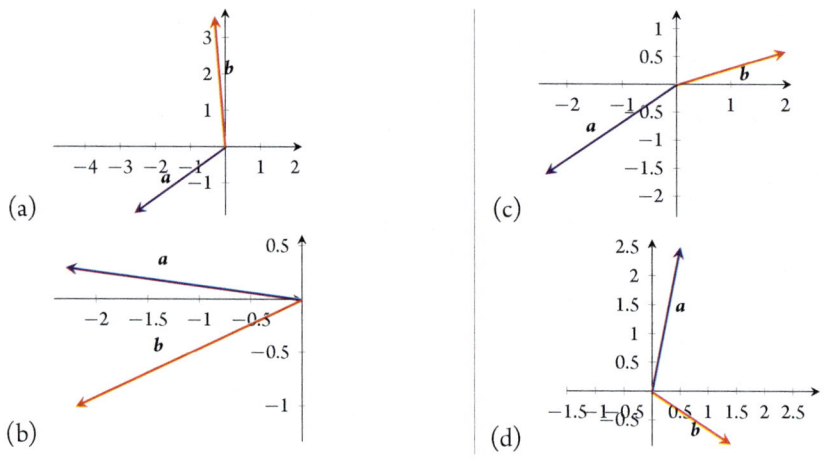

(a)

(b)

(c)

(d)

Exercise 1.2.3 For each of the following groups of vectors, use the distance between vectors to find which pair in the group are closest to each other, and which pair in the group are furthest from each other.

(a) $u = (-5, 0, 3)$, $v = (1, -6, 10)$, $w = (-4, 4, 11)$
(b) $u = 3i$, $v = 4i - 2j + 2k$, $w = 4i + 2j + 2k$
(c) $u = (-5, 3, 5, 6)$, $v = (-6, 1, 3, 10)$, $w = (-4, 6, 2, 15)$
(d) $u = 2e_1 + 4e_2 - e_3 + 5e_4$, $v = -2e_1 + 8e_2 - 6e_3 - 3e_4$, $w = -6e_3 + 11e_4$

Exercise 1.2.4 Find a parametric equation of the line through the given pairs of points.

(a) $(-11, 0, 3)$, $(-3, -2, 2)$
(b) $(-4, 1, -2)$, $(3, -5, 5)$

(c) $(2.2, 5.8, 4, 3, 2)$,
   $(-1.1, 2.2, -2.4, -3.2, 0.9)$
(d) $(1.8, -3.1, -1, -1.3, -3.3)$,
   $(-1.4, 0.8, -2.6, 3.1, -0.8)$

Exercise 1.2.5 Verify the algebraic properties of Theorem 1.2.19 for each of the following sets of vectors and scalars.

(a) $u = 2.4i - 0.3j$, $v = -1.9i + 0.5j$, $w = -3.5i - 1.8j$, $a = 0.4$, and $b = 1.4$.

(b) $u = -\frac{1}{2}j + \frac{3}{2}k$, $v = 2i - j$, $w = 2i - k$, $a = -3$, and $b = \frac{1}{2}$.

(c) $u = (2, 1, 4, -2)$, $v = (-3, -2, 0, -1)$, $w = (-6, 5, 4, 2)$, $a = -4$, and $b = 3$.

Exercise 1.2.6 Prove in detail some algebraic properties chosen from Theorem 1.2.19(b)–1.2.19(j) on vector addition and scalar multiplication.

Exercise 1.2.7 For each of the following vector equations, rearrange the equation to get vector x in terms of the other vectors. Give excruciating detail of the justification, using Theorem 1.2.19.

(a) $x + a = 0$.

(b) $2x - b = 3b$.

(c) $3(x + a) = x + (a - 2x)$.

(d) $-4b = x + 3(a - x)$.

Exercise 1.2.8 In a few sentences, answer/discuss each of the following.

(a) What empowers us to write every vector in terms of the standard unit vectors?

(b) We use the distance $|u - v|$ to measure how close the two vectors are to each other. Invent an alternative way to measure closeness of two vectors, and comment on why your invented alternative measures closeness.

(c) What is it about the parametric equation of a line that means it does indeed describe a line in space?

(d) Comment on why many of the properties of vector operations (Theorem 1.2.19) appear the same as those for operations with real numbers.

1.3 The dot product determines angles and lengths

The previous Section 1.2 discussed how to add, subtract, and stretch vectors. Question: can we multiply two vectors? The answer is that 'vector multiplication' has major differences to the multiplication of scalar numbers. There are at least four ways of multiplying vectors together: each way useful in appropriate circumstances. Often the angle between vectors is denoted by the Greek letter theta, θ. This section introduces one such multiplication, the so-called dot product of two vectors that, among other attributes, gives a valuable way to determine the angle between two vectors.

Example 1.3.1 Consider the two vectors $\boldsymbol{u} = (7, -1)$ and $\boldsymbol{v} = (2, 5)$ plotted first to the right. What is the angle θ between the two vectors?

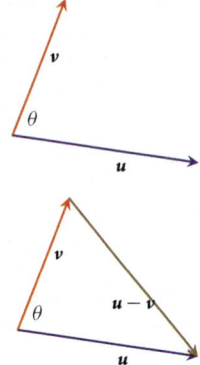

Solution: Form a triangle with the vector $\boldsymbol{u} - \boldsymbol{v} = (5, -6)$ going from the tip of $\boldsymbol{v}$ to the tip of $\boldsymbol{u}$, as plotted second. The sides of the triangles are of length $|\boldsymbol{u}| = \sqrt{7^2 + (-1)^2} = \sqrt{50} = 5\sqrt{2}$, $|\boldsymbol{v}| = \sqrt{2^2 + 5^2} = \sqrt{29}$, and $|\boldsymbol{u} - \boldsymbol{v}| = \sqrt{5^2 + (-6)^2} = \sqrt{61}$. By the cosine rule for triangles

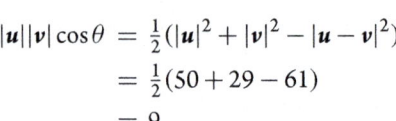

$$|\boldsymbol{u} - \boldsymbol{v}|^2 = |\boldsymbol{u}|^2 + |\boldsymbol{v}|^2 - 2|\boldsymbol{u}||\boldsymbol{v}| \cos\theta.$$

Here this rule rearranges to

$$
\begin{aligned}
|\boldsymbol{u}||\boldsymbol{v}| \cos\theta &= \tfrac{1}{2}(|\boldsymbol{u}|^2 + |\boldsymbol{v}|^2 - |\boldsymbol{u} - \boldsymbol{v}|^2) \\
&= \tfrac{1}{2}(50 + 29 - 61) \\
&= 9.
\end{aligned}
$$

(Recall that multiplication by $180/\pi$ converts an angle from radians to degrees: $1.3322 \cdot 180/\pi = 76.33°$.) Dividing by the product of the lengths then gives $\cos\theta = 9/(5\sqrt{58}) = 0.2364$, so the angle $\theta = \arccos(0.2364) = 1.3322 = 76.33°$ as is reasonable from the plots. $\square$

The interest in this Example 1.3.1 is the number 9 on the right-hand side of $|\boldsymbol{u}||\boldsymbol{v}| \cos\theta = 9$. The reason is that 9 just happens to be $14 - 5$, which in turn just happens to be $7 \cdot 2 + (-1) \cdot 5$, and it is no coincidence that this expression is the same as $u_1 v_1 + u_2 v_2$ in terms of vector components $\boldsymbol{u} = (u_1, u_2) = (7, -1)$ and $\boldsymbol{v} = (v_1, v_2) = (2, 5)$. Repeat this example for many pairs of vectors $\boldsymbol{u}$ and $\boldsymbol{v}$ to find that always $|\boldsymbol{u}||\boldsymbol{v}| \cos\theta = u_1 v_1 + u_2 v_2$ (Exercise 1.3.1). This equality suggests that the sum of products of corresponding components of $\boldsymbol{u}$ and $\boldsymbol{v}$ is closely connected to the angle between the vectors.

Definition 1.3.2 *For every two vectors in $\mathbb{R}^n$, $\boldsymbol{u} = (u_1, u_2, \ldots, u_n)$ and $\boldsymbol{v} = (v_1, v_2, \ldots, v_n)$, define the **dot product** (or **inner product**), denoted by a dot between the two vectors, as the scalar*

$$\boldsymbol{u} \cdot \boldsymbol{v} := u_1 v_1 + u_2 v_2 + \cdots + u_n v_n.$$

The dot product of two vectors gives a scalar result, a number, not a vector result.

When writing the vector dot product, the dot between the two vectors is essential. We sometimes also denote the scalar product by such a dot (to clarify a product) and sometimes omit the dot between the scalars, for example $a \cdot b = ab$ for scalars. But for the vector dot product, the dot must not be omitted: '$\boldsymbol{uv}$' is meaningless.

Example 1.3.3 Compute the dot product between the following pairs of vectors.

(a) $u = (-2,5,-2)$, $v = (3,3,-2)$

Solution: $u \cdot v = (-2)3 + 5 \cdot 3 + (-2)(-2) = 13$. Alternatively, $v \cdot u = 3(-2) + 3 \cdot 5 + (-2)(-2) = 13$. That these give the same result is a consequence of a general commutative law, Theorem 1.3.13(a), and so in the following we compute the dot product only one way around. □

(b) $u = (1,-3,0)$, $v = (1,2)$

Solution: There is no answer: a dot product cannot be computed here as the two vectors are of different sizes. □

(c) $a = (-7,3,0,2,2)$, $b = (-3,4,-4,2,0)$

Solution: $a \cdot b = (-7)(-3) + 3 \cdot 4 + 0(-4) + 2 \cdot 2 + 2 \cdot 0 = 37$. □

(d) $p = (-0.1,-2.5,-3.3,0.2)$, $q = (-1.6,1.1,-3.4,2.2)$

Solution: $p \cdot q = (-0.1)(-1.6) + (-2.5)1.1 + (-3.3)(-3.4) + 0.2 \cdot 2.2 = 9.07$. □

Activity 1.3.4 What is the dot product of the two vectors $u = 2i - j$ and $v = 3i + 4j$?

(a) 2   (b) 10   (c) 5   (d) 8

Theorem 1.3.5 *For every two nonzero vectors u and v in $\mathbb{R}^n$, the **angle** θ between the vectors is determined by*

$$\cos\theta = \frac{u \cdot v}{|u||v|}, \qquad 0 \le \theta \le \pi \quad (0 \le \theta \le 180°).$$

This picture illustrates the range of angles between two vectors: when they point in the same direction the angle is zero; when they are at right-angles to each other the angle is $\pi/2$, or equivalently 90°; when they point in opposite directions the angle is π, or equivalently 180°. Let's prove this theorem after some examples.

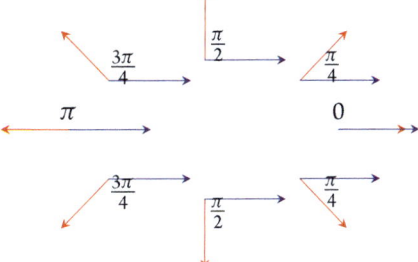

Example 1.3.6 Determine the angle between the following pairs of vectors.

(a) $(4,3)$ and $(5,12)$

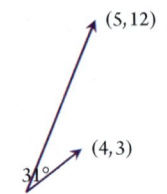

Solution: These vectors (shown to the right) have length $\sqrt{4^2 + 3^2} = \sqrt{25} = 5$ and $\sqrt{5^2 + 12^2} = \sqrt{169} = 13$, respectively. Their dot product $(4, 3) \cdot (5, 12) = 20 + 36 = 56$. Hence $\cos\theta = 56/(5 \cdot 13) = 0.8615$ and so angle $\theta = \arccos(0.8615) = 0.5325 = 30.51°$. ☐

(b) $(3,1)$ and $(-2,1)$

Solution: These vectors (shown to the right) have length $\sqrt{3^2 + 1^2} = \sqrt{10}$ and $\sqrt{(-2)^2 + 1^2} = \sqrt{5}$, respectively. Their dot product $(3, 1) \cdot (-2, 1) = -6 + 1 = -5$. Hence $\cos\theta = -5/(\sqrt{10} \cdot \sqrt{5}) = -1/\sqrt{2} = -0.7071$ and so angle $\theta = \arccos(-1/\sqrt{2}) = 2.3562 = \frac{3}{4}\pi = 135°$ (Table 1.1). ☐

(c) $(4,-2)$ and $(-1,-2)$

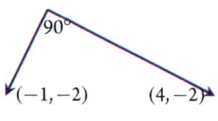

Solution: These vectors (shown to the right) have length $\sqrt{4^2 + (-2)^2} = \sqrt{20} = 2\sqrt{5}$ and $\sqrt{(-1)^2 + (-2)^2} = \sqrt{5}$, respectively. Their dot product $(4, -2) \cdot (-1, -2) = -4 + 4 = 0$. Hence $\cos\theta = 0/(2\sqrt{5} \cdot \sqrt{5}) = 0$ and so angle $\theta = \frac{1}{2}\pi = 90°$ (Table 1.1). ☐

Activity 1.3.7 What is the angle between the two vectors $(1, \sqrt{3})$ and $(\sqrt{3}, 1)$?

(a) $60°$ (b) $64.34°$ (c) $77.50°$ (d) $30°$

Example 1.3.8 In chemistry, one computes the angles between bonds in molecules and crystals. In engineering, one needs the angles between beams and struts in complex structures. The dot product determines such angles.

(a) Consider the cube drawn in stereo below, and compute the angle between the diagonals on two adjacent faces.

Table 1.1 When a cosine is one of these tabulated special values, then we know the corresponding angle exactly. In other cases, we usually use a calculator ($\texttt{arccos}$ or $\cos^{-1}$) or computer ($\texttt{acos()}$) to compute the angle numerically.

θ	θ	$\cos\theta$	$\cos\theta$
0	0°	1	1.
$\pi/6$	30°	$\sqrt{3}/2$	0.8660
$\pi/4$	45°	$1/\sqrt{2}$	0.7071
$\pi/3$	60°	$1/2$	0.5
$\pi/2$	90°	0	0.
$2\pi/3$	120°	$-1/2$	-0.5
$3\pi/4$	135°	$-1/\sqrt{2}$	-0.7071
$5\pi/6$	150°	$-\sqrt{3}/2$	-0.8660
π	180°	-1	$-1.$

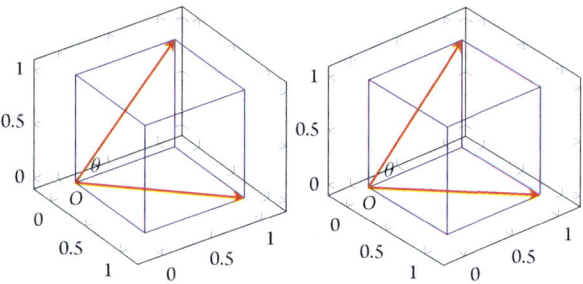

Solution: Draw two vectors along adjacent diagonals: the above pair of vectors are $(1,1,0)$ and $(0,1,1)$. They both have the same length as $|(1,1,0)| = \sqrt{1^2 + 1^2 + 0^2} = \sqrt{2}$ and $|(0,1,1)| = \sqrt{0^2 + 1^2 + 1^2} = \sqrt{2}$. The dot product is $(1,1,0) \cdot (0,1,1) = 0 + 1 + 0 = 1$. Hence the cosine $\cos\theta = 1/(\sqrt{2} \cdot \sqrt{2}) = 1/2$. Table 1.1 gives the angle $\theta = \frac{\pi}{3} = 60°$. □

(b) A body-centred cubic lattice (such as that formed by caesium chloride crystals) has one lattice point in the centre of the unit cell as well as the eight corner points. Consider the body-centred cube of atoms drawn in stereo below with the centre of the cube at the origin: what is the angle between the centre atom and any two adjacent corner atoms?

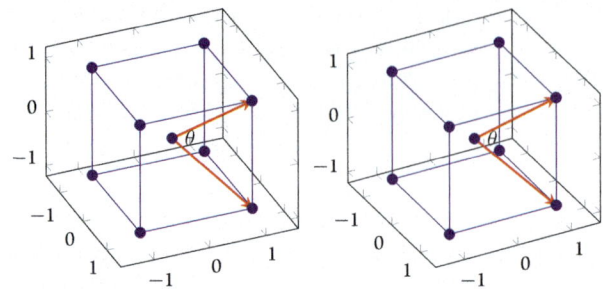

Solution: Draw two corresponding vectors from the centre atom: the above pair of vectors are $(1,1,1)$ and $(1,1,-1)$. These have the same length $|(1,1,1)| = \sqrt{1^2 + 1^2 + 1^2} = \sqrt{3}$ and $|(1,1,-1)| = \sqrt{1^2 + 1^2 + (-1)^2} = \sqrt{3}$. The dot product is $(1,1,1) \cdot (1,1,-1) = 1 + 1 - 1 = 1$. Hence $\cos\theta = 1/(\sqrt{3} \cdot \sqrt{3}) = 1/3 = 0.3333$. Then a calculator (or MATLAB/Octave, see Section 1.5) gives the angle $\theta = \arccos(1/3) = 1.2310 = 70.53°$.

$\square$

Example 1.3.9 *(semantic similarity)* Recall that Example 1.1.7 introduced the encoding of sentences and documents as word count vectors. In the example, a word vector has five components, $(N_{cat}, N_{dog}, N_{mat}, N_{sat}, N_{scratched})$ where the various N are the counts of each word in any sentence or document. For example,

(a) "The dog sat on the mat" has word vector $a = (0,1,1,1,0)$.

(b) "The cat scratched the dog" has word vector $b = (1,1,0,0,1)$.

(c) "The cat and dog sat on the mat" has word vector $c = (1,1,1,1,0)$.

Use the angle between these three word vectors to characterize the similarity of the sentences: a small angle means the sentences are somehow close; a large angle means the sentences are disparate.

Solution: First, these word vectors have lengths $|a| = |b| = \sqrt{3}$ and $|c| = \sqrt{4} = 2$. Second, the 'angles' between these sentences are the following.

• The angle θ_{ab} between "The dog sat on the mat" and "The cat scratched the dog" satisfies

$$\cos\theta_{ab} = \frac{a \cdot b}{|a||b|} = \frac{0+1+0+0+0}{\sqrt{3} \cdot \sqrt{3}} = \frac{1}{3}.$$

A calculator (or MATLAB/Octave, see Section 1.5) then gives the angle $\theta_{ab} = \arccos(1/3) = 1.2310 = 70.53°$ so the sentences are quite dissimilar.

• The angle θ_{ac} between "The dog sat on the mat" and "The cat and dog sat on the mat" satisfies

$$\cos\theta_{ac} = \frac{a \cdot c}{|a||c|} = \frac{0+1+1+1+0}{\sqrt{3} \cdot 2} = \frac{3}{2\sqrt{3}} = \frac{\sqrt{3}}{2}.$$

Table 1.1 gives the angle $\theta_{ac} = \frac{\pi}{6} = 30°$ so the sentences are roughly similar.

- The angle θ_{bc} between "The cat scratched the dog" and "The cat and dog sat on the mat" satisfies

$$\cos\theta_{bc} = \frac{\boldsymbol{b}\cdot\boldsymbol{c}}{|\boldsymbol{b}||\boldsymbol{c}|} = \frac{1+1+0+0+0}{\sqrt{3}\cdot 2} = \frac{2}{2\sqrt{3}} = \frac{1}{\sqrt{3}}.$$

A calculator (or MATLAB/Octave, see Section 1.5) then gives the angle $\theta_{bc} = \arccos(1/\sqrt{3}) = 0.9553 = 54.74°$ so the sentences are moderately dissimilar.

This stereo plot schematically draws these three vectors at the correct angles from each other, and with correct lengths, in some abstract coordinate system (Section 3.4 gives the techniques to do such plots systematically).

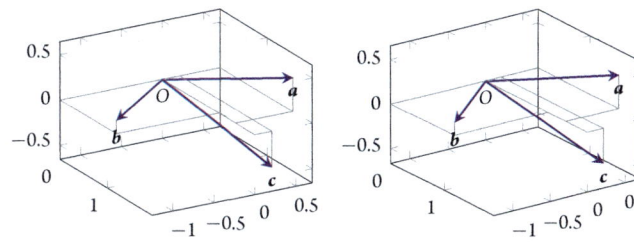

Proof. To prove the Angle Theorem 1.3.5, form a triangle from vectors $\boldsymbol{u}$, $\boldsymbol{v}$, and $\boldsymbol{u} - \boldsymbol{v}$ as illustrated to the right. Recall and apply the cosine rule for triangles

$$|\boldsymbol{u} - \boldsymbol{v}|^2 = |\boldsymbol{u}|^2 + |\boldsymbol{v}|^2 - 2|\boldsymbol{u}||\boldsymbol{v}|\cos\theta.$$

In $\mathbb{R}^n$ this rule rearranges to

$$\begin{aligned}
2|\boldsymbol{u}||\boldsymbol{v}|\cos\theta &= |\boldsymbol{u}|^2 + |\boldsymbol{v}|^2 - |\boldsymbol{u} - \boldsymbol{v}|^2 \\
&= u_1^2 + u_2^2 + \cdots + u_n^2 + v_1^2 + v_2^2 + \cdots + v_n^2 \\
&\quad - (u_1 - v_1)^2 - (u_2 - v_2)^2 - \cdots - (u_n - v_n)^2 \\
&= u_1^2 + u_2^2 + \cdots + u_n^2 + v_1^2 + v_2^2 + \cdots + v_n^2 \\
&\quad - u_1^2 + 2u_1 v_1 - v_1^2 - u_2^2 + 2u_2 v_2 - v_2^2 \\
&\quad - \cdots - u_n^2 + 2u_n v_n - v_n^2 \\
&= 2u_1 v_1 + 2u_2 v_2 + \cdots + 2u_n v_n \\
&= 2(u_1 v_1 + u_2 v_2 + \cdots + u_n v_n) \\
&= 2\boldsymbol{u}\cdot\boldsymbol{v}.
\end{aligned}$$

Dividing both sides by $2|\boldsymbol{u}||\boldsymbol{v}|$ gives $\cos\theta = \frac{\boldsymbol{u}\cdot\boldsymbol{v}}{|\boldsymbol{u}||\boldsymbol{v}|}$ as required.

1.3.1 Work done involves the dot product

In physics and engineering, "work" has a precise meaning related to energy: when a force of magnitude F acts on a body and that body moves a distance d, then the work done by the force is $W = Fd$. However, this formula applies only for one-dimensional force and displacement, the case when the force and the displacement are in the same direction. For example, if a 5 kg barbell drops downwards 2 m under the force of gravity (9.8 newtons/kg), then the work done by gravity on the barbell during the drop is the product

$$W = F \cdot d = (5 \cdot 9.8) \cdot 2 = 98 \text{ joules.}$$

This work done goes to the kinetic energy of the falling barbell. The kinetic energy dissipates when the barbell hits the floor.

In general, the applied force and the displacement are not in the same direction (as illustrated to the right). Consider the general case when a vector force F acts on a body which moves a displacement vector d. Then the work done by the force on the body is the length of the displacement times the component of the force in the direction of the displacement—the component of the force at right-angles to the displacement does no work.

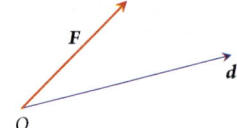

As illustrated to the right, draw a right-angled triangle to decompose the force F into the component F_0 in the direction of the displacement, and an unnamed component at right-angles. Then by the scalar formula, the work done is $W = F_0|d|$. As drawn, the force F makes an angle θ to the displacement d: the dot product determines

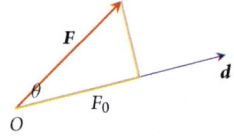

this angle via $\cos\theta = (F \cdot d)/(|F||d|)$ (Theorem 1.3.5). By basic trigonometry, the adjacent side of the force triangle has length $F_0 = |F|\cos\theta = |F|\frac{F \cdot d}{|F||d|} = \frac{F \cdot d}{|d|}$. Finally, the work done $W = F_0|d| = \frac{F \cdot d}{|d|}|d| = F \cdot d$: that is, the work done is the dot product of the vector force and vector displacement.

Example 1.3.10 A sailing boat travels a distance of 40 m east and 10 m north, as drawn to the right. The wind from abeam, of strength and direction $(1, -4)$ m/s, generates a force $F = (20, -10)$ (newtons) on the sail, as drawn. What is the work done by the wind?

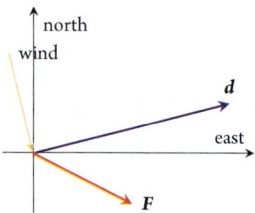

Solution: The direction of the wind is immaterial except for the force it generates. The displacement vector $d = (40, 10)$ m. Then the work done is $W = F \cdot d = (40, 10) \cdot (20, -10) = 800 - 100 = 700$ joules. ☐

Activity 1.3.11 Recall the force of gravity on an object is the mass of the object times the acceleration of gravity, $9.8 \, \text{m/s}^2$. A 3 kg ball is thrown horizontally from a height of 2 m and lands 10 m away on the ground: what is the total work done by gravity on the ball?

(a) 98 joules   (b) 29.4 joules   (c) 19.6 joules   (d) 58.8 joules

Finding components of vectors in various directions is called projection. Such projection is surprisingly common in applications and is developed much further by Section 3.5.3.

1.3.2 Algebraic properties of the dot product

To manipulate the dot product in algebraic expressions, we need to know its basic algebraic rules. The following rules of Theorem 1.3.13 are analogous to well known rules for scalar multiplication.

Example 1.3.12 Given vectors $u = (-2, 5, -2)$, $v = (3, 3, -2)$ and $w = (2, 0, -5)$, and scalar $a = 2$, verify that (Theorems 1.3.13(c) and 1.3.13(d))

- $a(u \cdot v) = (au) \cdot v = u \cdot (av)$ (a form of associativity);

- $(u + v) \cdot w = u \cdot w + v \cdot w$ (distributivity).

Solution:

- First, the following three evaluate to being equal:

$$a(u \cdot v) = 2\big((-2, 5, -2) \cdot (3, 3, -2)\big) = 2\big((-2)3 + 5 \cdot 3 + (-2)(-2)\big) = 2 \cdot 13 = 26;$$
$$(au) \cdot v = (-4, 10, -4) \cdot (3, 3, -2) = (-4)3 + 10 \cdot 3 + (-4)(-2) = 26;$$
$$u \cdot (av) = (-2, 5, -2) \cdot (6, 6, -4) = (-2)6 + 5 \cdot 6 + (-2)(-4) = 26.$$

- Second, the following two evaluate to being equal:

$$(u + v) \cdot w = (1, 8, -4) \cdot (2, 0, -5) = 1 \cdot 2 + 8 \cdot 0 + (-4)(-5) = 22;$$
$$u \cdot w + v \cdot w = (-2, 5, -2) \cdot (2, 0, -5) + (3, 3, -2) \cdot (2, 0, -5)$$
$$= [(-2)2 + 5 \cdot 0 + (-2)(-5)] + [3 \cdot 2 + 3 \cdot 0 + (-2)(-5)]$$
$$= 6 + 16 = 22. \qquad \square$$

Theorem 1.3.13 (*dot properties*) *For every three vectors* u, v, *and* w *in* $\mathbb{R}^n$, *and for every scalar* a, *the following properties hold:*

(a) $u \cdot v = v \cdot u$ (*commutative law*);

(b) $u \cdot 0 = 0 \cdot u = 0$;

(c) $a(u \cdot v) = (au) \cdot v = u \cdot (av)$;

(d) $(u + v) \cdot w = u \cdot w + v \cdot w$ (*distributive law*);

(e) $u \cdot u \geq 0$, *and moreover,* $u \cdot u = 0$ *if and only if* $u = 0$.

Proof. Here prove only the commutative law 1.3.13(a) and the inequality 1.3.13(e). Exercise 1.3.5 asks you to analogously prove the other properties. At the core of each proof is the definition of the dot product which empowers us to deduce a property via the corresponding property for scalars.

- To prove the commutative law 1.3.13(a) consider

$$\begin{aligned} \mathbf{u} \cdot \mathbf{v} &= u_1 v_1 + u_2 v_2 + \cdots + u_n v_n \quad \text{(by Definition 1.3.2)} \\ &= v_1 u_1 + v_2 u_2 + \cdots + v_n u_n \quad \text{(as each scalar multiplication commutes)} \\ &= \mathbf{v} \cdot \mathbf{u} \quad \text{(by Definition 1.3.2).} \end{aligned}$$

- To prove the inequality 1.3.13(e) consider

$$\begin{aligned} \mathbf{u} \cdot \mathbf{u} &= u_1 u_1 + u_2 u_2 + \cdots + u_n u_n \quad \text{(by Definition 1.3.2)} \\ &= u_1^2 + u_2^2 + \cdots + u_n^2 \\ &\geq 0 + 0 + \cdots + 0 \quad \text{(as each scalar term is } \geq 0) \\ &= 0. \end{aligned}$$

To prove the "moreover" part, first consider the zero vector. From Definition 1.3.2, in $\mathbb{R}^n$,

$$\mathbf{0} \cdot \mathbf{0} = 0^2 + 0^2 + \cdots + 0^2 = 0.$$

Second, suppose vector $\mathbf{u}$ in $\mathbb{R}^n$ satisfies $\mathbf{u} \cdot \mathbf{u} = 0$. Expanding the left-hand side in the components, $\mathbf{u} = (u_1, u_2, \ldots, u_n)$, gives that

$$\underbrace{u_1^2}_{\geq 0} + \underbrace{u_2^2}_{\geq 0} + \cdots + \underbrace{u_n^2}_{\geq 0} = 0.$$

Being squares, all terms on the left are non-negative, so the only way they can all add to zero is if they are all zero. That is, $u_1 = u_2 = \cdots = u_n = 0$. Hence, the vector $\mathbf{u}$ must be the zero vector $\mathbf{0}$.

Activity 1.3.14 For vectors $\mathbf{u}, \mathbf{v}, \mathbf{w}$ in $\mathbb{R}^n$, which of the following statements is not generally true?

(a) $\mathbf{u} \cdot \mathbf{v} - \mathbf{v} \cdot \mathbf{u} = 0$

(c) $\mathbf{u} \cdot (\mathbf{v} + \mathbf{w}) = \mathbf{u} \cdot \mathbf{v} + \mathbf{u} \cdot \mathbf{w}$

(b) $(\mathbf{u} - \mathbf{v}) \cdot (\mathbf{u} + \mathbf{v}) = \mathbf{u} \cdot \mathbf{u} - \mathbf{v} \cdot \mathbf{v}$

(d) $(2\mathbf{u}) \cdot (2\mathbf{v}) = 2(\mathbf{u} \cdot \mathbf{v})$

The above proof of Theorem 1.3.13(e), that $\mathbf{u} \cdot \mathbf{u} = 0$ if and only if $\mathbf{u} = \mathbf{0}$, may look uncannily familiar. The reason is that this last part is essentially the same as the proof of Theorem 1.1.13 that the zero vector is the only vector of length zero. The upcoming Theorem 1.3.17 establishes that this connection between dot products and lengths is no coincidence.

Example 1.3.15 For the two vectors $\mathbf{u} = (3, 4)$ and $\mathbf{v} = (2, 1)$ verify the following three properties:

(a) $\sqrt{u \cdot u} = |u|$, the length of u;

(b) $|u \cdot v| \leq |u||v|$ (Cauchy–Schwarz inequality);

(c) $|u + v| \leq |u| + |v|$ (triangle inequality).

Solution:

(a) Here $\sqrt{u \cdot u} = \sqrt{3 \cdot 3 + 4 \cdot 4} = \sqrt{25} = 5$, whereas the length $|u| = \sqrt{3^2 + 4^2} = \sqrt{25} = 5$ (Definition 1.1.9). These expressions are equal.

(b) Here $|u \cdot v| = |3 \cdot 2 + 4 \cdot 1| = 10$, whereas $|u||v| = 5\sqrt{2^2 + 1^2} = 5\sqrt{5} = 11.180$. Hence $|u \cdot v| = 10 \leq 11.180 = |u||v|$.

(c) Here $|u + v| = |(5, 5)| = \sqrt{5^2 + 5^2} = \sqrt{50} = 7.071$, whereas $|u| + |v| = 5 + \sqrt{5} = 7.236$. Hence $|u + v| = 7.071 \leq 7.236 = |u| + |v|$. This is called the triangle inequality because the vectors u, v, and $u + v$ may be viewed as forming a triangle, as illustrated to the right, and this inequality follows because the length of a side of a triangle must be less than the sum of the lengths of the other two sides.

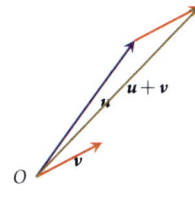

□

The Cauchy–Schwarz inequality is one point of distinction between this 'vector multiplication' and scalar multiplication: for scalars $|ab| = |a||b|$, whereas the dot product of vectors is typically less, $|u \cdot v| \leq |u||v|$.

Example 1.3.16 The general proof of the Cauchy–Schwarz inequality involves a trick, so let's introduce the trick using the vectors of Example 1.3.15. Let vectors $u = (3, 4)$ and $v = (2, 1)$. Then consider the line given parametrically (Definition 1.2.15) as the position vectors $x = u + tv = (3 + 2t, 4 + t)$ for scalar parameter t—illustrated to the right. The position vector x of any point on the line has length ℓ (Definition 1.1.9) where

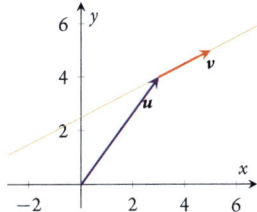

$$\ell^2 = (3 + 2t)^2 + (4 + t)^2$$
$$= 9 + 12t + 4t^2 + 16 + 8t + t^2$$
$$= \underbrace{25}_{c} + \underbrace{20}_{b} t + \underbrace{5}_{a} t^2,$$

a quadratic polynomial in t. We know that the length $\ell > 0$ (the line does not pass through the origin so no x is zero). Hence the quadratic in t cannot have any zeros. By the known properties of quadratic equations it follows that the discriminant $b^2 - 4ac < 0$. Indeed it is: here $b^2 - 4ac = 20^2 - 4 \cdot 5 \cdot 25 = 400 - 500 = -100 < 0$. Usefully, here $a = 5 = |v|^2$, $c = 25 = |u|^2$, and $b = 20 = 2 \cdot 10 = 2(u \cdot v)$. So $b^2 - 4ac < 0$, rewritten as $\frac{1}{4}b^2 < ac$, becomes the statement that $\frac{1}{4}[2(u \cdot v)]^2 = (u \cdot v)^2 < |v|^2|u|^2$. Taking the square-root of both sides verifies the Cauchy–Schwarz inequality. The proof of the next theorem establishes it in general.

Theorem 1.3.17 *For all vectors $\boldsymbol{u}$ and $\boldsymbol{v}$ in $\mathbb{R}^n$ the following properties hold:*

(a) $\sqrt{\boldsymbol{u} \cdot \boldsymbol{u}} = |\boldsymbol{u}|$, *the length of $\boldsymbol{u}$;*

(b) $|\boldsymbol{u} \cdot \boldsymbol{v}| \leq |\boldsymbol{u}||\boldsymbol{v}|$ (*Cauchy–Schwarz inequality*);

(c) $|\boldsymbol{u} \pm \boldsymbol{v}| \leq |\boldsymbol{u}| + |\boldsymbol{v}|$ (*triangle inequality*).

Proof. Except for the first, each property depends upon the previous.
1.3.17(a) Expand and rearrange

$$
\begin{aligned}
\sqrt{\boldsymbol{u} \cdot \boldsymbol{u}} &= \sqrt{u_1 u_1 + u_2 u_2 + \cdots + u_n u_n} \quad \text{(by Definition 1.3.2)} \\
&= \sqrt{u_1^2 + u_2^2 + \cdots + u_n^2} = |\boldsymbol{u}| \quad \text{(by Definition 1.1.9)}.
\end{aligned}
$$

1.3.17(b) To prove the Cauchy–Schwarz inequality between vectors $\boldsymbol{u}$ and $\boldsymbol{v}$ first consider the trivial case when $\boldsymbol{v} = \boldsymbol{0}$: then the left-hand side $|\boldsymbol{u} \cdot \boldsymbol{v}| = |\boldsymbol{u} \cdot \boldsymbol{0}| = |0| = 0$; whereas the right-hand side $|\boldsymbol{u}||\boldsymbol{v}| = |\boldsymbol{u}||\boldsymbol{0}| = |\boldsymbol{u}|0 = 0$; and so the inequality $|\boldsymbol{u} \cdot \boldsymbol{v}| \leq |\boldsymbol{u}||\boldsymbol{v}|$ is satisfied in this case.

Second, for the case when $\boldsymbol{v} \neq \boldsymbol{0}$, consider the line given parametrically by $\boldsymbol{x} = \boldsymbol{u} + t\boldsymbol{v}$ for (real) scalar parameter t (Definition 1.2.15), as illustrated to the right. The distance ℓ of a point on the line from the origin is the length of its position vector, and by property 1.3.17(a)

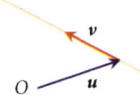

$$
\begin{aligned}
\ell^2 &= \boldsymbol{x} \cdot \boldsymbol{x} \\
&= (\boldsymbol{u} + t\boldsymbol{v}) \cdot (\boldsymbol{u} + t\boldsymbol{v}) \\
&= \boldsymbol{u} \cdot (\boldsymbol{u} + t\boldsymbol{v}) + (t\boldsymbol{v}) \cdot (\boldsymbol{u} + t\boldsymbol{v}) \quad \text{(using distributivity 1.3.13(d))} \\
&= \boldsymbol{u} \cdot \boldsymbol{u} + \boldsymbol{u} \cdot (t\boldsymbol{v}) + (t\boldsymbol{v}) \cdot \boldsymbol{u} + (t\boldsymbol{v}) \cdot (t\boldsymbol{v}) \quad \text{(again using distributivity 1.3.13(d))} \\
&= \boldsymbol{u} \cdot \boldsymbol{u} + t(\boldsymbol{u} \cdot \boldsymbol{v}) + t(\boldsymbol{v} \cdot \boldsymbol{u}) + t^2(\boldsymbol{v} \cdot \boldsymbol{v}) \quad \text{(using scalar mult. property 1.3.13(c))} \\
&= |\boldsymbol{u}|^2 + 2(\boldsymbol{u} \cdot \boldsymbol{v})t + |\boldsymbol{v}|^2 t^2 \quad \text{(using 1.3.17(a) and commutativity 1.3.13(a))} \\
&= at^2 + bt + c,
\end{aligned}
$$

a quadratic in t, with coefficients $a = |\boldsymbol{v}|^2 > 0$, $b = 2(\boldsymbol{u} \cdot \boldsymbol{v})$, and $c = |\boldsymbol{u}|^2$. Since $\ell^2 \geq 0$ (it may be zero if the line goes through the origin), then this quadratic in t has either no zeros or just one zero. By the properties of quadratic equations, the discriminant $b^2 - 4ac \leq 0$, that is, $\frac{1}{4}b^2 \leq ac$. Substituting the particular coefficients here gives $\frac{1}{4}[2(\boldsymbol{u} \cdot \boldsymbol{v})]^2 = (\boldsymbol{u} \cdot \boldsymbol{v})^2 \leq |\boldsymbol{v}|^2|\boldsymbol{u}|^2$. Taking the square-root of both sides then establishes the Cauchy–Schwarz inequality $|\boldsymbol{u} \cdot \boldsymbol{v}| \leq |\boldsymbol{u}||\boldsymbol{v}|$.

1.3.17(c) To prove the triangle inequality between vectors $\boldsymbol{u}$ and $\boldsymbol{v}$ first observe the Cauchy–Schwarz inequality implies $(\boldsymbol{u} \cdot \boldsymbol{v}) \leq |\boldsymbol{u}||\boldsymbol{v}|$ (since the left-hand side has magnitude $\leq$ the right-hand side). Then consider (analogous to the $t = 1$ case of the above)

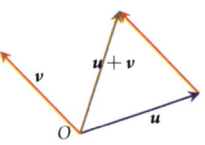

$$|u+v|^2 = (u+v) \cdot (u+v)$$
$$= u \cdot (u+v) + v \cdot (u+v) \quad \text{(using distributivity 1.3.13(d))}$$
$$= u \cdot u + u \cdot v + v \cdot u + v \cdot v \quad \text{(again using distributivity 1.3.13(d))}$$
$$= |u|^2 + 2(u \cdot v) + |v|^2 \quad \text{(using 1.3.17(a) and commutativity 1.3.13(a))}$$
$$\leq |u|^2 + 2|u||v| + |v|^2 \quad \text{(using Cauchy–Schwarz inequality)}$$
$$= (|u| + |v|)^2.$$

Take the square-root of both sides to establish the triangle inequality $|u+v| \leq |u| + |v|$. The minus case follows (illustrated) because $|u - v| = |u + (-v)| \leq |u| + |-v| = |u| + |v|$.

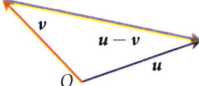

Example 1.3.18 Verify the Cauchy–Schwarz inequality and the triangle inequality ($+$ case) for the vectors $a = (-1, -2, 1, 3, -2)$ and $b = (-3, -2, 10, 2, 2)$.

Solution: We need the length of the vectors, and also the dot product:

$$|a| = \sqrt{(-1)^2 + (-2)^2 + 1^2 + 3^2 + (-2)^2} = \sqrt{19} = 4.3589,$$
$$|b| = \sqrt{(-3)^2 + (-2)^2 + 10^2 + 2^2 + 2^2} = \sqrt{121} = 11,$$
$$a \cdot b = (-1)(-3) + (-2)(-2) + 1 \cdot 10 + 3 \cdot 2 + (-2)2 = 19.$$

Hence $|a \cdot b| = 19 < 47.948 = |a||b|$, which verifies the Cauchy–Schwarz inequality.

Now, the length of the vector sum

$$|a + b| = |(-4, -4, 11, 5, 0)|$$
$$= \sqrt{(-4)^2 + (-4)^2 + 11^2 + 5^2 + 0^2}$$
$$= \sqrt{178} = 13.342.$$

Here $|a + b| = 13.342$ whereas $|a| + |b| = 11 + \sqrt{19} = 15.359$. Hence these indeed satisfy the triangle inequality $|a + b| \leq |a| + |b|$. $\square$

1.3.3 Orthogonal vectors are at right-angles

Of all the angles that vectors can make with each other, the two most important angles are when the vectors are aligned with each other, and when the vectors are at right-angles to each other. Recall Theorem 1.3.5 gives the angle θ between two vectors via $\cos\theta = \frac{u \cdot v}{|u||v|}$. For vectors at right-angles $\theta = 90°$, so $\cos\theta = 0$, and hence nonzero vectors are at right-angles only when the dot product $u \cdot v = 0$. We give a special name to vectors at right-angles.

Definition 1.3.19 *Two vectors* u *and* v *in* $\mathbb{R}^n$ *are termed* **orthogonal** *(or* **perpendicular***) if and only if their dot product* $u \cdot v = 0$.[7]

By convention the zero vector $\mathbf{0}$ is orthogonal to all other vectors. However, in practice, we almost always use the notion of orthogonality only in connection with *nonzero* vectors. Often the requirement that the orthogonal vectors are nonzero is explicitly made, but beware that sometimes the requirement may be implicit in the problem.

Example 1.3.20 The standard unit vectors (Definition 1.2.7) are orthogonal to each other. For example, consider the standard unit vectors $i, j,$ and k in $\mathbb{R}^3$:

- $i \cdot j = (1,0,0) \cdot (0,1,0) = 0 + 0 + 0 = 0$;
- $j \cdot k = (0,1,0) \cdot (0,0,1) = 0 + 0 + 0 = 0$;
- $k \cdot i = (0,0,1) \cdot (1,0,0) = 0 + 0 + 0 = 0$.

By Definition 1.3.19 these are orthogonal to each other.

Example 1.3.21 Which pairs of the following vectors, if any, are perpendicular to each other? $u = (-1,1,-3,0)$, $v = (2,4,2,-6)$, and $w = (-1,6,-2,3)$.

Solution: Is the dot product zero? or not?

- $u \cdot v = (-1,1,-3,0) \cdot (2,4,2,-6) = -2 + 4 - 6 + 0 = -4 \neq 0$ so this pair are not perpendicular.

- $u \cdot w = (-1,1,-3,0) \cdot (-1,6,-2,3) = 1 + 6 + 6 + 0 = 13 \neq 0$ so this pair are not perpendicular.

- $v \cdot w = (2,4,2,-6) \cdot (-1,6,-2,3) = -2 + 24 - 4 - 18 = 0$ so this pair of vectors are perpendicular to each other. $\square$

Activity 1.3.22 Which pair of the following three vectors are orthogonal to each other? $x = i - 2k$, $y = -3i - 4j$, $z = -i - 2j + 2k$

(a) no pair   (b) x, y (c) y, z (d) x, z

Example 1.3.23 Find the scalar number b such that vectors $a = i + 4j + 2k$ and $b = i + bj - 3k$ are at right-angles.

Solution: For vectors to be at right-angles, their dot product must be zero. Hence find b such that

---

[7] The term 'orthogonal' derives from the Greek for 'right-angled'.

$$0 = a \cdot b = (i + 4j + 2k) \cdot (i + bj - 3k) = 1 + 4b - 6 = 4b - 5.$$

Solving $0 = 4b - 5$ gives $b = 5/4$. That is, $i + \frac{5}{4}j - 3k$ is at right-angles to $i + 4j + 2k$. □

Key properties The next couple of innocuous looking theorems are vital keys to important results in subsequent chapters.

To introduce the first theorem, consider the 2D plane and try to draw a nonzero vector at right-angles to both the two standard unit vectors i and j. The red vectors to the right illustrate failed attempts to draw a nonzero vector at right-angles to both i and j. It cannot be done. No vector in the plane can be at right-angles to both the standard unit vectors in the plane.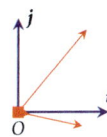

Theorem 1.3.24 *There is no nonzero vector orthogonal to all n standard unit vectors in $\mathbb{R}^n$.*

Proof. Let $u = (u_1, u_2, \ldots, u_n)$ be a vector in $\mathbb{R}^n$ that is orthogonal to all n standard unit vectors. Then by Definition 1.3.19 of orthogonality:

- $0 = u \cdot e_1 = (u_1, u_2, \ldots, u_n) \cdot (1,0,\ldots,0) = u_1 + 0 + \cdots + 0 = u_1,$ and so the first component must be zero;
- $0 = u \cdot e_2 = (u_1, u_2, \ldots, u_n) \cdot (0,1,\ldots,0) = 0 + u_2 + 0 + \cdots + 0 = u_2,$ and so the second component must be zero;
- and so on to
- $0 = u \cdot e_n = (u_1, u_2, \ldots, u_n) \cdot (0,0,\ldots,1) = 0 + 0 + \cdots + u_n = u_n,$ and so the last component must be zero.

Since $u_1 = u_2 = \cdots = u_n = 0$ the only vector that is orthogonal to all the standard unit vectors is $u = 0$, the zero vector.

To introduce the second theorem, imagine trying to draw three unit vectors in any orientation in the 2D plane such that all three are at right-angles to each other. The picture to the right illustrates one attempt. It cannot be done. There are at most two vectors in 2D that are all at right-angles to each other.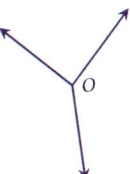

Theorem 1.3.25 (orthogonal completeness) *In a set of orthogonal unit vectors in $\mathbb{R}^n$, there can be no more than n vectors in the set.*[8]

Proof. Use contradiction. Suppose there are more than n orthogonal unit vectors in the set. Define a coordinate system for $\mathbb{R}^n$ using the first n of the given unit vectors as the n standard unit vectors (as illustrated for $\mathbb{R}^2$ to the right). Theorem 1.3.24 then says there cannot be any more nonzero vectors orthogonal than these n standard unit vectors. This contradicts there being more than n orthogonal unit vectors. To avoid this contradiction the supposition must be wrong; that is, there cannot be more than n orthogonal unit vectors in $\mathbb{R}^n$.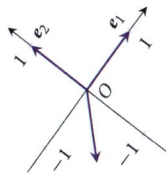

---

[8] For the pure at heart, this property is part of the definition of what we mean by $\mathbb{R}^n$. The representation of a vector in $\mathbb{R}^n$ by n components (here Definition 1.1.4) then follows as a consequence, instead of vice versa as here.

1.3.4 Normal vectors and equations of a plane

This section uses the dot product to find equations of a plane in 3D. The key is to write points in the plane as all those at right-angles to a certain direction. This direction is perpendicular to the required plane, and is called a normal. Let's start with an example of the idea in 2D.

Example 1.3.26 First find the equation of the line that is perpendicular to the vector $(2,3)$ and that passes through the origin. Second, find the equation of the line that passes through the point $(4,1)$ (instead of the origin).

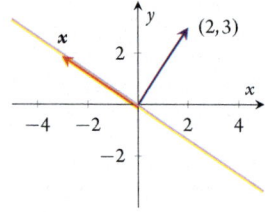

Solution: Recall that vectors at right-angles have a zero dot product (Section 1.3.3). Thus the position vector x of every point in the line satisfies the dot product $x \cdot (2,3) = 0$. For $x = (x,y)$, as illustrated to the above right, $x \cdot (2,3) = 2x + 3y$ so the equation of the line is $2x + 3y = 0$.

When the line goes through $(4,1)$ (instead of the origin), then it is the displacement vector $x - (4,1)$ that must be orthogonal to $(2,3)$, as illustrated. That is, the equation of the line is $(x - 4, y - 1) \cdot (2,3) = 0$. Evaluating the dot product gives $2(x-4) + 3(y-1) = 0$; that is, $2x + 3y = 2 \cdot 4 + 3 \cdot 1 = 11$ is an equation of the line.

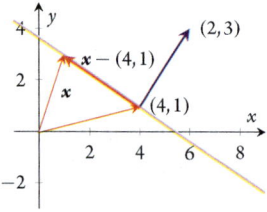

□

Activity 1.3.27 What is an equation of the line that is both through the point $(4,2)$, and at right-angles to the vector $(1,3)$?

(a) $2x + 3y = 11$ (b) $4x + y = 11$ (c) $x + 3y = 10$ (d) $4x + 2y = 10$

Now use the same approach to finding an equation of a plane in 3D. The problem is to find the equation of the plane that goes through a given point P and is perpendicular to a given vector n, called a **normal vector**. As illustrated in stereo, that means to find all points X such that $\overrightarrow{PX}$ is orthog-

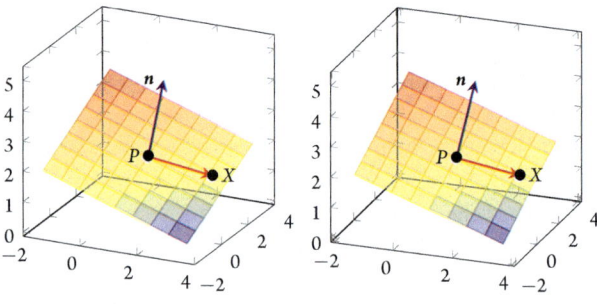

onal to n. Denote the position vector of P by $p = (x_0, y_0, z_0)$, the position vector of X by $x = (x,y,z)$, and let the normal vector be $n = (a,b,c)$. Then, as drawn below, the displacement vector $\overrightarrow{PX} = x - p = (x - x_0, y - y_0, z - z_0)$ and so for $\overrightarrow{PX}$ to be orthogonal to n requires $n \cdot (x - p) = 0$; that is, an **equation of the plane** is

$$a(x - x_0) + b(y - y_0) + c(z - z_0) = 0,$$

equivalently, an equation of the plane is

$$ax + by + cz = d \quad \text{for constant } d = ax_0 + by_0 + cz_0.$$

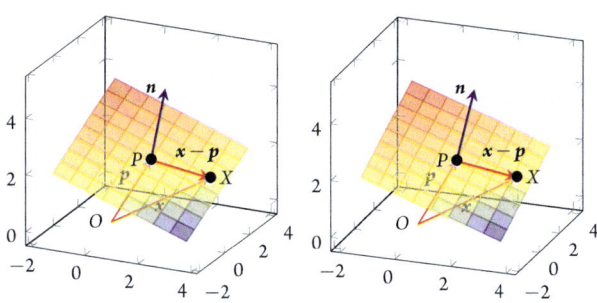

Example 1.3.28 Find an equation of the plane through point $P = (1,1,2)$ that has normal vector $n = (1,-1,3)$. (This is the case in the above illustrations.) Hence write down three distinct points on the plane.

Solution: Letting $x = (x,y,z)$ be the coordinates of a point in the plane, the above argument asserts an equation of the plane is $n \cdot (x - \overrightarrow{OP}) = 0$ which becomes $1(x-1) - 1(y-1) + 3(z-2) = 0$; that is, $x - 1 - y + 1 + 3z - 6 = 0$, which rearranged is $x - y + 3z = 6$.

To find some points in the plane, rearrange this equation to $z = 2 - x/3 + y/3$ and then substitute any values for x and y: $x = y = 0$ gives $z = 2$ so $(0,0,2)$ is on the plane; $x = 3$ and $y = 0$ gives $z = 1$ so $(3,0,1)$ is on the plane; $x = 2$ and $y = -2$ gives $z = 2/3$ so $(2,-2,\frac{2}{3})$ is on the plane; and so on. □

Example 1.3.29 Write down a normal vector to each of the following planes:

(a) $3x - 6y + 2z = 4$;

(b) $z = 0.2x - 3.3y - 1.9$.

Solution:

(a) In this standard form $3x - 6y + 2z = 4$ a normal vector is the coefficients of the variables, $n = (3,-6,2)$ (or any scalar multiple).

(b) Rearrange $z = 0.2x - 3.3y - 1.9$ to standard form $-0.2x + 3.3y + z = -1.9$ then a normal is $n = (-0.2,3.3,1)$ (or any scalar multiple).

□

Activity 1.3.30 Which vector is a normal vector to the plane $x_2 + 2x_3 + 4 = x_1$?

(a) none of these   (b) $(-1,1,2)$ (c) $(1,2,4)$ (d) $(1,2,1)$

Parametric equation of a plane An alternative way of describing a plane is via a parametric equation analogous to the parametric equation of a line (Section 1.2.3). Such a parametric representation generalizes to every dimension (Section 2.3).

The basic idea, as illustrated to the right, is that given any plane (through the origin for the moment), then choosing almost any two vectors in the plane allows us to write all points in the plane as a sum of multiples of the two vectors. With the given vectors u and v shown to the right, illustrated are the points $u + 2v$, $\frac{1}{2}u - 2v$, and $-2u + 3v$. Similarly, every point in the plane has a position vector in the form $su + tv$ for some scalar parameters s and t. The grid shown to the right illustrates the sum of integral and half-integral multiples. The formula $x = su + tv$ for parameters s and t is called a parametric equation of the plane.

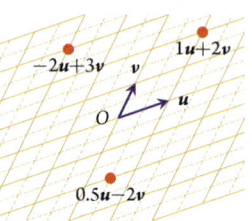

Example 1.3.31 Find a parametric equation of the plane that passes through the three points $P = (-1, 2, 3)$, $Q = (2, 3, 2)$, and $R = (0, 4, 5)$, drawn to the right in stereo.

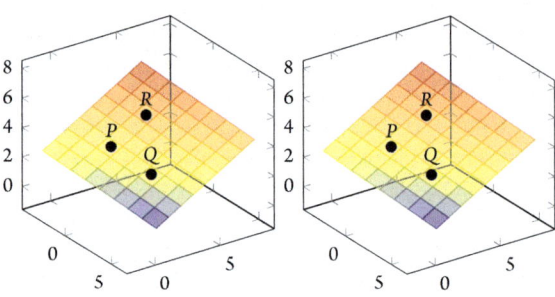

Solution: This plane does not pass through the origin, so we first choose a point and make the description relative to that point: say we choose the point P with position vector $p = \overrightarrow{OP} = -i + 2j + 3k$. Then, as illustrated below, two vectors parallel to the required plane are

$$
\begin{aligned}
u &= \overrightarrow{PQ} = \overrightarrow{OQ} - \overrightarrow{OP} \\
&= (2i + 3j + 2k) - (-i + 2j + 3k) \\
&= 3i + j - k, \\
v &= \overrightarrow{PR} = \overrightarrow{OR} - \overrightarrow{OP} \\
&= (4j + 5k) - (-i + 2j + 3k) \\
&= i + 2j + 2k.
\end{aligned}
$$

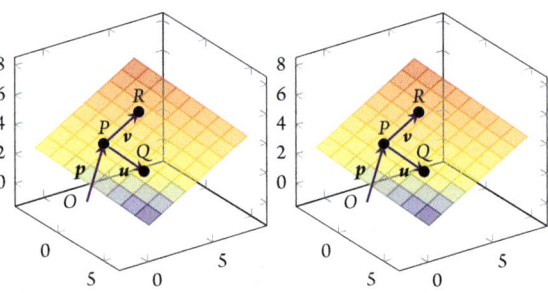

Lastly, every point in the plane is the sum of the displacement vector p and arbitrary multiples of the parallel vectors u and v. That is, a parametric equation of the plane is $x = p + su + tv$ which here is

$$
\begin{aligned}
x &= (-i + 2j + 3k) + s(3i + j - k) + t(i + 2j + 2k) \\
&= (-1 + 3s + t)i + (2 + s + 2t)j + (3 - s + 2t)k.
\end{aligned}
$$

□

Definition 1.3.32 *A **parametric equation** of a plane is $x = p + su + tv$ where p is the position vector of some point in the plane, the two vectors u and v are parallel to the plane ($u, v \neq 0$ and are at a nonzero/non-π angle to each other), and the scalar **parameters** s and t vary over all real values to give position vectors of all points in the plane.*

The beauty of this definition is that it applies for planes in any number of dimensions. To do so the parametric equations just use vectors with the corresponding number of components.

Example 1.3.33 Find a parametric equation of the plane that passes through the three points $P = (6, -4, 3)$, $Q = (-4, -18, 7)$, and $R = (11, 3, 1)$, drawn to the right in stereo.

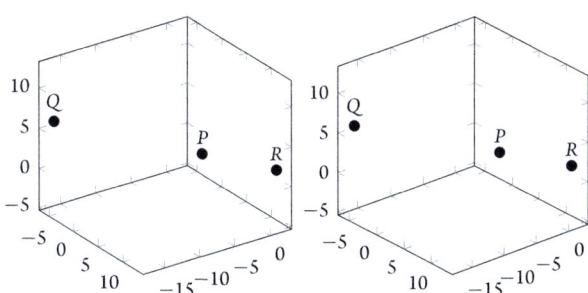

Solution: First choose a point and make the description relative to that point: say choose point P with position vector $p = \overrightarrow{OP} = 6i - 4j + 3k$. Then, as illustrated below, two vectors parallel to the required plane are

$$u = \overrightarrow{PQ} = \overrightarrow{OQ} - \overrightarrow{OP} = (-4i - 18j + 7k) - (6i - 4j + 3k) = -10i - 14j + 4k,$$
$$v = \overrightarrow{PR} = \overrightarrow{OR} - \overrightarrow{OP} = (11i + 3j + k) - (6i - 4j + 3k) = 5i + 7j - 2k.$$

Oops: notice that $u = -2v$ so the vectors u and v are not at a nontrivial angle; instead they are aligned along a line because the three points P, Q, and R are collinear. There are an infinite number of planes passing through such collinear points. Hence we cannot answer the question which requires "the plane". □

Example 1.3.34 Find a parametric equation of the plane that passes through the three points $A = (-1.2, 2.4, 0.8)$, $B = (1.6, 1.4, 2.4)$, and $C = (0.2, -0.4, -2.5)$, drawn to the right in stereo.

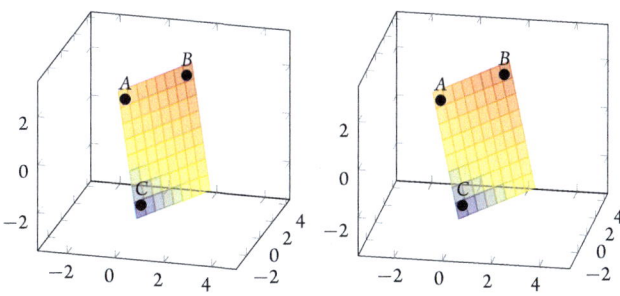

Solution: First choose a point and make the description relative to that point: say we choose the point A with position vector $a = \overrightarrow{OA} = -1.2i + 2.4j + 0.8k$. Then, as illustrated below, two vectors parallel to the required plane are

$$u = \overrightarrow{AB} = \overrightarrow{OB} - \overrightarrow{OA}$$
$$= (1.6i + 1.4j + 2.4k) - (-1.2i + 2.4j + 0.8k)$$
$$= 2.8i - j + 1.6k,$$
$$v = \overrightarrow{AC} = \overrightarrow{OC} - \overrightarrow{OA}$$

$$= (0.2i - 0.4j - 2.5k) - (-1.2i + 2.4j + 0.8k)$$
$$= 1.4i - 2.8j - 3.3k.$$

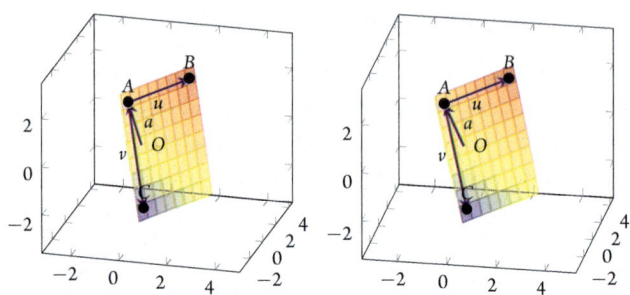

Lastly, every point in the plane is the sum of the displacement vector a, and some multiples of the 'parallel' vectors u and v. That is, a parametric equation of the plane is $x = a + su + tv$ which here is

$$x = \begin{bmatrix} -1.2 \\ 2.4 \\ 0.8 \end{bmatrix} + s \begin{bmatrix} 2.8 \\ -1 \\ 1.6 \end{bmatrix} + t \begin{bmatrix} 1.4 \\ -2.8 \\ -3.3 \end{bmatrix} = \begin{bmatrix} -1.2 + 2.8s + 1.4t \\ 2.4 - s - 2.8t \\ 0.8 + 1.6s - 3.3t \end{bmatrix}.$$

□

Activity 1.3.35 Which of the following is *not* a parametric equation of a plane?

(a) $i + sj + tk$

(b) $(-1, 1, -1)s + (4, 2, -1)t$

(c) $(3s + 2t, 4 + 2s + t, 4 + 3t)$

(d) $(4, 1, 4) + (3, 6, 3)s + (2, 4, 2)t$

1.3.5 Exercises

Exercise 1.3.1 Following Example 1.3.1, use the cosine rule for triangles to find the angle between the following pairs of vectors. Confirm that $|u||v| \cos\theta = u \cdot v$ in each case.

(a) $(6, 5)$ and $(-3, 1)$

(b) $(6, 2, 2)$ and $(-1, -2, 5)$

(c) $(2, 2.9)$ and $(-1.4, 0.8)$

(d) $(-3.6, 0, -0.7)$ and $(1.2, -0.9, -0.6)$

Exercise 1.3.2 Which of the following pairs of vectors appear orthogonal?

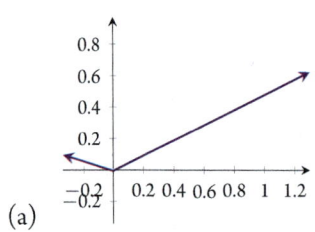

(a)

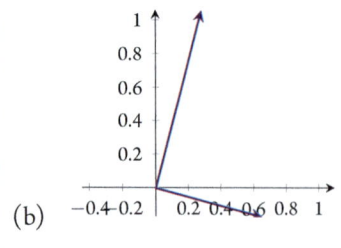

(b)

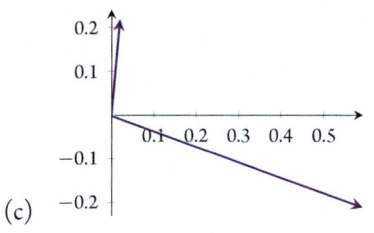

(c)

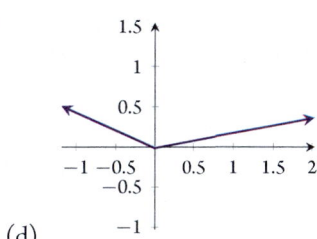

(d)

Exercise 1.3.3 Recall that Example 1.1.7 represented the following sentences by word vectors $w = (N_{cat}, N_{dog}, N_{mat}, N_{sat}, N_{scratched})$.

- "The cat and dog sat on the mat" is summarized by the vector $a = (1,1,1,1,0)$.
- "The dog scratched" is summarized by the vector $b = (0,1,0,0,1)$.
- "The dog sat on the mat; the cat scratched the dog." is summarized by the vector $c = (1,2,1,1,1)$.

Find the similarity between pairs of these sentences by calculating the angle between each pair of word vectors. What is the most similar pair of sentences?

Exercise 1.3.4 Suppose two nonzero word vectors are orthogonal. Explain what such orthogonality means in terms of the words of the original sentences.

Exercise 1.3.5 For the properties of the dot product, Theorem 1.3.13, prove some properties chosen from 1.3.13(b)–1.3.13(d).

Exercise 1.3.6 Verify the Cauchy–Schwarz inequality and also the triangle inequality (+ case) for the following pairs of vectors.

(a) $(2,-4,4)$ and $(6,7,6)$

(b) $(1,-2,2)$ and $(-3,6,-6)$

(c) $\begin{bmatrix} -0.2 \\ 0.8 \\ -3.8 \\ -0.3 \end{bmatrix}$ and $\begin{bmatrix} 2.4 \\ -5.2 \\ 5.0 \\ 1.9 \end{bmatrix}$

(d) $\begin{bmatrix} 0.8 \\ 0.8 \\ 6.6 \\ -1.5 \end{bmatrix}$ and $\begin{bmatrix} 4.4 \\ -0.6 \\ 2.1 \\ 2.2 \end{bmatrix}$

Exercise 1.3.7 Find an equation of the plane with the given normal vector n and through the given point P.

(a) $P = (1,2,-3)$, $n = (2,-5,-2)$.

(b) $P = (5,-4,-13)$, $n = (-1,0,-1)$.

(c) $P = (-7.3,-1.6,5.8)$, $n = (-2.8,-0.8,4.4)$.

(d) $P = (0,-1.2,2.2)$, $n = (-1.4,-8.1,-1.5)$.

Exercise 1.3.8 Write down a normal vector to the plane described by each of the following equations.

(a) $2x + 3y + 2z = 6$

(b) $-7x - 2y + 4 = -5z$

(c) $0.1x = 1.5y + 1.1z + 0.7$

(d) $-5.5x_1 + 1.6x_2 = 6.7x_3 - 1.3$

Exercise 1.3.9 For each case, find a parametric equation of the plane through the three given points.

(a) $(0, 5, -4)$, $(-3, -2, 2)$, $(5, 1, -3)$.

(b) $(0, -1, -1)$, $(-4, 1, -5)$, $(0, -3, -2)$.

(c) $(-5.6, -2.2, -6.8)$, $(-1.8, 4.3, -3.9)$, $(2.5, -3.5, -1.7)$.

(d) $(1.8, -0.2, -0.7)$, $(-1.6, 2, -3.7)$, $(1.4, -0.5, 0.5)$.

Exercise 1.3.10 For each case of Exercise 1.3.9 that you have done, find two other parametric equations of the plane.

Exercise 1.3.11 In a few sentences, answer/discuss each of the following.

(a) When using the dot product to determine the angle between a pair of vectors we only discuss angles between $0°$ and $180°$ (0 and π radians). Why do we not discuss larger angles, such as $246°$ or $315°$? nor negative angles?

(b) What properties of the dot product differ from that of the multiplication of scalar numbers?

(c) Describe a geometric reason for the Cauchy–Schwarz inequality.

(d) Why do we phrase an equation for a plane in terms of its perpendicular vector?

(e) Given that $x = p + td$ parametrizes a line, and that $x = p + sc + td$ parametrizes a plane, what would $x = p + rb + sc + td$ describe? why? are there any provisos?

1.4 The cross product

The dot product of Section 1.3 is not the only way to multiply vectors. In the three dimensions of the world we live in there is a second way to multiply vectors, called the cross product. But for more than three dimensions, qualitatively different techniques are developed in subsequent chapters.

This section is optional for us, but is vital in many topics of science and engineering.

Area of a parallelogram

Consider the parallelogram drawn in blue. It has sides given by vectors $v = (v_1, v_2)$ and $w = (w_1, w_2)$, as shown. Let's determine the area of the parallelogram. Its area is the containing rectangle less the two small rectangles and the four small triangles. The two small rectangles have the same area, namely $w_1 v_2$. The two small triangles on the left and the right also have the same area, namely $\frac{1}{2} w_1 w_2$. The two small triangles on the top and the bottom similarly have the same area, namely $\frac{1}{2} v_1 v_2$. Thus, the parallelogram has

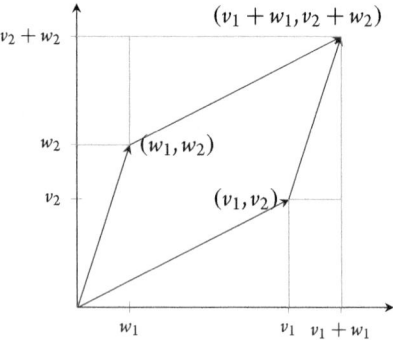

$$\begin{aligned} \text{area} &= (v_1 + w_1)(v_2 + w_2) - 2 w_1 v_2 - 2 \cdot \frac{1}{2} w_1 w_2 - 2 \cdot \frac{1}{2} v_1 v_2 \\ &= v_1 v_2 + v_1 w_2 + w_1 v_2 + w_1 w_2 - 2 w_1 v_2 - w_1 w_2 - v_1 v_2 \\ &= v_1 w_2 - v_2 w_1. \end{aligned}$$

In application, sometimes this right-hand side expression is negative because vectors v and w are the 'wrong way' around. Thus in general the parallelogram area $= |v_1 w_2 - v_2 w_1|$.

Example 1.4.1 What is the area of the parallelogram (illustrated to the right) whose edges are formed by the vectors $(3,2)$ and $(-1,4)$?

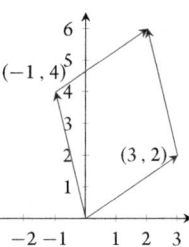

Solution: The parallelogram area $= |3 \cdot 4 - 2 \cdot (-1)| = |12 + 2| = 14$. The illustration indicates that this area must be about right, as with imagination one could cut the area and move the parts about to form a rectangle roughly 3 by 5, and hence the area should be roughly 15.

☐

Activity 1.4.2 What is the area of the parallelogram (illustrated to the right) whose edges are formed by the vectors $(5,3)$ and $(2,-2)$?

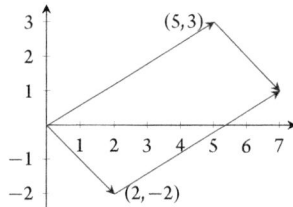

(a) 16   (b) 11   (c) 4   (d) 19

Interestingly, we meet this expression for area, $v_1 w_2 - v_2 w_1$, in another context: that of equations for a plane and its normal vector.

Normal vector to a plane

Recall Section 1.3.4 introduced that we describe planes either via an equation such as $x - y + 3z = 6$ or via a parametric description such as $x = (1,1,2) + (1,1,0)s + (0,3,1)t$. These determine the same plane; they are just different algebraic descriptions. One converts between these two descriptions using the cross product.

Example 1.4.3 Derive that the plane described parametrically by $x = (1,1,2) + (1,1,0)s + (0,3,1)t$ has normal equation $x - y + 3z = 6$.

Solution: The key to deriving the normal equation is to find that a normal vector to the plane is $(1,-1,3)$. This normal vector comes from the two vectors that multiply the parameters in the parametric form, $(1,1,0)$ and $(0,3,1)$. The following mysterious looking procedure may be a convenient way for you to remember an otherwise involved formula: if you prefer to remember the formula of Definition 1.4.5, then use that instead. (Those who have computed 3×3 determinants may recognize that the following has the same pattern—see Chapter 6.) Write the vectors as two consecutive columns, following a first column of the *symbols* of the standard unit vectors i, j, and k, in

$$n = \begin{vmatrix} i & 1 & 0 \\ j & 1 & 3 \\ k & 0 & 1 \end{vmatrix}$$

(then cross out 1st column and each row in turn, multiplying
each by common entry, with alternating sign)

$$= i\begin{vmatrix} 1 & 1 & 0 \\ 1 & 1 & 3 \\ k & 0 & 1 \end{vmatrix} - j\begin{vmatrix} i & 1 & 0 \\ j & 1 & 3 \\ k & 0 & 1 \end{vmatrix} + k\begin{vmatrix} i & 1 & 0 \\ j & 1 & 3 \\ k & 0 & 1 \end{vmatrix}$$

$$= i\begin{vmatrix} 1 & 3 \\ 0 & 1 \end{vmatrix} - j\begin{vmatrix} 1 & 0 \\ 0 & 1 \end{vmatrix} + k\begin{vmatrix} 1 & 0 \\ 1 & 3 \end{vmatrix}$$

(then draw diagonals, then subtract product of red
diagonal from product of the blue)

$$= i\begin{vmatrix} 1 & 3 \\ 0 & 1 \end{vmatrix} - j\begin{vmatrix} 1 & 0 \\ 0 & 1 \end{vmatrix} + k\begin{vmatrix} 1 & 0 \\ 1 & 3 \end{vmatrix}$$

$$= i(1 \cdot 1 - 0 \cdot 3) - j(1 \cdot 1 - 0 \cdot 0) + k(1 \cdot 3 - 1 \cdot 0)$$

$$= i - j + 3k.$$

Using this normal vector, the equation of the plane must be of the form $x - y + 3z = $ constant.
Since the plane goes through point $(1, 1, 2)$, the constant $= 1 - 1 + 3 \cdot 2 = 6$; that is, the plane is
$x - y + 3z = 6$ (as given). □

Activity 1.4.4 Use the procedure of Example 1.4.3 to derive a normal vector to the plane
described in parametric form as $x = (4, -1, -2) + (1, -2, 1)s + (2, -3, -2)t$. Which of the
following is your computed normal vector?

(a) $(7, 4, 1)$ (b) $(-4, 4, -10)$ (c) $(2, -2, 5)$ (d) $(5, 6, 7)$

Definition of a cross product

General formula The procedure used in Example 1.4.3 to derive a normal vector leads to
an algebraic formula. Let's apply the same procedure to two general vectors $v = (v_1, v_2, v_3)$ and
$w = (w_1, w_2, w_3)$. The procedure computes

$$n = \begin{vmatrix} i & v_1 & w_1 \\ j & v_2 & w_2 \\ k & v_3 & w_3 \end{vmatrix}$$

(then cross out 1st column and each row in turn, multiplying
each by common entry, with alternating sign)

$$= i\begin{vmatrix} i & v_1 & w_1 \\ j & v_2 & w_2 \\ k & v_3 & w_3 \end{vmatrix} - j\begin{vmatrix} i & v_1 & w_1 \\ j & v_2 & w_2 \\ k & v_3 & w_3 \end{vmatrix} + k\begin{vmatrix} i & v_1 & w_1 \\ j & v_2 & w_2 \\ k & v_3 & w_3 \end{vmatrix}$$

$$= i\begin{vmatrix} v_2 & w_2 \\ v_3 & w_3 \end{vmatrix} - j\begin{vmatrix} v_1 & w_1 \\ v_3 & w_3 \end{vmatrix} + k\begin{vmatrix} v_1 & w_1 \\ v_2 & w_2 \end{vmatrix}$$

(then draw diagonals, then subtract product of red
diagonal from product of the blue)

$$= i \begin{vmatrix} v_2 & w_2 \\ v_3 & w_3 \end{vmatrix} - j \begin{vmatrix} v_1 & w_1 \\ v_3 & w_3 \end{vmatrix} + k \begin{vmatrix} v_1 & w_1 \\ v_2 & w_2 \end{vmatrix}$$

$$= i(v_2 w_3 - v_3 w_2) - j(v_1 w_3 - v_3 w_1) + k(v_1 w_2 - v_2 w_1).$$

We use this formula to define the cross product algebraically, and then see what it means geometrically.

Definition 1.4.5   Let $v = (v_1, v_2, v_3)$ and $w = (w_1, w_2, w_3)$ be two vectors in $\mathbb{R}^3$. The **cross product** (or **vector product**) $v \times w$ is defined algebraically as

$$v \times w := i(v_2 w_3 - v_3 w_2) + j(v_3 w_1 - v_1 w_3) + k(v_1 w_2 - v_2 w_1).$$

Example 1.4.6   Among the standard unit vectors, derive that

(a) $i \times j = k$,

(b) $j \times i = -k$,

(c) $j \times k = i$,

(d) $k \times j = -i$,

(e) $k \times i = j$,

(f) $i \times k = -j$,

(g) $i \times i = j \times j = k \times k = 0.$

Solution: Using Definition 1.4.5:

$$i \times j = (1,0,0) \times (0,1,0)$$
$$= i(0 \cdot 0 - 0 \cdot 1) + j(0 \cdot 0 - 1 \cdot 0) + k(1 \cdot 1 - 0 \cdot 0)$$
$$= k;$$
$$j \times i = (0,1,0) \times (1,0,0)$$
$$= i(1 \cdot 0 - 0 \cdot 0) + j(0 \cdot 1 - 0 \cdot 0) + k(0 \cdot 0 - 1 \cdot 1)$$
$$= -k;$$
$$i \times i = (1,0,0) \times (1,0,0)$$
$$= i(0 \cdot 0 - 0 \cdot 0) + j(0 \cdot 1 - 1 \cdot 0) + k(1 \cdot 0 - 0 \cdot 1)$$
$$= 0.$$

Exercise 1.4.1 asks you to correspondingly establish the other six identities.   ◻

The cross products of this Example 1.4.6 most clearly demonstrate the orthogonality of a cross product to its two argument vectors (Theorem 1.4.10(a)), and that the direction is in the so-called right-hand sense (Theorem 1.4.10(b)).

Activity 1.4.7 Use Definition 1.4.5 to find the cross product of $(-4,1,-1)$ and $(-2,2,1)$ is which one of the following:

(a) $(-3,-6,6)$ (b) $(3,-6,-6)$ (c) $(-3,-6,6)$ (d) $(3,6,-6)$

Geometry of a cross product

Example 1.4.8 *(parallelogram area)* Let's revisit the introduction to this section. Consider the parallelogram in the x_1x_2-plane with edges formed by the $\mathbb{R}^3$ vectors $v = (v_1,v_2,0)$ and $w = (w_1,w_2,0)$. At the start of this Section 1.4 we derived that the parallelogram formed by these vectors has area $= |v_1w_2 - v_2w_1|$. Compare this area with the cross product

$$v \times w = i(v_2 \cdot 0 - 0 \cdot w_2) + j(0 \cdot w_1 - v_1 \cdot 0) + k(v_1w_2 - v_2w_1)$$
$$= i0 + j0 + k(v_1w_2 - v_2w_1)$$
$$= k(v_1w_2 - v_2w_1).$$

Consequently, the length of this cross product equals the area of the parallelogram formed by v and w (Theorem 1.4.10(d)). (Also the direction of the cross product, $\pm k$, is orthogonal to the x_1x_2-plane containing the two vectors—Theorem 1.4.10(a)).

Activity 1.4.9 Using property 1.4.10(b) of the next theorem, in which direction is the cross product $v \times w$ for the two vectors illustrated in stereo to the right?

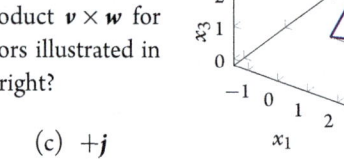 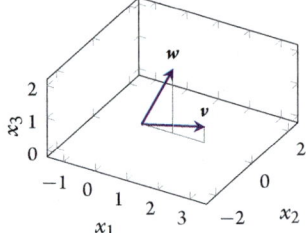

(a) $+i$ (c) $+j$
(b) $-j$ (d) $-i$

Theorem 1.4.10 *(cross product geometry)* Let v and w be two vectors in $\mathbb{R}^3$:

(a) the vector $v \times w$ is orthogonal to both v and w;

(b) the direction of $v \times w$ is in the **right-hand sense**, in that if v is in the direction of your thumb, and w is in the direction of your straight index finger, then $v \times w$ is in the direction of your bent second/longest finger—all on your right hand as illustrated to the right;

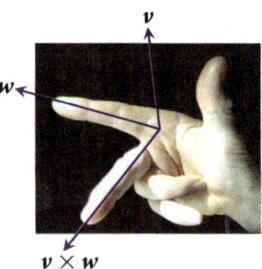

$v \times w$

(c) $|v \times w| = |v| \, |w| \sin \theta$ where θ is the angle between vectors v and w $(0 \le \theta \le \pi$, equivalently $0° \le \theta \le 180°)$; and

(d) the length $|v \times w|$ is the area of the parallelogram with edges v and w.

Proof. Let $v = (v_1, v_2, v_3)$ and $w = (w_1, w_2, w_3)$.

1.4.10(a) Recall that two vectors are orthogonal if their dot product is zero (Definition 1.3.19). To determine orthogonality between v and the cross product $v \times w$, consider

$$
\begin{aligned}
v \cdot (v \times w) &= (v_1 i + v_2 j + v_3 k) \cdot [i(v_2 w_3 - v_3 w_2) \\
&\quad + j(v_3 w_1 - v_1 w_3) + k(v_1 w_2 - v_2 w_1)] \\
&= v_1(v_2 w_3 - v_3 w_2) + v_2(v_3 w_1 - v_1 w_3) \\
&\quad + v_3(v_1 w_2 - v_2 w_1) \\
&= v_1 v_2 w_3 - v_1 v_3 w_2 + v_2 v_3 w_1 \\
&\quad - v_1 v_2 w_3 + v_1 v_3 w_2 - v_2 v_3 w_1 \quad = 0
\end{aligned}
$$

as each term in the penultimate line cancels with the term underneath in the last line. Since the dot product is zero, the cross product $v \times w$ is orthogonal to vector v. Similarly, $v \times w$ is orthogonal to w (Exercise 1.4.5).

1.4.10(b) This right-handed property follows from the convention that the standard unit vectors i, j, and k are right-handed: that if i is in the direction of your thumb, and j is in the direction of your straight index finger, then k is in the direction of your bent second/longest finger—all on your right hand.

We prove only for the case of vectors in the $x_1 x_2$-plane, in which case $v = (v_1, v_2, 0)$ and $w = (w_1, w_2, 0)$, and when both $v_1, w_1 > 0$. One example is in stereo below.

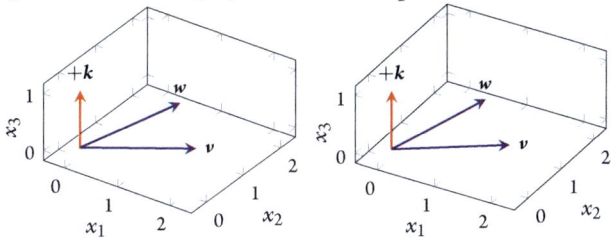

Example 1.4.8 derived the cross product $v \times w = k(v_1 w_2 - v_2 w_1)$. Consequently, this cross product is in the $+k$ direction only when $v_1 w_2 - v_2 w_1 > 0$ (it is in the $-k$ direction in the complementary case when $v_1 w_2 - v_2 w_1 < 0$). This inequality for $+k$ rearranges to $v_1 w_2 > v_2 w_1$. Dividing by the positive $v_1 w_1$ requires $\frac{w_2}{w_1} > \frac{v_2}{v_1}$. That is, in the $x_1 x_2$-plane the 'slope' of vector w must be greater than the 'slope' of vector v. In this case, if v is in the direction of your thumb on your right hand, and w is in the direction of your straight index finger, then your bent second/longest finger is in the direction $+k$ as required by the cross product $v \times w$.

1.4.10(c) Exercise 1.4.6 establishes the identity $|v \times w|^2 = |v|^2 |w|^2 - (v \cdot w)^2$. From Theorem 1.3.5 substitute $v \cdot w = |v||w|\cos\theta$ into this identity:

$$
\begin{aligned}
|v \times w|^2 &= |v|^2 |w|^2 - (v \cdot w)^2 \\
&= |v|^2 |w|^2 - (|v||w|\cos\theta)^2 \\
&= |v|^2 |w|^2 - |v|^2 |w|^2 \cos^2\theta \\
&= |v|^2 |w|^2 (1 - \cos^2\theta) \\
&= |v|^2 |w|^2 \sin^2\theta.
\end{aligned}
$$

Take the square-root of both sides to determine $|v \times w| = \pm |v||w| \sin \theta$. But $\sin \theta \geq 0$ since the angle $0 \leq \theta \leq \pi$, and all the lengths are also ≥ 0, so only the plus case applies. That is, the length $|v \times w| = |v||w| \sin \theta$ as required.

1.4.10(d) Consider the plane containing the vectors v and w, and hence containing the parallelogram formed by these vectors— as illustrated to the above right. Using vector v as the base of the parallelogram, with length $|v|$, by basic trigonometry the height of the parallelogram is then $|w| \sin \theta$. Hence the area of the parallelogram is the product base · height $= |v||w| \sin \theta =$ $|v \times w|$ by the previous part 1.4.10(c).

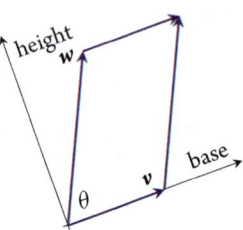

Example 1.4.11 Find the area of the parallelogram with edges formed by vectors $v = (-2, 0, 1)$ and $w = (2, 2, 1)$—as illustrated in stereo to the right.

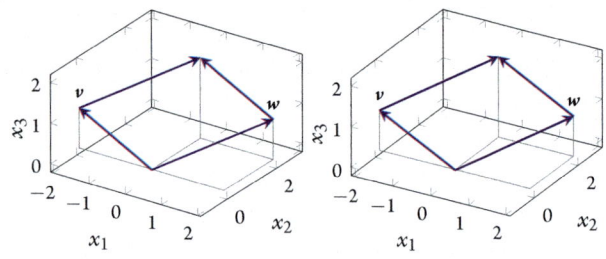

Solution: The area is the length of the cross product

$$v \times w = i(0 \cdot 1 - 1 \cdot 2) + j(1 \cdot 2 - (-2) \cdot 1) + k((-2) \cdot 2 - 0 \cdot 2)$$
$$= -2i + 4j - 4k.$$

Then the parallelogram area $|v \times w| = \sqrt{(-2)^2 + 4^2 + (-4)^2} = \sqrt{4 + 16 + 16} = \sqrt{36} = 6.$ □

Activity 1.4.12 What is the area of the parallelogram (in stereo to the right) with edges formed by vectors $v = (-2, 1, 0)$ and $w = (2, 0, -1)$?

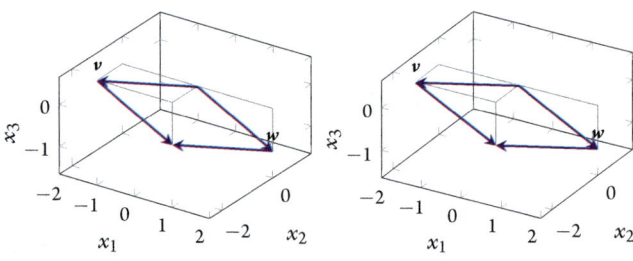

(a) 3   (c) 1
(b) $\sqrt{5}$ (d) 5

Example 1.4.13 Find a normal vector to the plane containing the two vectors $v = -2i + 3j + 2k$ and $w = 2i + 2j + 3k$ — illustrated to the right. Hence find an equation of the plane given parametrically as $x = -2i - j + 3k + (-2i + 3j + 2k)s + (2i + 2j + 3k)t$.

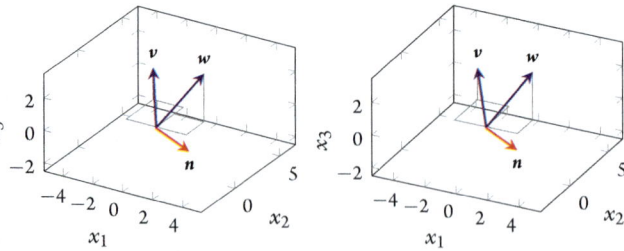

Solution: Use Definition 1.4.5 of the cross product to find a normal vector:

$$v \times w = i(3 \cdot 3 - 2 \cdot 2) + j(2 \cdot 2 - (-2) \cdot 3) + k((-2) \cdot 2 - 3 \cdot 2)$$
$$= 5i + 10j - 10k.$$

A normal vector is any vector proportional to this, so we could divide by five and choose normal vector $n = i + 2j - 2k$ (as illustrated above).

An equation of the plane through $-2i - j + 3k$ is then given by the dot product

$$(i + 2j - 2k) \cdot [(x+2)i + (y+1)j + (z-3)k] = 0,$$
$$\text{that is,} \quad x + 2 + 2y + 2 - 2z + 6 = 0,$$
$$\text{that is,} \quad x + 2y - 2z + 10 = 0$$

is the required normal equation of the plane. □

Algebraic properties of a cross product

Exercises 1.4.9 to 1.4.11 establish three of the following four useful algebraic properties of the cross product.

Theorem 1.4.14 (*cross product properties*) *Let u, v, and w be vectors in* $\mathbb{R}^3$, *and c be a scalar:*

(a) $v \times v = 0$;

(b) $w \times v = -(v \times w)$ (*not commutative*);

(c) $(cv) \times w = c(v \times w) = v \times (cw)$;

(d) $u \times (v + w) = u \times v + u \times w$ (*distributive law*).

Proof. Let's prove property 1.4.14(a) two ways—algebraically and geometrically. Exercises 1.4.9 to 1.4.11 ask you to prove the other properties.

- Algebraically: with vector $v = (v_1, v_2, v_3)$, Definition 1.4.5 gives

$$v \times v = i(v_2 v_3 - v_3 v_2) + j(v_3 v_1 - v_1 v_3) + k(v_1 v_2 - v_2 v_1)$$
$$= 0i + 0j + 0k = 0.$$

- Geometrically: from Theorem 1.4.10(d), $|v \times v|$ is the area of the parallelogram with edges v and v. But such a parallelogram has zero area, so $|v \times v| = 0$. Since the only vector of length zero is the zero vector (Theorem 1.1.13), $v \times v = 0$.

Example 1.4.15 As an example of Theorem 1.4.14(b), Example 1.4.6 shows that $i \times j = k$, whereas reversing the order of the cross product gives the negative $j \times i = -k$. Given Example 1.4.13 derived $v \times w = 5i + 10j - 10k$ in the case when $v = -2i + 3j + 2k$ and $w = 2i + 2j + 3k$, what is $w \times v$?

Solution: By 1.4.14(b), $w \times v = -(v \times w) = -5i - 10j + 10k$. □

Example 1.4.16 Given $(i+j+k) \times (-2i-j) = i - 2j + k$, what is $(3i + 3j + 3k) \times (-2i-j)$?

Solution: The first vector is $3(i+j+k)$ so by Theorem 1.4.14(c),

$$(3i+3j+3k) \times (-2i-j)$$
$$= [3(i+j+k)] \times (-2i-j)$$
$$= 3[(i+j+k) \times (-2i-j)]$$
$$= 3[i - 2j + k] = 3i - 6j + 3k.$$ □

Activity 1.4.17 For vectors $u = -i+3k$, $v = i+3j+5k$, and $w = -2i+j-k$ you are given that

$$u \times v = -9i + 8j - 3k,$$
$$u \times w = -3i - 7j - k,$$
$$v \times w = -8i - 9j + 7k.$$

Which is the cross product $(-i+3k) \times (-i+4j+4k)$?

(a) $i - 17j + 10k$ (c) $-12i + j - 4k$
(b) $-17i - j + 4k$ (d) $-11i - 16j + 6k$

Also, which is $(i+3j+5k) \times (-3i+j+2k)$?

Example 1.4.18 The properties of Theorem 1.4.14 empower algebraic manipulation. Use such algebraic manipulation, and the identities among standard unit vectors of Example 1.4.6, compute the cross product $(i-j) \times (4i+2k)$.

Solution: In full detail:

$$(i-j) \times (4i+2k)$$
$$= (i-j) \times (4i) + (i-j) \times (2k) \quad \text{(by 1.4.14(d))}$$
$$= 4(i-j) \times i + 2(i-j) \times k \quad \text{(by 1.4.14(c))}$$
$$= -4i \times (i-j) - 2k \times (i-j) \quad \text{(by 1.4.14(b))}$$
$$= -4[i \times i + i \times (-j)] - 2[k \times i + k \times (-j)] \quad \text{(by 1.4.14(d))}$$
$$= -4[i \times i - i \times j] - 2[k \times i - k \times j] \quad \text{(by 1.4.14(c))}$$
$$= -4[0 - k] - 2[j - (-i)] \quad \text{(by Example 1.4.6)}$$
$$= -2i - 2j + 4k.$$ □

Volume of a parallelepiped

Consider the paral-
lelepiped with edges
formed by three vectors u,
v, and w in $\mathbb{R}^3$, as illustrated in stereo to the right.
Our challenge is to derive

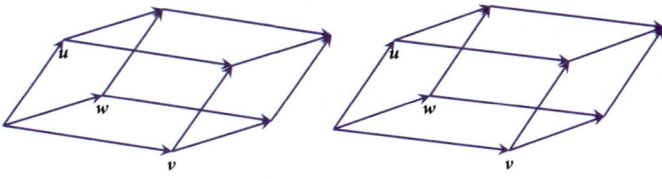

that the volume of the parallelepiped is $|u \cdot (v \times w)|$.

Let's use that we know the volume of the parallelepiped is the area of its base times its height.

- The base of the parallelepiped is the parallelogram formed with edges v and w. Hence the base has area $|v \times w|$ (Theorem 1.4.10(d)).

- The height of the parallelepiped is then that part
of u in the direction of a normal vector to v and w. We
know that $v \times w$ is orthogo-

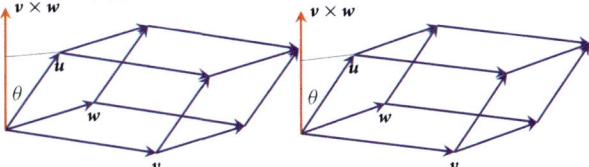

nal to both v and w (Theorem 1.4.10(a)), so by trigonometry the height must be $|u|\cos\theta$
for angle θ between u and $v \times w$, as illustrated.

To cater for cases where $v \times w$ points in the opposite direction to that shown, the height
is $|u||\cos\theta|$. The dot product determines this cosine (Theorem 1.3.5):

$$\cos\theta = \frac{u \cdot (v \times w)}{|u||v \times w|}.$$

The height of the parallelepiped is then

$$|u||\cos\theta| = |u|\frac{|u \cdot (v \times w)|}{|u||v \times w|} = \frac{|u \cdot (v \times w)|}{|v \times w|}.$$

Consequently, the volume of the parallelepiped equals

$$\text{base} \cdot \text{height} = |v \times w|\frac{|u \cdot (v \times w)|}{|v \times w|} = |u \cdot (v \times w)|.$$

Definition 1.4.19 *For every three vectors u, v, and w in $\mathbb{R}^3$, the **scalar triple product** is $u \cdot (v \times w)$.*

Example 1.4.20 Use the
scalar triple product to find the
volume of the parallelepiped
formed by vectors $u =$
$(0, 2, 1)$, $v = (-2, 0, 1)$ and
$w = (2, 2, 1)$—as illustrated in
stereo to the right.

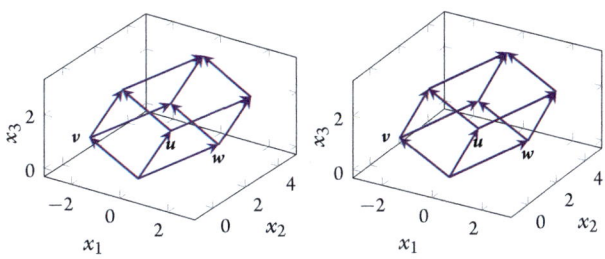

Solution: Example 1.4.11 found the cross product $v \times w = -2i + 4j - 4k$. So the scalar triple product $u \cdot (v \times w) = (2j + k) \cdot (-2i + 4j - 4k) = 8 - 4 = 4$. Hence the volume of the parallelepiped is 4 (cubic units).

The order of the vectors in a scalar triple product only affects the sign of the result. For example, we also find the volume of this parallelepiped via $v \cdot (u \times w)$. Returning to the procedure of Example 1.4.3 to find the cross product gives

$$u \times w = \begin{vmatrix} i & 0 & 2 \\ j & 2 & 2 \\ k & 1 & 1 \end{vmatrix}$$

$$= i \begin{vmatrix} \cancel{i} & \cancel{0} & \cancel{2} \\ j & 2 & 2 \\ \cancel{k} & 1 & 1 \end{vmatrix} - j \begin{vmatrix} i & 0 & 2 \\ \cancel{j} & \cancel{2} & \cancel{2} \\ k & 1 & 1 \end{vmatrix} + k \begin{vmatrix} i & 0 & 2 \\ j & 2 & 2 \\ \cancel{k} & \cancel{1} & \cancel{1} \end{vmatrix}$$

$$= i \begin{vmatrix} 2 & 2 \\ 1 & 1 \end{vmatrix} - j \begin{vmatrix} 0 & 2 \\ 1 & 1 \end{vmatrix} + k \begin{vmatrix} 0 & 2 \\ 2 & 2 \end{vmatrix}$$

$$= i \begin{vmatrix} 2 & 2 \\ 1 & 1 \end{vmatrix} - j \begin{vmatrix} 0 & 2 \\ 1 & 1 \end{vmatrix} + k \begin{vmatrix} 0 & 2 \\ 2 & 2 \end{vmatrix}$$

$$= i(2 \cdot 1 - 1 \cdot 2) - j(0 \cdot 1 - 1 \cdot 2) + k(0 \cdot 2 - 2 \cdot 2)$$

$$= 2j - 4k.$$

Then the triple product $v \cdot (u \times w) = (-2i + k) \cdot (2j - 4k) = 0 + 0 - 4 = -4$. Hence the volume of the parallelepiped is $|-4| = 4$ as before. $\square$

Using the procedure of Example 1.4.3 to find a scalar triple product establishes a strong connection to the matrix determinants of Chapter 6. In the second solution to the previous Example 1.4.20, in finding $u \times w$, the unit vectors i, j, and k just acted as place-holding symbols to eventually ensure a multiplication by the correct component of v in the dot product. We could seamlessly combine the two products by replacing the symbols i, j, and k directly with the corresponding component of v:

$$v \cdot (u \times w) = \begin{vmatrix} -2 & 0 & 2 \\ 0 & 2 & 2 \\ 1 & 1 & 1 \end{vmatrix}$$

$$= -2 \begin{vmatrix} \cancel{-2} & \cancel{0} & \cancel{2} \\ 0 & 2 & 2 \\ 1 & 1 & 1 \end{vmatrix} - 0 \begin{vmatrix} -2 & 0 & 2 \\ \cancel{0} & \cancel{2} & \cancel{2} \\ 1 & 1 & 1 \end{vmatrix} + 1 \begin{vmatrix} -2 & 0 & 2 \\ 0 & 2 & 2 \\ \cancel{1} & \cancel{1} & \cancel{1} \end{vmatrix}$$

$$= -2 \begin{vmatrix} 2 & 2 \\ 1 & 1 \end{vmatrix} - 0 \begin{vmatrix} 0 & 2 \\ 1 & 1 \end{vmatrix} + 1 \begin{vmatrix} 0 & 2 \\ 2 & 2 \end{vmatrix}$$

$$= -2 \begin{vmatrix} 2 & 2 \\ 1 & 1 \end{vmatrix} - 0 \begin{vmatrix} 0 & 2 \\ 1 & 1 \end{vmatrix} + 1 \begin{vmatrix} 0 & 2 \\ 2 & 2 \end{vmatrix}$$

$$= -2(2 \cdot 1 - 1 \cdot 2) - 0(0 \cdot 1 - 1 \cdot 2) + 1(0 \cdot 2 - 2 \cdot 2)$$

$$= -2 \cdot 0 - 0(-2) + 1(-4) = -4.$$

Hence the parallelepiped formed by u, v, and w has volume $|-4|$, as before. Here the volume follows from the above manipulations of the matrix of numbers formed with columns of the matrix being the vectors u, v, and w. Chapter 6 shows that this computation of volume generalizes to determining, via analogous matrices of vectors, the 'volume' of objects formed by vectors with any number of components.

1.4.1 Exercises

Exercise 1.4.1 Use Definition 1.4.5 to establish some of the standard unit vector identities in Example 1.4.6:

(a) $j \times k = i$, $\quad k \times j = -i$, $\quad j \times j = 0$;
(b) $k \times i = j$, $\quad i \times k = -j$, $\quad k \times k = 0$.

Exercise 1.4.2 Use Definition 1.4.5, perhaps via the procedure used in Example 1.4.3, to determine the following cross products. Confirm that each cross product is orthogonal to the two vectors in the given product. Show your details.

(a) $(3i + j) \times (3i - 3j - 2k)$
(b) $(3i + k) \times (5i + 6k)$
(c) $(4, 1, 3) \times (3, 2, -1)$
(d) $(3, -7, 3) \times (2, 1, 0)$

Exercise 1.4.3 For each of the stereo pictures below, estimate the area of the pictured parallelogram by estimating the edge vectors v and w (all components are integers), then computing their cross product.

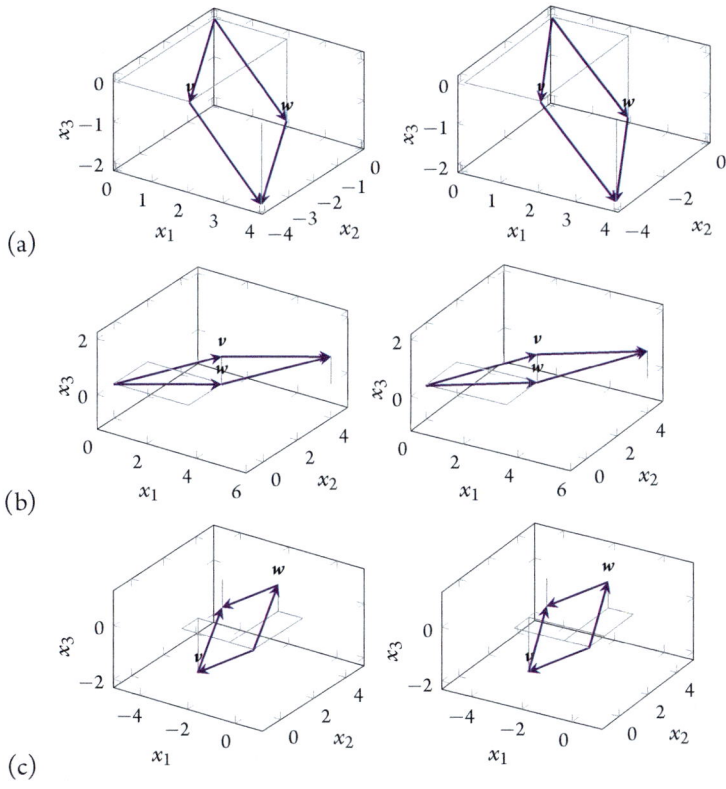

(a)

(b)

(c)

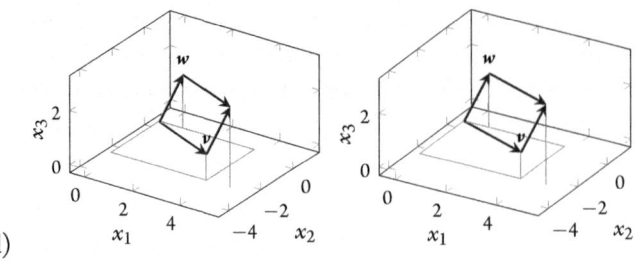

(d)

Exercise 1.4.4 Each of the following equations describes a plane in 3D. Find a normal vector to each of the planes.

(a) $x = (-1,0,1) + (-5,2,-1)s + (2,-4,0)t$

(b) $2x + 2y + 4z = 20$

(c) $x = j + 2k + (i - k)s + (-5i + j - 3k)t$

(d) $3y = x + 2z + 4$

Exercise 1.4.5 Use Definition 1.4.5 to prove that, for all vectors v, w in $\mathbb{R}^3$, the cross product $v \times w$ is orthogonal to w.

Exercise 1.4.6 Prove the identity that for every pair of vectors v, w in $\mathbb{R}^3$, $|v \times w|^2 = |v|^2|w|^2 - (v \cdot w)^2$ (an identity invoked in the proof of Theorem 1.4.10(c)). Use the algebraic Definitions 1.3.2 and 1.4.5 of the dot and cross products to expand both sides of the identity and show that both sides expand to the same complicated expression.

Exercise 1.4.7 Using Theorem 1.4.14, and the identities among standard unit vectors of Example 1.4.6, compute the following cross products. Record and justify each step in detail.

(a) $i \times (3j)$

(b) $(4j + 3k) \times k$

(c) $(2i + 2k) \times (i + j)$

(d) $(i - 5j) \times (-j + 3k)$

Exercise 1.4.8 You are given that three specific vectors u, v, and w in $\mathbb{R}^3$ have the following cross products:

$$u \times v = -j + k, \qquad u \times w = i - k, \qquad v \times w = -i + 2j.$$

Use Theorem 1.4.14 to compute the following cross products. Record and justify each step in detail.

(a) $(u + v) \times w$

(b) $(3u + w) \times (2u)$

(c) $(2v + 3w) \times (u + 2w)$

(d) $(u + 4v + 2w) \times w$

Exercise 1.4.9 Use Definition 1.4.5 to algebraically prove Theorem 1.4.14(b)—the property that $w \times v = -(v \times w)$. Explain how this property also follows from the basic geometry of the cross product (Theorem 1.4.10).

Exercise 1.4.10 Use Definition 1.4.5 to algebraically prove Theorem 1.4.14(c)—the property that $(c\boldsymbol{v}) \times \boldsymbol{w} = c(\boldsymbol{v} \times \boldsymbol{w}) = \boldsymbol{v} \times (c\boldsymbol{w})$. Explain how this property also follows from the basic geometry of the cross product (Theorem 1.4.10)—consider $c > 0$, $c = 0$, and $c < 0$ separately.

Exercise 1.4.11 Use Definition 1.4.5 to algebraically prove Theorem 1.4.14(d)—the distributive property that $\boldsymbol{u} \times (\boldsymbol{v} + \boldsymbol{w}) = \boldsymbol{u} \times \boldsymbol{v} + \boldsymbol{u} \times \boldsymbol{w}$.

Exercise 1.4.12 For each of the following illustrated parallelepipeds: estimate the edge vectors $\boldsymbol{u}$, $\boldsymbol{v}$, and $\boldsymbol{w}$ (all components are integers); then use the scalar triple product to estimate the volume of the parallelepiped.

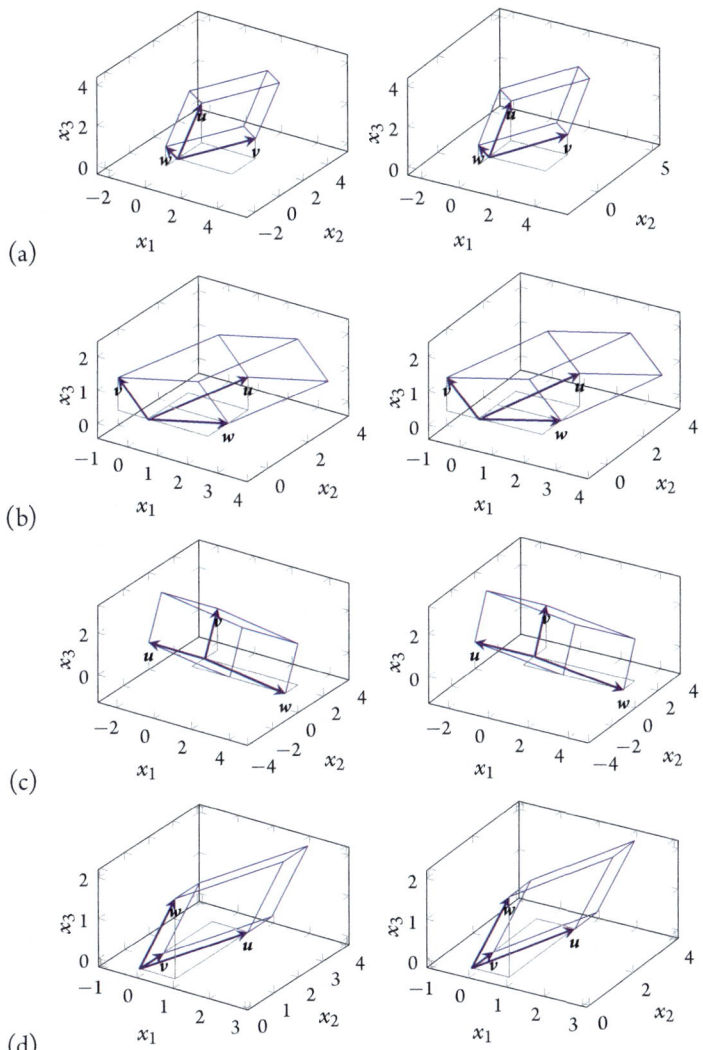

(a)

(b)

(c)

(d)

Exercise 1.4.13 In a few sentences, answer/discuss each of the following.

(a) What properties of the cross product differ from that of the multiplication of scalar numbers?

(b) How is the cross product useful in changing from a parametric equation of a plane to a normal equation of the plane?

(c) Given the properties $u \cdot v = |u||v| \cos \theta$ and $|u \times v| = |u||v| \sin \theta$, why is the dot product more useful for determining the angle θ between the vectors u and v?

1.5 Use MATLAB/Octave for vector computation

It is the science of *calculation*, which becomes continually more necessary at each step of our progress, and which must ultimately govern the whole of the applications of science to the arts of life.   *Charles Babbage, 1832*

Subsequent chapters invoke either of the computer packages MATLAB or Octave to perform calculations that would be tedious and error prone if done by hand. This section introduces MATLAB/Octave so that you can start to become familiar with it on small problems. You should directly compare the computed answer with your calculation by hand. The aim is to develop some basic confidence with MATLAB/Octave before later using it to save you considerable time in longer tasks.

- MATLAB is commercial software available from MathWorks.[9] It is also useable over the internet as MATLAB-Online or MATLAB-Mobile.

- Octave is free software, that for our purposes is almost identical to MATLAB, and download-able over the internet.[10] Octave is also freely useable over the internet.[11]

- Alternatively, your home institution may provide MATLAB/Octave via a web service that is useable via smart phones, tablets, and computers.

Example 1.5.1 Use the MATLAB/Octave command called norm() to compute the length/magnitude of the following vectors (Definition 1.1.9). Generally, computer commands and their output are typeset in the special fixed width font, such as this command norm().

(a) $(2, -1)$

Solution: Start MATLAB/Octave. After a prompt, ">>" in MATLAB or "octave:1>" in Octave, type a command, followed by the Return/Enter key to get it executed. As indicated by Table 1.2 the numbers with brackets separated by semicolons form a vector, and the = character assigns the result to a variable for subsequent use.

Assign the vector $(2, 1)$ to a variable a by the command a=[2;-1]. Then executing norm(a) reports ans = 2.2361 as shown in the dialogue to the right.

```
>> a=[2;-1]
a =
    2
   -1
>> norm(a)
ans =   2.2361
```

---

[9] http://mathworks.com
[10] https://www.gnu.org/software/octave/
[11] http://octave-online.net for example

Table 1.2 Use MATLAB/Octave to help compute vector results with the following basics. This and subsequent tables throughout the book summarize MATLAB/Octave for our use.

- Real numbers are limited to being zero or of magnitude from 10^{-323} to 10^{+308}, both positive and negative (called the **floating point** numbers). Real numbers are computed and stored to a maximum precision of nearly sixteen significant digits.[a]
- MATLAB/Octave potentially use complex numbers ($\mathbb{C}$), but mostly we stay within real numbers ($\mathbb{R}$).
- Each MATLAB/Octave command is usually typed on one line by itself.
- [. ; . ; .] where each dot denotes a number, forms vectors in $\mathbb{R}^3$ (or use newlines instead of the semicolons). Use n numbers separated by semicolons for vectors in $\mathbb{R}^n$.
- = assigns the result of the expression to the right of the = to the variable name on the left. If the result of an expression is not explicitly assigned to a variable, then by default it is assigned to the variable ans.
- Variable names are alphanumeric starting with a letter.
- size(v) returns the number of components of the vector (Definition 1.1.4): if the vector v is in $\mathbb{R}^m$, then size(v) returns $\begin{bmatrix} m & 1 \end{bmatrix}$.
- norm(v) computes the length/magnitude of the vector v (Definition 1.1.9).
- +, -, * is vector/scalar addition, subtraction, and scalar multiplication, but only provided the sizes of vectors are the same. Parentheses () control the order of operations.
- /x divides a vector/scalar by a scalar x. However, be warned that /v for a vector v typically gives a strange result as MATLAB/Octave interprets it to mean you want to approximately solve some linear equation.
- x^y for scalars x and y computes x^y.
- dot(u,v) computes the dot product of vectors u and v (Definition 1.3.2)—if they have the same size.
- acos(q) computes the arc-cos, the inverse cosine, of the scalar q in radians. To find the angle in degrees use acos(q)*180/pi (MATLAB/Octave knows $pi = \pi = 3.14159\cdots$).
- quit terminates the MATLAB/Octave session.

[a] If desired, 'computer algebra' software provides us with an arbitrary level of precision, even exact. Current computer algebra software includes the free Sage, Maxima, and Reduce, and the commercial Maple, Mathematica, and (via Matlab) MuPad.

This computes the answer $|(2,-1)| = \sqrt{2^2 + (-1)^2} = \sqrt{5} = 2.2361$ (to five significant digits, which we take to be practically exact). □

(b) $(-0.3, 4.3, -2.5, -2.8, 7, -1.9)$

Solution: In MATLAB/Octave:

– assign the vector with the command
c=[-0.3;4.3;-2.5;-2.8;7;-1.9]

– then executing norm(c) gives ans = 9.2347

Hence the length of vector $(-0.3, 4.3, -2.5, -2.8, 7, -1.9)$ is 9.2347 (to five significant digits). □

Example 1.5.2 Use MATLAB/Octave operators $+, -, *$ to compute the value of the expressions $u + v$, $u - v$, $3u$ for vectors $u = (-4.1, 1.7, 4.1)$ and $v = (2.9, 0.9, -2.4)$ (Definition 1.2.4).

Solution: In MATLAB/Octave type the commands, each followed by Return/Enter key.

Assign the named vectors with the commands u=[-4.1;1.7;4.1] and v=[2.9;0.9;-2.4] to see the two steps in the dialogue to the right.

```
>> u=[-4.1;1.7;4.1]
u =
   -4.1000
    1.7000
    4.1000
>> v=[2.9;0.9;-2.4]
v =
    2.9000
    0.9000
   -2.4000
```

Execute u+v to find from the dialogue on the right that the sum
$u + v = (-1.2, 2.6, 1.7)$.

```
>> u+v
ans =
   -1.2000
    2.6000
    1.7000
```

Execute u-v to find from the dialogue on the right that the difference
$u - v = (-7, 0.8, 6.5)$.

```
>> u-v
ans =
   -7.0000
    0.8000
    6.5000
```

Execute 3*u to find from the dialogue on the right that the scalar multiple
$3u = (-12.3, 5.1, 12.3)$ (the asterisk is essential to compute multiplication in MATLAB/Octave).

```
>> 3*u
ans =
  -12.3000
    5.1000
   12.3000
```
□

Example 1.5.3 Use MATLAB/Octave to confirm that $2(2p - 3q) + 6(q - p) = -2p$ for vectors $p = (1, 0, 2, -6)$ and $q = (2, 4, 3, 5)$.

Solution: In MATLAB/Octave

Assign the first vector with
p= [1;0;2;-6] as shown to the right.

```
>> p=[1;0;2;-6]
p =
    1
    0
    2
   -6
```

Assign the other vector with
q= [2;4;3;5].

```
>> q=[2;4;3;5]
q =
    2
    4
    3
    5
```

Compute $2(2\boldsymbol{p}-3\boldsymbol{q})+6(\boldsymbol{q}-\boldsymbol{p})$ with the
command 2*(2*p-3*q)+6*(q-p) as
shown to the right, and see the result is
evidently $-2\boldsymbol{p}$.

```
>> 2*(2*p-3*q)+6*(q-p)
ans =
   -2
    0
   -4
   12
```

Confirm that it is $-2\boldsymbol{p}$ by adding $2\boldsymbol{p}$ to the
above result with the command ans+2*p
as shown to the right, and see the zero vector
result.

```
>> ans+2*p
ans =
    0
    0
    0
    0
```

When using Octave, zeros such as the above may be reported as numbers such as 9.997e-17.
Such a number represents $9.997 \cdot 10^{-17}$ which is effectively zero when compared to the numbers
in the problem. Treat e-16 or e-17 as zero.   □

Example 1.5.4 Use MATLAB/Octave to confirm the commutative law (Theorem 1.2.19(a))
$\boldsymbol{u}+\boldsymbol{v}=\boldsymbol{v}+\boldsymbol{u}$ for vectors $\boldsymbol{u}=(8,-6,-4,-2)$ and $\boldsymbol{v}=(4,3,-1)$.

Solution: In MATLAB/Octave

Assign $u = (8, -6, -4, -2)$ with command u=[8;-6;-4;-2] as shown to the right.

```
>> u=[8;-6;-4;-2]
u =
     8
    -6
    -4
    -2
```

Assign $v = (4, 3, -1)$ with the command v=[4;3;-1].

```
>> v=[4;3;-1]
v =
     4
     3
    -1
```

Compute $u + v$ with the command u+v as shown to the right. MATLAB prints an error message because the vectors u and v are of different sizes and so cannot be added together.

```
>> u+v
Error using  +
Matrix dimensions must agree.
```

Check the sizes of the vectors in the sum using size(u) and size(v) to confirm u is in $\mathbb{R}^4$ whereas v is in $\mathbb{R}^3$. Hence the two vectors cannot be added (Definition 1.2.4).

```
>> size(u)
ans =
     4     1
>> size(v)
ans =
     3     1
```

Alternatively, Octave gives the following error message. (In such a message, "nonconformant arguments" means that the vectors are of a wrong size.)

```
error: operator +: nonconformant arguments
(op1 is 4x1, op2 is 3x1)
```
☐

Activity 1.5.5 You enter the two vectors into MATLAB by typing u=[1.1;3.7;-4.5] and v=[1.7;0.6;-2.6]. Which of the following is the result of typing the command u-v?

		(b)	Error			(d)	
	2.8000				2.2000		-0.6000
(a)	4.3000		using *	(c)	7.4000		3.1000
	-7.1000		Inner		-9.0000		-1.9000
			matrix				
			dimensions				
			must agree.				

Activity 1.5.6 For the vectors of the previous Activity 1.5.5: which is the result of typing the command 2*u? which is the result of typing the command u*v?

Example 1.5.7 Use MATLAB/Octave to compute the angles between the pair of vectors $(4,3)$ and $(5,12)$ (Theorem 1.3.5).

Solution: In MATLAB/Octave

Because each vector is used twice in the formula $\cos\theta = (\boldsymbol{u}\cdot\boldsymbol{v})/(|\boldsymbol{u}||\boldsymbol{v}|)$, give each a name as shown to the right.

```
>> u= [4;3]
u =
    4
    3
>> v= [5;12]
v =
    5
   12
```

Then compute the dot product, via dot (), in the formula for the cosine of the angle.

```
>> cost=dot (u,v)/norm (u)/norm (v)
cost =  0.8615
```

Lastly, invoke acos () for the arc-cosine, and convert the radians to degrees to find the angle $\theta = 30.510°$.

```
>> theta=acos (cost) *180/pi
theta =  30.510
```

□

Example 1.5.8 Verify the distributive law for the dot product $(\boldsymbol{u}+\boldsymbol{v})\cdot\boldsymbol{w}=\boldsymbol{u}\cdot\boldsymbol{w}+\boldsymbol{v}\cdot\boldsymbol{w}$ (Theorem 1.3.13(d)) for vectors $\boldsymbol{u}=(-0.1,-3.1,-2.9,-1.3)$, $\boldsymbol{v}=(-3,0.5,6.4,-0.9)$, and $\boldsymbol{w}=(-1.5,-0.2,0.4,-3.1)$.

Solution: In MATLAB/Octave

Assign vector
$u = (-0.1, -3.1, -2.9, -1.3)$ with the
command
`u=[-0.1;-3.1;-2.9;-1.3]` as
shown to the right.

```
>> u=[-0.1;-3.1;-2.9;-1.3]
u =
   -0.1000
   -3.1000
   -2.9000
   -1.3000
```

Assign vector $v = (-3, 0.5, 6.4, -0.9)$
with the command
`v=[-3;0.5;6.4;-0.9]` as shown to
the right.

```
>> v=[-3;0.5;6.4;-0.9]
v =
   -3.0000
    0.5000
    6.4000
   -0.9000
```

Assign vector $w = (-1.5, -0.2, 0.4, -3.1)$
with the command
`w=[-1.5;-0.2;0.4;-3.1]` as shown
to the right.

```
>> w=[-1.5;-0.2;0.4;-3.1]
w =
   -1.5000
   -0.2000
    0.4000
   -3.1000
```

Compute the dot product $(u + v) \cdot w$ with
the command `dot (u+v,w)` to find that
the answer is 13.390.

```
>> dot (u+v,w)
ans =   13.390
```

Compare this with the dot product
expression $u \cdot w + v \cdot w$ via the command
`dot (u,w)+dot (v,w)` to find that the
answer is 13.390.

```
>> dot (u,w)+dot (v,w)
ans =   13.390
```

That the two answers agree verifies the distributive law for dot products.   □

Activity 1.5.9 Given two vectors u and v that have already been typed into MATLAB/Octave, which of the following expressions could check the identity that $(u - 2v) \cdot (u + v) = u \cdot u - u \cdot v - 2v \cdot v$?

(a) `dot (u-2*v,u+v)-dot (u,u)+dot (u,v)+2*dot (v,v)`

(b) None of the others

(c) `dot (u-2v,u+v)-dot (u,u)+dot (u,v)+2dot (v,v)`

(d) `(u-2*v)*(u+v)-u*u+u*v+2*v*v`

Many other books (Quarteroni and Saleri, 2006, e.g.§§1.1–3) give more details about the basics than the essentials that are introduced here.

> On two occasions I have been asked [by members of Parliament!], "Pray, Mr. Babbage, if you put into the machine wrong figures, will the right answers come out?"
>
> I am not able rightly to apprehend the kind of confusion of ideas that could provoke such a question. *Charles Babbage*

1.5.1 Exercises

Exercise 1.5.1 Use MATLAB/Octave to compute the length of each of the following vectors.

(a) $\begin{bmatrix} 2 \\ 3 \\ 6 \end{bmatrix}$
(b) $\begin{bmatrix} 4 \\ -4 \\ 7 \end{bmatrix}$
(c) $\begin{bmatrix} 2.6 \\ -0.1 \\ 3.2 \\ -0.6 \\ -0.2 \end{bmatrix}$
(d) $\begin{bmatrix} 1.6 \\ -1.1 \\ -1.4 \\ 2.3 \\ -1.6 \end{bmatrix}$

Exercise 1.5.2 Use MATLAB/Octave to determine which are wrong out of the following identities and relations for vectors $p = (0.8, -0.3, 1.1, 2.6, 0.1)$ and $q = (1, 2.8, 1.2, 2.3, 2.3)$.

(a) $3(p-q) = 3p - 3q$

(b) $2(p - 3q) + 3(2q - p) = p$

(c) $|p - q| \le |p| + |q|$

(d) $|p \cdot q| \le |p||q|$

Exercise 1.5.3 Use MATLAB/Octave to find the angles between all pairs of vectors in each of the following groups.

(a) $p = (2, 3, 6), q = (6, 2, -3), r = (3, -6, 2)$

(b) $u = \begin{bmatrix} -1 \\ -7 \\ -1 \\ -7 \end{bmatrix}, v = \begin{bmatrix} -1 \\ 4 \\ 4 \\ 4 \end{bmatrix}, w = \begin{bmatrix} 1 \\ -4 \\ -4 \\ -4 \end{bmatrix}$

(c) $a = \begin{bmatrix} -0.5 \\ 2.0 \\ -3.4 \\ 1.8 \\ 0.1 \end{bmatrix}, b = \begin{bmatrix} 5.4 \\ 7.4 \\ 0.5 \\ 0.7 \\ 1.3 \end{bmatrix}, c = \begin{bmatrix} -0.2 \\ -1.5 \\ -0.3 \\ 1.1 \\ 2.5 \end{bmatrix}, d = \begin{bmatrix} 1.0 \\ 2.0 \\ -1.3 \\ -4.4 \\ -2.0 \end{bmatrix}$

Exercise 1.5.4 In a few sentences, answer/discuss each of the following.

(a) Explain the differences between `size (v)` and `norm (v)`.

(b) What are the different roles for parentheses, ( ), and brackets, [ ], in MATLAB/Octave?

(c) Why do computer languages use a symbol for multiplication, namely *? For example, if a=3.14 and v=[2;3;5], why do we need to type a*v instead of just av?

1.6 Summary of vectors

Vectors have magnitude and direction

★ Quantities that have the properties of both a magnitude and a direction are called vectors. Using a coordinate system, with n coordinate axes in $\mathbb{R}^n$, a **vector** is an ordered n-tuple of real numbers represented as a row in parentheses or as a column in brackets (Definition 1.1.4):

$$(v_1, v_2, \ldots, v_n) = \begin{bmatrix} v_1 \\ v_2 \\ \vdots \\ v_n \end{bmatrix}$$

In applications, the components of vectors have physical units such as metres, or km/hr, or numbers of words—usually the components all have the same units.

★ The set of all vectors with n components is denoted $\mathbb{R}^n$ (Definition 1.1.5). The vector with all components zero, $(0,0,\ldots,0)$, is called the **zero vector** and denoted by **0**.

★★ The **length** (or **magnitude**) of vector v is (Definition 1.1.9)

$$|v| := \sqrt{v_1^2 + v_2^2 + \cdots + v_n^2}.$$

A vector of length one is called a **unit vector**. The **zero vector 0** is the only vector of length zero (Theorem 1.1.13).

Adding and stretching vectors

★ The **sum** or **addition** of u and v, denoted $u + v$, is the vector obtained by joining v to u 'head-to-tail', and is computed as (Definition 1.2.4)

$$u + v := (u_1 + v_1, u_2 + v_2, \ldots, u_n + v_n).$$

The **scalar multiplication** of u by c is computed as

$$cu := (cu_1, cu_2, \ldots, cu_n),$$

and has length $|c||u|$ in the direction of u when $c > 0$ but in the opposite direction when $c < 0$.

★ The **standard unit vectors** in $\mathbb{R}^n$, $e_1, e_2, \ldots, e_n$, are the unit vectors in the direction of the corresponding coordinate axis (Definition 1.2.7). In $\mathbb{R}^2$ and $\mathbb{R}^3$ they are often denoted by $i = e_1, j = e_2$, and $k = e_3$.

- The **distance** between vectors u and v is the length of their difference, $|u - v|$ (Definition 1.2.10).
- A **parametric equation** of a line is $x = p + td$ where p is any point on the line, d is the **direction vector**, and the scalar **parameter** t varies over all real values (Definition 1.2.15).

★★ Addition and scalar multiplication of vectors satisfy the following familiar algebraic properties (Theorem 1.2.19):
- $u + v = v + u$ (commutative law);
- $(u + v) + w = u + (v + w)$ (associative law);
- $u + 0 = 0 + u = u$;
- $u + (-u) = (-u) + u = 0$;
- $a(u + v) = au + av$ (a distributive law);
- $(a + b)u = au + bu$ (a distributive law);
- $(ab)u = a(bu)$;
- $1u = u$;
- $0u = 0$;
- $|au| = |a| \cdot |u|$.

The dot product determines angles and lengths

★★ The **dot product** (or **inner product**) of two vectors u and v in $\mathbb{R}^n$ is the scalar (Definition 1.3.2)

$$u \cdot v := u_1 v_1 + u_2 v_2 + \cdots + u_n v_n.$$

★★ Determine the **angle** θ between the vectors by (Theorem 1.3.5)

$$\cos\theta = \frac{u \cdot v}{|u||v|}, \quad 0 \le \theta \le \pi \quad (0 \le \theta \le 180°).$$

In applications, the angle between two vectors tells us whether the vectors are in a similar direction, or not.

★★ The (nonzero) vectors are termed **orthogonal** (or **perpendicular**) if and only if their dot product $u \cdot v = 0$ (Definition 1.3.19).

- In mechanics the work done by a force F on a body that moves a distance d is the dot product $W = F \cdot d$.

★ The dot product (inner product) of vectors satisfies the following algebraic properties (Theorems 1.3.13 and 1.3.17):
- $u \cdot v = v \cdot u$ (commutative law);
- $u \cdot 0 = 0 \cdot u = 0$;
- $a(u \cdot v) = (au) \cdot v = u \cdot (av)$;
- $(u + v) \cdot w = u \cdot w + v \cdot w$ (distributive law);
- $u \cdot u \ge 0$, and moreover, $u \cdot u = 0$ if and only if $u = 0$.
- $\sqrt{u \cdot u} = |u|$, the length of u;

- $|\boldsymbol{u} \cdot \boldsymbol{v}| \leq |\boldsymbol{u}||\boldsymbol{v}|$ (Cauchy–Schwarz inequality);
- $|\boldsymbol{u} \pm \boldsymbol{v}| \leq |\boldsymbol{u}| + |\boldsymbol{v}|$ (triangle inequality).

- There is no nonzero vector orthogonal to all n standard unit vectors in $\mathbb{R}^n$ (Theorem 1.3.24). There can be no more than n orthogonal unit vectors in a set of vectors in $\mathbb{R}^n$ (Theorem 1.3.25).

- A **parametric equation** of a plane is $\boldsymbol{x} = \boldsymbol{p} + s\boldsymbol{u} + t\boldsymbol{v}$ for some point $\boldsymbol{p}$ in the plane, and two vectors $\boldsymbol{u}$ and $\boldsymbol{v}$ parallel to the plane, and where the scalar parameters s and t vary over all real values (Definition 1.3.32).

The cross product

★★ The **cross product** (or **vector product**) of vectors $\boldsymbol{v}$ and $\boldsymbol{w}$ in $\mathbb{R}^3$ is (Definition 1.4.5)

$$\boldsymbol{v} \times \boldsymbol{w} := \boldsymbol{i}(v_2 w_3 - v_3 w_2) + \boldsymbol{j}(v_3 w_1 - v_1 w_3) + \boldsymbol{k}(v_1 w_2 - v_2 w_1).$$

Theorem 1.4.10 gives the geometry:
  – the vector $\boldsymbol{v} \times \boldsymbol{w}$ is orthogonal to both $\boldsymbol{v}$ and $\boldsymbol{w}$;
  – the direction of $\boldsymbol{v} \times \boldsymbol{w}$ is in the right-hand sense;
  – $|\boldsymbol{v} \times \boldsymbol{w}| = |\boldsymbol{v}||\boldsymbol{w}|\sin\theta$ where θ is the angle between vectors $\boldsymbol{v}$ and $\boldsymbol{w}$ $(0 \leq \theta \leq \pi$, equivalently $0° \leq \theta \leq 180°)$; and
  – the length $|\boldsymbol{v} \times \boldsymbol{w}|$ is the area of the parallelogram with edges $\boldsymbol{v}$ and $\boldsymbol{w}$.

- The cross product has the following algebraic properties (Theorem 1.4.14):
  - $\boldsymbol{v} \times \boldsymbol{v} = \boldsymbol{0}$;
  - $\boldsymbol{w} \times \boldsymbol{v} = -(\boldsymbol{v} \times \boldsymbol{w})$ (not commutative);
  - $(c\boldsymbol{v}) \times \boldsymbol{w} = c(\boldsymbol{v} \times \boldsymbol{w}) = \boldsymbol{v} \times (c\boldsymbol{w})$;
  - $\boldsymbol{u} \times (\boldsymbol{v} + \boldsymbol{w}) = \boldsymbol{u} \times \boldsymbol{v} + \boldsymbol{u} \times \boldsymbol{w}$ (distributive law).

- The **scalar triple product** $\boldsymbol{u} \cdot (\boldsymbol{v} \times \boldsymbol{w})$ (Definition 1.4.19) is the volume of the parallelepiped with edges $\boldsymbol{u}$, $\boldsymbol{v}$, and $\boldsymbol{w}$.

Use MATLAB/*Octave for vector computation*

★★ [...] forms vectors: use n numbers separated by semicolons for vectors in $\mathbb{R}^n$ (or use newlines instead of the semicolons).

★★ = assigns the result of the expression to the right of the = to the variable name on the left.

★ norm(v) computes the length/magnitude of the vector v (Definition 1.1.9).

★ +, -, * is vector/scalar addition, subtraction, and multiplication. Parentheses () control the order of operations.

- /x divides a vector/scalar by a scalar x.

- x^y for scalars x and y computes x^y.

- dot(u,v) computes the dot product of vectors u and v (Definition 1.3.2).

- acos(q) computes the arc-cos, the inverse cosine, of the scalar q in radians. To find the angle in degrees use acos(q)*180/pi.

★★ quit terminates the MATLAB/Octave session.

Answers to selected activities

1.1.12d, 1.2.3b, 1.2.6a, 1.2.8a, 1.2.12b, 1.2.14c, 1.3.4a, 1.3.7d, 1.3.11d,
1.3.14d, 1.3.22a, 1.3.27c, 1.3.30b, 1.3.35d, 1.4.2a, 1.4.4a, 1.4.7d, 1.4.9b,
1.4.12a, 1.4.17c, 1.5.5d, 1.5.9a,

Answers to selected exercises

1.1.2b A$(2.6,-2.7)$ B$(3.9,2.7)$ C$(1.4,-2)$
     D$(0.2,0.6)$

1.1.2d A$(-0.2,2.4)$ B$(-2.9,-0.9)$
     C$(-1.2,-2.3)$ D$(-2.6,1.7)$

1.2.2b 1.3

1.2.2d 3.5

1.2.3b v and w are furthest; both the other
     pairs are equal closest.

1.2.3d u and w are closest; v and w are
     furthest.

1.2.4b One possibility is $x =$
     $(-4+7t)i + (1-6t)j + (-2+7t)k$

1.2.4d One possibility is $x = (1.8 - 3.2t)e_1 +$
     $(-3.1 + 3.9t)e_2 - (1 + 1.6t)e_3 +$
     $(-1.3 + 4.1t)e_4 + (-3.3 + 2.5t)e_5$

1.3.2b Orthogonal.

1.3.2d Not orthogonal.

1.3.9a $(-3, -2, 2) + (3, 7, -6)s + (8, 3,$
     $-5)t$

1.3.9c $(-1.8, 4.3, -3.9) + (-3.8, -6.5,$
     $-2.9)s + (4.3, -7.8, 2.2)t$

1.4.2b $-13j$

1.4.2d $(-3, 6, 17)$

1.4.3b $\sqrt{13} = 3.606$

1.4.3d $\sqrt{94} = 9.695$

1.4.4b $\propto (1, 1, 2)$

1.4.4d $\propto -i + 3j - 2k$

1.4.7b $4i$

1.4.7d $-15i - 3j - k$

1.4.8b $-2i + 2k$

1.4.8d $-3i + 8j - k$

1.4.12b 1

1.4.12d 2

1.5.1b 9

1.5.1d 3.6851

1.5.3b $\theta_{uv} = 147.44°, \theta_{uw} = 32.56°,$
     $\theta_{vw} = 180°$

2 Systems of linear equations

Linear relationships are commonly identified in science, engineering, and artificial intelligence. Such relationships are commonly expressed as linear equations. One of the reasons is that scientists and engineers can do amazingly powerful algebraic transformations with linear equations. Such transformations and their practical implications are the subject of this book.

One vital use in science and engineering is in the scientific task of taking scattered experimental data and inferring a general algebraic relation between the quantities measured. In computing science this task is often called 'data mining', 'knowledge discovery', or 'artificial intelligence'—although the algebraic relation is then typically discussed as a computational procedure. But always appearing within such tasks are linear equations to be solved.

Linear Algebra for the 21st Century. A. J. Roberts, Oxford University Press (2020). © A. J. Roberts.
DOI: 10.1093/oso/9780198856399.003.0002

Example 2.0.1 (*scientific inference*) Two colleagues, an American and a European, discuss the weather; in partic- ular, they discuss the temperature. (I am sure you can guess where we are going with this example, but let's pretend we do not know.) The American says "yesterday the tempera- ture was 80° but today is much cooler at 60°". The Euro- pean says, "that's not what I heard, I heard the temperature was 26° and today is 15°". (The graph to the right plots these two data points.) "Hmmmm, we must be using a different

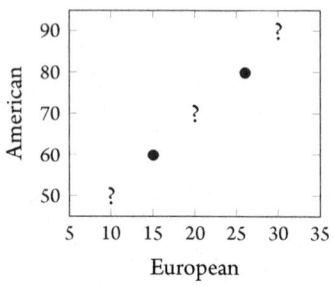

temperature scale", they say. Being scientists they start to use linear algebra to *infer*, from the two days of temperature data, a general relation between their temperature scales—a relationship valid over a wide range of temperatures (denoted by the question marks in the above-right figure). Let's assume that, in terms of the European temperature T_E, the American temperature $T_A = T_E \cdot c + d$ for some constants c and d that they and we aim to find. The two days of data then give that

$$80 = 26c + d \quad \text{and} \quad 60 = 15c + d.$$

To find the constants c and d:

- subtract the second equation from the first to deduce $80 - 60 = 26c + d - 15c - d$ which simplifies to $20 = 11c$, that is, $c = 20/11 = 1.82$ to two decimal places (2 d.p.);
- using this value of c in either equation, say the second, gives $60 = \frac{20}{11}15 + d$ which rear- ranges to $d = 360/11 = 32.73$ to two decimal places (2 d.p.).

We deduce that the temperature relationship is $T_A = 1.82 T_E + 32.73$ (as plotted to the right). The two col- leagues now *predict* that they will be able to use this formula to translate their temperature into that of the other, and vice versa.

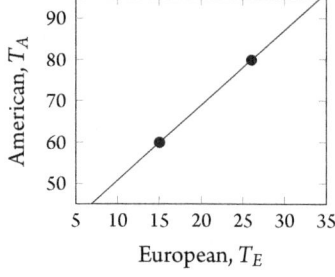

You may quite rightly object that the two colleagues *assumed* a linear relation; they do *not know* it is linear. You may also object that the predicted relation is erro- neous as it should be $T_A = \frac{9}{5}T_E + 32$ (the relation between Celsius and Fahrenheit). Absolutely, you should object. Scientifically, the deduced relation $T_A = 1.82 T_E + 32.73$ is only a conjecture that fits the known data. More data and more linear algebra together empower us to both confirm the linearity (or not as the case may be), and also to improve the accuracy of the coefficients. Such progressive refinement is fundamental scientific methodology—and central to it is the algebra of linear equations. ☐

Linear algebra and equations are also crucial for nonlinear relationships. Figure 2.1 shows four plots of the same nonlinear curve, but on successively smaller scales. Zooming in on the point $(0, 1)$ we see the curve looks straighter and straighter until on the microscale (bottom- right) it is effectively a straight line. The same is true for everywhere on every smooth curve: we may discover that every smooth curve looks like a straight line on its microscale. Thus we may view any smooth curve as roughly being made up of lots of microscale straight line segments.

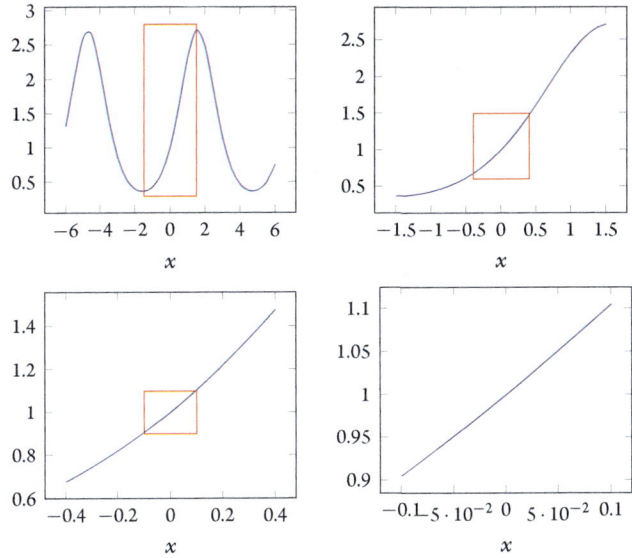

Figure 2.1 Zoom in anywhere on any smooth nonlinear curve, such as the plotted $f(x)$, and we discover that the curve looks like a straight line on the microscale. The (red) rectangles show the region plotted in the next graph in the sequence.

Linear equations and their algebra on this microscale empower our understanding of nonlinear relationships—for example, microscale linearity underwrites all of calculus.

2.1 Introduction to systems of linear equations

The great aspect of linear equations is that we straightforwardly manipulate them algebraically to deduce results: some results are immensely useful not only in applications but also in further theory.

Example 2.1.1 *(simple algebraic manipulation)* Following Example 2.0.1, recall that the temperature in Fahrenheit $T_F = \frac{9}{5}T_C + 32$ in terms of the temperature in Celsius, T_C. Straightforward algebra answers the following questions.

- What is a formula for the Celsius temperature as a function of the temperature in Fahrenheit? Answer by rearranging the equation: subtract 32 from both sides, $T_F - 32 = \frac{9}{5}T_C$; multiply both sides by $\frac{5}{9}$, then $\frac{5}{9}(T_F - 32) = T_C$; that is, $T_C = \frac{5}{9}T_F - \frac{160}{9}$.
- What temperature has the same *numerical value* in the two scales? That is, when is $T_F = T_C$? Answer by algebra: we want $T_C = T_F = \frac{9}{5}T_C + 32$; subtract $\frac{9}{5}T_C$ from both sides to give $-\frac{4}{5}T_C = 32$; multiply both sides by $-\frac{5}{4}$, then $T_C = -\frac{5}{4} \times 32 = -40$. This algebra discovers that $-40°$C is the same temperature as $-40°$F. ☐

Table 2.1 Examples of linear equations, and equations that are not linear (called nonlinear equations).

linear	nonlinear
$-3x + 2 = 0$	$x^2 - 3x + 2 = 0$
$2x - 3y = -1$	$2xy = 3$
$-1.2x_1 + 3.4x_2 - x_3 = 5.6$	$x_1^2 + 2x_2^2 = 4$
$r - 5s = 2 - 3s + 2t$	$r/s = 2 + t$
$\sqrt{3}t_1 + \frac{\pi}{2}t_2 - t_3 = 0$	$3\sqrt{t_1} + t_2^3/t_3 = 0$
$(\cos\frac{\pi}{6})x + e^2 y = 1.23$	$x + e^{2y} = 1.23$

Linear equations are characterized by each unknown never being multiplied or divided by another unknown, or itself, or inside 'curvaceous' functions. Table 2.1 lists examples of both. Generally, problems have many unknown variables. The power of linear algebra is especially important for large numbers of unknown variables. The number n of unknown variables may be two or three as in many examples herein, or may be thousands or millions in many modern applications.

Definition 2.1.2 *A **linear equation** in the n variables $x_1, x_2, \ldots, x_n$ is an equation that can be written in the form*

$$a_1 x_1 + a_2 x_2 + \cdots + a_n x_n = b$$

*where the **coefficients** a_1, a_2, $\ldots$, a_n and the **constant term** b are given scalar constants. An equation that cannot be written in this form is called a **nonlinear equation**. A **system** of linear equations is a set of one or more linear equations in one or more variables (usually more than one).*

Example 2.1.3 *(two equations in two variables)* Graphically and algebraically solve each of the following systems.

(a) $x + y = 3$
$\quad 2x - 4y = 0$

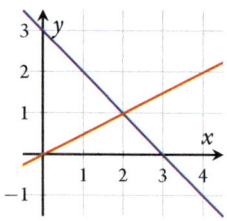

Solution: To draw the graphs seen in the plot to the right, rearrange the linear equations as $y = 3 - x$ and $y = x/2$. From the graph, they intersect only at the point $(2, 1)$ so $x = 2$ and $y = 1$ is the unique solution.

Algebraically, one could add twice the first equation to half of the second equation: $2(x + y) + \frac{1}{2}(2x - 4y) = 2 \cdot 3 + \frac{1}{2} \cdot 0$ which simplifies to $3x = 6$ as the y terms cancel; hence $x = 2$. Then, consider the second equation, $2x - 4y = 0$, which now becomes $2 \cdot 2 - 4y = 0$, that is, $y = 1$. This algebra gives the same solution $(x, y) = (2, 1)$ as graphically. □

(b) $2x - 3y = 2$
$-4x + 6y = 3$

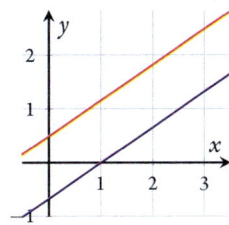

Solution: To draw the graphs seen in the plot to the right, rearrange the linear equations as $y = \frac{2}{3}x - \frac{2}{3}$ and $y = \frac{2}{3}x + \frac{1}{2}$. Evidently these lines never intersect: they appear parallel so there appears to be no solution.

Algebraically, one could add twice the first equation to the second equation: $2(2x - 3y) + (-4x + 6y) = 2 \cdot 2 + 3$ which, as all the x and y terms cancel, simplifies to $0 = 7$. This equation is a contradiction, as zero cannot be equal to seven. Thus there are no solutions to the system. □

(c) $x + 2y = 4$
$2x + 4y = 8$

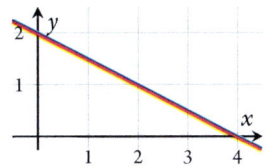

Solution: To draw the graphs seen in the plot to the right, rearrange the linear equations as $y = 2 - x/2$ and $y = 2 - x/2$. They are the same line so every point on this line is a solution of the system. There are an infinite number of possible solutions.

Algebraically, the rearrangement of both equations to exactly the same $y = 2 - x/2$ establishes an infinite number of solutions, here parametrized by x. □

Activity 2.1.4 Solve the system $x + 5y = 9$ and $x + 2y = 3$ to find that the only solution is which of the following?

(a) $(1,1)$ (b) $(1,2)$ (c) $(-1,2)$ (d) $(-1,1)$

Example 2.1.5 *(Global Positioning System)* The Global Positioning System (GPS) is a network of 24 satellites orbiting the Earth. Each satellite knows very accurately its position at all times, and broadcasts this position by radio. A receiver, say your smart-phone, picks up these signals and, from the time taken for the signals to arrive, knows the distance to all those satellites within 'sight'. Your smart-phone then solves a system of equations and informs you of its precise position.

Let's solve a definite example problem, but in two dimensions for simplicity. Suppose you and your smart-phone are at some unknown location (x, y) in the 2D-plane, on the Earth's surface where the Earth has radius about 6 Mm (here all distances are measured in units of megametres, Mm, equivalently thousands of km). But your smart-phone picks up the broadcast from three GPS satellites, and then determines their distance from you.

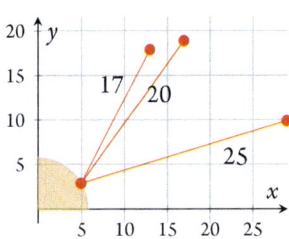

From the broadcast and the timing, suppose you then know that a satellite at $(29, 10)$ is 25 away (all in Mm), one at $(17, 19)$ is 20 away, and one at $(13, 18)$ is 17 away (as drawn to the right). Find your location (x, y).

Solution: From these three sources of information, Pythagoras and the length of displacement vectors (Definition 1.1.9) gives the three equations

$$(x-29)^2 + (y-10)^2 = 25^2,$$
$$(x-17)^2 + (y-19)^2 = 20^2,$$
$$(x-13)^2 + (y-18)^2 = 17^2.$$

These three equations constrain the unknown variables of your, as yet unknown, location (x,y). Expanding the squares in these equations gives the equivalent system of equations

$$x^2 - 58x + 841 + y^2 - 20y + 100 = 625,$$
$$x^2 - 34x + 289 + y^2 - 38y + 361 = 400,$$
$$x^2 - 26x + 169 + y^2 - 36y + 324 = 289.$$

Involving squares of the unknowns, x^2 and y^2, these form a *nonlinear* system of equations and so appear to lie outside the remit of this book. However, straightforward algebra transforms these three nonlinear equations into a system of two linear equations which we solve.

Let's subtract the third equation from each of the other two, then the nonlinear squared terms cancel, giving a system of two linear equations in two variables:

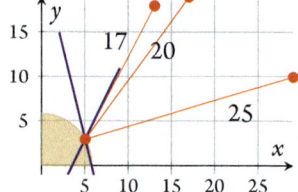

$$-32x + 672 + 16y - 224 = 336 \iff -2x + y = -7;$$
$$-8x + 120 - 2y + 37 = 111 \iff -4x - y = -23.$$

Graphically, include these two lines to the picture (in blue), namely $y = -7 + 2x$ and $y = 23 - 4x$, and then their intersection gives your location.

Algebraically, one could add the two equations together: $(-2x + y) + (-4x - y) = -7 - 23$ which reduces to $-6x = -30$, that is, $x = 5$. Then either equation, say the first, determines $y = -7 + 2x = -7 + 2 \cdot 5 = 3$. That is, your location is $(x,y) = (5,3)$ (in Mm), as drawn. □

If the x-axis is a line through the equator, and the y-axis goes through the North Pole, then trigonometry gives that your location would be at latitude $\tan^{-1}\frac{3}{5} = 0.5404 = 30.96°$N.

Example 2.1.6 *(three equations in three variables)* Graph the surfaces and algebraically solve the system

$$x_1 + x_2 - x_3 = -2,$$
$$x_1 + 3x_2 + 5x_3 = 8,$$
$$x_1 + 2x_2 + x_3 = 1.$$

Solution: The plot to the right shows the three planes represented by the given equations (in the order blue, brown, red), and plots the (black) point of intersection of all three planes. The black intersection point is where all three equations are satisfied. Let's find this point.

Algebraically we combine and manipulate the equations in a sequence of steps designed to simplify the form of the system. *By doing the same manipulation to the whole of each of the equations, we ensure that the intersection point remains the same throughout.*

(a) Subtract the first equation from each of the other two equations to deduce (as illustrated)

$$x_1 + x_2 - x_3 = -2,$$
$$2x_2 + 6x_3 = 10,$$
$$x_2 + 2x_3 = 3.$$

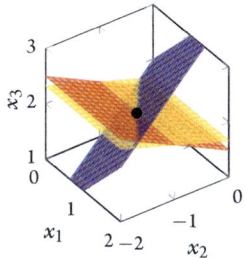

(b) Divide the new second equation by two:

$$x_1 + x_2 - x_3 = -2,$$
$$x_2 + 3x_3 = 5,$$
$$x_2 + 2x_3 = 3.$$

(c) Subtract the new second equation from each of the other two (as illustrated):

$$x_1 \qquad - 4x_3 = -7,$$
$$x_2 + 3x_3 = 5,$$
$$-x_3 = -2.$$

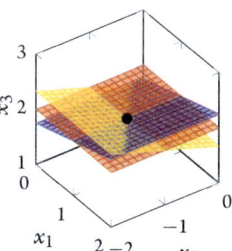

(d) Multiply the new third equation by (-1):

$$x_1 \qquad - 4x_3 = -7,$$
$$x_2 + 3x_3 = 5,$$
$$x_3 = 2.$$

(e) Add four times the new third equation to the first, and subtract three times it from the second (as illustrated):

$$x_1 \qquad\qquad = 1,$$
$$x_2 \qquad = -1,$$
$$x_3 = 2.$$

Thus the only solution to this system of three linear equations in three variables is $(x_1, x_2, x_3) = (1, -1, 2)$. ☐

The sequence of graphs in the previous Example 2.1.6 illustrates the equations at each main step in the algebraic manipulations. Apart from keeping the solution intersection point fixed, the sequence of graphs looks rather chaotic. Indeed, for each of these algebraic steps there is no particular geometric pattern or interpretation. In contrast, one feature of the upcoming Section 3.3 is that we discover how the so-called 'singular value decomposition' solves linear equations via a great method with a strong geometric interpretation. This geometric interpretation then empowers further methods useful in applications.

Transform into abstract setting Linear algebra has an important aspect crucial in applications. A crucial skill in applying linear algebra is that it takes an application problem and transforms it into an abstract setting. Example 2.0.1 transformed the problem of inferring a line through two data points into solving two linear equations. The next Example 2.1.7 similarly transforms the problem of inferring a plane through three data points into solving three linear equations. The original application is often not easily recognizable in the abstract version. Nonetheless, it is the abstraction by linear algebra that empowers wonderful results for applications.

Example 2.1.7 (infer a surface through three points) This example illustrates the previous paragraph. Given a geometric problem of inferring what plane passes through three given points, we transform this problem into the linear algebra task of finding the intersection point of three specific planes (in a different space). This task we then do.

Table 2.2 In some artificial units, this table lists measured temperature, humidity, and rainfall.

temp	humid	rain
1	−1	−2
3	5	8
2	1	1

Suppose we observe that at some given temperature and humidity we get some rainfall: let's find a formula that predicts the rainfall from temperature and humidity measurements. In some *completely artificial units*, Table 2.2 lists measured temperature ('temp'), humidity ('humid'), and rainfall ('rain').

Solution: To infer a relation to hold generally—to fill in the gaps between the known measurements, seek 'rainfall' to be predicted by the linear formula

$$(\text{'rain'}) = x_1 + x_2(\text{'temp'}) + x_3(\text{'humid'}),$$

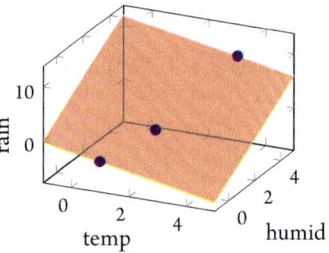

for some coefficients x_1, x_2, and x_3 to be determined. The measured data of Table 2.2 constrains and determines these coefficients: substitute each triple of measurements to require

$$
\begin{aligned}
-2 &= x_1 + x_2(1) + x_3(-1), \\
8 &= x_1 + x_2(3) + x_3(5), \\
1 &= x_1 + x_2(2) + x_3(1),
\end{aligned}
\quad\Longleftrightarrow\quad
\begin{aligned}
x_1 + x_2 - x_3 &= -2, \\
x_1 + 3x_2 + 5x_3 &= 8, \\
x_1 + 2x_2 + x_3 &= 1.
\end{aligned}
$$

The previous Example 2.1.6 solves this set of three linear equations in three unknowns to determine that the only solution is that the coefficients $(x_1, x_2, x_3) = (1, -1, 2)$. That is, the requisite formula to infer rain from any given temperature and humidity is

$$(\text{'rain'}) = 1 - (\text{'temp'}) + 2(\text{'humid'}).$$

This example illustrates that the geometry of fitting a plane to three points (as plotted) translates into the abstract geometry of finding the intersection of three planes (plotted in the previous example). The linear algebra procedure for this latter abstract problem then gives the required 'physical' solution. □

The solution of three linear equations in three variables leads to finding the intersection point of three planes. Figure 2.2 illustrates the three general possibilities: a unique solution (as in Example 2.1.6), or infinitely many solutions, or no solution. The solution of two linear equations in two variables also has the same three possibilities—as deduced and illustrated in Example 2.1.3. The next Section 2.2 establishes the general key property of a system of any number of linear equations in any number of variables: the system has either

- a unique solution (a consistent system), or
- infinitely many solutions (a consistent system), or
- no solutions (an inconsistent system).

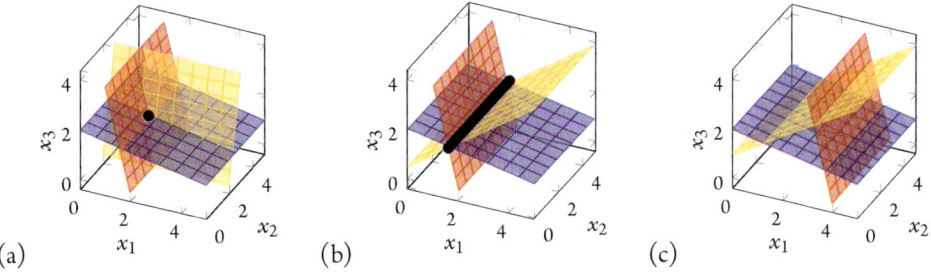

(a) (b) (c)

Figure 2.2 Solving three linear equations in three variables finds the intersection point(s) of three planes. The only three possibilities are: (a) a unique solution; (b) infinitely many solutions; or (c) no solution.

2.1.1 Exercises

Exercise 2.1.1 Graphically and algebraically solve each of the following systems.

(a) $\begin{aligned} x - 2y &= -3 \\ -4x &= -4 \end{aligned}$

(b) $\begin{aligned} 3x - 2y &= 2 \\ -3x + 2y &= -2 \end{aligned}$

(c) $\begin{aligned} 3x - 2y &= 1 \\ 6x - 4y &= -2 \end{aligned}$

(d) $\begin{aligned} 4x - 3y &= -1 \\ -5x + 4y &= 1 \end{aligned}$

(e) $\begin{aligned} p + q &= 3 \\ -p - q &= 2 \end{aligned}$

(f) $\begin{aligned} 4u + 4v &= -2 \\ -u - v &= 1 \end{aligned}$

(g) $\begin{aligned} -3s + 4t &= 0 \\ -3s + 3t &= -\tfrac{3}{2} \end{aligned}$

(h) $\begin{aligned} -4s + t &= -2 \\ 4s - t &= 2 \end{aligned}$

Exercise 2.1.2 For each of the following graphs: estimate the equations of the pair of lines; solve the pair of equations algebraically; and confirm that the algebraic solution is reasonably close to the intersection of the pair of lines.

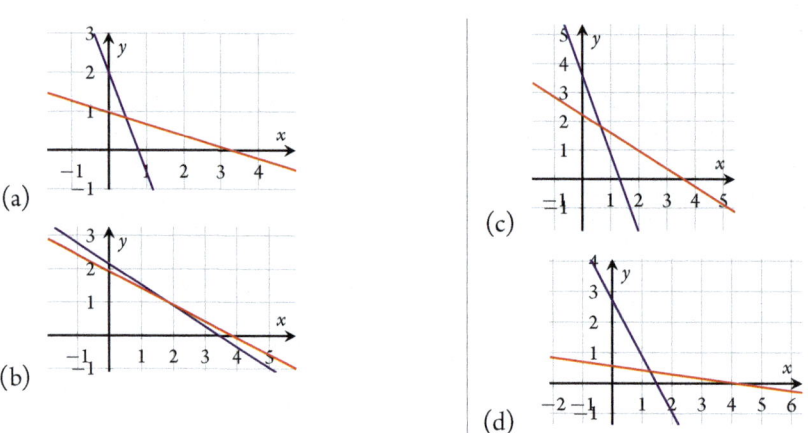

(a)

(b)

(c)

(d)

Exercise 2.1.3 Graphically and algebraically solve each of the following systems of three equations for the two unknowns.

(a) $\begin{aligned} 4x + y &= 8 \\ 3x - 3y &= -\tfrac{3}{2} \\ -4x + 2y &= -2 \end{aligned}$

(b) $\begin{aligned} -4x + 3y &= \tfrac{7}{2} \\ 7x + y &= -3 \\ x - 2y &= \tfrac{3}{2} \end{aligned}$

(c) $\begin{aligned} 3x + 2y &= 4 \\ -2x - 4y &= -4 \\ 4x + 2y &= 5 \end{aligned}$

(d) $\begin{aligned} -2x + 3y &= -3 \\ -5x + 2y &= -9 \\ 3x + 3y &= 6 \end{aligned}$

Exercise 2.1.4 *(Global Positioning System in 2D)* For each case below, and in two dimensions, suppose you know from three GPS satellites that you and your GPS receiver are given distances away from the given locations of each of the three satellites (all locations and distances are in Mm). Following Example 2.1.5, determine your position.

	25 from $(7,30)$		20 from $(22,12)$
(a)	26 from $(10,30)$	(c)	26 from $(16,24)$
	29 from $(20,27)$		29 from $(26,21)$
	25 from $(11,29)$		17 from $(12,21)$
(b)	26 from $(28,15)$	(d)	25 from $(10,29)$
	20 from $(16,21)$		26 from $(27,15)$

In which of these cases: are you at the 'North Pole'? flying high above the Earth? the measurement data is surely in error?

Exercise 2.1.5 In a few sentences, answer/discuss each of the following.

(a) Is the equation $x = 5/(3 + 4y/x)$ linear? or nonlinear? Explain.
(b) Recall that Example 2.1.5 determined your position in 2D from the GPS readings of three satellites. Can you determine your position in 2D from just two GPS satellites? Using linear algebra?

2.2 Directly solve linear systems

The previous Section 2.1 solved some example systems of linear equations by hand algebraic manipulation. We continue to do so for small systems. However, such by-hand solutions are tedious for systems bigger than say four equations in four unknowns. For bigger systems with anything from tens to millions of equations—which are typical in applications—we use computers to find solutions because computers are ideal for tedious repetitive calculations.

2.2.1 Compute a system's solution

It is unworthy of excellent persons to lose hours like slaves in the labour of calculation.
Gottfried Wilhelm von Leibniz

Computers primarily deal with numbers, not algebraic equations, so we have to abstract the coefficients of a system into a numerical data structure. We use matrices and vectors.

Example 2.2.1 The first system of Example 2.1.3(a)

$$x + y = 3 \qquad \text{is written} \qquad \underbrace{\begin{bmatrix} 1 & 1 \\ 2 & -4 \end{bmatrix}}_{A} \underbrace{\begin{bmatrix} x \\ y \end{bmatrix}}_{x} = \underbrace{\begin{bmatrix} 3 \\ 0 \end{bmatrix}}_{b}.$$

That is, the system $\begin{cases} x + y = 3 \\ 2x - 4y = 0 \end{cases}$ is equivalent to $Ax = b$ for

- the so-called coefficient matrix $A = \begin{bmatrix} 1 & 1 \\ 2 & -4 \end{bmatrix}$,
- the right-hand side vector $\boldsymbol{b} = (3,0)$, and
- the vector of unknown variables $\boldsymbol{x} = (x,y)$. ☐

The beauty of the form $A\boldsymbol{x} = \boldsymbol{b}$ is that the numbers involved in the system are abstracted into the matrix A and vector $\boldsymbol{b}$: MATLAB/Octave handles such numerical matrices and vectors.[1] For some of you, writing a system in this matrix-vector form $A\boldsymbol{x} = \boldsymbol{b}$ (Definition 2.2.2 below) may appear to be just some mystic rearrangement of symbols—such an interpretation is sufficient for this chapter. However, those of you who have met matrix multiplication will recognise that $A\boldsymbol{x} = \boldsymbol{b}$ is an expression involving natural operations for matrices and vectors: Section 3.1 defines and explores such useful operations.

Definition 2.2.2 *(matrix-vector form) For every given system of m linear equations in n variables*

$$a_{11}x_1 + a_{12}x_2 + \cdots + a_{1n}x_n = b_1,$$
$$a_{21}x_1 + a_{22}x_2 + \cdots + a_{2n}x_n = b_2,$$
$$\vdots$$
$$a_{m1}x_1 + a_{m2}x_2 + \cdots + a_{mn}x_n = b_m,$$

its ***matrix-vector form*** *is* $A\boldsymbol{x} = \boldsymbol{b}$ *for the* $m \times n$ ***matrix*** *of coefficients*

$$A = \begin{bmatrix} a_{11} & a_{12} & \cdots & a_{1n} \\ a_{21} & a_{22} & \cdots & a_{2n} \\ \vdots & \vdots & \ddots & \vdots \\ a_{m1} & a_{m2} & \cdots & a_{mn} \end{bmatrix},$$

and vectors $\boldsymbol{x} = (x_1, x_2, \ldots, x_n)$ *and* $\boldsymbol{b} = (b_1, b_2, \ldots, b_m)$. *If* $m = n$ *(the number of equations is the same as the number of variables), then A is called a* ***square matrix*** *(the number of rows is the same as the number of columns).*

Example 2.2.3 *(matrix-vector form)* Write the following systems in matrix-vector form.

(a)
$$x_1 + x_2 - x_3 = -2,$$
$$x_1 + 3x_2 + 5x_3 = 8,$$
$$x_1 + 2x_2 + x_3 = 1.$$

(b)
$$-2r + 3s = 6,$$
$$s - 4t = -\pi.$$

---

[1] In this chapter, the two-character symbol '$A\boldsymbol{x}$' is just a shorthand for all the left-hand sides of the linear equations in a system. Section 3.1 then defines a crucial multiplicative meaning to the composite symbol '$A\boldsymbol{x}$'.

Solution:

(a) The first system, that of Example 2.1.6, is of three equations in three variables $(m = n = 3)$ and is written in the form $A\boldsymbol{x} = \boldsymbol{b}$, for square matrix A, as

$$\underbrace{\begin{bmatrix} 1 & 1 & -1 \\ 1 & 3 & 5 \\ 1 & 2 & 1 \end{bmatrix}}_{A} \underbrace{\begin{bmatrix} x_1 \\ x_2 \\ x_3 \end{bmatrix}}_{\boldsymbol{x}} = \underbrace{\begin{bmatrix} -2 \\ 8 \\ 1 \end{bmatrix}}_{\boldsymbol{b}}.$$

(b) The second system has three variables called r, s, and t and two equations. Variables 'missing' from an equation are represented as zero times that variable, thus the system, for 2×3 matrix A,

$$\begin{matrix} -2r + 3s + 0t = 6, \\ 0r + s - 4t = -\pi, \end{matrix} \quad \text{is} \quad \underbrace{\begin{bmatrix} -2 & 3 & 0 \\ 0 & 1 & -4 \end{bmatrix}}_{A} \underbrace{\begin{bmatrix} r \\ s \\ t \end{bmatrix}}_{\boldsymbol{x}} = \underbrace{\begin{bmatrix} 6 \\ -\pi \end{bmatrix}}_{\boldsymbol{b}} \qquad \square$$

Activity 2.2.4 Which of the following systems correspond to the matrix-vector equation

$$\begin{bmatrix} -1 & 3 \\ 1 & 2 \end{bmatrix} \begin{bmatrix} u \\ w \end{bmatrix} = \begin{bmatrix} 1 \\ 0 \end{bmatrix}?$$

(a) $\begin{matrix} -x + y = 1 \\ 3x + 2y = 0 \end{matrix}$
(b) $\begin{matrix} -u + w = 1 \\ 3u + 2w = 0 \end{matrix}$
(c) $\begin{matrix} -x + 3y = 1 \\ x + 2y = 0 \end{matrix}$
(d) $\begin{matrix} -u + 3w = 1 \\ u + 2w = 0 \end{matrix}$

Procedure 2.2.5 (unique solution) *In* MATLAB/Octave, *to solve the matrix-vector system* $A\boldsymbol{x} = \boldsymbol{b}$ *for a square matrix A, use commands listed in Tables 1.2 and 2.3 to:*

1. *form matrix A and column vector* $\boldsymbol{b}$;

2. *check* rcond (A) *exists and is not too small,* $1 \geq good > 10^{-2} > poor > 10^{-4} > bad > 10^{-8} > terrible,$ (rcond (A) *is always between zero and one, inclusive);*

3. *if* rcond (A) *both exists and is acceptable, then execute* x=A\b *to compute the solution vector* $\boldsymbol{x}$.

Checking rcond (A) avoids gross mistakes. Section 3.3.2 discovers what rcond () is, and why rcond () avoids mistakes.[2] In practice, decisions about acceptability are rarely black and white, and so the qualitative ranges of rcond () in Procedure 2.2.5 reflect practical realities.

> In theory, there is no difference between theory and practice. But, in practice, there is.
> Jan L. A. van de Snepscheut

[2] Interestingly, there are incredibly rare pathological matrices for which even rcond () and A\ do fail us (Driscoll and Maki, 2007). For example, among 32×32 matrices the probability is about 10^{-22} of encountering a matrix for which rcond () misleads us by a factor of more than a hundred in using A\.

Table 2.3 To realize Procedure 2.2.5, and other procedures, we need these basics of MATLAB/Octave as well as that of Table 1.2.

- The floating point numbers are extended by `Inf`, denoting 'infinity', and `NaN`, denoting 'not a number' such as the indeterminate $0/0$.
- `[ ... ; ... ; ... ]` forms both matrices and vectors, or use newlines instead of the semicolons.
- `rcond(A)` of a square matrix A *estimates* the reciprocal of the so-called condition number of A (defined precisely by Definition 3.3.16).
- `x=A\b` computes an 'answer' to $Ax = b$ —but to be a solution requires `rcond(A)` to both exist and be not small.
- Change one element of an array or vector by assigning a new value with assignments `A(i,j) = ...` or `b(i) = ...` where `i` and `j` denote some indices.
- For a vector (or matrix) `t` and an exponent `p`, the operation `t.^p` computes the pth power of each element in the vector; for example, if `t= [1;2;3;4;5]` then `t.^2` results in `[1;4;9;16;25]`.
- The function `ones(m,1)` gives a (column) vector of m ones, $(1, 1, \ldots, 1)$.
- Lastly, always remember that 'the answer' by a computer is not necessarily 'the solution' of your problem.

Example 2.2.6   Use MATLAB/Octave to solve the system (from Example 2.1.6)

$$x_1 + x_2 - x_3 = -2,$$
$$x_1 + 3x_2 + 5x_3 = 8,$$
$$x_1 + 2x_2 + x_3 = 1.$$

Solution: Begin by writing the system in the abstract matrix-vector form $Ax = b$ as already done by Example 2.2.3. Then the three steps of Procedure 2.2.5 are the following.

(a) Form matrix A and column vector b with the MATLAB/Octave assignments (Remember that the symbol "=" in MATLAB/Octave is a procedural assignment of a value—very different in nature to the "=" in algebra which denotes equality.)

```
A= [1 1 -1; 1 3 5; 1 2 1]
b= [-2;8;1]
```

Table 2.3 summarizes that, in MATLAB/Octave, each line is one command; the = symbol assigns the value of the right-hand expression to the variable name of the left-hand side; and the brackets [ ] construct both matrices and vectors.

(b) Check the value of `rcond(A)` : here it is 0.018 which is in the good range.

(c) Since `rcond(A)` is acceptable, execute `x = A\b` to compute the solution vector $x = (1, -1, 2)$ (and assign it to the variable x, see Table 2.3).

All together, that is the four commands

```
A=[1 1 -1; 1 3 5; 1 2 1]
b=[-2;8;1]
rcond(A)
x=A\b
```
☐

Activity 2.2.7 Use MATLAB/Octave to solve the system $7x + 8y = 42$ and $32x + 38y = 57$, to find the answer for (x, y) is which of the following?

(a) $\begin{bmatrix} -94.5 \\ 114 \end{bmatrix}$

(c) $\begin{bmatrix} 73.5 \\ 342 \end{bmatrix}$

(b) $\begin{bmatrix} 114 \\ -94.5 \end{bmatrix}$

(d) $\begin{bmatrix} 342 \\ 73.5 \end{bmatrix}$

Example 2.2.8 Following the previous Example 2.2.6, solve each of the two systems:

$$\text{(a)} \quad \begin{aligned} x_1 + x_2 - x_3 &= -2, \\ x_1 + 3x_2 + 5x_3 &= 5, \\ x_1 - 3x_2 + x_3 &= 1; \end{aligned} \qquad \text{(b)} \quad \begin{aligned} x_1 + x_2 - x_3 &= -2, \\ x_1 + 3x_2 - 2x_3 &= 5, \\ x_1 - 3x_2 + x_3 &= 1. \end{aligned}$$

Solution: Begin by writing, or at least by imagining, each system in matrix-vector form:

$$\underbrace{\begin{bmatrix} 1 & 1 & -1 \\ 1 & 3 & 5 \\ 1 & -3 & 1 \end{bmatrix}}_{A} \underbrace{\begin{bmatrix} x_1 \\ x_2 \\ x_3 \end{bmatrix}}_{x} = \underbrace{\begin{bmatrix} -2 \\ 5 \\ 1 \end{bmatrix}}_{b} ; \qquad \underbrace{\begin{bmatrix} 1 & 1 & -1 \\ 1 & 3 & -2 \\ 1 & -3 & 1 \end{bmatrix}}_{A} \underbrace{\begin{bmatrix} x_1 \\ x_2 \\ x_3 \end{bmatrix}}_{x} = \underbrace{\begin{bmatrix} -2 \\ 5 \\ 1 \end{bmatrix}}_{b} .$$

As the matrices and vectors are modifications of the previous Example 2.2.6, we reduce typing by modifying the matrix and vector of the previous example (using the ability to change an element in a matrix, see Table 2.3).

(a) For the first system execute A (3 , 2) =- 3 and b (2) = 5 to see the matrix and vector are now

```
A =
     1     1    -1
     1     3     5
     1    -3     1

b =
    -2
     5
     1
```

Check the value of rcond (A) ; here 0.14 is good. Then obtain the solution from x=A\b with the result

```
x =
    -0.6429
    -0.1429
     1.2143
```

That is, the solution $x = (-0.64, -0.14, 1.21)$ to two decimal places (2 d.p.).[3]

(b) For the second system now execute A (2 , 3) =-2 to see the new matrix is the required

```
A =
     1      1     -1
     1      3     -2
     1     -3      1
```

Check: find that rcond (A) is zero, which is classified as terrible. Consequently we cannot compute a solution of this second system of linear equations (as in Figure 2.2(c)).

If we were to try x=A\b in this second system, then MATLAB/Octave would report[4]

Warning: Matrix is singular to working precision.

However, we cannot rely on MATLAB/Octave producing such useful messages: we must use rcond () to avoid mistakes. □

Example 2.2.9 Use MATLAB/Octave to solve the system

$$
\begin{aligned}
x_1 - 2x_2 + 3x_3 + x_4 + 2x_5 &= 7, \\
-2x_1 - 6x_2 - 3x_3 - 2x_4 + 2x_5 &= -1, \\
2x_1 + 3x_2 - 2x_5 &= -9, \\
-2x_1 + x_2 &= -3, \\
-2x_1 - 2x_2 + x_3 + x_4 - 2x_5 &= 5.
\end{aligned}
$$

Solution: Following Procedure 2.2.5, form the corresponding matrix and vector, with appropriate zeros, as

---

[3] The four or five significant digits printed by MATLAB/Octave is effectively exact for most practical purposes. This book often reports two significant digits, because two is enough for most human readable purposes. When a numerical result is reported to two decimal places, the book indicates this truncation with "(2 d.p.)".

[4] Section 3.2 introduces that the term 'singular' means that the matrix does not have a so-called inverse. The 'working precision' is the sixteen significant digits mentioned in Table 1.2.

```
A= [1    -2    3    1    2
    -2   -6   -3   -2    2
     2    3    0    0   -2
    -2    1    0    0    0
    -2   -2    1    1   -2 ]
b= [7;-1;-9;-3;5]
```

Check: first find that rcond (A) is acceptably 0.020, and so then compute the solution via x=A\b to obtain the result

```
x =
       0.8163
      -1.3673
      -6.7551
      17.1837
       3.2653
```

that is, the solution $x = (0.82, -1.37, -6.76, 17.18, 3.27)$ (2 d.p.). □

Example 2.2.10 What system of linear equations is represented by the following matrix-vector expression? and what is the result of using Procedure 2.2.5 for this system?

$$\begin{bmatrix} -7 & 3 \\ 7 & -5 \\ 1 & -2 \end{bmatrix} \begin{bmatrix} y \\ z \end{bmatrix} = \begin{bmatrix} 3 \\ -2 \\ 1 \end{bmatrix}.$$

Solution: The corresponding system of linear equations is

$$\begin{aligned} -7y + 3z &= 3, \\ 7y - 5z &= -2, \\ y - 2z &= 1. \end{aligned}$$

Invoking Procedure 2.2.5:

(a) form matrix A and column vector b with

```
A= [-7 3; 7 -5; 1 -2]
b= [3;-2;1]
```

(b) check rcond (A) : MATLAB/Octave gives the message

```
Error using rcond
Input must be a square matrix.
```

As rcond (A) does not exist, the procedure cannot give a solution.

The reason for the procedure not leading to a solution is that a system of three equations in two variables, as here, generally does not have a solution.[5]   □

Example 2.2.11 (rcond *avoids disaster*) In Example 2.0.1 an American and a European compared temperatures and, using measurements from two days, discovered the approximation that the American temperature $T_A = 1.82\,T_E + 32.73$ where T_E denotes the European temperature. Continuing the story, three days later they again meet and compare the temperatures they experienced: the American reports that "for the last three days it has been 51°, 74° and 81°", whereas the European reports "why, I recorded it as 11°, 23° and 27°". The graph to the right plots this data with the original two data points, apparently confirming a reasonable linear relationship between the two temperature scales.

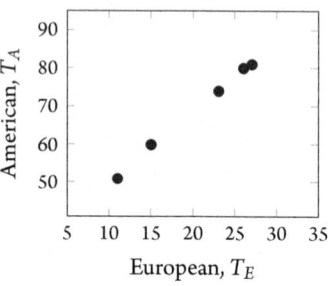

Let's fit a polynomial to this temperature data.

Solution: There are five data points. Each data point gives us an equation to be satisfied. This suggests we use linear algebra to determine five coefficients in a formula. Let's fit the data with the quartic polynomial

$$T_A = c_1 + c_2 T_E + c_3 T_E^2 + c_4 T_E^3 + c_5 T_E^4, \tag{2.1}$$

and use the data to determine the coefficients c_1, c_2, ..., c_5. Substituting each of the five pairs of T_E and T_A into this equation gives the five linear equations

$$60 = c_1 + 15c_2 + 225c_3 + 3375c_4 + 50625c_5,$$

$$\vdots$$

$$81 = c_1 + 27c_2 + 729c_3 + 19683c_4 + 531441c_5.$$

Form these into the matrix-vector equation $A\mathbf{c} = \mathbf{t}_A$ for the unknown coefficients $\mathbf{c} = (c_1, c_2, c_3, c_4, c_5)$. In MATLAB/Octave the vectors of American temperatures $\mathbf{t}_A$, and the 5×5 matrix A are constructed below (recall from Table 2.3 that te.^p computes the pth power of each element in the column vector te).

```
te=[15;26;11;23;27]
ta=[60;80;51;74;81]
plot(te,ta,'o')
A=[ones(5,1) te te.^2 te.^3 te.^4]
```

---

[5] If you were to execute x=A\b, then you would find MATLAB/Octave gives the 'answer' $\mathbf{x} = (-0.77, -0.73)$ (2 d.p.). But this answer is not a solution. Instead, this answer has another meaning, often sensibly useful, which is explained by Section 3.5. Using rcond() helps us to avoid confusing such an answer with a solution.

Then solve for the coefficients using c=A\ta to get

A =

1	15	225	3375	50625
1	26	676	17576	456976
1	11	121	1331	14641
1	23	529	12167	279841
1	27	729	19683	531441

c =
   -163.5469
     46.5194
     -3.6920
      0.1310
     -0.0017

Job done—or is it? To check, let's plot the predictions of the quartic polynomial (2.1) with these coefficients. In MATLAB/Octave we may plot a graph with the following

```
t=linspace(5,35);
plot(t,c(1)+c(2)*t+c(3)*t.^2+c(4)
   *t.^3+c(5)*t.^4)
```

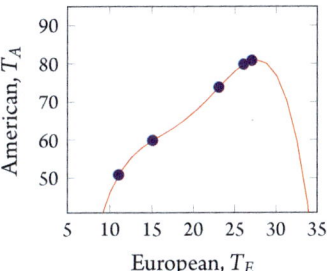

and see a graph like the one to the right. Disaster: the quartic polynomial relationship is clearly terrible as it is too wavy and nothing like the straight line we know it should be ($T_A = \frac{9}{5}T_E + 32$).

The problem is we forgot rcond. In MATLAB/Octave execute rcond(A) and discover rcond is $3 \cdot 10^{-9}$. This value is in the 'terrible' range classified by Procedure 2.2.5. Thus the solution of the linear equations must not be used: here the plot shown above-right shows the solution coefficients are not acceptable. Always use rcond to check for bad systems of linear equations. ☐

The previous Example 2.2.11 also illustrates one of the 'rules of thumb' in science and engineering: *for data fitting, avoid using polynomials of degree higher than cubic.*[6]

Example 2.2.12 (Global Positioning System in space-time) Now recall Example 2.1.5. Consider the GPS receiver in your smart-phone. The phone's clock is generally in error; it may only be by a second but the GPS needs microsecond precision. Because of such a timing unknown, five satellites determine our precise position in space *and* time.

Suppose at some time (according to our smart-phone) the phone receives from a GPS satellite that it is at 3D location (6, 12, 23) Mm (megametres) and that the signal was sent at a true time 0.04 s (seconds) before the phone's time. But the phone's time is different from the true time

---

[6] In finance, a 'rule of thumb' is to avoid fitting data with anything high-order than straight lines!

by some unknown amount, say t. Consequently, the travel time of the signal from the satellite to the phone is actually $t + 0.04$ s. Given the speed of light is $c = 300$ Mm/s, this is a distance of $300(t + 0.04) = 300t + 12$ —linear in the discrepancy of the phone's clock to the GPS clock. Let (x, y, z) be you and your phone's position in 3D space, then the distance to the satellite is also $\sqrt{(x-6)^2 + (y-12)^2 + (z-23)^2}$. Equating the squares of these two gives one equation

$$(x-6)^2 + (y-12)^2 + (z-23)^2 = (300t + 12)^2.$$

Similarly other satellites give other equations that help determine our position. But writing "300t" all the time is a bit tedious, so replace it with the new unknown $w = 300t$.

Given that your phone also detects that four other satellites broadcast the following position and time information: $(13, 20, 12)$ time shift 0.04 s before; $(17, 14, 10)$ time shift $0.033 \cdots$ s before; $(8, 21, 10)$ time shift $0.033 \cdots$ s before; and $(22, 9, 8)$ time shift 0.04 s before. Adapting the approach of Example 2.1.5, use linear algebra to determine your phone's location in space.

Solution: Let your unknown position be (x, y, z) and the unknown time shift to the phone's clock t be found from $w = 300t$. Then the five equations from the five satellites are, respectively,

$$(x-6)^2 + (y-12)^2 + (z-23)^2 = (300t + 12)^2 = (w+12)^2,$$
$$(x-13)^2 + (y-20)^2 + (z-12)^2 = (300t + 12)^2 = (w+12)^2,$$
$$(x-17)^2 + (y-14)^2 + (z-10)^2 = (300t + 10)^2 = (w+10)^2,$$
$$(x-8)^2 + (y-21)^2 + (z-10)^2 = (300t + 10)^2 = (w+10)^2,$$
$$(x-22)^2 + (y-9)^2 + (z-8)^2 = (300t + 12)^2 = (w+12)^2.$$

Expand all the squares in these equations:

$$x^2 - 12x + 36 + y^2 - 24y + 144 + z^2 - 46z + 529 = w^2 + 24w + 144,$$
$$x^2 - 26x + 169 + y^2 - 40y + 400 + z^2 - 24z + 144 = w^2 + 24w + 144,$$
$$x^2 - 34x + 289 + y^2 - 28y + 196 + z^2 - 20z + 100 = w^2 + 20w + 100,$$
$$x^2 - 16x + 64 + y^2 - 42y + 441 + z^2 - 20z + 100 = w^2 + 20w + 100,$$
$$x^2 - 44x + 484 + y^2 - 18y + 81 + z^2 - 16z + 64 = w^2 + 24w + 144.$$

As before, these form a system of nonlinear equations and so appear outside the remit of the course, but a little algebra brings them within. Subtract the last equation, say, from each of the first four equations: then *all* of the nonlinear squares of variables cancel leaving a linear system. Combining the constants on the right-hand side, and moving the w terms to the left gives this system of four linear equations:

$$32x - 6y - 30z + 0w = -80,$$
$$18x - 22y - 8z + 0w = -84,$$
$$10x - 10y - 4z + 4w = 0,$$
$$28x - 24y - 4z + 4w = -20.$$

Following Procedure 2.2.5, solve this system by forming the corresponding matrix and vector as

```
A= [32    -6  -30   0
    18  -22   -8   0
    10  -10   -4   4
    28  -24   -4   4 ]
b= [-80;-84;0;-20]
```

Check `rcond (A)` : it is acceptably 0.023 so compute the solution via `x=A\b` to find

```
x  =
    2
    4
    4
    9
```

Hence your phone is at location $(x,y,z) = (2,4,4)$ Mm. Further, the time discrepancy between your phone and the GPS satellites' time is proportional to $w = 9$ Mm. Since $w = 300t$, where 300 Mm/s is the speed of light, the time discrepancy is $t = \frac{9}{300} = 0.03$ s. □

2.2.2 Algebraic manipulation solves systems

A variant of GE [Gaussian Elimination] was used by the Chinese around the first century AD; the *Jiu Zhang Suanshu* (Nine Chapters of the Mathematical Art) contains a worked example for a system of five equations in five unknowns   *Higham (1996) [p.195]*

To solve linear equations with non-square matrices, or with poorly conditioned matrices we need to know many more details about linear algebra.

This and the next subsection are not essential, but many further courses currently assume knowledge of the content. Theorems 2.2.26 and 2.2.30 are convenient to establish in the next subsection, but could alternatively be established using Procedure 3.3.15.

This subsection systematizes the algebraic working of Examples 2.1.3 and 2.1.6. The systematic approach empowers by-hand solution of systems of linear equations, together with two general properties on the number of solutions possible. The algebraic methodology invoked here also reinforces algebraic skills that will help in further courses.

In hand calculations we often want to minimize writing, so the discussion here uses two forms side-by-side for the linear equations: one form with all symbols recorded for best clarity; and beside it, a form where only coefficients are recorded for quickest writing. Translating from one to the other is crucial, even in a computing era as the computer also primarily deals with arrays of numbers, and we must interpret what those arrays of numbers mean in terms of linear equations.

Example 2.2.13 Recall the system of linear equations of Example 2.1.6:

$$x_1 + x_2 - x_3 = -2,$$
$$x_1 + 3x_2 + 5x_3 = 8,$$
$$x_1 + 2x_2 + x_3 = 1.$$

The first crucial level of abstraction is to write this in the matrix-vector form, Example 2.2.3,

$$\underbrace{\begin{bmatrix} 1 & 1 & -1 \\ 1 & 3 & 5 \\ 1 & 2 & 1 \end{bmatrix}}_{A} \underbrace{\begin{bmatrix} x_1 \\ x_2 \\ x_3 \end{bmatrix}}_{x} = \underbrace{\begin{bmatrix} -2 \\ 8 \\ 1 \end{bmatrix}}_{b}$$

A second step of abstraction omits the symbols "]$x =$ ["—often we draw a vertical (dotted) line to show where the symbols "]$x =$ [" were, but this (dotted) line is not essential and the theoretical statements ignore such a drawn line. Here this second step of abstraction represents this linear system by the so-called augmented matrix

$$\begin{bmatrix} 1 & 1 & -1 & \vdots & -2 \\ 1 & 3 & 5 & \vdots & 8 \\ 1 & 2 & 1 & \vdots & 1 \end{bmatrix}$$

☐

Definition 2.2.14 *The **augmented matrix** of the system of linear equations* $Ax = b$ *is the matrix* $[A \vdots b]$.

Example 2.2.15 Write down augmented matrices for the two following systems:

(a) $\begin{aligned} -2r + 3s &= 6, \\ s - 4t &= -\pi, \end{aligned}$

(b) $\begin{aligned} -7y + 3z &= 3, \\ 7y - 5z &= -2, \\ y - 2z &= 1. \end{aligned}$

Solution:

$$\begin{cases} -2r + 3s = 6 \\ s - 4t = -\pi \end{cases} \iff \begin{bmatrix} -2 & 3 & 0 & \vdots & 6 \\ 0 & 1 & -4 & \vdots & -\pi \end{bmatrix}$$

$$\begin{cases} -7y + 3z = 3 \\ 7y - 5z = -2 \\ y - 2z = 1 \end{cases} \iff \begin{bmatrix} -7 & 3 & \vdots & 3 \\ 7 & -5 & \vdots & -2 \\ 1 & -2 & \vdots & 1 \end{bmatrix}$$

An augmented matrix is not unique: it depends upon the order of the equations, and also upon the order you choose for the variables in x. The first example implicitly chose $x = (r, s, t)$; if instead we choose to order the variables as $x = (s, t, r)$, then

$$\begin{cases} 3s - 2r = 6 \\ s - 4t = -\pi \end{cases} \iff \begin{bmatrix} 3 & 0 & -2 & \vdots & 6 \\ 1 & -4 & 0 & \vdots & -\pi \end{bmatrix}$$

Such variations to the augmented matrix are valid, but you must remember your corresponding chosen order of the variables. ☐

Activity 2.2.16 Which of the following *cannot* be an augmented matrix for the system $p + 4q = 3$ and $-p + 2q = -2$?

(a) $\begin{bmatrix} -1 & 2 & \vdots & -2 \\ 1 & 4 & \vdots & 3 \end{bmatrix}$

(c) $\begin{bmatrix} 1 & 4 & \vdots & 3 \\ -1 & 2 & \vdots & -2 \end{bmatrix}$

(b) $\begin{bmatrix} 4 & 1 & \vdots & 3 \\ 2 & -1 & \vdots & -2 \end{bmatrix}$

(d) $\begin{bmatrix} 2 & -1 & \vdots & 3 \\ 4 & 1 & \vdots & -2 \end{bmatrix}$

Recall that Examples 2.1.3 and 2.1.6 manipulate the linear equations to deduce solution(s) to systems of linear equations. The following theorem validates such manipulations in general, and gives the basic operations a collective name.

Theorem 2.2.17 *The following **elementary row operations** can be performed either on a system of linear equations or on its corresponding augmented matrix without changing the solutions:*

(a) *interchange two equations/rows; or*

(b) *multiply an equation/row by a nonzero constant; or*

(c) *add a multiple of an equation/row to another.*

Proof. We just address the system of equations form because the augmented matrix form is equivalent but more abstract.

(a) Swapping the order of two equations does not change the system of equations, the set of relations between the variables, so does not change the solution.

(b) Let vector x satisfy $a_1 x_1 + a_2 x_2 + \cdots + a_n x_n = b$. Then $c a_1 x_1 + c a_2 x_2 + \cdots + c a_n x_n = c(a_1 x_1 + a_2 x_2 + \cdots + a_n x_n) = cb$ and so x satisfies c times the equation. When the constant c is nonzero, the above can be reversed through dividing by c. Hence, multiplying an equation by a nonzero constant c does not change the possible solutions.

(c) Let vector x satisfy both $a_1 x_1 + a_2 x_2 + \cdots + a_n x_n = b$ and $a_1' x_1 + a_2' x_2 + \cdots + a_n' x_n = b'$. Then

$$(a_1' + c a_1) x_1 + (a_2' + c a_2) x_2 + \cdots + (a_n' + c a_n) x_n$$
$$= a_1' x_1 + c a_1 x_1 + a_2' x_2 + c a_2 x_2 + \cdots + a_n' x_n + c a_n x_n$$
$$= a_1' x_1 + a_2' x_2 + \cdots + a_n' x_n + c(a_1 x_1 + a_2 x_2 + \cdots + a_n x_n)$$
$$= b' + cb.$$

That is, x also satisfies the equation formed by adding c times the first to the second. Conversely, every vector x that satisfies both $a_1 x_1 + a_2 x_2 + \cdots + a_n x_n = b$ and $(a_1' + c a_1) x_1 + (a_2' + c a_2) x_2 + \cdots + (a_n' + c a_n) x_n = b' + cb$, by adding $(-c)$ times the first to the second as above, also satisfies $a_1' x_1 + a_2' x_2 + \cdots + a_n' x_n = b'$. Hence adding a multiple of an equation to another does not change the possible solutions. □

Example 2.2.18 Use elementary row operations to find the only solution of the following system of linear equations:

$$x + 2y + z = 1,$$
$$2x - 3y = 2,$$
$$-3y - z = 2.$$

Confirm with MATLAB/Octave.

Solution: In order to know what the row operations should find, let's first solve the system with MATLAB/Octave via Procedure 2.2.5. In matrix-vector form the system is

$$\begin{bmatrix} 1 & 2 & 1 \\ 2 & -3 & 0 \\ 0 & -3 & -1 \end{bmatrix} \begin{bmatrix} x \\ y \\ z \end{bmatrix} = \begin{bmatrix} 1 \\ 2 \\ 2 \end{bmatrix};$$

hence in MATLAB/Octave execute

```
A= [1 2 1;2 -3 0;0 -3 -1]
b= [1;2;2]
rcond(A)
x=A\b
```

`rcond(A)` is just good, 0.0104, so the computed answer $x = (x,y,z) = (7,4,-14)$ is the solution.

Second, use elementary row operations. Let's write the working in both full symbolic equations and in augmented matrix form, in order to see the correspondence between the two—you would not have to do both, either one would suffice.

$$\begin{cases} x + 2y + z = 1 \\ 2x - 3y + 0z = 2 \\ 0x - 3y - z = 2 \end{cases} \iff \left[\begin{array}{ccc:c} 1 & 2 & 1 & 1 \\ 2 & -3 & 0 & 2 \\ 0 & -3 & -1 & 2 \end{array}\right]$$

Add (-2) times the first equation/row to the second.

$$\begin{cases} x + 2y + z = 1 \\ 0x - 7y - 2z = 0 \\ 0x - 3y - z = 2 \end{cases} \iff \left[\begin{array}{ccc:c} 1 & 2 & 1 & 1 \\ 0 & -7 & -2 & 0 \\ 0 & -3 & -1 & 2 \end{array}\right]$$

This makes the first column have a leading one (Definition 2.2.19). Start on the second column by dividing the second equation/row by (-7).

$$\begin{cases} x + 2y + z = 1 \\ 0x + y + \frac{2}{7}z = 0 \\ 0x - 3y - z = 2 \end{cases} \iff \left[\begin{array}{ccc:c} 1 & 2 & 1 & 1 \\ 0 & 1 & \frac{2}{7} & 0 \\ 0 & -3 & -1 & 2 \end{array}\right]$$

Now subtract twice the second equation/row from the first, and add three times the second to the third.

$$\begin{cases} x + 0y + \frac{3}{7}z = 1 \\ 0x + y + \frac{2}{7}z = 0 \\ 0x + 0y - \frac{1}{7}z = 2 \end{cases} \iff \left[\begin{array}{ccc:c} 1 & 0 & \frac{3}{7} & 1 \\ 0 & 1 & \frac{2}{7} & 0 \\ 0 & 0 & -\frac{1}{7} & 2 \end{array}\right]$$

This makes the second column have the second leading one (Definition 2.2.19). Start on the third column by multiplying the third equation/row by (-7).

$$\begin{cases} x + 0y + \frac{3}{7}z = 1 \\ 0x + y + \frac{2}{7}z = 0 \\ 0x + 0y + z = -14 \end{cases} \iff \left[\begin{array}{ccc:c} 1 & 0 & \frac{3}{7} & 1 \\ 0 & 1 & \frac{2}{7} & 0 \\ 0 & 0 & 1 & -14 \end{array}\right]$$

Now subtract 3/7 of the third equation/row from the first, and 2/7 from the second.

$$\begin{cases} x + 0y + 0z = 7 \\ 0x + y + 0z = 4 \\ 0x + 0y + z = -14 \end{cases} \quad \Longleftrightarrow \quad \left[\begin{array}{ccc:c} 1 & 0 & 0 & 7 \\ 0 & 1 & 0 & 4 \\ 0 & 0 & 1 & -14 \end{array}\right]$$

This completes the transformation of the equations/augmented matrix into a so-called reduced row echelon form (Definition 2.2.19). From this form we read off the solution: the system of equations on the left directly gives $x = 7$, $y = 4$, and $z = -14$, that is, the solution vector $x = (x,y,z) = (7,4,-14)$ (as computed by MATLAB/Octave); the transformed augmented matrix on the right tells us exactly the same thing because (Definition 2.2.14) it means the same as the matrix-vector

$$\begin{bmatrix} 1 & 0 & 0 \\ 0 & 1 & 0 \\ 0 & 0 & 1 \end{bmatrix} \begin{bmatrix} x \\ y \\ z \end{bmatrix} = \begin{bmatrix} 7 \\ 4 \\ -14 \end{bmatrix},$$

which is the same as the system on the above-left and tells us the solution $x = (x,y,z) = (7,4,-14)$. □

Definition 2.2.19 *A system of linear equations or (augmented) matrix is in **reduced row echelon form** (RREF) if:*

(a) *any equations with all zero coefficients, or rows of the matrix consisting entirely of zeros, are at the bottom;*

(b) *in each nonzero equation/row, the first nonzero coefficient/entry is a one (called the **leading one**), and is in a variable/column to the left of any leading ones below it; and*

(c) *each variable/column containing a leading one has zero coefficients/entries in every other equation/row.*

*A **free variable** is any variable which is **not** multiplied by a leading one when the reduced row echelon form is translated to its corresponding algebraic equations.*

Example 2.2.20 *(reduced row echelon form)* Which of the following are in reduced row echelon form (RREF)? For those that are, identify the leading ones, and treating other variables as free variables write down the most general solution of the system of linear equations.

(a) $\begin{cases} x_1 + x_2 + 0x_3 - 2x_4 = -2 \\ 0x_1 + 0x_2 + x_3 + 4x_4 = 5 \end{cases}$

Solution: This is in RREF, with leading ones on the variables x_1 and x_3. Let the other variables be free by, say, setting $x_2 = s$ and $x_4 = t$ for arbitrary parameters s and t. Then the two equations give $x_1 = -2 - s + 2t$ and $x_3 = 5 - 4t$. Consequently, the most general solution is $x = (x_1,x_2,x_3,x_4) = (-2 - s + 2t, s, 5 - 4t, t)$ for arbitrary s and t. □

(b) $\left[\begin{array}{ccc:c} 1 & 0 & -1 & 1 \\ 0 & 1 & -1 & -2 \\ 0 & 0 & 0 & 4 \end{array}\right]$

Solution: This augmented matrix is in RREF with leading ones in the first and second columns. To find solutions, explicitly write down the corresponding system of linear equations. But we do not know the variables! If the context does not give variable names, then use the generic x_1, x_2, ..., x_n. Thus here the corresponding system is

$$x_1 - x_3 = 1, \quad x_2 - x_3 = -2, \quad 0 = 4.$$

The first two equations are valid, but the last is contradictory as $0 \neq 4$. Hence there are no solutions to the system. □

(c) $\begin{cases} x + 2y = 3 \\ 0x + y = -2 \end{cases}$

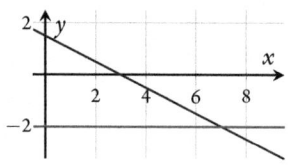

Solution: This system is not in RREF: although there are two leading ones multiplying x and y in the first and the second equation respectively, the variable y does not have zero coefficients in the first equation. (A solution to this system exists, shown to the right, but the question does not ask for it.) □

(d) $\begin{bmatrix} -1 & 4 & 1 & 6 & \vdots & -1 \\ 3 & 0 & 1 & -2 & \vdots & -2 \end{bmatrix}$

Solution: This augmented matrix is not in RREF as there are no leading ones. □

Activity 2.2.21 Which one of the following augmented matrices is *not* in reduced row echelon form?

(a) $\begin{bmatrix} 0 & 1 & 0 & \vdots & 1 \\ 0 & 0 & 1 & \vdots & 2 \end{bmatrix}$

(c) $\begin{bmatrix} 1 & 0 & 1 & \vdots & 2 \\ 0 & 1 & 0 & \vdots & 1 \end{bmatrix}$

(b) $\begin{bmatrix} 0 & 1 & 1 & \vdots & -1 \\ 0 & 0 & 0 & \vdots & 0 \end{bmatrix}$

(d) $\begin{bmatrix} 1 & 1 & 0 & \vdots & 0 \\ 0 & -1 & 1 & \vdots & -1 \end{bmatrix}$

Activity 2.2.22 Which one of the following is a general solution to the system with augmented matrix in reduced row echelon form of

$$\begin{bmatrix} 1 & 0 & -0.2 & \vdots & 0.4 \\ 0 & 1 & -1.2 & \vdots & -0.6 \end{bmatrix}?$$

(a) solution does not exist

(c) $(0.2t + 0.4, 1.2t - 0.6, t)$

(b) $(0.4, -0.6, 0)$

(d) $(0.2 + 0.4t, 1.2 + 0.6t, t)$

The previous Example 2.2.20 shows that, given a system of linear equations in reduced row echelon form, we can either immediately write down all solutions, or immediately determine if none exists. Generalizing Example 2.2.18, the following Gauss–Jordan procedure uses elementary row operations (Theorem 2.2.17) to find an equivalent system of equations in reduced row echelon form. From such a form we then write down a general solution.

Procedure 2.2.23 (Gauss–Jordan elimination)[7]

1. Write down either the full symbolic form of the system of linear equations, or the augmented matrix of the system of linear equations.

2. Use elementary row operations to reduce the system/augmented matrix to reduced row echelon form.

3. If the resulting system is consistent, then solve for the leading variables in terms of any remaining free variables to obtain a **general solution**.

Example 2.2.24 Use Gauss–Jordan elimination, Procedure 2.2.23, to find all possible solutions to the system

$$-x - y = -3,$$
$$x + 4y = -1,$$
$$2x + 4y = c,$$

depending upon the parameter c.

Solution: Here write both the full symbolic equations and the augmented matrix form—you would only have to do one.

$$\begin{cases} -x - y = -3 \\ x + 4y = -1 \\ 2x + 4y = c \end{cases} \iff \begin{bmatrix} -1 & -1 & \vdots & -3 \\ 1 & 4 & \vdots & -1 \\ 2 & 4 & \vdots & c \end{bmatrix}$$

Multiply the first by (-1).

$$\begin{cases} x + y = 3 \\ x + 4y = -1 \\ 2x + 4y = c \end{cases} \iff \begin{bmatrix} 1 & 1 & \vdots & 3 \\ 1 & 4 & \vdots & -1 \\ 2 & 4 & \vdots & c \end{bmatrix}$$

Subtract the first from the second, and twice the first from the third.

$$\begin{cases} x + y = 3 \\ 0x + 3y = -4 \\ 0x + 2y = c - 6 \end{cases} \iff \begin{bmatrix} 1 & 1 & \vdots & 3 \\ 0 & 3 & \vdots & -4 \\ 0 & 2 & \vdots & c - 6 \end{bmatrix}$$

Divide the second by three.

$$\begin{cases} x + y = 3 \\ 0x + y = -\frac{4}{3} \\ 0x + 2y = c - 6 \end{cases} \iff \begin{bmatrix} 1 & 1 & \vdots & 3 \\ 0 & 1 & \vdots & -\frac{4}{3} \\ 0 & 2 & \vdots & c - 6 \end{bmatrix}$$

---

[7] Computers and graphics calculators perform Gauss–Jordan elimination for you; for example, A\ in MATLAB/Octave. However, when rcond indicates A\ is inappropriate, then the singular value decomposition of Section 3.3 is a far better choice than such Gauss–Jordan elimination.

Subtract the second from the first, and twice the second from the third.

$$\begin{cases} x + 0y = \frac{13}{3} \\ 0x + y = -\frac{4}{3} \\ 0x + 0y = c - \frac{10}{3} \end{cases} \iff \left[\begin{array}{cc:c} 1 & 0 & \frac{13}{3} \\ 0 & 1 & -\frac{4}{3} \\ 0 & 0 & c - \frac{10}{3} \end{array}\right]$$

The system is now in reduced row echelon form. The last row immediately tells us that there is no solution for parameter $c \neq \frac{10}{3}$ as the equation would then be inconsistent. If parameter $c = \frac{10}{3}$, then the system is consistent and the first two rows give that the only solution is $(x,y) = (\frac{13}{3}, -\frac{4}{3})$. □

Example 2.2.25 Use Gauss–Jordan elimination, Procedure 2.2.23, to find all possible solutions to the system

$$\begin{cases} -2v + 3w = -1, \\ 2u + v - w = -1. \end{cases}$$

Solution: Here write both the full symbolic equations and the augmented matrix form—you would choose one or the other representations.

$$\begin{cases} 0u - 2v + 3w = -1 \\ 2u + v - w = -1 \end{cases} \iff \left[\begin{array}{ccc:c} 0 & -2 & 3 & -1 \\ 2 & 1 & 1 & -1 \end{array}\right]$$

Swap the two rows to get a nonzero top-left entry.

$$\begin{cases} 2u + v - w = -1 \\ 0u - 2v + 3w = -1 \end{cases} \iff \left[\begin{array}{ccc:c} 2 & 1 & -1 & -1 \\ 0 & -2 & 3 & -1 \end{array}\right]$$

Divide the first row by two.

$$\begin{cases} u + \frac{1}{2}v - \frac{1}{2}w = -\frac{1}{2} \\ 0u - 2v + 3w = -1 \end{cases} \iff \left[\begin{array}{ccc:c} 1 & \frac{1}{2} & -\frac{1}{2} & -\frac{1}{2} \\ 0 & -2 & 3 & -1 \end{array}\right]$$

Divide the second row by (-2).

$$\begin{cases} u + \frac{1}{2}v - \frac{1}{2}w = -\frac{1}{2} \\ 0u + v - \frac{3}{2}w = \frac{1}{2} \end{cases} \iff \left[\begin{array}{ccc:c} 1 & \frac{1}{2} & -\frac{1}{2} & -\frac{1}{2} \\ 0 & 1 & -\frac{3}{2} & \frac{1}{2} \end{array}\right]$$

Subtract half the second row from the first.

$$\begin{cases} u + 0v + \frac{1}{4}w = -\frac{3}{4} \\ 0u + v - \frac{3}{2}w = \frac{1}{2} \end{cases} \iff \left[\begin{array}{ccc:c} 1 & 0 & \frac{1}{4} & -\frac{3}{4} \\ 0 & 1 & -\frac{3}{2} & \frac{1}{2} \end{array}\right]$$

The system is now in reduced row echelon form. The third column is that of a free variable so set the third component $w = t$ for arbitrary t. Then the first row gives $u = -\frac{3}{4} - \frac{1}{4}t$, and the second row gives $v = \frac{1}{2} + \frac{3}{2}t$. That is, the solutions are $(u,v,w) = (-\frac{3}{4} - \frac{1}{4}t, \frac{1}{2} + \frac{3}{2}t, t)$ for arbitrary t. □

2.2.3 Three possible numbers of solutions

The number of possible solutions to a system of equations is fundamental. We need to know all the possibilities. As seen in previous examples, the following theorem says there are only three possibilities for linear equations.

Theorem 2.2.26 *For every system of linear equations $Ax = b$, exactly one of the following is true:*

- *there is no solution;*
- *there is a unique solution;*
- *there are infinitely many solutions.*

Proof. First, if there is exactly none or one solution to $Ax = b$, then the theorem holds. Second, suppose there are two distinct solutions; let them be y and z so $Ay = b$ and $Az = b$. Then consider $x = ty + (1 - t)z$ for every t (a parametric description of the line through y and z, Section 1.2.3). Consider the first row of Ax: by Definition 2.2.2 it is

$$a_{11}x_1 + a_{12}x_2 + \cdots + a_{1n}x_n$$
$$= a_{11}[ty_1 + (1 - t)z_1] + a_{12}[ty_2 + (1 - t)z_2] + \cdots + a_{1n}[ty_n + (1 - t)z_n]$$

(then rearrange this scalar expression)

$$= t[a_{11}y_1 + a_{12}y_2 + \cdots + a_{1n}y_n] + (1 - t)[a_{11}z_1 + a_{12}z_2 + \cdots + a_{1n}z_n]$$
$$= t[\text{first row of } Ay] + (1 - t)[\text{first row of } Az]$$
$$= tb_1 + (1 - t)b_1 \quad (\text{as } Ay = b \text{ and } Az = b)$$
$$= b_1.$$

Similarly for all rows of Ax: that is, each row in Ax equals the corresponding element of b. Consequently, $Ax = b$ for every t. Hence, if there are ever two distinct solutions, then there are an infinite number of solutions: $x = ty + (1 - t)z$ for every t. □

An important class of linear equations always has at least one solution, never none. For example, modify Example 2.2.24 to

$$-x - y = 0,$$
$$x + 4y = 0,$$
$$2x + 4y = 0,$$

and then $x = y = 0$ is immediately a solution. The reason is that the right-hand side is all zeros and so $x = y = 0$ makes the left-hand sides also zero.

Definition 2.2.27 *A system of linear equations is called **homogeneous** if the (right-hand side) constant term in each equation is zero; that is, when the system may be written $Ax = 0$. Otherwise the system is termed **non-homogeneous**.*

Example 2.2.28

(a) $\begin{cases} 3x_1 - 3x_2 = 0 \\ -x_1 - 7x_2 = 0 \end{cases}$ is homogeneous. Solving, the first equation gives $x_1 = x_2$ and substituting in the second then gives $-x_2 - 7x_2 = 0$ so that $x_1 = x_2 = 0$ is the only solution. It must have $x = 0$ as a solution as the system is homogeneous.

(b) $\begin{cases} -2+y+3z=0 \\ 2x+y+2z=0 \end{cases}$ is not homogeneous because there is a nonzero constant in the first equation, the (-2), even though it is here sneakily written on the left-hand side.

(c) $\begin{cases} x_1+2x_2+4x_3-3x_4=0 \\ x_1+2x_2-3x_3+6x_4=0 \end{cases}$ is homogeneous. Use Gauss–Jordan elimination, Procedure 2.2.23, to solve:

$$\begin{cases} x_1+2x_2+4x_3-3x_4=0 \\ x_1+2x_2-3x_3+6x_4=0 \end{cases} \iff \begin{bmatrix} 1 & 2 & 4 & -3 & \vdots & 0 \\ 1 & 2 & -3 & 6 & \vdots & 0 \end{bmatrix}$$

Subtract the first row from the second. $\begin{cases} x_1+2x_2+4x_3-3x_4=0 \\ 0x_1+0x_2-7x_3+9x_4=0 \end{cases} \iff \begin{bmatrix} 1 & 2 & 4 & -3 & \vdots & 0 \\ 0 & 0 & -7 & 9 & \vdots & 0 \end{bmatrix}$

Divide the second row by (-7). $\begin{cases} x_1+2x_2+4x_3-3x_4=0 \\ 0x_1+0x_2+x_3-\frac{9}{7}x_4=0 \end{cases} \iff \begin{bmatrix} 1 & 2 & 4 & -3 & \vdots & 0 \\ 0 & 0 & 1 & -\frac{9}{7} & \vdots & 0 \end{bmatrix}$

Subtract four times the second row from the first. $\begin{cases} x_1+2x_2+0x_3+\frac{15}{7}x_4=0 \\ 0x_1+0x_2+x_3-\frac{9}{7}x_4=0 \end{cases} \iff \begin{bmatrix} 1 & 2 & 0 & \frac{15}{7} & \vdots & 0 \\ 0 & 0 & 1 & -\frac{9}{7} & \vdots & 0 \end{bmatrix}$

The system is now in reduced row echelon form. The second and fourth columns are those of free variables so set the second and fourth component $x_2=s$ and $x_4=t$ for arbitrary s and t. Then the first row gives $x_1=-2s-\frac{15}{7}t$, and the second row gives $x_3=\frac{9}{7}t$. That is, the solutions are $x=(-2s-\frac{15}{7}t,s,\frac{9}{7}t,t)=(-2,1,0,0)s+(-\frac{15}{7},0,\frac{9}{7},1)t$ for arbitrary s and t. These solutions include $x=0$ via the choice $s=t=0$. $\square$

Activity 2.2.29 Which one of the following systems of equations for x and y is homogeneous?

(a) $3x+1=0$ and $-x-y=0$

(b) $-3x-y=0$ and $7x+5y=3$

(c) $-2x+y-3=0$ and $x+4=2y$

(d) $5y=3x$ and $4x=2y$

As Example 2.2.28(c) illustrates, a further subclass of homogeneous systems is immediately known to have an infinite number of solutions. Namely, if the number of equations is less than the number of unknowns (two is less than four in Example 2.2.28(c)), then a homogeneous system always has an infinite number of solutions.

Theorem 2.2.30 If $Ax=0$ is a homogeneous system of m linear equations with n variables where $m<n$, then the system has infinitely many solutions.

Remember that this theorem says nothing about the cases where there are at least as many equations as variables ($m\geq n$), when there may or may not be an infinite number of solutions.

Proof. The zero vector, $x=0$ in $\mathbb{R}^n$, is a solution of $Ax=0$, so a homogeneous system is always consistent. In the reduced row echelon form there are at most m leading variables—one for each row. Here $n > m$ and so the number of free variables is at least $n - m > 0$. Hence there is at least one free variable and consequently an infinite number of solutions. □

Prefer a matrix/vector level

Working at the element level in this way leads to a profusion of symbols, superscripts, and subscripts that tend to obscure the mathematical structure and hinder insights being drawn into the underlying process. One of the key developments in the last century was the recognition that it is much more profitable to work at the matrix level.

Higham (2015b) [§2]

A large part of this and preceding sections is devoted to arithmetic and algebraic manipulations on the individual coefficients and variables in the system. This is working at the 'element level' commented on by Higham. But as Higham also comments, we need to work more at a whole matrix level. This means we need to discuss and manipulate matrices as a whole, not get enmeshed in the intricacies of the element operations. This has close intellectual parallels in computing where abstract data structures empower us to encode complex tasks: here the analogous abstract data structures are matrices and vectors, and working with matrices and vectors as objects in their own right empowers linear algebra. The next chapter proceeds to develop linear algebra at the matrix level.

But first, the next Section 2.3 establishes some necessary fundamental aspects at the vector level.

2.2.4 Exercises

Exercise 2.2.1   For each of the following systems, write down two different matrix-vector forms of the equations. For each system: how many different possible matrix-vector forms could be written down?

(a)
$$-3x + 6y = -6$$
$$-x - 3y = 4$$

(b)
$$-2p - q + 1 = 0$$
$$p - 6q = 2$$

(c)
$$7.9x - 4.7y = -1.7$$
$$2.4x - 0.1y = 1$$
$$-3.1x + 2.7y = 2.3$$

(d)
$$3a + 4b - \tfrac{5}{2}c = 0$$
$$-\tfrac{7}{2}a + b - \tfrac{9}{2}c - \tfrac{9}{2} = 0$$

(e)
$$u + v - 2w = -1$$
$$-2u - v + 2w = 3$$
$$u + v + 5w = 2$$

Exercise 2.2.2   Use Procedure 2.2.5 in MATLAB/Octave to try to solve each of the systems of Exercise 2.2.1.

Exercise 2.2.3   Use Procedure 2.2.5 in MATLAB/Octave to try to solve each of the following systems.

$$-5x - 3y + 5z = -3$$
(a) $\quad 2x + 3y - z = -5$
$$-2x + 3y + 4z = -3$$

$$-p + 2q - r = -2$$
(b) $\quad -2p + q + 2r = 1$
$$-3p + 4q = 4$$

$$u + 3v + 2w = -1$$
(c) $\quad 3v + 5w = 1$
$$-u + 3w = 2$$

Exercise 2.2.4 Use elementary row operations (Theorem 2.2.17) to solve the systems in Exercise 2.2.3.

Exercise 2.2.5 Use Procedure 2.2.5 in MATLAB/Octave to try to solve each of the following systems.

(a)
$$2.2x_1 - 2.2x_2 - 3.5x_3 - 2.2x_4 = 2.9$$
$$4.8x_1 + 1.8x_2 - 3.1x_3 - 4.8x_4 = -1.6$$
$$-0.8x_1 + 1.9x_2 - 3.2x_3 + 4.1x_4 = -5.1$$
$$-9x_1 + 3.5x_2 - 0.7x_3 + 1.6x_4 = -3.3$$

(b)
$$0.7c_1 + 0.7c_2 + 4.1c_3 - 4.2c_4 = -0.70$$
$$c_1 + c_2 + 2.1c_3 - 5.1c_4 = -2.8$$
$$4.3c_1 + 5.4c_2 + 0.5c_3 + 5.5c_4 = -6.1$$
$$-0.6c_1 + 7.2c_2 + 1.9c_3 - 0.6c_4 = -0.3$$

Exercise 2.2.6 Each of the following show some MATLAB/Octave commands and their results. Write down a possible problem that these commands aim to solve, and interpret what the results mean for the problem.

(a)
```
>> A=[0.1 -0.3; 2.2 0.8]
A =
     0.1000   -0.3000
     2.2000    0.8000
>> b=[-1.2; 0.6]
b =
    -1.2000
     0.6000
>> rcond(A)
ans =    0.1072
>> x=A\b
x =
    -1.054
     3.649
```

(b)
```
>> A=[0.7 1.4; -0.5 -0.9; 1.9 0.7]
A =
     0.7000    1.4000
    -0.5000   -0.9000
     1.9000    0.7000
```

```
>> b=[1.1; -0.2; -0.6]
b =
   1.1000
  -0.2000
  -0.6000
>> rcond(A)
error: rcond: matrix must be square
>> x=A\b
x =
  -0.6808
   0.9751
```

(c)
```
>> A=[-2 1.2 -0.8; 1.2 -0.8 1.1; 0 0.1 -1]
A =
  -2.0000    1.2000   -0.8000
   1.2000   -0.8000    1.1000
   0.0000    0.1000   -1.0000
>> b=[0.8; -0.4; -2.4]
b =
   0.8000
  -0.4000
  -2.4000
>> rcond(A)
ans =   0.003389
>> x=A\b
x =
   42.44
   78.22
   10.22
```

(d)
```
>> A=[0.3 0.6 1.7 -0.3
-0.2 -1 0.2 1.5
0.2 -0.8 1 1.3
1.2 0.8 -1.1 -0.9]
A =
   0.3000    0.6000    1.7000   -0.3000
  -0.2000   -1.0000    0.2000    1.5000
   0.2000   -0.8000    1.0000    1.3000
   1.2000    0.8000   -1.1000   -0.9000
>> b=[-1.5; -1.3; -2; 1.2]
b =
  -1.500
  -1.300
  -2.000
   1.200
```

```
>> rcond(A)
ans =   0.02162
>> x=A\b
x =
   -0.3350
   -0.3771
   -0.8747
   -1.0461
```

(e)
```
>> A=[1.4 0.9 1.9; -0.9 -0.2 0.4]
A =
     1.4000     0.9000     1.9000
    -0.9000    -0.2000     0.4000
>> b=[-2.3; -0.6]
b =
    -2.3000
    -0.6000
>> rcond(A)
error: rcond: matrix must be square
>> x=A\b
x =
     0.1721
    -0.2306
    -1.2281
```

(f)
```
>> A=[0.3 0.3 0.3 -0.5
   1.5 -0.2 -1 1.5
   -0.6 1.1 -0.9 -0.4
   1.8 1.1 -0.9 0.2]
A =
     0.3000     0.3000     0.3000    -0.5000
     1.5000    -0.2000    -1.0000     1.5000
    -0.6000     1.1000    -0.9000    -0.4000
     1.8000     1.1000    -0.9000     0.2000
>> b=[-1.1; -0.7; 0; 0.3]
b =
    -1.1000
    -0.7000
     0.0000
     0.3000
>> rcond(A)
ans = 5.879e-05
>> x=A\b
x =
  -501.5000
```

```
1979.7500
1862.2500
2006.5000
```

Exercise 2.2.7 Which of the following systems are in reduced row echelon form? For those that are, determine all solutions, if any.

(a)
$$x_1 = -194$$
$$x_2 = 564$$
$$x_3 = -38$$
$$x_4 = 275$$

(b)
$$y_1 - 13.3y_4 = -13.1$$
$$y_2 + 6.1y_4 = 5.7$$
$$y_3 + 3.3y_4 = 3.1$$

(c)
$$z_1 - 13.3z_3 = -13.1$$
$$z_2 + 6.1z_3 = 5.7$$
$$3.3z_3 + z_4 = 3.1$$

(d)
$$a - d = -4$$
$$b - \tfrac{7}{2}d = -29$$
$$c - \tfrac{1}{4}d = -\tfrac{7}{2}$$

Exercise 2.2.8 For each of the following tables of data, use a system of linear equations to determine the nominated polynomial that finds the second column as a function of the first column. Sketch a graph of your fitted polynomial and the data points. Record your working.

(a) linear

x	y
2	-4
3	4

(b) quadratic

x	y
-2	-1
1	0
2	5

(c) quadratic

p	q
0	-1
2	3
3	4

(d) cubic

r	t
-3	-4
-2	0
-1	-3
0	-6

Exercise 2.2.9 In three consecutive years a company sells goods to the value of $51M, $81M and $92M (millions of dollars). Find a quadratic that fits this data, and use the quadratic to predict the value of sales in the fourth year.

Exercise 2.2.10 In 2011, there were 98 wolves in Yellowstone National Park; in 2012 there were 83 wolves; and in 2013 there were 95 wolves. Find a quadratic that fits this data, and use the quadratic to predict the number of wolves in 2014. To keep the coefficients manageable, write the quadratic in terms of the number of years from the starting year of 2011.

Exercise 2.2.11 Table 2.4 lists the time taken by a planet to orbit the Sun and a typical distance of the planet from the Sun. Analogous to Example 2.2.11, fit a quadratic polynomial $T = c_1 + c_2 R + c_3 R^2$ for the period T as a function of distance R. Use the data for Mercury, Venus, and Earth. Then use the quadratic to predict the period of Mars; what is the error in your prediction? (Example 3.5.11 shows that a power law fit is better, and that the power law agrees with Kepler's law.)

Table 2.4 Orbital periods for four planets of the solar system: the periods are in (Earth) days; the distance is the length of the semi-major axis of the orbits (Wikipedia, 2014). https://en.wikipedia.org/wiki/Orbital_period

planet	distance (gigametres)	period (days)
Mercury	57.91	87.97
Venus	108.21	224.70
Earth	149.60	365.26
Mars	227.94	686.97

Exercise 2.2.12 *(Global Positioning System in space-time)* For each case below, and in space-time, suppose you know from five GPS satellites that you and your GPS receiver are the given measured time shifts away from the given locations of each of the three satellites (locations are in Mm). Following Example 2.2.12, determine both your position and the discrepancy in time between your GPS receiver and the satellites' GPS time. Which case needs another satellite?

(a) $\begin{cases} 0.03\,s & \text{time shift before } (17,11,17) \\ 0.03\,s & \text{time shift before } (11,20,14) \\ 0.0233\cdots s & \text{time shift before } (20,10,9) \\ 0.03\,s & \text{time shift before } (9,13,21) \\ 0.03\,s & \text{time shift before } (7,24,8) \end{cases}$

(b) $\begin{cases} 0.1\,s & \text{time shift before } (11,12,18) \\ 0.1066\cdots s & \text{time shift before } (18,6,19) \\ 0.1\,s & \text{time shift before } (11,19,9) \\ 0.1066\cdots s & \text{time shift before } (9,10,22) \\ 0.1\,s & \text{time shift before } (23,3,9) \end{cases}$

(c) $\begin{cases} 0.03\,s & \text{time shift before } (17,11,17) \\ 0.03\,s & \text{time shift before } (19,12,14) \\ 0.0233\cdots s & \text{time shift before } (20,10,9) \\ 0.03\,s & \text{time shift before } (9,13,21) \\ 0.03\,s & \text{time shift before } (7,24,8) \end{cases}$

Exercise 2.2.13 Formulate the following two-thousand-year-old Chinese puzzle as a system of linear equations. Use algebraic manipulation to solve the system.

There are three classes of grain, of which three bundles of the first class, two of the second, and one of the third make 39 measures. Two of the first, three of the second, and one of the third make 34 measures. And one of the first, two of the second, and three of the third make 26 measures. How many measures of grain are contained in one bundle of each class?

Jiuzhang Suanshu, 200BC (Chartier 2015, p.3)

Exercise 2.2.14 Suppose you are given data at n points, equi-spaced in x. Say the known data points are $(1, y_1), (2, y_2), \ldots, (n, y_n)$ for some given $y_1, y_2, \ldots, y_n$. Seek a polynomial fit to the data of degree $(n-1)$; that is, seek the fit $y = c_1 + c_2 x + c_3 x^2 + \cdots + c_n x^{n-1}$. In MATLAB/Octave, form the matrix of the linear equations that need to be solved for the coefficients $c_1, c_2, \ldots, c_n$. According to Procedure 2.2.5, for what number n of data points is $\mathtt{rcond}$ good? poor? bad? terrible?

Exercise 2.2.15 In a few sentences, answer/discuss each of the following.

(a) What is the role of the MATLAB/Octave function $\mathtt{rcond()}$?

(b) How does the $=$ symbol in MATLAB/Octave compare with the $=$ symbol in mathematics?

(c) How can there be several different augmented matrices for a given system of linear equations? How many different augmented matrices may there be for a system of three linear equations in three unknowns?

(d) Why is the reduced row echelon form important?

(e) Why cannot there be exactly three solutions to a system of linear equations?

2.3 Linear combinations span sets

A common feature in the solution to linear equations is the appearance of combinations of several vectors. For example, the general solution to Example 2.2.28(c) is

$$x = (-2s - \tfrac{15}{7}t, s, \tfrac{9}{7}t, t)$$
$$= \underbrace{s(-2, 1, 0, 0) + t(-\tfrac{15}{7}, 0, \tfrac{9}{7}, 1)}_{\text{linear combination}}.$$

The general solution to Example 2.2.20(a) is

$$x = (-2 - s + 2t, s, 5 - 4t, t)$$
$$= \underbrace{1 \cdot (-2, 0, 5, 0) + s(-1, 1, 0, 0) + t(2, 0, -4, 1)}_{\text{linear combination}}.$$

Such so-called linear combinations occur in many other contexts. Recall that the standard unit vectors in $\mathbb{R}^3$ are $e_1 = (1, 0, 0)$, $e_2 = (0, 1, 0)$ and $e_3 = (0, 0, 1)$ (Definition 1.2.7); so any other vector in $\mathbb{R}^3$ may be written as

$$x = (x_1, x_2, x_3)$$
$$= x_1(1, 0, 0) + x_2(0, 1, 0) + x_3(0, 0, 1)$$
$$= \underbrace{x_1 e_1 + x_2 e_2 + x_3 e_3}_{\text{linear combination}}.$$

The widespread appearance of such combinations calls for the following definition.

Definition 2.3.1 *A vector v is a **linear combination** of vectors v_1, v_2, ..., v_k if there are scalars c_1, c_2, ..., c_k (called the **coefficients**) such that $v = c_1 v_1 + c_2 v_2 + \cdots + c_k v_k$.*

Example 2.3.2 Estimate roughly each of the blue vectors as a linear combination of the given red vectors in the following graphs (estimate coefficients to say roughly 10% error).

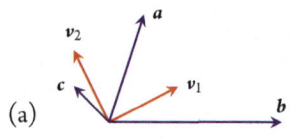

(a)

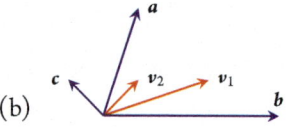

(b)

Solution: By visualizing various combinations:

$a \approx v_1 + v_2 = 1v_1 + 1v_2$;

$b \approx 2v_1 - v_2 = 2v_1 + (-1)v_2$;

$c \approx -0.2v_1 + 0.6v_2$. □

Solution: By visualizing various combinations: $a \approx -1v_1 + 4v_2$;

$b \approx 2v_1 - 2v_2$; $c \approx -1v_1 + 2v_2$. □

Activity 2.3.3 Choose any one of these linear combinations:

$$2v_1 - 0.5v_2 ; \quad 0v_1 - v_2 ; \quad -0.5v_1 + 0.5v_2 ; \quad v_1 + v_2.$$

Then in the plot to the right, which vector, a, b, c, or d, corresponds to your chosen linear combination?

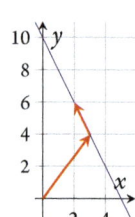

Example 2.3.4 Parametric descriptions of lines and planes involve linear combinations (Sections 1.2 and 1.3).

(a) For each value of t, the expression $(3, 4) + t(-1, 2)$ is a linear combination of the two vectors $(3, 4)$ and $(-1, 2)$. Over all values of parameter t, it describes the line illustrated to the right. (The line is alternatively described as $2x + y = 10$.)

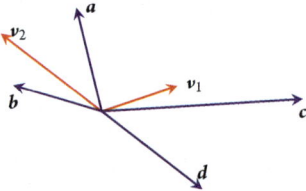

(b) For each value of s and t, the expression $2(1, 0, 1) + s(-1, -\frac{1}{2}, \frac{1}{2}) + t(1, -1, 0)$ is a linear combination of the three vectors

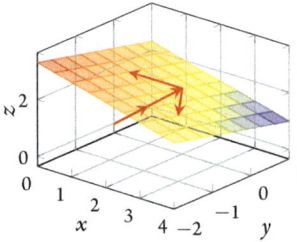
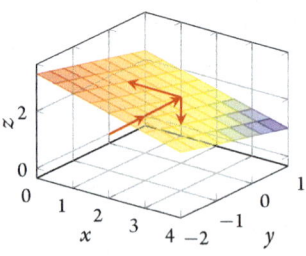

$(1, 0, 1)$, $(-1, -\frac{1}{2}, \frac{1}{2})$, and $(1, -1, 0)$. Over all values of the parameters s and t it describes the plane illustrated above-right in stereo. (Alternatively the plane could be described as $x + y + 3z = 8$).

(c) The expression $t(-1, 2, 0) + t^2(0, 2, 1)$ is a linear combination of the two vectors $(-1, 2, 0)$ and $(0, 2, 1)$ as the vectors are multiplied by scalars and then

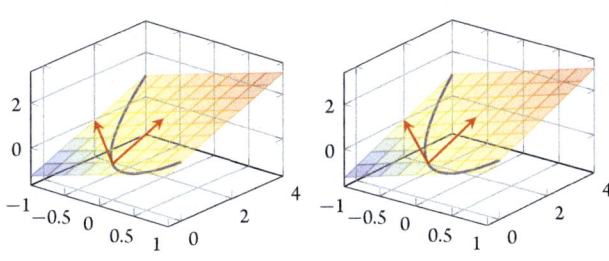

added. That a coefficient is a nonlinear function of some parameter is irrelevant to the property of linear combination. This expression is a parametric description of a parabola in $\mathbb{R}^3$, as illustrated above in stereo, and very soon we will be able to say it is a parabola in the plane spanned by $(-1, 2, 0)$ and $(0, 2, 1)$. □

The matrix-vector form $A\boldsymbol{x} = \boldsymbol{b}$ of a system of linear equations involves a linear combination on the left-hand side.

Example 2.3.5 Recall from Definition 2.2.2 that $\begin{bmatrix} -5 & 4 \\ 3 & 2 \end{bmatrix}\begin{bmatrix} x \\ y \end{bmatrix} = \begin{bmatrix} 1 \\ -2 \end{bmatrix}$ is our matrix-vector representation of the system of the two equations $-5x + 4y = 1$, and $3x + 2y = -2$. Form both sides into a vector so that

$$\begin{bmatrix} -5x + 4y \\ 3x + 2y \end{bmatrix} = \begin{bmatrix} 1 \\ -2 \end{bmatrix}.$$

Write the left-hand side as the sum of two vectors:

$$\begin{bmatrix} -5x \\ 3x \end{bmatrix} + \begin{bmatrix} 4y \\ 2y \end{bmatrix} = \begin{bmatrix} 1 \\ -2 \end{bmatrix}.$$

By scalar multiplication the system becomes

$$\begin{bmatrix} -5 \\ 3 \end{bmatrix} x + \begin{bmatrix} 4 \\ 2 \end{bmatrix} y = \begin{bmatrix} 1 \\ -2 \end{bmatrix}.$$

That is, the left-hand side is a linear combination of $(-5, 3)$ and $(4, 2)$, the two columns of the matrix. □

Example 2.3.6 Let's repeat the previous example in general. Recall from Definition 2.2.2 that $A\boldsymbol{x} = \boldsymbol{b}$ is our matrix-vector representation for the system of m equations

$$a_{11}x_1 + a_{12}x_2 + \cdots + a_{1n}x_n = b_1,$$
$$a_{21}x_1 + a_{22}x_2 + \cdots + a_{2n}x_n = b_2,$$
$$\vdots$$
$$a_{m1}x_1 + a_{m2}x_2 + \cdots + a_{mn}x_n = b_m.$$

Form both sides into a vector so that

$$
\begin{bmatrix} a_{11}x_1 + a_{12}x_2 + \cdots + a_{1n}x_n \\ a_{21}x_1 + a_{22}x_2 + \cdots + a_{2n}x_n \\ \vdots \\ a_{m1}x_1 + a_{m2}x_2 + \cdots + a_{mn}x_n \end{bmatrix} = \begin{bmatrix} b_1 \\ b_2 \\ \vdots \\ b_m \end{bmatrix}.
$$

Then use addition and scalar multiplication of vectors (Definition 1.2.4) to rewrite the left-hand side vector as

$$
\begin{bmatrix} a_{11} \\ a_{21} \\ \vdots \\ a_{m1} \end{bmatrix} x_1 + \begin{bmatrix} a_{12} \\ a_{22} \\ \vdots \\ a_{m2} \end{bmatrix} x_2 + \cdots + \begin{bmatrix} a_{1n} \\ a_{2n} \\ \vdots \\ a_{mn} \end{bmatrix} x_n = \begin{bmatrix} b_1 \\ b_2 \\ \vdots \\ b_m \end{bmatrix}.
$$

This left-hand side is a linear combination of the columns of matrix A: define from the columns of A the n vectors, $a_1 = (a_{11}, a_{21}, \ldots, a_{m1})$, $a_2 = (a_{12}, a_{22}, \ldots, a_{m2})$, ..., $a_n = (a_{1n}, a_{2n}, \ldots, a_{mn})$, then the left-hand side is a linear combination of these vectors, with the coefficients of the linear combination being x_1, x_2, ..., x_n. That is, the system $Ax = b$ is identical to the linear combination $x_1 a_1 + x_2 a_2 + \cdots + x_n a_n = b$. $\qquad \square$

Theorem 2.3.7 *A system of linear equations $Ax = b$ is consistent (Procedure 2.2.23) if and only if the right-hand side vector b is a linear combination of the columns of A.*

Be aware of a subtle twist going on in this theorem: for the general Example 2.3.6 this theorem turns a question about the existence of an n variable solution x, into a question about vectors with m components, and vice versa.

Proof. Example 2.3.6 establishes that if a solution x exists, then b is a linear combination of the columns. Conversely, if vector b is a linear combination of the columns, then a solution x exists with components of x set to the coefficients in the linear combination. $\qquad \square$

Example 2.3.8 This first example considers the simplest cases when the matrix has only one column, and so any linear combination is only a scalar multiple of that column. Compare the consistency of the equations with the right-hand side being a linear combination of the column of the matrix.

(a) $\begin{bmatrix} -1 \\ 2 \end{bmatrix} x = \begin{bmatrix} -2 \\ 4 \end{bmatrix}$.

Solution: The system is consistent because $x = 2$ is a solution (Procedure 2.2.23). Also, the right-hand side $b = (-2, 4)$ is the linear combination $2(-1, 2)$ of the column of the matrix. $\qquad \square$

(b) $\begin{bmatrix} -1 \\ 2 \end{bmatrix} x = \begin{bmatrix} 2 \\ 3 \end{bmatrix}.$

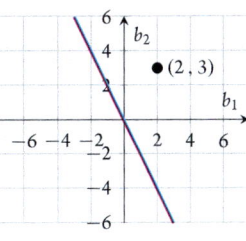

Solution: The system is inconsistent as the first equation requires $x = -2$ whereas the second requires $x = \frac{3}{2}$ and these cannot hold simultaneously (Procedure 2.2.23). Also, there is no multiple of $(-1, 2)$ that gives the right-hand side $\boldsymbol{b} = (2, 3)$ so the right-hand side cannot be a linear combination of the column of the matrix—as illustrated to the right. □

(c) $\begin{bmatrix} 1 \\ a \end{bmatrix} x = \begin{bmatrix} 3 \\ -6 \end{bmatrix}$ depending upon parameter a.

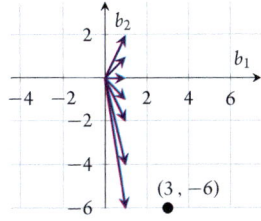

Solution: The first equation requires $x = 3$ whereas the second equation requires $ax = -6$; that is, $a \cdot 3 = -6$, that is, $a = -2$. Thus it is only for $a = -2$ that the system is consistent; for $a \neq -2$ the system is inconsistent. Also, plotted to the right are vectors $(1, a)$ for various a. It is only for $a = -2$ that the vector is aligned towards the given $(3, -6)$. Hence it is only for $a = -2$ that a linear combination of $(1, a)$ can give the required $(3, -6)$. □

Activity 2.3.9 For what value of a is the system $\begin{bmatrix} 3-a \\ -2a \end{bmatrix} x = \begin{bmatrix} 1 \\ 1 \end{bmatrix}$ consistent?

(a) $a = 1$ (b) $a = -\frac{1}{2}$ (c) $a = 2$ (d) $a = -3$

In the Examples 2.3.4 and 2.3.6 of linear combination, the coefficients mostly are a variable parameter or unknown. Consequently, mostly we are interested in the range of possibilities encompassed by a given set of vectors.

Definition 2.3.10 *Let a set of k vectors in $\mathbb{R}^n$ be $S = \{v_1, v_2, \ldots, v_k\}$, then the set of all linear combinations of $v_1, v_2, \ldots, v_k$ is called the **span** of $v_1, v_2, \ldots, v_k$, and is denoted by $\mathrm{span}\{v_1, v_2, \ldots, v_k\}$ or span S.*[8]

Example 2.3.11

(a) Let the set $S = \{(-1, 2)\}$ with just one vector. Then span $S = \mathrm{span}\{(-1, 2)\}$ is the set of all vectors encompassed by the form $t(-1, 2)$. From the parametric equation of a line (Definition 1.2.15), span S is all vectors in the line $y = -2x$, as shown to the right.

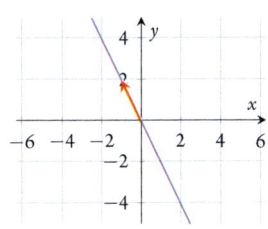

[8] In the degenerate case of the set S being the empty set, we take its span to be just the zero vector; that is, by convention span$\{\} = \{0\}$. But we rarely need this degenerate case.

(b) With two vectors in the set, $\text{span}\{(-1,2),(3,4)\} = \mathbb{R}^2$ is the entire 2D plane. To see this, recall that any point in the span must be of the form $s(-1,2)+t(3,4)$. Given any vector (x_1,x_2) in $\mathbb{R}^2$ we choose $s = (-4x_1+3x_2)/10$ and $t = (2x_1+x_2)/10$ and then the linear combination

$$s\begin{bmatrix}-1\\2\end{bmatrix} + t\begin{bmatrix}3\\4\end{bmatrix} = \frac{-4x_1+3x_2}{10}\begin{bmatrix}-1\\2\end{bmatrix} + \frac{2x_1+x_2}{10}\begin{bmatrix}3\\4\end{bmatrix}$$

$$= x_1\left(\frac{-4}{10}\begin{bmatrix}-1\\2\end{bmatrix} + \frac{2}{10}\begin{bmatrix}3\\4\end{bmatrix}\right)$$

$$+ x_2\left(\frac{3}{10}\begin{bmatrix}-1\\2\end{bmatrix} + \frac{1}{10}\begin{bmatrix}3\\4\end{bmatrix}\right)$$

$$= x_1\begin{bmatrix}1\\0\end{bmatrix} + x_2\begin{bmatrix}0\\1\end{bmatrix} = \begin{bmatrix}x_1\\x_2\end{bmatrix}.$$

Since every vector in $\mathbb{R}^2$ can be expressed as $s(-1,2)+t(3,4)$, it follows that then $\mathbb{R}^2 = \text{span}\{(-1,2),(3,4)\}$.

(c) But if two vectors are proportional to each other then their span is a line. For example, $\text{span}\{(-1,2),(2,-4)\}$ is the set of all vectors of the form $r(-1,2)+s(2,-4) = r(-1,2)+(-2s)(-1,2) = (r-2s)(-1,2) = t(-1,2)$ for $t = r-2s$. That is, $\text{span}\{(-1,2),(2,-4)\} = \text{span}\{(-1,2)\}$ as illustrated to the right.

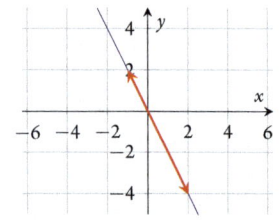

(d) In 3D, $\text{span}\{(-1,2,0),(0,2,1)\}$ is the set of all linear combinations $s(-1,2,0)+t(0,2,1)$ which here is a parametric form of the plane illust-

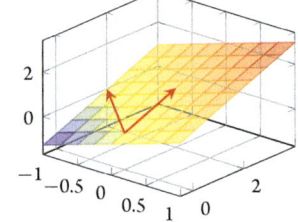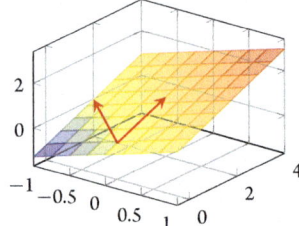

rated here (Definition 1.3.32). The plane passes through the origin $\mathbf{0}$, obtained when $s = t = 0$. We could also check that the vector $(2,1,-2)$ is orthogonal to these two vectors, hence is a normal to the plane, and so the plane may also be expressed as $2x+y-2z = 0$.

(e) For the complete set of n standard unit vectors in $\mathbb{R}^n$ (Definition 1.2.7), $\text{span}\{\mathbf{e}_1, \mathbf{e}_2, \ldots, \mathbf{e}_n\} = \mathbb{R}^n$. This is because every vector $\mathbf{x} = (x_1, x_2, \ldots, x_n)$ in $\mathbb{R}^n$ may be written as the linear combination $\mathbf{x} = x_1\mathbf{e}_1 + x_2\mathbf{e}_2 + \cdots + x_n\mathbf{e}_n$, and hence every vector is in $\text{span}\{\mathbf{e}_1, \mathbf{e}_2, \ldots, \mathbf{e}_n\}$.

(f) The homogeneous system (Definition 2.2.27) of linear equations from Example 2.2.28(c) has solutions $\mathbf{x} = (-2s - \frac{15}{7}t, s, \frac{9}{7}t, t) = (-2,1,0,0)s + (-\frac{15}{7},0,\frac{9}{7},1)t$ for arbitrary s and t. That is, the set of solutions is $\text{span}\{(-2,1,0,0),(-\frac{15}{7},0,\frac{9}{7},1)\}$, a subset of $\mathbb{R}^4$.

Generally, the set of solutions to a homogeneous system is the span of some set.

(g) However, the set of solutions to a non-homogeneous system is generally not the span of some set. For example, the solutions to Example 2.2.25 are all of the form $(u,v,w) = (-\frac{3}{4} - \frac{1}{4}t, \frac{1}{2} + \frac{3}{2}t, t) = (-\frac{3}{4}, \frac{1}{2}, 0) + t(-\frac{1}{4}, \frac{3}{2}, 1)$ for arbitrary t. True, each of these solutions is a linear combination of vectors $(-\frac{3}{4}, \frac{1}{2}, 0)$ and $(-\frac{1}{4}, \frac{3}{2}, 1)$. But the multiple of $(-\frac{3}{4}, \frac{1}{2}, 0)$ is always fixed, whereas the span invokes *all* multiples. Consequently, all the possible solutions cannot be the same as the span of a set of vectors.

Activity 2.3.12 To the right is drawn a line: for which one of the following vectors u is span$\{u\}$ *not* the drawn line?

(a) $(4,2)$

(b) $(-1,-0.5)$

(c) $(-1,-2)$

(d) $(2,1)$

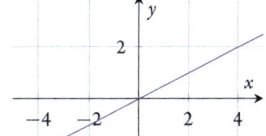

Example 2.3.13 Describe in other words span$\{i,k\}$ in $\mathbb{R}^3$.

Solution: All vectors in span$\{i,k\}$ are of the form $c_1 i + c_2 k = c_1(1,0,0) + c_2(0,0,1) = (c_1, 0, c_2)$. Hence the span is all vectors with second component zero—it is the plane $y = 0$ in (x,y,z) coordinates. □

Example 2.3.14 Find a set S such that span $S = \{(3b, a+b, -2a-4b) : a,b \text{ scalars}\}$. Similarly, find a set T such that span $T = \{(-a-2b-2, -b+1, -3b-1) : a,b \text{ scalars}\}$.

Solution: Because vectors $(3b, a+b, -2a-4b) = a(0,1,-2) + b(3,1,-4)$ for all scalars a and b, a suitable set is $S = \{(0,1,-2),(3,1,-4)\}$.

Second, vectors $(-a-2b-2, -b+1, 3b-1) = a(-1,0,0) + b(-2,-1,3) + (-2,1,-1)$ which are linear combinations for all a and b. However, the vectors cannot form a span due to the constant vector $(-2,1,-1)$ because a span requires *all* linear combinations of its component vectors. The given set cannot be expressed as a span. □

Geometrically, the span of a set of vectors is always all vectors lying in either a line, a plane, or a higher dimensional hyper-plane, that passes *through the origin* (discussed further by Section 3.4).

2.3.1 Exercises

Exercise 2.3.1 For each of the following, express vectors a and b as a linear combination of vectors v_1 and v_2. Estimate the coefficients roughly (to say 10%).

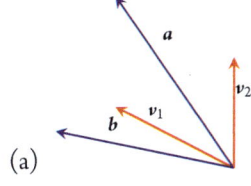

(a)

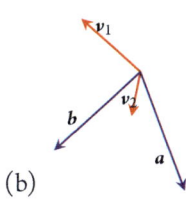

(b)

(c)　　　　　　　　　　　　　　　　(d)

Exercise 2.3.2 For each of the following lines in 2D, write down a parametric equation of the line as a linear combination of two vectors, one of which is multiplied by the parameter.

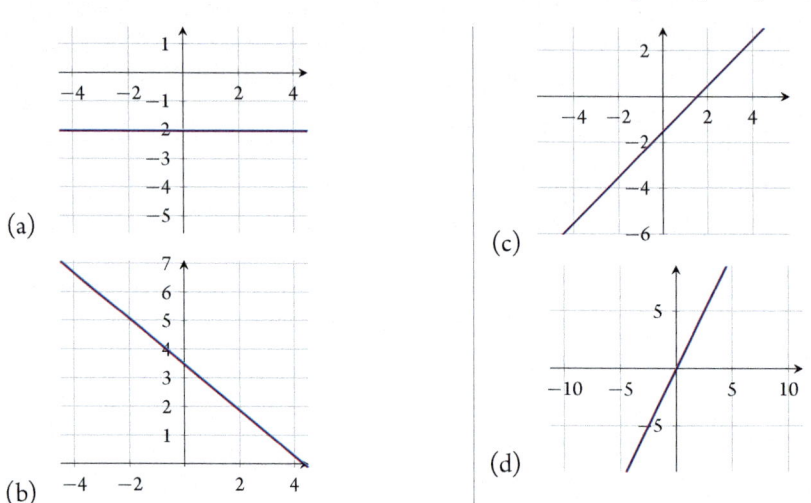

(a)

(b)

(c)

(d)

Exercise 2.3.3 Write each of the following systems of linear equations as one vector equation involving a linear combination of vectors.

(a) $\begin{aligned} -2x + y - 2z &= -2 \\ -4x + 2y - z &= 2 \end{aligned}$

(b) $\begin{aligned} -3x + 2y - 3z &= 0 \\ y - z &= 0 \\ x - 3y &= 0 \end{aligned}$

(c) $\begin{aligned} x_1 + 3x_2 + x_3 - 2x_4 &= 2 \\ 2x_1 + x_2 + 4x_3 - 2x_4 &= -1 \\ -x_1 + 2x_2 - 2x_3 - x_4 &= 3 \end{aligned}$

(d) $\begin{aligned} -2p - 2q &= -1 \\ q &= 2 \\ 3p - q &= 1 \end{aligned}$

Exercise 2.3.4 For each of the cases in Exercise 2.3.3, by attempting to solve the system, determine if the right-hand side vector is in the span of the vectors on the left-hand side.

Exercise 2.3.5 For each of the following sets, write the set as a span, if possible. Give reasons.

(a) $\{(p - 4q, p + 2q, p + 2q) : p, q \text{ scalars}\}$
(b) $\{(-p + 2r, 2p - 2q, p + 2q + r, -q - 3r) : p, q, r \text{ scalars}\}$
(c) The line $y = 2x + 1$ in $\mathbb{R}^2$.
(d) The line $x = y = z$ in $\mathbb{R}^3$.
(e) The set of vectors x in $\mathbb{R}^4$ with component $x_3 = 0$.

Exercise 2.3.6 Show the following identities hold for any given vectors u, v and w:

(a) $\operatorname{span}\{u, v\} = \operatorname{span}\{u - v, u + v\}$;

(b) $\operatorname{span}\{u, v, w\} = \operatorname{span}\{u, u - v, u + v + w\}$.

Exercise 2.3.7 Suppose u_1, u_2, ..., u_s are any s vectors in $\mathbb{R}^n$. Let the set $R = \{u_1, u_2, \ldots, u_r\}$ for some $r < s$, and the set $S = \{u_1, u_2, \ldots, u_r, u_{r+1}, \ldots, u_s\}$.

(a) Prove that $\operatorname{span} R \subseteq \operatorname{span} S$; that is, that every vector in $\operatorname{span} R$ is also in $\operatorname{span} S$.

(b) Hence deduce that if $\operatorname{span} R = \mathbb{R}^n$, then $\operatorname{span} S = \mathbb{R}^n$.

Exercise 2.3.8 Suppose u_1, u_2, ..., u_r and v_1, v_2, ..., v_s are all vectors in $\mathbb{R}^n$.

(a) Prove that if every vector u_j is a linear combination of v_1, v_2, ..., v_s, then $\operatorname{span}\{u_1, u_2, \ldots, u_r\} \subseteq \operatorname{span}\{v_1, v_2, \ldots, v_s\}$; that is, that every vector in $\operatorname{span}\{u_1, u_2, \ldots, u_r\}$ is also in $\operatorname{span}\{v_1, v_2, \ldots, v_s\}$.

(b) Prove that if, additionally, every vector v_j is a linear combination of u_1, u_2, ..., u_r, then $\operatorname{span}\{u_1, u_2, \ldots, u_r\} = \operatorname{span}\{v_1, v_2, \ldots, v_s\}$.

Exercise 2.3.9 Let $S = \{v_1, v_2, \ldots, v_s\}$ be a set of vectors in $\mathbb{R}^n$ such that vector v_1 is a linear combination of v_2, v_3, ..., v_s. Prove that $\operatorname{span} S = \operatorname{span}\{v_2, v_3, \ldots, v_s\}$.

Exercise 2.3.10 In a few sentences, answer/discuss each of the following.

(a) In what circumstances have you encountered linear combinations of vectors—even though they may not have been identified as linear combinations. Justify classifying the circumstances as involving linear combinations.

(b) Why is the zero vector always in the span of a set of vectors?

(c) Suppose three nonzero vectors are $u, v, w \in \mathbb{R}^3$. Describe all the possibilities for the span of these three vectors.

(d) How does the span of a set of vectors arise in linear equations?

2.4 Summary of linear equations

Introduction to systems of linear equations

⋆⋆ A **linear equation** in the n variables $x_1, x_2, \ldots, x_n$ is an equation that can be written in the form (Definition 2.1.2)

$$a_1 x_1 + a_2 x_2 + \cdots + a_n x_n = b.$$

A **system** of linear equations is a set of one or more linear equations in one or more variables.

- Algebraic manipulation, such as that for the GPS, can sometimes extract a tractable system of linear equations from an intractable nonlinear problem. Often the algebraic manipulation forms equations in an abstract setting where it is difficult to interpret the mathematical quantities—but the effort is worthwhile.

Directly solve linear systems

★ The **matrix-vector form** of a given system is $Ax = b$ (Definition 2.2.2) for the $m \times n$ **matrix** of coefficients

$$
A = \begin{bmatrix}
a_{11} & a_{12} & \cdots & a_{1n} \\
a_{21} & a_{22} & \cdots & a_{2n} \\
\vdots & \vdots & \ddots & \vdots \\
a_{m1} & a_{m2} & \cdots & a_{mn}
\end{bmatrix},
$$

and vectors $x = (x_1, x_2, \ldots, x_n)$ and $b = (b_1, b_2, \ldots, b_m)$. If $m = n$, then A is called a **square matrix**.

★★ Procedure 2.2.5 uses MATLAB/Octave to solve the matrix-vector system $Ax = b$, for a square matrix A:

1. form matrix A and column vector b;

2. check rcond(A) exists and is not too small, $1 \geq$ good $> 10^{-2} >$ poor $> 10^{-4} >$ bad $> 10^{-8} >$ terrible;

3. if rcond(A) is acceptable, then execute x=A\b to compute the solution vector x.

Checking rcond() avoids many mistakes in applications.

- In MATLAB/Octave you may need the following.
    - ★★ [... ; ... ; ...] forms both matrices and vectors, or use newlines instead of the semicolons.
    - ★★ rcond(A) of a square matrix A *estimates* the reciprocal of the so-called condition number.
    - ★★ x=A\b computes an 'answer' to $Ax = b$ —but it may not be a solution unless rcond(A) exists and is not small;
    - — Change an element of an array or vector by assigning a new value with assignments A(i,j)=... or b(i)=... where i and j denote some indices.
    - — For a vector (or matrix) t and an exponent p, the operation t.^p computes the pth power of each element in the vector.
    - ★ The function ones(m,1) gives a (column) vector of m ones, $(1,1,\ldots,1)$.
- For fitting data, generally avoid using polynomials of degree higher than cubic.
- To solve systems by hand we need several more notions. The **augmented matrix** of the system $Ax = b$ is the matrix $[A \vdots b]$ (Definition 2.2.14).
- **Elementary row operations** either on a system of linear equations or on its corresponding augmented matrix do not change the solutions (Theorem 2.2.17):
    - — interchange two equations/rows; or
    - — multiply an equation/row by a nonzero constant; or
    - — add a multiple of an equation/row to another.
- A system of linear equations or (augmented) matrix is in **reduced row echelon form** (RREF) when (Definition 2.2.19) all the following hold:
    - — any equations with all zero coefficients, or rows of the matrix consisting entirely of zeros, are at the bottom;

- in each nonzero equation/row, the first nonzero coefficient/entry is a one (called the **leading one**), and is in a variable/column to the left of any leading ones below it; and
- each variable/column containing a leading one has zero coefficients/entries in every other equation/row.

A **free variable** is any variable which is *not* multiplied by a leading one in the algebraic equations.

★ Gauss–Jordan elimination solves systems by hand (Procedure 2.2.23):

1. Write down either the full symbolic form of the system of linear equations, or the augmented matrix of the system of linear equations.

2. Use elementary row operations to reduce the system/augmented matrix to reduced row echelon form.

3. If the resulting system is consistent, then solve for the leading variables in terms of any remaining free variables.

★★ For every system of linear equations $Ax = b$, exactly one of the following holds (Theorem 2.2.26): there is either no solution, or a unique solution, or infinitely many solutions.

- A system of linear equations is called **homogeneous** when the system may be written $Ax = 0$ (Definition 2.2.27), otherwise the system is termed **non-homogeneous**.

★ If $Ax = 0$ is a homogeneous system of m linear equations with n variables where $m < n$, then the system has infinitely many solutions (Theorem 2.2.30).

Linear combinations span sets

★★ A vector v is a **linear combination** of vectors v_1, v_2, ..., v_k if there are scalars c_1, c_2, ..., c_k (called the **coefficients**) such that $v = c_1 v_1 + c_2 v_2 + \cdots + c_k v_k$ (Definition 2.3.1).

- A system of linear equations $Ax = b$ is **consistent** if and only if the right-hand side vector b is a linear combination of the columns of A (Theorem 2.3.7).

★★ For any given set of vectors $S = \{v_1, v_2, \ldots, v_k\}$, the set of all linear combinations of v_1, v_2, ..., v_k is called the **span** of v_1, v_2, ..., v_k, and is denoted by span$\{v_1, v_2, \ldots, v_k\}$ or span S (Definition 2.3.10).

Answers to selected activities

2.1.4c, 2.2.4d, 2.2.7b, 2.2.16d, 2.2.21d, 2.2.22c, 2.2.29d, 2.3.9d, 2.3.12c,

Answers to selected exercises

2.1.1b line $y = \frac{3}{2}x - 1$

2.1.1d $(x, y) = (-1, -1)$

2.1.1f no solution

2.1.1h line $t = 4s - 2$

2.1.2b (1.8, 1.04)

2.1.2d (1.25, 0.41)

2.1.3b no solution

2.1.3d no solution

2.1.4b (4, 5)

2.1.4d $(-1, 11)/9$ surely indicates an error.

2.2.1b e.g. $\begin{bmatrix} -2 & -1 \\ 1 & -6 \end{bmatrix}\begin{bmatrix} p \\ q \end{bmatrix} = \begin{bmatrix} -1 \\ 2 \end{bmatrix}$,

$\begin{bmatrix} 1 & -6 \\ -2 & -1 \end{bmatrix}\begin{bmatrix} p \\ q \end{bmatrix} = \begin{bmatrix} 2 \\ -1 \end{bmatrix}$. Four: two orderings of rows, and two orderings of the variables.

2.2.1d Twelve possibilities: two orderings of rows, and $3! = 6$ orderings of the variables.

2.2.2 (a) $(x, y) = (-0.4, -1.2)$

(b) $(p, q) = (0.6154, -0.2308)$

(c) No solution as rcond requires a square matrix.

(d) No solution as rcond requires a square matrix.

(e) $(u, v, w) =$ $(-2, 1.8571, 0.4286)$

2.2.3b $(p, q, r) = (32, 25, 20)$

2.2.4 As before, except $(u, v, w) = (-2 + 3t, \frac{1}{3} - \frac{5}{3}t, t)$ for all t

2.2.5b $c = (-1.64, -0.30, 0.58, 0.41)$ (2 d.p.)

2.2.6b Fails to solve the system
$0.7x_1 + 1.4x_2 = 1.1$,
$-0.5x_1 - 0.9x_2 = -0.2$,
$1.9x_1 + 0.7x_2 = -0.6$ because the matrix is not square as reported by rcond. The 'answer' x is not relevant (yet).

2.2.6d Solve the system
$0.3x_1 + 0.6x_2 + 1.7x_3 - 0.3x_4 = -1.5$,
$-0.2x_1 - x_2 + 0.2x_3 + 1.5x_4 = -1.3$,
$0.2x_1 - 0.8x_2 + x_3 + 1.3x_4 = -2$,
$1.2x_1 + 0.8x_2 - 1.1x_3 - 0.9x_4 = 1.2$.
Since rcond is good, the solution is
$x_1 = -0.34, x_2 = -0.38, x_3 = -0.87$
and $x_4 = -1.05$ (2 d.p.).

2.2.6f Solves the system
$0.3x_1 + 0.3x_2 + 0.3x_3 - 0.5x_4 = -1.1$,

$1.5x_1 - 0.2x_2 - x_3 + 1.5x_4 = -0.7$,
$-0.6x_1 + 1.1x_2 - 0.9x_3 - 0.4x_4 = 0$,
$1.8x_1 + 1.1x_2 - 0.9x_3 + 0.2x_4 = 0.3$.
But since rcond is bad, the 'answer' x is likely to be meaningless: here the large values $x =$
$(-501.5, 1979.75, 1862.25, 2006.5)$
also suggest a bad answer.

2.2.7b Yes, $y = (-13.1 + 13.3t, 5.7 - 6.1t, 3.1 - 3.3t, t)$ for all t

2.2.7d Yes. $(a, b, c, d) =$ $(-4 + t, -29 + \frac{7}{2}t, -\frac{7}{2} + \frac{1}{4}t, t)$ for all t

2.2.8b $y = -\frac{8}{3} + \frac{3}{2}x + \frac{7}{6}x^2$

2.2.8d $t = -6 - \frac{2}{3}t + \frac{7}{2}t^2 + \frac{7}{6}t^3$

2.2.10 134 wolves

2.2.12a rcond $= 0.014$, $(3, 4, 3)$, shift $= 12/300 = 0.04$ s

2.2.12c rcond $= 0$, singular, needs another satellite.

2.2.14 Good, $\{1, 2\}$; poor, $\{3, 4\}$; bad, $\{5, 6, 7\}$; terrible, $\{8, 9, \ldots\}$.

2.3.1b $a = -1v_1 + 1.5v_2$, $b = 1v_1 + 3v_2$

2.3.1d $a = -1.8v_1 + 1.5v_2$, $b = -1.5v_1 - 0.6v_2$

2.3.2b e.g. $(0, 3.5) + t(-2.5, 2)$

2.3.2d e.g. $(0.5, 1) + t(0.5, 1)$

2.3.3b $\begin{bmatrix} -3 \\ 0 \\ 1 \end{bmatrix} x + \begin{bmatrix} 2 \\ 1 \\ -3 \end{bmatrix} y + \begin{bmatrix} -3 \\ 1 \\ 0 \end{bmatrix} z = \begin{bmatrix} 0 \\ 0 \\ 0 \end{bmatrix}$

2.3.3d $\begin{bmatrix} -2 \\ 0 \\ 3 \end{bmatrix} p + \begin{bmatrix} -2 \\ 1 \\ -1 \end{bmatrix} q = \begin{bmatrix} -1 \\ 2 \\ 1 \end{bmatrix}$

2.3.5a e.g. span$\{(1, 1, 1), (-4, 2, 2)\}$

2.3.5c Not a span.

2.3.5e e.g. span$\{e_1, e_2, e_4\}$.

3 Matrices encode system interactions

Linear Algebra for the 21st Century. A. J. Roberts, Oxford University Press (2020). © A. J. Roberts.
DOI: 10.1093/oso/9780198856399.003.0003

Section 2.2 introduced matrices in the matrix-vector form $Ax = b$ of a system of linear equations. This chapter starts with Sections 3.1 and 3.2 developing the basic operations on matrices that make them so useful in applications and theory—including making sense of the 'product' Ax. Section 3.3 then explores how the so-called "singular value decomposition (SVD)" of a matrix empowers us to understand how to solve general linear systems of equations, and a graphical meaning of a matrix in terms of rotations and stretching. The structures discovered by an SVD lead to further conceptual development (Section 3.4) that underlies the, at first paradoxical, solution of inconsistent equations (Section 3.5). Finally, Section 3.6 unifies the geometric views invoked.

> the language of mathematics reveals itself unreasonably effective in the natural sciences ...
> a wonderful gift which we neither understand nor deserve. We should be grateful for it and
> hope that it will remain valid in future research and that it will extend, for better or for worse,
> to our pleasure even though perhaps also to our bafflement, to wide branches of learning
>
> *Wigner, 1960 (Mandelbrot, 1982, p.3)*

3.1 Matrix operations and algebra

This section introduces basic matrix concepts, operations, and algebra. You may have met some of it in previous study.

3.1.1 Basic matrix terminology

Let's start with some basic definitions of terminology.

- As already introduced by Section 2.2, a **matrix** is a rectangular array of real numbers, scalars, written inside **brackets** $[\cdots]$, such as these six examples:[1]

$$\begin{bmatrix} -2 & -5 & 4 \\ 1 & -3 & 0 \\ 2 & 4 & 0 \end{bmatrix}, \quad \begin{bmatrix} -2.33 & 3.66 \\ -4.17 & -0.36 \end{bmatrix}, \quad \begin{bmatrix} 0.56 \\ 3.99 \\ -5.22 \end{bmatrix},$$

$$\begin{bmatrix} 1 & -\sqrt{3} & \pi \\ -5/3 & \sqrt{5} & -1 \end{bmatrix}, \quad \begin{bmatrix} 1 & \frac{10}{3} & \frac{\pi^2}{4} \end{bmatrix}, \quad \begin{bmatrix} 0.35 \end{bmatrix}. \tag{3.1}$$

- The **size** of a matrix is its number of rows and columns—written $m \times n$ where m is the number of rows and n is the number of columns. The six example matrices of (3.1) are of size, respectively, $3 \times 3, 2 \times 2, 3 \times 1, 2 \times 3, 1 \times 3,$ and 1×1.
  Recall from Definition 2.2.2 that if the number of rows equals the number of columns, $m = n$, then it is called a square matrix. For example, the first, second, and last matrices in (3.1) are square; the others are not.

---

[1] Chapter 7 starts using complex numbers in a matrix, but until then we stay within the realm of real numbers. Some books use parentheses, $(\cdot)$, around matrices: we do not as here parentheses denote a vector when we write the components horizontally on the page (most often used when written in text).

- To correspond with vectors, we often invoke the term **column vector** which means a matrix with only one column; that is, a matrix of size $m \times 1$ for some m. For convenience and compatibility with vectors, we often write a column vector horizontally within **parentheses** $(\cdots)$. The third matrix of (3.1) is an example, and may also be written as $(0.56, 3.99, -5.22)$.

  Occasionally we refer to a **row vector** to mean a matrix with one row; that is, a $1 \times n$ matrix for some n, such as the fifth matrix of (3.1). Remember the distinction: a row of numbers written within brackets, $[\cdots]$, is a row vector, whereas a row of numbers written within parentheses, $(\cdots)$, is a column vector.

- The numbers appearing in a matrix are called the **entries**, **elements**, or **components** of the matrix. For example, the first matrix in (3.1) has entries/elements/components of the numbers $-5, -3, -2, 0, 1, 2$, and 4.

- But it is important to identify where the numbers appear in a matrix: the **double subscript** notation identifies the location of an entry. For a matrix A, the entry in row i and column j is denoted by a_{ij}: by convention we use capital (uppercase) letters for a matrix, and the corresponding lowercase letter subscripted for its entries.[2] For example, let matrix

$$A = \begin{bmatrix} -2 & -5 & 4 \\ 1 & -3 & 0 \\ 2 & 4 & 0 \end{bmatrix},$$

then entries $a_{12} = -5$, $a_{22} = -3$ and $a_{31} = 2$.

- The first of two special matrices is a **zero matrix** of all zeros and of any size: the symbol $O_{m \times n}$ denotes the $m \times n$ zero matrix, such as

$$O_{2 \times 4} = \begin{bmatrix} 0 & 0 & 0 & 0 \\ 0 & 0 & 0 & 0 \end{bmatrix}.$$

  The symbol O_n denotes the square zero matrix of size $n \times n$, whereas the plain symbol O denotes a zero matrix whose size is apparent from the context.

- Arising from the nature of matrix multiplication (Section 3.1.2), the second special matrix is the **identity matrix**: the symbol I_n denotes an $n \times n$ square matrix which has zero entries except for the diagonal from the top-left to the bottom-right which are all ones. Occasionally we invoke non-square 'identity' matrices denoted by $I_{m \times n}$. Three examples are

---

[2] Some books use the capital letter subscripted for its entries: that is, some use A_{ij} to denote the entry in the ith row and jth column of matrix A. However, we use A_{ij} to mean something else, the so-called 'minor' (Theorem 6.2.11).

$$I_3 = \begin{bmatrix} 1 & 0 & 0 \\ 0 & 1 & 0 \\ 0 & 0 & 1 \end{bmatrix}, \quad I_{2\times 3} = \begin{bmatrix} 1 & 0 & 0 \\ 0 & 1 & 0 \end{bmatrix}, \quad I_{4\times 2} = \begin{bmatrix} 1 & 0 \\ 0 & 1 \\ 0 & 0 \\ 0 & 0 \end{bmatrix}.$$

The plain symbol I denotes an identity matrix whose size is apparent from the context.

- Using the double subscript notation, and as already used in Definition 2.2.2, a general $m \times n$ matrix

$$A = \begin{bmatrix} a_{11} & a_{12} & \cdots & a_{1n} \\ a_{21} & a_{22} & \cdots & a_{2n} \\ \vdots & \vdots & \ddots & \vdots \\ a_{m1} & a_{m2} & \cdots & a_{mn} \end{bmatrix}.$$

Often, as already seen in Example 2.3.6, it is useful to write a matrix A in terms of its n column vectors a_j, $A = \begin{bmatrix} a_1 & a_2 & \cdots & a_n \end{bmatrix}$. For example, matrix

$$B = \begin{bmatrix} 1 & -\sqrt{3} & \pi \\ -5/3 & \sqrt{5} & -1 \end{bmatrix} = \begin{bmatrix} b_1 & b_2 & b_3 \end{bmatrix}$$

for the three column vectors

$$b_1 = \begin{bmatrix} 1 \\ -5/3 \end{bmatrix}, \quad b_2 = \begin{bmatrix} -\sqrt{3} \\ \sqrt{5} \end{bmatrix}, \quad b_3 = \begin{bmatrix} \pi \\ -1 \end{bmatrix}.$$

Alternatively these column vectors may be written as $b_1 = (1, -5/3)$, $b_2 = (-\sqrt{3}, \sqrt{5})$, and $b_3 = (\pi, -1)$.

- Lastly, two matrices are **equal** ($=$) if they both have the same size *and* their corresponding entries are equal. Otherwise the matrices are not equal. For example, consider matrices

$$A = \begin{bmatrix} 2 & \pi \\ 3 & 9 \end{bmatrix}, \quad B = \begin{bmatrix} \sqrt{4} & \pi \\ 2+1 & 3^2 \end{bmatrix}, \quad C = \begin{bmatrix} 2 & \pi \end{bmatrix}, \quad D = \begin{bmatrix} 2 \\ \pi \end{bmatrix} = (2, \pi).$$

The matrices $A = B$ because they are the same size and their corresponding entries are equal, such as $a_{11} = 2 = \sqrt{4} = b_{11}$. Matrix A cannot be equal to C because their sizes are different. Matrices C and D are not equal, despite having the same elements in the same order, because they have different sizes: 1×2 and 2×1 respectively.

Activity 3.1.1 Which of the following matrices equals $\begin{bmatrix} 3 & -1 & 4 \\ -2 & 0 & 1 \end{bmatrix}$?

(a) $\begin{bmatrix} 3 & -2 \\ -1 & 0 \\ 4 & 1 \end{bmatrix}$

(c) $\begin{bmatrix} 3 & 1-2 & \sqrt{16} \\ 3-2 & 0 & e^0 \end{bmatrix}$

(b) $\begin{bmatrix} \sqrt{9} & -1 & 2^2 \\ -2 & 0 & \cos 0 \end{bmatrix}$

(d) $\begin{bmatrix} 3 & -1 \\ 4 & -2 \\ 0 & 1 \end{bmatrix}$

3.1.2 Addition, subtraction, and multiplication with matrices

A matrix is not just an array of numbers: associated with a matrix is a suite of operations that empower a matrix to be useful in applications. We start with addition and multiplication; 'division' is addressed by Section 3.2 and others.

An analogue in computing science is the concept of object orientated programming. In object oriented programming one defines not just data structures, but also the functions that operate on those structures. Analogously, an array is just a group of numbers, but a matrix is an array together with many operations explicitly available. The power and beauty of matrices results from the ramifications of its associated operations.

Matrix addition and subtraction

Corresponding to vector addition and subtraction (Definition 1.2.4), matrix addition and subtraction is done component wise, but only between matrices of the same size.

Example 3.1.2 Let matrices

$$A = \begin{bmatrix} 4 & 0 \\ -5 & -4 \\ 0 & -3 \end{bmatrix}, \quad B = \begin{bmatrix} 1 & 0 & 2 \\ -3 & 0 & 3 \end{bmatrix}, \quad C = \begin{bmatrix} -4 & -1 \\ -4 & -1 \\ 1 & 4 \end{bmatrix},$$

$$D = \begin{bmatrix} -2 & -1 & -3 \\ 1 & 3 & 0 \end{bmatrix}, \quad E = \begin{bmatrix} 5 & -2 & -2 \\ 0 & -3 & 2 \\ -4 & 7 & -1 \end{bmatrix}.$$

Then the addition and subtraction

$$A + C = \begin{bmatrix} 4 & 0 \\ -5 & -4 \\ 0 & -3 \end{bmatrix} + \begin{bmatrix} -4 & -1 \\ -4 & -1 \\ 1 & 4 \end{bmatrix}$$

$$= \begin{bmatrix} 4 + (-4) & 0 + (-1) \\ -5 + (-4) & -4 + (-1) \\ 0 + 1 & -3 + 4 \end{bmatrix} = \begin{bmatrix} 0 & -1 \\ -9 & -5 \\ 1 & 1 \end{bmatrix},$$

$$B - D = \begin{bmatrix} 1 & 0 & 2 \\ -3 & 0 & 3 \end{bmatrix} - \begin{bmatrix} -2 & -1 & -3 \\ 1 & 3 & 0 \end{bmatrix}$$

$$= \begin{bmatrix} 1 - (-2) & 0 - (-1) & 2 - (-3) \\ -3 - 1 & 0 - 3 & 3 - 0 \end{bmatrix} = \begin{bmatrix} 3 & 1 & 5 \\ -4 & -3 & 3 \end{bmatrix}.$$

But because the matrices are of different sizes, the following are not defined and must not be attempted: $A + B$, $A - D$, $E - A$, $B + C$, $E - C$, for example. □

In general, when A and B are both $m \times n$ matrices, with entries a_{ij} and b_{ij} respectively, then we define their **sum** or **addition**, $A + B$, as the $m \times n$ matrix whose (i,j)th entry is $a_{ij} + b_{ij}$. Similarly, define the **difference** or **subtraction** $A - B$ as the $m \times n$ matrix whose (i,j)th entry is $a_{ij} - b_{ij}$. That is,

$$A + B = \begin{bmatrix} a_{11} + b_{11} & a_{12} + b_{12} & \cdots & a_{1n} + b_{1n} \\ a_{21} + b_{21} & a_{22} + b_{22} & \cdots & a_{2n} + b_{2n} \\ \vdots & \vdots & \ddots & \vdots \\ a_{m1} + b_{m1} & a_{m2} + b_{m2} & \cdots & a_{mn} + b_{mn} \end{bmatrix},$$

$$A - B = \begin{bmatrix} a_{11} - b_{11} & a_{12} - b_{12} & \cdots & a_{1n} - b_{1n} \\ a_{21} - b_{21} & a_{22} - b_{22} & \cdots & a_{2n} - b_{2n} \\ \vdots & \vdots & \ddots & \vdots \\ a_{m1} - b_{m1} & a_{m2} - b_{m2} & \cdots & a_{mn} - b_{mn} \end{bmatrix}.$$

Consequently, letting O denote the zero matrix of the appropriate size,

$$A \pm O = A, \quad O + A = A, \quad \text{and} \quad A - A = O.$$

Activity 3.1.3 Given the two matrices $A = \begin{bmatrix} 3 & -2 \\ 1 & -1 \end{bmatrix}$ and $B = \begin{bmatrix} 2 & 1 \\ 3 & 2 \end{bmatrix}$, which of the following is the matrix $\begin{bmatrix} 5 & -1 \\ -2 & -3 \end{bmatrix}$?

(a) $A + B$

(b) $B - A$

(c) none of the others

(d) $A - B$

Scalar multiplication of matrices

Corresponding to multiplication of a vector by a scalar (Definition 1.2.4), multiplication of a matrix by a scalar means that every entry of the matrix is multiplied by the scalar.

Example 3.1.4 Let the three matrices

$$A = \begin{bmatrix} 5 & 2 \\ -2 & 3 \end{bmatrix}, \quad B = \begin{bmatrix} 1 \\ 0 \\ -6 \end{bmatrix}, \quad C = \begin{bmatrix} 5 & -6 & 4 \\ -1 & -3 & -3 \end{bmatrix}.$$

Then the scalar multiplications

$$3A = \begin{bmatrix} 3 \cdot 5 & 3 \cdot 2 \\ 3 \cdot (-2) & 3 \cdot 3 \end{bmatrix} = \begin{bmatrix} 15 & 6 \\ -6 & 9 \end{bmatrix},$$

$$-B = (-1)B = \begin{bmatrix} (-1) \cdot 1 \\ (-1) \cdot 0 \\ (-1) \cdot (-6) \end{bmatrix} = \begin{bmatrix} -1 \\ 0 \\ 6 \end{bmatrix},$$

$$-\pi C = (-\pi)C = \begin{bmatrix} -5\pi & 6\pi & -4\pi \\ \pi & 3\pi & 3\pi \end{bmatrix}.$$

□

In general, when A is an $m \times n$ matrix, with entries a_{ij}, then we define the **scalar product** by c, denoted by either cA or Ac, as the $m \times n$ matrix whose (i,j)th entry is ca_{ij}.[3] That is,

$$cA = Ac = \begin{bmatrix} ca_{11} & ca_{12} & \cdots & ca_{1n} \\ ca_{21} & ca_{22} & \cdots & ca_{2n} \\ \vdots & \vdots & \ddots & \vdots \\ ca_{m1} & ca_{m2} & \cdots & ca_{mn} \end{bmatrix}.$$

Matrix-vector multiplication transforms

Recall that the matrix-vector form of a system of linear equations, Definition 2.2.2, is written $Ax = b$. In this form, Ax denotes a matrix-vector product. As implied by Definition 2.2.2, we define the general **matrix-vector product**

$$Ax := \begin{bmatrix} a_{11}x_1 + a_{12}x_2 + \cdots + a_{1n}x_n \\ a_{21}x_1 + a_{22}x_2 + \cdots + a_{2n}x_n \\ \vdots \\ a_{m1}x_1 + a_{m2}x_2 + \cdots + a_{mn}x_n \end{bmatrix}$$

for $m \times n$ matrix A and vector x in $\mathbb{R}^n$ with entries/components

$$A = \begin{bmatrix} a_{11} & a_{12} & \cdots & a_{1n} \\ a_{21} & a_{22} & \cdots & a_{2n} \\ \vdots & \vdots & \ddots & \vdots \\ a_{m1} & a_{m2} & \cdots & a_{mn} \end{bmatrix} \quad \text{and} \quad x = \begin{bmatrix} x_1 \\ x_2 \\ \vdots \\ x_n \end{bmatrix}.$$

This product is only defined when the number of columns of matrix A is the same as the number of components of vector x.[4] If not, then the product cannot be used.

Example 3.1.5 Let matrices

$$A = \begin{bmatrix} 3 & 2 \\ -2 & 1 \end{bmatrix}, \quad B = \begin{bmatrix} 5 & -6 & 4 \\ -1 & -3 & -3 \end{bmatrix},$$

[3] Be aware that MATLAB/Octave reasonably treats multiplication by a '1×1 matrix' as a scalar multiplication. Strictly speaking products such as '$[0.35]A$' are not defined because strictly speaking $[0.35]$ is not a scalar but is a 1×1 matrix. However, MATLAB/Octave do not make the distinction.

[4] Some of you who have studied calculus may wonder about what might be called 'continuous matrices' $A(x, y)$ which multiply a function $f(x)$ according to the integral $\int_a^b A(x, y)f(y)\, dy$. Then you might wonder about solving problems such as find the unknown $f(x)$ such that $\int_0^1 A(x, y)f(y)\, dy = \sin \pi x$ for given 'continuous matrix' $A(x, y) := \min(x, y)$ $[1 - \max(x, y)]$; you may check that here the solution is $f = \pi^2 \sin \pi x$. Such notions are a useful generalization of our linear algebra: they are called integral equations; the main structures and patterns developed by this course also apply to such integral equations.

and vectors $x = (2, -3)$ and $b = (1, 0, 4)$. Then the matrix-vector products

$$Ax = \begin{bmatrix} 3 & 2 \\ -2 & 1 \end{bmatrix} \begin{bmatrix} 2 \\ -3 \end{bmatrix} = \begin{bmatrix} 3 \cdot 2 + 2 \cdot (-3) \\ (-2) \cdot 2 + 1 \cdot (-3) \end{bmatrix} = \begin{bmatrix} 0 \\ -7 \end{bmatrix},$$

$$Bb = \begin{bmatrix} 5 & -6 & 4 \\ -1 & -3 & -3 \end{bmatrix} \begin{bmatrix} 1 \\ 0 \\ 4 \end{bmatrix}$$

$$= \begin{bmatrix} 5 \cdot 1 + (-6) \cdot 0 + 4 \cdot 4 \\ (-1) \cdot 1 + (-3) \cdot 0 + (-3) \cdot 4 \end{bmatrix} = \begin{bmatrix} 21 \\ -13 \end{bmatrix}.$$

The combinations Ab and Bx are not defined because the number of columns of each matrix is not equal to the number of components in the multiplying vector.

Further, we do not here define vector-matrix products such as xA or bB: the order of multiplication matters with matrices and so these are not in the scope of this definition. □

Activity 3.1.6 Which of the following is the result of the matrix-vector product $\begin{bmatrix} 4 & 1 \\ 3 & -2 \end{bmatrix} \begin{bmatrix} 3 \\ 2 \end{bmatrix}$?

(a) $\begin{bmatrix} 21 \\ -2 \end{bmatrix}$
(b) $\begin{bmatrix} 15 \\ 2 \end{bmatrix}$
(c) $\begin{bmatrix} 14 \\ 5 \end{bmatrix}$
(d) $\begin{bmatrix} 18 \\ -1 \end{bmatrix}$

Geometric interpretation Multiplication of a vector by a *square matrix* transforms the vector into another in the same space. The right graph shows the example of Ax from Example 3.1.5.

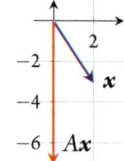

For another vector $y = (1, 1)$ and the same matrix A the product

$$Ay = \begin{bmatrix} 3 & 2 \\ -2 & 1 \end{bmatrix} \begin{bmatrix} 1 \\ 1 \end{bmatrix} = \begin{bmatrix} 3 \cdot 1 + 2 \cdot 1 \\ (-2) \cdot 1 + 1 \cdot 1 \end{bmatrix} = \begin{bmatrix} 5 \\ -1 \end{bmatrix},$$

as illustrated in the second right-hand picture.

Similarly, for the vector $z = (-1, 2)$ and the same matrix A the product

$$Az = \begin{bmatrix} 3 & 2 \\ -2 & 1 \end{bmatrix} \begin{bmatrix} -1 \\ 2 \end{bmatrix} = \begin{bmatrix} 3 \cdot (-1) + 2 \cdot 2 \\ (-2) \cdot (-1) + 1 \cdot 2 \end{bmatrix} = \begin{bmatrix} 1 \\ 4 \end{bmatrix},$$

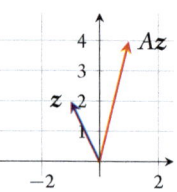

as illustrated in the third right-hand picture. Such a geometric interpretation underlies the use of matrix multiplication in video and picture processing, for example. Such video/picture processing employs stretching and shrinking (Section 3.2.2), rotations (Section 3.2.3), among more general transformations (Section 3.6).

Example 3.1.7 Recall I_n is the $n \times n$ identity matrix. Then the products

$$I_2 x = \begin{bmatrix} 1 & 0 \\ 0 & 1 \end{bmatrix} \begin{bmatrix} 2 \\ -3 \end{bmatrix} = \begin{bmatrix} 1 \cdot 2 + 0 \cdot (-3) \\ 0 \cdot 2 + 1 \cdot (-3) \end{bmatrix} = \begin{bmatrix} 2 \\ -3 \end{bmatrix},$$

$$I_3 b = \begin{bmatrix} 1 & 0 & 0 \\ 0 & 1 & 0 \\ 0 & 0 & 1 \end{bmatrix} \begin{bmatrix} 1 \\ 0 \\ 4 \end{bmatrix} = \begin{bmatrix} 1 \cdot 1 + 0 \cdot 0 + 0 \cdot 4 \\ 0 \cdot 1 + 1 \cdot 0 + 0 \cdot 4 \\ 0 \cdot 1 + 0 \cdot 0 + 1 \cdot 4 \end{bmatrix} = \begin{bmatrix} 1 \\ 0 \\ 4 \end{bmatrix}.$$

That is, and justifying its name of "identity", the products with an identity matrix give the result that is the vector itself: $I_2 x = x$ and $I_3 b = b$. Multiplication by the identity matrix leaves the vector unchanged (Theorem 3.1.26(e)).

Example 3.1.8 *(rabbits multiply)* In 1202, Fibonacci[5] famously considered the breeding of rabbits—such as the following question. One pair of rabbits can give birth to another pair of rabbits (called kittens) every month, say. Each kitten becomes fertile after it has aged a month, when it becomes adult and is called a buck (male) or doe (female). The new bucks and does then also start breeding. How many rabbits are there after six months?

Let's count just the females, the does, and the female kittens. At the start of any month let there be x_1 kittens (female) and x_2 does. Then at the end of the month:

- because all the female kittens grow up to be does, the number of does is now $x_2' = x_2 + x_1$;

- and because all the does at the start month have bred another pair of kittens, of which we expect one to be female, the new number of female kittens just born is $x_1' = x_2$.

Then x_1' and x_2' is the number of kittens and does at the start of the next month. Write this as a matrix-vector system. Let the female population be $x = (x_1, x_2)$ and the population one month later be $x' = (x_1', x_2')$. Then our model is that

$$x' = \begin{bmatrix} x_1' \\ x_2' \end{bmatrix} = \begin{bmatrix} x_2 \\ x_1 + x_2 \end{bmatrix} = \begin{bmatrix} 0 & 1 \\ 1 & 1 \end{bmatrix} \begin{bmatrix} x_1 \\ x_2 \end{bmatrix} = L x \quad \text{for } L = \begin{bmatrix} 0 & 1 \\ 1 & 1 \end{bmatrix},$$

called a Leslie matrix.[6]

- At the start there is one adult pair, one doe, and no female kittens, so the initial population is $x = (0, 1)$.

- After one month, females $x' = L x = \begin{bmatrix} 0 & 1 \\ 1 & 1 \end{bmatrix} \begin{bmatrix} 0 \\ 1 \end{bmatrix} = \begin{bmatrix} 1 \\ 1 \end{bmatrix}$.

- After two months, females $x'' = L x' = \begin{bmatrix} 0 & 1 \\ 1 & 1 \end{bmatrix} \begin{bmatrix} 1 \\ 1 \end{bmatrix} = \begin{bmatrix} 1 \\ 2 \end{bmatrix}$.

- After three months, females $x''' = L x'' = \begin{bmatrix} 0 & 1 \\ 1 & 1 \end{bmatrix} \begin{bmatrix} 1 \\ 2 \end{bmatrix} = \begin{bmatrix} 2 \\ 3 \end{bmatrix}$.

---

[5] Fibonacci's real name is Leonardo Bonacci. He lived circa 1175 to 1250, travelled extensively from Pisa, and is considered to be one of the most talented Western mathematicians of the Middle Ages.

[6] Named after Patrick H. Leslie, Leslie matrices are widely used in ecology to model the evolution over time of a population of organisms.

- After four months, females $x^{iv} = Lx''' = \begin{bmatrix} 0 & 1 \\ 1 & 1 \end{bmatrix}\begin{bmatrix} 2 \\ 3 \end{bmatrix} = \begin{bmatrix} 3 \\ 5 \end{bmatrix}$.

- After five months, females $x^v = Lx^{iv} = \begin{bmatrix} 0 & 1 \\ 1 & 1 \end{bmatrix}\begin{bmatrix} 3 \\ 5 \end{bmatrix} = \begin{bmatrix} 5 \\ 8 \end{bmatrix}$.

- After six months, females $x^{vi} = Lx^v = \begin{bmatrix} 0 & 1 \\ 1 & 1 \end{bmatrix}\begin{bmatrix} 5 \\ 8 \end{bmatrix} = \begin{bmatrix} 8 \\ 13 \end{bmatrix}$.

Fibonacci's model predicts the rabbit population grows rapidly according to the famous Fibonacci numbers $1, 2, 3, 5, 8, 13, 21, 34, 55, 89, \ldots$. $\square$

Example 3.1.9 *(age structured population)* An ecologist studies an isolated population of a species of animal. The growth of the population depends primarily upon the females so it is only these that are counted. The females are grouped into three ages: female pups (in their first year), juvenile females (one year old), and adult females (two years or older). During the study, the ecologist observes the following happens over the period of a year:

- half of the female pups survive and become juvenile females;
- one-third of the juvenile females survive and become adult females;
- each adult female breeds and produces four female pups;
- one-third of the adult females survive to breed in the following year;
- female pups and juvenile females do not breed.

(a) Let x_1, x_2, and x_3 be the number of females at the start of a year, of ages zero, one and two+ respectively, and let x_1', x_2', and x_3' be their number at the start of the next year. Use the ecologist's observations to write x_1', x_2', and x_3' as a function of x_1, x_2, and x_3 (this function is called a Markov chain).

(b) Letting vectors $x = (x_1, x_2, x_3)$ and $x' = (x_1', x_2', x_3')$ write down your function as the matrix-vector product $x' = Lx$ for some matrix L (called a Leslie matrix).

(c) Suppose the ecologist observes the numbers of females at the start of a given year is $x = (60, 70, 20)$, use your matrix to predict the numbers x' at the start of the next year. Continue similarly to predict the numbers after two years (x''), and three years (x''').

Solution:

(a) Since adult females breed and produce four female pups, $x_1' = 4x_3$. Since half of the female pups survive and become juvenile females, $x_2' = \frac{1}{2}x_1$. Since one-third of the juvenile females survive and become adult females, $\frac{1}{3}x_2$ contributes to x_3', but additionally one-third of the adult females survive to breed in the following year, so $x_3' = \frac{1}{3}x_2 + \frac{1}{3}x_3$.

(b) Writing these equations into vector form

$$x' = \begin{bmatrix} x_1' \\ x_2' \\ x_3' \end{bmatrix} = \begin{bmatrix} 4x_3 \\ \frac{1}{2}x_1 \\ \frac{1}{3}x_2 + \frac{1}{3}x_3 \end{bmatrix} = \underbrace{\begin{bmatrix} 0 & 0 & 4 \\ \frac{1}{2} & 0 & 0 \\ 0 & \frac{1}{3} & \frac{1}{3} \end{bmatrix}}_{L} x.$$

(c) Given the initial numbers of female animals is $x = (60, 70, 20)$, the number of females after one year is then predicted by the matrix-vector product

$$x' = Lx = \begin{bmatrix} 0 & 0 & 4 \\ \frac{1}{2} & 0 & 0 \\ 0 & \frac{1}{3} & \frac{1}{3} \end{bmatrix} \begin{bmatrix} 60 \\ 70 \\ 20 \end{bmatrix} = \begin{bmatrix} 80 \\ 30 \\ 30 \end{bmatrix}.$$

That is, the predicted numbers of females are 80 pups, 30 juveniles, and 30 adults.

After a second year the number of females is then predicted by the matrix-vector product $x'' = Lx'$. Here

$$x'' = Lx' = \begin{bmatrix} 0 & 0 & 4 \\ \frac{1}{2} & 0 & 0 \\ 0 & \frac{1}{3} & \frac{1}{3} \end{bmatrix} \begin{bmatrix} 80 \\ 30 \\ 30 \end{bmatrix} = \begin{bmatrix} 120 \\ 40 \\ 20 \end{bmatrix}.$$

After a third year the number of females is predicted by the matrix-vector product $x''' = Lx''$. Here

$$x''' = Lx'' = \begin{bmatrix} 0 & 0 & 4 \\ \frac{1}{2} & 0 & 0 \\ 0 & \frac{1}{3} & \frac{1}{3} \end{bmatrix} \begin{bmatrix} 120 \\ 40 \\ 20 \end{bmatrix} = \begin{bmatrix} 80 \\ 60 \\ 20 \end{bmatrix}.$$

□

Matrix-matrix multiplication

Matrix-vector multiplication explicitly uses the vector in its equivalent form as an $n \times 1$ matrix—a matrix with one column. Such multiplication immediately generalizes to the case of a right-hand matrix with multiple columns.

Example 3.1.10 Let two matrices

$$A = \begin{bmatrix} 3 & 2 \\ -2 & 1 \end{bmatrix}, \quad B = \begin{bmatrix} 5 & -6 & 4 \\ -1 & -3 & -3 \end{bmatrix},$$

then the matrix multiplication AB may be done as the matrix A multiplying each of the three columns in B. That is, in detail write

$$AB = A \begin{bmatrix} 5 & -6 & 4 \\ -1 & -3 & -3 \end{bmatrix} = A \begin{bmatrix} 5 & \vdots & -6 & \vdots & 4 \\ -1 & \vdots & -3 & \vdots & -3 \end{bmatrix} = \begin{bmatrix} A \begin{bmatrix} 5 \\ -1 \end{bmatrix} & \vdots & A \begin{bmatrix} -6 \\ -3 \end{bmatrix} & \vdots & A \begin{bmatrix} 4 \\ -3 \end{bmatrix} \end{bmatrix}$$

$$= \begin{bmatrix} \begin{bmatrix} 13 \\ -11 \end{bmatrix} & \vdots & \begin{bmatrix} -24 \\ 9 \end{bmatrix} & \vdots & \begin{bmatrix} 6 \\ -11 \end{bmatrix} \end{bmatrix} = \begin{bmatrix} 13 & -24 & 6 \\ -11 & 9 & -11 \end{bmatrix}.$$

Conversely, the product BA cannot be done because if we try the same procedure then

$$BA = B \begin{bmatrix} 3 & 2 \\ -2 & 1 \end{bmatrix} = B \left(\begin{bmatrix} 3 \\ -2 \end{bmatrix} : \begin{bmatrix} 2 \\ 1 \end{bmatrix} \right) = \begin{bmatrix} B \begin{bmatrix} 3 \\ -2 \end{bmatrix} : B \begin{bmatrix} 2 \\ 1 \end{bmatrix} \end{bmatrix},$$

and neither of these matrix-vector products can be done because, for example,

$$B \begin{bmatrix} 3 \\ -2 \end{bmatrix} = \begin{bmatrix} 5 & -6 & 4 \\ -1 & -3 & -3 \end{bmatrix} \begin{bmatrix} 3 \\ -2 \end{bmatrix}$$

the number of columns of the left matrix is not equal to the number of elements of the vector on the right. Hence the product BA is not defined. □

Example 3.1.11 For the following two matrices, compute, if possible, CD and DC, and compare these products:

$$C = \begin{bmatrix} -4 & -1 \\ -4 & -1 \\ 1 & 4 \end{bmatrix}, \quad D = \begin{bmatrix} -2 & -1 & -3 \\ 1 & 3 & 0 \end{bmatrix}.$$

Solution:

- On the one hand,

$$CD = C \begin{bmatrix} -2 & -1 & -3 \\ 1 & 3 & 0 \end{bmatrix} = \begin{bmatrix} C \begin{bmatrix} -2 \\ 1 \end{bmatrix} : C \begin{bmatrix} -1 \\ 3 \end{bmatrix} : C \begin{bmatrix} -3 \\ 0 \end{bmatrix} \end{bmatrix} = \begin{bmatrix} 7 & 1 & 12 \\ 7 & 1 & 12 \\ 2 & 11 & -3 \end{bmatrix}.$$

- Conversely,

$$DC = D \begin{bmatrix} -4 & -1 \\ -4 & -1 \\ 1 & 4 \end{bmatrix} = \begin{bmatrix} D \begin{bmatrix} -4 \\ -4 \\ 1 \end{bmatrix} : D \begin{bmatrix} -1 \\ -1 \\ 4 \end{bmatrix} \end{bmatrix} = \begin{bmatrix} 9 & -9 \\ -16 & -4 \end{bmatrix}.$$

Interestingly, $CD \neq DC$ —they are not even of the same size! □

Definition 3.1.12 *(matrix product)* Let matrix A be $m \times n$, and matrix B be $n \times p$, then the ***matrix product*** $C = AB$, or ***matrix multiplication***, is the $m \times p$ matrix whose (i,j)th entry is

$$c_{ij} = a_{i1} b_{1j} + a_{i2} b_{2j} + \cdots + a_{in} b_{nj}.$$

This formula looks like a dot product (Definition 1.3.2) of two vectors: indeed, we do use that the expression for the (i,j)th entry is the dot product of the ith row of A and the jth column of B as illustrated by

$$\begin{bmatrix} a_{11} & a_{12} & \cdots & a_{1n} \\ \vdots & \vdots & & \vdots \\ \boxed{a_{i1} \quad a_{i2} \quad \cdots \quad a_{in}} \\ \vdots & \vdots & & \vdots \\ a_{m1} & a_{m2} & \cdots & a_{mn} \end{bmatrix} \begin{bmatrix} b_{11} & \cdots & \boxed{b_{1j}} & \cdots & b_{1p} \\ b_{21} & \cdots & \boxed{b_{2j}} & \cdots & b_{2p} \\ \vdots & & \vdots & & \vdots \\ b_{n1} & \cdots & \boxed{b_{nj}} & \cdots & b_{np} \end{bmatrix}.$$

As seen in the examples, although the two matrices A and B may be of different sizes, the number of columns of A must equal the number of rows of B in order for the product AB to be defined.

Activity 3.1.13 Which one of the following matrix products is not defined?

(a) $\begin{bmatrix} 8 & 9 & 3 \\ 2 & 5 & 1 \end{bmatrix} \begin{bmatrix} -2 & 8 \\ 3 & -2 \end{bmatrix}$

(c) $\begin{bmatrix} 3 & -1 \end{bmatrix} \begin{bmatrix} -3 & 1 \\ 7 & -3 \end{bmatrix}$

(b) $\begin{bmatrix} 2 & 5 & -3 \end{bmatrix} \begin{bmatrix} -3 & 1 \\ -5 & -1 \\ 2 & -2 \end{bmatrix}$

(d) $\begin{bmatrix} -1 & 2 \\ 1 & -3 \end{bmatrix} \begin{bmatrix} -3 & -1 & -1 \\ 0 & -4 & -1 \end{bmatrix}$

Example 3.1.14 Matrix multiplication leads to powers of a square matrix. Let matrix $A = \begin{bmatrix} 3 & 2 \\ -2 & 1 \end{bmatrix}$, then by A^2 we mean the product

$$AA = \begin{bmatrix} 3 & 2 \\ -2 & 1 \end{bmatrix} \begin{bmatrix} 3 & 2 \\ -2 & 1 \end{bmatrix} = \begin{bmatrix} 5 & 8 \\ -8 & -3 \end{bmatrix},$$

and by A^3 we mean the product

$$AAA = AA^2 = \begin{bmatrix} 3 & 2 \\ -2 & 1 \end{bmatrix} \begin{bmatrix} 5 & 8 \\ -8 & -3 \end{bmatrix} = \begin{bmatrix} -1 & 18 \\ -18 & -19 \end{bmatrix},$$

and so on. ☐

In general, for an $n \times n$ square matrix A and a positive integer exponent p we define the **matrix power**

$$A^p = \underbrace{AA \cdots A}_{p \text{ factors}}.$$

The matrix powers A^p are also $n \times n$ square matrices.

Example 3.1.15 *(age structured population)* Matrix powers occur naturally in modelling populations by ecologists such as the animals of Example 3.1.9. Recall that given the numbers of female pups, juveniles, and adults formed into a vector $x = (x_1, x_2, x_3)$, the number in each age one year later (indicated here by a dash) is $x' = Lx$ for Leslie matrix

$$L = \begin{bmatrix} 0 & 0 & 4 \\ \frac{1}{2} & 0 & 0 \\ 0 & \frac{1}{3} & \frac{1}{3} \end{bmatrix}.$$

Hence the number in each age category two years later (indicated here by two dashes) is

$$\boldsymbol{x}'' = L\boldsymbol{x}' = L(L\boldsymbol{x}) = (LL)\boldsymbol{x} = L^2\boldsymbol{x},$$

provided that matrix multiplication is associative (established by Theorem 3.1.26(c)) to enable us to write $L(L\boldsymbol{x}) = (LL)\boldsymbol{x}$. Then the matrix square

$$L^2 = \begin{bmatrix} 0 & 0 & 4 \\ \frac{1}{2} & 0 & 0 \\ 0 & \frac{1}{3} & \frac{1}{3} \end{bmatrix} \begin{bmatrix} 0 & 0 & 4 \\ \frac{1}{2} & 0 & 0 \\ 0 & \frac{1}{3} & \frac{1}{3} \end{bmatrix} = \begin{bmatrix} 0 & \frac{4}{3} & \frac{4}{3} \\ 0 & 0 & 2 \\ \frac{1}{6} & \frac{1}{9} & \frac{1}{9} \end{bmatrix}.$$

Continuing to use such associativity, the number in each age category three years later (indicated here by threes dashes) is

$$\boldsymbol{x}''' = L\boldsymbol{x}'' = L(L^2\boldsymbol{x}) = (LL^2)\boldsymbol{x} = L^3\boldsymbol{x},$$

where the matrix cube

$$L^3 = LL^2 = \begin{bmatrix} 0 & 0 & 4 \\ \frac{1}{2} & 0 & 0 \\ 0 & \frac{1}{3} & \frac{1}{3} \end{bmatrix} \begin{bmatrix} 0 & \frac{4}{3} & \frac{4}{3} \\ 0 & 0 & 2 \\ \frac{1}{6} & \frac{1}{9} & \frac{1}{9} \end{bmatrix} = \begin{bmatrix} \frac{2}{3} & \frac{4}{9} & \frac{4}{9} \\ 0 & \frac{2}{3} & \frac{2}{3} \\ \frac{1}{18} & \frac{1}{27} & \frac{19}{27} \end{bmatrix}.$$

That is, the powers of the Leslie matrix help predict what happens two, three, or more years into the future. $\square$

The transpose of a matrix

The operations so far defined for matrices correspond directly to analogous operations for scalars. The transpose has no corresponding analogue. At first mysterious, the transpose occurs frequently—often due to it linking the dot product of vectors with matrix multiplication. The transpose also reflects symmetry in applications (Chapter 4), such as Newton's law that every action has an equal and opposite reaction.

Example 3.1.16 Let matrices

$$A = \begin{bmatrix} -4 & 2 \\ -3 & 4 \\ -1 & -7 \end{bmatrix}, \quad B = \begin{bmatrix} 2 & 0 & -1 \end{bmatrix}, \quad C = \begin{bmatrix} 1 & 1 & 1 \\ -1 & -3 & 0 \\ 2 & 3 & 2 \end{bmatrix}.$$

Then obtain the transpose of each of these three matrices by writing each of their rows as columns, in order:

$$A^T = \begin{bmatrix} -4 & -3 & -1 \\ 2 & 4 & -7 \end{bmatrix}, \quad B^T = \begin{bmatrix} 2 \\ 0 \\ -1 \end{bmatrix}, \quad C^T = \begin{bmatrix} 1 & -1 & 2 \\ 1 & -3 & 3 \\ 1 & 0 & 2 \end{bmatrix}.$$

☐

These examples illustrate the following definition.

Definition 3.1.17 *(transpose)* *The **transpose** of an $m \times n$ matrix A is the $n \times m$ matrix, denoted A^T, obtained by writing the ith row of A as the ith column of A^T, or equivalently by writing the jth column of A to be the jth row of A^T. That is, if $B = A^T$, then $b_{ij} = a_{ji}$.*

Activity 3.1.18 Which of the following matrices is the transpose of the matrix

$$\begin{bmatrix} 1 & -0.5 & 2.9 \\ -1.4 & -1.4 & -0.2 \\ 0.9 & -2.3 & 1.6 \end{bmatrix} ?$$

(a) $\begin{bmatrix} 2.9 & -0.5 & 1 \\ -0.2 & -1.4 & -1.4 \\ 1.6 & -2.3 & 0.9 \end{bmatrix}$
(c) $\begin{bmatrix} 1.6 & -2.3 & 0.9 \\ -0.2 & -1.4 & -1.4 \\ 2.9 & -0.5 & 1 \end{bmatrix}$

(b) $\begin{bmatrix} 1 & -1.4 & 0.9 \\ -0.5 & -1.4 & -2.3 \\ 2.9 & -0.2 & 1.6 \end{bmatrix}$
(d) $\begin{bmatrix} 0.9 & -2.3 & 1.6 \\ -1.4 & -1.4 & -0.2 \\ 1 & -0.5 & 2.9 \end{bmatrix}$

Example 3.1.19 *(transpose and dot product)* Consider two vectors in $\mathbb{R}^n$, say $u = (u_1, u_2, \ldots, u_n)$ and $v = (v_1, v_2, \ldots, v_n)$; that is,

$$u = \begin{bmatrix} u_1 \\ u_2 \\ \vdots \\ u_n \end{bmatrix}, \quad v = \begin{bmatrix} v_1 \\ v_2 \\ \vdots \\ v_n \end{bmatrix}.$$

Then the dot product between the two vectors

$$u \cdot v = u_1 v_1 + u_2 v_2 + \cdots + u_n v_n \quad \text{(Definition 1.3.2 of dot)}$$

$$= \begin{bmatrix} u_1 & u_2 & \cdots & u_n \end{bmatrix} \begin{bmatrix} v_1 \\ v_2 \\ \vdots \\ v_n \end{bmatrix} \qquad \text{(Definition 3.1.12 of mult.)}$$

$$= \begin{bmatrix} u_1 \\ u_2 \\ \vdots \\ u_n \end{bmatrix}^{\mathrm{T}} \begin{bmatrix} v_1 \\ v_2 \\ \vdots \\ v_n \end{bmatrix} \qquad \text{(transpose Definition 3.1.17)}$$

$$= u^{\mathrm{T}} v.$$

Subsequent sections and chapters often use this identity, that the dot product $u \cdot v = u^{\mathrm{T}} v$. ☐

Definition 3.1.20 (symmetry) *A (real) matrix A is a **symmetric matrix** if $A^T = A$; that is, if the matrix is equal to its transpose.*

A symmetric matrix must be a square matrix—as otherwise the sizes of A and A^{T} would be different and so the matrices could not be equal.

Example 3.1.21 None of the three matrices in Example 3.1.16 are symmetric: the first two matrices are not square so cannot be symmetric, and the third matrix $C \neq C^{\mathrm{T}}$. The following matrix is symmetric:

$$D = \begin{bmatrix} 2 & 0 & 1 \\ 0 & -6 & 3 \\ 1 & 3 & 4 \end{bmatrix} = D^{\mathrm{T}}.$$

When is the following general 2×2 matrix symmetric?

$$E = \begin{bmatrix} a & b \\ c & d \end{bmatrix}.$$

Solution: Consider the transpose

$$E^{\mathrm{T}} = \begin{bmatrix} a & c \\ b & d \end{bmatrix} \quad \text{compared with } E = \begin{bmatrix} a & b \\ c & d \end{bmatrix}.$$

The top-left and bottom-right elements are always the same. The top-right and bottom-left elements are the same if and only if $b = c$. That is, the 2×2 matrix E is symmetric if and only if $b = c$. ☐

Symmetric matrices of note are the $n \times n$ identity matrix and $n \times n$ zero matrix, I_n and O_n.

Activity 3.1.22 Which one of the following matrices is a symmetric matrix?

(a) $\begin{bmatrix} 0 & -3.2 & -0.8 \\ 3.2 & 0 & 3.2 \\ 0.8 & -3.2 & 0 \end{bmatrix}$

(c) $\begin{bmatrix} -2.6 & 0.3 & -1.3 \\ 0.3 & -0.2 & 0 \\ -1.3 & 0 & -2 \end{bmatrix}$

(b) $\begin{bmatrix} 2.2 & -0.9 & -1.2 \\ -0.9 & -1.2 & -3.1 \end{bmatrix}$

(d) $\begin{bmatrix} 2.3 & -1.3 & -2 \\ -3.2 & -1 & -1.3 \\ -3 & -3.2 & 2.3 \end{bmatrix}$

Compute in MATLAB/Octave

MATLAB/Octave empowers us to compute all these operations quickly, especially for the large problems found in applications: after all, MATLAB is an abbreviation of *Matrix Laboratory*. Table 3.1 summarizes the MATLAB/Octave version of the operations introduced so far, and used in the rest of this book.

Table 3.1 As well as the basics of MATLAB/Octave listed in Tables 1.2 and 2.3, we need these matrix operations.

- `size(A)` returns the number of rows and columns of matrix A: if A is $m \times n$, then `size(A)` returns $\begin{bmatrix} m & n \end{bmatrix}$.
- `A(i,j)` is the (i,j)th entry of a matrix A, `A(:,j)` is the jth column, `A(i,:)` is the ith row; either to use the value(s) or to assign value(s).
- `+,-,*` is matrix/vector/scalar addition, subtraction, and multiplication, but only provided the sizes of the two operands are compatible.
- `A^p` for scalar p computes the pth power of square matrix A (in contrast to `A.^p` which computes the pth power of *each element* of A, Table 2.3).
- The character single **prime/quote/dash**, `A'`, transposes the matrix A. But when using complex numbers be wary: `A'` is the complex conjugate transpose (which is what we usually want); whereas `A.'` is the transpose without complex conjugation.
- Predefined matrices include:
  - `zeros(m,n)` is the zero matrix $O_{m \times n}$;
  - `eye(m,n)` is $m \times n$ 'identity matrix' $I_{m \times n}$;
  - `ones(m,n)` is the $m \times n$ matrix where all entries are one;
  - `randn(m,n)` is a $m \times n$ matrix with random entries (independent, distributed Normally, mean zero, standard deviation one).

  A single argument gives the square matrix version:
  - `zeros(n)` is $O_n = O_{n \times n}$;
  - `eye(n)` is the $n \times n$ identity matrix $I_n = I_{n \times n}$;
  - `ones(n)` is the $n \times n$ matrix of all ones;
  - `randn(n)` is an $n \times n$ matrix with random entries.

  With no argument, these functions return the corresponding scalar: for example, `randn` computes a single random number.
- Very large and small magnitude numbers are printed in MATLAB/Octave like the following:
  - `4.852e+08` denotes the large $4.852 \cdot 10^8$; whereas
  - `3.469e-16` denotes the small $3.469 \cdot 10^{-16}$.

Matrix size and elements   Let the matrix

$$A = \begin{bmatrix} 0 & 0 & -2 & -11 & 5 \\ 0 & 1 & -1 & 11 & -8 \\ -4 & 2 & 10 & 2 & -3 \end{bmatrix}.$$

We readily see this is a 3×5 matrix, but to check that MATLAB/Octave agrees, execute the following in MATLAB/Octave:

```
A= [0 0 -2 -11   5
    0 1 -1   11 -8
   -4 2 10    2 -3]
size (A)
```

The answer, "3   5", confirms A is 3×5. MATLAB/Octave accesses individual elements, rows and columns. For example, execute each of the following:

- A(2,4) gives a_{24} which here results in 11;

- A(:,5) is the fifth column vector, here $\begin{bmatrix} 5 \\ -8 \\ -3 \end{bmatrix}$;

- A(1,:) is the first row, here $\begin{bmatrix} 0 & 0 & -2 & -11 & 5 \end{bmatrix}$.

One may also use these constructs to change the elements in matrix A: for example, executing A(2,4)=9 changes matrix A to

```
A =
      0      0    -2   -11      5
      0      1    -1     9     -8
     -4      2    10     2     -3
```

then A(:,5) = [2;-3;1] changes matrix A to

```
A =
      0      0    -2   -11      2
      0      1    -1     9     -3
     -4      2    10     2      1
```

whereas A(1,:) = [1 2 3 4 5] changes matrix A to

```
A =
      1      2     3     4      5
      0      1    -1     9     -3
     -4      2    10     2      1
```

Matrix addition and subtraction   To illustrate further operations let's use some random matrices generated by MATLAB/Octave: you will generate different matrices to the following, but the operations will work the same. Table 3.1 mentions that randn (m) and randn (m, n) generate random matrices so execute, say,

```
A=randn(4)
B=randn(4)
C=randn(4,2)
```

and obtain matrices such as (2 d.p.)

```
A =
   -1.31    2.07    0.08    2.05
    1.25   -1.35   -1.00    1.94
    1.08    1.79   -0.99    0.93
    1.34   -0.99   -0.23   -0.22
B =
    1.21   -0.46    0.09    0.58
    1.67   -1.96    1.26    1.93
    0.24   -0.46    2.77   -0.59
    0.03   -0.28   -0.76    0.13

C =
    1.14    0.85
   -0.48    0.17
    0.37   -0.64
    0.62   -1.17
```

Then A+B gives here the sum

```
ans =
   -0.10    1.62    0.17    2.63
    2.92   -3.31    0.26    3.87
    1.31    1.33    1.78    0.34
    1.37   -1.27   -0.99   -0.09
```

and A-B the difference

```
ans =
   -2.52    2.53   -0.01    1.46
   -0.41    0.62   -2.25    0.01
    0.84    2.26   -3.76    1.52
    1.31   -0.71    0.53   -0.35
```

You could check that B+A gives the same matrix as A+B (Theorem 3.1.24(a)) by seeing that their difference is the 4×4 zero matrix: execute (A+B) - (B+A) (the parentheses control the order of evaluation). However, expressions such as B+C and A-C give an error, because the matrices are of incompatible sizes, reported by MATLAB as

```
Error using  +
Matrix dimensions must agree.
```

or reported by Octave as

```
error: operator +: nonconformant arguments
```

Scalar multiplication of matrices   In MATLAB/Octave the asterisk indicates multiplication. Scalar multiplication can be done either way around. For example, generate a random 4×3 matrix A and compute $2A$ and $A\frac{1}{10}$. These commands

```
A=randn(4,3)
2*A
A*0.1
```

might give the following (2 d.p.)

```
A =
    0.82    2.54   -0.98
    2.30    0.05    2.63
   -1.45    2.15    0.89
   -2.58   -0.09   -0.55

>> 2*A
ans =
    1.64    5.07   -1.97
    4.61    0.10    5.25
   -2.90    4.30    1.77
   -5.16   -0.18   -1.11

>> A*0.1
ans =
    0.08    0.25   -0.10
    0.23    0.00    0.26
   -0.15    0.21    0.09
   -0.26   -0.01   -0.06
```

Division by a scalar is also defined in MATLAB/Octave and means multiplication by the reciprocal; for example, the product `A*0.1` could equally well be computed as `A/10`.

In mathematical algebra we would not normally accept statements such as $A + 3$ or $2A - 5$ because addition and subtraction with matrices has only been defined between matrices of the same size.[7] However, MATLAB/Octave usefully extends addition and subtraction so that `A+3` and `2*A-5` mean add three to *every* element of A and subtract five from *every* element of $2A$. For example, with the above random 4×3 matrix A,

```
>> A+3
ans =
    3.82    5.54    2.02
    5.30    3.05    5.63
    1.55    5.15    3.89
    0.42    2.91    2.45
```

---

[7] Although in some contexts such mathematical expressions are routinely accepted, be careful of their meaning.

```
>> 2*A-5
ans =
   -3.36     0.07    -6.97
   -0.39    -4.90     0.25
   -7.90    -0.70    -3.23
  -10.16    -5.18    -6.11
```

This last computation illustrates that, in any expression, the operations of multiplication and division are performed before additions and subtractions—as normal in mathematics.

Matrix multiplication In MATLAB/Octave the asterisk also invokes matrix-matrix and matrix-vector multiplication. For example, generate and multiply two random matrices say of size 3×4 and 4×2 with

```
A=randn(3,4)
B=randn(4,2)
C=A*B
```

might give the following result (2 d.p.)

```
A =
   -0.02     1.31    -0.74    -0.49
   -0.36    -1.30    -0.23     0.41
   -0.88    -0.34     0.28    -0.99


B =
   -1.32    -0.79
    0.71     1.48
   -0.48     2.79
    1.40    -0.41

>> C=A*B
C =
    0.62     0.10
    0.24    -2.44
   -0.60     1.38
```

Without going into excruciating arithmetic detail this product is hard to check. However, we can check several things such as c_{11} comes from the first row of A times the first column of B by computing A(1,:)*B(:,1) and seeing it does give 0.62 as required. Also check that the two columns of C may be viewed as the two matrix-vector products $A\boldsymbol{b}_1$ and $A\boldsymbol{b}_2$ by comparing C with [A*B(:,1) A*B(:,2)] and seeing they are the same.

Recall that in a matrix product the number of columns of the left matrix has to be the same as the number of rows of the right matrix. MATLAB/Octave gives an error message if this is not the case, such as occurs upon asking it to compute B*A when MATLAB reports

```
Error using  *
Inner matrix dimensions must agree.
```

and Octave reports

```
error: operator *: nonconformant arguments
```

The caret symbol, ^, computes matrix powers in MATLAB/Octave, such as the cube A^3. But such matrix powers only make sense and work for square matrices A.[8] For example, if matrix A was 3×4, then $A^2 = AA$ would involve multiplying a 3×4 matrix by a 3×4 matrix: since the number of columns of the left A, 4, is not the same as the number of rows of the right A, 3, such a multiplication is not allowed.

The transpose and symmetry   In MATLAB/Octave the single apostrophe denotes matrix transpose. For example, see it transpose a couple of random matrices with

```
A=randn(3,4)
B=randn(4,2)
A'
B'
```

giving here for example (2 d.p.)

```
A =
      0.80     0.30   -0.12   -0.57
      0.07    -0.51   -0.81    1.95
      0.29    -0.10    0.17    0.70
B =
     -0.71    -0.34
     -0.33    -0.73
      1.11    -0.21
      0.41     0.33

>> A'
ans =
      0.80     0.07    0.29
      0.30    -0.51   -0.10
     -0.12    -0.81    0.17
     -0.57     1.95    0.70

>> B'
ans =
     -0.71    -0.33    1.11    0.41
     -0.34    -0.73   -0.21    0.33
```

One can do further operations after the transposition, such as checking the multiplication rule that $(AB)^T = B^T A^T$ (Theorem 3.1.29(d)) by verifying the result of (A*B)'-B'*A' is the zero matrix, here $O_{2 \times 3}$.

---

[8] Here we define matrix powers for only integer power. MATLAB/Octave computes the power of a square matrix for any real/complex exponent, but its meaning involves matrix exponentials and matrix logarithms that we do not explore here.

You can generate a symmetric matrix by adding a square matrix to its transpose (Theorem 3.1.29(f)): for example, generate a random square matrix by first C=randn(3) then C=C+C' makes a random symmetric matrix such as the following (2 d.p.)

```
>> C=randn(3)
C =
   -0.33     0.65    -0.62
   -0.43    -2.18    -0.28
    1.86    -1.00    -0.52

>> C=C+C'
C =
   -0.65     0.22     1.24
    0.22    -4.36    -1.28
    1.24    -1.28    -1.04

>> C-C'
ans =
    0.00     0.00     0.00
    0.00     0.00     0.00
    0.00     0.00     0.00
```

That the resulting matrix C is symmetric is checked by this last step which computes the difference between C and C^T and confirming the difference is zero. Hence C and C^T must be equal.

3.1.3 Familiar algebraic properties of matrix operations

Almost all of the familiar algebraic properties of scalar addition, subtraction, and multiplication—namely commutativity, associativity, and distributivity—hold for matrix addition, subtraction, and multiplication.

The one outstanding exception is that matrix multiplication is *not* commutative: for matrices A and B the products AB and BA are usually not equal.

Example 3.1.23 Let matrices $A = \begin{bmatrix} 0 & 1 \\ 1 & 1 \end{bmatrix}$ and $B = \begin{bmatrix} 1 & -1 \\ 1 & 0 \end{bmatrix}$. Show that the two products AB and BA are not equal.

Solution: Compute

$$AB = \begin{bmatrix} 0 & 1 \\ 1 & 1 \end{bmatrix} \begin{bmatrix} 1 & -1 \\ 1 & 0 \end{bmatrix} = \begin{bmatrix} 1 & 0 \\ 2 & -1 \end{bmatrix},$$

$$BA = \begin{bmatrix} 1 & -1 \\ 1 & 0 \end{bmatrix} \begin{bmatrix} 0 & 1 \\ 1 & 1 \end{bmatrix} = \begin{bmatrix} -1 & 0 \\ 0 & 1 \end{bmatrix}.$$

These two products are not equal, $AB \neq BA$. ☐

This Example 3.1.23 illustrates that matrix multiplication is not commutative. We are used to such non-commutativity in life. For example, when you go home, to enter your house you first open the door, second walk in, and third close the door. You cannot swap the order and try to walk in before opening the door—these operations do not commute. Similarly, for another example, I often teach classes on the third floor of a building next to my office: after finishing classes, first I walk downstairs to ground level, and second I cross the road to my office. If I try to cross the road before going downstairs, then the force of gravity has something very painful to say about the outcome—the operations do not commute. Similar to these analogues, the result of a matrix multiplication depends upon the order of the matrices in the multiplication.

Theorem 3.1.24 (*Properties of addition and scalar multiplication*) *Let matrices A, B, and C be of the same size, and let c and d be scalars. Then:*

(a) $A + B = B + A$ *(commutativity of addition);*

(b) $(A + B) + C = A + (B + C)$ *(associativity of addition);*

(c) $A \pm O = A = O + A;$

(d) $c(A \pm B) = cA \pm cB$ *(distributivity over matrix addition);*

(e) $(c \pm d)A = cA \pm dA$ *(distributivity over scalar addition);*

(f) $c(dA) = (cd)A$ *(associativity of scalar multiplication);*

(g) $1A = A$*; and*

(h) $0A = O.$

Proof. The proofs directly match those of the corresponding vector properties and are set as exercises.

Example 3.1.25 (*geometry of associativity*) Many properties of matrix multiplication have a useful geometric interpretation such as that discussed for matrix-vector products. Recall the earlier Example 3.1.15 invoked the associativity Theorem 3.1.26(c). For another example, consider the two matrices and vector

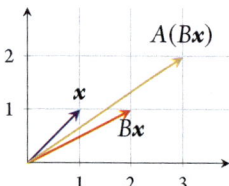

$$A = \begin{bmatrix} 1 & 1 \\ 1 & 0 \end{bmatrix}, \quad B = \begin{bmatrix} 2 & 0 \\ 2 & -1 \end{bmatrix}, \quad x = \begin{bmatrix} 1 \\ 1 \end{bmatrix}.$$

Now the transform $x' = Bx = (2, 1)$, and then transforming with A gives $x'' = Ax' = A(Bx) = (3, 2)$, as illustrated to the above-right.

This is the same results as forming the product

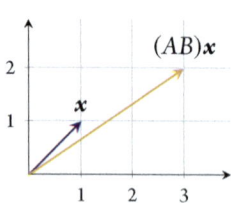

$$AB = \begin{bmatrix} 1 & 1 \\ 1 & 0 \end{bmatrix} \begin{bmatrix} 2 & 0 \\ 2 & -1 \end{bmatrix} = \begin{bmatrix} 4 & -1 \\ 2 & 0 \end{bmatrix}$$

and then computing $(AB)x = (3, 2)$ as also illustrated to the right. Such associativity asserts that $A(Bx) = (AB)x$: that is, the geometric transform of x by matrix B followed by the transform of matrix A is the same result as just transforming by the matrix formed from the product AB—as assured by Theorem 3.1.26(c). $\square$

Theorem 3.1.26 (*properties of matrix multiplication*) *Let matrices A, B, and C be of sizes such that the following expressions are defined, and let c be a scalar, then:*

(a) $A(B \pm C) = AB \pm AC$ (*distributivity of matrix multiplication*);

(b) $(A \pm B)C = AC \pm BC$ (*distributivity of matrix multiplication*);

(c) $A(BC) = (AB)C$ (*associativity of matrix multiplication*);

(d) $c(AB) = (cA)B = A(cB)$;

(e) $I_m A = A = AI_n$ *for $m \times n$ matrix A* (*multiplicative identity*);

(f) $O_m A = O_{m \times n} = AO_n$ *for $m \times n$ matrix A*;

(g) $A^p A^q = A^{p+q}$, $(A^p)^q = A^{pq}$ *and* $(cA)^p = c^p A^p$ *for square matrices A and for positive integers p and q.*[9]

Proof. Let's document a few proofs, others are exercises.

3.1.26(a) This direct proof involves some long expressions involving the entries of $m \times n$ matrix A, and $n \times p$ matrices B and C. Let $(\cdot)_{ij}$ denote the (i,j)th entry of whatever matrix expression is inside the parentheses. By Definition 3.1.12 of matrix multiplication

$$(A(B \pm C))_{ij}$$
$$= a_{i1}(B \pm C)_{1j} + a_{i2}(B \pm C)_{2j} + \cdots + a_{in}(B \pm C)_{nj}$$

(then by definition of matrix addition)

$$= a_{i1}(b_{1j} \pm c_{1j}) + a_{i2}(b_{2j} \pm c_{2j}) + \cdots + a_{in}(b_{nj} \pm c_{nj})$$

(then distributing the scalar multiplications)

$$= a_{i1}b_{1j} \pm a_{i1}c_{1j} + a_{i2}b_{2j} \pm a_{i2}c_{2j} + \cdots + a_{in}b_{nj} \pm a_{in}c_{nj}$$

(then upon reordering terms in the sum)

$$= a_{i1}b_{1j} + a_{i2}b_{2j} + \cdots + a_{in}b_{nj}$$
$$\pm (a_{i1}c_{1j} + a_{i2}c_{2j} + \cdots + a_{in}c_{nj})$$

(then using Definition 3.1.12 for matrix products)

$$= (AB)_{ij} \pm (AC)_{ij}.$$

Since this identity holds for all indices i and j, the matrix identity $A(B \pm C) = AB \pm AC$ holds, proving Theorem 3.1.26(a).

3.1.26(c) Associativity involves some longer expressions involving the entries of $m \times n$ matrix A, $n \times p$ matrix B, and $p \times q$ matrix C. By Definition 3.1.12 of matrix multiplication

---

[9] Generally these exponent properties hold for all scalar p and q, although one has to be very careful with non-integer exponents.

$$(A(BC))_{ij} = a_{i1}(BC)_{1j} + a_{i2}(BC)_{2j} + \cdots + a_{in}(BC)_{nj}$$

(then using Definition 3.1.12 for BC)

$$= a_{i1}(b_{11}c_{1j} + b_{12}c_{2j} + \cdots + b_{1p}c_{pj})$$
$$+ a_{i2}(b_{21}c_{1j} + b_{22}c_{2j} + \cdots + b_{2p}c_{pj})$$
$$+ \cdots$$
$$+ a_{in}(b_{n1}c_{1j} + b_{n2}c_{2j} + \cdots + b_{np}c_{pj})$$

(distributing the scalar multiplications)

$$= a_{i1}b_{11}c_{1j} + a_{i1}b_{12}c_{2j} + \cdots + a_{i1}b_{1p}c_{pj}$$
$$+ a_{i2}b_{21}c_{1j} + a_{i2}b_{22}c_{2j} + \cdots + a_{i2}b_{2p}c_{pj}$$
$$+ \cdots$$
$$+ a_{in}b_{n1}c_{1j} + a_{in}b_{n2}c_{2j} + \cdots + a_{in}b_{np}c_{pj}$$

(then reordering the terms—transpose)

$$= a_{i1}b_{11}c_{1j} + a_{i2}b_{21}c_{1j} + \cdots + a_{in}b_{n1}c_{1j}$$
$$+ a_{i1}b_{12}c_{2j} + a_{i2}b_{22}c_{2j} + \cdots + a_{in}b_{n2}c_{2j}$$
$$+ \cdots$$
$$+ a_{i1}b_{1p}c_{pj} + a_{i2}b_{2p}c_{pj} + \cdots + a_{in}b_{np}c_{pj}$$

(then factoring $c_{1j}, c_{2j}, \ldots c_{pj}$)

$$= (a_{i1}b_{11} + a_{i2}b_{21} + \cdots + a_{in}b_{n1})c_{1j}$$
$$+ (a_{i1}b_{12} + a_{i2}b_{22} + \cdots + a_{in}b_{n2})c_{2j}$$
$$+ \cdots$$
$$+ (a_{i1}b_{1p} + a_{i2}b_{2p} + \cdots + a_{in}b_{np})c_{pj}$$

(then recognizing the entries for $(AB)_{ik}$)

$$= (AB)_{i1}c_{1j} + (AB)_{i2}c_{2j} + \cdots + (AB)_{ip}c_{pj}$$

(then again using Definition 3.1.12)

$$= ((AB)C)_{ij}.$$

Since this identity holds for all indices i and j, the matrix identity $A(BC) = (AB)C$ holds, proving Theorem 3.1.26(c).

3.1.26(g) Other proofs develop from previous parts of the theorem. For example, to establish $A^p A^q = A^{p+q}$ start from the definition of matrix powers:

$$A^p A^q = \underbrace{(AA \cdots A)}_{p \text{ times}} \underbrace{(AA \cdots A)}_{q \text{ times}}$$

(then using associativity, Theorem 3.1.26(c))

$$= \underbrace{AA \cdots A}_{p+q \text{ times}} = A^{p+q}.$$

Example 3.1.27 Show that $(A+B)^2 \neq A^2 + 2AB + B^2$ in general.

Solution: Consider

$$
\begin{aligned}
(A+B)^2 &= (A+B)(A+B) \quad \text{(matrix power)} \\
&= A(A+B) + B(A+B) \quad \text{(Theorem 3.1.26(b))} \\
&= AA + AB + BA + BB \quad \text{(Theorem 3.1.26(a))} \\
&= A^2 + AB + BA + B^2 \quad \text{(matrix power).}
\end{aligned}
$$

This expression is only equal to $A^2 + 2AB + B^2$ if we can replace BA by AB. But this requires $BA = AB$ which is generally not true (e.g., see Example 3.1.23). That is, $(A+B)^2 = A^2 + 2AB + B^2$ *only if $BA = AB$.*   □

Example 3.1.28 Show that the matrix $J = \begin{bmatrix} 0 & 0 & 1 \\ 0 & 1 & 0 \\ 1 & 0 & 0 \end{bmatrix}$ is not a multiplicative identity (despite having ones down a diagonal, this diagonal is the wrong one for an identity).

Solution: Among many other ways to show J is not a multiplicative identity, let's invoke a general 3×3 matrix

$$
A = \begin{bmatrix} a & b & c \\ d & e & f \\ g & h & i \end{bmatrix},
$$

and evaluate the product

$$
JA = \begin{bmatrix} 0 & 0 & 1 \\ 0 & 1 & 0 \\ 1 & 0 & 0 \end{bmatrix} \begin{bmatrix} a & b & c \\ d & e & f \\ g & h & i \end{bmatrix} = \cdots = \begin{bmatrix} g & h & i \\ d & e & f \\ a & b & c \end{bmatrix} \neq A.
$$

Since $JA \neq A$ then matrix J cannot be a multiplicative identity (the multiplicative identity is only when the ones are along the diagonal from top-left to bottom-right).   □

Theorem 3.1.29 *(properties of transpose) Let matrices A and B be of sizes such that the following expressions are defined, then:*

(a) $(A^{\mathrm{T}})^{\mathrm{T}} = A$;

(b) $(A \pm B)^{\mathrm{T}} = A^{\mathrm{T}} \pm B^{\mathrm{T}}$;

(c) $(cA)^{\mathrm{T}} = c(A^{\mathrm{T}})$ *for every scalar c;*

(d) $(AB)^{\mathrm{T}} = B^{\mathrm{T}}A^{\mathrm{T}}$ —*remember the reversed order in this identity;*

(e) $(A^p)^T = (A^T)^p$ for every positive integer exponent p;[10]

(f) $A + A^T$, $A^T A$ and $A A^T$ are symmetric matrices.

Proof. Let's document a few proofs, others are exercises. Some proofs use primitive definitions—usually using $(\cdot)_{ij}$ to denote the (i,j)th entry of whatever matrix expression is inside the parentheses—others invoke earlier proved parts.

3.1.29(b) Recall from Definition 3.1.17 of the transpose that

$$((A \pm B)^T)_{ij} = (A \pm B)_{ji}$$

(then by definition of addition)

$$= a_{ji} \pm b_{ji}$$

(then by Definition 3.1.17 of transpose)

$$= (A^T)_{ij} \pm (B^T)_{ij}.$$

Since this identity holds for all indices i and j, then $(A \pm B)^T = A^T \pm B^T$.

3.1.29(d) The transpose of matrix multiplication is more involved. Let matrices A and B be of sizes $m \times n$ and $n \times p$ respectively. Then from Definition 3.1.17 of the transpose

$$((AB)^T)_{ij} = (AB)_{ji}$$

(then by Definition 3.1.12 of multiplication)

$$= a_{j1}b_{1i} + a_{j2}b_{2i} + \cdots + a_{jn}b_{ni}$$

(then commuting the scalar products)

$$= b_{1i}a_{j1} + b_{2i}a_{j2} + \cdots + b_{ni}a_{jn}$$

(then by Definition 3.1.17 of transpose)

$$= (B^T)_{i1}(A^T)_{1j} + (B^T)_{i2}(A^T)_{2j} + \cdots + (B^T)_{in}(A^T)_{nj}$$

(then by Definition 3.1.12 of multiplication)

$$= (B^T A^T)_{ij}.$$

Since this identity holds for all indices i and j, then $(AB)^T = B^T A^T$.

3.1.29(f) To prove the second, that $A^T A$ is equal to its transpose, we invoke earlier parts. Consider the transpose

$$(A^T A)^T = (A)^T (A^T)^T \quad \text{(by Theorem 3.1.29(d))}$$
$$= A^T A \quad \text{(by Theorem 3.1.29(a))}.$$

Since $A^T A$ equals its transpose, it is symmetric.

---

[10] With care, this property also holds for all scalar exponents p—although this book only considers integer exponents p, see Definition 3.2.15.

3.1.4 Exercises

Exercise 3.1.1 Consider the following six matrices: $A = \begin{bmatrix} -1 & 3 \\ 0 & -5 \\ 0 & -7 \end{bmatrix}$;

$B = \begin{bmatrix} -4 & -3 & -3 & 1 \\ -3 & -2 & 0 & -1 \end{bmatrix}$; $C = \begin{bmatrix} -3 & 1 \end{bmatrix}$; $D = \begin{bmatrix} 0 & 6 & 6 & 3 \\ 2 & 2 & 0 & -5 \end{bmatrix}$; $E = \begin{bmatrix} 0 & 1 & 1 & -2 \\ -1 & 5 & 4 & -1 \\ 1 & -3 & 7 & 3 \\ -6 & -3 & 0 & 2 \end{bmatrix}$;

$F = \begin{bmatrix} 4 & 1 & 0 \\ -1 & 1 & 6 \\ -4 & 5 & -2 \end{bmatrix}$.

(a) What is the size of each of these matrices?

(b) Which pairs of matrices may be added or subtracted?

(c) Which matrix multiplications can be performed between two of the matrices?

Exercise 3.1.2 Given the matrix

$$A = \begin{bmatrix} -0.3 & 2.1 & -4.8 \\ -5.9 & 3.6 & -1.3 \end{bmatrix}:$$

write down its column vectors; what are the values of elements a_{13} and a_{21}?

Exercise 3.1.3 Given the matrix

$$B = \begin{bmatrix} 7.6 & -1.1 & -0.7 & -4.5 \\ -1.1 & -9.3 & 0.1 & 8.2 \\ 2.6 & 6.9 & 1.2 & -3.6 \\ -1.5 & -7.5 & 3.7 & 2.6 \\ -0.2 & 5.5 & -0.9 & 2.4 \end{bmatrix}:$$

write down its column vectors; what are the values of entries b_{13}, b_{31}, b_{42}?

Exercise 3.1.4 Write down the column vectors of the identity I_4. What do we call these column vectors?

Exercise 3.1.5 For the following pairs of matrices, calculate their sum and difference.

(a) $A = \begin{bmatrix} 2 & 1 & -1 \\ -4 & 1 & -3 \\ -2 & 2 & -1 \end{bmatrix}, B = \begin{bmatrix} 1 & 1 & 0 \\ 4 & -6 & -6 \\ -6 & 4 & 0 \end{bmatrix}$

(b) $C = \begin{bmatrix} -2 & -2 & -7 \end{bmatrix}, D = \begin{bmatrix} 4 & 2 & -2 \end{bmatrix}$

(c) $R = \begin{bmatrix} -2.5 & -0.4 \\ -1.0 & -3.5 \\ -3.3 & 1.8 \end{bmatrix}, S = \begin{bmatrix} -0.9 & 4.9 \\ -1.2 & -0.7 \\ -4.0 & -5.4 \end{bmatrix}$

Exercise 3.1.6 For the given matrix, evaluate the following matrix-scalar products.

(a) $A = \begin{bmatrix} -3 & -2 \\ 4 & -2 \\ 2 & -4 \end{bmatrix}$: $-2A$, $2A$, and $3A$.

(b) $B = \begin{bmatrix} 4 & 0 \\ -1 & -1 \end{bmatrix}$: $1.9B$, $2.6B$, and $-6.9B$.

(c) $V = \begin{bmatrix} -2.6 & -3.2 \\ 3.3 & -0.8 \\ -0.3 & 0.3 \end{bmatrix}$: $1.3V$, $-3.7V$, and $2.5V$.

Exercise 3.1.7 Use MATLAB/Octave to generate some random matrices of a suitable size of your choice, and some random scalars (see Table 3.1). Then confirm the addition and scalar multiplication properties of Theorem 3.1.24. Record all your commands and the output from MATLAB/Octave.

Exercise 3.1.8 Use the definition of matrix addition and scalar multiplication to prove the basic properties of Theorem 3.1.24.

Exercise 3.1.9 For each of the given matrices, calculate the specified matrix-vector products.

(a) For $A = \begin{bmatrix} 4 & -3 \\ -2 & 5 \end{bmatrix}$ and vectors $p = \begin{bmatrix} -6 \\ -5 \end{bmatrix}$, $q = \begin{bmatrix} -2 \\ -4 \end{bmatrix}$, and $r = \begin{bmatrix} -3 \\ 1 \end{bmatrix}$, calculate Ap, Aq, and Ar.

(b) For $B = \begin{bmatrix} 1 & 6 \\ 4 & -5 \end{bmatrix}$ and vectors $p = \begin{bmatrix} -3 \\ -3 \end{bmatrix}$, $q = \begin{bmatrix} 2 \\ 1 \end{bmatrix}$, and $r = \begin{bmatrix} -5 \\ 2 \end{bmatrix}$, calculate Bp, Bq, and Br.

(c) For $C = \begin{bmatrix} -3 & 0 & -3 \\ -1 & -1 & 1 \end{bmatrix}$ and vectors $u = \begin{bmatrix} -4 \\ 3 \\ 2 \end{bmatrix}$, $v = \begin{bmatrix} -3 \\ 1 \\ 2 \end{bmatrix}$, and $w = \begin{bmatrix} -4 \\ 5 \\ -4 \end{bmatrix}$, calculate Cu, Cv and Cw.

Exercise 3.1.10 For each of the given matrices and vectors, calculate the matrix-vector products. Plot in 2D, and label, the vectors and the specified matrix-vector products.

(a) $A = \begin{bmatrix} 3 & 2 \\ -3 & -1 \end{bmatrix}$, $u = \begin{bmatrix} 1 \\ 2 \end{bmatrix}$, $v = \begin{bmatrix} 0 \\ -3 \end{bmatrix}$, and $w = \begin{bmatrix} 1 \\ 3 \end{bmatrix}$.

(b) $B = \begin{bmatrix} 3 & -2 \\ 3 & 2 \end{bmatrix}$, $p = \begin{bmatrix} 0 \\ 1 \end{bmatrix}$, $q = \begin{bmatrix} -1 \\ 2 \end{bmatrix}$, and $r = \begin{bmatrix} -2 \\ 1 \end{bmatrix}$.

(c) $C = \begin{bmatrix} -2.1 & 1.1 \\ 4.6 & -1 \end{bmatrix}$, $x_1 = \begin{bmatrix} 2.1 \\ 0 \end{bmatrix}$, $x_2 = \begin{bmatrix} -0.1 \\ 1.1 \end{bmatrix}$, and $x_3 = \begin{bmatrix} -0.3 \\ -1 \end{bmatrix}$.

Exercise 3.1.11 For each of the given matrices and vectors, calculate the matrix-vector products. Plot in 2D, and label, the vectors and the specified matrix-vector products. For each of the matrices, interpret the matrix multiplication of the vectors as either a rotation, a reflection, a stretch, or none of these.

(a) $P = \begin{bmatrix} 1 & 0 \\ 0 & -1 \end{bmatrix}$, $u = \begin{bmatrix} 1 \\ -1.4 \end{bmatrix}$, $v = \begin{bmatrix} -3.6 \\ -1.7 \end{bmatrix}$, and $w = \begin{bmatrix} 0.1 \\ 2.3 \end{bmatrix}$.

(b) $Q = \begin{bmatrix} 2 & 0 \\ 0 & 2 \end{bmatrix}$, $p = \begin{bmatrix} 2.1 \\ 1.9 \end{bmatrix}$, $q = \begin{bmatrix} 2.8 \\ -1.1 \end{bmatrix}$, and $r = \begin{bmatrix} 0.8 \\ 3.3 \end{bmatrix}$.

(c) $R = \begin{bmatrix} 0.8 & -0.6 \\ 0.6 & 0.8 \end{bmatrix}$, $x_1 = \begin{bmatrix} -4 \\ 2 \end{bmatrix}$, $x_2 = \begin{bmatrix} 4 \\ -3 \end{bmatrix}$, and $x_3 = \begin{bmatrix} 2 \\ 3 \end{bmatrix}$.

Exercise 3.1.12 Using the matrix-vector products you calculated for Exercise 3.1.9, write down the results of the following matrix-matrix products.

(a) For $A = \begin{bmatrix} 4 & -3 \\ -2 & 5 \end{bmatrix}$, write down the matrix products

   i. $A \begin{bmatrix} -6 & -2 \\ -5 & -4 \end{bmatrix}$,
                    iii. $A \begin{bmatrix} -2 & -3 \\ -4 & 1 \end{bmatrix}$,

   ii. $A \begin{bmatrix} -6 & -3 \\ -5 & 1 \end{bmatrix}$,
                    iv. $A \begin{bmatrix} -6 & -2 & -3 \\ -5 & -4 & 1 \end{bmatrix}$.

(b) For $B = \begin{bmatrix} 1 & 6 \\ 4 & -5 \end{bmatrix}$, write down the matrix products

   i. $B \begin{bmatrix} -3 & 2 \\ -3 & 1 \end{bmatrix}$,
                    iii. $B \begin{bmatrix} -5 & -3 \\ 2 & -3 \end{bmatrix}$,

   ii. $B \begin{bmatrix} -5 & 2 \\ 2 & 1 \end{bmatrix}$,
                    iv. $B \begin{bmatrix} -5 & 2 & -3 \\ 2 & 1 & -3 \end{bmatrix}$.

(c) For $C = \begin{bmatrix} -3 & 0 & -3 \\ -1 & -1 & 1 \end{bmatrix}$, write down the matrix products

   i. $C \begin{bmatrix} -4 & -3 \\ 3 & 1 \\ 2 & 2 \end{bmatrix}$,
                    iii. $C \begin{bmatrix} -4 & -4 \\ 5 & 3 \\ -4 & 2 \end{bmatrix}$,

   ii. $C \begin{bmatrix} -4 & -3 \\ 5 & 1 \\ -4 & 2 \end{bmatrix}$,
                    iv. $C \begin{bmatrix} -4 & -3 & -4 \\ 5 & 1 & 3 \\ -4 & 2 & 2 \end{bmatrix}$.

Exercise 3.1.13 Use MATLAB/Octave to generate some random matrices of a suitable size of your choice, and some random scalars (see Table 3.1). Choose some suitable exponents. Then confirm the matrix multiplication properties of Theorem 3.1.26. Record all your commands and the output from MATLAB/Octave.

In checking some properties, you may get matrices with elements such as 2.2204e-16: recall from Table 3.1 that this denotes the very small number $2.2204 \cdot 10^{-16}$. When adding and subtracting numbers of magnitude one or so, the result 2.2204e-16 is effectively zero (due to the sixteen digit precision of MATLAB/Octave, Table 1.2).

Exercise 3.1.14 Use Definition 3.1.12 of matrix-matrix multiplication to prove multiplication properties of Theorem 3.1.26. Prove parts: 3.1.26(b), distributivity; 3.1.26(d), scalar associativity; 3.1.26(e), identity; 3.1.26(f), zeros.

Exercise 3.1.15 Use the other parts of Theorem 3.1.26 to prove part 3.1.26(g) that $(A^p)^q = A^{pq}$ and $(cA)^p = c^p A^p$ for square matrix A, scalar c, and for positive integer exponents p and q.

Exercise 3.1.16 *(Tasmanian Devils)* Ecologists studying a colony of Tasmanian Devils, an Australian marsupial, observed the following: two-thirds of the female newborns survive to be one year old; two-thirds of female one year olds survive to be two years old; one-half of female two year olds survive to be three years old; each year, each female aged two or three years gives birth to two female offspring; female Tasmanian Devils survive for four years, at most.

Tasmanian Devil

Analogous to Example 3.1.9 define a vector x in $\mathbb{R}^4$ to be the number of females of specified ages. Use the above information to write down the Leslie matrix L that predicts the number in the next year, x', from the number in any year, x. Given the observed initial female numbers of 18 newborns, 9 one year olds, 18 two year olds, and 18 three year olds, use matrix multiplication to predict the numbers of female Tasmanian Devils one, two, and three years later. Does the population appear to be increasing? or decreasing?

Exercise 3.1.17 Write down the transpose of each of the following matrices. Which of the following matrices are a symmetric matrix?

(a) $\begin{bmatrix} -2 & 3 \\ 3 & 0 \\ -8 & 2 \\ -2 & -4 \end{bmatrix}$

(b) $\begin{bmatrix} 3 & -4 & -2 & 2 \\ -5 & 2 & -3 & 3 \end{bmatrix}$

(c) $\begin{bmatrix} 14 & 5 & 3 & 2 \\ 5 & 0 & -1 & 1 \\ 3 & -1 & -6 & -4 \\ 2 & 1 & -4 & 4 \end{bmatrix}$

(d) $\begin{bmatrix} -4 & -5.1 & 0.3 \\ -5.1 & -7.4 & -3. \\ 0.3 & -3 & 2.6 \end{bmatrix}$

(e) $\begin{bmatrix} -1.5 & -0.6 & -1.7 \\ -1 & -0.4 & -5.6 \end{bmatrix}$

(f) $\begin{bmatrix} 1.7 & -0.2 & -0.4 \\ 0.7 & -0.3 & -0.4 \\ 0.6 & 3 & -2.2 \end{bmatrix}$

Exercise 3.1.18 For each of the following matrices, are they symmetric? I_4, $I_{3\times4}$, O_3, and $O_{3\times1}$.

Exercise 3.1.19 Use MATLAB/Octave to generate some random matrices of a suitable size of your choice, and some random scalars (see Table 3.1). Choose some suitable exponents. Recalling that in MATLAB/Octave the quote/dash ' performs the transpose, confirm the matrix transpose properties of Theorem 3.1.29. Record all your commands and the output from MATLAB/Octave.

Exercise 3.1.20 Use Definition 3.1.17 of the matrix transpose to prove Theorems 3.1.29(a) and 3.1.29(c).

Exercise 3.1.21 Use the other parts of Theorem 3.1.29 to prove parts 3.1.29(e) and 3.1.29(f).

Exercise 3.1.22 In a few sentences, answer/discuss each of the following.

(a) Why is the size of a matrix important?

(b) What causes these identities to hold? $A \pm O = A$ and $A - A = O$.

(c) In the matrix product AB, why do the number of columns of A have to be the same as the number of rows of B?

(d) What can you say about the sizes of matrices A and B if both products AB and BA are computable?

(e) What causes multiplication of a vector by a square matrix to be viewed as a transformation?

(f) What causes the identity matrix to be zero except for ones along the diagonal from the top-left to the bottom-right?

(g) How does multiplication by a square matrix arise in studying the age structure of populations?

(h) Why is it impossible to compute powers of a non-square matrix?

(i) Why did we invoke random matrices?

(j) Among all the properties for matrix addition and multiplication operations, which ones are different from the analogous properties for scalars? why?

(k) What constraint must be put on matrix A in order for $A + A^{\mathsf{T}}$ to be defined? What constraint must be put on matrix B in order for BB^{T} to be defined? Give reasons.

3.2 The inverse of a matrix

The previous Section 3.1 introduced addition, subtraction, multiplication, and other operations of matrices. Conspicuously missing from the list is 'division' by a matrix—missing because division is complicated. This section develops 'division' by a matrix as multiplication by the inverse of a matrix. The analogue in ordinary arithmetic is that division by ten is the same as multiplying by its reciprocal, one-tenth. But the inverse of a matrix looks almost nothing like a reciprocal.

3.2.1 Introducing the unique inverse

Let's start with an example that illustrates an analogy with the reciprocal/inverse of a scalar number.

Example 3.2.1 Recall that a crucial property is that a number multiplied by its reciprocal/ inverse is one: for example, $2 \times 0.5 = 1$ so 0.5 is the reciprocal/inverse of 2. Similarly, show that matrix

$$B = \begin{bmatrix} -3 & 1 \\ -4 & 1 \end{bmatrix} \text{ is an inverse of } A = \begin{bmatrix} 1 & -1 \\ 4 & -3 \end{bmatrix}$$

by showing their product is the 2×2 identity matrix I_2.

Solution: Multiply

$$AB = \begin{bmatrix} 1 & -1 \\ 4 & -3 \end{bmatrix}\begin{bmatrix} -3 & 1 \\ -4 & 1 \end{bmatrix} = \begin{bmatrix} 1 & 0 \\ 0 & 1 \end{bmatrix} = I_2$$

the multiplicative identity. But matrix multiplication is generally not commutative (Section 3.1.3), so also consider

$$BA = \begin{bmatrix} -3 & 1 \\ -4 & 1 \end{bmatrix}\begin{bmatrix} 1 & -1 \\ 4 & -3 \end{bmatrix} = \begin{bmatrix} 1 & 0 \\ 0 & 1 \end{bmatrix} = I_2.$$

Since the identity matrix is analogous to the number one in scalar arithmetic, that both these products are the identity means that the matrix A has the same relation to the matrix B as a number has to its reciprocal/inverse.

Being the inverse, matrix B 'undoes' the action of matrix A—as illustrated to the right. The first picture shows that multiplication by A transforms the vector $(2,1)$ to the vector $(1,5)$: $A\begin{bmatrix} 2 \\ 1 \end{bmatrix} = \begin{bmatrix} 1 \\ 5 \end{bmatrix}$. The second picture shows that multiplication by B undoes the transform by A because $B\begin{bmatrix} 1 \\ 5 \end{bmatrix} = \begin{bmatrix} 2 \\ 1 \end{bmatrix}$ the original vector.

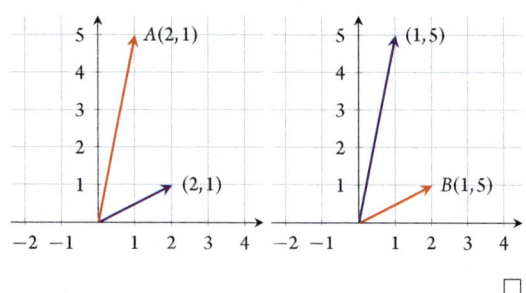

The previous Example 3.2.1 shows at least one case when we can do some sort of matrix 'division': that is, multiplying by B is equivalent to 'dividing' by A. One restriction is that a clearly defined 'division' only works for square matrices. Part of the reason for this restriction is because we need to be able to compute both AB and BA.

Definition 3.2.2 *(inverse) For every $n \times n$ square matrix A, an **inverse** of A is an $n \times n$ matrix B such that both $AB = I_n$ and $BA = I_n$. If such a matrix B exists, then matrix A is called **invertible**.*

By saying "an inverse", this definition allows for the possibility of many inverses, but Theorem 3.2.6 establishes that the inverse is unique. Further, an inverse may not exist for a given matrix A.

Example 3.2.3 Show that matrix

$$B = \begin{bmatrix} 0 & -\frac{1}{4} & -\frac{1}{8} \\ \frac{3}{2} & 1 & \frac{7}{8} \\ \frac{1}{2} & \frac{1}{4} & \frac{3}{8} \end{bmatrix} \text{ is an inverse of } A = \begin{bmatrix} 1 & -1 & 5 \\ -5 & -1 & 3 \\ 2 & 2 & -6 \end{bmatrix}.$$

Solution: First compute

$$AB = \begin{bmatrix} 1 & -1 & 5 \\ -5 & -1 & 3 \\ 2 & 2 & -6 \end{bmatrix} \begin{bmatrix} 0 & -\frac{1}{4} & -\frac{1}{8} \\ \frac{3}{2} & 1 & \frac{7}{8} \\ \frac{1}{2} & \frac{1}{4} & \frac{3}{8} \end{bmatrix}$$

$$= \begin{bmatrix} 1\cdot 0 - 1\cdot\frac{3}{2}+5\cdot\frac{1}{2} & 1\cdot(-\frac{1}{4})-1\cdot 1+5\cdot\frac{1}{4} & 1\cdot(-\frac{1}{8})-1\cdot\frac{7}{8}+5\cdot\frac{3}{8} \\ -5\cdot 0 - 1\cdot\frac{3}{2}+3\cdot\frac{1}{2} & -5\cdot(-\frac{1}{4})-1\cdot 1+3\cdot\frac{1}{4} & -5\cdot(-\frac{1}{8})-1\cdot\frac{7}{8}+3\cdot\frac{3}{8} \\ 2\cdot 0 + 2\cdot\frac{3}{2}-6\cdot\frac{1}{2} & 2\cdot(-\frac{1}{4})+2\cdot 1-6\cdot\frac{1}{4} & 2\cdot(-\frac{1}{8})+2\cdot\frac{7}{8}-6\cdot\frac{3}{8} \end{bmatrix}$$

$$= \begin{bmatrix} 1 & 0 & 0 \\ 0 & 1 & 0 \\ 0 & 0 & 1 \end{bmatrix} = I_3.$$

Second compute

$$BA = \begin{bmatrix} 0 & -\frac{1}{4} & -\frac{1}{8} \\ \frac{3}{2} & 1 & \frac{7}{8} \\ \frac{1}{2} & \frac{1}{4} & \frac{3}{8} \end{bmatrix} \begin{bmatrix} 1 & -1 & 5 \\ -5 & -1 & 3 \\ 2 & 2 & -6 \end{bmatrix}$$

$$= \begin{bmatrix} 0\cdot 1 - \frac{1}{4}\cdot(-5)-\frac{1}{8}\cdot 2 & 0\cdot(-1)-\frac{1}{4}\cdot(-1)-\frac{1}{8}\cdot 2 & 0\cdot 5 - \frac{1}{4}\cdot 3 - \frac{1}{8}\cdot(-6) \\ \frac{3}{2}\cdot 1+1\cdot(-5)+\frac{7}{8}\cdot 2 & \frac{3}{2}\cdot(-1)+1\cdot(-1)+\frac{7}{8}\cdot 2 & \frac{3}{2}\cdot 5+1\cdot 3+\frac{7}{8}\cdot(-6) \\ \frac{1}{2}\cdot 1+\frac{1}{4}\cdot(-5)+\frac{3}{8}\cdot 2 & \frac{1}{2}\cdot(-1)+\frac{1}{4}\cdot(-1)+\frac{3}{8}\cdot 2 & \frac{1}{2}\cdot 5+\frac{1}{4}\cdot 3+\frac{3}{8}\cdot(-6) \end{bmatrix}$$

$$= \begin{bmatrix} 1 & 0 & 0 \\ 0 & 1 & 0 \\ 0 & 0 & 1 \end{bmatrix} = I_3.$$

Since both of these products are the identity, then matrix A is invertible, and B is an inverse of A.
□

Activity 3.2.4 What value of b makes the matrix $\begin{bmatrix} -1 & b \\ 1 & 2 \end{bmatrix}$ to be an inverse of $\begin{bmatrix} 2 & 3 \\ -1 & -1 \end{bmatrix}$?

(a) -2 (b) -3 (c) 3 (d) 1

But even among square matrices, there are many nonzero matrices which do not have an inverse! A matrix which is not invertible is sometimes called a **singular matrix**. The next Section 3.3 further explores why some matrices do not have an inverse: the reason is associated with both rcond being zero (Procedure 2.2.5) and/or the so-called determinant being zero (Chapter 6).

Example 3.2.5 *(no inverse)* Prove that the matrix $A = \begin{bmatrix} 1 & -2 \\ -3 & 6 \end{bmatrix}$ does not have an inverse.

Solution: Assume that there is an inverse matrix $B = \begin{bmatrix} a & b \\ c & d \end{bmatrix}$. Then by Definition 3.2.2 the product $AB = I_2$; that is,

$$AB = \begin{bmatrix} 1 & -2 \\ -3 & 6 \end{bmatrix}\begin{bmatrix} a & b \\ c & d \end{bmatrix}$$

$$= \begin{bmatrix} a - 2c & b - 2d \\ -3a + 6c & -3b + 6d \end{bmatrix} = \begin{bmatrix} 1 & 0 \\ 0 & 1 \end{bmatrix}.$$

The bottom-left entry in this matrix equality asserts $-3a + 6c = 0$ which is $-3(a - 2c) = 0$, that is, $a - 2c = 0$. But the top-left entry in the matrix equality asserts $a - 2c = 1$. Both of these equations involving a and c cannot be true simultaneously, therefore the assumption of an inverse must be incorrect. This matrix A does not have an inverse. $\quad\square$

Theorem 3.2.6 (*unique inverse*) *If A is an invertible matrix, then its inverse is unique (and denoted by A^{-1}).*

Proof. We suppose there are two inverses, say B_1 and B_2, and proceed to show they must be the same. Since they are inverses, by Definition 3.2.2 both $AB_1 = B_1A = I_n$ and $AB_2 = B_2A = I_n$. Consequently, using associativity of matrix multiplication (Theorem 3.1.26(c)),

$$B_1 = B_1 I_n = B_1(AB_2) = (B_1A)B_2 = I_n B_2 = B_2.$$

That is, $B_1 = B_2$, and so the inverse is unique. $\quad\square$

In the elementary case of 1×1 matrices, that is $A = \begin{bmatrix} a_{11} \end{bmatrix}$, the inverse is simply the reciprocal of the entry, that is $A^{-1} = \begin{bmatrix} 1/a_{11} \end{bmatrix}$ provided a_{11} is nonzero. The reason is that $AA^{-1} = \begin{bmatrix} a_{11} \cdot \frac{1}{a_{11}} \end{bmatrix} = \begin{bmatrix} 1 \end{bmatrix} = I_1$ and $A^{-1}A = \begin{bmatrix} \frac{1}{a_{11}} \cdot a_{11} \end{bmatrix} = \begin{bmatrix} 1 \end{bmatrix} = I_1$.

In the case of 2×2 matrices the inverse is a little more complicated, but should be remembered. (For larger sized matrices, any direct general formulas for an inverse are too complicated to remember.)

Theorem 3.2.7 (*2 × 2 inverse*) *Let 2×2 matrix $A = \begin{bmatrix} a & b \\ c & d \end{bmatrix}$. Then A is invertible if and only if the **determinant** $ad - bc \neq 0$, in which case*

$$A^{-1} = \frac{1}{ad - bc}\begin{bmatrix} d & -b \\ -c & a \end{bmatrix}. \tag{3.2}$$

If the determinant $ad - bc = 0$, then A is not invertible (it is a singular matrix).

Example 3.2.8

(a) Recall that Example 3.2.1 verified that

$$B = \begin{bmatrix} -3 & 1 \\ -4 & 1 \end{bmatrix} \text{ is an inverse of } A = \begin{bmatrix} 1 & -1 \\ 4 & -3 \end{bmatrix}.$$

Formula (3.2) gives this inverse from the matrix A: its elements are $a = 1$, $b = -1$, $c = 4$ and $d = -3$ so the determinant $ad - bc = 1 \cdot (-3) - (-1) \cdot 4 = 1$ and hence formula (3.2) derives the inverse

$$A^{-1} = \frac{1}{1}\begin{bmatrix} -3 & -(-1) \\ -4 & 1 \end{bmatrix} = \begin{bmatrix} -3 & 1 \\ -4 & 1 \end{bmatrix} = B.$$

(b) Further, recall Example 3.2.5 proved that there is no inverse for matrix

$$A = \begin{bmatrix} 1 & -2 \\ -3 & 6 \end{bmatrix}.$$

Theorem 3.2.7 also establishes that this matrix is not invertible because the matrix determinant $ad - bc = 1 \cdot 6 - (-2) \cdot (-3) = 6 - 6 = 0$. □

Activity 3.2.9 Which of the following matrices is invertible?

(a) $\begin{bmatrix} -2 & -1 & 4 \\ 3 & 1 & 1 \end{bmatrix}$

(b) $\begin{bmatrix} 0 & -3 \\ 4 & -2 \end{bmatrix}$

(c) $\begin{bmatrix} -4 & -2 \\ 2 & 2 \\ -3 & 1 \end{bmatrix}$

(d) $\begin{bmatrix} -2 & 1 \\ 4 & -2 \end{bmatrix}$

Proof. To prove Theorem 3.2.7, first show the given A^{-1} satisfies Definition 3.2.2 when the determinant $ad - bc \neq 0$ (and using associativity of scalar-matrix multiplication, Theorem 3.1.26(d)). For the proposed A^{-1}, on the one hand,

$$A^{-1}A = \frac{1}{ad - bc}\begin{bmatrix} d & -b \\ -c & a \end{bmatrix}\begin{bmatrix} a & b \\ c & d \end{bmatrix}$$

$$= \frac{1}{ad - bc}\begin{bmatrix} da - bc & db - bd \\ -ca + ac & -cb + ad \end{bmatrix}$$

$$= \begin{bmatrix} 1 & 0 \\ 0 & 1 \end{bmatrix} = I_2.$$

On the other hand,

$$AA^{-1} = \begin{bmatrix} a & b \\ c & d \end{bmatrix}\begin{bmatrix} d & -b \\ -c & a \end{bmatrix}\frac{1}{ad - bc}$$

$$= \begin{bmatrix} ad - bc & -ab + ba \\ cd - dc & -cb + da \end{bmatrix}\frac{1}{ad - bc}$$

$$= \begin{bmatrix} 1 & 0 \\ 0 & 1 \end{bmatrix} = I_2.$$

By uniqueness (Theorem 3.2.6), formula (3.2) is the only inverse when $ad - bc \neq 0$.

Now eliminate the case when $ad - bc = 0$. If an inverse exists, say X, then it must satisfy $AX = I_2$. The top-left entry of this matrix equality requires $ax_{11} + bx_{21} = 1$, whereas the bottom-left equality requires $cx_{11} + dx_{21} = 0$. Regard these as a system of two linear equations for the as yet unknowns x_{11} and x_{21}: from $d\times$ the first subtract $b\times$ the second to deduce that an inverse requires $dax_{11} + dbx_{21} - bcx_{11} - bdx_{21} = d \cdot 1 - b \cdot 0$. By cancellation and factorizing x_{11}, this equation then requires $(ad - bc)x_{11} = d$. But the determinant is zero, so this equation requires $0 \cdot x_{11} = d$.

- If element d is nonzero, then this equation cannot be satisfied and hence no inverse can be found.
- Otherwise, if d is zero, then any x_{11} satisfies this equation, so if there is an inverse then there would be an infinite number of inverses (through the free variable x_{11}). But this contradicts the uniqueness Theorem 3.2.6, so an inverse cannot exist in this case either.

That is, if the determinant $ad - bc = 0$, then the 2×2 matrix is not invertible. $\qquad\square$

Computational considerations   Except for easy cases such as 2×2 matrices, we rarely explicitly compute the inverse of a matrix. Computationally there are (almost) always better ways such as the MATLAB/Octave operation $A\backslash b$ of Procedure 2.2.5. The inverse is a crucial theoretical device, but rarely a practical computational tool.

Almost anything you can do with A^{-1} can be done without it.
G. E. Forsythe and C. B. Moler, 1967 (Higham, 1996, p.261)

The following Theorem 3.2.10 is an example: for a system of linear equations the theorem connects the existence of a unique solution to the invertibility of the matrix of coefficients. Further, Section 3.3.2 connects solutions to the rcond invoked by Procedure 2.2.5. Although in theoretical statements we write expressions like $x = A^{-1}b$, practically, once we know a solution exists (rcond is acceptable), we almost always compute a solution without ever constructing the inverse A^{-1}.

Theorem 3.2.10   *If A is an invertible $n \times n$ matrix, then the system of linear equations $Ax = b$ has the unique solution $x = A^{-1}b$ for every b in $\mathbb{R}^n$. Equivalently, if there is either no solution or an infinite number of solutions (Theorem 2.2.26), then the matrix A is not invertible (it is singular).*

The forthcoming Theorem 3.3.27 strengthens this theorem to "if and only if": that is, A is invertible if and only if $Ax = b$ has a unique solution.

Proof.   The proof has two parts: first showing $x = A^{-1}b$ is a solution, and second showing that there are no others. First, try $x = A^{-1}b$ and use associativity (Theorem 3.1.26(c)) and the inverse Definition 3.2.2:

$$Ax = A(A^{-1}b) = (AA^{-1})b = I_n b = b.$$

Second, suppose y is any solution, that is, $Ay = b$. Multiply both sides by the inverse A^{-1}, and again use associativity and the definition of the inverse, to deduce

$$A^{-1}(Ay) = A^{-1}b \implies (A^{-1}A)y = x$$
$$\implies I_ny = x$$
$$\implies y = x.$$

Since any solution y has to be the same as x, $x = A^{-1}b$ is the unique solution.

The equivalent statement follows because if there is not a unique solution then the precondition, that the matrix is invertible, cannot hold. ☐

Example 3.2.11 Use the matrices of Examples 3.2.1, 3.2.3 and 3.2.5 to decide whether each of the following systems have a unique solution, or not.

(a) $\begin{cases} x - y = 4, \\ 4x - 3y = 3. \end{cases}$
(c) $\begin{cases} r - 2s = -1, \\ -3r + 6s = 3. \end{cases}$

(b) $\begin{cases} u - v + 5w = 2, \\ -5u - v + 3w = 5, \\ 2u + 2v - 6w = 1. \end{cases}$

Solution:

(a) A matrix for this system is $\begin{bmatrix} 1 & -1 \\ 4 & -3 \end{bmatrix}$ which Example 3.2.1 shows has an inverse. Theorem 3.2.10 then assures us the system has a unique solution.

(b) A matrix for this system is $\begin{bmatrix} 1 & -1 & 5 \\ -5 & -1 & 3 \\ 2 & 2 & -6 \end{bmatrix}$ which Example 3.2.3 shows has an inverse. Theorem 3.2.10 then assures us that the system has a unique solution.

(c) A matrix for this system is $\begin{bmatrix} 1 & -2 \\ -3 & 6 \end{bmatrix}$ which Example 3.2.5 shows is not invertible. Thus the precondition of Theorem 3.2.10 is not satisfied and so this theorem tells us nothing in this scenario.

However, the more complete forthcoming Theorem 3.3.27 establishes for us that the system does not have a unique solution (either no solution or an infinite number of solutions—the matrix alone does not tell us which).

☐

Example 3.2.12 Given the following information about solutions of systems of linear equations, write down if the matrix associated with each system is invertible, or not, or there is not enough given information to decide. Give reasons.

(a) A general solution is $(1, -5, 0, 3)$ (and no other).

(b) A general solution is $(3, -5 + 3t, 3 - t, -1)$ (and no other).

(c) A solution of a system is $(-3/2, -2, -\pi, 2, -4)$ (but there may be others).

(d) A solution of a homogeneous system is $(1, 2, -8)$ (but there may be others).

Solution:

(a) Since the solution is unique, the preconditions of Theorem 3.2.10 do not hold and so we cannot (as yet) decide. (We need Theorem 3.3.27 to certify that the matrix in the system must be invertible.)

(b) This solution has an apparent free parameter, t, and so there are many solutions, which implies that the matrix is not invertible.

(c) Not enough information is given as we do not know whether there are any more solutions.

(d) Since a homogeneous system always has $\mathbf{0}$ as a solution (Section 2.2.3), then we know that there are at least two solutions to the system, and hence the matrix is not invertible.

□

Recall from Section 3.1 the properties of scalar multiplication, matrix powers, transpose, and their computation (Table 3.1). The next theorem incorporates the inverse into this suite of properties.

Theorem 3.2.13 (*properties of the inverse*) *Let A and B be invertible matrices of the same size, then:*

(a) *matrix A^{-1} is invertible and $(A^{-1})^{-1} = A$;*

(b) *if scalar $c \neq 0$, then matrix cA is invertible and $(cA)^{-1} = \frac{1}{c}A^{-1}$;*

(c) *matrix AB is invertible and $(AB)^{-1} = B^{-1}A^{-1}$ —remember the reversed order in this identity;*

(d) *matrix A^{T} is invertible and $(A^{\mathrm{T}})^{-1} = (A^{-1})^{\mathrm{T}}$;*

(e) *for every $p = 1, 2, 3, \ldots$, matrix A^p is invertible and $(A^p)^{-1} = (A^{-1})^p$.*

Proof. Three parts are proved, and two are left as exercises.

3.2.13(a) By Definition 3.2.2 the matrix A^{-1} satisfies $A^{-1}A = AA^{-1} = I$. But also by Definition 3.2.2 these are exactly the identities we need to assert that matrix A is the inverse of matrix (A^{-1}). Hence $A = (A^{-1})^{-1}$.

3.2.13(c) Test that $B^{-1}A^{-1}$ has the required properties for the inverse of AB. First, by associativity (Theorem 3.1.26(c)) and multiplication by the identity (Theorem 3.1.26(e))

$$(B^{-1}A^{-1})(AB) = B^{-1}(A^{-1}A)B = B^{-1}IB = B^{-1}B = I.$$

Second, and similarly

$$(AB)(B^{-1}A^{-1}) = A(BB^{-1})A^{-1} = AIA^{-1} = AA^{-1} = I.$$

Hence by Definition 3.2.2 and the uniqueness Theorem 3.2.6, matrix AB is invertible and $B^{-1}A^{-1}$ is the inverse.

3.2.13(e) Prove by induction and use Theorem 3.2.13(c). First, for the case of exponent $p = 1$, $(A^1)^{-1} = (A)^{-1} = A^{-1} = (A^{-1})^1$ and so the identity holds.

Then, for any integer exponent $p \geq 2$, assume the identity $(A^{p-1})^{-1} = (A^{-1})^{p-1}$. Consider

$$
\begin{aligned}
(A^p)^{-1} &= (AA^{p-1})^{-1} && \text{(by power law Theorem 3.1.26(g))} \\
&= (A^{p-1})^{-1}A^{-1} && \text{(Theorem 3.2.13(c), } B = A^{p-1}) \\
&= (A^{-1})^{p-1}A^{-1} && \text{(by inductive assumption)} \\
&= (A^{-1})^p && \text{(by power law Theorem 3.1.26(g)).}
\end{aligned}
$$

By induction, the identity $(A^p)^{-1} = (A^{-1})^p$ holds for every exponent $p = 1, 2, 3, \ldots$.

☐

Activity 3.2.14 The matrix $\begin{bmatrix} 3 & -5 \\ 4 & -7 \end{bmatrix}$ has inverse $\begin{bmatrix} 7 & -5 \\ 4 & -3 \end{bmatrix}$.

- What then is the inverse of the matrix $\begin{bmatrix} 6 & -10 \\ 8 & -14 \end{bmatrix}$?

(a) $\begin{bmatrix} 14 & -10 \\ 8 & -3 \end{bmatrix}$

(b) $\begin{bmatrix} 7 & 4 \\ -5 & -3 \end{bmatrix}$

(c) $\begin{bmatrix} 3.5 & -2.5 \\ 2 & -1.5 \end{bmatrix}$

(d) $\begin{bmatrix} 3.5 & 2 \\ -2.5 & -1.5 \end{bmatrix}$

- Further, which of the above is the inverse of $\begin{bmatrix} 3 & 4 \\ -5 & -7 \end{bmatrix}$?

Definition 3.2.15 *(non-positive powers)* For every invertible matrix A, define $A^0 := I$ and for every positive integer p define $A^{-p} := (A^{-1})^p$ (or by Theorem 3.2.13(e) equivalently as $(A^p)^{-1}$).

Example 3.2.16 Recall from Example 3.2.1 that matrix

$$A = \begin{bmatrix} 1 & -1 \\ 4 & -3 \end{bmatrix} \text{ has inverse } A^{-1} = \begin{bmatrix} -3 & 1 \\ -4 & 1 \end{bmatrix}.$$

Compute A^{-2} and A^{-4}.

Solution: From Definition 3.2.15,

$$A^{-2} = (A^{-1})^2 = \begin{bmatrix} -3 & 1 \\ -4 & 1 \end{bmatrix}\begin{bmatrix} -3 & 1 \\ -4 & 1 \end{bmatrix} = \begin{bmatrix} 5 & -2 \\ 8 & -3 \end{bmatrix},$$

$$A^{-4} = (A^{-1})^4 = [(A^{-1})^2]^2$$

$$= \begin{bmatrix} 5 & -2 \\ 8 & -3 \end{bmatrix}\begin{bmatrix} 5 & -2 \\ 8 & -3 \end{bmatrix} = \begin{bmatrix} 9 & -4 \\ 16 & -7 \end{bmatrix},$$

upon using one of the power laws of Theorem 3.1.26(g).   □

Activity 3.2.17 The previous Example 3.2.16 gives the inverse of a matrix A and determines A^{-2}: what then is A^{-3}?

(a) $\begin{bmatrix} -7 & 3 \\ -12 & 5 \end{bmatrix}$
(c) $\begin{bmatrix} 3 & 5 \\ -7 & -12 \end{bmatrix}$

(b) $\begin{bmatrix} 3 & -7 \\ 5 & -12 \end{bmatrix}$
(d) $\begin{bmatrix} -7 & -12 \\ 3 & 5 \end{bmatrix}$

Example 3.2.18 (predict the past) Recall Example 3.1.9 introduced how to use a Leslie matrix to predict the future population of an animal. If $x = (60, 70, 20)$ is the current number of pups, juveniles, and mature females respectively, then our modelling predicts the population numbers after a year is $x' = Lx$, after two years is $x'' = Lx' = L^2 x$, and so on. In these formulas, and for this example, the Leslie matrix

$$L = \begin{bmatrix} 0 & 0 & 4 \\ \frac{1}{2} & 0 & 0 \\ 0 & \frac{1}{3} & \frac{1}{3} \end{bmatrix}, \quad \text{which has inverse } L^{-1} = \begin{bmatrix} 0 & 2 & 0 \\ -\frac{1}{4} & 0 & 3 \\ \frac{1}{4} & 0 & 0 \end{bmatrix}.$$

Assume the same rule applies for earlier years.

- Letting the population numbers a year ago be denoted by x^- then according to the model the current population $x = Lx^-$. Multiply by the inverse of L: $L^{-1}x = L^{-1}Lx^- = x^-$; that is, the population a year before the current is $x^- = L^{-1}x$.

- Similarly, letting the population numbers two years ago be denoted by $x^=$ then according to the model $x^- = Lx^=$. So multiplication by L^{-1} gives $x^= = L^{-1}x^- = L^{-1}L^{-1}x = L^{-2}x$.

- One more year earlier, letting the population numbers three years ago be denoted by $x^{\equiv}$, then according to the model $x^= = Lx^{\equiv}$. So multiplication by L^{-1} gives $x^{\equiv} = L^{-1}x^= = L^{-1}L^{-2}x = L^{-3}x$.

Hence use the inverse powers of L to predict the earlier history of the population of female animals in the given example; but first verify the given inverse is correct.

Solution: Verify the given inverse by evaluating (showing only nonzero terms in a sum)

$$LL^{-1} = \begin{bmatrix} 0 & 0 & 4 \\ \frac{1}{2} & 0 & 0 \\ 0 & \frac{1}{3} & \frac{1}{3} \end{bmatrix} \begin{bmatrix} 0 & 2 & 0 \\ -\frac{1}{4} & 0 & 3 \\ \frac{1}{4} & 0 & 0 \end{bmatrix}$$

$$= \begin{bmatrix} 4 \cdot \frac{1}{4} & 0 & 0 \\ 0 & \frac{1}{2} \cdot 2 & 0 \\ \frac{1}{3} \cdot (-\frac{1}{4}) + \frac{1}{3} \cdot \frac{1}{4} & 0 & \frac{1}{3} \cdot 3 \end{bmatrix}$$

$$= \begin{bmatrix} 1 & 0 & 0 \\ 0 & 1 & 0 \\ 0 & 0 & 1 \end{bmatrix} = I_3,$$

$$L^{-1}L = \begin{bmatrix} 0 & 2 & 0 \\ -\frac{1}{4} & 0 & 3 \\ \frac{1}{4} & 0 & 0 \end{bmatrix} \begin{bmatrix} 0 & 0 & 4 \\ \frac{1}{2} & 0 & 0 \\ 0 & \frac{1}{3} & \frac{1}{3} \end{bmatrix}$$

$$= \begin{bmatrix} 2\frac{1}{2} & 0 & 0 \\ 0 & 3 \cdot \frac{1}{3} & -\frac{1}{4} \cdot 4 + 3 \cdot \frac{1}{3} \\ 0 & 0 & \frac{1}{4} \cdot 4 \end{bmatrix}$$

$$= \begin{bmatrix} 1 & 0 & 0 \\ 0 & 1 & 0 \\ 0 & 0 & 1 \end{bmatrix} = I_3.$$

Hence the given L^{-1} is indeed the inverse. For the current population $x = (60, 70, 20)$, now use the inverse to compute earlier populations.

- The population of females one year ago was

$$x^- = L^{-1}x = \begin{bmatrix} 0 & 2 & 0 \\ -\frac{1}{4} & 0 & 3 \\ \frac{1}{4} & 0 & 0 \end{bmatrix} \begin{bmatrix} 60 \\ 70 \\ 20 \end{bmatrix} = \begin{bmatrix} 140 \\ 45 \\ 15 \end{bmatrix}.$$

That is, there were 140 pups, 45 juveniles, and 15 mature females.

- Computing the square of the inverse

$$L^{-2} = (L^{-1})^2 = \begin{bmatrix} 0 & 2 & 0 \\ -\frac{1}{4} & 0 & 3 \\ \frac{1}{4} & 0 & 0 \end{bmatrix}^2 = \begin{bmatrix} -\frac{1}{2} & 0 & 6 \\ \frac{3}{4} & -\frac{1}{2} & 0 \\ 0 & \frac{1}{2} & 0 \end{bmatrix},$$

we predict that the population of females two years ago was

$$x^= = L^{-2}x = \begin{bmatrix} -\frac{1}{2} & 0 & 6 \\ \frac{3}{4} & -\frac{1}{2} & 0 \\ 0 & \frac{1}{2} & 0 \end{bmatrix} \begin{bmatrix} 60 \\ 70 \\ 20 \end{bmatrix} = \begin{bmatrix} 90 \\ 10 \\ 35 \end{bmatrix}.$$

- Similarly, computing the cube of the inverse

$$L^{-3} = L^{-2}L^{-1} = \cdots = \begin{bmatrix} \frac{3}{2} & -1 & 0 \\ \frac{1}{8} & \frac{3}{2} & -\frac{3}{2} \\ -\frac{1}{8} & 0 & \frac{3}{2} \end{bmatrix},$$

we predict the population of females three years ago was

$$x^{\equiv} = L^{-3}x = \begin{bmatrix} \frac{3}{2} & -1 & 0 \\ \frac{1}{8} & \frac{3}{2} & -\frac{3}{2} \\ -\frac{1}{8} & 0 & \frac{3}{2} \end{bmatrix} \begin{bmatrix} 60 \\ 70 \\ 20 \end{bmatrix} = \begin{bmatrix} 20 \\ 82.5 \\ 22.5 \end{bmatrix}.$$

(Predicting half animals in this last calculation is because the modelling only deals with average numbers, not exact numbers.) □

Example 3.2.19 As an alternative to the hand calculations of Example 3.2.18, predict earlier populations by computing in MATLAB/Octave without ever explicitly finding the inverse or powers of the inverse. The procedure is to solve the linear system $Lx^- = x$ for the population x^- a year ago, and then similarly solve $Lx^= = x^-$, $Lx^{\equiv} = x^=$, and so on.

Solution: Execute

```
L= [0  0  4;1/2  0  0;0  1/3  1/3]
x= [60;70;20]
rcond (L)
xm=L\x
xmm=L\xm
xmmm=L\xmm
```

Since rcond is 0.08 (good by Procedure 2.2.5), this code uses L\ to solve the linear systems and confirm the population of females in previous years is as determined by Example 3.2.18, namely

$$xm = \begin{bmatrix} 140 \\ 45 \\ 15 \end{bmatrix}, \quad xmm = \begin{bmatrix} 90 \\ 10 \\ 35 \end{bmatrix}, \quad xmmm = \begin{bmatrix} 20 \\ 82.5 \\ 22.5 \end{bmatrix}.$$

□

3.2.2 Diagonal matrices stretch and shrink

This section and the next Section 3.2.3 introduce two classes of matrices whose inverses we can easily write down and use. Importantly, the subsequent Section 3.3 shows that both classes lie at the heart of every matrix via the singular value decomposition of the matrix!

Recall that identity matrices are zero except for a diagonal of ones from the top-left to the bottom-right of the matrix. Because of the nature of matrix multiplication it is this diagonal that is special. Because of the special nature of this diagonal, this section explores matrices which are zero except for the numbers (not generally ones) in the top-left to bottom-right diagonal.

Example 3.2.20 That is, this section explores the nature of so-called diagonal matrices such as

$$
\begin{bmatrix} 3 & 0 \\ 0 & 2 \end{bmatrix}, \quad
\begin{bmatrix} 0.58 & 0 & 0 \\ 0 & -1.61 & 0 \\ 0 & 0 & 2.17 \end{bmatrix}, \quad
\begin{bmatrix} \pi & 0 & 0 \\ 0 & \sqrt{3} & 0 \\ 0 & 0 & 0 \end{bmatrix}.
$$

We use the term diagonal matrix to also include non-square matrices such as

$$
\begin{bmatrix} -\sqrt{2} & 0 \\ 0 & \frac{1}{2} \\ 0 & 0 \end{bmatrix}, \quad
\begin{bmatrix} 1 & 0 & 0 & 0 & 0 \\ 0 & \pi & 0 & 0 & 0 \\ 0 & 0 & e & 0 & 0 \end{bmatrix}.
$$

The term diagonal matrix does *not* describe matrices such as

$$
\begin{bmatrix} 0 & 0 & 1 \\ 0 & 2 & 0 \\ -\frac{1}{2} & 0 & 0 \end{bmatrix}, \quad \text{and} \quad
\begin{bmatrix} -0.17 & 0 & 0 & 0 \\ 0 & -4.22 & 0 & 0 \\ 0 & 0 & 0 & 3.05 \end{bmatrix}.
$$

Amazingly, the singular value decomposition of Section 3.3 proves that diagonal matrices lie at the very heart of the action of *every* matrix.

Definition 3.2.21 *(diagonal matrix)* *For every $m \times n$ matrix A, the **diagonal entries** of A are $a_{11}, a_{22}, \ldots, a_{pp}$ where $p = \min(m,n)$. A matrix whose off-diagonal entries are all zero is called a **diagonal matrix**.*

For brevity, sometimes we write $\mathrm{diag}(v_1, v_2, \ldots, v_n)$ to denote the $n \times n$ square matrix with diagonal entries $v_1, v_2, \ldots, v_n$, or $\mathrm{diag}_{m \times n}(v_1, v_2, \ldots, v_p)$ for an $m \times n$ matrix with diagonal entries $v_1, v_2, \ldots, v_p$.

Example 3.2.22 The five diagonal matrices of Example 3.2.20 could equivalently be written as $\mathrm{diag}(3,2)$, $\mathrm{diag}(0.58, -1.61, 2.17)$, $\mathrm{diag}(\pi, \sqrt{3}, 0)$, $\mathrm{diag}_{3 \times 2}(-\sqrt{2}, \frac{1}{2})$, and $\mathrm{diag}_{3 \times 5}(1, \pi, e)$, respectively.

Diagonal matrices may also have zeros on the diagonal, as well as the required zeros for the off-diagonal entries.

Table 3.2 As well as the basics of MATLAB/Octave listed in Tables 1.2, 2.3 and 3.1, we need these matrix operations.

- diag (v) where v is a row/column vector of size p generates the $p \times p$ matrix

$$\mathrm{diag}(v_1, v_2, \ldots, v_p) = \begin{bmatrix} v_1 & 0 & \cdots & 0 \\ 0 & v_2 & & \vdots \\ \vdots & & \ddots & \\ 0 & \cdots & & v_p \end{bmatrix}.$$

- In MATLAB/Octave (but not usually in algebra), diag also does the opposite: for an $m \times n$ matrix A such that both m, $n \geq 2$, diag (A) returns the (column) vector $(a_{11}, a_{22}, \ldots, a_{pp})$ of diagonal entries where the result vector size $p = \min(m, n)$.
- The dot operators ./ and .* do element-by-element division and multiplication of two matrices/vectors of the same size. For example,

  [5 14 33]./[5 7 3] = [1 2 11]

- Section 3.5 also needs to compute the logarithm of data: log10 (v) finds the logarithm to base 10 of each component of v and returns the results in a vector of the same size; log (v) does the same but for the natural logarithm (not ln (v)).

Activity 3.2.23 Which of the following matrices are not diagonal?

(a) $\begin{bmatrix} 1 & 0 & 0 \\ 0 & 0 & 0 \\ 0 & 2 & 0 \\ 0 & 0 & 0 \end{bmatrix}$ (b) $\begin{bmatrix} 1 & 0 & 0 & 0 \\ 0 & 2 & 0 & 0 \\ 0 & 0 & 0 & 0 \end{bmatrix}$ (c) O_n
(d) I_n

Solve systems whose matrix is diagonal

Solving a system of linear equations (Definition 2.1.2) is particularly straightforward when the matrix of the system is diagonal. Indeed, much mathematics in both theory and applications is devoted to transforming a given problem so that the matrix appearing in the system is diagonal (e.g., Sections 2.2.2 and 3.3.2 and Chapters 4 and 7).

Example 3.2.24 Solve $\begin{bmatrix} 3 & 0 \\ 0 & 2 \end{bmatrix} \begin{bmatrix} x_1 \\ x_2 \end{bmatrix} = \begin{bmatrix} 2 \\ -5 \end{bmatrix}$.

Solution: Algebraically this matrix-vector equation means

$$\begin{cases} 3x_1 + 0x_2 = 2 \\ 0x_1 + 2x_2 = -5 \end{cases} \iff \begin{cases} 3x_1 = 2 \\ 2x_2 = -5 \end{cases} \iff \begin{cases} x_1 = 2/3 \\ x_2 = -5/2 \end{cases}$$

The solution is $x = (2/3, -5/2)$. Interestingly, the two components of this solution are firstly the 2 on the right-hand side divided by the 3 in the matrix, and secondly the -5 on the right-hand side divided by the 2 in the matrix. $\square$

Example 3.2.25 For any given numbers b_1, b_2, b_3, solve

$$\begin{bmatrix} 2 & 0 & 0 \\ 0 & \frac{2}{3} & 0 \\ 0 & 0 & -1 \end{bmatrix} \begin{bmatrix} x_1 \\ x_2 \\ x_3 \end{bmatrix} = \begin{bmatrix} b_1 \\ b_2 \\ b_3 \end{bmatrix}$$

Solution: Algebraically this equation means

$$\begin{cases} 2x_1 + 0x_2 + 0x_3 = b_1 \\ 0x_1 + \frac{2}{3}x_2 + 0x_3 = b_2 \\ 0x_1 + 0x_2 - 1x_3 = b_3 \end{cases} \Longleftrightarrow \begin{cases} 2x_1 = b_1 \\ \frac{2}{3}x_2 = b_2 \\ -x_3 = b_3 \end{cases} \Longleftrightarrow \begin{cases} x_1 = \frac{1}{2}b_1 \\ x_2 = \frac{3}{2}b_2 \\ x_3 = -b_3 \end{cases}.$$

The solution is $x = (\frac{1}{2}b_1, \frac{3}{2}b_2, -b_3)$.

This solution may be rewritten as $x = \begin{bmatrix} 1/2 & 0 & 0 \\ 0 & 3/2 & 0 \\ 0 & 0 & -1 \end{bmatrix} \begin{bmatrix} b_1 \\ b_2 \\ b_3 \end{bmatrix}$. Consequently, by its uniqueness

(Theorem 3.2.6), the inverse of the given diagonal matrix must be

$$\begin{bmatrix} 2 & 0 & 0 \\ 0 & \frac{2}{3} & 0 \\ 0 & 0 & -1 \end{bmatrix}^{-1} = \begin{bmatrix} \frac{1}{2} & 0 & 0 \\ 0 & \frac{3}{2} & 0 \\ 0 & 0 & -1 \end{bmatrix},$$

which interestingly is simply the diagonal of reciprocals of the given matrix's diagonal. □

Activity 3.2.26 What is the solution to $\begin{bmatrix} 0.4 & 0 \\ 0 & 0.1 \end{bmatrix} x = \begin{bmatrix} 0.1 \\ -0.2 \end{bmatrix}$?

(a) $(4, -2)$ (b) $(4, -\frac{1}{2})$ (c) $(\frac{1}{4}, -2)$ (d) $(\frac{1}{4}, -\frac{1}{2})$

Theorem 3.2.27 (*inverse of diagonal matrix*) *For every $n \times n$ diagonal matrix $D = \text{diag}(d_1, d_2, \ldots, d_n)$, if all the diagonal entries are nonzero, $d_i \neq 0$ for every $i = 1, 2, \ldots, n$, then D is invertible and the inverse $D^{-1} = \text{diag}(1/d_1, 1/d_2, \ldots, 1/d_n)$. Conversely, if a diagonal matrix is invertible, then all its diagonal entries are nonzero.*

Proof. Consider the matrix product

$$\begin{bmatrix} d_1 & 0 & \cdots & 0 \\ 0 & d_2 & & 0 \\ \vdots & & \ddots & \vdots \\ 0 & 0 & \cdots & d_n \end{bmatrix} \begin{bmatrix} \frac{1}{d_1} & 0 & \cdots & 0 \\ 0 & \frac{1}{d_2} & & 0 \\ \vdots & & \ddots & \vdots \\ 0 & 0 & \cdots & \frac{1}{d_n} \end{bmatrix}$$

$$
\begin{aligned}
&= \begin{bmatrix}
d_1\frac{1}{d_1}+0+\cdots+0 & d_10+0\frac{1}{d_2}+\cdots+0 & \cdots & d_10+0+\cdots+0\frac{1}{d_n} \\
0\frac{1}{d_1}+d_20+\cdots+0 & 0+d_2\frac{1}{d_2}+\cdots+0 & & 0+d_20+\cdots+0\frac{1}{d_n} \\
\vdots & & \ddots & \vdots \\
0\frac{1}{d_1}+0+\cdots+d_n0 & 0+0\frac{1}{d_2}+\cdots+d_n0 & \cdots & 0+0+\cdots+d_n\frac{1}{d_n}
\end{bmatrix} \\[2mm]
&= \begin{bmatrix}
1 & 0 & \cdots & 0 \\
0 & 1 & & 0 \\
\vdots & & \ddots & \vdots \\
0 & 0 & \cdots & 1
\end{bmatrix} = I_n.
\end{aligned}
$$

Similarly for the reverse product. By Definition 3.2.2, D is invertible with the given inverse. Exercise 3.2.14 asks you to use Theorem 3.2.10 to establish the last, converse, part of this theorem. ☐

Example 3.2.28 The previous Example 3.2.25 gives the inverse of a 3×3 matrix. For the 2×2 matrix $D = \mathrm{diag}(3,2) = \begin{bmatrix} 3 & 0 \\ 0 & 2 \end{bmatrix}$ the inverse is $D^{-1} = \mathrm{diag}(\frac{1}{3},\frac{1}{2}) = \begin{bmatrix} 1/3 & 0 \\ 0 & 1/2 \end{bmatrix}$. Then the solution to

$$
\begin{bmatrix} 3 & 0 \\ 0 & 2 \end{bmatrix} x = \begin{bmatrix} 2 \\ -5 \end{bmatrix} \quad \text{is } x = \begin{bmatrix} \frac{1}{3} & 0 \\ 0 & \frac{1}{2} \end{bmatrix} \begin{bmatrix} 2 \\ -5 \end{bmatrix} = \begin{bmatrix} 2/3 \\ -5/2 \end{bmatrix}.
$$

Compute in MATLAB/Octave To solve the matrix-vector equation $Dx = b$ recognize that this equation means

$$
\begin{bmatrix}
d_1 & 0 & \cdots & 0 \\
0 & d_2 & & 0 \\
\vdots & & \ddots & \vdots \\
0 & 0 & \cdots & d_n
\end{bmatrix}
\begin{bmatrix} x_1 \\ x_2 \\ \vdots \\ x_n \end{bmatrix}
=
\begin{bmatrix} d_1x_1 \\ d_2x_2 \\ \vdots \\ d_nx_n \end{bmatrix}
=
\begin{bmatrix} b_1 \\ b_2 \\ \vdots \\ b_n \end{bmatrix}
$$

$$
\Longleftrightarrow \quad
\begin{cases}
d_1x_1 = b_1 \\
d_2x_2 = b_2 \\
\vdots \\
d_nx_n = b_n
\end{cases}
\quad \Longleftrightarrow \quad
\begin{cases}
x_1 = b_1/d_1 \\
x_2 = b_2/d_2 \\
\vdots \\
x_n = b_n/d_n
\end{cases}
\tag{3.3}
$$

- Suppose you have a column vector d of the diagonal entries of D and a column vector b of the RHS; then compute a solution by, for example,

```
d= [2;2/3;-1]
b= [1;2;3]
x=b./d
```

to find the answer [0.5;3;-3]. Here the MATLAB/Octave operation ./ does element-by-element division (Table 3.2).

- When you have the diagonal matrix in full: extract the diagonal elements into a column vector with `diag()` (Table 3.2); then execute the element-by-element division; for example,

```
D= [2 0 0;0 2/3 0;0 0 -1]
b= [1;2;3]
x=b./diag(D)
```

But do not divide by zero

Dividing by zero is almost always nonsense. Instead use reasoning. Consider solving $Dx = b$ for diagonal $D = \text{diag}(d_1, d_2, \ldots, d_n)$ where $d_n = 0$ (and similarly if any others are zero). From (3.3) we need to solve $d_n x_n = b_n$, which here is $0 \cdot x_n = b_n$, that is, $0 = b_n$. There are two cases:

- if $b_n \neq 0$, then there is no solution; alternatively
- if $b_n = 0$, then there are an infinite number of solutions as every x_n satisfies $0 \cdot x_n = 0$.

Example 3.2.29 Solve the two systems (the only difference is the last component on the RHS)

$$\begin{bmatrix} 2 & 0 & 0 \\ 0 & \frac{2}{3} & 0 \\ 0 & 0 & 0 \end{bmatrix} \begin{bmatrix} x_1 \\ x_2 \\ x_3 \end{bmatrix} = \begin{bmatrix} 1 \\ 2 \\ 3 \end{bmatrix} \quad \text{and} \quad \begin{bmatrix} 2 & 0 & 0 \\ 0 & \frac{2}{3} & 0 \\ 0 & 0 & 0 \end{bmatrix} \begin{bmatrix} x_1 \\ x_2 \\ x_3 \end{bmatrix} = \begin{bmatrix} 1 \\ 2 \\ 0 \end{bmatrix}$$

Solution: Algebraically, the first system means

$$\begin{aligned} 2x_1 + 0x_2 + 0x_3 &= 1 \\ 0x_1 + \tfrac{2}{3}x_2 + 0x_3 &= 2 \\ 0x_1 + 0x_2 + 0x_3 &= 3 \end{aligned} \quad \Longleftrightarrow \quad \begin{aligned} 2x_1 &= 1 \\ \tfrac{2}{3}x_2 &= 2 . \\ 0x_3 &= 3 \end{aligned}$$

There is no solution in this first case as there is no choice of x_3 such that $0x_3 = 3$.

Algebraically, the second system means

$$\begin{aligned} 2x_1 + 0x_2 + 0x_3 &= 1 \\ 0x_1 + \tfrac{2}{3}x_2 + 0x_3 &= 2 \\ 0x_1 + 0x_2 + 0x_3 &= 0 \end{aligned} \quad \Longleftrightarrow \quad \begin{aligned} 2x_1 &= 1 \\ \tfrac{2}{3}x_2 &= 2 . \\ 0x_3 &= 0 \end{aligned}$$

In this second case we satisfy the equation $0x_3 = 0$ with any x_3. Hence there are an infinite number of solutions, namely $x = (\tfrac{1}{2}, 3, t)$ for every t—a free variable just as in Gauss–Jordan elimination (Procedure 2.2.23). □

Stretch or squash the unit square

> Equations are just the boring part of mathematics. I attempt to see things in terms of geometry.
>
> Stephen Hawking, 2005

Multiplication by matrices transforms shapes: multiplication by diagonal matrices just stretches or squashes and/or reflects in the direction of the coordinate axes. The next Section 3.2.3 introduces matrices that rotate.

Example 3.2.30 Consider $A = \text{diag}(3,2) = \begin{bmatrix} 3 & 0 \\ 0 & 2 \end{bmatrix}$. The picture to the right shows this matrix stretches the (blue) unit square (drawn with a 'roof') by a factor of three horizontally and two vertically (to the red). Recall that (x_1, x_2) denotes the corresponding column vector. As seen in the corner points of the graphic to the right, $A(1,0) = (3,0)$, $A(0,1) = (0,2)$, and $A(0,0) = (0,0)$, and $A(1,1) = (3,2)$. The 'roof' just helps us to track which corner goes where.

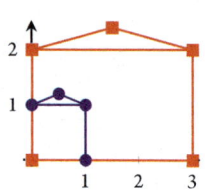

The inverse $A^{-1} = \text{diag}(\frac{1}{3}, \frac{1}{2}) = \begin{bmatrix} 1/3 & 0 \\ 0 & 1/2 \end{bmatrix}$ undoes the stretching of the matrix A by squashing in both the horizontal and vertical directions (from blue to red).

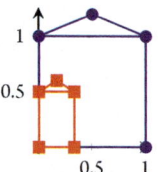

□

Example 3.2.31 Consider $\text{diag}(2, \frac{2}{3}, -1) = \begin{bmatrix} 2 & 0 & 0 \\ 0 & 2/3 & 0 \\ 0 & 0 & -1 \end{bmatrix}$: the stereo pair to the right illustrates how this diagonal matrix stretches in one direction, squashes in another, and reflects in the vertical. By multiplying the matrix by the eight corner vectors, $(1,0,0)$, $(0,1,0)$, $(0,0,1)$, and so on, we see that the blue unit cube (with 'roof' and 'door') maps to the red.

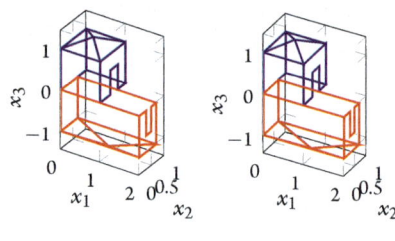

□

One great aspect of a diagonal matrix is that it is easy to separate its effects into each coordinate direction. For example, the above 3×3 matrix is the same as the combined effects of the following three.

$\begin{bmatrix} 2 & 0 & 0 \\ 0 & 1 & 0 \\ 0 & 0 & 1 \end{bmatrix}$. Stretch by a factor of two in the x_1 direction.

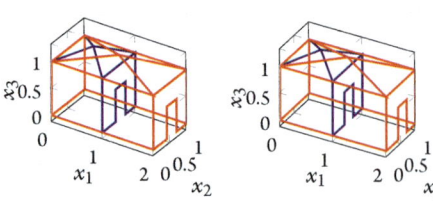

$\begin{bmatrix} 1 & 0 & 0 \\ 0 & 2/3 & 0 \\ 0 & 0 & 1 \end{bmatrix}$. Squash through 'stretching' by a factor of $2/3$ in the x_2 direction.

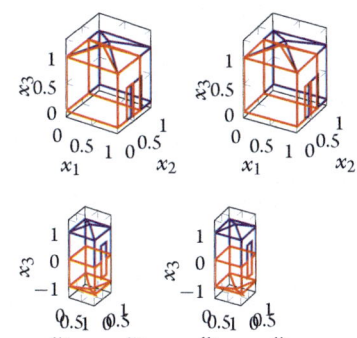

$\begin{bmatrix} 1 & 0 & 0 \\ 0 & 1 & 0 \\ 0 & 0 & -1 \end{bmatrix}$. Reflect in the vertical x_3 direction.

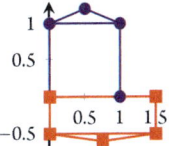

Example 3.2.32 From the illustration to the right, estimate the diagonal matrix that transforms the blue unit square to the red rectangle.

Solution: In the illustration, the horizontal is stretched by a factor of $3/2$, whereas the vertical is squashed by 'stretching' with a factor of $1/2$, as well as being reflected (minus sign). Hence the matrix is $\text{diag}(\frac{3}{2}, -\frac{1}{2}) = \begin{bmatrix} 3/2 & 0 \\ 0 & -1/2 \end{bmatrix}$. ☐

Activity 3.2.33 Which of the following diagrams represents the transformation from the (blue) unit square to the (red) rectangle by the matrix $\text{diag}(-1.3, 0.7)$?

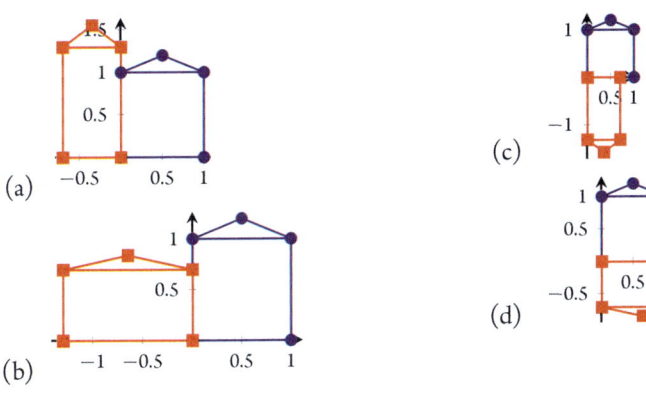

Some diagonal matrices rotate Now consider the transformation of multiplying by matrix $\begin{bmatrix} -1 & 0 \\ 0 & -1 \end{bmatrix}$: the two reflections of this diagonal matrix, the two -1s, have the same effect as one rotation, here by $180°$, as shown to the right. Matrices that rotate are incredibly useful, and this is the topic of the next Section 3.2.3.

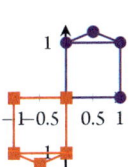

Sketch convenient coordinates

This optional subsubsection is a preliminary to diagonalization.

One of the fundamental principles of applying mathematics in science and engineering is that the real world—nature—does its thing irrespective of our mathematical description. Hence we often simplify our mathematical description of real world applications by choosing a coordinate system to suit its nature. That is, although this book (almost) always draws the x or x_1 axis horizontally, and the y or x_2 axis vertically, in applications it is often better to draw the axes in some other directions—directions that are convenient for the application. This example illustrates the principle.

Example 3.2.34 Consider the transformation shown to the right (it might arise from the deformation of some material and we need to know the internal stretching and shrinking to predict failure). The drawing has no coordinate axes shown because it is supposed to be some transformation in nature. Now we impose on nature our mathematical description. Draw approximate coordinate axes, with origin at the common point at the lower-left corner, so the transformation becomes that of the diagonal matrix $\mathrm{diag}(\frac{1}{2}, 2) = \begin{bmatrix} 1/2 & 0 \\ 0 & 2 \end{bmatrix}$.

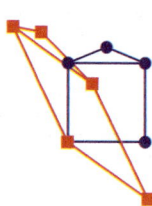

Solution: From the diagonal matrix we first look for a direction in which the transformation squashes—'stretching' by a factor of $1/2$: from the marginal graph, this direction must be towards the top-right (or bottom-left). Second, from the diagonal matrix we look for a direction in which the transformation stretches by a factor of two: from the above right picture this direction must be aligned to the top-left (or bottom-right). Because the top-right corner of the square is stretched a little in this second direction, the first direction must be aimed a little lower than this corner. Hence, coordinate axes that make the transformation the given diagonal matrix are as shown to the right. □

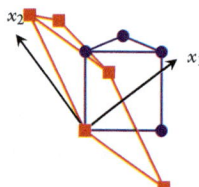

Example 3.2.35 Consider the transformation shown to the right. It has no coordinate axes shown because it is supposed to be some transformation in nature. Now impose on nature our mathematical description. Draw approximate coordinate axes, with origin at the common corner point, so the transformation becomes that of the diagonal matrix $\mathrm{diag}(3, -1) = \begin{bmatrix} 3 & 0 \\ 0 & -1 \end{bmatrix}$.

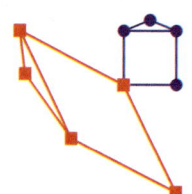

Solution: From the diagonal matrix we first look for a direction in which the transformation stretches by a factor of three: from the above right graph, this direction must be aligned along the diagonal top-left to bottom-right. Second, from the diagonal matrix we look for a direction in which the transformation reflects: from the above right picture this direction must be aligned along the top-right to bottom-left. Hence, coordinate axes that make the transformation the given diagonal matrix are as shown to the right. □

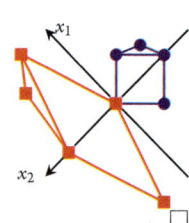

Finding such coordinate systems in which a given real world transformation is diagonal is important in science, engineering, and computer science. Systematic methods for such diagonalization are developed in Section 3.3 and Chapters 4 and 7. These rely on understanding the algebra and geometry of not only diagonal matrices, but also rotations, which is our next topic.

3.2.3 Orthogonal matrices rotate

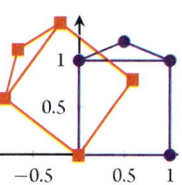

Whereas diagonal matrices stretch and squash, the so-called 'orthogonal matrices' represent just rotations (and/or reflection). This section starts by showing that multiplying by the 'orthogonal matrix' $\begin{bmatrix} 3/5 & -4/5 \\ 4/5 & 3/5 \end{bmatrix}$ rotates by $53.13°$ as shown to the right. Such orthogonal matrices are the best to compute with, such as in solving linear equations, since they all have $\texttt{rcond} = 1$. To see these and related marvellous properties, we must explore the effect of matrix multiplication on the geometry of lengths and angles.

Recall that the dot product determines lengths and angles   Section 1.3 introduced the dot product $\boldsymbol{u} \cdot \boldsymbol{v}$ between two vectors (Definition 1.3.2). The dot product is the same as the matrix product (Example 3.1.19): $\boldsymbol{u} \cdot \boldsymbol{v} = \boldsymbol{u}^T \boldsymbol{v} = \boldsymbol{v}^T \boldsymbol{u}$. Further (Theorem 1.3.17(a)), the length of a vector is $|\boldsymbol{v}| = \sqrt{\boldsymbol{v} \cdot \boldsymbol{v}}$. For two nonzero vectors, Theorem 1.3.5 defines the angle θ between the vectors via

$$\cos\theta = \frac{\boldsymbol{u} \cdot \boldsymbol{v}}{|\boldsymbol{u}||\boldsymbol{v}|}, \quad 0 \le \theta \le \pi.$$

If the two vectors are at right-angles, then the dot product is zero and the two vectors are termed orthogonal (Definition 1.3.19).

Orthogonal set of vectors

We need sets of orthogonal vectors (nonzero vectors that are all at right-angles to each other). (Recall that a **set** of objects is denoted by a list of the objects enclosed within braces, $\{\ldots\}$.) One example is the set of standard unit vectors $\{\boldsymbol{e}_1, \boldsymbol{e}_2, \ldots, \boldsymbol{e}_n\}$ that are aligned with the coordinate axes in $\mathbb{R}^n$.

Example 3.2.36   The set of two vectors $\{(3,4), (-8,6)\}$ shown to the right is an orthogonal set as the two vectors have dot product $= 3 \cdot (-8) + 4 \cdot 6 = -24 + 24 = 0$ and hence are orthogonal.

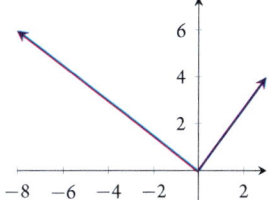

Example 3.2.37   Let vectors $\boldsymbol{q}_1 = (1,-2,2)$, $\boldsymbol{q}_2 = (2,2,1)$ and $\boldsymbol{q}_3 = (-2,1,2)$, illustrated in stereo to the right. Is $\{\boldsymbol{q}_1, \boldsymbol{q}_2, \boldsymbol{q}_3\}$ an orthogonal set?

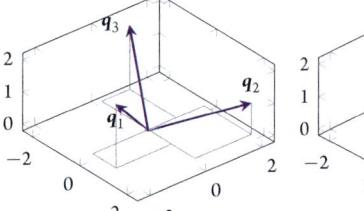
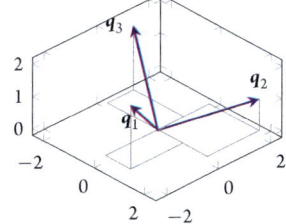

Solution: Yes, because all the pairwise dot products are zero: $q_1 \cdot q_2 = 2 - 4 + 2 = 0$; $q_1 \cdot q_3 = -2 - 2 + 4 = 0$; $q_2 \cdot q_3 = -4 + 2 + 2 = 0$. Hence every pair of vectors is orthogonal.

□

Definition 3.2.38 *A set of nonzero vectors* $\{q_1, q_2, \ldots, q_k\}$ *in* $\mathbb{R}^n$ *is called an* **orthogonal set** *if all pairs of distinct vectors in the set are orthogonal: that is,* $q_i \cdot q_j = 0$ *whenever* $i \neq j$ *for* $i, j = 1, 2, \ldots, k$. *A set of vectors in* $\mathbb{R}^n$ *is called an* **orthonormal set** *if it is an orthogonal set of unit vectors.*

A single nonzero vector always forms an orthogonal set. A single unit vector always forms an orthonormal set.

Example 3.2.39 Any set, or subset, of standard unit vectors in $\mathbb{R}^n$ (Definition 1.2.7) are an orthonormal set as they are all at right-angles (orthogonal), and all of length one.

Example 3.2.40 Let vectors $q_1 = (\frac{1}{3}, -\frac{2}{3}, \frac{2}{3})$, $q_2 = (\frac{2}{3}, \frac{2}{3}, \frac{1}{3})$, $q_3 = (-\frac{2}{3}, \frac{1}{3}, \frac{2}{3})$. Show that the set $\{q_1, q_2, q_3\}$ is an orthonormal set.

Solution: These vectors are all $1/3$ of the vectors in Example 3.2.37 and so are orthogonal. They all have length one: $|q_1|^2 = \frac{1}{9} + \frac{4}{9} + \frac{4}{9} = 1$; $|q_2|^2 = \frac{4}{9} + \frac{4}{9} + \frac{1}{9} = 1$; $|q_3|^2 = \frac{4}{9} + \frac{1}{9} + \frac{4}{9} = 1$. Hence $\{q_1, q_2, q_3\}$ is an orthonormal set in $\mathbb{R}^3$.

□

Activity 3.2.41 Which one of the following sets of vectors is *not* an orthogonal set?

(a) $\{(-2, 3), (6, 4)\}$

(b) $\{i, k\}$

(c) $\{(-5, 4)\}$

(d) $\{(2, 3), (4, -1)\}$

Orthogonal matrices

Now let's see how orthonormal vectors form a so-called orthogonal matrix and underlie its marvellous properties.

Example 3.2.42 Example 3.2.36 showed $\{(3, 4), (-8, 6)\}$ is an orthogonal set. The vectors have lengths 5 and 10, respectively, so dividing each by their length means that $\{(\frac{3}{5}, \frac{4}{5}), (-\frac{4}{5}, \frac{3}{5})\}$ is an orthonormal set. Form the matrix with these two vectors as its columns: $Q = \begin{bmatrix} 3/5 & -4/5 \\ 4/5 & 3/5 \end{bmatrix}$.
Then consider

$$Q^T Q = \begin{bmatrix} \frac{3}{5} & \frac{4}{5} \\ -\frac{4}{5} & \frac{3}{5} \end{bmatrix} \begin{bmatrix} \frac{3}{5} & -\frac{4}{5} \\ \frac{4}{5} & \frac{3}{5} \end{bmatrix} = \begin{bmatrix} \frac{9+16}{25} & \frac{-12+12}{25} \\ \frac{-12+12}{25} & \frac{16+9}{25} \end{bmatrix} = \begin{bmatrix} 1 & 0 \\ 0 & 1 \end{bmatrix}.$$

Similarly $QQ^T = I_2$. Consequently, the transpose Q^T is here the inverse of Q (Definition 3.2.2). The transpose being the inverse is no accident here.

Also no accident is that multiplication by this Q gives the rotation illustrated at the start of this section (Section 3.2.3).

Definition 3.2.43 *(orthogonal matrices)* *A square $n \times n$ matrix Q is called an* **orthogonal** **matrix** *if $Q^T Q = I_n$. Because of its special properties (Theorem 3.2.48), multiplication by an orthogonal matrix is called a* **rotation and/or reflection;**[11] *for brevity, and depending upon the circumstances, it may be called just a* **rotation** *or just a* **reflection.**

Activity 3.2.44 For which of the following values of p is the matrix $Q = \begin{bmatrix} \frac{1}{2} & p \\ -p & \frac{1}{2} \end{bmatrix}$ orthogonal?

(a) $p = \sqrt{3}/2$

(b) some other value

(c) $p = -1/2$

(d) $p = 3/4$

Example 3.2.45 In the following equation, check that the matrix is orthogonal. Use its orthogonality to solve the equation $Qx = b$:

$$\begin{bmatrix} \frac{1}{3} & -\frac{2}{3} & \frac{2}{3} \\ \frac{2}{3} & \frac{2}{3} & \frac{1}{3} \\ -\frac{2}{3} & \frac{1}{3} & \frac{2}{3} \end{bmatrix} x = \begin{bmatrix} 0 \\ 2 \\ -1 \end{bmatrix}.$$

The stereo pair to the right illustrates the rotation of the unit cube under multiplication by this matrix Q: every point x in the (blue) unit cube (with 'roof and door'), is mapped to the point Qx to form the (red) result.

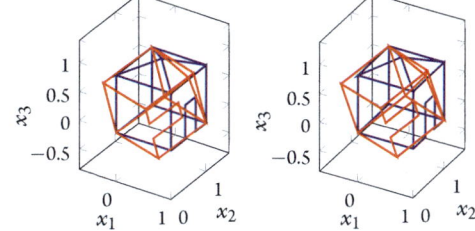

Solution: In MATLAB/Octave (recall the single quote (prime) gives the transpose, Table 3.1),

```
Q= [1,-2,2;2,2,1;-2,1,2]/3
Q'*Q
```

Since the product $Q^T Q$ is I_3, the matrix is orthogonal. Multiplying by Q^T both sides of the equation $Qx = b$ gives $Q^T Qx = Q^T b$; that is, $I_3 x = Q^T b$, equivalently, $x = Q^T b$. Here,

```
x=Q' * [0;2;-1]
```

gives the solution $x = (2, 1, 0)$. □

---

[11] Although herein we term multiplication by an orthogonal matrix as a 'rotation', in three or more dimensions it generally is not a single rotation about a single axis. Instead, generally the 'rotation' characterized by any one orthogonal matrix may be composed of a sequence of several rotations about different axes—each axis with a different orientation.

Example 3.2.46 Given the matrix is orthogonal, solve the linear equation

$$
\begin{bmatrix}
\frac{1}{2} & \frac{1}{2} & \frac{1}{2} & \frac{1}{2} \\
\frac{1}{2} & -\frac{1}{2} & \frac{1}{2} & -\frac{1}{2} \\
\frac{1}{2} & -\frac{1}{2} & -\frac{1}{2} & \frac{1}{2} \\
\frac{1}{2} & \frac{1}{2} & -\frac{1}{2} & -\frac{1}{2}
\end{bmatrix} \boldsymbol{x} =
\begin{bmatrix} 1 \\ -1 \\ 1 \\ 3 \end{bmatrix}.
$$

Solution: Denote the matrix by Q. Given the matrix Q is orthogonal, we know $Q^{\mathsf{T}}Q = I_4$. Just multiply the equation $Q\boldsymbol{x} = \boldsymbol{b}$ by the transpose Q^{T} to deduce $Q^{\mathsf{T}}Q\boldsymbol{x} = Q^{\mathsf{T}}\boldsymbol{b}$, that is, $\boldsymbol{x} = Q^{\mathsf{T}}\boldsymbol{b}$. Here this equation gives the solution to be

$$
\boldsymbol{x} =
\begin{bmatrix}
\frac{1}{2} & \frac{1}{2} & \frac{1}{2} & \frac{1}{2} \\
\frac{1}{2} & -\frac{1}{2} & -\frac{1}{2} & \frac{1}{2} \\
\frac{1}{2} & \frac{1}{2} & -\frac{1}{2} & -\frac{1}{2} \\
\frac{1}{2} & -\frac{1}{2} & \frac{1}{2} & -\frac{1}{2}
\end{bmatrix}
\begin{bmatrix} 1 \\ -1 \\ 1 \\ 3 \end{bmatrix} =
\begin{bmatrix} 2 \\ 2 \\ -2 \\ 0 \end{bmatrix}.
$$

Example 3.2.47 The graph to the right shows a rotation of the unit square. From the graph, estimate roughly the matrix Q such that multiplication by Q performs the rotation. Confirm that your estimated matrix is orthogonal (approximately).

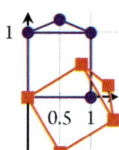

Solution: Consider what happens to the standard unit vectors as shown in the graph: multiplying Q by $(1,0)$ gives $(0.5, -0.9)$ roughly (to one decimal place); whereas multiplying Q by $(0,1)$ gives $(0.9, 0.5)$ roughly. To do this, the matrix Q must have these two vectors as its two columns, that is,

$$
Q \approx \begin{bmatrix} 0.5 & 0.9 \\ -0.9 & 0.5 \end{bmatrix}.
$$

Let's check what happens to the corner point $(1,1)$: $Q(1,1) \approx (1.4, -0.4)$, which looks approximately correct. To confirm orthogonality of Q, find

$$
Q^{\mathsf{T}}Q = \begin{bmatrix} 0.5 & -0.9 \\ 0.9 & 0.5 \end{bmatrix}\begin{bmatrix} 0.5 & 0.9 \\ -0.9 & 0.5 \end{bmatrix} = \begin{bmatrix} 1.06 & 0 \\ 0 & 1.06 \end{bmatrix} \approx I_2,
$$

and so Q is approximately orthogonal.

Because orthogonal matrices represent rotations, they arise frequently in engineering and scientific mechanics of bodies. Also, the ease in solving equations with orthogonal matrices puts orthogonal matrices at the heart of coding and decoding photographs (jpeg), videos (mpeg), signals (Fourier transforms), and so on. Furthermore, an extension of orthogonal matrices to complex valued matrices, the so-called unitary matrices, is at the core of quantum physics and quantum computing. Moreover, the next Section 3.3 establishes that orthogonal matrices express the orientation of the action of *every* matrix and hence are a vital component of solving linear equations in general. But to utilize orthogonal matrices across the wide range of applications we need to establish the following properties.

Theorem 3.2.48 *For every square matrix Q, the following statements are equivalent:*

(a) *Q is an orthogonal matrix;*

(b) *the column vectors of Q form an orthonormal set;*

(c) *Q is invertible and $Q^{-1} = Q^T$;*

(d) *Q^T is an orthogonal matrix;*

(e) *the row vectors of Q form an orthonormal set;*

(f) *multiplication by Q preserves all lengths and angles (and hence corresponds to our intuition of a rotation and/or reflection).*

Proof. Write $n \times n$ matrix $Q = \begin{bmatrix} q_1 & q_2 & \cdots & q_n \end{bmatrix}$ in terms of its n columns $q_1, q_2, \ldots, q_n$. So then the transpose Q^T has the vectors $q_1, q_2, \ldots, q_n$ as its rows—the ith row being the row vector q_i^T.

3.2.48(a) $\Longleftrightarrow$ 3.2.48(b) Consider

$$
Q^T Q = \begin{bmatrix} q_1^T \\ q_2^T \\ \vdots \\ q_n^T \end{bmatrix} \begin{bmatrix} q_1 & q_2 & \cdots & q_n \end{bmatrix}
$$

$$
= \begin{bmatrix} q_1^T q_1 & q_1^T q_2 & \cdots & q_1^T q_n \\ q_2^T q_1 & q_2^T q_2 & \cdots & q_2^T q_n \\ \vdots & \vdots & \ddots & \vdots \\ q_n^T q_1 & q_n^T q_2 & \cdots & q_n^T q_n \end{bmatrix}
$$

$$
= \begin{bmatrix} q_1 \cdot q_1 & q_1 \cdot q_2 & \cdots & q_1 \cdot q_n \\ q_2 \cdot q_1 & q_2 \cdot q_2 & \cdots & q_2 \cdot q_n \\ \vdots & \vdots & \ddots & \vdots \\ q_n \cdot q_1 & q_n \cdot q_2 & \cdots & q_n \cdot q_n \end{bmatrix}
$$

Matrix Q is orthogonal if and only if this product is the identity (Definition 3.2.43) which is if and only if $q_i \cdot q_j = 0$ for $i \neq j$ and $|q_i|^2 = q_i \cdot q_i = 1$, that is, if and only if the columns q_i are orthonormal (Definition 3.2.38).

3.2.48(b) $\Longrightarrow$ 3.2.48(c) First, consider the homogeneous system $Q^T x = 0$ for x in $\mathbb{R}^n$. We establish $x = 0$ is the only solution. The system $Q^T x = 0$, written in terms of the orthonormal columns of Q, is

$$
\begin{bmatrix} q_1^T \\ q_2^T \\ \vdots \\ q_n^T \end{bmatrix} x = \begin{bmatrix} q_1^T x \\ q_2^T x \\ \vdots \\ q_n^T x \end{bmatrix} = \begin{bmatrix} q_1 \cdot x \\ q_2 \cdot x \\ \vdots \\ q_n \cdot x \end{bmatrix} = \begin{bmatrix} 0 \\ 0 \\ \vdots \\ 0 \end{bmatrix}.
$$

Since the dot products are all zero, either $x = 0$ or x is orthogonal (at right-angles) to all of the n orthonormal vectors $\{q_1, q_2, \ldots, q_n\}$. In $\mathbb{R}^n$ we cannot have $(n+1)$ nonzero vectors all at right-angles to each other (Theorem 1.3.25), consequently $x = 0$ is the only possibility as the solution of $Q^T x = 0$.

Second, let $n \times n$ matrix $X = I_n - QQ^T$. Pre-multiply by Q^T: $Q^T X = Q^T(I_n - QQ^T) = Q^T I_n - (Q^T Q)Q^T = Q^T I_n - I_n Q^T = Q^T - Q^T = O_n$. That is, $Q^T X = O_n$. But each column of $Q^T X = O_n$ is of the form $Q^T x = 0$ which we know requires $x = 0$, hence each column of X is zero, and so $X = O_n$. Then $X = I_n - QQ^T = O_n$ which rearranged gives $QQ^T = I_n$. Put $QQ^T = I_n$ together with $Q^T Q = I_n$ (Definition 3.2.43), then by Definition 3.2.2 Q is invertible with inverse Q^T.

3.2.48(c) $\implies$ **3.2.48(a), 3.2.48(d)** Part 3.2.48(c) asserts Q is invertible with inverse Q^T: by Definition 3.2.2 of the inverse, $Q^T Q = QQ^T = I_n$. Since $Q^T Q = I_n$, matrix Q is orthogonal.
Since $I_n = QQ^T = (Q^T)^T Q^T$, by Definition 3.2.43 Q^T is orthogonal.

3.2.48(d) $\iff$ **3.2.48(e)** The proof is similar to that for 3.2.48(a) $\iff$ 3.2.48(b), but for the rows of Q and $QQ^T = I_n$.

3.2.48(e) $\implies$ **3.2.48(c)** Similar to that for 3.2.48(b) $\implies$ 3.2.48(c), but for the rows of Q, $Qx = 0$ and $X = I_n - Q^T Q$.

3.2.48(a) $\implies$ **3.2.48(f)** We prove that multiplication by orthogonal Q preserves all lengths and angles, as illustrated in Examples 3.2.45 and 3.2.47, by comparing the properties of transformed vectors Qu with the properties of the original u. For any vectors u, v in $\mathbb{R}^n$, consider the dot product $(Qu) \cdot (Qv) = (Qu)^T Qv = u^T Q^T Qv = u^T I_n v = u^T v = u \cdot v$. Now let's use this identity that $(Qu) \cdot (Qv) = u \cdot v$.

- Firstly, the length $|Qu| = \sqrt{(Qu) \cdot (Qu)} = \sqrt{u \cdot u}$ by the identity. But $\sqrt{u \cdot u} = |u|$ so the length $|u| = |Qu|$ is preserved. Similarly for v.

- Secondly, let θ be the angle between u and v and θ' be the angle between Qu and Qv (recall $0 \le$ angle $\le \pi$), then, via the identity,

$$\cos \theta' = \frac{(Qu) \cdot (Qv)}{|Qu||Qv|} = \frac{u \cdot v}{|u||v|} = \cos \theta.$$

Since $\cos \theta' = \cos \theta$ and the cosine is a one-one function for $0 \le$ angles $\le \pi$, $\theta' = \theta$ so all angles are preserved.

3.2.48(f) $\implies$ **3.2.48(b)** Look at the consequences of matrix Q preserving all lengths and angles when applied to the standard unit vectors $e_1, e_2, \ldots, e_n$. Observe $Qe_j = q_j$, the jth column of matrix Q. Then for all j, the length of the jth column $|q_j| = |Qe_j| = |e_j| = 1$ by the preservation of the length of the standard unit vector. Also, for all $i \ne j$ the dot product of columns $q_i \cdot q_j = |q_i||q_j| \cos \theta' = 1 \cdot 1 \cdot \cos \frac{\pi}{2} = 0$ where θ' is the angle between q_i and q_j, which is the angle between e_i and e_j by preservation, namely the angle $\frac{\pi}{2}$. That is, the columns of Q form an orthonormal set. $\qquad\square$

Another important property, proved by Exercise 3.2.21, is that the product of orthogonal matrices is also an orthogonal matrix.

Example 3.2.49 Show that these matrices are orthogonal and hence write down their inverses:

$$\begin{bmatrix} 0 & 0 & 1 \\ 1 & 0 & 0 \\ 0 & 1 & 0 \end{bmatrix}, \quad \begin{bmatrix} \cos\theta & -\sin\theta \\ \sin\theta & \cos\theta \end{bmatrix}.$$

Solution: For the first matrix, by inspection each column is a unit vector, and each column is orthogonal to each other. Since the matrix has orthonormal columns, the matrix is orthogonal (Theorem 3.2.48(b)). Its inverse is then the transpose (Theorem 3.2.48(c))

$$\begin{bmatrix} 0 & 1 & 0 \\ 0 & 0 & 1 \\ 1 & 0 & 0 \end{bmatrix}.$$

For the second matrix the two columns are unit vectors as $|(\cos\theta, \sin\theta)|^2 = \cos^2\theta + \sin^2\theta = 1$ and $|(-\sin\theta, \cos\theta)|^2 = \sin^2\theta + \cos^2\theta = 1$. The two columns are orthogonal as the dot product $(\cos\theta, \sin\theta) \cdot (-\sin\theta, \cos\theta) = -\cos\theta \sin\theta + \sin\theta \cos\theta = 0$. Since the matrix has orthonormal columns, the matrix is orthogonal (Theorem 3.2.48(b)). Its inverse is then the transpose (Theorem 3.2.48(c))

$$\begin{bmatrix} \cos\theta & \sin\theta \\ -\sin\theta & \cos\theta \end{bmatrix}. \qquad \square$$

Example 3.2.50 The following graphs illustrate the transformation of the unit square through multiplying by some different matrices. Using Theorem 3.2.48(f), which transformations appear to be that of multiplying by an orthogonal matrix?

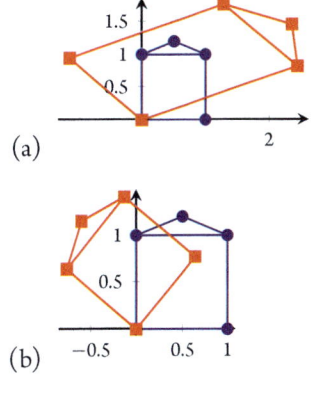

(a)

(b)

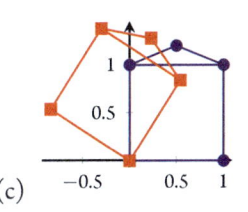

(c)

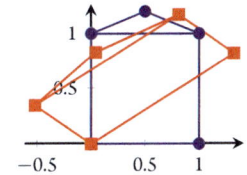

(d)

Solution:

(a) No—the square is stretched and angles changed.

(b) Yes—the square is just rotated.

(c) Yes—the square is rotated and reflected.

(d) No—the square is squashed and angles changed. $\qquad \square$

Activity 3.2.51 The following graphs illustrate the transformation of the unit square through multiplying by some different matrices.

- Which transformation appears to be that of multiplying by an orthogonal matrix?

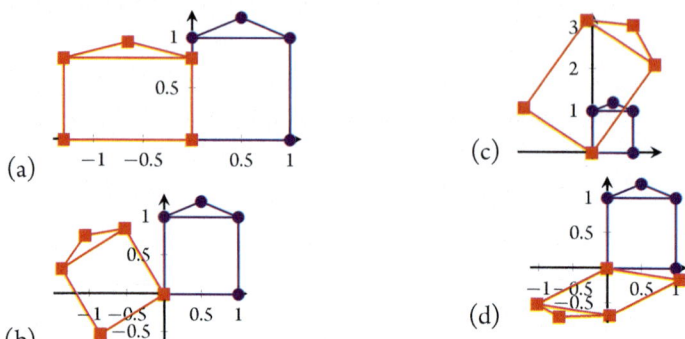

- Further, which of the above transformations appears to be that of multiplying by a diagonal matrix?

Example 3.2.52 The following stereo pairs illustrate the transformation of the unit cube through multiplying by some different matrices: using Theorem 3.2.48(f), which transformations appear to be that of multiplying by an orthogonal matrix?

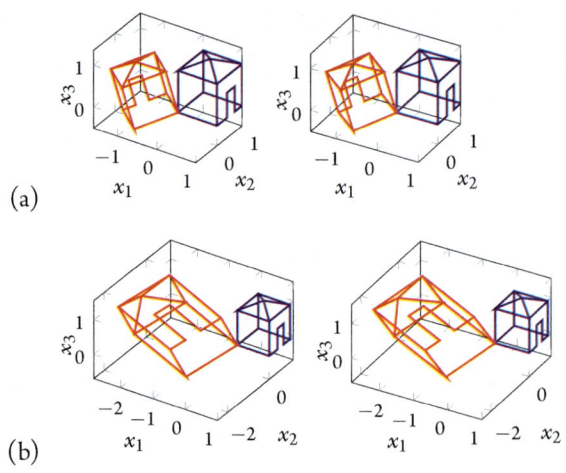

Solution:

    (a) Yes—the cube is just rotated.
    (b) No—the cube is stretched and angles changed.   □

3.2.4 Exercises

Exercise 3.2.1 By direct multiplication, both ways, for each of the following pairs of matrices, confirm whether matrix B is an inverse of matrix A, or not.

(a) $A = \begin{bmatrix} 0 & -4 \\ 4 & 4 \end{bmatrix}, B = \begin{bmatrix} 1/4 & 1/4 \\ -1/4 & 0 \end{bmatrix}$

(b) $A = \begin{bmatrix} -1 & 1 \\ -5 & -1 \end{bmatrix}, B = \begin{bmatrix} -1/6 & -1/6 \\ 5/6 & -1/6 \end{bmatrix}$

(c) $A = \begin{bmatrix} -2 & 4 & 2 \\ 1 & 1 & -1 \\ 1 & -1 & 1 \end{bmatrix}, B = \begin{bmatrix} 0 & 1/2 & 1/2 \\ 1/6 & 1/3 & 0 \\ 1/6 & -1/6 & 1/2 \end{bmatrix}$

(d) $A = \begin{bmatrix} -3 & -3 & 3 \\ 4 & 3 & -3 \\ -2 & -1 & 4 \end{bmatrix}, B = \begin{bmatrix} 1 & 1 & 0 \\ -1 & -2/3 & 1/3 \\ 2/9 & 1/3 & 1/3 \end{bmatrix}$

(e) $A = \begin{bmatrix} -1 & 3 & 4 & -1 \\ 2 & 2 & -2 & 0 \\ 1 & 2 & -2 & 0 \\ 4 & 2 & 4 & -1 \end{bmatrix}, B = \begin{bmatrix} 0 & 1 & -1 & 0 \\ 1 & 5 & -5 & -1 \\ 1 & 11/2 & -6 & -1 \\ 6 & 36 & -38 & -7 \end{bmatrix}$, use MATLAB/Octave

(f) $A = \begin{bmatrix} 1 & -7 & -4 & 3 \\ 0 & 2 & 1 & -1 \\ 0 & -2 & -3 & 2 \\ 3 & -2 & -4 & 1 \end{bmatrix}, B = \begin{bmatrix} -2 & -7 & -1 & 1 \\ -1 & -8/3 & 0 & 1/3 \\ -2 & -22/3 & -1 & 2/3 \\ -4 & -41/3 & -1 & 4/3 \end{bmatrix}$, use MATLAB/Octave

Exercise 3.2.2 Use the direct formula of Theorem 3.2.7 to calculate the inverse, when it exists, of each of the following 2×2 matrices.

(a) $\begin{bmatrix} -2 & 2 \\ -1 & 4 \end{bmatrix}$

(b) $\begin{bmatrix} -5 & -10 \\ -1 & -2 \end{bmatrix}$

(c) $\begin{bmatrix} -2 & -4 \\ 5 & 2 \end{bmatrix}$

(d) $\begin{bmatrix} -3 & 2 \\ -1 & -2 \end{bmatrix}$

Exercise 3.2.3 Given the inverses of Exercise 3.2.1, solve each of the following systems of linear equations with a matrix-vector multiplication (Theorem 3.2.10).

(a) $\begin{cases} -4y = 1 \\ 4x + 4y = -5 \end{cases}$

(b) $\begin{cases} m - x = 1 \\ -m - 5x = -1 \end{cases}$

(c) $\begin{cases} 2p - 2q + 4r = -1 \\ -p + q + r = -2 \\ p + q - r = 1 \end{cases}$

(d) $\begin{cases} -x_1 + 3x_2 + 4x_3 - x_4 = 0 \\ 2x_1 + 2x_2 - 2x_3 = -1 \\ x_1 + 2x_2 - 2x_3 = 3 \\ 4x_1 + 2x_2 + 4x_3 - x_4 = -5 \end{cases}$

(e) $\begin{cases} p - 7q - 4r + 3s = -1 \\ 2q + r - s = -5 \\ -2q - 3r + 2s = 3 \\ 3p - 2q - 4r + s = -1 \end{cases}$

Exercise 3.2.4 Given the following information about solutions of systems of linear equations, write down if the matrix associated with each system is invertible, or not, or there is not enough given information to decide. Give reasons.

(a) The general solution is $(-2, 1, 2, 0, 2)$.

(b) A solution of a system is $(2.4, -2.8, -3.6, -2.2, -3.8)$.

(c) A solution of a homogeneous system is $(0.8, 0.4, -2.3, 2.5)$.

(d) The general solution of a system is $(4, 1, 0, 2)t$ for every t.

(e) The general solution of a homogeneous system is $(0, 0, 0, 0)$.

(f) A solution of a homogeneous system is $(0, 0, 0, 0, 0)$.

Exercise 3.2.5 Use MATLAB/Octave to generate some random matrices of a suitable size of your choice (see Table 3.1). Also record some 'random' integer exponents of your choice. Then confirm the properties of inverse matrices given by Theorem 3.2.13. For the purposes of this exercise, use the MATLAB/Octave function inv(A), which computes the inverse of the matrix A if it exists (as commented, remember that computing the inverse of a matrix is generally inappropriate—the inverse is primarily a theoretical device—this exercise only computes the inverse for educational purposes). Record all your commands and the output from MATLAB/Octave.

Exercise 3.2.6 Consider Theorem 3.2.13 on the properties of the inverse. Invoking properties of matrix operations from Section 3.1.3,

(a) prove property 3.2.13(b) using associativity, and

(b) prove property 3.2.13(d) using the transpose.

Exercise 3.2.7 Recall the properties of the inverse, Theorem 3.2.13, and matrix operations, Section 3.1.3. Use these properties to prove that for every $n \times n$ invertible matrix A, and every pair of $n \times m$ matrices U and V such that $I_m + V^{\mathsf{T}} A^{-1} U$ is invertible, then the matrix $A + UV^{\mathsf{T}}$ is invertible and[12]

$$(A + UV^{\mathsf{T}})^{-1} = A^{-1} - (A^{-1}U)(I_m + V^{\mathsf{T}} A^{-1} U)^{-1}(V^{\mathsf{T}} A^{-1}).$$

Hint: directly check Definition 3.2.2.

Exercise 3.2.8 Using the inverses identified in Exercise 3.2.1, and matrix multiplication, calculate the following matrix powers.

(a) $\begin{bmatrix} 0 & -4 \\ 4 & 4 \end{bmatrix}^{-2}$

(b) $\begin{bmatrix} -1/6 & -1/6 \\ 5/6 & -1/6 \end{bmatrix}^{-4}$

(c) $\begin{bmatrix} 0 & 1/2 & 1/2 \\ 1/6 & 1/3 & 0 \\ 1/6 & -1/6 & 1/2 \end{bmatrix}^{-2}$

(d) $\begin{bmatrix} -1 & 3 & 4 & -1 \\ 2 & 2 & -2 & 0 \\ 1 & 2 & -2 & 0 \\ 4 & 2 & 4 & -1 \end{bmatrix}^{-2}$ use MATLAB/Octave

---

[12] This is the Woodbury generalization of the Sherman–Morrison formula. It is used to update matrices while searching for optima and solutions.

Exercise 3.2.9 Which of the following matrices are diagonal? For those that are diagonal, write down how they may be represented with the diag function (algebraic, not MATLAB/Octave).

(a) $\begin{bmatrix} 9 & 0 & 0 \\ 0 & -5 & 0 \\ 0 & 0 & 4 \end{bmatrix}$

(b) $\begin{bmatrix} 0 & 0 & 1 \\ 0 & 2 & 0 \\ -2 & 0 & 0 \end{bmatrix}$

(c) $\begin{bmatrix} -5 & 0 & 0 & 0 & 0 \\ 0 & 1 & 0 & 0 & 0 \\ 0 & 0 & 9 & 0 & 0 \\ 0 & 0 & 0 & 1 & 0 \\ 0 & 0 & 0 & 0 & 0 \end{bmatrix}$

(d) $\begin{bmatrix} 6 & 0 & 0 & 0 \\ 0 & 1 & 0 & -9 \\ 0 & 0 & 0 & 0 \\ 0 & 0 & 0 & 0 \end{bmatrix}$

(e) $\begin{bmatrix} 0 & 0 & 0 \\ 0 & -1 & 0 \\ 0 & 0 & -5 \\ 0 & 0 & 0 \\ 0 & 0 & 0 \end{bmatrix}$

(f) $\begin{bmatrix} 1 & 0 \\ 0 & 1 \\ 0 & 0 \\ 2 & 0 \\ 0 & 0 \end{bmatrix}$

Exercise 3.2.10 Write down the individual algebraic equations represented by each of the following diagonal matrix-vector equations. Hence, where possible solve each system.

(a) $\begin{bmatrix} -4 & 0 & 0 & 0 & 0 \\ 0 & 1 & 0 & 0 & 0 \\ 0 & 0 & -2 & 0 & 0 \\ 0 & 0 & 0 & 1 & 0 \\ 0 & 0 & 0 & 0 & -3 \end{bmatrix} \begin{bmatrix} x_1 \\ x_2 \\ x_3 \\ x_4 \\ x_5 \end{bmatrix} = \begin{bmatrix} 6 \\ -1 \\ -4 \\ -2 \\ 4 \end{bmatrix}$

(b) $\begin{bmatrix} 0 & 0 & 0 & 0 \\ 0 & 4 & 0 & 0 \\ 0 & 0 & -1 & 0 \end{bmatrix} \begin{bmatrix} x_1 \\ x_2 \\ x_3 \\ x_4 \end{bmatrix} = \begin{bmatrix} -8 \\ 3 \\ -1 \end{bmatrix}$

(c) $\begin{bmatrix} 1 & 0 & 0 & 0 \\ 0 & -3 & 0 & 0 \\ 0 & 0 & -3 & 0 \\ 0 & 0 & 0 & 0 \end{bmatrix} \begin{bmatrix} p \\ q \\ r \\ s \end{bmatrix} = \begin{bmatrix} -2 \\ -6 \\ 8 \\ 0 \end{bmatrix}$

(d) $\begin{bmatrix} -1 & 0 & 0 & 0 \\ 0 & 3 & 0 & 0 \\ 0 & 0 & 0 & 0 \end{bmatrix} \begin{bmatrix} p \\ q \\ r \\ s \end{bmatrix} = \begin{bmatrix} -3 \\ 0 \\ 3 \end{bmatrix}$

Exercise 3.2.11 In each of the following illustrations, the unit square (blue, with 'roof') is transformed by a matrix multiplication to some shape (red). Which of these transformations appear to correspond to multiplication by a diagonal matrix? For those that are, estimate the elements of the diagonal matrix.

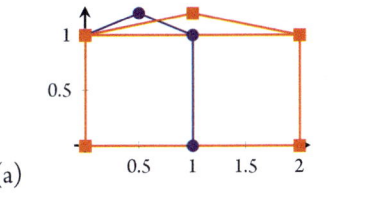

(a)

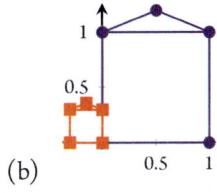

(b)

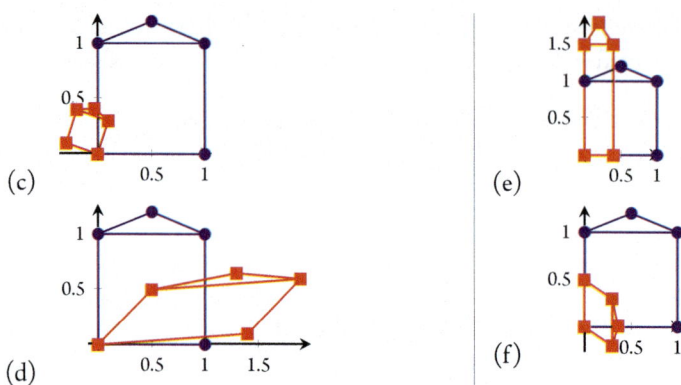

(c)

(d)

(e)

(f)

Exercise 3.2.12 In each of the following stereo illustrations, the unit cube (blue) is transformed by a matrix multiplication to some shape (red). Which of these transformations appear to correspond to multiplication by a diagonal matrix? For those that are, estimate the elements of the diagonal matrix.

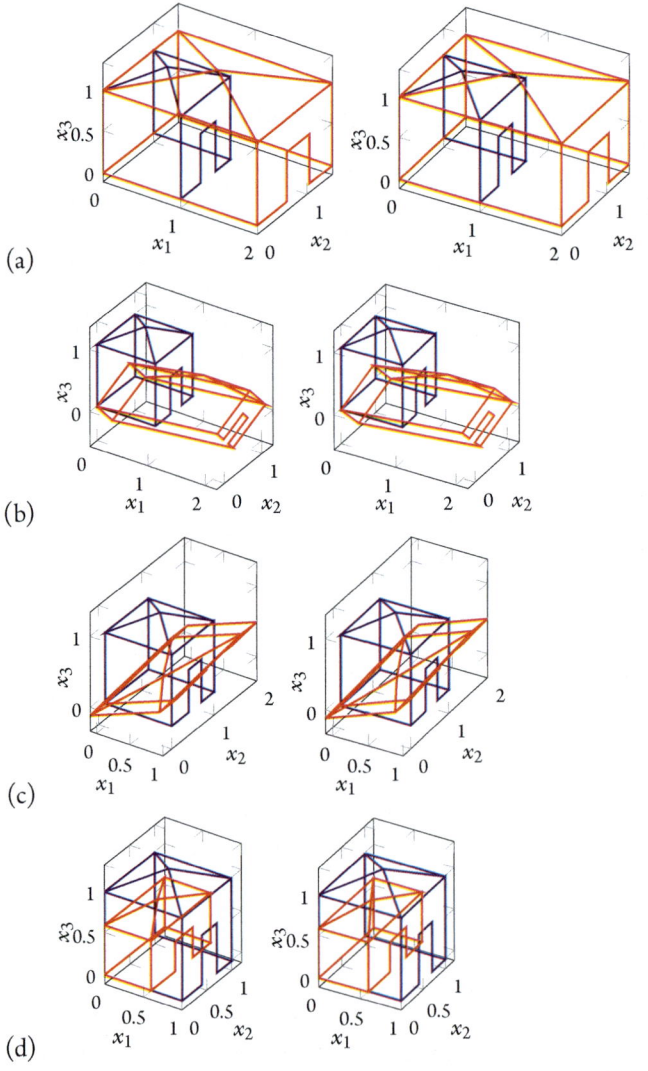

(a)

(b)

(c)

(d)

Exercise 3.2.13 Consider each of the transformations shown below that transform the from blue unit square to the red parallelogram. They each have no coordinate axes shown because it is supposed to be some transformation in nature. Now impose on nature our mathematical description. Draw approximate orthogonal coordinate axes, with origin at the common corner point, so the transformation becomes that of multiplication by the specified diagonal matrix.

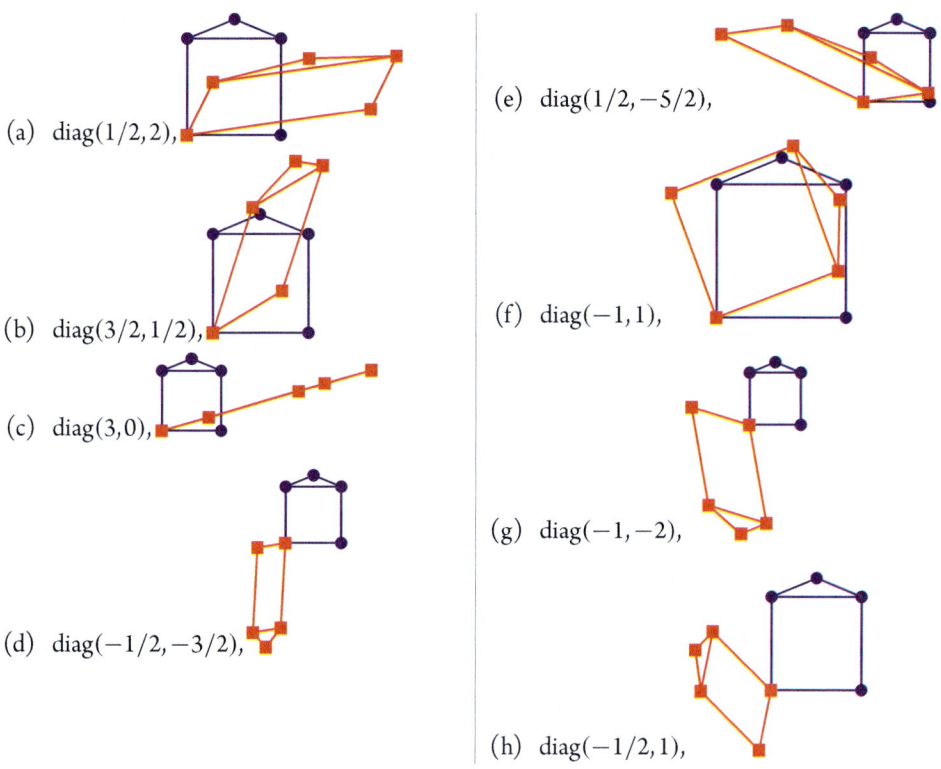

(a) $\mathrm{diag}(1/2,2)$,

(b) $\mathrm{diag}(3/2,1/2)$,

(c) $\mathrm{diag}(3,0)$,

(d) $\mathrm{diag}(-1/2,-3/2)$,

(e) $\mathrm{diag}(1/2,-5/2)$,

(f) $\mathrm{diag}(-1,1)$,

(g) $\mathrm{diag}(-1,-2)$,

(h) $\mathrm{diag}(-1/2,1)$,

Exercise 3.2.14 Use Theorem 3.2.10 to establish the converse part of Theorem 3.2.27: namely, that if a diagonal matrix is invertible, then all its diagonal entries are nonzero.

Exercise 3.2.15 Which of the following pairs of vectors appear to form an orthogonal set of two vectors? Which appear to form an orthonormal set of two vectors?

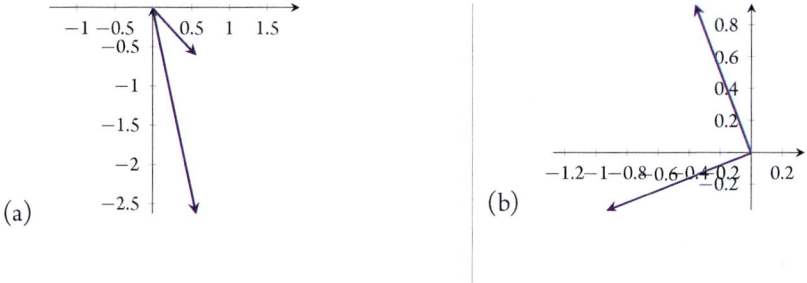

(a)

(b)

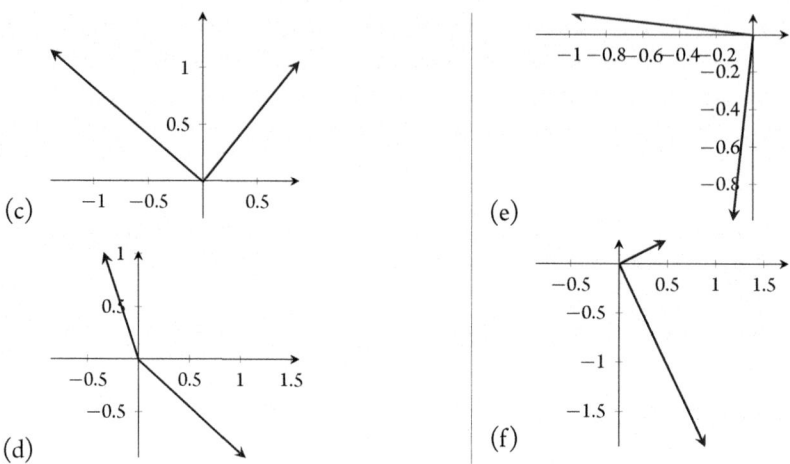

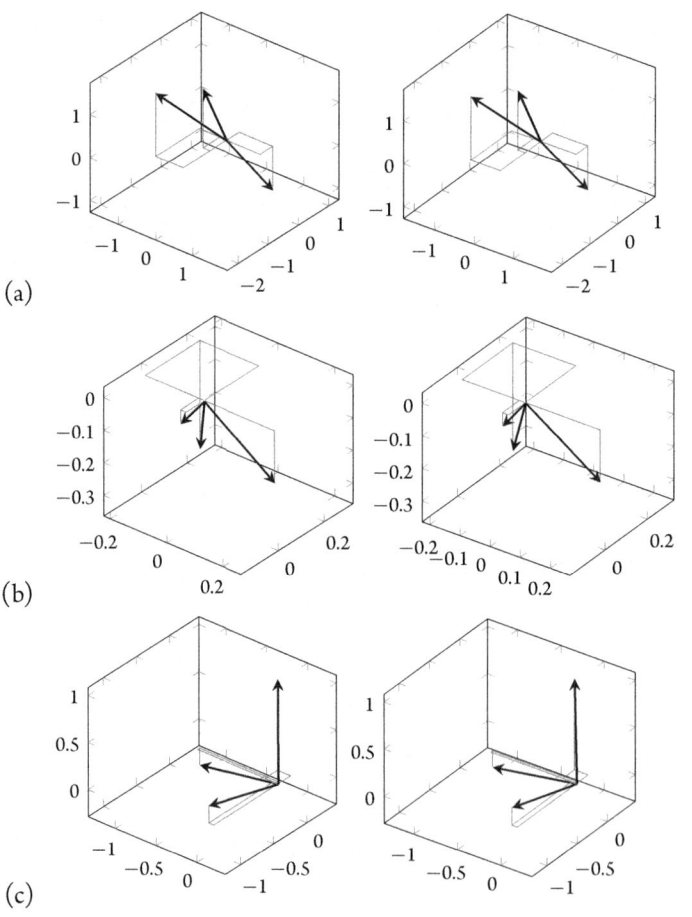

Exercise 3.2.16 Which of the following sets of vectors, drawn as stereo pairs, appear to form an orthogonal set? Which appear to form an orthonormal set?

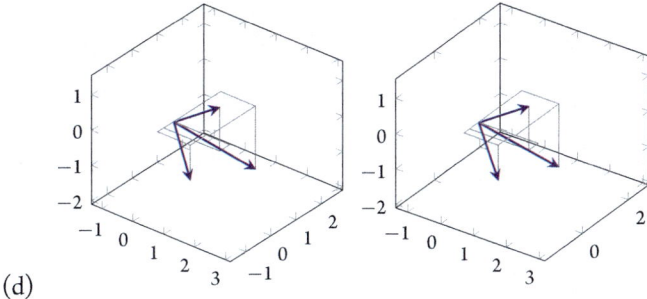

(d)

Exercise 3.2.17 Use dot products to determine which of the following sets of vectors are orthogonal sets. For the orthogonal sets, scale the vectors to form an orthonormal set.

(a) $\{(2,3,6), (3,-6,2), (6,2,-3)\}$

(b) $\{(6,3,2), (3,-6,2), (2,-3,6)\}$

(c) $\{(1,1,1,1), (1,1,-1,-1), (1,-1,-1,1), (1,-1,1,-1)\}$

(d) $\{(5,6,2,4), (-2,6,-5,-4), (6,-5,-4,2)\}$

Exercise 3.2.18 Using Definition 3.2.43 determine which of the following matrices are orthogonal matrices. For those matrices which are orthogonal, confirm Theorem 3.2.48(c).

(a) $\begin{bmatrix} \frac{5}{13} & \frac{12}{13} \\ -\frac{12}{13} & \frac{5}{13} \end{bmatrix}$

(b) $\begin{bmatrix} -3 & 4 \\ 4 & 3 \end{bmatrix}$

(c) $\begin{bmatrix} \frac{2}{7} & \frac{3}{7} & \frac{6}{7} \\ \frac{3}{7} & -\frac{6}{7} & \frac{2}{7} \\ \frac{6}{7} & \frac{2}{7} & -\frac{3}{7} \end{bmatrix}$

(d) $\begin{bmatrix} \frac{2}{11} & \frac{9}{11} \\ \frac{6}{11} & -\frac{6}{11} \\ \frac{9}{11} & \frac{2}{11} \end{bmatrix}$

(e) $\frac{1}{7} \begin{bmatrix} 1 & 4 & 4 & 4 \\ 4 & -5 & 2 & 2 \\ 4 & 2 & -5 & 2 \\ 4 & 2 & 2 & -5 \end{bmatrix}$

(f) $\begin{bmatrix} 0.2 & 0.4 & 0.4 \\ 0.4 & -0.2 & 0.8 \\ 0.4 & -0.8 & -0.2 \\ 0.8 & 0.4 & -0.4 \end{bmatrix}$

(g) $\frac{1}{6} \begin{bmatrix} 1 & 1 & 3 & 5 \\ -5 & -3 & 1 & 1 \\ 3 & -5 & -1 & 1 \\ 1 & -1 & 5 & 3 \end{bmatrix}$

(h) $\begin{bmatrix} 0.1 & 0.5 & 0.5 & 0.7 \\ 0.5 & -0.1 & -0.7 & 0.5 \\ 0.5 & 0.7 & -0.1 & -0.5 \\ 0.7 & -0.5 & 0.5 & -0.1 \end{bmatrix}$

Exercise 3.2.19 Each part gives an orthogonal matrix Q and two vectors u and v. For each part calculate the lengths of u and v, and the angle between u and v. Confirm that these are the same as the lengths of Qu and Qv, and the angle between Qu and Qv, respectively.

(a) $Q = \begin{bmatrix} 0 & -1 \\ 1 & 0 \end{bmatrix}, u = \begin{bmatrix} 3 \\ 4 \end{bmatrix}, v = \begin{bmatrix} 12 \\ 5 \end{bmatrix}$

(b) $Q = \begin{bmatrix} \frac{3}{5} & \frac{4}{5} \\ -\frac{4}{5} & \frac{3}{5} \end{bmatrix}, u = \begin{bmatrix} 7 \\ -4 \end{bmatrix}, v = \begin{bmatrix} 1 \\ 2 \end{bmatrix}$

(c) $Q = \frac{1}{11} \begin{bmatrix} 7 & 6 & 6 \\ 6 & 2 & -9 \\ 6 & -9 & 2 \end{bmatrix}, u = \begin{bmatrix} 0 \\ 1 \\ 1 \end{bmatrix}, v = \begin{bmatrix} 1 \\ 0 \\ 1 \end{bmatrix}$

(d) $Q = \begin{bmatrix} 0.1 & 0.3 & 0.3 & 0.9 \\ 0.3 & -0.1 & -0.9 & 0.3 \\ 0.3 & 0.9 & -0.1 & -0.3 \\ 0.9 & -0.3 & 0.3 & -0.1 \end{bmatrix}, u = \begin{bmatrix} 3 \\ 1 \\ 2 \\ 1 \end{bmatrix}, v = \begin{bmatrix} -2 \\ -1 \\ 0 \\ 2 \end{bmatrix}$

Exercise 3.2.20 Using one or other of the orthogonal matrices appearing in Exercise 3.2.19, solve each of the following systems of linear equations by a single matrix-vector multiplication.

(a) $\begin{cases} \frac{3}{5}x + \frac{4}{5}y = 5 \\ -\frac{4}{5}x + \frac{3}{5}y = 2 \end{cases}$

(b) $\begin{cases} 3x + 4y = 20 \\ -4x + 3y = 5 \end{cases}$

(c) $\begin{cases} 7a + 6b + 6c = 22 \\ 6a + 2b - 9c = 11 \\ 6a - 9b + 2c = -22 \end{cases}$

(d) $\begin{cases} z_1 + 3z_2 + 3z_3 + 9z_4 = 5 \\ 3z_1 - z_2 - 9z_3 + 3z_4 = 0 \\ 3z_1 + 9z_2 - z_3 - 3z_4 = -1 \\ 9z_1 - 3z_2 + 3z_3 - z_4 = -3 \end{cases}$

Exercise 3.2.21 *(product of orthogonal matrices)* Use Definition 3.2.43 to prove that if Q_1 and Q_2 are orthogonal matrices of the same size, then so is the product $Q_1 Q_2$. Consider $(Q_1 Q_2)^{\mathsf{T}}(Q_1 Q_2)$.

Exercise 3.2.22 Fill in details of the proof for Theorem 3.2.48 to establish that a matrix Q^{T} is orthogonal if and only if the row vectors of Q form an orthonormal set.

Exercise 3.2.23 Fill in details of the proof for Theorem 3.2.48 to establish that if the row vectors of Q form an orthonormal set, then Q is invertible and $Q^{-1} = Q^{\mathsf{T}}$.

Exercise 3.2.24 The following graphs illustrate the transformation of the unit square through multiplying by some different matrices. Using Theorem 3.2.48(f), which transformations appear to be that of multiplying by an orthogonal matrix?

(a)

(b)

(c)

(d)

Exercise 3.2.25 The following stereo pairs illustrate the transformation of the unit cube through multiplying by some different matrices. Using Theorem 3.2.48(f), which transformations appear to be that of multiplying by an orthogonal matrix?

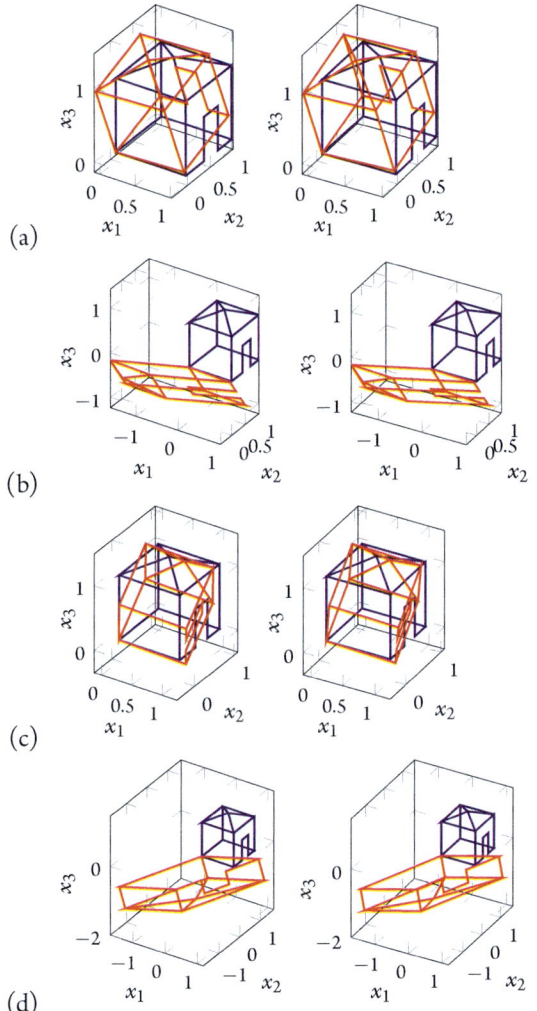

(a)

(b)

(c)

(d)

Exercise 3.2.26 In a few sentences, answer/discuss each of the following.

(a) How is the notion of a matrix inverse related to that of the reciprocal of a number?

(b) Given the formula for the inverse of a 2×2 matrix, Theorem 3.2.7, what is one way to choose a matrix with integer components whose inverse also has integer components?

(c) What happens if we try to find an inverse of a non-square matrix? Explore trying to find the inverse of the 2×1 matrix $\begin{bmatrix} a \\ b \end{bmatrix}$.

(d) Why is the concept of an inverse important?

(e) What enables negative powers of a matrix to be defined?

(f) How do negative powers arise in modelling populations?

(g) Why are diagonal matrices important for solving equations?

(h) Compare and contrast an orthogonal set with an orthonormal set.

(i) What causes an orthogonal matrix to have its transpose as its inverse?

(j) Why is multiplication by an orthogonal matrix referred to as a 'rotation and/or reflection'?

3.3 Factorize to the singular value decomposition

The singular value decomposition (SVD) is sometimes called the jewel in the crown of linear algebra.[13] The SVD's importance is certified by the many names by which it is invoked in scientific and engineering applications: principal component analysis, singular spectrum analysis, proper orthogonal decomposition, latent semantic indexing, Schmidt decomposition, correspondence analysis, Lanczos methods, dimension reduction, and so on. Let's start seeing what it can do for us.

3.3.1 Introductory examples

Let's introduce an analogous problem, so the SVD procedure follows more easily.

You are a contestant in a quiz show. The final million dollar question is:

in your head, without a calculator, solve $42x = 1554$ within twenty seconds,

your time starts now . . . .

Solution: Long division is hopeless in the time available. However, recognize that 42 factors, $42 = 2 \cdot 3 \cdot 7$, and so divide 1554 by 2 to get 777, divide 777 by 3 to get 259, and divide 259 by 7 to get the answer $x = 37$. You win the prize!   □

Activity 3.3.1   Given $154 = 2 \cdot 7 \cdot 11$, solve in your head $154x = 8008$ or 9856 or 12628 or 13090 or 14322 (teacher to choose): first to answer wins!

Such examples show that factorization can turn a hard problem into several easy problems. We adopt an analogous matrix factorization to solve and understand general linear equations.

[13] Beltrami first derived the SVD in 1873 (Stewart, 1993). The first reliable method for computing an SVD was developed by Golub and Kahan in 1965, and only thereafter did applications proliferate.

To illustrate the procedure to come, let's write the above solution steps in detail: we solve $42x = 1554$.

(a) Factorize the coefficient $42 = 2 \cdot 3 \cdot 7$ so the equation becomes

$$2 \cdot 3 \cdot \underbrace{7 \cdot x}_{=z} = 1554,$$

where $=y$ is indicated over $\overbrace{7 \cdot x}$

and introduce two intermediate unknowns y and z as indicated above (that is, $z = 3y = 21x$ and $y = 7x$): these are as yet unknown as x is as yet unknown.

(b) Solve $2z = 1554$ to determine $z = 777$.

(c) Solve $3y = z = 777$ to determine $y = 259$.

(d) Solve $7x = y = 259$ to determine $x = 37$ —the answer.

Now let's proceed to small matrix examples. Each example follows analogous solution steps to those above. The following examples introduce the general matrix procedure empowered by a factorization of a matrix. Specifically, we use the matrix factorization called a singular value decomposition (SVD).

Example 3.3.2 Solve the 2×2 system

$$\begin{bmatrix} 10 & 2 \\ 5 & 11 \end{bmatrix} x = \begin{bmatrix} 18 \\ -1 \end{bmatrix}$$

for x given the matrix factorization

$$\begin{bmatrix} 10 & 2 \\ 5 & 11 \end{bmatrix} = \begin{bmatrix} \frac{3}{5} & -\frac{4}{5} \\ \frac{4}{5} & \frac{3}{5} \end{bmatrix} \begin{bmatrix} 10\sqrt{2} & 0 \\ 0 & 5\sqrt{2} \end{bmatrix} \begin{bmatrix} \frac{1}{\sqrt{2}} & -\frac{1}{\sqrt{2}} \\ \frac{1}{\sqrt{2}} & \frac{1}{\sqrt{2}} \end{bmatrix}^{\mathrm{T}}$$

(remember the transpose on the last matrix).

Solution: Optionally check the factorization if you like:

$$\begin{bmatrix} \frac{3}{5} & -\frac{4}{5} \\ \frac{4}{5} & \frac{3}{5} \end{bmatrix} \begin{bmatrix} 10\sqrt{2} & 0 \\ 0 & 5\sqrt{2} \end{bmatrix} = \begin{bmatrix} 6\sqrt{2} & -4\sqrt{2} \\ 8\sqrt{2} & 3\sqrt{2} \end{bmatrix};$$

then

$$\begin{bmatrix} 6\sqrt{2} & -4\sqrt{2} \\ 8\sqrt{2} & 3\sqrt{2} \end{bmatrix} \begin{bmatrix} \frac{1}{\sqrt{2}} & \frac{1}{\sqrt{2}} \\ -\frac{1}{\sqrt{2}} & \frac{1}{\sqrt{2}} \end{bmatrix} = \begin{bmatrix} 10 & 2 \\ 5 & 11 \end{bmatrix}.$$

Given the factorization, the following four steps form the general procedure.

(a) Write the system using the factorization, and with two intermediate unknowns y and z as indicated below:

$$\underbrace{\begin{bmatrix} \frac{3}{5} & -\frac{4}{5} \\ \frac{4}{5} & \frac{3}{5} \end{bmatrix} \begin{bmatrix} 10\sqrt{2} & 0 \\ 0 & 5\sqrt{2} \end{bmatrix} \overbrace{\begin{bmatrix} \frac{1}{\sqrt{2}} & -\frac{1}{\sqrt{2}} \\ \frac{1}{\sqrt{2}} & \frac{1}{\sqrt{2}} \end{bmatrix}^{\text{T}}}^{=y}}_{=z} x = \begin{bmatrix} 18 \\ -1 \end{bmatrix}.$$

(b) Solve $\begin{bmatrix} 3/5 & -4/5 \\ 4/5 & 3/5 \end{bmatrix} z = \begin{bmatrix} 18 \\ -1 \end{bmatrix}$: recall that the matrix appearing here is orthogonal (and this orthogonality is no accident), so multiplying by its transpose gives the intermediary

$$z = \begin{bmatrix} \frac{3}{5} & \frac{4}{5} \\ -\frac{4}{5} & \frac{3}{5} \end{bmatrix} \begin{bmatrix} 18 \\ -1 \end{bmatrix} = \begin{bmatrix} 10 \\ -15 \end{bmatrix}.$$

(c) Now solve $\begin{bmatrix} 10\sqrt{2} & 0 \\ 0 & 5\sqrt{2} \end{bmatrix} y = z = \begin{bmatrix} 10 \\ -15 \end{bmatrix}$: the matrix appearing here is diagonal (and this is no accident), so dividing by the respective diagonal elements gives the intermediary

$$y = \begin{bmatrix} 10\sqrt{2} & 0 \\ 0 & 5\sqrt{2} \end{bmatrix}^{-1} \begin{bmatrix} 10 \\ -15 \end{bmatrix} = \begin{bmatrix} 1/\sqrt{2} \\ -3/\sqrt{2} \end{bmatrix}.$$

(d) Finally solve $\begin{bmatrix} 1/\sqrt{2} & -1/\sqrt{2} \\ 1/\sqrt{2} & 1/\sqrt{2} \end{bmatrix}^{\text{T}} x = y = \begin{bmatrix} 1/\sqrt{2} \\ -3/\sqrt{2} \end{bmatrix}$: now the matrix appearing here is also orthogonal (this orthogonality is also no accident), so multiplying by itself (the transpose of the transpose, Theorem 3.1.29(a)) gives the solution

$$x = \begin{bmatrix} \frac{1}{\sqrt{2}} & -\frac{1}{\sqrt{2}} \\ \frac{1}{\sqrt{2}} & \frac{1}{\sqrt{2}} \end{bmatrix} \begin{bmatrix} \frac{1}{\sqrt{2}} \\ -\frac{3}{\sqrt{2}} \end{bmatrix} = \begin{bmatrix} \frac{1}{2} + \frac{3}{2} \\ \frac{1}{2} - \frac{3}{2} \end{bmatrix} = \begin{bmatrix} 2 \\ -1 \end{bmatrix}.$$

That is, we obtain the solution of the matrix-vector system via two orthogonal matrices and a diagonal matrix. □

Activity 3.3.3 Let's solve the system $\begin{bmatrix} 12 & -41 \\ 34 & -12 \end{bmatrix} x = \begin{bmatrix} 94 \\ 58 \end{bmatrix}$ using the factorization $\begin{bmatrix} 12 & -41 \\ 34 & -12 \end{bmatrix} =$

$\begin{bmatrix} \frac{4}{5} & -\frac{3}{5} \\ \frac{3}{5} & \frac{4}{5} \end{bmatrix} \begin{bmatrix} 50 & 0 \\ 0 & 25 \end{bmatrix} \begin{bmatrix} \frac{3}{5} & \frac{4}{5} \\ -\frac{4}{5} & \frac{3}{5} \end{bmatrix}^{\text{T}}$ in which the first and third matrices on the right-hand side are

orthogonal. After solving $\begin{bmatrix} \frac{4}{5} & -\frac{3}{5} \\ \frac{3}{5} & \frac{4}{5} \end{bmatrix} z = \begin{bmatrix} 94 \\ 58 \end{bmatrix}$, the next step is to solve which of the following?

(a) $\begin{bmatrix} 50 & 0 \\ 0 & 25 \end{bmatrix} y = \begin{bmatrix} 10 \\ 110 \end{bmatrix}$

(c) $\begin{bmatrix} 50 & 0 \\ 0 & 25 \end{bmatrix} y = \begin{bmatrix} \frac{202}{5} \\ \frac{514}{5} \end{bmatrix}$

(b) $\begin{bmatrix} 50 & 0 \\ 0 & 25 \end{bmatrix} y = \begin{bmatrix} \frac{514}{5} \\ -\frac{202}{5} \end{bmatrix}$

(d) $\begin{bmatrix} 50 & 0 \\ 0 & 25 \end{bmatrix} y = \begin{bmatrix} 110 \\ -10 \end{bmatrix}$

Example 3.3.4 Solve the 3×3 system

$$Ax = \begin{bmatrix} 10 \\ 2 \\ -2 \end{bmatrix} \quad \text{for matrix } A = \begin{bmatrix} -4 & -2 & 4 \\ -8 & -1 & -4 \\ 6 & 6 & 0 \end{bmatrix}$$

using the following given matrix factorization (note the last matrix is transposed)

$$A = \begin{bmatrix} \frac{1}{3} & -\frac{2}{3} & \frac{2}{3} \\ \frac{2}{3} & \frac{2}{3} & \frac{1}{3} \\ -\frac{2}{3} & \frac{1}{3} & \frac{2}{3} \end{bmatrix} \begin{bmatrix} 12 & 0 & 0 \\ 0 & 6 & 0 \\ 0 & 0 & 3 \end{bmatrix} \begin{bmatrix} -\frac{8}{9} & -\frac{1}{9} & -\frac{4}{9} \\ -\frac{4}{9} & \frac{4}{9} & \frac{7}{9} \\ -\frac{1}{9} & -\frac{8}{9} & \frac{4}{9} \end{bmatrix}^{\text{T}}.$$

Solution: Use MATLAB/Octave. Enter the matrices and the right-hand side, and check the factorization (and the typing):

```
U=[1,-2,2;2,2,1;-2,1,2]/3
S=[12,0,0;0,6,0;0,0,3]
V=[-8,-1,-4;-4,4,7;-1,-8,4]/9
b=[10;2;-2]
A=U*S*V'
```

(a) Write the system $Ax = b$ using the factorization, and with two intermediate unknowns y and z:

$$\underbrace{\begin{bmatrix} \frac{1}{3} & -\frac{2}{3} & \frac{2}{3} \\ \frac{2}{3} & \frac{2}{3} & \frac{1}{3} \\ -\frac{2}{3} & \frac{1}{3} & \frac{2}{3} \end{bmatrix} \underbrace{\begin{bmatrix} 12 & 0 & 0 \\ 0 & 6 & 0 \\ 0 & 0 & 3 \end{bmatrix} \overbrace{\begin{bmatrix} -\frac{8}{9} & -\frac{1}{9} & -\frac{4}{9} \\ -\frac{4}{9} & \frac{4}{9} & \frac{7}{9} \\ -\frac{1}{9} & -\frac{8}{9} & \frac{4}{9} \end{bmatrix}^{\text{T}} x}^{=y}}_{=z} = \begin{bmatrix} 10 \\ 2 \\ -2 \end{bmatrix}.$$

(b) Solve $\begin{bmatrix} 1/3 & -2/3 & 2/3 \\ 2/3 & 2/3 & 1/3 \\ -2/3 & 1/3 & 2/3 \end{bmatrix} z = \begin{bmatrix} 10 \\ 2 \\ -2 \end{bmatrix}$. Now the matrix on the left, called U, is orthogonal—
check by computing U'*U—so multiplying by the transpose gives the intermediary:
z=U'*b = $(6, -6, 6)$.

(c) Then solve $\begin{bmatrix} 12 & 0 & 0 \\ 0 & 6 & 0 \\ 0 & 0 & 3 \end{bmatrix} y = z = \begin{bmatrix} 6 \\ -6 \\ 6 \end{bmatrix}$: this matrix, called S, is diagonal, so dividing by the
respective diagonal elements gives the intermediary y=z./diag(S) $= (\frac{1}{2}, -1, 2)$.

(d) Finally solve $\begin{bmatrix} -8/9 & -1/9 & -4/9 \\ -4/9 & 4/9 & 7/9 \\ -1/9 & -8/9 & 4/9 \end{bmatrix}^{\mathsf{T}} x = y = \begin{bmatrix} 1/2 \\ -1 \\ 2 \end{bmatrix}$. This matrix, called V, is also orthogonal—check by computing $V' * V$—so multiplying by itself (the transpose of the transpose) gives the final solution $x = V*y = (-\frac{11}{9}, \frac{8}{9}, \frac{31}{18})$. □

Warning: do *not* solve in reverse order

Example 3.3.5 Reconsider Example 3.3.2 wrongly.

(a) After writing the system using the SVD as

$$\begin{bmatrix} \frac{3}{5} & -\frac{4}{5} \\ \frac{4}{5} & \frac{3}{5} \end{bmatrix} \overbrace{\begin{bmatrix} 10\sqrt{2} & 0 \\ 0 & 5\sqrt{2} \end{bmatrix} \begin{bmatrix} \frac{1}{\sqrt{2}} & -\frac{1}{\sqrt{2}} \\ \frac{1}{\sqrt{2}} & \frac{1}{\sqrt{2}} \end{bmatrix}^{\mathsf{T}}}^{=y} x = \begin{bmatrix} 18 \\ -1 \end{bmatrix},$$

$$=z$$

one might be inadvertently tempted to 'solve' the system by using the matrices in reverse order as in the following: *do not do this.*

(b) First solve $\begin{bmatrix} 1/\sqrt{2} & -1/\sqrt{2} \\ 1/\sqrt{2} & 1/\sqrt{2} \end{bmatrix}^{\mathsf{T}} x = \begin{bmatrix} 18 \\ -1 \end{bmatrix}$: this matrix is orthogonal, so multiplying by itself (the transpose of the transpose) gives

$$x = \begin{bmatrix} \frac{1}{\sqrt{2}} & -\frac{1}{\sqrt{2}} \\ \frac{1}{\sqrt{2}} & \frac{1}{\sqrt{2}} \end{bmatrix} \begin{bmatrix} 18 \\ -1 \end{bmatrix} = \begin{bmatrix} \frac{19}{\sqrt{2}} \\ \frac{17}{\sqrt{2}} \end{bmatrix}.$$

(c) Inappropriately 'solve' $\begin{bmatrix} 10\sqrt{2} & 0 \\ 0 & 5\sqrt{2} \end{bmatrix} y = \begin{bmatrix} 19/\sqrt{2} \\ 17/\sqrt{2} \end{bmatrix}$: this matrix is diagonal, so dividing by the diagonal elements gives

$$y = \begin{bmatrix} 10\sqrt{2} & 0 \\ 0 & 5\sqrt{2} \end{bmatrix}^{-1} \begin{bmatrix} 19/\sqrt{2} \\ 17/\sqrt{2} \end{bmatrix} = \begin{bmatrix} \frac{19}{20} \\ \frac{17}{10} \end{bmatrix}.$$

(d) Inappropriately 'solve' $\begin{bmatrix} 3/5 & -4/5 \\ 4/5 & 3/5 \end{bmatrix} z = \begin{bmatrix} 19/20 \\ 17/10 \end{bmatrix}$: this matrix is orthogonal, so multiplying by the transpose gives

$$z = \begin{bmatrix} \frac{3}{5} & \frac{4}{5} \\ -\frac{4}{5} & \frac{3}{5} \end{bmatrix} \begin{bmatrix} \frac{19}{20} \\ \frac{17}{10} \end{bmatrix} = \begin{bmatrix} 1.93 \\ 0.26 \end{bmatrix}.$$

And then, since the solution is to be called x, we might inappropriately call what we just calculated as the solution $x = (1.93, 0.26)$. □

Avoid this reverse process as it is wrong. Matrix multiplication is *not* commutative (Section 3.1.3). We must use matrix factorization in the correct order: to solve linear equations use the matrices in a factorization from left to right.

3.3.2 The SVD solves general systems

The previous examples depended upon a matrix being factored into a product of three matrices: two orthogonal and one diagonal. Amazingly, such factorization is always possible. (http://www. youtube.com/watch?v=JEYLfIVvR9I is an entertaining prelude—their *D* is our *S*.)

Theorem 3.3.6 (SVD *factorization*) *Every $m \times n$ real matrix A can be factored into a product of three matrices*

$$A = USV^{\mathrm{T}}, \tag{3.4}$$

called a **singular value decomposition** (SVD), where
- $m \times m$ matrix $U = \begin{bmatrix} u_1 & u_2 & \cdots & u_m \end{bmatrix}$ is orthogonal,
- $n \times n$ matrix $V = \begin{bmatrix} v_1 & v_2 & \cdots & v_n \end{bmatrix}$ is orthogonal, and
- $m \times n$ diagonal matrix S is zero except for non-negative diagonal elements called **singular values** $\sigma_1 , \sigma_2 , \ldots , \sigma_{\min(m,n)}$ (the symbol σ is the Greek letter sigma, and denotes singular values), which are unique when ordered from largest to smallest so that $\sigma_1 \geq \sigma_2 \geq \cdots \geq \sigma_{\min(m,n)} \geq 0$. The two orthonormal sets of vectors $\{u_j\}$ and $\{v_j\}$ are called **singular vectors**.[14]

Proof. Detailed in Section 3.3.3.

Importantly, the singular values are unique (when ordered), although the orthogonal matrices U and V are not unique (e.g., one may change the sign of any column in U together with its corresponding column in V). Nonetheless, although there are many possible matrices U and V in the SVDs of a matrix, all such SVDs are equivalent in application.

Some people may be disturbed by the non-uniqueness of an SVD. But the non-uniqueness is analogous to the non-uniqueness of Gauss–Jordan elimination upon reordering of equations, and/or reordering the variables in the equations (Section 2.2.2). Do not be disturbed by any non-uniqueness in U and V.

Example 3.3.7 Example 3.3.2 invoked the SVD

$$\begin{bmatrix} 10 & 2 \\ 5 & 11 \end{bmatrix} = \begin{bmatrix} \frac{3}{5} & -\frac{4}{5} \\ \frac{4}{5} & \frac{3}{5} \end{bmatrix} \begin{bmatrix} 10\sqrt{2} & 0 \\ 0 & 5\sqrt{2} \end{bmatrix} \begin{bmatrix} \frac{1}{\sqrt{2}} & -\frac{1}{\sqrt{2}} \\ \frac{1}{\sqrt{2}} & \frac{1}{\sqrt{2}} \end{bmatrix}^{\mathrm{T}} ,$$

where the two outer matrices are orthogonal (check), so the singular values of this matrix are $\sigma_1 = 10\sqrt{2}$ and $\sigma_2 = 5\sqrt{2}$.

---

[14] This enormously useful theorem also generalizes from $m \times n$ real matrices to complex matrices, and also to analogues in 'infinite' dimensions (Stewart, 1993, §5): an SVD exists for all compact linear operators (Kress, 2015, §7).

Example 3.3.4 invoked the SVD

$$
\begin{bmatrix} -4 & -2 & 4 \\ -8 & -1 & -4 \\ 6 & 6 & 0 \end{bmatrix} = \begin{bmatrix} \frac{1}{3} & -\frac{2}{3} & \frac{2}{3} \\ \frac{2}{3} & \frac{2}{3} & \frac{1}{3} \\ -\frac{2}{3} & \frac{1}{3} & \frac{2}{3} \end{bmatrix} \begin{bmatrix} 12 & 0 & 0 \\ 0 & 6 & 0 \\ 0 & 0 & 3 \end{bmatrix} \begin{bmatrix} -\frac{8}{9} & -\frac{1}{9} & -\frac{4}{9} \\ -\frac{4}{9} & \frac{4}{9} & \frac{7}{9} \\ -\frac{1}{9} & -\frac{8}{9} & \frac{4}{9} \end{bmatrix}^{\mathrm{T}},
$$

where the two outer matrices are orthogonal (check), so the singular values of this matrix are $\sigma_1 = 12, \sigma_2 = 6$, and $\sigma_3 = 3$. ☐

Example 3.3.8 Any orthogonal matrix Q, say $n \times n$, has an SVD $Q = QI_nI_n^{\mathrm{T}}$; that is, $U = Q$, $S = V = I_n$. The S-matrix is the identity. Hence every $n \times n$ orthogonal matrix has singular values $\sigma_1 = \sigma_2 = \cdots = \sigma_n = 1$.

Example 3.3.9 *(some non-uniqueness)*

- An identity matrix, say I_n, has an SVD $I_n = I_nI_nI_n^{\mathrm{T}}$.
- Additionally, for *every* $n \times n$ orthogonal matrix Q, the identity I_n also has the SVD $I_n = QI_nQ^{\mathrm{T}}$—as this right-hand side $QI_nQ^{\mathrm{T}} = QQ^{\mathrm{T}} = I_n$.
- Further, any constant multiple of an identity, say $sI_n = \mathrm{diag}(s, s, \ldots, s)$, has the same non-uniqueness: an SVD is $sI_n = USV^{\mathrm{T}}$ for matrices $U = Q$, $S = sI_n$, and $V = Q$ for every $n \times n$ orthogonal Q (provided $s \geq 0$).

The matrices in this example are characterized by all their singular values having an identical value. In general, analogous non-uniqueness in U and V occurs whenever two or more singular values are identical in value.

Activity 3.3.10 Example 3.3.8 commented that $QI_nI_n^{\mathrm{T}}$ is an SVD of an orthogonal matrix Q. Which of the following is also an SVD of a given $n \times n$ orthogonal matrix Q?

(a) $I_nI_n(Q^{\mathrm{T}})^{\mathrm{T}}$
(b) $I_nQI_n^{\mathrm{T}}$
(c) $Q(-I_n)(-I_n)^{\mathrm{T}}$
(d) $I_nI_nQ^{\mathrm{T}}$

Example 3.3.11 *(positive ordering)* Find an SVD of the diagonal matrix

$$
D = \begin{bmatrix} 2.7 & 0 & 0 \\ 0 & -3.9 & 0 \\ 0 & 0 & -0.9 \end{bmatrix}.
$$

Solution: Singular values cannot be negative, so a factorization is

$$
D = \begin{bmatrix} 1 & 0 & 0 \\ 0 & -1 & 0 \\ 0 & 0 & -1 \end{bmatrix} \begin{bmatrix} 2.7 & 0 & 0 \\ 0 & 3.9 & 0 \\ 0 & 0 & 0.9 \end{bmatrix} \begin{bmatrix} 1 & 0 & 0 \\ 0 & 1 & 0 \\ 0 & 0 & 1 \end{bmatrix}^{\mathrm{T}},
$$

where the (-1)s in the first matrix encode the signs of the corresponding diagonal elements (one could alternatively use the rightmost matrix to encode the pattern of signs). However, Theorem 3.3.6 requires that singular values be ordered in decreasing magnitude, so sort the diagonal of the middle matrix into order and correspondingly permute the columns of the outer two matrices to obtain the following SVD:

$$
D = \begin{bmatrix} 0 & 1 & 0 \\ -1 & 0 & 0 \\ 0 & 0 & -1 \end{bmatrix} \begin{bmatrix} 3.9 & 0 & 0 \\ 0 & 2.7 & 0 \\ 0 & 0 & 0.9 \end{bmatrix} \begin{bmatrix} 0 & 1 & 0 \\ 1 & 0 & 0 \\ 0 & 0 & 1 \end{bmatrix}^{\mathrm{T}}.
$$

□

Computers empower use of the SVD

Except for simple cases such as 2×2 matrices (Example 3.3.33), constructing an SVD is usually far too laborious by hand.[15] Typically, this book either gives an SVD (as in the earlier two examples) or asks you to compute an SVD in MATLAB/Octave with [U,S,V]=svd(A) (Table 3.3).

The SVD Theorem 3.3.6 asserts that every matrix is the product of two orthogonal matrices and a diagonal matrix. Because, in a matrix's SVD factorization, the rotations (and/or reflection) by the two orthogonal matrices are so 'nice', any 'badness' or 'trickiness' in the matrix is represented in the diagonal matrix S of the singular values.

This and following examples illustrate the cases of either no or infinite solutions, to complement the case of unique solutions of the first two examples.

Example 3.3.12 *(rate sport teams/players)* Consider three table tennis players, Anne, Bob, and Chris: Anne beat Bob 3 games to 2 games; Anne beat Chris 3-1; Bob beat Chris 3-2. How good

Table 3.3 As well as the MATLAB/Octave commands and operations listed in Tables 1.2, 2.3, 3.1 and 3.2, we need these matrix operations.

- [U,S,V]=svd(A) computes the three matrices U, S, and V in a singular value decomposition (SVD) of the $m \times n$ matrix: $A = USV^{\mathrm{T}}$ for $m \times m$ orthogonal matrix U, $n \times n$ orthogonal matrix V, and $m \times n$ non-negative diagonal matrix S (Theorem 3.3.6). svd(A) just reports (in a vector) the singular values.
- Complementing information of Table 3.1, to extract and compute with a subset of rows/columns of a matrix, specify the vector of indices. For example:
  - V(:,1:r) selects the first r columns of V;
  - A([2 3 5],:) selects the second, third, and fifth row of matrix A;
  - B(4:6,1:3) selects the 3×3 submatrix of the first three columns of the fourth, fifth, and sixth rows.

---

[15] For those interested advanced students, Trefethen and Bau (1997) [p.234] discusses how the standard method of numerically computing an SVD is based upon first transforming a matrix to bidiagonal form, and then using an iteration based upon a so-called QR factorization. See https://www.youtube.com/watch?v=R9UoFyqJca8 for a visualization from 1976.

are they all? What are their ratings? That is, we seek to assign a number to each player, called their rating, indicating how good they are at playing: the higher the rating the better the player.

Solution: Denote Anne's rating by x_1, Bob's rating by x_2, and Chris' rating by x_3. The ratings should predict the results of matches, so from the above three match results, surely

- Anne beat Bob 3 games to 2 $\leftrightarrow$ $x_1 - x_2 = 3 - 2 = 1$;

- Anne beat Chris 3-1 $\leftrightarrow$ $x_1 - x_3 = 3 - 1 = 2$; and

- Bob beat Chris 3-2 $\leftrightarrow$ $x_2 - x_3 = 3 - 2 = 1$.

In matrix-vector form, $Ax = b$,

$$\begin{bmatrix} 1 & -1 & 0 \\ 1 & 0 & -1 \\ 0 & 1 & -1 \end{bmatrix} x = \begin{bmatrix} 1 \\ 2 \\ 1 \end{bmatrix}.$$

In MATLAB/Octave, we might try Procedure 2.2.5:

```
A= [1,-1,0;1,0,-1;0,1,-1]
b= [1;2;1]
rcond(A)
```

but find `rcond=0`, which is extremely terrible so we cannot use A\b to solve the system $Ax = b$. *Whenever difficulties arise, use an* SVD.

(a) Compute an SVD $A = USV^T$ with `[U,S,V] =svd(A)` (Table 3.3): here

```
U =
      0.4082    -0.7071     0.5774
     -0.4082    -0.7071    -0.5774
     -0.8165    -0.0000     0.5774
S =
      1.7321          0          0
           0     1.7321          0
           0          0     0.0000
V =
      0.0000    -0.8165     0.5774
     -0.7071     0.4082     0.5774
      0.7071     0.4082     0.5774
```

so the three singular values are $\sigma_1 = \sigma_2 = 1.7321 = \sqrt{3}$ and $\sigma_3 = 0$ (different computers may give different U and V, but any deductions are equivalent). The system of equations for the ratings then become

$$Ax = U \underbrace{S \overbrace{V^T x}^{=y}}_{=z} = b = \begin{bmatrix} 1 \\ 2 \\ 1 \end{bmatrix}.$$

(b) As U is orthogonal, $Uz = b$ has unique solution $z = U^T b$ computed by z=U' *b:

```
z =
    -1.2247
    -2.1213
         0
```

(c) Now solve $Sy = z$. But S has a troublesome zero on the diagonal. So interpret the equation $Sy = z$ in detail as

$$\begin{bmatrix} 1.7321 & 0 & 0 \\ 0 & 1.7321 & 0 \\ 0 & 0 & 0 \end{bmatrix} y = \begin{bmatrix} -1.2247 \\ -2.1213 \\ 0 \end{bmatrix} :$$

  i. the first line implies $y_1 = -1.2247/1.7321$;
  ii. the second line implies $y_2 = -2.1213/1.7321$;
  iii. the third line is $0y_3 = 0$, which is satisfied for all y_3.

In using MATLAB/Octave you must notice $\sigma_3 = 0$, check that the corresponding $z_3 = 0$, and then compute a *particular solution* from the first two components to give the first two components of y:

```
y=z(1:2)./diag(S(1:2,1:2))
y =
   -0.7071
   -1.2247
```

The third component, involving the free variable y_3, we omit from this numerical computation.

(d) Finally, as V is orthogonal, $V^T x = y$ has the solution $x = Vy$ (unique for each valid y): in MATLAB/Octave, compute a particular solution with x=V(:,1:2)*y

```
x =
    1.0000
    0.0000
   -1.0000
```

Then for a general solution remember to add an arbitrary multiple, y_3, of V(:,3) = $(0.5774, 0.5774, 0.5774) = (1, 1, 1)/\sqrt{3}$.

Thus the three player ratings may be any one from the general solution

$$(x_1, x_2, x_3) = (1, 0, -1) + y_3(1, 1, 1)/\sqrt{3}.$$

In this application we only care about relative ratings, not absolute ratings, so here adding any multiple of $(1, 1, 1)$ is immaterial. This solution for the ratings indicates that Anne is the best player, and Chris the worst. □

Compute in MATLAB/Octave   As seen in the previous example, often we need to compute with a subset of the components of matrices, a submatrix (Table 3.3):

- b(1:r) selects the first r entries of vector b;
- S(1:r,1:r) selects the top-left $r \times r$ submatrix of S;
- V(:,1:r) selects the first r columns of matrix V.

Example 3.3.13 But what if Bob beat Chris 3-1?

Solution: The only change to the problem is the new right-hand side $b = (1,2,2)$.

(a) An SVD of matrix A remains the same.

(b) $Uz = b$ has unique solution z=U'*b of

```
z =
       -2.0412
       -2.1213
        0.5774
```

(c) We need to interpret $Sy = z$,

$$\begin{bmatrix} 1.7321 & 0 & 0 \\ 0 & 1.7321 & 0 \\ 0 & 0 & 0 \end{bmatrix} y = \begin{bmatrix} -2.0412 \\ -2.1213 \\ 0.5774 \end{bmatrix}.$$

The third line of this system says $0y_3 = 0.5774$, which is impossible for any y_3.

In this case there is no solution of the system of equations. It would appear that we cannot assign ratings to the players!   □

Section 3.5 further explores systems with no solution and uses the SVD to determine a best approximate solution (Example 3.5.3).

Example 3.3.14 Find the value(s) of the parameter c such that the following system has a solution, and find a general solution for that (those) parameter value(s):

$$\begin{bmatrix} -9 & -15 & -9 & -15 \\ -10 & 2 & -10 & 2 \\ 8 & 4 & 8 & 4 \end{bmatrix} x = \begin{bmatrix} c \\ 8 \\ -5 \end{bmatrix}.$$

Solution: Because the matrix is not square, we cannot use Procedure 2.2.5: instead use an SVD.

(a) In MATLAB/Octave, compute an SVD of this 3×4 matrix with

```
A=[-9 -15 -9 -15; -10 2 -10 2; 8 4 8 4]
[U,S,V]=svd(A)

U =
```

$$
\begin{array}{ccc}
0.8571 & 0.4286 & 0.2857 \\
0.2857 & -0.8571 & 0.4286 \\
-0.4286 & 0.2857 & 0.8571
\end{array}
$$

$$
S =
\begin{array}{cccc}
28.0000 & 0 & 0 & 0 \\
0 & 14.0000 & 0 & 0 \\
0 & 0 & 0.0000 & 0
\end{array}
$$

$$
V =
\begin{array}{cccc}
-0.5000 & 0.5000 & -0.1900 & -0.6811 \\
-0.5000 & -0.5000 & 0.6811 & -0.1900 \\
-0.5000 & 0.5000 & 0.1900 & 0.6811 \\
-0.5000 & -0.5000 & -0.6811 & 0.1900
\end{array}
$$

(Depending upon MATLAB/Octave, you may get different alternatives for the last two columns for V, and different signs for columns of U and V—adjust accordingly.) The singular values are $\sigma_1 = 28$, $\sigma_2 = 14$ and the problematic $\sigma_3 = 0$ (it is computed as the negligible 10^{-15}).

(b) To solve $Uz = b$ we compute $z = U^{\mathsf{T}} b$. But for the next step we must have the third component of z to be zero as otherwise there is no solution. Now $z_3 = u_3^{\mathsf{T}} b$ (where u_3 is the third column of U); that is, $z_3 = 0.2857 \times c + 0.4286 \times 8 + 0.8571 \times (-5)$ needs to be zero, which requires $c = -(0.4286 \times 8 + 0.8571 \times (-5))/0.2857$. Recognize that this expression is equivalent to $c = -(0.2857 \times 0 + 0.4286 \times 8 + 0.8571 \times (-5))/0.2857 = u_3 \cdot (0, 8, -5)/0.2857$ and so compute

```
c=-U(:,3)'*[0;8;-5]/U(1,3)
```

Having found that $c = 3$, compute z from `z=U' * [3;8;-5]` to find $z = (7, -7, 0)$.

(c) Find a general solution of the diagonal system $Sy = z$:

$$
\begin{bmatrix}
28 & 0 & 0 & 0 \\
0 & 14 & 0 & 0 \\
0 & 0 & 0 & 0
\end{bmatrix}
y =
\begin{bmatrix}
7 \\
-7 \\
0
\end{bmatrix}.
$$

The first line gives $y_1 = 7/28 = 1/4$, the second line gives $y_2 = -7/14 = -1/2$, and the third line is $0y_3 + 0y_4 = 0$, which is satisfied for every y_3 and y_4 (because we chose c correctly). Thus $y = (\frac{1}{4}, -\frac{1}{2}, y_3, y_4)$ is a general solution for this intermediary. Compute the particular solution with $y_3 = y_4 = 0$ via

```
y=z(1:2)./diag(S(1:2,1:2))
```

(d) Finally, solve $V^{\mathsf{T}} x = y$ as $x = V y$, namely

$$
x =
\begin{bmatrix}
-0.5 & 0.5 & -0.1900 & -0.6811 \\
-0.5 & -0.5 & 0.6811 & -0.1900 \\
-0.5 & 0.5 & 0.1900 & 0.6811 \\
-0.5 & -0.5 & -0.6811 & 0.1900
\end{bmatrix}
\begin{bmatrix}
1/4 \\
-1/2 \\
y_3 \\
y_4
\end{bmatrix}
$$

Obtain a particular solution with x=V(:,1:2)*y of $x = (-3, 1, -3, 1)/8$, and then add the free components:

$$x = \begin{bmatrix} -\frac{3}{8} \\ \frac{1}{8} \\ -\frac{3}{8} \\ \frac{1}{8} \end{bmatrix} + \begin{bmatrix} -0.1900 \\ 0.6811 \\ 0.1900 \\ -0.6811 \end{bmatrix} y_3 + \begin{bmatrix} -0.6811 \\ -0.1900 \\ 0.6811 \\ 0.1900 \end{bmatrix} y_4.$$

This formula for x is a most general solution to the given system. $\square$

Procedure 3.3.15 (general solution) *Obtain a general solution of the system $Ax = b$ using an SVD and via intermediate unknowns.*

1. *Obtain an SVD factorization $A = USV^\mathsf{T}$.*

2. *Solve $Uz = b$ by $z = U^\mathsf{T}b$ (unique given U).*

3. *When possible, solve $Sy = z$ as follows.[16] Identify the nonzero and the zero singular values: suppose $\sigma_1 \geq \sigma_2 \geq \cdots \geq \sigma_r > 0$ and $\sigma_{r+1} = \cdots = \sigma_{\min(m,n)} = 0$:*

   - *if $z_i \neq 0$ for any one (or more) $i = r+1, \ldots, m$, then there is no solution (the equations are **inconsistent**);*

   - *otherwise (when $z_i = 0$ for every $i = r+1, \ldots, m$) determine the ith component of y by $y_i = z_i/\sigma_i$ for $i = 1, \ldots, r$ (for which $\sigma_i > 0$), whereas for $i = r+1, \ldots, n$ let y_i be a free variable.*

4. *Solve $V^\mathsf{T}x = y$ (unique given V and for each y) to derive that a general solution is $x = Vy$.*

Proof. Given an SVD $A = USV^\mathsf{T}$ (Theorem 3.3.6), consider each and every solution of $Ax = b$:

$$Ax = b \iff USV^\mathsf{T}x = b \quad \text{(by Step 1)}$$
$$\iff S(V^\mathsf{T}x) = U^\mathsf{T}b$$
$$\iff Sy = z \quad \text{(by Step 2 and 4)},$$

and Step 3 determines all possible y satisfying $Sy = z$. Hence Procedure 3.3.15 determines all possible solutions of $Ax = b$.[17]

This Procedure 3.3.15 determines for us that there is either none, one, or an infinite number of solutions, as Theorem 2.2.26 requires.

However, MATLAB/Octave's "A\" gives one 'answer' for all of these cases, even when there is no solution or an infinite number of solutions. The function rcond(A) indicates whether the 'answer' is a good unique solution of $Ax = b$ (Procedure 2.2.5). Section 3.5 addresses what the 'answer' by MATLAB/Octave means in the other cases of no or infinite solutions.

---

[16] Being diagonal, S is in a special type of reduced row echelon form (Definition 2.2.19).
[17] Any non-uniqueness in the orthogonal U and V just gives rise to equivalent different algebraic expressions for the same set of possibilities.

Condition number and rank determine the possibilities

> The expression 'ill-conditioned' is sometimes used merely as a term of abuse ... It is charac-
> teristic of ill-conditioned sets of equations that small percentage errors in the coefficients given
> may lead to large percentage errors in the solution.   *Alan Turing, 1934 (Higham, 1996, p.131)*

The MATLAB/Octave function rcond () roughly estimates the reciprocal of what is called the
condition number (estimates it to within a factor of two or three, usually).

Definition 3.3.16 *For every $m \times n$ matrix A, the **condition number** of A is the ratio of the largest
to smallest of its singular values: $\operatorname{cond} A := \sigma_1/\sigma_{\min(m,n)}$. By convention: if $\sigma_{\min(m,n)} = 0$, then
$\operatorname{cond} A := \infty$ (infinity); also, for zero matrices $\operatorname{cond} O_{m \times n} := \infty$.*

Example 3.3.17 Example 3.3.7 gives the singular values of two matrices: for the 2×2 matrix
the condition number $\sigma_1/\sigma_2 = (10\sqrt{2})/(5\sqrt{2}) = 2$ (for which rcond $= 0.5$); for the 3×3
matrix the condition number $\sigma_1/\sigma_3 = 12/3 = 4$ (for which rcond $= 0.25$). Example 3.3.8
comments that every $n \times n$ orthogonal matrix has singular values $\sigma_1 = \cdots = \sigma_n = 1$; hence
every orthogonal matrix has condition number one (rcond $= 1$). Such condition numbers less
than 100 (non-small rcond) indicate that all these matrices are "good" matrices (as classified by
Procedure 2.2.5).

However, the matrix in the sports ranking Example 3.3.12 has singular values $\sigma_1 = \sigma_2 = \sqrt{3}$
and $\sigma_3 = 0$ so its condition number $\sigma_1/\sigma_3 = \sqrt{3}/0 = \infty$ (correspondingly, rcond $= 0$) which
indicates that the equations are likely to be unsolvable. (In MATLAB/Octave, see that $\sigma_3 = 2 \cdot
10^{-17}$ so a numerical calculation would give condition number $1.7321/\sigma_3 = 7 \cdot 10^{16}$ which is
effectively infinite.) ☐

Activity 3.3.18 What is the condition number of the matrix of Example 3.3.14,

$$\begin{bmatrix} -9 & -15 & -9 & -15 \\ -10 & 2 & -10 & 2 \\ 8 & 4 & 8 & 4 \end{bmatrix},$$

given it has an SVD (2 d.p.)

$$\begin{bmatrix} -0.86 & 0.43 & 0.29 \\ -0.29 & -0.86 & 0.43 \\ 0.43 & 0.29 & 0.86 \end{bmatrix} \begin{bmatrix} 28 & 0 & 0 & 0 \\ 0 & 14 & 0 & 0 \\ 0 & 0 & 0 & 0 \end{bmatrix} \begin{bmatrix} 0.50 & 0.50 & -0.19 & -0.68 \\ 0.50 & -0.50 & 0.68 & -0.19 \\ 0.50 & 0.50 & 0.19 & 0.68 \\ 0.50 & -0.50 & -0.68 & 0.19 \end{bmatrix}^{T}$$

(a) 0   (b) ∞ (c) 0.5   (d) 2

   In practice, a condition number $> 10^8$ is effectively infinite (equivalently rcond $< 10^{-8}$ is
effectively zero, and hence called "terrible" by Procedure 2.2.5). The closely related important
property of a matrix is the *number* of singular values that are nonzero. When applying the following
definition in practical computation (e.g., MATLAB/Octave), any singular values less than $10^{-8}\sigma_1$
are effectively zero.

Definition 3.3.19 *The **rank** of a matrix A is the number of nonzero singular values in an* SVD, $A = USV^{\mathsf{T}}$: *letting* $r = \text{rank}\,A$,

$$S = \begin{bmatrix} \sigma_1 & \cdots & 0 & & \\ \vdots & \ddots & \vdots & O_{r \times (n-r)} & \\ 0 & \cdots & \sigma_r & & \\ & & & & \\ O_{(m-r) \times r} & & O_{(m-r) \times (n-r)} & \end{bmatrix},$$

equivalently $S = \text{diag}_{m \times n}(\sigma_1, \sigma_2, \ldots, \sigma_r, 0, \ldots, 0)$.

Example 3.3.20 In the four matrices of Example 3.3.17, the respective ranks are 2, 3, n, and 2.

Theorem 3.3.6 asserts that the singular values are unique for a given matrix, so the rank of a matrix is independent of its different SVDs.

Activity 3.3.21 What is the rank of the matrix of Example 3.3.14,

$$\begin{bmatrix} -9 & -15 & -9 & -15 \\ -10 & 2 & -10 & 2 \\ 8 & 4 & 8 & 4 \end{bmatrix},$$

given it has an SVD (2 d.p.)

$$\begin{bmatrix} -0.86 & 0.43 & 0.29 \\ -0.29 & -0.86 & 0.43 \\ 0.43 & 0.29 & 0.86 \end{bmatrix} \begin{bmatrix} 28 & 0 & 0 & 0 \\ 0 & 14 & 0 & 0 \\ 0 & 0 & 0 & 0 \end{bmatrix} \begin{bmatrix} 0.50 & 0.50 & -0.19 & -0.68 \\ 0.50 & -0.50 & 0.68 & -0.19 \\ 0.50 & 0.50 & 0.19 & 0.68 \\ 0.50 & -0.50 & -0.68 & 0.19 \end{bmatrix}^{\mathsf{T}} ?$$

(a) 4   (b) 2   (c) 3   (d) 1

Example 3.3.22 Use MATLAB/Octave to find the rank of the matrix $\begin{bmatrix} 0 & 1 & 0 \\ 1 & 1 & -1 \\ 1 & 0 & -1 \\ 2 & 0 & -2 \end{bmatrix}$

Solution: Enter the matrix into MATLAB/Octave and compute its singular values with svd (A):

```
A= [0  1  0
    1  1  -1
    1  0  -1
    2  0  -2 ]
svd (A)
```

The singular values are 3.49, 1.34 (2 d.p.), and $1.55 \cdot 10^{-16} \approx 0$. Since two singular values are nonzero, the rank of the matrix is two. □

Example 3.3.23 Use MATLAB/Octave to find the rank of the matrix $\begin{bmatrix} 1 & -2 & -1 & 2 & 1 \\ -2 & -2 & -0 & 2 & -0 \\ -2 & -3 & 1 & -1 & 1 \\ -3 & 0 & 1 & -0 & -1 \\ 2 & 1 & 1 & 2 & -1 \end{bmatrix}$

Solution: Enter the matrix into MATLAB/Octave and compute its singular values with svd (A) :[18]

```
A= [1 -2 -1 2 1
  -2 -2 -0 2 -0
  -2 -3 1 -1 1
  -3 0 1 -0 -1
  2 1 1 2 -1 ]
svd (A)
```

The singular values are 5.58, 4.17, 3.13, 1.63 (2 d.p.), and $2.99 \cdot 10^{-16} \approx 0$. Since four singular values are nonzero, the rank of the matrix is four. □

Theorem 3.3.24 *For every matrix A, let an SVD of A be USV^T, then the transpose A^T has an SVD of $V(S^\mathsf{T})U^\mathsf{T}$. Further, $\operatorname{rank}(A^\mathsf{T}) = \operatorname{rank} A$.*

Proof. Let $m \times n$ matrix A have SVD USV^T. Using the properties of the matrix transpose (Theorem 3.1.29),

$$A^\mathsf{T} = (USV^\mathsf{T})^\mathsf{T} = (V^\mathsf{T})^\mathsf{T} S^\mathsf{T} U^\mathsf{T} = V(S^\mathsf{T})U^\mathsf{T}$$

which is an SVD for A^T since U and V are orthogonal, and S^T has the necessary diagonal structure. Further, since the number of nonzero values along the diagonal of S^T is precisely the same as that of the diagonal of S, $\operatorname{rank}(A^\mathsf{T}) = \operatorname{rank} A$.

Example 3.3.25 From earlier examples, write down an SVD of the matrices

$$\begin{bmatrix} 10 & 5 \\ 2 & 11 \end{bmatrix} \quad \text{and} \quad \begin{bmatrix} -4 & -8 & 6 \\ -2 & -1 & 6 \\ 4 & -4 & 0 \end{bmatrix}.$$

Solution: These matrices are the transpose of the two matrices whose SVDs are given in Example 3.3.7. Hence their SVDs are the transpose of the SVDs in that example (remembering that the transpose of a product is the product of the transposes but in reverse order, Theorem 3.1.29(d)):

---

[18] Some advanced students may know that MATLAB/Octave provides the rank () function to directly compute the rank. However, this example is to reinforce its meaning in terms of singular values.

- $$\begin{bmatrix} 10 & 5 \\ 2 & 11 \end{bmatrix} = \begin{bmatrix} \frac{1}{\sqrt{2}} & -\frac{1}{\sqrt{2}} \\ \frac{1}{\sqrt{2}} & \frac{1}{\sqrt{2}} \end{bmatrix} \begin{bmatrix} 10\sqrt{2} & 0 \\ 0 & 5\sqrt{2} \end{bmatrix} \begin{bmatrix} \frac{3}{5} & -\frac{4}{5} \\ \frac{4}{5} & \frac{3}{5} \end{bmatrix}^{\mathrm{T}},$$

- $$\begin{bmatrix} -4 & -8 & 6 \\ -2 & -1 & 6 \\ 4 & -4 & 0 \end{bmatrix} = \begin{bmatrix} -\frac{8}{9} & -\frac{1}{9} & -\frac{4}{9} \\ -\frac{4}{9} & \frac{4}{9} & \frac{7}{9} \\ -\frac{1}{9} & -\frac{8}{9} & \frac{4}{9} \end{bmatrix} \begin{bmatrix} 12 & 0 & 0 \\ 0 & 6 & 0 \\ 0 & 0 & 3 \end{bmatrix} \begin{bmatrix} \frac{1}{3} & -\frac{2}{3} & \frac{2}{3} \\ \frac{2}{3} & \frac{2}{3} & \frac{1}{3} \\ -\frac{2}{3} & \frac{1}{3} & \frac{2}{3} \end{bmatrix}^{\mathrm{T}}.$$

$\square$

Activity 3.3.26 Recall that

$$\begin{bmatrix} -0.86 & 0.43 & 0.29 \\ -0.29 & -0.86 & 0.43 \\ 0.43 & 0.29 & 0.86 \end{bmatrix} \begin{bmatrix} 28 & 0 & 0 & 0 \\ 0 & 14 & 0 & 0 \\ 0 & 0 & 0 & 0 \end{bmatrix} \begin{bmatrix} 0.50 & 0.50 & -0.19 & -0.68 \\ 0.50 & -0.50 & 0.68 & -0.19 \\ 0.50 & 0.50 & 0.19 & 0.68 \\ 0.50 & -0.50 & -0.68 & 0.19 \end{bmatrix}^{\mathrm{T}}$$

is an SVD (2 d.p.) of the matrix of Example 3.3.14,

$$\begin{bmatrix} -9 & -15 & -9 & -15 \\ -10 & 2 & -10 & 2 \\ 8 & 4 & 8 & 4 \end{bmatrix}.$$

Which of the following is an SVD of the transpose of this matrix?

(a) $$\begin{bmatrix} -0.86 & -0.29 & 0.43 \\ 0.43 & -0.86 & 0.29 \\ 0.29 & 0.43 & 0.86 \end{bmatrix} \begin{bmatrix} 28 & 0 & 0 \\ 0 & 14 & 0 \\ 0 & 0 & 0 \\ 0 & 0 & 0 \end{bmatrix} \begin{bmatrix} 0.50 & 0.50 & 0.50 & 0.50 \\ 0.50 & -0.50 & 0.50 & -0.50 \\ -0.19 & 0.68 & 0.19 & -0.68 \\ -0.68 & -0.19 & 0.68 & 0.19 \end{bmatrix}^{\mathrm{T}}$$

(b) $$\begin{bmatrix} -0.86 & 0.43 & 0.29 \\ -0.29 & -0.86 & 0.43 \\ 0.43 & 0.29 & 0.86 \end{bmatrix} \begin{bmatrix} 28 & 0 & 0 \\ 0 & 14 & 0 \\ 0 & 0 & 0 \\ 0 & 0 & 0 \end{bmatrix} \begin{bmatrix} 0.50 & 0.50 & -0.19 & -0.68 \\ 0.50 & -0.50 & 0.68 & -0.19 \\ 0.50 & 0.50 & 0.19 & 0.68 \\ 0.50 & -0.50 & -0.68 & 0.19 \end{bmatrix}^{\mathrm{T}}$$

(c) $$\begin{bmatrix} 0.50 & 0.50 & 0.50 & 0.50 \\ 0.50 & -0.50 & 0.50 & -0.50 \\ -0.19 & 0.68 & 0.19 & -0.68 \\ -0.68 & -0.19 & 0.68 & 0.19 \end{bmatrix} \begin{bmatrix} 28 & 0 & 0 \\ 0 & 14 & 0 \\ 0 & 0 & 0 \\ 0 & 0 & 0 \end{bmatrix} \begin{bmatrix} -0.86 & -0.29 & 0.43 \\ 0.43 & -0.86 & 0.29 \\ 0.29 & 0.43 & 0.86 \end{bmatrix}^{\mathrm{T}}$$

(d) $$\begin{bmatrix} 0.50 & 0.50 & -0.19 & -0.68 \\ 0.50 & -0.50 & 0.68 & -0.19 \\ 0.50 & 0.50 & 0.19 & 0.68 \\ 0.50 & -0.50 & -0.68 & 0.19 \end{bmatrix} \begin{bmatrix} 28 & 0 & 0 \\ 0 & 14 & 0 \\ 0 & 0 & 0 \\ 0 & 0 & 0 \end{bmatrix} \begin{bmatrix} -0.86 & 0.43 & 0.29 \\ -0.29 & -0.86 & 0.43 \\ 0.43 & 0.29 & 0.86 \end{bmatrix}^{\mathrm{T}}$$

Let's now return to the topic of linear equations and connect new concepts to the task of solving linear equations. In particular, the following theorem addresses when a unique solution exists to a system of linear equations. Concepts developed in subsequent sections extend this theorem further (Theorems 3.4.43 and 7.2.41).

Theorem 3.3.27 (*Unique Solutions: version 1*) For every $n \times n$ square matrix A, the following statements are equivalent:

(a) A is invertible;

(b) $Ax = b$ has a **unique solution** for every b in $\mathbb{R}^n$;

(c) $Ax = 0$ has only the zero solution;

(d) all n singular values of A are nonzero;

(e) the condition number of A is finite ($\texttt{rcond} > 0$);

(f) $\operatorname{rank} A = n$.

Proof. Prove a circular chain of implications, with one 'side' equivalence in addition.

3.3.27(a) $\Longrightarrow$ **3.3.27(b)** Established by Theorem 3.2.10.

3.3.27(b) $\Longrightarrow$ **3.3.27(c)** Now $x = 0$ is always a solution of $Ax = 0$. If property 3.3.27(b) holds, then this is the only solution.

3.3.27(c) $\Longrightarrow$ **3.3.27(d)** Use contradiction. Assume a singular value is zero. Then Procedure 3.3.15 finds an infinite number of solutions to the homogeneous system $Ax = 0$, which contradicts 3.3.27(c). Hence the assumption is wrong, so all singular values are nonzero.

3.3.27(d) $\Longleftrightarrow$ **3.3.27(e)** By Definition 3.3.16, the condition number is finite if and only if the smallest singular value is > 0, and hence if and only if all singular values are nonzero.

3.3.27(d) $\Longrightarrow$ **3.3.27(f)** Property 3.3.27(f) is direct from Definition 3.3.19.

3.3.27(f) $\Longrightarrow$ **3.3.27(a)** Find an svd $A = USV^{\mathsf{T}}$. The inverse of the $n \times n$ diagonal matrix S exists as its diagonal elements are the n nonzero singular values (Theorem 3.2.27). Let $B = VS^{-1}U^{\mathsf{T}}$. Then $AB = USV^{\mathsf{T}}VS^{-1}U^{\mathsf{T}} = USS^{-1}U^{\mathsf{T}} = UU^{\mathsf{T}} = I_n$. Similarly $BA = I_n$. From Definition 3.2.2, A is invertible (with $B = VS^{-1}U^{\mathsf{T}}$ as its inverse). $\qquad\square$

Practical shades of grey The preceding Unique Solution Theorem 3.3.27 is 'black-and-white': either a solution exists, or it does not. This is a great theory. But in applications, problems arise in 'all shades of grey'. Practical issues in applications are better phrased in terms of reliability, uncertainty, and error estimates. For example, suppose in an experiment you measure quantities b to three significant digits, then solve the linear equations $Ax = b$ to estimate quantities of interest x: how accurate are your estimates of the interesting quantities x? or are your estimates complete nonsense?

This (optional) discussion and theorem reinforces why we must check condition numbers in computation.

Example 3.3.28 Consider the following innocuous looking system of linear equations

$$\begin{cases} -2q + r = 3 \\ p - 5q + r = 8 \\ -3p + 2q + 3r = -5 \end{cases}$$

Solve by hand (Procedure 2.2.23) to find the unique solution is $(p, q, r) = (2, -1, 1)$.

But—and it is a big but in practical applications—what happens if the right-hand side comes from experimental measurements with a relative error of 1%? (Recall that the *relative error* of an approximation, compared to the exact, is $(|(\text{approx}) - (\text{exact})|)/|\text{exact}|$.) Let's explore by writing the system in matrix-vector form and using MATLAB/Octave to solve with various example errors.

(a) First solve the system as stated. Denoting the unknowns by vector $x = (p,q,r)$, write the system as $Ax = b$ for matrix

$$A = \begin{bmatrix} 0 & -2 & 1 \\ 1 & -5 & 1 \\ -3 & 2 & 3 \end{bmatrix}, \quad \text{and right-hand side } b = \begin{bmatrix} 3 \\ 8 \\ -5 \end{bmatrix}.$$

Use Procedure 2.2.5 to solve the system in MATLAB/Octave:

   i. enter the matrix and vector with

```
A= [0 -2 1; 1 -5 1; -3 2 3]
b= [3;8;-5]
```

   ii. find rcond (A) is 0.0031, which is poor, but we proceed anyway;

   iii. then x=A\b gives the solution $x = (2,-1,1)$ as before.

(b) Now let's recognize that the right-hand side comes from experimental measurements with a 1% error. In MATLAB/Octave, norm (b) computes the length $|b| = 9.90$ (2 d.p.). Thus a 1% error corresponds to changing b by $0.01 \times 9.90 \approx 0.1$. Let's say the first component of b is in error by this amount and see what the new solution would be:

   i. executing x1=A\ (b+ [0.1;0;0]) adds the 1% error $(0.1,0,0)$ to b and then solves the new system to find $x' = (3.7,-0.4,2.3)$. This solution is very different to the original solution $x = (2,-1,1)$!

   ii. relerr1=norm (x-x1) /norm (x) computes its relative error $|x - x'|/|x|$ to be 0.91, that is, 91%—a grossly large relative error.

As illustrated to the right, the large difference between x and x' indicates that 'the solution' x is almost complete nonsense. How can a 1% error in b turn into the astonishingly

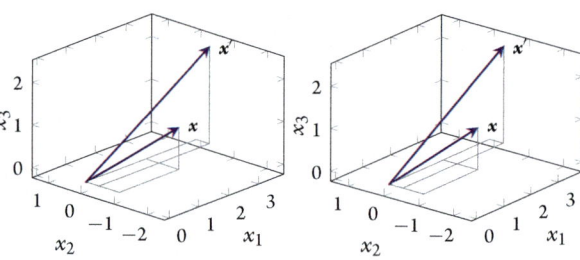

large 91% error in solution x? Theorem 3.3.30 below shows that it is no accident that the magnification of the error by a factor of 91 (from 1% to 91%)) is of the same order of magnitude as the condition number $= 152.27$ computed via s=svd (A) and then condA=s (1) /s (3).

(c) To explore further, let's say the second component of b is in error by 1% of b, that is, by 0.1. As in the previous case, add $(0, 0.1, 0)$ to the right-hand side and solve to find now

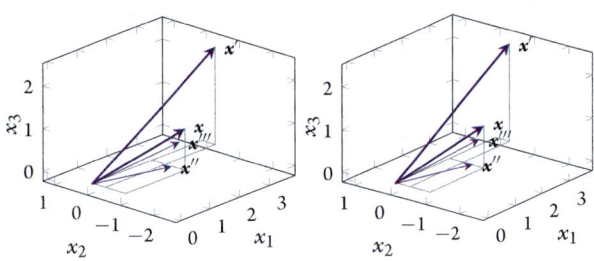

$x'' = (1.2, -1.3, 0.4)$ which is quite different to both x and x', as illustrated to the right. Compute its relative error $|x - x''|/|x| = 0.43$. At 43%, the relative error in solution x'' is also much larger than the 1% error in b.

(d) Lastly, let's say the third component of b is in error by 1% of b, that is, by 0.1. As in the previous cases, add $(0,0,0.1)$ to the right-hand side and solve to find now $x''' = (1.7, -1.1, 0.8)$ which, as illustrated above, is at least roughly x. Compute its relative error $|x - x'''|/|x| = 0.15$. At 15%, the relative error in solution x''' is significantly larger than the 1% error in b.

This example shows that the apparently innocuous matrix A variously multiplies measurement errors in b by factors of at least up to 91 when finding 'the solution' x to $Ax = b$. The matrix A must, after all, be a bad matrix. Theorem 3.3.30 shows that this badness is quantified by its condition number 152.27, and its poor (estimated) reciprocal $\mathtt{rcond\,(A)} = 0.0031$. □

Example 3.3.29 Consider solving the system of linear equations

$$\begin{bmatrix} 0.4 & 0.4 & -0.2 & 0.8 \\ -0.2 & 0.8 & -0.4 & -0.4 \\ 0.4 & -0.4 & -0.8 & -0.2 \\ -0.8 & -0.2 & -0.4 & 0.4 \end{bmatrix} x = \begin{bmatrix} -3 \\ 3 \\ -9 \\ -1 \end{bmatrix}.$$

Use MATLAB/Octave to explore the effect on the solution x of 1% errors in the right-hand side vector.

Solution: Enter the matrix and right-hand side vector into MATLAB/Octave, then solve with Procedure 2.2.5:

```
Q=[0.4 0.4 -0.2 0.8
   -0.2 0.8 -0.4 -0.4
   0.4 -0.4 -0.8 -0.2
   -0.8 -0.2 -0.4 0.4]
b=[-3;3;-9;-1]
rcond(Q)
x=Q\b
```

to find the solution $x = (-4.6, 5, 7, -2.2)$.

Now see the effect on this solution of 1% errors in b. Since the length $|b| = \text{norm}(b) = 10$ we find the solution for various changes to b of magnitude 0.1.

- For example, adding the 1% error $(0.1, 0, 0, 0)$ to b, the MATLAB/Octave commands

  ```
  x1=Q\(b+[0.1;0;0;0])
  relerr1=norm(x-x1)/norm(x)
  ```

  show the changed solution is $x' = (-4.56, 5.04, 6.98, -2.12)$ which here is reasonably close to x. Indeed, its relative error $|x - x'|/|x|$ is computed to be $0.0100 = 1\%$. Here the relative error in solution x is exactly the same as the relative error in b.

- Exploring further, upon adding the 1% error $(0, 0.1, 0, 0)$ to b, analogous commands show the changed solution is $x'' = (-4.62, 5.08, 6.96, -2.24)$ which has relative error $|x - x''|/|x| = 0.0100 = 1\%$ again.

- Whereas, upon adding 1% error $(0, 0, 0.1, 0)$ to b, analogous commands show the changed solution is $x''' = (-4.56, 4.96, 6.92, -2.22)$ which has relative error $|x - x'''|/|x| = 0.0100 = 1\%$ again.

- Lastly, upon adding 1% error $(0, 0, 0, 0.1)$ to b, analogous commands show the changed solution is $x'''' = (-4.68, 4.98, 6.96, -2.16)$ which has relative error $|x - x''''|/|x| = 0.0100 = 1\%$ yet again.

In this example, and in contrast to the previous Example 3.3.28, throughout, the relative error in the solution x is exactly the same as the relative error in b. The reason is that here the matrix Q is an orthogonal matrix—check by computing $Q' * Q$ (Definition 3.2.43). Being orthogonal, multiplication by Q only rotates or reflects, and never stretches or distorts (Theorem 3.2.48(c)). Consequently, errors remain the same magnitude when multiplied by such orthogonal matrices, as seen in this example, and as reflected in the condition number of Q being one (as computed via $s=\text{svd}(Q)$ and then $\text{condQ}=s(1)/s(4)$). □

The condition number determines the reliability of the solution of a system of linear equations. This is why we should always precede the computation of a solution with an estimate of the condition number such as that provided by the reciprocal $\text{rcond}()$ (Procedure 2.2.5). The next theorem establishes that the condition number characterizes the amplification of errors that occurs in solving a linear system. Hence solving a system of linear equations with a large condition number (small rcond) means that errors are amplified by a large factor as happens in Example 3.3.28.

Theorem 3.3.30 (*error magnification*) *Consider solving $Ax = b$ for $n \times n$ matrix A with full rank $A = n$. Suppose the right-hand side b has relative error of magnitude ϵ (the symbol ϵ is the Greek letter epsilon, and often denotes errors), then the solution x has relative error $\leq \epsilon \, \text{cond} A$, with equality in the worst case.*

Proof. Let the length of the right-hand side vector be $b = |b|$. Then the error in b has magnitude ϵb since ϵ is the relative error. Following Procedure 3.3.15, let $A = USV^{\mathsf{T}}$ be an

SVD for matrix A. Compute $z = U^T b$: recall that multiplication by orthogonal U preserves lengths (Theorem 3.2.48), so not only is $|z| = b$, but also z will be in error by an amount ϵb since b has this error. Consider solving $Sy = z$: the diagonals of S stretch and shrink both the 'signal' and the 'noise'. The *worst case* is when $z = (b, 0, \ldots, 0, \epsilon b)$; that is, when all the 'signal' happens to be in the first component of z, and all the 'noise', the error, is in the last component. Then the intermediary $y = (b/\sigma_1, 0, \ldots, \epsilon b/\sigma_n)$. Consequently, the intermediary has relative error $(\epsilon b/\sigma_n)/(b/\sigma_1) = \epsilon(\sigma_1/\sigma_n) = \epsilon \operatorname{cond} A$. Again because multiplication by orthogonal V preserves lengths, the solution $x = Vy$ has the same relative error: in the worst case this relative error is $\epsilon \operatorname{cond} A$.

Example 3.3.31 Each of the following cases involves solving a linear system $Ax = b$ to determine quantities of interest x from some measured quantities b. From the given information, estimate the maximum relative error in x, if possible, otherwise say so.

(a) Quantities b are measured to a relative error 0.001, and matrix A has condition number of 10.

(b) Quantities b are measured to three significant digits and $\texttt{rcond (A)} = 0.025$.

(c) Measurements are accurate to two decimal places, and matrix A has condition number of 20.

(d) Measurements are correct to two significant digits and $\texttt{rcond (A)} = 0.002$.

Solution:

(a) The relative error in x could be as big as $0.001 \times 10 = 0.01$.

(b) Recall that three significant digits means numbers such as 123 or 0.123 with an error of less than a half of the least significant digit, here 0.5 or 0.0005 respectively. So measuring to three significant digits means the relative error could be as large as 0.005. Then here with $\texttt{rcond (A)} = 0.025$, matrix A has condition number of roughly 40, so the relative error of x is less than $0.005 \times 40 = 0.2$; that is, up to 20%.

(c) There is not enough information as we cannot determine the *relative* error in measurements b.

(d) Two significant digits means the relative error could be as large as 0.05, whereas matrix A has condition number of roughly $1/0.002 = 500$ so the relative error of x could be as big as $0.05 \times 500 = 25$; that is, the estimated solution x is likely to be complete rubbish.

$\square$

Activity 3.3.32 In some experiment the components of b, $|b| = 5$, are measured to two decimal places. We compute a vector x by solving $Ax = b$. For matrix A, we compute $\texttt{rcond (A)} = 0.02$. What is our estimate of the largest possible relative error in x?

(a) 20%   (b) 2%   (c) 0.1%   (d) 5%

This issue of the amplification of errors occurs in other contexts. The eminent mathematician Henri Poincaré (1854–1912) was the first to detect possible chaos in the orbits of the planets.

If we knew exactly the laws of nature and the situation of the universe at the initial moment, we could predict exactly the situation of that same universe at a succeeding moment. But even if it were the case that the natural laws had no longer any secret for us, we could still only know the initial situation approximately. If that enabled us to predict the succeeding situation with the same approximation, that is all we require, and we should say that the phenomenon had been predicted, that it is governed by laws. But it is not always so; it may happen that small differences in the initial conditions produce very great ones in the final phenomena. A small error in the former will produce an enormous error in the latter. Prediction becomes impossible, and we have the fortuitous phenomenon.   *Poincaré, 1903*

The analogue for us in solving linear equations such as $Ax = b$ is the following: it may happen that a small error in the elements of b produces an enormous error in the final x. The condition number warns when this happens by characterizing the amplification.

3.3.3 Prove the SVD Theorem 3.3.6

When doing maths there's this great feeling. You start with a problem that just mystifies you. You can't understand it, it's so complicated, you just can't make head nor tail of it. But then when you finally resolve it, you have this incredible feeling of how beautiful it is, how it all fits together so elegantly.   *Andrew Wiles, C1993*

Two preliminary examples introduce the structure of the general proof that an SVD exists.[19] As in this example prelude, the proof of a general singular value decomposition is similarly constructive.

Prelude to the proof

These first two examples are optional: their purpose is to introduce two key parts of the general proof in a definite setting.

Example 3.3.33 *(a 2 × 2 case)* Recall Example 3.3.2 factorized the matrix

$$A = \begin{bmatrix} 10 & 2 \\ 5 & 11 \end{bmatrix} = \begin{bmatrix} \frac{3}{5} & -\frac{4}{5} \\ \frac{4}{5} & \frac{3}{5} \end{bmatrix} \begin{bmatrix} 10\sqrt{2} & 0 \\ 0 & 5\sqrt{2} \end{bmatrix} \begin{bmatrix} \frac{1}{\sqrt{2}} & -\frac{1}{\sqrt{2}} \\ \frac{1}{\sqrt{2}} & \frac{1}{\sqrt{2}} \end{bmatrix}^{\mathrm{T}}.$$

Here we find this factorization, $A = USV^{\mathrm{T}}$, by maximizing $|Av|$ over all unit vectors v (all vectors of length one).

Solution: In 2D, all unit vectors are of the form $v = (\cos t, \sin t)$ for $-\pi < t \leq \pi$. The picture to the right plots these unit vectors v in blue for 32 angles t.[20] Plotted in red from the end of each v is the vector Av (scaled down by a factor of ten for clarity). Our aim is to find the v that maximizes the length of the corresponding adjoined Av. By inspection,

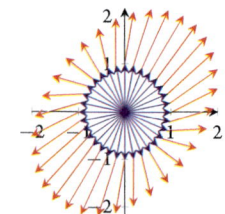

---

[19] This proof may be delayed until the last week of a semester. It may be given together with the closely related classic proof of Theorem 4.2.16 on the eigenvectors of symmetric matrices.

[20] The MATLAB function `eigshow(A)` (download for R2017b and later) provides an interactive alternative to this static view—click on the `eig/`(`svd`) button to make `eigshow(A)` show `svd/`(`eig`).

the longest red vectors $A\boldsymbol{v}$ occur towards the top-right or the bottom-left, either of these directions $\boldsymbol{v}$ are what we first find.

Maximizing $|A\boldsymbol{v}|$ is the same as maximizing $|A\boldsymbol{v}|^2$ which is what the following considers: since

$$Av = \begin{bmatrix} 10 & 2 \\ 5 & 11 \end{bmatrix} \begin{bmatrix} \cos t \\ \sin t \end{bmatrix} = \begin{bmatrix} 10\cos t + 2\sin t \\ 5\cos t + 11\sin t \end{bmatrix},$$

$$\begin{aligned} |Av|^2 &= (10\cos t + 2\sin t)^2 + (5\cos t + 11\sin t)^2 \\ &= 100\cos^2 t + 40\cos t \sin t + 4\sin^2 t \\ &\quad + 25\cos^2 t + 110\cos t \sin t + 121\sin^2 t \\ &= 125(\cos^2 t + \sin^2 t) + 150\sin t \cos t \\ &= 125 + 75\sin 2t \qquad \text{(shown to the below-right).} \end{aligned}$$

Since the sine function has maximum of one at angle $\frac{\pi}{2}$ (90°), the maximum of $|A\boldsymbol{v}|^2$ is $125 + 75 = 200$ for $2t = \frac{\pi}{2}$, that is, for $t = \frac{\pi}{4}$ corresponding to unit vector $\boldsymbol{v}_1 = (\cos\frac{\pi}{4}, \sin\frac{\pi}{4}) = (\frac{1}{\sqrt{2}}, \frac{1}{\sqrt{2}})$—this vector points to the top-right as identified from the previous right-hand figure. This vector is the first column of V.

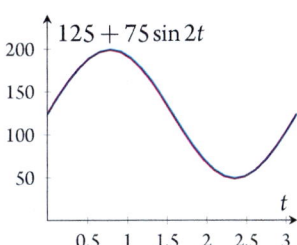

Now multiply to find $A\boldsymbol{v}_1 = (6\sqrt{2}, 8\sqrt{2})$. The length of this vector is $\sqrt{72 + 128} = \sqrt{200} = 10\sqrt{2} = \sigma_1$ the first singular value. Normalize the vector $A\boldsymbol{v}_1$ by $A\boldsymbol{v}_1/\sigma_1 = (6\sqrt{2}, 8\sqrt{2})/(10\sqrt{2}) = (\frac{3}{5}, \frac{4}{5}) = \boldsymbol{u}_1$, the first column of U.

The other column of V must be orthogonal (at right-angles) to $\boldsymbol{v}_1$ in order for matrix V to be orthogonal. Thus set $\boldsymbol{v}_2 = (-\frac{1}{\sqrt{2}}, \frac{1}{\sqrt{2}})$, as shown in the plot to the right. Now multiply to find $A\boldsymbol{v}_2 = (-4\sqrt{2}, 3\sqrt{2})$: magically, and a crucial part of the general proof, the vector $A\boldsymbol{v}_2$ is orthogonal to $\boldsymbol{u}_1$. The length of $A\boldsymbol{v}_2 = (-4\sqrt{2}, 3\sqrt{2})$ is $\sqrt{32 + 18} = \sqrt{50} = 5\sqrt{2} = \sigma_2$, the other singular value. Normalize the vector to $A\boldsymbol{v}_2/\sigma_2 = (-4\sqrt{2}, 3\sqrt{2})/(2\sqrt{2}) = (-\frac{4}{5}, \frac{3}{5}) = \boldsymbol{u}_2$, the second column of U.

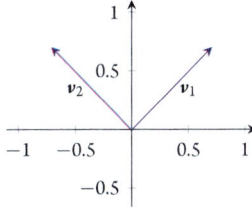

This construction establishes that here $AV = US$. Then post-multiply each side by V^{T} to find an SVD is $A = USV^{\mathsf{T}}$.

In this example, we could have chosen the negative of $\boldsymbol{v}_1$ (angle $t = -\frac{3\pi}{4}$), and/or chosen the negative of $\boldsymbol{v}_2$. The result would still be a valid SVD of the matrix A. The orthogonal matrices in an SVD are not unique, and need not be. The ordered singular values are unique. ☐

Example 3.3.34 (*a* 3×1 *case*) Find the following SVD for the 3×1 matrix

$$A = \begin{bmatrix} 1 \\ 1 \\ 1 \end{bmatrix} = \begin{bmatrix} \frac{1}{\sqrt{3}} & \cdot & \cdot \\ \frac{1}{\sqrt{3}} & \cdot & \cdot \\ \frac{1}{\sqrt{3}} & \cdot & \cdot \end{bmatrix} \begin{bmatrix} \sqrt{3} \\ 0 \\ 0 \end{bmatrix} \begin{bmatrix} 1 \end{bmatrix}^{\mathsf{T}} = USV^{\mathsf{T}},$$

where we do not worry about the elements denoted by dots as they are only multiplied by the zeros in the 'diagonal' $S = (\sqrt{3}, 0, 0)$.

Solution: We seek to maximize $|Av|^2$ but, since matrix A is 3×1, here vector v is in $\mathbb{R}^1$. Being of unit magnitude, there are two alternatives: $v = (\pm 1)$. Each alternative gives the same $|Av|^2 = |(\pm 1, \pm 1, \pm 1)| = 3$. Choosing one alternative, say $v_1 = (1)$, then fixes the matrix $V = \begin{bmatrix} 1 \end{bmatrix}$.

Then $Av_1 = (1, 1, 1)$, which is of length $\sqrt{3}$. This length is the singular value $\sigma_1 = \sqrt{3}$. Dividing Av_1 by its length gives the unit vector $u_1 = (\frac{1}{\sqrt{3}}, \frac{1}{\sqrt{3}}, \frac{1}{\sqrt{3}})$, the first column of U. To find the other columns of U, consider the three standard unit

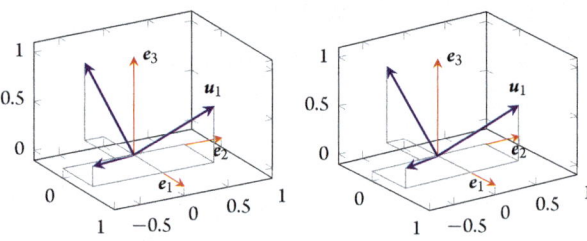

vectors in $\mathbb{R}^3$ (red in the illustration to the right), rotate them all together so that one lines up with u_1, and then the other two rotated unit vectors form the other two columns of U (blue vectors in the illustration). Since the columns of U are then orthonormal, U is an orthogonal matrix (Theorem 3.2.48). □

Outline of the general proof   We use induction on the size $m \times n$ of the matrix.

- First, zero matrices have a trivial SVD, and, further, $m \times 1$ and $1 \times n$ matrices have a straightforward SVD (as in Example 3.3.34).

- For any given $m \times n$ matrix A, choose v_1 to maximize $|Av|^2$ among all unit vectors v in $\mathbb{R}^n$.

- Crucially, we then establish that for every vector v orthogonal to v_1, the vector Av is orthogonal to Av_1.

- Then rotate the standard unit vectors to align one with v_1. Similarly for Av_1.

- This rotation transforms the matrix A to strip off the leading singular value, and to effectively leave an $(m-1) \times (n-1)$ matrix.

- By induction on the size, an SVD exists for all sizes.

This proof corresponds closely to the proof of the spectral Theorem 4.2.16 for symmetric matrices of Section 4.2.

Detailed proof of the SVD Theorem 3.3.6

Use induction on the size $m \times n$ of the matrix A: we assume an SVD exists for all $(m-1) \times (n-1)$ matrices, and prove that consequently an SVD must exist for all $m \times n$ matrices. There are three base cases to establish: one for $m \leq n$, one for $m \geq n$, and one for matrix $A = O$; then the induction extends to all sized matrices.

Case $A = O_{m \times n}$: When $m \times n$ matrix $A = O_{m \times n}$ then choose $U = I_m$ (orthogonal), $S = O_{m \times n}$ (diagonal), and $V = I_n$ (orthogonal), so then $USV^\mathsf{T} = I_m O_{m \times n} I_n^\mathsf{T} = O_{m \times n} = A$.

Consequently, the rest of the proof only considers the non-trivial cases when the matrix A is not all zero.

Case $m \times 1$ ($n = 1$): Here the $m \times 1$ nonzero matrix $A = \begin{bmatrix} a_1 \end{bmatrix}$ for $a_1 = (a_{11}, a_{21}, \ldots, a_{m1})$. Set the singular value $\sigma_1 = |a_1| = \sqrt{a_{11}^2 + a_{21}^2 + \cdots + a_{m1}^2}$ and unit vector $u_1 = a_1/\sigma_1$. Set 1×1 orthogonal matrix $V = \begin{bmatrix} 1 \end{bmatrix}$; $m \times 1$ diagonal matrix $S = (\sigma_1, 0, \ldots, 0)$; and $m \times m$ orthogonal matrix $U = \begin{bmatrix} u_1 & u_2 & \cdots & u_m \end{bmatrix}$. Matrix U exists because we can take the orthonormal set of standard unit vectors in $\mathbb{R}^m$ and rotate them all together so that the first lines up with u_1; the other $(m-1)$ unit vectors then become the other u_j. Then an SVD for the $m \times 1$ matrix A is

$$USV^\mathsf{T} = \begin{bmatrix} u_1 & u_2 & \cdots & u_m \end{bmatrix} \begin{bmatrix} \sigma_1 \\ 0 \\ \vdots \\ 0 \end{bmatrix} 1^\mathsf{T} = \sigma_1 u_1 = \begin{bmatrix} a_1 \end{bmatrix} = A.$$

Case $1 \times n$ ($m = 1$): use an exactly complementary argument to the preceding $m \times 1$ case.

Induction Assume an SVD exists for all $(m-1) \times (n-1)$ matrices: we proceed to prove that consequently an SVD must exist for all $m \times n$ matrices. Consider any $m \times n$ nonzero matrix A with $m, n \geq 2$. Set vector v_1 in $\mathbb{R}^n$ to be a unit vector that maximizes $|Av|^2$ for unit vectors v in $\mathbb{R}^n$; that is, vector v_1 achieves the maximum in $\max_{|v|=1} |Av|^2$.

1. *Such a maximum exists by the Extreme Value Theorem in calculus.* This theorem is proved in higher level analysis.

   As matrix A is nonzero, there exists v such that $|Av| > 0$. Since v_1 maximizes $|Av|$ it follows that $|Av_1| > 0$.

   The vector v_1 is not unique: for example, the negative $-v_1$ is another unit vector that achieves the maximum value. Sometimes there are other unit vectors that achieve the maximum value. Choose any one of them.

   Nonetheless, the maximum value of $|Av|^2$ is unique, and so the following singular value σ_1 is unique.

2. *Set the singular value $\sigma_1 := |Av_1| > 0$ and unit vector $u_1 := (Av_1)/\sigma_1$ in $\mathbb{R}^m$. For every unit vector v orthogonal to v_1 we now prove that the vector Av is orthogonal to u_1.* Let $u := Av$ in $\mathbb{R}^m$ and consider $f(t) := |A(v_1 \cos t + v \sin t)|^2$. Since v_1 achieves the maximum, and $v_1 \cos t + v \sin t$ is a unit vector for all t (Exercise 3.3.15), then $f(t)$ must have a maximum at $t = 0$ (maybe at other t as well), and so $f'(0) = 0$ (from the calculus of a maximum). On the other hand,

$$f(t) = |A\mathbf{v}_1 \cos t + A\mathbf{v} \sin t|^2$$
$$= |\sigma_1 \mathbf{u}_1 \cos t + \mathbf{u} \sin t|^2$$
$$= (\sigma_1 \mathbf{u}_1 \cos t + \mathbf{u} \sin t) \cdot (\sigma_1 \mathbf{u}_1 \cos t + \mathbf{u} \sin t)$$
$$= \sigma_1^2 \cos^2 t + \sigma_1 \mathbf{u} \cdot \mathbf{u}_1 2 \sin t \cos t + |\mathbf{u}|^2 \sin^2 t;$$

differentiating $f(t)$ and evaluating at zero gives $f'(0) = \sigma_1 \mathbf{u} \cdot \mathbf{u}_1$. As $t = 0$ is a maximum, this derivative is zero, so $\sigma_1 \mathbf{u} \cdot \mathbf{u}_1 = 0$. Since the singular value $\sigma_1 > 0$, we must have $\mathbf{u} \cdot \mathbf{u}_1 = 0$ and so $\mathbf{u}_1$ and $\mathbf{u}$ are orthogonal (Definition 1.3.19).

3. Consider the orthonormal set of standard unit vectors in $\mathbb{R}^n$: rotate them so that the first unit vector lines up with $\mathbf{v}_1$, and let the other $(n-1)$ rotated unit vectors become the columns of the $n \times (n-1)$ matrix $\bar{V}$. Then set the $n \times n$ matrix $V_1 := \begin{bmatrix} \mathbf{v}_1 & \bar{V} \end{bmatrix}$ which is orthogonal as its columns are orthonormal (Theorem 3.2.48(b)). Similarly, set an $m \times m$ orthogonal matrix $U_1 := \begin{bmatrix} \mathbf{u}_1 & \bar{U} \end{bmatrix}$. Compute the $m \times n$ matrix

$$A_1 := U_1^T A V_1 = \begin{bmatrix} \mathbf{u}_1^T \\ \bar{U}^T \end{bmatrix} A \begin{bmatrix} \mathbf{v}_1 & \bar{V} \end{bmatrix} = \begin{bmatrix} \mathbf{u}_1^T A \mathbf{v}_1 & \mathbf{u}_1^T A \bar{V} \\ \bar{U}^T A \mathbf{v}_1 & \bar{U}^T A \bar{V} \end{bmatrix}$$

where

- the top-left entry $\mathbf{u}_1^T A \mathbf{v}_1 = \mathbf{u}_1^T \sigma_1 \mathbf{u}_1 = \sigma_1 |\mathbf{u}_1|^2 = \sigma_1$,
- the bottom-left column $\bar{U}^T A \mathbf{v}_1 = \bar{U}^T \sigma_1 \mathbf{u}_1 = O_{m-1 \times 1}$ as the columns of $\bar{U}$ are orthogonal to $\mathbf{u}_1$,
- the top-right row $\mathbf{u}_1^T A \bar{V} = O_{1 \times n-1}$ as each column of $\bar{V}$ is orthogonal to $\mathbf{v}_1$ and hence each column of $A \bar{V}$ is orthogonal to $\mathbf{u}_1$,
- and set the bottom-right block $B := \bar{U}^T A \bar{V}$ which is an $(m-1) \times (n-1)$ matrix as $\bar{U}^T$ is $(m-1) \times m$ and $\bar{V}$ is $n \times (n-1)$.

Consequently,

$$A_1 = \begin{bmatrix} \sigma_1 & O_{1 \times n-1} \\ O_{m-1 \times 1} & B \end{bmatrix}.$$

Note: rearranging $A_1 := U_1^T A V_1$ gives $A V_1 = U_1 A_1$.

4. *By induction assumption, $(m-1) \times (n-1)$ matrix B has an SVD, and so we now construct an SVD for $m \times n$ matrix A.* Let $B = \hat{U} \hat{S} \hat{V}^T$ be an SVD for B. Then construct (for appropriately sized zero matrices O)

$$U := U_1 \begin{bmatrix} 1 & O \\ O & \hat{U} \end{bmatrix}, \quad V := V_1 \begin{bmatrix} 1 & O \\ O & \hat{V} \end{bmatrix}, \quad S := \begin{bmatrix} \sigma_1 & O \\ O & \hat{S} \end{bmatrix}.$$

Matrices U and V are orthogonal as each is the product of two orthogonal matrices (Exercise 3.2.21). Also, matrix S is diagonal. These form an SVD for matrix A since

$$AV = AV_1 \begin{bmatrix} 1 & O \\ O & \hat{V} \end{bmatrix} = U_1 A_1 \begin{bmatrix} 1 & O \\ O & \hat{V} \end{bmatrix} = U_1 \begin{bmatrix} \sigma_1 & O \\ O & B \end{bmatrix} \begin{bmatrix} 1 & O \\ O & \hat{V} \end{bmatrix}$$

$$= U_1 \begin{bmatrix} \sigma_1 & O \\ O & B\hat{V} \end{bmatrix} = U_1 \begin{bmatrix} \sigma_1 & O \\ O & \hat{U}\hat{S} \end{bmatrix} = U_1 \begin{bmatrix} 1 & O \\ O & \hat{U} \end{bmatrix} \begin{bmatrix} \sigma_1 & O \\ O & \hat{S} \end{bmatrix} = US.$$

Hence $A = USV^{\mathsf{T}}$ is an SVD for A.

By induction, an SVD exists for all $m \times n$ matrices. This argument establishes the SVD Theorem 3.3.6.

3.3.4 Exercises

Exercise 3.3.1 Using a factorization of the left-hand side coefficient, quickly solve by hand the following equations.

(a) $18x = 1134$

(b) $70x = 3150$

(c) $99x = 8118$

(d) $242x = 20086$

(e) $245x = 12495$

(f) $539x = 28028$

Exercise 3.3.2 Find a general solution, if a solution exists, of each of the following systems of linear equations using Procedure 3.3.15. Calculate by hand using the given SVD factorization; record your working.

(a) $\underbrace{\begin{bmatrix} -\frac{9}{5} & \frac{12}{5} \\ -4 & -3 \end{bmatrix}}_{=A} x = \begin{bmatrix} -\frac{9}{5} \\ \frac{17}{2} \end{bmatrix}$ given the SVD

$$A = \begin{bmatrix} 0 & 1 \\ 1 & 0 \end{bmatrix} \begin{bmatrix} 5 & 0 \\ 0 & 3 \end{bmatrix} \begin{bmatrix} -\frac{4}{5} & -\frac{3}{5} \\ -\frac{3}{5} & \frac{4}{5} \end{bmatrix}^{\mathsf{T}}$$

(b) $\underbrace{\begin{bmatrix} -\frac{5}{26} & -\frac{6}{13} \\ -\frac{12}{13} & \frac{5}{13} \end{bmatrix}}_{=B} x = \begin{bmatrix} -\frac{7}{13} \\ \frac{34}{13} \end{bmatrix}$ given the SVD

$$B = \begin{bmatrix} 0 & 1 \\ 1 & 0 \end{bmatrix} \begin{bmatrix} 1 & 0 \\ 0 & \frac{1}{2} \end{bmatrix} \begin{bmatrix} -\frac{12}{13} & -\frac{5}{13} \\ \frac{5}{13} & -\frac{12}{13} \end{bmatrix}^{\mathsf{T}}$$

(c) $\underbrace{\begin{bmatrix} -\frac{2}{3} & \frac{23}{51} & \frac{22}{51} \\ \frac{1}{6} & \frac{7}{51} & -\frac{31}{51} \end{bmatrix}}_{=C} x = \begin{bmatrix} -\frac{115}{102} \\ -\frac{35}{102} \end{bmatrix}$ given the SVD

$$C = \begin{bmatrix} \frac{15}{17} & -\frac{8}{17} \\ -\frac{8}{17} & -\frac{15}{17} \end{bmatrix} \begin{bmatrix} 1 & 0 & 0 \\ 0 & \frac{1}{2} & 0 \end{bmatrix} \begin{bmatrix} -\frac{2}{3} & \frac{1}{3} & -\frac{2}{3} \\ \frac{1}{3} & -\frac{2}{3} & -\frac{2}{3} \\ \frac{2}{3} & \frac{2}{3} & -\frac{1}{3} \end{bmatrix}^{\mathrm{T}}$$

(d) $\underbrace{\begin{bmatrix} \frac{36}{119} & -\frac{11}{17} \\ \frac{164}{119} & -\frac{18}{17} \\ -\frac{138}{119} & -\frac{6}{17} \end{bmatrix}}_{=D} x = \begin{bmatrix} \frac{11}{17} \\ \frac{9}{17} \\ \frac{3}{17} \end{bmatrix}$ given the SVD

$$D = \begin{bmatrix} \frac{2}{7} & -\frac{3}{7} & -\frac{6}{7} \\ \frac{6}{7} & -\frac{2}{7} & \frac{3}{7} \\ -\frac{3}{7} & -\frac{6}{7} & \frac{2}{7} \end{bmatrix} \begin{bmatrix} 2 & 0 \\ 0 & 1 \\ 0 & 0 \end{bmatrix} \begin{bmatrix} \frac{15}{17} & \frac{8}{17} \\ -\frac{8}{17} & \frac{15}{17} \end{bmatrix}^{\mathrm{T}}$$

(e) $\underbrace{\begin{bmatrix} -\frac{17}{18} & -\frac{8}{9} & -\frac{8}{9} \\ 1 & \frac{2}{3} & -\frac{2}{3} \\ -\frac{11}{9} & \frac{8}{9} & -\frac{7}{9} \end{bmatrix}}_{=E} x = \begin{bmatrix} -\frac{17}{18} \\ \frac{5}{3} \\ -\frac{7}{18} \end{bmatrix}$ given the SVD

$$E = \begin{bmatrix} -\frac{2}{3} & -\frac{1}{3} & -\frac{2}{3} \\ \frac{1}{3} & \frac{2}{3} & -\frac{2}{3} \\ -\frac{2}{3} & \frac{2}{3} & \frac{1}{3} \end{bmatrix} \begin{bmatrix} 2 & 0 & 0 \\ 0 & \frac{3}{2} & 0 \\ 0 & 0 & 1 \end{bmatrix} \begin{bmatrix} \frac{8}{9} & \frac{1}{9} & -\frac{4}{9} \\ \frac{1}{9} & \frac{8}{9} & \frac{4}{9} \\ \frac{4}{9} & -\frac{4}{9} & \frac{7}{9} \end{bmatrix}^{\mathrm{T}}$$

(f) $\underbrace{\begin{bmatrix} -\frac{6}{11} & -\frac{1}{11} & \frac{81}{22} \\ \frac{7}{11} & \frac{3}{11} & \frac{27}{11} \\ -\frac{6}{11} & \frac{9}{22} & -\frac{9}{11} \end{bmatrix}}_{=F} x = \begin{bmatrix} -\frac{35}{2} \\ -\frac{41}{4} \\ \frac{15}{8} \end{bmatrix}$ given the SVD

$$F = \begin{bmatrix} \frac{9}{11} & \frac{6}{11} & \frac{2}{11} \\ \frac{6}{11} & -\frac{7}{11} & -\frac{6}{11} \\ -\frac{2}{11} & \frac{6}{11} & -\frac{9}{11} \end{bmatrix} \begin{bmatrix} \frac{9}{2} & 0 & 0 \\ 0 & 1 & 0 \\ 0 & 0 & \frac{1}{2} \end{bmatrix} \begin{bmatrix} 0 & -1 & 0 \\ 0 & 0 & -1 \\ 1 & 0 & 0 \end{bmatrix}^{\mathrm{T}}$$

Exercise 3.3.3 Find a general solution, if a solution exists, of each of the following systems of linear equations. Calculate by hand using the given SVD factorization; check the SVD and confirm your calculations with MATLAB/Octave (Procedure 3.3.15); then compare and contrast the two methods.

(a)
$$
\underbrace{\begin{bmatrix}
\frac{7}{180} & \frac{8}{45} & \frac{41}{180} \\
\frac{19}{180} & -\frac{22}{45} & \frac{101}{180} \\
-\frac{19}{180} & \frac{4}{45} & \frac{133}{180} \\
\frac{59}{60} & -\frac{2}{15} & \frac{91}{60}
\end{bmatrix}}_{=A}
x =
\begin{bmatrix}
-\frac{13}{40} \\
-\frac{9}{8} \\
-\frac{17}{20} \\
-\frac{15}{4}
\end{bmatrix}
\text{ given the SVD}
$$

$$
A =
\begin{bmatrix}
-\frac{1}{10} & \frac{3}{10} & \frac{3}{10} & -\frac{9}{10} \\
-\frac{3}{10} & -\frac{9}{10} & -\frac{1}{10} & -\frac{3}{10} \\
-\frac{3}{10} & -\frac{1}{10} & \frac{9}{10} & \frac{3}{10} \\
-\frac{9}{10} & \frac{3}{10} & -\frac{3}{10} & \frac{1}{10}
\end{bmatrix}
\begin{bmatrix}
2 & 0 & 0 \\
0 & \frac{1}{2} & 0 \\
0 & 0 & \frac{1}{2} \\
0 & 0 & 0
\end{bmatrix}
\begin{bmatrix}
-\frac{4}{9} & \frac{4}{9} & -\frac{7}{9} \\
\frac{1}{9} & \frac{8}{9} & \frac{4}{9} \\
-\frac{8}{9} & -\frac{1}{9} & \frac{4}{9}
\end{bmatrix}^{\mathrm{T}}
$$

(b)
$$
\underbrace{\begin{bmatrix}
\frac{57}{22} & -\frac{3}{22} & -\frac{45}{22} & -\frac{9}{22} \\
-\frac{14}{11} & \frac{32}{11} & -\frac{4}{11} & -\frac{14}{11} \\
-\frac{9}{22} & -\frac{3}{22} & \frac{45}{22} & \frac{57}{22}
\end{bmatrix}}_{=B}
x =
\begin{bmatrix}
117 \\
-72 \\
63
\end{bmatrix}
\text{ given the SVD}
$$

$$
B =
\begin{bmatrix}
-\frac{6}{11} & \frac{9}{11} & \frac{2}{11} \\
\frac{7}{11} & \frac{6}{11} & -\frac{6}{11} \\
-\frac{6}{11} & -\frac{2}{11} & -\frac{9}{11}
\end{bmatrix}
\begin{bmatrix}
4 & 0 & 0 & 0 \\
0 & 3 & 0 & 0 \\
0 & 0 & 3 & 0
\end{bmatrix}
\begin{bmatrix}
-\frac{1}{2} & \frac{1}{2} & \frac{1}{2} & \frac{1}{2} \\
\frac{1}{2} & \frac{1}{2} & -\frac{1}{2} & \frac{1}{2} \\
\frac{1}{2} & -\frac{1}{2} & \frac{1}{2} & \frac{1}{2} \\
-\frac{1}{2} & -\frac{1}{2} & -\frac{1}{2} & \frac{1}{2}
\end{bmatrix}^{\mathrm{T}}
$$

(c)
$$
\underbrace{\begin{bmatrix}
-\frac{2}{5} & -\frac{2}{5} & -\frac{26}{45} & \frac{26}{45} \\
\frac{11}{9} & \frac{11}{9} & -\frac{1}{3} & \frac{1}{3} \\
\frac{31}{90} & \frac{31}{90} & \frac{17}{90} & -\frac{17}{90} \\
\frac{4}{9} & \frac{4}{9} & -\frac{2}{9} & \frac{2}{9}
\end{bmatrix}}_{=C}
x =
\begin{bmatrix}
3 \\
-6 \\
-3 \\
-2
\end{bmatrix}
\text{ given the SVD}
$$

$$
C =
\begin{bmatrix}
\frac{2}{9} & \frac{8}{9} & \frac{1}{3} & \frac{2}{9} \\
-\frac{8}{9} & \frac{2}{9} & -\frac{2}{9} & \frac{1}{3} \\
-\frac{2}{9} & -\frac{1}{3} & \frac{8}{9} & \frac{2}{9} \\
-\frac{1}{3} & \frac{2}{9} & \frac{2}{9} & -\frac{8}{9}
\end{bmatrix}
\begin{bmatrix}
2 & 0 & 0 & 0 \\
0 & 1 & 0 & 0 \\
0 & 0 & 0 & 0 \\
0 & 0 & 0 & 0
\end{bmatrix}
\begin{bmatrix}
-\frac{7}{10} & -\frac{1}{10} & \frac{1}{10} & -\frac{7}{10} \\
-\frac{7}{10} & \frac{1}{10} & \frac{1}{10} & \frac{7}{10} \\
\frac{1}{10} & -\frac{7}{10} & \frac{7}{10} & \frac{1}{10} \\
\frac{1}{10} & \frac{7}{10} & \frac{7}{10} & -\frac{1}{10}
\end{bmatrix}^{\mathrm{T}}
$$

Exercise 3.3.4 Find a general solution, if possible, of each of the following systems of linear equations with MATLAB/Octave and using Procedure 3.3.15.

(a) $\begin{bmatrix} 2.4 & 1.6 & 1 & -0.8 \\ -1.2 & 3.2 & -2 & -0.4 \\ -1.2 & -0.8 & 2 & -1.6 \\ 0.6 & -1.6 & -4 & -0.8 \end{bmatrix} x = \begin{bmatrix} -29.4 \\ -12.4 \\ 13.2 \\ -0.8 \end{bmatrix}$

(b) $\begin{bmatrix} 1.38 & 0.50 & 3.30 & 0.34 \\ -0.66 & -0.70 & 1.50 & -2.38 \\ -0.90 & 2.78 & -0.54 & 0.10 \\ 0.00 & 1.04 & -0.72 & -1.60 \end{bmatrix} x = \begin{bmatrix} -7.64 \\ -7.72 \\ -20.72 \\ -20.56 \end{bmatrix}$

(c) $\begin{bmatrix} 1.32 & 1.40 & 1.24 & -0.20 \\ 1.24 & 3.00 & 2.68 & 1.00 \\ 1.90 & -1.06 & -1.70 & 2.58 \\ -1.30 & 0.58 & 0.90 & -0.94 \end{bmatrix} x = \begin{bmatrix} -5.28 \\ 2.04 \\ 6.30 \\ 2.50 \end{bmatrix}$

(d) $\begin{bmatrix} 7 & 1 & -1 & 4 \\ 2 & 4 & -4 & 0 \\ 0 & 4 & 0 & -1 \\ -4 & 1 & 1 & -1 \\ -1 & 0 & -1 & 3 \end{bmatrix} x = \begin{bmatrix} 22.4 \\ 11.2 \\ -6.1 \\ -8.3 \\ 17.8 \end{bmatrix}$

(e) $\begin{bmatrix} 7 & 1 & -1 & 4 \\ 2 & 4 & -4 & 0 \\ 0 & 4 & 0 & -1 \\ -4 & 1 & 1 & -1 \\ -1 & 0 & -1 & 3 \end{bmatrix} x = \begin{bmatrix} -2.1 \\ 2.2 \\ 4.6 \\ -0.7 \\ 5.5 \end{bmatrix}$

Exercise 3.3.5 Recall Theorems 2.2.26 and 2.2.30 on the existence of none, one, or an infinite number of solutions to linear equations. Use Procedure 3.3.15 to provide an alternative proof to each of these two theorems.

Exercise 3.3.6 Write down the condition number and the rank of each of the matrices in Exercise 3.3.2 using the given svds.

Exercise 3.3.7 In MATLAB/Octave, use randn() to generate some random matrices A of chosen sizes, and some correspondingly sized random right-hand side vectors b. For each, find a general solution, if possible, of the system $Ax = b$ with MATLAB/Octave and using Procedure 3.3.15. Record each step, the condition number and rank of A, and comment on what is interesting about the sizes you choose.

Exercise 3.3.8 Let $m \times n$ matrix A have the svd $A = USV^{\mathsf{T}}$. Derive that the matrix $A^{\mathsf{T}}A$ has an svd $A^{\mathsf{T}}A = V\bar{S}V^{\mathsf{T}}$, for what matrix $\bar{S}$? Derive that the matrix AA^{T} has an svd $AA^{\mathsf{T}} = U\tilde{S}U^{\mathsf{T}}$, for what matrix $\tilde{S}$?

Exercise 3.3.9 Consider the problems in Exercises 3.3.2 and 3.3.3. For each of these problems comment on the applicability of the Unique Solution Theorem 3.3.27, and comment on how the solution(s) illustrate the theorem.

Exercise 3.3.10 Recall that Definition 3.2.2 says that a square matrix A is invertible if there exists a matrix B such that *both* $AB = I$ *and* $BA = I$. We now see that we need only one of these to ensure that the matrix is invertible.

(a) Use Theorem 3.3.27(c) to now prove that a square matrix A is invertible if there exists a matrix B such that $BA = I$.

(b) Use the transpose and Theorem 3.3.27(f) and Theorem 3.3.24 to then prove that a square matrix A is invertible if there exists a matrix B such that $AB = I$.

Exercise 3.3.11 For each of the following systems, explore the effect on the solution of 1% errors in the right-hand side, and comment on the relation to the given condition number of the matrix.

(a) $\begin{bmatrix} 2 & -4 \\ -2 & -1 \end{bmatrix} x = \begin{bmatrix} 10 \\ 0 \end{bmatrix}$, cond $= 2$

(c) $\begin{bmatrix} -1 & 1 \\ 4 & -5 \end{bmatrix} x = \begin{bmatrix} -2 \\ 10 \end{bmatrix}$, cond $= 42.98$

(b) $\begin{bmatrix} -3 & 1 \\ -4 & 2 \end{bmatrix} x = \begin{bmatrix} 6 \\ 8 \end{bmatrix}$, cond $= 14.93$

Exercise 3.3.12 For each of the following systems, use MATLAB/Octave to explore the effect on the solution of 0.1% errors in the right-hand side. Record your commands and output, and comment on the relation to the condition number of the matrix.

(a) $\begin{bmatrix} 1 & 2 & 2 \\ -1 & -1 & 0 \\ 0 & 3 & 1 \end{bmatrix} x = \begin{bmatrix} -7 \\ 2 \\ -7 \end{bmatrix}$

(b) $\begin{bmatrix} -3 & -2 & -2 & -2 \\ 2 & 1 & -5 & -7 \\ 2 & 4 & 3 & 3 \\ 2 & 1 & 1 & 1 \end{bmatrix} x = \begin{bmatrix} -3 \\ -8 \\ 5 \\ 2 \end{bmatrix}$

(c) $\begin{bmatrix} -1 & 6 & -6 & 2 & 7 \\ -7 & 4 & 3 & 1 & -8 \\ 7 & 6 & 4 & 0 & 5 \\ -8 & 3 & 3 & 2 & 4 \\ 2 & 0 & -3 & 1 & 0 \end{bmatrix} x = \begin{bmatrix} 5 \\ -7 \\ 5 \\ -2 \\ 1 \end{bmatrix}$

Exercise 3.3.13 For any $m \times n$ matrix A, use an SVD $A = USV^T$ to prove that $\operatorname{rank}(A^T A) = \operatorname{rank} A$ and that $\operatorname{cond}(A^T A) = \operatorname{cond}(A)^2$ (see Exercise 3.3.8).

Exercise 3.3.14 Recall how Example 3.3.33 introduced that finding a singular vector and singular value of a matrix A came from maximizing $|Av|$. Each of the following matrices, say A for discussion, has plotted Av (red) adjoined the corresponding unit vector v (blue). For each case:

 i. by inspection of the plot, estimate a singular vector v_1 that appears to maximize $|Av_1|$ (to one decimal place say);

 ii. estimate the corresponding singular value σ_1 by measuring $|Av_1|$ on the plot;

 iii. set the second singular vector v_2 to be orthogonal to v_1 by swapping components, and making one negative;

 iv. estimate the corresponding singular value σ_2 by measuring $|Av_2|$ on the plot;

 v. compute the matrix-vector products Av_1 and Av_2, and confirm that they are orthogonal (approximately).

(a) $A = \begin{bmatrix} 1 & 1 \\ 0.2 & 1.4 \end{bmatrix}$

(c) $C = \begin{bmatrix} 1.3 & 0.9 \\ 1.4 & 0.9 \end{bmatrix}$

(b) $B = \begin{bmatrix} 0 & -1.3 \\ 0.4 & 1.1 \end{bmatrix}$

(d) $D = \begin{bmatrix} 1.4 & -0.4 \\ -1.6 & 0.9 \end{bmatrix}$

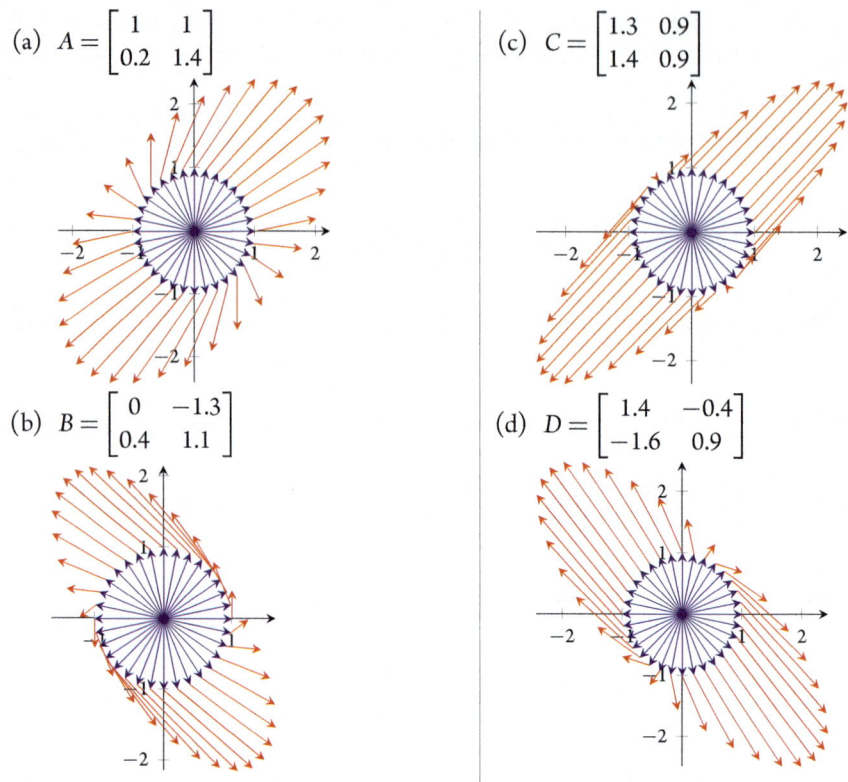

Exercise 3.3.15 Use properties of the dot product to prove that when v_1 and v are orthogonal unit vectors the vector $v_1 \cos t + v \sin t$ is also a unit vector for all t (used in the proof of the SVD in Section 3.3.3).

Exercise 3.3.16 In a few sentences, answer/discuss each of the following.

(a) Why can factorization be useful in solving equations?

(b) What is it about the matrices in an SVD that makes the factorization useful?

(c) In solving linear equations, how does the SVD show that non-unique solutions arise in two ways?

(d) Why is every condition number greater than or equal to one?

(e) When using a computer, why do we often treat computed numbers of magnitude about 10^{-16} as effectively zero? and computed numbers of magnitude about 10^{16} as effectively infinity?

(f) In estimating the relative error in the solution x of $Ax = b$ in terms of the relative error of b, how does the worst case error arise?

(g) Why does Procedure 2.2.5 check the value of `rcond (A)` before computing a solution to $Ax = b$?

3.4 Subspaces, basis and dimension

[Nature] is written in that great book which ever lies before our eyes — I mean the universe — but we cannot understand it if we do not first learn the language and grasp the symbols in which it is written. The book is written in the mathematical language, and the symbols are triangles, circles, and other geometric figures, without whose help it is impossible to comprehend a single word of it; without which one wanders in vain through a dark labyrinth.

Galileo Galilei, 1610

Some of the most fundamental geometric structures in mathematics, especially linear algebra, are the lines or planes through the origin, and higher dimensional analogues. For example, a general solution of linear equations often involves linear combinations such as $(-2,1,0,0)s +$ $(-\frac{15}{7},0,\frac{9}{7},1)t$ (Example 2.2.28(c)) and $y_3 v_3 + y_4 v_4$ (Example 3.3.14): such combinations for all values of the free variables form a plane through the origin (Section 1.3.4). The aim of this section is to connect geometric structures, such as lines and planes, to the information in a singular value decomposition. The structures are called subspaces.

3.4.1 Subspaces are lines, planes, and so on

Let's introduce graphically the concept of a "subspace". Definition 3.4.3 then gives the precise algebraic description.

Example 3.4.1   The following graphs illustrate the concept of subspaces through examples (imagine the graphs extend to infinity, as appropriate).

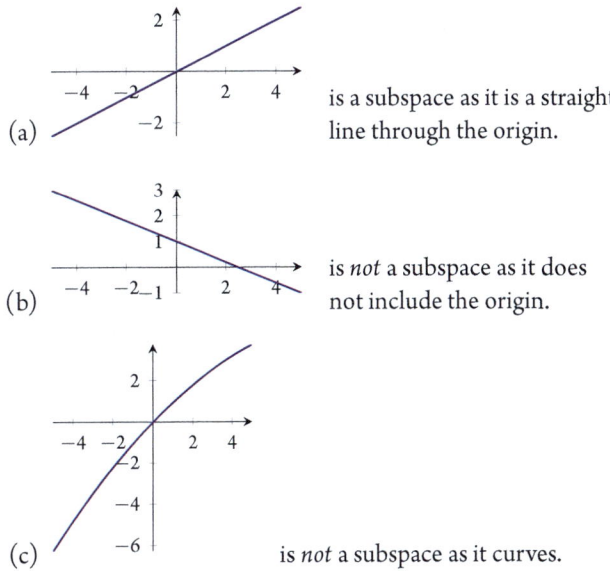

(a) is a subspace as it is a straight line through the origin.

(b) is *not* a subspace as it does not include the origin.

(c) is *not* a subspace as it curves.

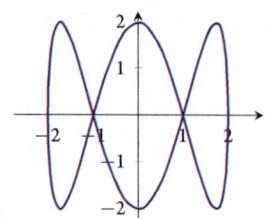

(d)

is *not* a subspace as it not only curves, but also does not include the origin.

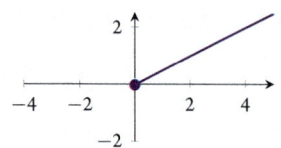

(e)

where the disc indicates an end to the line, is *not* a subspace as it does not extend infinitely in both directions.

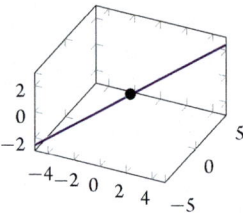

(f)

is a subspace as it is a line through the origin (marked ● in these 3D plots).

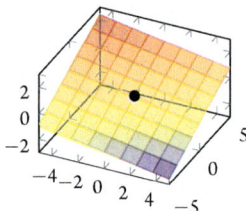

(g)

is a subspace as it is a plane through the origin.

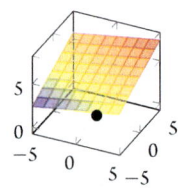

(h)

is *not* a subspace as it does not go through the origin.

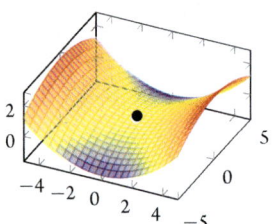

(i)

is *not* a subspace as it curves.

Activity 3.4.2 Given the examples and comments of Example 3.4.1, which of the following is a subspace?

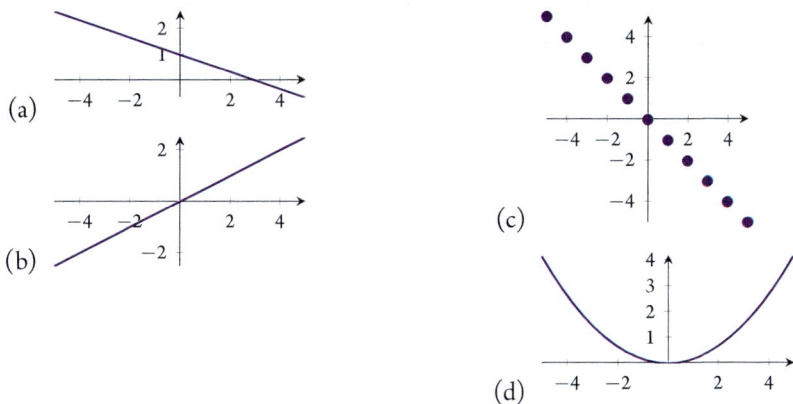

(a)

(b)

(c)

(d)

The following definition expresses precisely in algebra the concept of a subspace. This book uses the 'blackboard bold' font, such as $\mathbb{W}$ and $\mathbb{R}$, for names of spaces and subspaces.

Recall that the mathematical symbol "$\in$" means "in" or "in the set" or "is an element of the set". For two examples: "$c \in \mathbb{R}$" means "c is a real number"; whereas "$v \in \mathbb{R}^3$" means "v is a vector with three components". For conciseness, hereafter this book uses "$\in$" extensively.

Definition 3.4.3 *A **subspace** $\mathbb{W}$ of $\mathbb{R}^n$ is a set of vectors with $\mathbf{0} \in \mathbb{W}$, and such that for every $c \in \mathbb{R}$ and for every $\mathbf{u}, \mathbf{v} \in \mathbb{W}$, then both $\mathbf{u} + \mathbf{v} \in \mathbb{W}$ and $c\mathbf{u} \in \mathbb{W}$. (When these properties hold we often say that $\mathbb{W}$ is **closed** under addition and scalar multiplication.)*

Example 3.4.4 Use Definition 3.4.3 to determine whether each of the following are subspaces, or not.

(a) All vectors (x, y) in the line $y = x/2$ (Example 3.4.1(a)).

Solution: The origin, the zero vector $\mathbf{0}$, is in the line $y = x/2$ as $x = y = 0$ satisfies the equation. The line $y = x/2$ is composed of vectors in the form $\mathbf{u} = (1, \frac{1}{2})t$ for some parameter t. Then for any $c \in \mathbb{R}$, $c\mathbf{u} = c(1, \frac{1}{2})t = (1, \frac{1}{2})(ct) = (1, \frac{1}{2})t'$ for new parameter $t' = ct$; hence $c\mathbf{u}$ is in the line. Let $\mathbf{v} = (1, \frac{1}{2})s$ be another vector in the line for some parameter s, then $\mathbf{u} + \mathbf{v} = (1, \frac{1}{2})t + (1, \frac{1}{2})s = (1, \frac{1}{2})(t + s) = (1, \frac{1}{2})t'$ for new parameter $t' = t + s$; hence $\mathbf{u} + \mathbf{v}$ is in the line. The three requirements of Definition 3.4.3 are met, and so this line is a subspace. □

(b) All vectors (x, y) with end-points on the curve $y = x - x^2/20$ (Example 3.4.1(c)).

Solution: A vector is 'in the set' when its end-point lies on a plot of the set, as to the right. To show something is not a subspace, we only need to give one instance when 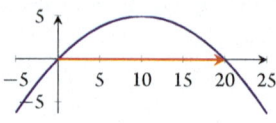 any one of the properties fail. One instance is that the vector $(20, 0)$ is in the curve as $20 - 20^2/20 = 0$, but the scalar multiple of half of this vector $\frac{1}{2}(20, 0) = (10, 0)$ is not as $10 - 10^2/20 = 5 \neq 0$. That is, the curve is not closed under scalar multiplication and hence is not a subspace.   □

(c) All vectors (x,y) in the line $y = x/2$ for $x, y \geq 0$ (Example 3.4.1(e)).

Solution: A vector is 'in the set' when its end-point lies on a plot of the set, as to the right. Although vectors (x, y) in the line $y = x/2$ for $x, y \geq 0$ includes the origin and is 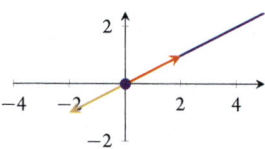 closed under addition, it fails the scalar multiplication test. For example, $u = (2, 1)$ is in the line, but the scalar multiple $(-1)u = (-2, -1)$ is not. Hence it is not a subspace.   □

(d) All vectors (x,y,z) in the plane $z = -x/6 + y/3$ (Example 3.4.1(g)).

Solution: The   origin, the   zero   vector   **0**,   is   in the   plane $z = -x/6 + y/3$ as $x = y = z = 0$ satisfies the   equation.   A   vector 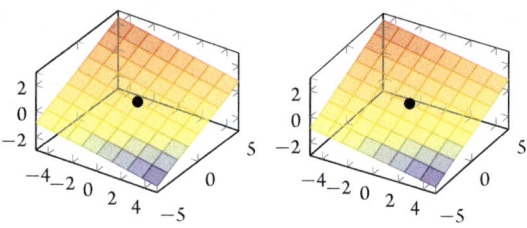 $u = (u_1, u_2, u_3)$ is   in   the plane provided $-u_1 + 2u_2 - 6u_3 = 0$. Consider $cu = (cu_1, cu_2, cu_3)$ for which $-(cu_1) + 2(cu_2) - 6(cu_3) = c(-u_1 + 2u_2 - 6u_3) = c \times 0 = 0$ and hence must also be in the plane. Also let vector $v = (v_1, v_2, v_3)$ be in the plane and consider $u + v = (u_1 + v_1, u_2 + v_2, u_3 + v_3)$ for which $-(u_1 + v_1) + 2(u_2 + v_2) - 6(u_3 + v_3) = -u_1 - v_1 + 2u_2 + 2v_2 - 6u_3 - 6v_3 = (-u_1 + 2u_2 - 6u_3) + (-v_1 + 2v_2 - 6v_3) = 0 + 0 = 0$ and hence must also be in the plane. The three requirements of Definition 3.4.3 are met, and so this plane is a subspace.   □

(e) All vectors (x,y,z) in the plane $z = 5 + x/6 + y/3$ (Example 3.4.1(h)).

Solution: The origin is not in the plane as $u = (0,0,0)$ does not satisfy the equation. Hence this plane is not a subspace.   □

(f) $\{\mathbf{0}\}$ (the set of the zero vector in some $\mathbb{R}^n$).

Solution: The zero vector forms a trivial subspace, $\mathbb{W} = \{\mathbf{0}\}$: firstly, $\mathbf{0} \in \mathbb{W}$; secondly, the only vector in $\mathbb{W}$ is $\mathbf{u} = \mathbf{0}$ for which every scalar multiple $c\mathbf{u} = c\mathbf{0} = \mathbf{0} \in \mathbb{W}$; and thirdly, a second vector $\mathbf{v}$ in $\mathbb{W}$ can only be $\mathbf{v} = \mathbf{0}$ so $\mathbf{u} + \mathbf{v} = \mathbf{0} + \mathbf{0} = \mathbf{0} \in \mathbb{W}$. The three requirements of Definition 3.4.3 are met, and so $\{\mathbf{0}\}$ is always a subspace. $\quad\square$

(g) $\mathbb{R}^n$.

Solution: Lastly, $\mathbb{R}^n$ also is a subspace: firstly, $\mathbf{0} = (0, 0, \ldots, 0) \in \mathbb{R}^n$; secondly, for $\mathbf{u} = (u_1, u_2, \ldots, u_n) \in \mathbb{R}^n$, the scalar multiplication $c\mathbf{u} = c(u_1, u_2, \ldots, u_n) = (cu_1, cu_2, \ldots, cu_n) \in \mathbb{R}^n$; and thirdly, for $\mathbf{v} = (v_1, v_2, \ldots, v_n) \in \mathbb{R}^n$, the vector addition $\mathbf{u} + \mathbf{v} = (u_1, u_2, \ldots, u_n) + (v_1, v_2, \ldots, v_n) = (u_1 + v_1, u_2 + v_2, \ldots, u_n + v_n) \in \mathbb{R}^n$. The three requirements of Definition 3.4.3 are met, and so $\mathbb{R}^n$ is always a subspace. $\quad\square$

Activity 3.4.5 The following pairs of vectors are all in the set shown to the right (in the sense that their end-points lie on the plotted curve). The sum of which pair proves that the curve plotted to the right is not a subspace?

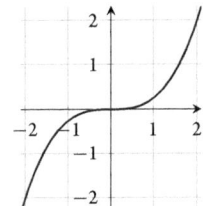

(a) $(2,2)$, $(-2,-2)$

(b) $(1, \frac{1}{4})$, $(0,0)$

(c) $(0,0)$, $(2,2)$

(d) $(-1, -\frac{1}{4})$, $(2,2)$

In summary:

- in two dimensions (denoted $\mathbb{R}^2$), subspaces are the origin $\{\mathbf{0}\}$, any line through $\mathbf{0}$, or the entire plane $\mathbb{R}^2$;
- in three dimensions (denoted $\mathbb{R}^3$), subspaces are the origin $\{\mathbf{0}\}$, any line through $\mathbf{0}$, any plane through $\mathbf{0}$, or the entire space $\mathbb{R}^3$;
- and analogously for higher dimensions (denoted $\mathbb{R}^n$).

Recall that the set of all linear combinations of a set of vectors, such as $(-2,1,0,0)s + (-\frac{15}{7}, 0, \frac{9}{7}, 1)t$ (Example 2.2.28(c)), is called the span of that set (Definition 2.3.10).

Theorem 3.4.6 *Let $\mathbf{w}_1, \mathbf{w}_2, \ldots, \mathbf{w}_k$ be k vectors in $\mathbb{R}^n$, then $\mathrm{span}\{\mathbf{w}_1, \mathbf{w}_2, \ldots, \mathbf{w}_k\}$ is a subspace of $\mathbb{R}^n$.*

Proof. Denote $\mathrm{span}\{\mathbf{w}_1, \mathbf{w}_2, \ldots, \mathbf{w}_k\}$ by $\mathbb{W}$; we prove it is a subspace (Definition 3.4.3). First, $\mathbf{0} = 0\mathbf{w}_1 + 0\mathbf{w}_2 + \cdots + 0\mathbf{w}_k$ which is a linear combination of $\mathbf{w}_1, \mathbf{w}_2, \ldots, \mathbf{w}_k$, and so the zero vector $\mathbf{0} \in \mathbb{W}$. Now let $\mathbf{u}, \mathbf{v} \in \mathbb{W}$ then by Definition 2.3.10 there are coefficients $a_1, a_2, \ldots, a_n$ and $b_1, b_2, \ldots, b_n$ such that

$$\mathbf{u} = a_1\mathbf{w}_1 + a_2\mathbf{w}_2 + \cdots + a_k\mathbf{w}_k,$$
$$\mathbf{v} = b_1\mathbf{w}_1 + b_2\mathbf{w}_2 + \cdots + b_k\mathbf{w}_k.$$

Secondly, consequently

$$\begin{aligned}
\boldsymbol{u} + \boldsymbol{v} &= a_1\boldsymbol{w}_1 + a_2\boldsymbol{w}_2 + \cdots + a_k\boldsymbol{w}_k \\
&\quad + b_1\boldsymbol{w}_1 + b_2\boldsymbol{w}_2 + \cdots + b_k\boldsymbol{w}_k \\
&= (a_1 + b_1)\boldsymbol{w}_1 + (a_2 + b_2)\boldsymbol{w}_2 + \cdots + (a_k + b_k)\boldsymbol{w}_k,
\end{aligned}$$

which is a linear combination of $\boldsymbol{w}_1$, $\boldsymbol{w}_2$, ..., $\boldsymbol{w}_k$, and so is in $\mathbb{W}$. Thirdly, for every scalar c,

$$cu = c(a_1\boldsymbol{w}_1 + a_2\boldsymbol{w}_2 + \cdots + a_k\boldsymbol{w}_k) = ca_1\boldsymbol{w}_1 + ca_2\boldsymbol{w}_2 + \cdots + ca_k\boldsymbol{w}_k,$$

which is a linear combination of $\boldsymbol{w}_1$, $\boldsymbol{w}_2$, ..., $\boldsymbol{w}_k$, and so is in $\mathbb{W}$. Hence $\mathbb{W} = \text{span}\{\boldsymbol{w}_1, \boldsymbol{w}_2, \ldots, \boldsymbol{w}_k\}$ is a subspace.

Example 3.4.7 $\text{span}\{(1, \frac{1}{2})\}$ is the subspace $y = x/2$. The reason is that a vector $\boldsymbol{u} \in \text{span}\{(1, \frac{1}{2})\}$ only if there is some constant a_1 such that $\boldsymbol{u} = a_1(1, \frac{1}{2}) = (a_1, a_1/2)$. That is, the y-component is half the x-component and hence it lies on the line $y = x/2$.

$\text{span}\{(1, \frac{1}{2}), (-2, -1)\}$ is also the subspace $y = x/2$ since every linear combination $a_1(1, \frac{1}{2}) + a_2(-2, -1) = (a_1 - 2a_2, a_1/2 - a_2)$ satisfies that the y-component is half the x-component and hence the linear combination lies on the line $y = x/2$.

Example 3.4.8 The plane $z = -x/6 + y/3$ may be written as $\text{span}\{(3, 3, 1/2), (0, 3, 1)\}$, as illustrated in stereo to the right, since every linear combination of these two vectors fills out the plane: $a_1(3, 3, 1/2) + a_2(0, 3, 1) = (3a_1, 3a_1 + 3a_2, a_1/2 + a_2)$ and so lies in the plane as $-x/6 + y/3 - z = -\frac{1}{6}3a_1 + \frac{1}{3}(3a_1 + 3a_2) - (a_1/2 + a_2) = -\frac{1}{2}a_1 + a_1 + a_2 - \frac{1}{2}a_1 - a_2 = 0$ for all a_1 and a_2 (although such arguments do not establish that the linear combinations cover the whole plane—we need Theorem 3.4.14).

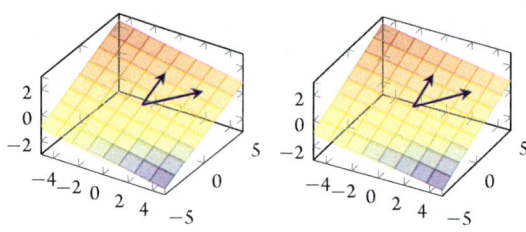

Also, $\text{span}\{(5, 1, -1/2), (0, -3, -1), (-4, 1, 1)\}$ is the plane $z = -x/6 + y/3$, as illustrated to the right. The reason is that every linear combination of these three vectors fills out the plane: $a_1(5, 1, -1/2) + a_2(0, -3, -1) + a_3(-4, 1, 1) = (5a_1 - 4a_3, a_1 - 3a_2 + a_3, -a_1/2 - a_2 + a_3)$ and so lies in the plane as $-x/6 + y/3 - z = -\frac{1}{6}(5a_1 - 4a_3) + \frac{1}{3}(a_1 - 3a_2 + a_3) - (-a_1/2 - a_2 + a_3) = -\frac{5}{6}a_1 + \frac{2}{3}a_3 + \frac{1}{3}a_1 - a_2 + \frac{1}{3}a_3 + \frac{1}{2}a_1 + a_2 - a_3 = 0$ for all a_1, a_2 and a_3. $\square$

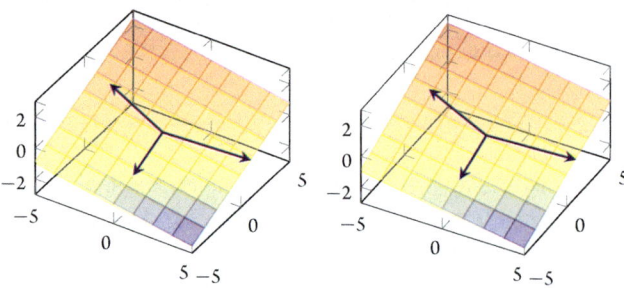

Example 3.4.9 Find a set of two vectors that spans the plane $x - 2y + 3z = 0$.

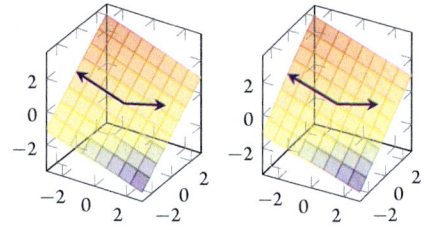

Solution: Write the equation for this plane as $x = 2y - 3z$, say, then vectors in the plane are all of the form $\boldsymbol{u} = (x,y,z) = (2y - 3z, y, z) = (2,1,0)y + (-3,0,1)z$. That is, all vectors in the plane may be written as a linear combination of the two vectors $(2,1,0)$ and $(-3,0,1)$, hence the plane is span$\{(2,1,0), (-3,0,1)\}$ as illustrated in stereo to the right. □

Such subspaces connect with matrices. The connection is via a matrix whose columns are the vectors appearing within the span although sometimes we also use the rows of the matrix to be the vectors in the span.

Definition 3.4.10 *For every $m \times n$ matrix A:*

(a) *the **column space** of A is the subspace of $\mathbb{R}^m$ spanned by the n column vectors of A;*[21]

(b) *the **row space** of A is the subspace of $\mathbb{R}^n$ spanned by the m row vectors (transposed) of A.*

Example 3.4.11 Examples 3.4.7 to 3.4.9 provide some cases.

- From Example 3.4.7, the column space of $A = \begin{bmatrix} 1 & -2 \\ 1/2 & -1 \end{bmatrix}$ is the line $y = x/2$. The row space of this matrix A is span$\{(1, -2), (\frac{1}{2}, -1)\}$. This row space is the set of all vectors of the form $(1, -2)s + (\frac{1}{2}, -1)t = (s + t/2, -2s - t) = (1, -2)(s + t/2) = (1, -2)t'$ is the line $y = -2x$ as illustrated to the right. That the row space and the column space are both lines, albeit different lines, is not a coincidence (Theorem 3.4.32).

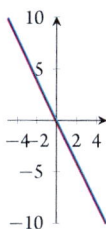

- Example 3.4.8 shows that the column space of matrix

$$B = \begin{bmatrix} 3 & 0 \\ 3 & 3 \\ \frac{1}{2} & 1 \end{bmatrix}$$

is the plane $z = -x/6 + y/3$ in $\mathbb{R}^3$.

---

[21] Some of you may know that the column space is also called the range, but for the moment we just use the term column space.

The row space of matrix B is span$\{(3,0),(3,3),(\frac{1}{2},1)\}$ which is a subspace of $\mathbb{R}^2$—the right-hand plot shows the three vectors. Whereas the column space is a subspace of $\mathbb{R}^3$. Here the row space is all of $\mathbb{R}^2$ as for each $(x,y) \in \mathbb{R}^2$ choose the linear combination $\frac{x-y}{3}(3,0)+\frac{y}{3}(3,3)+0(\frac{1}{2},1)=(x-y+y+0,0+y+0)=(x,y)$ so each (x,y) is in the span, and hence all of the $\mathbb{R}^2$ plane is the span. That the column space and the row space are both planes is no coincidence (Theorem 3.4.32).

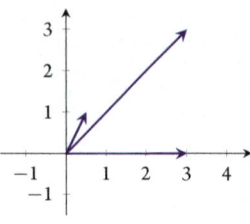

- Example 3.4.8 also shows that the column space of matrix

$$C = \begin{bmatrix} 5 & 0 & -4 \\ 1 & -3 & 1 \\ -\frac{1}{2} & -1 & 1 \end{bmatrix}$$

is also the plane $z = -x/6 + y/3$ in $\mathbb{R}^3$.

Now, span$\{(5,0,-4),(1,-3,1),(-\frac{1}{2},-1,1)\}$ is the row space of matrix C. It is not readily apparent, but we can check that this space is the plane $4x+3y+5z=0$, as illustrated to the right in stereo. To see this, consider all linear combinations $a_1(5,0,-4)+a_2(1,-3,1)+a_3(-\frac{1}{2},-1,1)=(5a_1+a_2-a_3/2,-3a_2-a_3,-4a_1+a_2+a_3)$ satisfy $4x+3y+5z=4(5a_1+a_2-a_3/2)+3(-3a_2-a_3)+5(-4a_1+a_2+a_3)=20a_1+4a_2-2a_3-9a_2-3a_3-20a_1+5a_2+5a_3=0$. Again, it is no coincidence that the row and column spaces of C are both planes (Theorem 3.4.32). □

Activity 3.4.12 Which one of the following vectors is in the column space of the matrix

$$\begin{bmatrix} 6 & 2 \\ -3 & 5 \\ -2 & -1 \end{bmatrix} ?$$

(a) $\begin{bmatrix} 2 \\ -3 \\ -3 \end{bmatrix}$ (b) $\begin{bmatrix} 8 \\ 5 \\ -2 \end{bmatrix}$ (c) $\begin{bmatrix} 2 \\ 2 \\ -3 \end{bmatrix}$ (d) $\begin{bmatrix} 8 \\ 2 \\ -3 \end{bmatrix}$

Example 3.4.13 Is vector $b = (-0.6, 0, -2.1, 1.9, 1.2)$ in the column space of matrix

$$A = \begin{bmatrix} 2.8 & -3.1 & 3.4 \\ 4.0 & 1.7 & 0.8 \\ -0.4 & -0.1 & 4.4 \\ 1.0 & -0.4 & -4.7 \\ -0.3 & 1.9 & 0.7 \end{bmatrix} ?$$

What about vector $c = (15.2, 5.4, 3.8, -1.9, -3.7)$?

Solution: The question is: can we find a linear combination of the columns of A which equals vector $\boldsymbol{b}$? That is, can we find some vector $\boldsymbol{x}$ of the coefficients in the linear combination such that $A\boldsymbol{x} = \boldsymbol{b}$? Answer using our knowledge of linear equations.

Let's use Procedure 3.3.15 in MATLAB/Octave.

(a) Compute an SVD of this 5×3 matrix with

```
A= [2.8 -3.1   3.4
    4.0   1.7   0.8
   -0.4 -0.1   4.4
    1.0 -0.4 -4.7
   -0.3   1.9   0.7]
[U,S,V] =svd(A)
```

to find (2 d.p.)

```
U =
   -0.58    0.49    0.53   -0.07    0.37
   -0.17    0.69   -0.65   -0.04   -0.25
   -0.56   -0.28   -0.10    0.74   -0.22
    0.57    0.43    0.21    0.66    0.14
   -0.04   -0.15   -0.49    0.10    0.85
S =
    7.52       0       0
       0    4.91       0
       0       0    3.86
       0       0       0
       0       0       0
V = ...
```

(b) Then solve $U\boldsymbol{z} = \boldsymbol{b}$ with `z=U' * [-0.6;0;-2.1;1.9;1.2]` to find (2 d.p.) $\boldsymbol{z} = (2.55, 0.92, -0.29, -0.15, 1.54)$.

(c) Now the diagonal matrix S has three nonzero singular values, and the last two rows are zero. So to be able to solve $S\boldsymbol{y} = \boldsymbol{z}$ we need the last two components of $\boldsymbol{z}$ to be zero. But, at -0.15 and 1.54, they are not zero, so the system is not solvable. Hence there is no linear combination of the columns of A that gives us vector $\boldsymbol{b}$. Consequently, vector $\boldsymbol{b}$ is not in the column space of A.

(d) For the case of vector $\boldsymbol{c} = (15.2, 5.4, 3.8, -1.9, -3.7)$ solve $U\boldsymbol{z} = \boldsymbol{c}$ with

```
z=U' * [15.2;5.4;3.8;-1.9;-3.7]
```

to find $\boldsymbol{z} = (-12.800, 9.876, 5.533, 0.000, 0.000)$. Since the last two entries in vector $\boldsymbol{z}$ are zero, corresponding to the zero rows of S, a solution exists to $S\boldsymbol{y} = \boldsymbol{z}$. Hence a solution exists to $A\boldsymbol{x} = \boldsymbol{c}$. Consequently, vector $\boldsymbol{c}$ is in the column space of A.

(Incidentally, you may check that $\boldsymbol{c} = 2\boldsymbol{a}_1 - 2\boldsymbol{a}_2 + \boldsymbol{a}_3$.)   □

Another subspace associated with matrices is the set of possible solutions to a homogeneous system of linear equations.

Theorem 3.4.14 *For any $m \times n$ matrix A, define the set $\mathrm{null}(A)$ to be all the solutions x of the homogeneous system $Ax = 0$. The set $\mathrm{null}(A)$ is a subspace of $\mathbb{R}^n$ called the **nullspace** of A.*

Proof. First, $A0 = 0$ so $0 \in \mathrm{null}\,A$. Let $u, v \in \mathrm{null}\,A$; that is, $Au = 0$ and $Av = 0$. Second, by the distributivity of matrix-vector multiplication (Theorem 3.1.26), $A(u + v) = Au + Av = 0 + 0 = 0$ and so $u + v \in \mathrm{null}\,A$. Third, by the associativity and commutativity of *scalar* multiplication (Theorem 3.1.24), for every $c \in \mathbb{R}$, $A(cu) = Acu = cAu = c(Au) = c0 = 0$ and so $cu \in \mathrm{null}\,A$. Hence $\mathrm{null}\,A$ is a subspace (Definition 3.4.3). ☐

Example 3.4.15

- Example 2.2.28(a) showed that the only solution of the homogeneous system $\begin{cases} 3x_1 - 3x_2 = 0 \\ -x_1 - 7x_2 = 0 \end{cases}$ is $x = 0$. Thus its set of solutions is $\{0\}$ which is a subspace (Example 3.4.4(f)). Thus $\{0\}$ is the nullspace of matrix $\begin{bmatrix} 3 & -3 \\ -1 & -7 \end{bmatrix}$.

- Recall that the homogeneous system of linear equations from Example 2.2.28(c) has solutions $x = (-2s - \frac{15}{7}t, s, \frac{9}{7}t, t) = (-2, 1, 0, 0)s + (-\frac{15}{7}, 0, \frac{9}{7}, 1)t$ for arbitrary s and t. That is, the set of solutions is $\mathrm{span}\{(-2, 1, 0, 0), (-\frac{15}{7}, 0, \frac{9}{7}, 1)\}$. Since the set is a span (Theorem 3.4.6), the set of solutions is a subspace of $\mathbb{R}^4$. Thus this set of solutions is the nullspace of the matrix $\begin{bmatrix} 1 & 2 & 4 & -3 \\ 1 & 2 & -3 & 6 \end{bmatrix}$.

- In contrast, Example 2.2.25 shows that the set of solutions of the *non*-homogeneous system $\begin{cases} -2v + 3w = -1, \\ 2u + v + w = -1. \end{cases}$ is $(u, v, w) = (-\frac{3}{4} - \frac{1}{4}t, \frac{1}{2} + \frac{3}{2}t, t) = (-\frac{3}{4}, \frac{1}{2}, 0) + (-\frac{1}{4}, \frac{3}{2}, 1)t$ over all values of parameter t. But there is no value of parameter t giving 0 as a solution: for the last component to be zero requires $t = 0$, but when $t = 0$ neither of the other components are zero, so they cannot all be zero. Since the origin 0 is not in the set of solutions, the set does not form a subspace. A *non*-homogeneous system does not form a subspace of solutions. ☐

Example 3.4.16 Is the vector $v = (-2, 6, 1)$ in the null space of $A = \begin{bmatrix} 3 & 1 & 0 \\ -5 & -1 & -4 \end{bmatrix}$? What about vector $w = (1, -3, 2)$?

Solution: To test a given vector, just multiply by the matrix and see if the result is zero.

- $Av = \begin{bmatrix} 3 \cdot (-2) + 1 \cdot 6 + 0 \cdot 1 \\ -5 \cdot (-2) - 1 \cdot 6 - 4 \cdot 1 \end{bmatrix} = \begin{bmatrix} 0 \\ 0 \end{bmatrix} = 0$, so $v \in \mathrm{null}\,A$.

- $Aw = \begin{bmatrix} 3 \cdot 1 + 1 \cdot (-3) + 0 \cdot 2 \\ -5 \cdot 1 - 1 \cdot (-3) - 4 \cdot 2 \end{bmatrix} = \begin{bmatrix} 0 \\ -10 \end{bmatrix} \neq 0$, so w is not in the nullspace. ☐

Activity 3.4.17 Which vector is in the nullspace of the matrix

$$\begin{bmatrix} 4 & 5 & 1 \\ 4 & 3 & -1 \\ 4 & 2 & -2 \end{bmatrix}?$$

(a) $\begin{bmatrix} 2 \\ -2 \\ 2 \end{bmatrix}$ (b) $\begin{bmatrix} 3 \\ -4 \\ 0 \end{bmatrix}$ (c) $\begin{bmatrix} 0 \\ 1 \\ 3 \end{bmatrix}$ (d) $\begin{bmatrix} -1 \\ 0 \\ 4 \end{bmatrix}$

Summary Three common ways that subspaces arise from a matrix are as the column space, row space, and nullspace.

3.4.2 Orthonormal bases form a foundation

The importance of orthogonal basis functions in interpolation and approximation cannot be overstated.

Cuyt (2015)

Given that subspaces arise frequently in linear algebra, and that there are many ways of representing the same subspace (as seen in some previous examples), is there a 'best' way of representing subspaces? The next definition and theorems largely answer this challenge.

We prefer to use an orthonormal set of vectors to span a subspace. The virtue is that orthonormal sets have many practically useful properties. Because of their beautiful properties, orthonormal sets underpin JPEG images, our understanding of vibrations, reliable weather forecasting, and much more. Recall that an orthonormal set (Definition 3.2.38) is composed of vectors that are both all at right-angles to each other (their dot products are zero) and all of unit length.

Definition 3.4.18 *An **orthonormal basis** for a subspace $\mathbb{W}$ of $\mathbb{R}^n$ is an orthonormal set of vectors that span $\mathbb{W}$.*

Example 3.4.19 Recall that $\mathbb{R}^n$ is itself a subspace of $\mathbb{R}^n$ (Example 3.4.4(g)).

(a) The n standard unit vectors e_1, e_2, $\ldots$, e_n in $\mathbb{R}^n$ form a set of n orthonormal vectors. They span the subspace $\mathbb{R}^n$, as every vector in $\mathbb{R}^n$ can be written as a linear combination $x = (x_1, x_2, \ldots, x_n) = x_1 e_1 + x_2 e_2 + \cdots + x_n e_n$. Hence the set of standard unit vectors in $\mathbb{R}^n$ is an orthonormal basis for the subspace $\mathbb{R}^n$.

(b) The n columns q_1, q_2, $\ldots$, q_n of an $n \times n$ orthogonal matrix Q also form an orthonormal basis for the subspace $\mathbb{R}^n$. The reasons are: first, Theorem 3.2.48(b) establishes that the column vectors of Q are orthonormal; and second they span the subspace $\mathbb{R}^n$, as for every vector $x \in \mathbb{R}^n$ there exists a linear combination $x = c_1 q_1 + c_2 q_2 + \cdots + c_n q_n$ obtained by solving $Qc = x$ through calculating $c = Q^\mathsf{T} x$ since Q^T is the inverse of an orthogonal matrix Q (Theorem 3.2.48(c)).

This example also illustrates that generally there are many different orthonormal bases for a given subspace. □

Activity 3.4.20 Which of the following sets is an orthonormal basis for $\mathbb{R}^2$?

(a) $\{\frac{1}{5}(3,-4), \frac{1}{13}(12,5)\}$

(c) $\{0, i, j\}$

(b) $\{(1,1), (1,-1)\}$

(d) $\{\frac{1}{2}(1,\sqrt{3}), \frac{1}{2}(-\sqrt{3},1)\}$

Example 3.4.21 Find an orthonormal basis for the line $x = y = z$ in $\mathbb{R}^3$.

Solution: This line is a subspace as it passes through 0. A parametric description of the line is $x = (x,y,z) = (t,t,t) = (1,1,1)t$ for every t. So the subspace is spanned by $\{(1,1,1)\}$. But this is not an orthonormal basis as it is not of unit length, so divide by its length $|(1,1,1)| = \sqrt{1^2 + 1^2 + 1^2} = \sqrt{3}$. That is, $\{(1/\sqrt{3}, 1/\sqrt{3}, 1/\sqrt{3})\}$

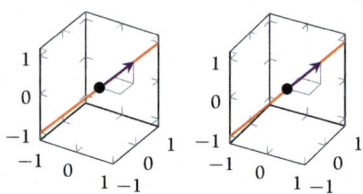

is an orthonormal basis for the subspace, as illustrated in stereo to the right. The only other orthonormal basis is the unit vector in the opposite direction, $\{(-1/\sqrt{3}, -1/\sqrt{3}, -1/\sqrt{3})\}$. ☐

For subspaces that are planes in $\mathbb{R}^n$, orthonormal bases have more details to confirm as in the next example. The SVD then empowers us to find such bases as in the next Procedure 3.4.23.

Example 3.4.22 Confirm that the plane $-x + 2y - 2z = 0$ has an orthonormal basis $\{u_1, u_2\}$ where $u_1 = (-\frac{2}{3}, \frac{1}{3}, \frac{2}{3})$, and $u_2 = (\frac{2}{3}, \frac{2}{3}, \frac{1}{3})\}$ as illustrated in stereo to the right.

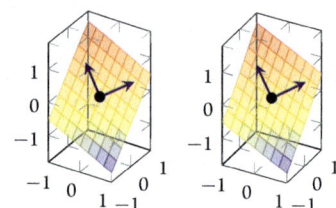

Solution: First, the given set is of unit vectors as the lengths are $|u_1| = \sqrt{\frac{4}{9} + \frac{1}{9} + \frac{4}{9}} = 1$ and $|u_2| = \sqrt{\frac{4}{9} + \frac{4}{9} + \frac{1}{9}} = 1$. Second, the set is orthonormal as their dot product is zero: $u_1 \cdot u_2 = -\frac{4}{9} + \frac{2}{9} + \frac{2}{9} = 0$. Third, they both lie in the plane as we check by substituting their components in the equation: for u_1, $-x + 2y - 2z = \frac{2}{3} + 2(\frac{1}{3}) - 2(\frac{2}{3}) = \frac{2}{3} + \frac{2}{3} - \frac{4}{3} = 0$; and for u_2, $-x + 2y - 2z = -\frac{2}{3} + 2(\frac{2}{3}) - 2(\frac{1}{3}) = -\frac{2}{3} + \frac{4}{3} - \frac{2}{3} = 0$. Lastly, from the parametric form of an equation for a plane (Section 1.3.4) we know that all linear combinations of u_1 and u_2 span the plane. ☐

Procedure 3.4.23 (orthonormal basis for a span) *Let* $\{a_1, a_2, \ldots, a_n\}$ *be a set of n vectors in* $\mathbb{R}^m$, *then the following procedure finds an orthonormal basis for the subspace* span$\{a_1, a_2, \ldots, a_n\}$.

1. *Form $m \times n$ matrix* $A := \begin{bmatrix} a_1 & a_2 & \cdots & a_n \end{bmatrix}$.

2. *Factorize A into an SVD,* $A = USV^{\mathsf{T}}$. *Let* $u_1, u_2, \ldots, u_m$ *denote the m columns of U (singular vectors), and let $r = \text{rank} A$ be the number of nonzero singular values (Definition 3.3.19).*

3. *Then* $\{u_1, u_2, \ldots, u_r\}$ *is an orthonormal basis for the subspace* span$\{a_1, a_2, \ldots, a_n\}$.

Proof. The argument corresponds to that for Procedure 3.3.15. Consider any point $b \in \text{span}\{a_1, a_2, \ldots, a_n\}$. Because b is in the span, there exist coefficients $x_1, x_2, \ldots, x_n$ such that

$$
\begin{aligned}
b &= a_1 x_1 + a_2 x_2 + \cdots + a_n x_n \\
&= Ax \quad \text{(by matrix-vector product Section 3.1.2)} \\
&= USV^{\mathsf{T}} x \quad \text{(by the SVD of } A) \\
&= USy \quad (\text{for } y = V^{\mathsf{T}} x) \\
&= Uz \quad (\text{for } z = (z_1, z_2, \ldots, z_r, 0, \ldots, 0) = Sy) \\
&= u_1 z_1 + u_2 z_2 + \cdots + u_r z_r \quad \text{(by matrix-vector product)} \\
&\in \text{span}\{u_1, u_2, \ldots, u_r\}.
\end{aligned}
$$

These equalities also hold in reverse, due to the invertibility of U and V, and with $y_i = z_i/\sigma_i$ for $i = 1, 2, \ldots, r$. Hence a point is in $\text{span}\{a_1, a_2, \ldots, a_n\}$ if and only if it is in $\text{span}\{u_1, u_2, \ldots, u_r\}$. Lastly, U is an orthogonal matrix and so the set of columns $\{u_1, u_2, \ldots, u_r\}$ is an orthonormal set. Hence $\{u_1, u_2, \ldots, u_r\}$ forms an orthonormal basis for $\text{span}\{a_1, a_2, \ldots, a_n\}$. ☐

Example 3.4.24 Compute an orthonormal basis for $\text{span}\{(1, \frac{1}{2}), (-2, -1)\}$.

Solution: Form the matrix whose columns are the given vectors $A = \begin{bmatrix} 1 & -2 \\ 1/2 & -1 \end{bmatrix}$, then ask MATLAB/Octave for an SVD and interpret.

```
A= [1 -2; 1/2 -1]
[U,S,V] =svd (A)
```

The computed SVD is (V is immaterial here)

```
U =
    -0.8944    -0.4472
    -0.4472     0.8944

S =
     2.5000          0
          0     0.0000
V = ...
```

There is one nonzero singular value—the matrix has rank one—so an orthonormal basis for the span is the first column of matrix U, namely the set $\{(-0.89, -0.45)\}$ (2 d.p.). That is, every vector in $\text{span}\{(1, \frac{1}{2}), (-2, -1)\}$ can be written as $(-0.89, -0.45)t$ for some t: hence the span is a line. ☐

Example 3.4.25 Recall that Example 3.4.8 found the plane $z = -x/6 + y/3$ could be written as span$\{(3,3,1/2), (0,3,1)\}$ or as span$\{(5,1,-1/2), (0,-3,-1), (-4,1,1)\}$. Use each of these spans to find two different orthonormal bases for the plane.

Solution:

- Form the matrix whose columns are the given vectors $A = \begin{bmatrix} 3 & 0 \\ 3 & 3 \\ 1/2 & 1 \end{bmatrix}$, then ask MAT-LAB/Octave for the SVD and interpret. In MATLAB/Octave it is often easier to form the matrix by entering the vectors as rows and then transposing (with the dash):

```
A= [3 3 1/2;0 3 1]'
[U,S,V]=svd(A)
```

The computed SVD is (2 d.p.)

```
U =
   -0.51    0.85    0.16
   -0.84   -0.44   -0.31
   -0.20   -0.29    0.94
S =
    4.95       0
       0    1.94
       0       0
V = ...
```

There are two nonzero singular values—the matrix has rank two—so an orthonormal basis for the plane is the set of the first two columns of matrix U, namely

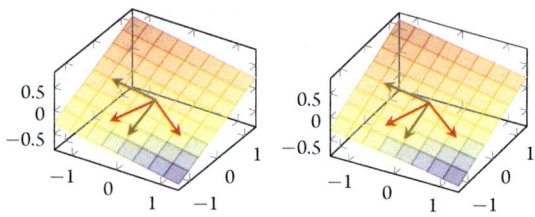

$(-0.51, -0.84, -0.20)$ and $(0.85, -0.44, -0.29)$. These basis vectors are illustrated as the pair of red vectors in the stereo to the above-right.

- Similarly, form the matrix $B = \begin{bmatrix} 5 & 0 & -4 \\ 1 & -3 & 1 \\ -1/2 & -1 & 1 \end{bmatrix}$, then ask MATLAB/Octave for the SVD and interpret. Form the matrix in MATLAB/Octave by entering the vectors as rows and then transposing:

```
B= [5 1 -1/2; 0 -3 -1; -4 1 1]'
[U,S,V]=svd(B)
```

The computed SVD is (2 d.p.)

```
U =
   -0.99   -0.04    0.16
   -0.01   -0.95   -0.31
    0.16   -0.31    0.94
S =
    6.49       0       0
       0    3.49       0
       0       0    0.00
V = ...
```

There are two nonzero singular values—the matrix has rank two—so an orthonormal basis for the plane spanned by the three vectors is the set of the first two columns of matrix U, namely the vectors $(-0.99, -0.01, 0.16)$ and $(-0.04, -0.95, -0.31)$. These are the pair of brown vectors in the above stereo illustration. ☐

Activity 3.4.26 The matrix

$$A = \begin{bmatrix} 4 & 5 & 1 \\ 4 & 3 & -1 \\ 4 & 2 & -2 \end{bmatrix}$$

has the following SVD computed by $[U, S, V] = \text{svd}(A)$ in MATLAB/Octave: what is an orthonormal basis for the column space of the matrix A (2 d.p.)?

```
U =
   -0.67    0.69    0.27
   -0.55   -0.23   -0.80
   -0.49   -0.69    0.53
S =
    9.17       0       0
       0    2.83       0
       0       0    0.00
V =
   -0.75   -0.32   -0.58
   -0.66    0.49    0.58
    0.09    0.81   -0.58
```

(a) $\{(-0.67, 0.69, 0.27), (-0.55, -0.23, -0.80)\}$
(b) $\{(-0.75, -0.32, -0.58), (-0.66, 0.49, 0.58)\}$
(c) $\{(-0.67, -0.55, -0.49), (0.69, -0.23, -0.69)\}$
(d) $\{(-0.75, -0.66, 0.09), (-0.32, 0.49, 0.81)\}$

Extension: recalling Theorem 3.3.24, which of the above is an orthonormal basis for the row space of A?

Example 3.4.27 *(data reduction)* Every four or five years the phenomenon of El Niño makes a large impact on the world's weather: from drought in Australia to floods in South America. We would like to predict El Niño in advance to save lives and businesses. El Niño is correlated significantly with the difference in atmospheric pressure between Darwin and Tahiti—the so-called Southern Oscillation Index (SOI). This example seeks patterns in the SOI in order to be able to predict the SOI and hence predict El Niño.

Figure 3.1 plots the yearly average SOI each year for fifty years up to 1993. A strong regular structure is apparent, but there are significant variations and complexities in the year-to-year signal. The challenge of this example is to explore the full details of this signal.

Let's use a general technique called Singular Spectrum Analysis. The figure shows that the SOI oscillates, to and fro, a couple of times every ten years. This suggests that if we analyse ten year 'snapshots', or 'windows', of the SOI data then there should be some common pattern of oscillations apparent—somehow. Consider a window of ten years of the SOI, and let the window 'slide' across the data to give us many 'local' pictures of the evolution in time. For example, Figure 3.2 plots six windows (each displaced vertically for clarity) each of length ten years. As the 'window' slides across the fifty year data of Figure 3.1 there are 41 such local views of the data of length ten years. Let's invoke the concept of subspaces to detect regularity in the data via these windows.

The fundamental property is that if the data has regularities, then it should lie in some subspace. We detect such subspaces using the SVD of a matrix.

- First, form the 41 data windows of length ten into a matrix of size 10×41. The numerical values of the SOI data of Figure 3.1 are the following:

```
year=(1944:1993)'
```

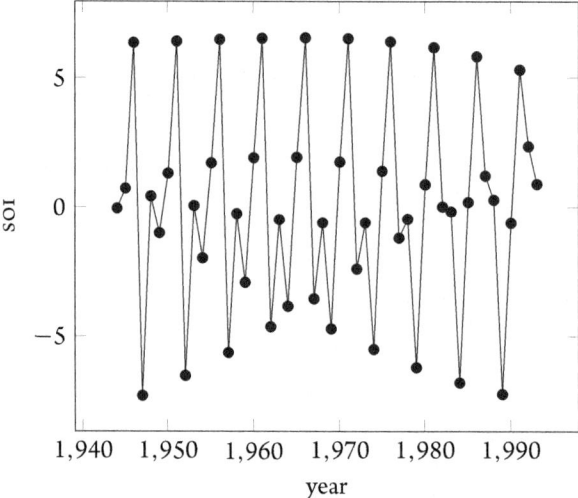

Figure 3.1 Yearly average SOI over 50 years ('smoothed' somewhat for the purposes of the example). The nearly regular behaviour suggests that it should be predictable.

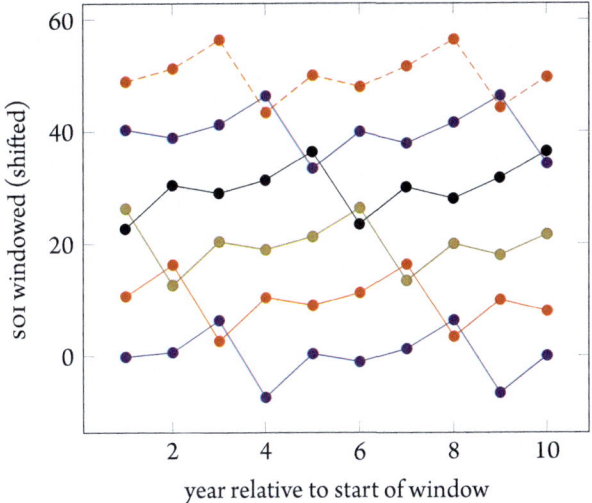

Figure 3.2 The first six windows of the soi data of Figure 3.1—displaced vertically for clarity. Each window is of length ten years: lowest, the first window is data 1944–1953; second lowest, covers 1945–1954; third lowest, covers 1946–1955; and so on to the 41st window which is data 1984–1993, not shown.

```
soi=[-0.03; 0.74; 6.37; -7.28; 0.44; -0.99; 1.32
6.42; -6.51; 0.07; -1.96; 1.72; 6.49; -5.61
-0.24; -2.90; 1.92; 6.54; -4.61; -0.47; -3.82
1.94; 6.56; -3.53; -0.59; -4.69; 1.76; 6.53
-2.38; -0.59; -5.48; 1.41; 6.41; -1.18; -0.45
-6.19; 0.89; 6.19; 0.03; -0.16; -6.78; 0.21; 5.84
1.23; 0.30; -7.22; -0.60; 5.33; 2.36; 0.91 ]
```

- Second, form the 10 × 41 matrix of the windows of the data, the first seven columns being

```
A =
   Columns 1 through 7
 -0.03    0.74    6.37   -7.28    0.44   -0.99    1.32
  0.74    6.37   -7.28    0.44   -0.99    1.32    6.42
  6.37   -7.28    0.44   -0.99    1.32    6.42   -6.51
 -7.28    0.44   -0.99    1.32    6.42   -6.51    0.07
  0.44   -0.99    1.32    6.42   -6.51    0.07   -1.96
 -0.99    1.32    6.42   -6.51    0.07   -1.96    1.72
  1.32    6.42   -6.51    0.07   -1.96    1.72    6.49
  6.42   -6.51    0.07   -1.96    1.72    6.49   -5.61
 -6.51    0.07   -1.96    1.72    6.49   -5.61   -0.24
  0.07   -1.96    1.72    6.49   -5.61   -0.24   -2.90
```

Figure 3.2 plots the first six of these columns. The simplest way to form this matrix in MATLAB/Octave—useful for all such shifting windows of data—is to invoke the hankel() function:

```
A=hankel(soi(1:10),soi(10:50))
```

In MATLAB/Octave the command hankel(s(1:w),s(w:n)) forms the $w \times (n - w + 1)$ so-called Hankel matrix

$$
\begin{bmatrix}
s_1 & s_2 & s_3 & \cdots & s_{n-w} & s_{n-w+1} \\
s_2 & s_3 & \vdots & & s_{n-w+1} & \vdots \\
s_3 & \vdots & s_w & & \vdots & \vdots \\
\vdots & s_w & s_{w+1} & & \vdots & s_{n-1} \\
s_w & s_{w+1} & s_{w+2} & \cdots & s_{n-1} & s_n
\end{bmatrix}
$$

- Lastly, compute the SVD of the matrix of these windows:

```
[U,S,V]=svd(A);
singValues=diag(S)
plot(U(:,1:4))
```

The computed singular values are 44.63, 43.01, 39.37, 36.69, 0.03, 0.03, 0.02, 0.02, 0.02, 0.01. In practice, treat the six small singular values as zero. Since there are four 'nonzero' singular values, the windows of data lie in a subspace spanned by the first four columns of U.

That is, all the structure seen in the fifty year SOI data of Figure 3.1 can be expressed in terms of the orthonormal basis of the four ten-year vectors plotted in Figure 3.3. This analysis implies that the SOI data is composed of two cycles of two different frequencies.[22] □

Example 3.4.25 obtained two different orthonormal bases for the one plane. Although the bases are different, they both had the same number of vectors. The next theorem establishes that this same number always occurs.

Theorem 3.4.28 *For every given subspace, any two orthonormal bases have the same number of vectors.*

---

[22] However, I 'smoothed' the SOI data for the purposes of this example. The real SOI data is much noisier. Also we would use 600 monthly averages not 50 yearly averages: so a ten-year window would be a window of 120 months, and the matrix would be considerably larger 120×481. Nonetheless, the conclusions with the real data—and justified by Chapter 5—are much the same.

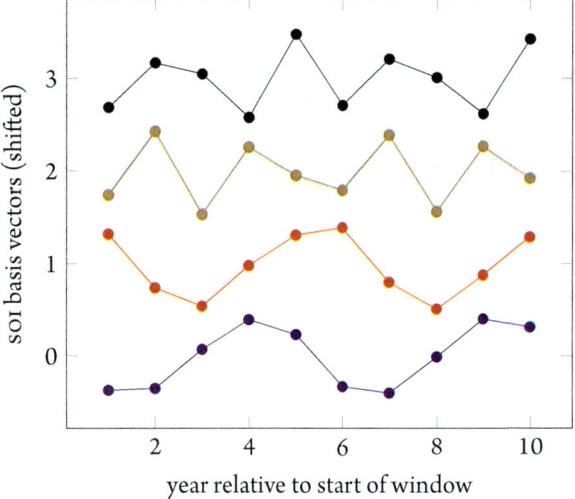

Figure 3.3 The first four singular vectors of the soi data—displaced vertically for clarity. The bottom two form a pair to show a five-year cycle. The top two are a pair that show a two–three-year cycle. The combination of these two cycles leads to the structure of the soi in Figure 3.1.

Proof. Let $\mathcal{U} = \{u_1, u_2, \ldots, u_r\}$, with r vectors, and $\mathcal{V} = \{v_1, v_2, \ldots, v_s\}$, with s vectors, be any two orthonormal bases for a given subspace in $\mathbb{R}^n$. Prove that the number of vectors $r = s$ by contradiction. First assume $r < s$ ($\mathcal{U}$ has less vectors than $\mathcal{V}$). Since $\mathcal{U}$ is an orthonormal basis for the subspace every vector in $\mathcal{V}$ can be written as a linear combination of vectors in $\mathcal{U}$ with some coefficients a_{ij}:

$$v_1 = u_1 a_{11} + u_2 a_{21} + \cdots + u_r a_{r1},$$
$$v_2 = u_1 a_{12} + u_2 a_{22} + \cdots + u_r a_{r2},$$
$$\vdots$$
$$v_s = u_1 a_{1s} + u_2 a_{2s} + \cdots + u_r a_{rs}.$$

Write each of these equations, such as the first one, in the form

$$v_1 = \begin{bmatrix} u_1 & u_2 & \cdots & u_r \end{bmatrix} \begin{bmatrix} a_{11} \\ a_{21} \\ \vdots \\ a_{r1} \end{bmatrix} = U a_1,$$

upon setting matrix $U := \begin{bmatrix} u_1 & u_2 & \cdots & u_r \end{bmatrix}$. Then setting the $n \times s$ matrix

$$V := \begin{bmatrix} v_1 & v_2 & \cdots & v_s \end{bmatrix}$$
$$= \begin{bmatrix} Ua_1 & Ua_2 & \cdots & Ua_s \end{bmatrix}$$
$$= U \begin{bmatrix} a_1 & a_2 & \cdots & a_s \end{bmatrix} = UA$$

for the $r \times s$ matrix $A := \begin{bmatrix} a_1 & a_2 & \cdots & a_s \end{bmatrix}$. By assumption, $r < s$ and so Theorem 2.2.30 assures us that the homogeneous system $Ax = 0$ has infinitely many solutions, choose any one non-trivial solution $x \neq 0$. Consider

$$Vx = UAx \quad \text{(from above)}$$
$$= U0 \quad \text{(since } Ax = 0\text{)}$$
$$= 0.$$

Premultiplying $Vx = 0$ by V^{T} gives $V^{\mathsf{T}}Vx = V^{\mathsf{T}}0$, but as the columns of V are orthonormal this simplifies to $I_s x = 0$; that is, $x = 0$. But $x = 0$ contradicts $x \neq 0$, so the assumption cannot be correct, that is, we cannot have $r < s$.

Second, a corresponding argument establishes that we cannot have $s < r$. Hence $r = s$, and so all orthonormal bases of a given subspace must have the same number of vectors. □

An existential issue   How do we know that every subspace has an orthonormal basis? We know that many subspaces, such as row and column spaces, have an orthonormal basis because they are the span of rows and columns of a matrix, and then Procedure 3.4.23 assures us they have an orthonormal basis. But do all subspaces have an orthonormal basis? The following theorem certifies that they do.

Theorem 3.4.29 (*existence of basis*) *Let* $\mathbb{W}$ *be a subspace of* $\mathbb{R}^n$, *then there exists an orthonormal basis for* $\mathbb{W}$.

Proof.   In the trivial case when the subspace is $\mathbb{W} = \{0\}$, then $\mathbb{W} = \text{span}\{\}$ gives a basis and this trivial case is done.

For every other subspace $\mathbb{W} \neq \{0\}$ there exists a nonzero vector $w \in \mathbb{W}$. Normalize the vector by setting $u_1 = w/|w|$. Then all scalar multiples $cu_1 =$

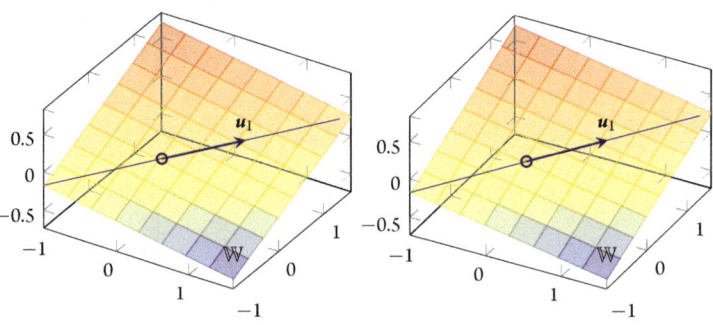

$cw/|w| = (c/|w|)w \in \mathbb{W}$ by closure of $\mathbb{W}$ under scalar multiplication (Definition 3.4.3). Hence span$\{u_1\} \subseteq \mathbb{W}$ (as illustrated to the above-right for example). Consequently, either span$\{u_1\} = \mathbb{W}$ and we are done, or we repeat the following step until the space $\mathbb{W}$ is spanned.

Given orthonormal vectors u_1, u_2, ..., u_k such that the set span$\{u_1, u_2, ..., u_k\} \subset \mathbb{W}$, so span$\{u_1, u_2, ..., u_k\} \neq \mathbb{W}$. Then there must

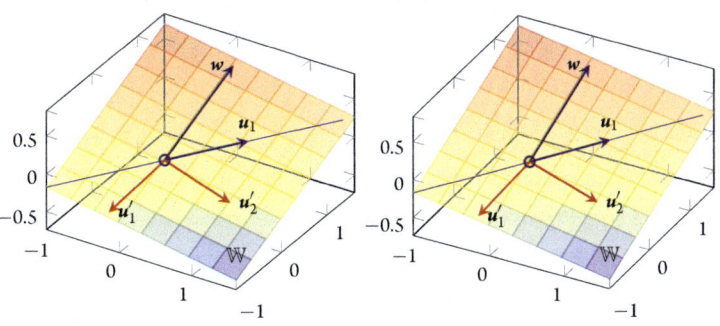

exist a vector $w \in \mathbb{W}$ which is not in the set span$\{u_1, u_2, ..., u_k\}$. By the closure of subspace $\mathbb{W}$ under addition and scalar multiplication (Definition 3.4.3), the set span$\{u_1, u_2, ..., u_k, w\} \subseteq \mathbb{W}$. Procedure 3.4.23, on the orthonormal basis for a span, then assures us that an SVD gives an orthonormal basis $\{u_1', u_2', ..., u_{k+1}'\}$ for the set span$\{u_1, u_2, ..., u_k, w\} \subseteq \mathbb{W}$ (as illustrated for an example). Consequently, either span$\{u_1', u_2', ..., u_{k+1}'\} = \mathbb{W}$ and we are done, or we repeat the process of this paragraph with k bigger by one.

The process must terminate because $\mathbb{W} \subseteq \mathbb{R}^n$. If the process repeats until $k = n$, then we know $\mathbb{W} = \mathbb{R}^n$ and we are done as $\mathbb{R}^n$ is spanned by the n standard unit vectors. □

Ensemble simulation makes better weather forecasts Near the end of the twentieth century weather forecasts were becoming amazingly good at predicting the chaotic weather days in advance. However, there were notable failures: occasionally the weather forecast would give no hint of storms that developed (such as the severe 1999 storm in Sydney[23]). Why?

Occasionally the weather is both near a 'tipping point' where small changes may cause a storm, and where the errors in measuring the current weather are of the magnitude of the necessary changes. Then the storm would be within the possibilities. But the storm would not be forecast if the measurements were, by chance error, the 'other side' of the tipping point (as happened in 1999). Meteorologists now mostly overcome this problem by executing on their computers an ensemble of simulations, perhaps an ensemble of a hundred different forecast simulations (Roulstone and Norbury, 2013, pp.274–80, e.g.). Such a set of 100 simulations essentially lie in a subspace spanned by 100 vectors in the vastly larger space, say $\mathbb{R}^{1,000,000,000}$, of the maybe billion variables in the weather model. But what happens in the computational simulations is that the ensemble of simulations degenerate in time. To avoid such degeneracy, the meteorologists continuously 'renormalize' the ensemble of simulations by rewriting the ensemble in terms of an *orthonormal basis* of 100 vectors. Such an orthonormal basis for the ensemble reasonably ensures that unusual storms are retained in the range of possibilities explored by the ensemble forecast, and hence make weather forecasting much more complete.

3.4.3 Is it a line? a plane? The dimension answers

> ... *physical dimension.* It is an intuitive notion that appears to go back to an archaic state before Greek geometry, yet deserves to be taken up again. *Mandelbrot (1982)*

[23] http://en.wikipedia.org/wiki/1999_Sydney_hailstorm [Oct 2019]

One of the beauties of an orthonormal basis is that, being orthonormal, they look just like a rotated
version of the standard unit vectors. For example, in a *plane* any
two orthonormal basis vectors of the
plane could form the two 'standard

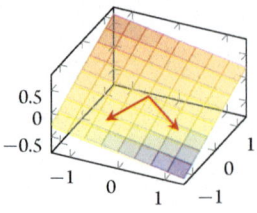 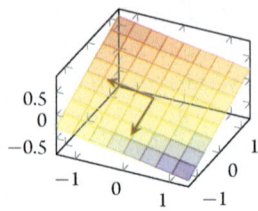

unit vectors' of a coordinate system in that plane—as suggested by the illustration. Example 3.4.25
found that the plane $z = -x/6 + y/3$ could have the following two orthonormal bases: either
of these orthonormal bases, or indeed any other pair of orthonormal vectors, could act as a pair
of 'standard unit vectors' of the given planar subspace. Similarly in other dimensions for other
subspaces. Just as $\mathbb{R}^n$ is called n-dimensional and has n standard unit vectors, so we analogously
define the dimension of any subspace.

Definition 3.4.30 *Let $\mathbb{W}$ be a subspace of $\mathbb{R}^n$. The number of vectors in an orthonormal basis for $\mathbb{W}$
is called the **dimension** of $\mathbb{W}$, denoted $\dim \mathbb{W}$. By convention, $\dim\{\mathbf{0}\} = 0$.*

Example 3.4.31

- Example 3.4.21 finds that the linear subspace $x = y = z$ is spanned by the orthonormal
  basis $\{(1/\sqrt{3}, 1/\sqrt{3}, 1/\sqrt{3})\}$. With one vector in the basis, the line is one-dimensional.
- Example 3.4.22 finds that the planar subspace $-x + 2y - 2z = 0$ is spanned by the
  orthonormal basis $\{\mathbf{u}_1, \mathbf{u}_2\}$ where $\mathbf{u}_1 = (-\frac{2}{3}, \frac{1}{3}, \frac{2}{3})$, and $\mathbf{u}_2 = (\frac{2}{3}, \frac{2}{3}, \frac{1}{3})$. With two vectors
  in the basis, the plane is two-dimensional.
- Subspace $\mathbb{W} = \mathrm{span}\{(5, 1, -1/2), (0, -3, -1), (-4, 1, 1)\}$ of Example 3.4.25 is found to
  have an orthonormal basis of vectors $(-0.99, -0.01, 0.16)$ and $(-0.04, -0.95, -0.31)$.
  With two vectors in the basis, the subspace is two-dimensional; that is, $\dim \mathbb{W} = 2$.
- Since the subspace $\mathbb{R}^n$ (Example 3.4.4(g)) has an orthonormal basis of the n standard unit
  vectors, $\{\mathbf{e}_1, \mathbf{e}_2, \ldots, \mathbf{e}_n\}$, then $\dim \mathbb{R}^n = n$.
- The El Niño windowed data of Example 3.4.27 is effectively spanned by four orthonormal
  vectors. Despite the apparent complexity of the signal, the data effectively lies in a subspace
  of dimension four (that of two oscillators). ☐

Theorem 3.4.32 *The row space and column space of a matrix A have the same dimension. Further,
given an SVD of the matrix, say $A = USV^T$ and setting $r = \mathrm{rank}\, A$, an orthonormal basis for the
column space is the first r columns of U, and that for the row space is the first r columns of V.*

Proof. From Definition 3.1.17 of the transpose, the rows of A are the same as the columns of A^T,
and so the row space of A is the same as the column space of its transpose, A^T. Hence,

dimension of the row space of A

$= $ dimension of the column space of A^T

$= \mathrm{rank}(A^T)$ (by Procedure 3.4.23)

$= \operatorname{rank} A$ (by Theorem 3.3.24)

$=$ dimension of the column space of A (by Procedure 3.4.23).

Let $m \times n$ matrix A have an SVD $A = USV^{\mathsf{T}}$ and $r = \operatorname{rank} A$. Then Procedure 3.4.23 establishes that an orthonormal basis for the column space of A is the first r columns of U. Recall that $A^{\mathsf{T}} = (USV^{\mathsf{T}})^{\mathsf{T}} = VS^{\mathsf{T}}U^{\mathsf{T}}$ is an SVD for A^{T} (Theorem 3.3.24), and so an orthonormal basis for the column space of A^{T} is the first r columns of V (Procedure 3.4.23). Since the row space of A is the column space of A^{T}, an orthonormal basis for the row space of A is the first r columns of V. □

Example 3.4.33 Find an SVD of the matrix $A = \begin{bmatrix} 1 & -4 \\ 1/2 & -2 \end{bmatrix}$ and compare the column space and the row space of the matrix.

Solution: Ask MATLAB/Octave for an SVD and interpret:

```
A= [1 -4; 1/2 -4]
[U,S,V] =svd (A)
```

computes the SVD

```
U =
   -0.8944   -0.4472
   -0.4472    0.8944
S =
    4.6098         0
         0    0.0000
V =
   -0.2425    0.9701
    0.9701    0.2425
```

There is one nonzero singular value—the matrix has rank one—so an orthonormal basis for the column space is the first column of matrix U, namely $(-0.89, -0.45)$ (2 d.p.).

Complementing this, as there is one nonzero singular value—the matrix has rank one—so an orthonormal basis for the row space is the first column of matrix V, namely $(-0.24, 0.97)$. As illustrated to the right, the two subspaces, the row space (red), and the column space (blue), are different but of the same dimension. (As in general, here the row and column spaces are not orthogonal.)

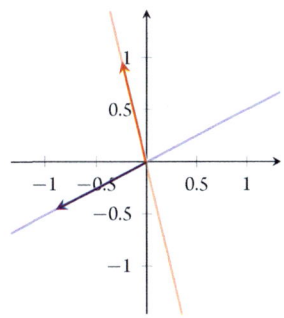

□

Activity 3.4.34 Using the SVD of Example 3.4.33, what is the dimension of the nullspace of the matrix $\begin{bmatrix} 1 & -4 \\ 1/2 & -2 \end{bmatrix}$?

(a) 1   (b) 2   (c) 3   (d) 0

Example 3.4.35 Use the SVD of the matrix B in Example 3.4.25 to compare the column space and the row space of matrix B.

Solution: Recall that there are two nonzero singular values—the matrix has rank two—so an orthonormal basis for the column space is the first two columns of matrix U, namely the vectors $(-0.99, -0.01, 0.16)$ and $(-0.04, -0.95, -0.31)$.

Complementing this, as there are two nonzero singular values—the matrix has rank two—so an orthonormal basis for the row space is the set of the first two columns of matrix V, namely the vectors $(-0.78, -0.02, 0.63)$ and $(-0.28, 0.91, -0.32)$. As illustrated to the right in stereo, the two subspaces, the row space (red) and the column space (blue), are different but of the same dimension.

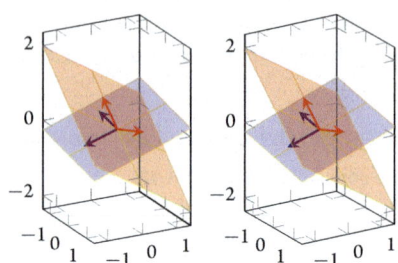

☐

Definition 3.4.36 The **nullity** of a matrix A is the dimension of its nullspace (defined in Theorem 3.4.14), and is denoted by nullity(A).

Example 3.4.37 Example 3.4.15 finds the nullspace of the two matrices

$$\begin{bmatrix} 3 & -3 \\ -1 & -7 \end{bmatrix} \quad \text{and} \quad \begin{bmatrix} 1 & 2 & 4 & -3 \\ 1 & 2 & -3 & 6 \end{bmatrix}.$$

- The first matrix has nullspace $\{0\}$ which has dimension zero and hence the nullity of the matrix is zero.

- The second matrix, 2×4, has nullspace written as span$\{(-2, 1, 0, 0), (-\frac{15}{7}, 0, \frac{9}{7}, 1)\}$. Being spanned by two vectors not proportional to each other, we expect the dimension of the nullspace, the nullity, to be two. To check, compute the singular values of the matrix whose columns are these vectors: calling the matrix N for nullspace,

```
N=[-2 1 0 0; -15/7 0 9/7 1]'
svd(N)
```

which computes the singular values

```
    3.2485
    1.3008
```

Since there are two nonzero singular values, there are two orthonormal vectors spanning the subspace, the nullspace, hence its dimension, the nullity, is two. ☐

Example 3.4.38 For the matrix

$$C = \begin{bmatrix} -1 & -2 & 2 & 1 \\ -3 & 3 & 1 & 0 \\ 2 & -5 & 1 & 1 \end{bmatrix},$$

find an orthonormal basis for its nullspace and hence determine its nullity.

Solution: To find the nullspace construct a general solution to the homogeneous system $Cx = 0$ with Procedure 3.3.15.

(a) Enter into MATLAB/Octave the matrix C and compute an SVD via $[U, S, V] = svd(C)$ to find (2 d.p.)

```
U =
    0.24    0.78   -0.58
   -0.55    0.60    0.58
    0.80    0.18    0.58
S =
    6.95       0       0       0
       0    3.43       0       0
       0       0    0.00       0
V =
    0.43   -0.65    0.63   -0.02
   -0.88   -0.19    0.42    0.10
    0.11    0.68    0.62   -0.37
    0.15    0.28    0.21    0.92
```

(b) Since the right-hand side of $Cx = 0$ is zero the solution to $Uz = 0$ is $z = 0$.

(c) Then, because the rank of the matrix is two, the solution to $Sy = z = 0$ is $y = (0,0,y_3,y_4)$ for free variables y_3 and y_4.

(d) The solution to $V^Tx = y$ is $x = Vy = v_3 y_3 + v_4 y_4 = y_3(0.63, 0.42, 0.62, 0.21) + y_4(-0.02, 0.10, -0.37, 0.92)$.

Hence span$\{(0.63,0.42,0.62,0.21), (-0.02,0.10,-0.37,0.92)\}$ is the nullspace of the matrix C (2 d.p.). Because the columns of V are orthonormal, the *two* vectors appearing in this span are orthonormal and so form an orthonormal basis for the nullspace. Hence the nullity $C = 2$. □

This Example 3.4.38 indicates that the nullity is determined by the number of zero columns in the diagonal matrix S of an SVD. Conversely, the rank of a matrix is determined by the number of nonzero columns in the diagonal matrix S of an SVD. Put these two facts together in general and we get the following theorem that helps characterize solutions of linear equations.

Theorem 3.4.39 (*rank theorem*) *For every $m \times n$ matrix A, rank A + nullity $A = n$, the number of columns of A.*

Proof. Set $r = \operatorname{rank} A$. By Procedure 3.3.15 a general solution to the homogeneous system $Ax = 0$ involves $n - r$ free variables $y_{r+1}, \ldots, y_n$ in the linear combination form $v_{r+1}y_{r+1} + \cdots + v_ny_n$. Hence the nullspace is $\operatorname{span}\{v_{r+1}, \ldots, v_n\}$. Because matrix V is orthogonal, the vectors $v_{r+1}, \ldots, v_n$ are orthonormal; that is, they form an orthonormal basis for the nullspace, and so the nullspace is of dimension $n - r$. Consequently, $\operatorname{rank} A + \text{nullity} \, A = r + (n - r) = n$.

Example 3.4.40 Compute SVDs to determine the rank and nullity of each of the given matrices.

(a) $\begin{bmatrix} 1 & -1 & 2 \\ 2 & -2 & 4 \end{bmatrix}$

Solution: Enter the matrix into MATLAB/Octave and compute the singular values:

```
A= [1  -1  2
2  -2  4]
svd (A)
```

The resultant singular values are

```
5.4772
0.0000
```

The one nonzero singular value indicates rank $A = 1$. Since the matrix has three columns, the nullity—the dimension of the nullspace—is $3 - 1 = 2$. □

(b) $\begin{bmatrix} 1 & -1 & -1 \\ 1 & 0 & -1 \\ -1 & 3 & 1 \end{bmatrix}$

Solution: Enter the matrix into MATLAB/Octave and compute the singular values:

```
B= [1  -1  -1
1  0  -1
-1  3  1]
svd (B)
```

The resultant singular values are

```
3.7417
1.4142
0.0000
```

The two nonzero singular values indicate rank $B = 2$. Since the matrix has three columns, the nullity—the dimension of the nullspace—is $3 - 2 = 1$. □

(c) $\begin{bmatrix} 0 & 0 & -1 & -3 & 2 \\ -2 & -2 & 1 & 0 & 1 \\ 1 & -1 & 2 & 8 & -2 \\ -1 & 1 & 0 & -2 & -2 \\ -3 & -1 & 0 & -5 & 1 \end{bmatrix}$

Solution: Enter the matrix into MATLAB/Octave and compute the singular values:

```
C= [0  0  -1  -3  2
-2  -2  1  -0  1
1  -1  2  8  -2
-1  1  -0  -2  -2
-3  -1  -0  -5  1]
svd (C)
```

The resultant singular values are

```
10.8422
4.0625
3.1532
0.0000
0.0000
```

Three nonzero singular values indicate rank $C = 3$. Since the matrix has five columns, the nullity—the dimension of the nullspace—is $5 - 3 = 2$. ☐

Activity 3.4.41 The matrix

$$\begin{bmatrix} -2 & 1 & 4 & 0 & -4 \\ -1 & 1 & 0 & -2 & 0 \\ -3 & 1 & 3 & 2 & -3 \\ 0 & 0 & 1 & 0 & -1 \end{bmatrix} \quad \text{has singular values} \quad \begin{matrix} 8.1975 \\ 2.6561 \\ 1.6572 \\ 0.0000 \end{matrix}$$

computed with svd (). What is its nullity?

(a) 3   (b) 0   (c) 1   (d) 2

Example 3.4.42 Each of the following graphs plots all the column vectors of a matrix. What is the nullity of each of the matrices? Give reasons.

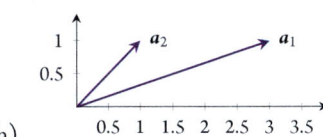

(a)

Solution: Zero. These two column vectors in the plane must come from a 2×2 matrix A. Since the two columns are at a non-trivial angle, every point in the plane may be written as a linear combination of a_1 and a_2, hence the column space of A is $\mathbb{R}^2$. Consequently, rank $A = 2$. From the Rank Theorem 3.4.39: nullity $A = n - \text{rank} A = 2 - 2 = 0$. ☐

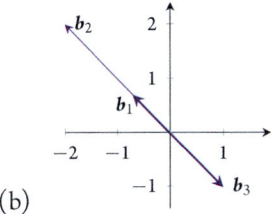

(b)

Solution: Two. These three column vectors in the plane must come from a 2×3 matrix B. The three vectors are all in a line, so the column space of matrix B is a line. Consequently, rank $B = 1$. From the Rank Theorem 3.4.39: nullity $B = n - \text{rank} B = 3 - 1 = 2$. ☐

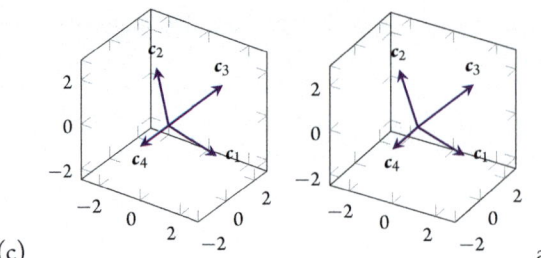

(c)   a stereo pair.

Solution: One. These four column vectors in 3D space must come from a 3×4 matrix C. Since the four columns do not all lie in a line or plane, every point in space may be written as a linear combination of c_1, c_2, ..., c_4, hence the column space of C is $\mathbb{R}^3$. Consequently, rank $C = 3$. From the Rank Theorem 3.4.39: nullity $C = n - \text{rank}\, C = 4 - 3 = 1$. □

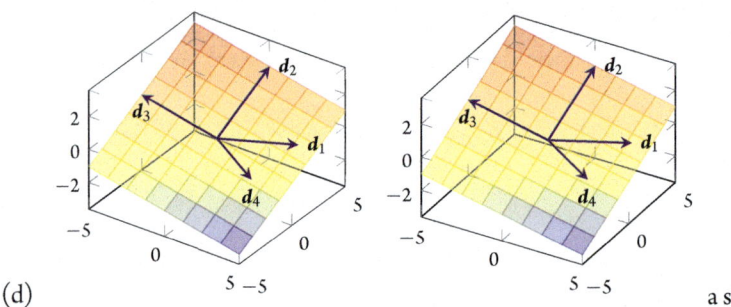

(d)   a stereo pair.

Solution: Two. These four column vectors in 3D space must come from a 3×4 matrix D. Since the four columns all lie in a plane (as indicated by the drawn plane), and linear combinations can give every point in the plane, hence the column space of D has dimension two. Consequently, rank $D = 2$. The Rank Theorem 3.4.39 gives nullity $D = n - \text{rank}\, D = 4 - 2 = 2$. □

Recall the list of exact properties that ensure a system of linear equations has a unique solution, Theorem 3.3.27. The recognition of these new concepts associated with matrices and linear equations lead us to extend this list.

Theorem 3.4.43 (*Unique Solutions: version 2*) *For every $n \times n$ square matrix A, and extending Theorem 3.3.27, the following statements are equivalent:*

(a) *A is invertible;*

(b) *$Ax = b$ has a **unique solution** for every $b \in \mathbb{R}^n$;*

(c) *$Ax = 0$ has only the zero solution;*

(d) *all n singular values of A are nonzero;*

(e) *the condition number of A is finite ($\text{rcond} > 0$);*

(f) rank $A = n$;

(g) nullity $A = 0$;

(h) the column vectors of A span $\mathbb{R}^n$;

(i) the row vectors of A span $\mathbb{R}^n$.

Proof. Theorem 3.3.27 establishes the equivalence of the statements 3.4.43(a)–3.4.43(f). We here prove the equivalence of these with the statements 3.4.43(g)–3.4.43(i).

3.4.43(f) ⟺ 3.4.43(g) The Rank Theorem 3.4.39 assures us that nullity $A = 0$ if and only if rank $A = n$.

3.4.43(b) ⟹ 3.4.43(h) By 3.4.43(b) every $b \in \mathbb{R}^n$ can be written as $b = Ax$ for some x. But Ax is a linear combination of the columns of A and so b is in the span of the columns. Hence the column vectors of A span $\mathbb{R}^n$.

3.4.43(h) ⟹ 3.4.43(f) Suppose rank $A = r$ reflecting r nonzero singular values in an SVD $A = USV^T$. Procedure 3.4.23 assures us the column space of A has orthonormal basis $\{u_1, u_2, \ldots, u_r\}$. But the column space is $\mathbb{R}^n$ (statement 3.4.43(h)) which also has the orthonormal basis of the n standard unit vectors. Theorem 3.4.28 assures us that the number of basis vectors must be the same; that is, rank $A = r = n$.

3.4.43(f) ⟺ 3.4.43(i) Theorem 3.3.24 asserts rank$(A^T) = $ rank A, so the statement 3.4.43(f) implies rank$(A^T) = n$, and so statement 3.4.43(h) asserts the columns of A^T span $\mathbb{R}^n$. But the columns of A^T are the rows of A so the rows of A span $\mathbb{R}^n$. Conversely, if the rows of A span $\mathbb{R}^n$, then so do the columns of A^T, hence rank$(A^T) = n$ which by Theorem 3.3.24 implies rank $A = n$.

This completes the extended theorem.

3.4.4 Exercises

Exercise 3.4.1 Use your intuitive notion of a subspace to decide whether each of the following drawn sets (3D in stereo pair) is a subspace, or not.

(a)

(b)
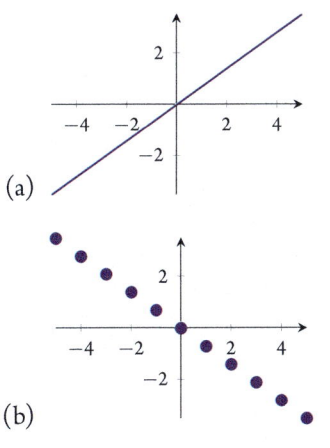

(c)

(d)
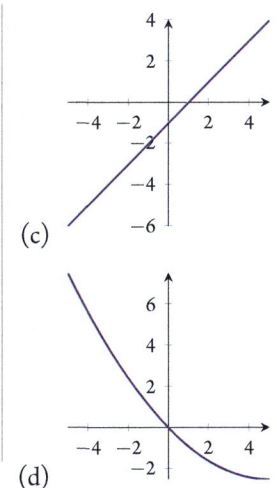

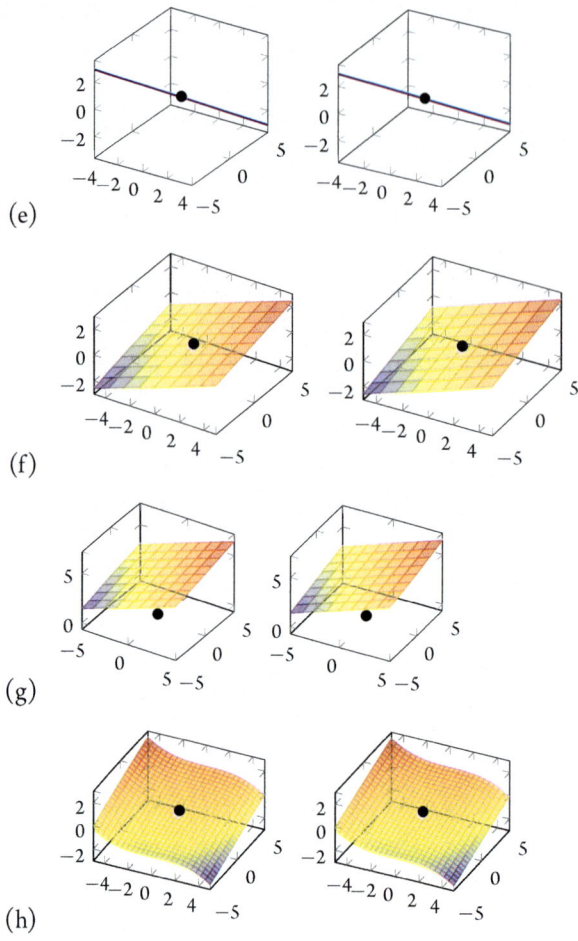

(e)

(f)

(g)

(h)

Exercise 3.4.2 Use Definition 3.4.3 to decide whether each of the following is a subspace, or not. Give reasons.

(a) All vectors in the line $y = 2x$.

(b) All vectors $(1.3n, -3.4n)$ for all integer n.

(c) All vectors $(x, y) = (-3.3 - 0.3t, 2.4 - 1.8t)$ for all real t.

(d) The vectors $(2, 1, -3)t + (5, -\frac{1}{2}, 2)s$ for all real s, t.

(e) The vectors $(0.9, 2.4, 1)t - (0.2, 0.6, 0.3)s$ for all real s, t.

(f) The vectors $(t, n, 2t + 3n)$ for all real t and integer n.

(g) The vectors $(1.4, 2.3, 1.5, 4) + (1.2, -0.8, -1.2, 2)t$ for all real t.

(h) The vectors $(t^3, 2t^3)$ for all real t (tricky).

(i) The vectors $(t^2, 3t^2)$ for all real t (tricky).

Exercise 3.4.3 Let W_1 and W_2 be any two subspaces of $\mathbb{R}^n$ (Definition 3.4.3).

(a) Use the definition to prove that the intersection of W_1 and W_2 is also a subspace of $\mathbb{R}^n$.

(b) Give an example to prove that the union of W_1 and W_2 is not necessarily a subspace of $\mathbb{R}^n$.

Exercise 3.4.4 For each of the following matrices, partially solve linear equations to determine whether the given vector b_j is in the column space, and to determine if the given vector r_j is in the row space of the matrix. Work small problems by hand, and address larger problems with MATLAB/Octave. Record your working or MATLAB/Octave commands and output.

(a) $A = \begin{bmatrix} 2 & 1 \\ 5 & 4 \end{bmatrix}$, $b_1 = \begin{bmatrix} 3 \\ -2 \end{bmatrix}$, $r_1 = \begin{bmatrix} 1 \\ 0 \end{bmatrix}$

(b) $B = \begin{bmatrix} -2 & -4 & -5 \\ -6 & -2 & 1 \end{bmatrix}$, $b_2 = \begin{bmatrix} 3 \\ 1 \end{bmatrix}$, $r_2 = \begin{bmatrix} 2 \\ -6 \\ -11 \end{bmatrix}$

(c) $C = \begin{bmatrix} 3 & 2 & 4 \\ 1 & 6 & 0 \\ 1 & -2 & 2 \end{bmatrix}$, $b_3 = \begin{bmatrix} 10 \\ 2 \\ 4 \end{bmatrix}$, $r_3 = \begin{bmatrix} 0 \\ -1 \\ 2 \end{bmatrix}$

(d) $D = \begin{bmatrix} -2 & 1 & 1 & -1 \\ 2 & -2 & -1 & 0 \\ -2 & 1 & 1 & -2 \\ 2 & 5 & -1 & -1 \end{bmatrix}$, $b_4 = \begin{bmatrix} 2 \\ -1 \\ 3 \\ 0 \end{bmatrix}$, $r_4 = \begin{bmatrix} -1 \\ -2 \\ 0 \\ -1 \end{bmatrix}$

(e) $E = \begin{bmatrix} 1.0 & 0.8 & 2.1 & 1.4 \\ 0.6 & -0.1 & 2.1 & 1.8 \\ 0.1 & -0.1 & 2.1 & 1.2 \\ 1.7 & -1.1 & -1.9 & 2.9 \end{bmatrix}$, $b_5 = \begin{bmatrix} 4.3 \\ 1.2 \\ 0.5 \\ 0.3 \end{bmatrix}$, $r_5 = \begin{bmatrix} 0.0 \\ 1.5 \\ -0.5 \\ 1.0 \end{bmatrix}$

Exercise 3.4.5 In each of the following, is the given vector in the nullspace of the given matrix?

(a) $A = \begin{bmatrix} -11 & -2 & 5 \\ -1 & 1 & 1 \end{bmatrix}$, $p = \begin{bmatrix} -7 \\ 6 \\ -13 \end{bmatrix}$

(b) $B = \begin{bmatrix} 3 & -3 & 2 \\ 1 & 1 & -3 \end{bmatrix}$, $q = \begin{bmatrix} 1 \\ 1 \\ 1 \end{bmatrix}$

(c) $C = \begin{bmatrix} -3 & 2 & 3 & 1 \\ -3 & -2 & -1 & 4 \\ 6 & 1 & -1 & -1 \end{bmatrix}$, $r = \begin{bmatrix} 2 \\ -4 \\ 1 \\ -2 \end{bmatrix}$

(d) $D = \begin{bmatrix} -4 & -2 & 2 & -2 \\ 2 & -1 & -2 & 1 \\ 0 & 2 & 1 & 0 \\ 0 & 0 & -8 & -2 \end{bmatrix}$, $s = \begin{bmatrix} 11 \\ -2 \\ 4 \\ -16 \end{bmatrix}$

Exercise 3.4.6 Given the SVDs of Exercises 3.3.2 and 3.3.3, write down an orthonormal basis for the span of the following sets of vectors.

(a) $(-\frac{9}{5}, -4)$, $(\frac{12}{5}, -3)$

(b) $(-\frac{2}{5}, \frac{11}{9}, \frac{31}{90}, \frac{4}{9})$, $(-\frac{2}{5}, \frac{11}{9}, \frac{31}{90}, \frac{4}{9})$, $(-\frac{26}{45}, -\frac{1}{3}, \frac{17}{90}, -\frac{2}{9})$,
$(\frac{26}{45}, \frac{1}{3}, -\frac{17}{90}, \frac{2}{9})$

Exercise 3.4.7 Given any $m \times n$ matrix A.

(a) Explain how Procedure 3.4.23 uses an SVD to find an orthonormal basis for the column space of A.

(b) How does the same SVD give the orthonormal basis $\{v_1 , v_2 , \dots , v_r\}$ for the row space of A? Justify your answer.

(c) Why does the same SVD also give the orthonormal basis $\{v_{r+1}, \dots , v_n\}$ for the nullspace of A? Justify.

Exercise 3.4.8 For each of the following matrices, compute an SVD with MATLAB/Octave, and then use the properties of Exercise 3.4.7 to write down an orthonormal basis for the column space, the row space, and the nullspace of the matrix. (The bases, especially for the nullspace, may differ in detail depending upon your version of MATLAB/Octave.)

(a) $\begin{bmatrix} 19 & -36 & -18 \\ -3 & 12 & 6 \\ -17 & 48 & 24 \end{bmatrix}$

(b) $\begin{bmatrix} -13 & 9 & 10 & -4 & -6 \\ -7 & 27 & -2 & 4 & -10 \\ -4 & 0 & 4 & 4 & -4 \\ -4 & -18 & 10 & -8 & 5 \end{bmatrix}$

(c) $\begin{bmatrix} 1 & -2 & 3 & 9 \\ -1 & 5 & 0 & 0 \\ 0 & 3 & 3 & 9 \\ 2 & -9 & 1 & 3 \\ 1 & -7 & -2 & -6 \end{bmatrix}$

(d) $\begin{bmatrix} -128 & 6 & 55 & -28 & -1 \\ 20 & 12 & -31 & 18 & -3 \\ -12 & -30 & 39 & -24 & 7 \\ -1 & 6 & -1 & 7 & -3 \end{bmatrix}$

Exercise 3.4.9 For each of the matrices in Exercise 3.4.8, from your computed bases write down the dimension of the column space, the row space, and the nullspace. Comment on how these confirm the Rank Theorem 3.4.39.

Exercise 3.4.10 What are the possible values for nullity(A) in the following cases?

(a) A is a 2×5 matrix.

(b) A is a 3×3 matrix.

(c) A is a 3×2 matrix.

(d) A is a 4×6 matrix.

(e) A is a 4×4 matrix.

(f) A is a 6×5 matrix.

Exercise 3.4.11 *(Cowen (1997))* Alice and Bob are taking a course on linear algebra. One of the problems in their homework assignment is to find the nullspace of a 4×5 matrix A. In each of the following cases: are their answers consistent with each other? Give reasons.

(a) Alice's answer is that the nullspace is spanned by $(-2, -2, 0, 2, -6)$, $(1, 5, 4, -3, 11)$, $(3, 5, 2, -4, 13)$, and $(0, -2, -2, 1, -4)$. Bob's answer is that the nullspace is spanned by $(1, 1, 0, -1, 3)$, $(-2, 0, 2, 1, -2)$, and $(-1, 3, 4, 1, 5)$.

(b) Alice's answer is that the nullspace is spanned by $(2, -3, 1, -2, -5)$, $(2, -7, 2, -1, -6)$, $(1, -2, 1, 1, 0)$, $(3, -6, 3, 3, 0)$. Bob's answer is that the nullspace is spanned by $(1, -2, 1, 1, 0)$, $(0, 4, -1, -1, 1)$, $(1, -1, 0, -3, -5)$.

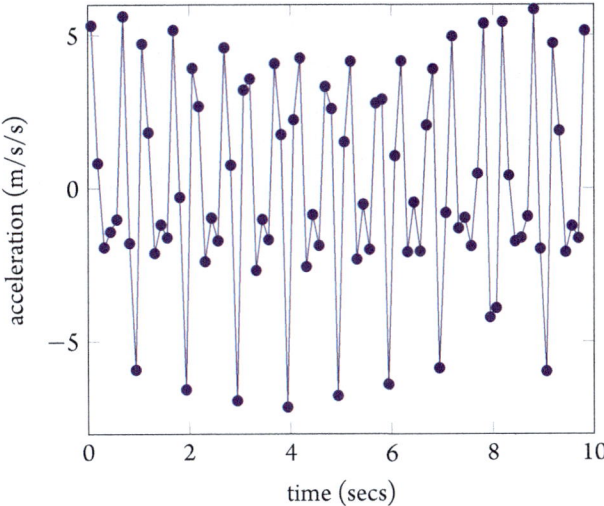

Figure 3.4 Vertical acceleration of the ankle of a person walking normally, a person who has Parkinson's disease. The data is recorded 0.125 s apart (here subsampled and smoothed for the purposes of the exercise).

(c) Alice's answer is that the nullspace is spanned by $(-2, 0, -2, 4, -5)$, $(0, 2, -2, 2, -2)$, $(0, -2, 2, -2, 2)$, $(-4, -12, 8, -4, 2)$. Bob's answer is that the nullspace is spanned by $(0, 2, -2, 2, -2)$, $(-2, -4, 2, 0, -1)$, $(1, 0, -1, 1, -3)$.

(d) Alice's answer is that the nullspace is spanned by $(-1, 0, 0, 0, 0)$, $(5, 3, -2, 5, 1)$, $(-5, 1, 0, -6, -2)$, $(4, -2, 0, 1, 8)$. Bob's answer is that the nullspace is spanned by $(1, -2, 0, -3, 4)$, $(2, -1, 1, 3, 2)$, $(3, 0, -1, 2, 3)$.

Exercise 3.4.12 Prove that if the columns of a matrix A are orthonormal, then the columns must form an orthonormal basis for the column space of A.

Exercise 3.4.13 Let A be any $m \times n$ matrix. Use an SVD to prove that every vector in the row space of A is orthogonal to every vector in the nullspace.

Exercise 3.4.14 Bachlin et al. [*IEEE Transactions on Information Technology in Biomedicine,* 14(2), 2010] explored the walking gait of people with Parkinson's disease. Among many measurements, they measured the vertical ankle acceleration of the people when they walked. Figure 3.4 shows ten seconds of just one example: use the so-called Singular Spectrum Analysis to find the regular structures in this complex data.

Following Example 3.4.27:

(a) enter the data into MATLAB/Octave;

```
time=(0.0625:0.125:9.85)'
acc=[5.34;  0.85;  -1.90;  -1.39;  -0.99;  5.64;
-1.76;  -5.90;  4.74;  1.85;  -2.09;  -1.16;  -1.58;
5.19;  -0.27;  -6.54;  3.94;  2.70;  -2.36;  -0.94;
-1.68;  4.61;  0.79;  -6.90;  3.23;  3.59;  -2.65;
```

```
-0.99; -1.65; 4.09; 1.78; -7.11; 2.26; 4.27;
-2.53; -0.84; -1.84; 3.34; 2.62; -6.74; 1.54;
4.16; -2.29; -0.50; -1.97; 2.80; 2.92; -6.37;
1.09; 4.17; -2.05; -0.44; -2.03; 2.08; 3.91;
-5.84; -0.78; 4.98; -1.28; -0.94; -1.86; 0.50;
5.40; -4.19; -3.88; 5.45; 0.44; -1.71; -1.59;
-0.90; 5.86; -1.95; -5.95; 4.75; 1.90; -2.06;
-1.21; -1.61; 5.16]
```

(b) use hankel() to form a matrix whose 66 columns are 66 'windows' of accelerations, each of length fourteen data points (of length 1.75 s);

(c) compute an SVD of the matrix, and explain why the windows of measured accelerations are close to lying in a four-dimensional subspace;

(d) plot orthonormal basis vectors for the four-dimensional subspace of the windowed accelerations.

Exercise 3.4.15 Consider $m \times n$ matrices bordered by zeros in the block form

$$E = \begin{bmatrix} F & O_{k \times n - \ell} \\ O_{m-k \times \ell} & O_{m-k \times n-\ell} \end{bmatrix}$$

where F is some $k \times \ell$ matrix. Given that matrix F has an SVD, find an SVD of matrix E, and hence prove that rank E = rank F.

Exercise 3.4.16 For compatibly sized matrices A and B, use their SVDs, and the result of the previous Exercise 3.4.15 (applied to the matrix $S_A V_A^T U_B S_B$), to prove that rank$(AB) \leq$ rank A and also that rank$(AB) \leq$ rank B.

Exercise 3.4.17 In a few sentences, answer/discuss each of the following.

(a) In Definition 3.4.3 what causes a subspace to be a line, plane, ..., through the origin?

(b) What is the key feature of the concept of the span of a set that causes the span to always be a subspace?

(c) How does the column space of a matrix relate to its row space?

(d) Why is an orthonormal basis important?

(e) How does the concept of the dimension occur?

3.5 Project to solve inconsistent equations

Agreement with experiment is the sole criterion of truth for a physical theory.

Pierre Duhem, 1906

The scientific method is to infer general laws from data and then validate the laws. This section addresses some aspects of the inference of general laws from data. A huge challenge is that data is typically corrupted by noise and errors. So this section shows how the singular value decomposition (SVD) leads to understanding 'least square methods' for handling noisy errors.

As well as being fundamental to engineering, scientific, and computational inference, approximately solving inconsistent equations also introduces the linear transformation of "projection".

3.5.1 Make a minimal change to the problem

Example 3.5.1 *(rationalize contradictions)* I weighed myself the other day. I weighed myself four times, each time separated by a few minutes: the scales reported my weight in kilograms (kg) as 84.8, 84.1, 84.7 and 84.4. The measurements give four different weights! What sense can we make of this apparently contradictory data? Traditionally we just average and say my weight is $x \approx (84.8 + 84.1 + 84.7 + 84.4)/4 = 84.5$ kg. Let's see this same answer from a new linear algebra justification.

In the linear algebra, view my weight x as an unknown. The four experimental measurements give four equations for this one unknown:

$$x = 84.8, \quad x = 84.1, \quad x = 84.7, \quad x = 84.4.$$

Despite being manifestly impossible to satisfy all four equations, let's see what linear algebra can do for us. Linear algebra writes these four equations as the matrix-vector system

$$Ax = b, \quad \text{namely} \quad \begin{bmatrix} 1 \\ 1 \\ 1 \\ 1 \end{bmatrix} x = \begin{bmatrix} 84.8 \\ 84.1 \\ 84.7 \\ 84.4 \end{bmatrix}.$$

The linear algebra Procedure 3.3.15 is to 'solve' this system, despite its contradictions, via an SVD and some intermediaries:

$$Ax = U \underbrace{S \overbrace{V^{\mathsf{T}} x}^{=y}}_{=z} = b.$$

(a) We are given that this particular matrix A of a column of ones has an SVD of

$$A = \begin{bmatrix} 1 \\ 1 \\ 1 \\ 1 \end{bmatrix} = \begin{bmatrix} \frac{1}{2} & \frac{1}{2} & \frac{1}{2} & \frac{1}{2} \\ \frac{1}{2} & \frac{1}{2} & -\frac{1}{2} & -\frac{1}{2} \\ \frac{1}{2} & -\frac{1}{2} & -\frac{1}{2} & \frac{1}{2} \\ \frac{1}{2} & -\frac{1}{2} & \frac{1}{2} & -\frac{1}{2} \end{bmatrix} \begin{bmatrix} 2 \\ 0 \\ 0 \\ 0 \end{bmatrix} [1]^{\mathsf{T}} = USV^{\mathsf{T}}$$

(perhaps check the columns of U are orthonormal).

(b) Solve $Uz = b$ by computing

$$z = U^T b = \begin{bmatrix} \frac{1}{2} & \frac{1}{2} & \frac{1}{2} & \frac{1}{2} \\ \frac{1}{2} & \frac{1}{2} & -\frac{1}{2} & -\frac{1}{2} \\ \frac{1}{2} & -\frac{1}{2} & -\frac{1}{2} & \frac{1}{2} \\ \frac{1}{2} & -\frac{1}{2} & \frac{1}{2} & -\frac{1}{2} \end{bmatrix} \begin{bmatrix} 84.8 \\ 84.1 \\ 84.7 \\ 84.4 \end{bmatrix} = \begin{bmatrix} 169 \\ -0.1 \\ 0.2 \\ 0.5 \end{bmatrix}.$$

(c) Now try to solve $Sy = z$, that is,

$$\begin{bmatrix} 2 \\ 0 \\ 0 \\ 0 \end{bmatrix} y = \begin{bmatrix} 169 \\ -0.1 \\ 0.2 \\ 0.5 \end{bmatrix}.$$

We cannot, because the last three components in the equation are impossible: we cannot satisfy any of

$$0y = -0.1, \quad 0y = 0.2, \quad 0y = 0.5.$$

Instead of seeking an *exact* solution, ask what is the *smallest change* we can make to $z = (169, -0.1, 0.2, 0.5)$ so that we can report a solution to a slightly different problem? Answer: we *have to* adjust the last three components to zero. Moreover, any adjustment to the first component is not needed, it would make the change to z bigger than necessary, and so we do not adjust the first component. Hence we solve a slightly different problem, that of

$$\begin{bmatrix} 2 \\ 0 \\ 0 \\ 0 \end{bmatrix} y = \begin{bmatrix} 169 \\ 0 \\ 0 \\ 0 \end{bmatrix},$$

with solution $y = 84.5$. *Let's treat this exact solution to a slightly different problem as an approximate solution to the original problem.*

(d) Lastly, solve $V^T x = y$ by computing $x = Vy = 1y = y = 84.5 \, \text{kg}$ (upon including the physical units). That is, this linear algebra procedure gives my weight as $x = 84.5 \, \text{kg}$ (approximately).

This linear algebra procedure recovers the traditional answer of averaging measurements. □

The methodology of the previous Example 3.5.1 illustrates how traditional averaging emerges from trying to make sense of apparently inconsistent information. Importantly, the principle of making the smallest possible change to the intermediary z is equivalent to making the smallest possible change to the original data vector b. The reason is that $b = Uz$ for an orthogonal matrix U: since U is an orthogonal matrix, multiplication by U preserves distances and angles (Theorem 3.2.48) and so the smallest possible change to $\boldsymbol{b}$ is precisely the same magnitude as the smallest possible change to $\boldsymbol{z}$. Scientists and engineers implicitly use this same 'smallest change' approach to approximately solve many sorts of inconsistent linear equations.

Activity 3.5.2 Consider the inconsistent equations $3x = 1$ and $4x = 3$ formed as the system (illustrated to the right)

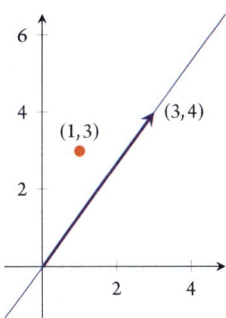

$$\begin{bmatrix} 3 \\ 4 \end{bmatrix} x = \begin{bmatrix} 1 \\ 3 \end{bmatrix}, \quad \text{and given} \quad \begin{bmatrix} 3 \\ 4 \end{bmatrix} = \begin{bmatrix} \frac{3}{5} & \frac{4}{5} \\ \frac{4}{5} & -\frac{3}{5} \end{bmatrix} \begin{bmatrix} 5 \\ 0 \end{bmatrix} [1]^\mathsf{T}$$

is an SVD factorization of the 2×1 matrix. Following the procedure of the previous Example 3.5.1, what is the 'best' approximate solution to these inconsistent equations?

(a) $x = 3/4$ (b) $x = 3/5$ (c) $x = 4/7$ (d) $x = 1/3$

Example 3.5.3 Recall the table tennis player rating Example 3.3.13. There we found that we could not solve the equations to find some ratings because the equations were inconsistent. In our new terminology of the previous Section 3.4, the right-hand side vector $\boldsymbol{b}$ is not in the column space of the matrix A (Definition 3.4.10):

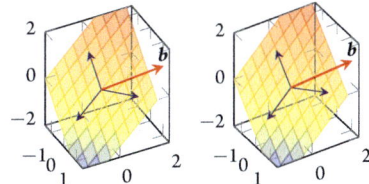

the stereo picture to the right illustrates the 2D column space spanned by the three columns of A and that the vector $\boldsymbol{b}$, of the results, lies outside the column space.

Now reconsider Step 3 in Example 3.3.13.

(a) We need to interpret and 'solve' $S\boldsymbol{y} = \boldsymbol{z}$ which here is

$$\begin{bmatrix} 1.7321 & 0 & 0 \\ 0 & 1.7321 & 0 \\ 0 & 0 & 0 \end{bmatrix} \boldsymbol{y} = \begin{bmatrix} -2.0412 \\ -2.1213 \\ 0.5774 \end{bmatrix}.$$

The third line of this system says $0y_3 = 0.5774$ which is impossible for any y_3: we cannot have zero on the left-hand side equalling 0.5774 on the right-hand side. Instead of seeking an *exact* solution, ask what is the *smallest change* we can make to $\boldsymbol{z} = (-2.0412, -2.1213, 0.5774)$ so that we can report a solution, albeit to a slightly different problem? Answer: we *must* change the last component of $\boldsymbol{z}$ to zero. Moreover, any change to the first two components is not needed, it would make the change bigger than necessary, and so we do not change the first two components. Hence we find an approximate solution to the player ratings via solving

$$\begin{bmatrix} 1.7321 & 0 & 0 \\ 0 & 1.7321 & 0 \\ 0 & 0 & 0 \end{bmatrix} y = \begin{bmatrix} -2.0412 \\ -2.1213 \\ 0 \end{bmatrix}.$$

Here, via `y=z(1:2)./diag(S(1:2,1:2))`, a general solution is that vector $y = (-1.1785, -1.2247, y_3)$. Varying the free variable y_3 gives equally good approximate solutions.

(b) Lastly, solve $V^\mathsf{T} x = y$, via computing `x=V(:,1:2)*y`, to determine

$$x = Vy = \begin{bmatrix} 0.0000 & -0.8165 & 0.5774 \\ -0.7071 & 0.4082 & 0.5774 \\ 0.7071 & 0.4082 & 0.5774 \end{bmatrix} \begin{bmatrix} -1.1785 \\ -1.2247 \\ y_3 \end{bmatrix}$$

$$= \begin{bmatrix} 1 \\ \frac{1}{3} \\ -\frac{4}{3} \end{bmatrix} + \frac{y_3}{\sqrt{3}} \begin{bmatrix} 1 \\ 1 \\ 1 \end{bmatrix}.$$

As before, it is only the relative ratings that are important so we choose any particular (approximate) solution by setting y_3 to anything we like, such as zero. The predicted ratings are then $x = (1, \frac{1}{3}, -\frac{4}{3})$ for Anne, Bob, and Chris, respectively. □

The reliability and likely error of such approximate solutions are the province of statistics courses. We focus on the geometry and linear algebra of obtaining such a 'best' approximate solution.

Procedure 3.5.4 (approximate solution) *Obtain the so-called 'least square' approximate solution(s) of inconsistent equations $Ax = b$ using an SVD and via intermediate unknowns:*

1. *factorize $A = USV^\mathsf{T}$ and set $r = \operatorname{rank} A$ (remembering that relatively small singular values are effectively zero);*

2. *solve $Uz = b$ by $z = U^\mathsf{T} b$;*

3. *disregard the inconsistent equations for $i = r+1, \ldots, m$ as errors, set $y_i = z_i / \sigma_i$ for $i = 1, \ldots, r$ (as these $\sigma_i > 0$), and otherwise y_i is free for $i = r+1, \ldots, n$;*

4. *solve $V^\mathsf{T} x = y$ to obtain a general approximate solution as $x = Vy$.*

Example 3.5.5 You are given the choice of two different types of concrete mix. One type contains 40% cement, 40% gravel, and 20% sand; whereas the other type contains 20% cement, 10% gravel, and 70% sand. How many kilograms of each type should you mix together to obtain a concrete mix as close as possible to 3 kg of cement, 2 kg of gravel, and 4 kg of sand.

Solution: Let variables x_1 and x_2 be the as yet unknown amounts, in kg, of each type of concrete mix. Then for the cement component we want $0.4x_1 + 0.2x_2 = 3$, whereas for the gravel component we want $0.4x_1 + 0.1x_2 = 2$, and for the sand component $0.2x_1 + 0.7x_2 = 4$. These form the matrix-vector system $Ax = b$ for matrix and vector

$$A = \begin{bmatrix} 0.4 & 0.2 \\ 0.4 & 0.1 \\ 0.2 & 0.7 \end{bmatrix}, \quad b = \begin{bmatrix} 3 \\ 2 \\ 4 \end{bmatrix}.$$

Apply Procedure 3.5.4.

(a) Enter the matrix A and vector b into MATLAB/Octave with

```
A= [0.4 0.2; 0.4 0.1; 0.2 0.7]
b= [3;2;4]
```

Then factorize matrix $A = USV^T$ with $[U, S, V] = svd(A)$:

```
U =
   -0.4638   -0.5018   -0.7302
   -0.3681   -0.6405    0.6740
   -0.8058    0.5814    0.1123
S =
    0.8515         0
         0    0.4182
         0         0
V =
   -0.5800   -0.8146
   -0.8146    0.5800
```

The system of equations $Ax = b$ for the mix becomes

$$U S \overbrace{V^T x}^{=y} = b.$$
$$\underbrace{}_{=z}$$

(b) Solve $Uz = b$ by $z = U^T b$ via computing z=U' *b to get

```
z =
   -5.3510
   -0.4608
   -0.3932
```

(c) Now solve $Sy = z$. But the last (third) row of the diagonal matrix S is zero, whereas the last component of z is nonzero: hence there is no exact solution. Instead, we approximate by setting the last component of z to zero. This approximation is the *smallest change* we can make to the required mix that is possible.

Table 3.4 The results of six games played in a round robin: the scores are games/goals/points scored by each when playing the others. For example, Dee beat Anne 3 to 1.

	Anne	Bob	Chris	Dee
Anne	—	3	3	1
Bob	2	—	2	4
Chris	0	1	—	2
Dee	3	0	3	—

That is, since rank $A = 2$ from the two nonzero singular values, so we approximately solve the system in MATLAB/Octave by y=z(1:2)./diag(S) (there are no free variables here):

```
y =
   -6.284
   -1.102
```

(d) Lastly solve $V^T x = y$ as $x = V y$ by computing x=V*y:

```
x =
    4.543
    4.479
```

Then interpret: from this solution $x \approx (4.5, 4.5)$ we need to mix close to 4.5 kg of both the types of concrete to get as close as possible to the desired mix. Multiplication, Ax or A*x, tells us that the resultant mix is about 2.7 kg cement, 2.3 kg gravel, and 4.0 kg of sand.

Compute x=A\b and find that it directly gives exactly the same answer: Section 3.5.2 discusses why A\b gives exactly the same 'best' approximate solution. ☐

Example 3.5.6 *(round robin tournament)* Consider four players (or teams) that play in a round robin sporting event: Anne, Bob, Chris, and Dee. Table 3.4 summarizes the results of the six games played. From these results estimate the relative player ratings of the four players. As in many real-life situations, the information appears contradictory such as Anne beats Bob, who beats Dee, who in turn beats Anne. Assume that the rating x_i of player i is to reflect, as best we can, the difference in scores upon playing player j: that is, pose the difference in ratings, $x_i - x_j$, should equal the difference in the scores when they play.

Solution: The first stage is to model the results by idealised mathematical equations. From Table 3.4 six games were played with the following scores. Each game then generates the shown ideal equation for the difference between two ratings.

- Anne beats Bob 3-2, so $x_1 - x_2 = 3 - 2 = 1$.

- Anne beats Chris 3-0, so $x_1 - x_3 = 3 - 0 = 3$.

- Bob beats Chris 2-1, so $x_2 - x_3 = 2 - 1 = 1$.

- Anne is beaten by Dee 1-3, so $x_1 - x_4 = 1 - 3 = -2$.

- Bob beats Dee 4-0, so $x_2 - x_4 = 4 - 0 = 4$.

- Chris is beaten by Dee 2-3, so $x_3 - x_4 = 2 - 3 = -1$.

These six equations form the linear system $Ax = b$ where

$$
A = \begin{bmatrix}
1 & -1 & 0 & 0 \\
1 & 0 & -1 & 0 \\
0 & 1 & -1 & 0 \\
1 & 0 & 0 & -1 \\
0 & 1 & 0 & -1 \\
0 & 0 & 1 & -1
\end{bmatrix}, \quad
b = \begin{bmatrix}
1 \\
3 \\
1 \\
-2 \\
4 \\
-1
\end{bmatrix}.
$$

We cannot satisfy all these equations exactly, so we have to accept an approximate solution that estimates the ratings as best we can. The second stage uses an SVD and Procedure 3.5.4 to 'best' solve the equations.

(a) Enter the matrix A and vector b into MATLAB/Octave with

```
A= [1   -1    0    0
     1    0   -1    0
     0    1   -1    0
     1    0    0   -1
     0    1    0   -1
     0    0    1   -1 ]
b= [1;3;1;-2;4;-1]
```

Then factorize matrix $A = USV^\mathsf{T}$ with [U, S, V] =svd (A) (2 d.p.):

```
U =
    0.31  -0.26  -0.58  -0.26   0.64  -0.15
    0.07   0.40  -0.58   0.06  -0.49  -0.51
   -0.24   0.67   0.00  -0.64   0.19   0.24
   -0.38  -0.14  -0.58   0.21  -0.15   0.66
   -0.70   0.13   0.00   0.37   0.45  -0.40
   -0.46  -0.54  -0.00  -0.58  -0.30  -0.26
S =
    2.00      0      0      0
       0   2.00      0      0
       0      0   2.00      0
       0      0      0   0.00
       0      0      0      0
       0      0      0      0
V =
    0.00   0.00  -0.87  -0.50
   -0.62   0.53   0.29  -0.50
```

```
-0.14 -0.80   0.29 -0.50
 0.77  0.28   0.29 -0.50
```

Although the first three columns of U and V may be different for you (because the first three singular values are all the same), the eventual solution is the same. The system of equations $Ax = b$ for the ratings becomes

$$\underbrace{US\underbrace{\overbrace{V^\mathsf{T}x}^{=y}}_{=z}} = b.$$

(b) Solve $Uz = b$ by $z = U^\mathsf{T}b$ via computing z=U'*b to get the $\mathbb{R}^6$ vector

```
z =
    -1.27
     2.92
    -1.15
     0.93
     1.76
    -4.07
```

(c) Now solve $Sy = z$. But the last three rows of the diagonal matrix S are zero, whereas the last three components of z are nonzero: hence there is no exact solution. Instead, we approximate by setting the last three components of z to zero. This approximation is the *smallest change* we can make to the data of the game results that makes the results consistent.

That is, since rank $A = 3$ from the three nonzero singular values, we approximately solve the system in MATLAB/Octave by y=z(1:3)./diag(S(1:3,1:3)):

```
y =
    -0.63
     1.46
    -0.58
```

The fourth component y_4 is arbitrary.

(d) Lastly, solve $V^\mathsf{T}x = y$ as $x = Vy$. Obtain a particular solution in MATLAB/Octave by computing x=V(:,1:3)*y:

```
x =
     0.50
     1.00
    -1.25
    -0.25
```

Be aware of Kenneth Arrow's Impossibility Theorem (Arrow, 1950)—one of the great theorems of the 20th century: *all 1D ranking systems are flawed!* Wikipedia[a] (2014) described the theorem this way (in the context of voting systems): that among

> three or more distinct alternatives (options), no rank order voting system can convert the ranked preferences of individuals into a community-wide (complete and transitive) ranking while also meeting [four sensible] criteria ... called unrestricted domain, non-dictatorship, Pareto efficiency, and independence of irrelevant alternatives.

In rating sport players/teams:

- the "distinct alternatives" are the players/teams;
- the "ranked preferences of individuals" are the individual results of each game played; and
- the "community-wide ranking" is the assumption that we can rate each player/team by a one-dimensional numerical rating.

Arrow's theorem assures us that every such scheme must violate at least one of four sensible criteria. Every ranking scheme is thus open to criticism. But every alternative scheme would also be open to criticism by also violating at least one of the criteria.

[a] https://en.wikipedia.org/wiki/Arrow%27s_impossibility_theorem

Add an arbitrary multiple of the fourth column of V to get a general solution

$$
x = \begin{bmatrix} \frac{1}{2} \\ 1 \\ -\frac{5}{4} \\ -\frac{1}{4} \end{bmatrix} + y_4 \begin{bmatrix} -\frac{1}{2} \\ -\frac{1}{2} \\ -\frac{1}{2} \\ -\frac{1}{2} \end{bmatrix}.
$$

The final stage is to interpret the solution for the application. In this application, the absolute ratings are not important, so we ignore y_4 (consider it zero). From the game results of Table 3.4 this analysis indicates that the players' rankings are, in decreasing order, Bob, Anne, Dee, and Chris.

□

When rating players or teams based upon results, be clear of the purpose. For example, is the purpose to summarize past performance? or to predict future contests? If the latter, then my limited experience suggests that one should fit the win-loss record instead of the scores. Explore the alternatives for your favourite sport.

Activity 3.5.7 Listed below are four approximate solutions to the system $Ax = b$,

$$
\begin{bmatrix} 5 & 3 \\ 3 & -1 \\ 1 & 1 \end{bmatrix} \begin{bmatrix} x \\ y \end{bmatrix} = \begin{bmatrix} 9 \\ 2 \\ 10 \end{bmatrix}.
$$

Setting vector $b' = Ax$ for each, which one minimizes the distance between the original right-hand side $b = (9, 2, 10)$ and the approximate b'?

(a) $x = \begin{bmatrix} 2 \\ 1 \end{bmatrix}$ (b) $x = \begin{bmatrix} 2 \\ 2 \end{bmatrix}$ (c) $x = \begin{bmatrix} 1 \\ 2 \end{bmatrix}$ (d) $x = \begin{bmatrix} 1 \\ 1 \end{bmatrix}$

Theorem 3.5.8 (*smallest change*) *All approximate solutions obtained by Procedure 3.5.4 solve the linear system $Ax = b'$ for the unique consistent right-hand side vector b' that minimizes the distance $|b' - b|$.*

(The dash on b' is to suggest an approximation to b.)

Proof. Find an SVD $A = USV^{\mathsf{T}}$ of $m \times n$ matrix A. Then Procedure 3.5.4 computes $z = U^{\mathsf{T}}b \in \mathbb{R}^m$, that is, $b = Uz$ as U is orthogonal. For any $b' \in \mathbb{R}^m$ let $z' = U^{\mathsf{T}}b' \in \mathbb{R}^m$, that is, $b' = Uz'$. Then $|b' - b| = |Uz' - Uz| = |U(z' - z)| = |z' - z|$ as multiplication by orthogonal U preserves distances (Theorem 3.2.48). Thus minimizing $|b' - b|$ is equivalent to minimizing $|z' - z|$. Procedure 3.5.4 seeks to solve the diagonal system $Sy = z$ for $y \in \mathbb{R}^n$. That is, for a matrix of rank $A = r$

$$\begin{bmatrix} \sigma_1 & \cdots & 0 & & \\ \vdots & \ddots & \vdots & & O_{r\times(n-r)} \\ 0 & \cdots & \sigma_r & & \\ & & & & \\ O_{(m-r)\times r} & & O_{(m-r)\times(n-r)} & \end{bmatrix} \, y = \begin{bmatrix} z_1 \\ \vdots \\ z_r \\ z_{r+1} \\ \vdots \\ z_m \end{bmatrix}.$$

Procedure 3.5.4 approximately solves this inconsistent system by adjusting the right-hand side to $z' = (z_1, \ldots, z_r, 0, \ldots, 0) \in \mathbb{R}^m$. This change makes $|z - z'|$ as small as possible because we *must* zero the last $(m - r)$ components of z in order to obtain a consistent set of equations, and because any adjustment to the first r components of z would only increase $|z - z'|$. Further, it is the only change to z that does so, so z' is the unique minimizer. Hence the solution computed by Procedure 3.5.4 solves the consistent system $Ax = b'$ (with the unique $b' = Uz'$) such that $|b - b'|$ is minimized. □

Example 3.5.9 (*life expectancy*) Table 3.5 lists life expectancies of people born in a given year; Figure 3.5 plots the data points. Over the decades, the life expectancies have increased. Let's quantify the overall trend to be able to draw, as in Figure 3.5, the best straight line to the female life expectancy. Solve the approximation problem with an SVD and confirm that it gives the same solution as A\b in MATLAB/Octave.

Table 3.5 Life expectancy in years of (white) females and males born in the given years [http://www.infoplease.com/ipa/A0005140.html, 2014]. Used by Example 3.5.9.

year	1951	1961	1971	1981	1991	2001	2011
female	72.0	74.2	75.5	78.2	79.6	80.2	81.1
male	66.3	67.5	67.9	70.8	72.9	75.0	76.3

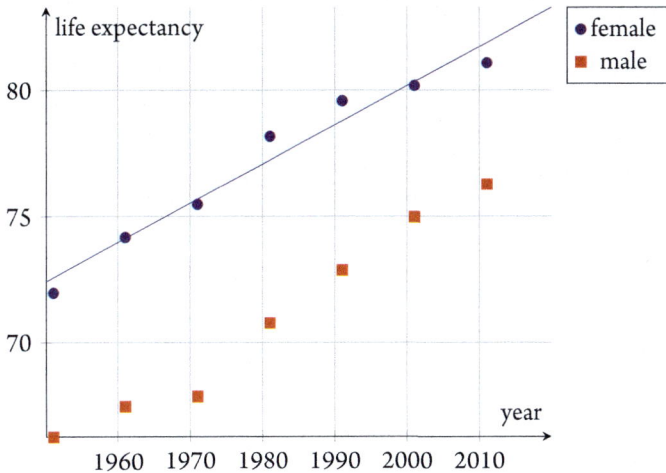

Figure 3.5 The life expectancies in years of females and males born in the given years (Table 3.5). Also plotted is the best straight line fit to the female data obtained by Example 3.5.9.

Solution: Start by posing a mathematical model: let's suppose that the life expectancy ℓ is a straight line function of year of birth: $\ell = x_1 + x_2 t$, where we need to find the coefficients x_1 and x_2, and where t counts the number of decades since 1951, the start of the data. Table 3.5 then gives seven ideal equations to solve for x_1 and x_2:

$$(1951) \quad x_1 + 0x_2 = 72.0,$$
$$(1961) \quad x_1 + 1x_2 = 74.2,$$
$$(1971) \quad x_1 + 2x_2 = 75.5,$$
$$(1981) \quad x_1 + 3x_2 = 78.2,$$
$$(1991) \quad x_1 + 4x_2 = 79.6,$$
$$(2001) \quad x_1 + 5x_2 = 80.2,$$
$$(2011) \quad x_1 + 6x_2 = 81.1.$$

Form these into the matrix-vector system $Ax = b$ where

$$A = \begin{bmatrix} 1 & 0 \\ 1 & 1 \\ 1 & 2 \\ 1 & 3 \\ 1 & 4 \\ 1 & 5 \\ 1 & 6 \end{bmatrix}, \quad b = \begin{bmatrix} 72.0 \\ 74.2 \\ 75.5 \\ 78.2 \\ 79.6 \\ 80.2 \\ 81.1 \end{bmatrix}.$$

Procedure 3.5.4 then determines a best approximate solution.

(a) Enter the matrix A and vector b into MATLAB/Octave, and compute an SVD of $A = USV^{\mathsf{T}}$
via [U,S,V]=svd(A) (2 d.p.):

```
U =
    0.02   0.68 -0.38 -0.35 -0.32 -0.30 -0.27
    0.12   0.52 -0.14   0.06   0.26   0.45   0.65
    0.22   0.36   0.89 -0.09 -0.08 -0.07 -0.05
    0.32   0.20 -0.10   0.88 -0.13 -0.15 -0.16
    0.42   0.04 -0.10 -0.14   0.81 -0.23 -0.28
    0.52 -0.12 -0.09 -0.16 -0.24   0.69 -0.39
    0.62 -0.28 -0.09 -0.19 -0.29 -0.40   0.50
S =
    9.80      0
       0   1.43
       0      0
       0      0
       0      0
       0      0
       0      0
V =
    0.23   0.97
    0.97 -0.23
```

(b) Solve $Uz = b$ to give this first intermediary $z = U^{\mathsf{T}}b$ via the command z=U'*b:

```
z =
   178.19
   100.48
    -0.05
     1.14
     1.02
     0.10
    -0.52
```

(c) Now solve approximately $Sy = z$. From the two nonzero singular values in S the matrix A
has rank 2. So the approximation is to discard/zero (as 'errors') all but the first two elements of z and find the best approximate y via y=z(1:2)./diag(S(1:2,1:2)):

```
y =
    18.19
    70.31
```

(d) Solve $V^{\mathsf{T}}x = y$ by $x = Vy$ via x=V*y:

```
x =
    72.61
     1.55
```

Compute x=A\b to find that it gives exactly the same answer: Section 3.5.2 discusses why A\b gives exactly the same 'best' approximate solution.

Lastly, interpret the answer. The approximation gives $x_1 = 72.61$ and $x_2 = 1.55$ (2 d.p.). Since the ideal model was life expectancy $\ell = x_1 + x_2 t$ we determine a 'best' approximate model is $\ell \approx 72.61 + 1.55 t$ years where t is the number of decades since 1951: this is the straight line drawn in Figure 3.5. That is, females tend to live an extra 1.55 years for every decade born after 1951. For example, for females born in 2021, some seven decades after 1951, this model predicts a life expectancy of $\ell \approx 72.61 + 1.55 \times 7 = 83.46$ years. □

Activity 3.5.10 In calibrating a vortex flowmeter the following flow rates were obtained for various applied voltages.

voltage (V)	1.18	1.85	2.43	2.81
flow rate (litre/s)	0.18	0.57	0.93	1.27

Letting v_i be the voltages and f_i the flow rates, which of the following is a reasonable model to seek? (for coefficients x_1, x_2, x_3)

(a) $f_i = x_1$

(b) $f_i = x_1 + x_2 v_i$

(c) $v_i = x_1 + x_2 f_i$

(d) $v_i = x_1 + x_2 f_i + x_3 f_i^2$

Example 3.5.11 *(planetary orbital periods)* Table 3.6 lists each orbital period of the planets of the solar system; Figure 3.6 plots the data points as a function of the distance of the planets from the sun. Let's infer Kepler's law that the period grows as the distance to the power $3/2$: shown by the best straight line fit in Figure 3.6. Use the data for the planets from Mercury to Uranus to infer the law with an SVD, confirm that it gives the same solution as A\b in MATLAB/Octave, and use the fit to predict Neptune's period from its distance.

Solution: Start by posing a mathematical model: Kepler's law is a power law that the ith period $p_i = c_1 d_i^{c_2}$ for some unknown coefficient c_1 and exponent c_2. Take logarithms (to any base so let's use base 10) and seek that $\log_{10} p_i = \log_{10} c_1 + c_2 \log_{10} d_i$; that is, seek unknowns x_1 and x_2 such that $\log_{10} p_i = x_1 + x_2 \log_{10} d_i$. The first seven rows of Table 3.6 then gives seven ideal linear equations to solve for x_1 and x_2:

$$x_1 + \log_{10} 57.91 \, x_2 = \log_{10} 87.97,$$
$$x_1 + \log_{10} 108.21 \, x_2 = \log_{10} 224.70,$$
$$x_1 + \log_{10} 149.60 \, x_2 = \log_{10} 365.26,$$
$$x_1 + \log_{10} 227.94 \, x_2 = \log_{10} 686.97,$$
$$x_1 + \log_{10} 778.55 \, x_2 = \log_{10} 4332.59,$$
$$x_1 + \log_{10} 1433.45 \, x_2 = \log_{10} 10759.22,$$
$$x_1 + \log_{10} 2870.67 \, x_2 = \log_{10} 30687.15.$$

Power laws and the log-log plot Hundreds of power laws have been identified in engineering, physics, biology, and the social sciences. These laws are typically detected via log-log plots. A log-log plot is a two-dimensional graph of the numerical data that uses a logarithmic scale on both the horizontal and vertical axes, as in Figure 3.6. Then curvaceous relationships of the form $y = cx^a$ between the vertical variable, y, and the horizontal variable, x, appear as straight lines on a log-log plot. For example, below-left is a plot of the three curves $y \propto x^2$, $y \propto x^3$, and $y \propto x^4$. It is hard to tell which is which.

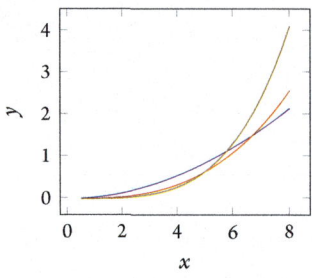 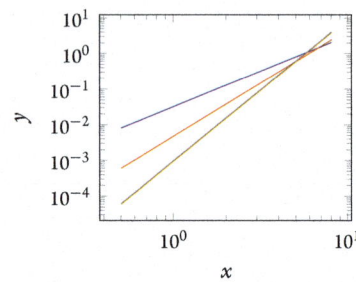

However, plot the same curves on the above-right log-log plot and it distinguishes the curves as different straight lines: the steepest line is the curve with the largest exponent, $y \propto x^4$, whereas the least steep line is the curve with the smallest exponent, $y \propto x^2$.

For example, suppose you make three measurements that at $x = 1.8$, 3.3, 6.7 the value of $y = 0.9$, 4.6, 29.1, respectively. The graph below-left show the three data points (x, y). Find the power law curve $y = cx^a$ that explains these points.

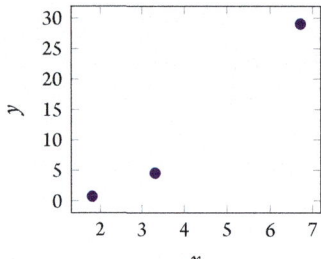 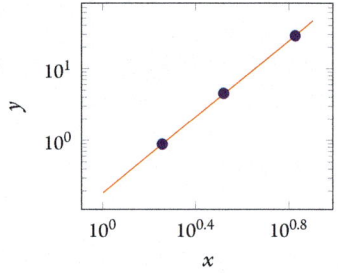

Take the logarithm (to any base, so let's choose base 10) of both sides of $y = cx^a$ to get $\log_{10} y = (\log_{10} c) + a(\log_{10} x)$, equivalently, $(\log_{10} y) = a(\log_{10} x) + b$ for constant $b = \log_{10} c$. That is, there is a straight line relationship between $(\log_{10} y)$ and $(\log_{10} x)$, as illustrated above-right. Here $\log_{10} x = 0.26$, 0.52, 0.83 and $\log_{10} y = -0.04$, 0.66, 1.46, respectively (2 d.p.). Using the end-points to estimate the slope gives $a = 2.63$, the exponent in the power law. Then the constant $b = -0.04 - 2.63 \cdot 0.26 = -0.72$ so the coefficient $c = 10^b = 0.19$. That is, via the log-log plot, the power law $y = 0.19 \cdot 2.63^x$ explains the data. Such log-log plots are not only used in Example 3.5.11, they are endemic in science and engineering.

Table 3.6 Orbital periods for the eight planets of the solar system: the periods are in (Earth) days; the distance is the length of the semi-major axis of the orbits (Wikipedia, 2014). Used by Example 3.5.11. [https://en.wikipedia.org/wiki/Orbital_period]

planet	distance (gigametres)	period (days)
Mercury	57.91	87.97
Venus	108.21	224.70
Earth	149.60	365.26
Mars	227.94	686.97
Jupiter	778.55	4332.59
Saturn	1433.45	10759.22
Uranus	2870.67	30687.15
Neptune	4498.54	60190.03

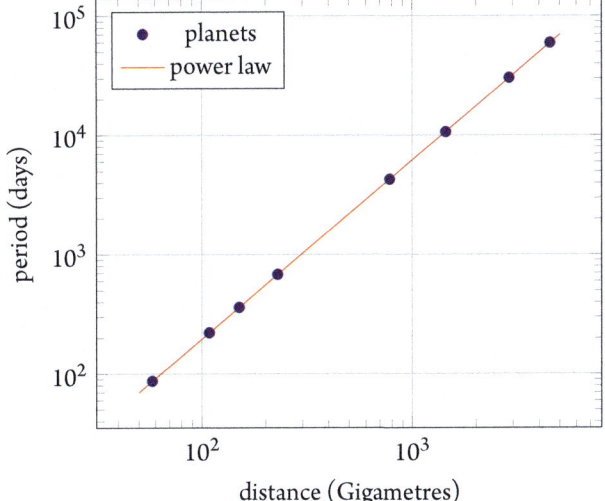

Figure 3.6 The planetary periods as a function of the distance, from the data of Table 3.6: the graph is a log-log plot to show the excellent power law. Also plotted is the power law fit computed by Example 3.5.11.

Form these into the matrix-vector system $Ax = b$: for simplicity recorded here to two decimal places, albeit computed more accurately,

$$A = \begin{bmatrix} 1 & 1.76 \\ 1 & 2.03 \\ 1 & 2.17 \\ 1 & 2.36 \\ 1 & 2.89 \\ 1 & 3.16 \\ 1 & 3.46 \end{bmatrix}, \quad b = \begin{bmatrix} 1.94 \\ 2.35 \\ 2.56 \\ 2.84 \\ 3.64 \\ 4.03 \\ 4.49 \end{bmatrix}.$$

Procedure 3.5.4 then determines a best approximate solution.

(a) Enter these matrices in MATLAB/Octave by the commands, for example,

```
d= [    57.91
       108.21
       149.60
       227.94
       778.55
      1433.45
      2870.67] ;
p= [ 87.97
     224.70
     365.26
     686.97
    4332.59
   10759.22
   30687.15] ;
A= [ones (7,1)  log10 (d)]
b=log10 (p)
```

since the MATLAB/Octave function log10 () computes the logarithm to base 10 of each component in its argument (Table 3.2). Then compute an SVD of $A = USV^{\mathsf{T}}$ via [U, S, V] =svd (A) (2 d.p.):

```
U =
  -0.27 -0.57 -0.39 -0.38 -0.34 -0.32 -0.30
  -0.31 -0.40 -0.21 -0.09  0.27  0.45  0.65
  -0.32 -0.31  0.88 -0.10 -0.06 -0.04 -0.02
  -0.35 -0.19 -0.11  0.90 -0.10 -0.09 -0.09
  -0.41  0.14 -0.08 -0.11  0.80 -0.25 -0.30
  -0.45  0.31 -0.06 -0.11 -0.26  0.67 -0.41
  -0.49  0.51 -0.04 -0.11 -0.31 -0.41  0.47
S =
   7.38      0
      0   0.55
      0      0
      0      0
      0      0
      0      0
      0      0
V =
  -0.35 -0.94
  -0.94  0.35
```

(b) Solve $Uz = b$ to give this first intermediary $z = U^{\mathsf{T}}b$ via the command z=U'*b:

```
z =
   -8.5507
    0.6514
    0.0002
    0.0004
    0.0005
   -0.0018
    0.0012
```

(c) Now solve approximately $Sy = z$. From the two nonzero singular values in S the matrix A has rank two. So the approximation is to discard/zero all but the first two elements of z (as an error, here all small in value). Then find the best approximate y via the command y=z(1:2)./diag(S(1:2,1:2)) :

```
y =
   -1.1581
    1.1803
```

(d) Solve $V^{\mathsf{T}}x = y$ by $x = Vy$ via x=V*y :

```
x =
   -0.6980
    1.4991
```

Also check that computing x=A\b gives exactly the same 'best' approximate solution.

Lastly, interpret the answer. The approximation gives $x_1 = -0.6980$ and $x_2 = 1.4991$. Since the ideal model was the log of the period $\log_{10} p = x_1 + x_2 \log_{10} d$, we determine that a 'best' approximate model is $\log_{10} p \approx -0.6980 + 1.4991 \log_{10} d$. Raising ten to the power of both sides gives the power law that the period $p \approx 0.2005\, d^{1.4991}$ days: this is the straight line drawn in Figure 3.6. The exponent 1.4991 is within 0.1% of the exponent $3/2$ that is Kepler's law.

For example, for Neptune with a semi-major axis distance of 4498.542 Gm, using the 'best' model predicts Neptune's period

$$10^{-0.6980+1.4991\log_{10} 4498.542} = 60019 \text{ days}.$$

This prediction is pleasingly close to the observed period of 60190 days. □

Compute in MATLAB/Octave   There are two separate important computational issues.

- Many books approximate solutions of $Ax = b$ by solving the associated normal equation $(A^{\mathsf{T}}A)x = (A^{\mathsf{T}}b)$. For *theoretical purposes* this normal equation is very useful. However, in practical computation avoid the normal equation because forming $A^{\mathsf{T}}A$, and then manipulating it, is both expensive and error-enhancing (especially in large problems). For example, $\text{cond}(A^{\mathsf{T}}A) = (\text{cond}A)^2$ (Exercise 3.3.13) so matrix $A^{\mathsf{T}}A$ typically has a much worse condition number than matrix A (Procedure 2.2.5). To paraphrase Cleve Molar: Almost anything you can do with $A^{\mathsf{T}}A$ can be done without it [via the SVD].

- The last two examples observe that A\b gives an answer that was identical to what the SVD procedure gives. Thus A\b can serve as a very useful shortcut to finding a best approximate solution. For non-square matrices with more rows than columns (more equations than variables), A\b generally does this (without comment as MATLAB/Octave assumes you know what you are doing). For other scenarios A\b does something different, so be wary.

3.5.2 Compute the smallest appropriate solution

I'm thinking of two numbers. Their average is three. What are the numbers?

Cleve Moler, The world's simplest impossible problem *(1990)*

The MATLAB/Octave operation A\b   Examples 3.5.9 and 3.5.11 observe that A\b gives an answer identical to the best approximate solution given by the SVD Procedure 3.5.4. But there are just as many circumstances when A\b is not 'the approximate answer' that you want. Beware.

Example 3.5.12 Use x=A\b to 'solve' the problems of Examples 3.5.1, 3.5.3 and 3.5.6.

- With Octave, observe that the answer returned is the *particular* solution determined by the SVD Procedure 3.5.4 (whether approximate or exact): respectively 84.5 kg; ratings $(1, \frac{1}{3}, -\frac{4}{3})$; and ratings $(\frac{1}{2}, 1, -\frac{5}{4}, -\frac{1}{4})$.

- With MATLAB (R2013b), the computed answers are often different: respectively 84.5 kg (the same); ratings $(\mathrm{NaN}, \mathrm{Inf}, \mathrm{Inf})$ with a warning; and ratings $(\frac{3}{4}, \frac{5}{4}, -1, 0)$ with a warning.

How do we make sense of such differences in computed answers?   □

Recall that systems of linear equations may not have unique solutions (as in the rating examples): what does A\b compute when there are an infinite number of solutions?

- For systems of equations with the number of equations not equal to the number of variables, $m \neq n$, the Octave operation A\b computes for you the *smallest solution* (that is, a solution that is of least magnitude, smallest norm) of all valid solutions (Theorem 3.5.13): often 'exact' when $m < n$, or approximate when $m > n$ (Theorem 3.5.8). Using A\b is the most efficient computationally, but using the SVD helps us understand what it does.

- MATLAB (R2013b etc.) does something different with A\b in the case of fewer equations than variables, $m < n$. MATLAB's different 'answer' does reinforce that a choice of one solution among many is a subjective decision. But Octave's choice of the smallest valid solution is often more appealing.

Theorem 3.5.13 (*smallest solution*) *Obtain the smallest solution, whether exact or as an approximation, to a system of linear equations by invoking Procedure 3.3.15 or Procedure 3.5.4, respectively, and then setting to zero the free variables, that is, $y_{r+1} = \cdots = y_n = 0$.*

Proof. We obtain all possible solutions, whether exact (Procedure 3.3.15) or approximate (Procedure 3.5.4), from solving $x = Vy$. Since multiplication by orthogonal V preserves lengths (Theorem 3.2.48), the lengths of x and y are the same: consequently, $|x|^2 = |y|^2 = y_1^2 + \cdots + y_r^2 + y_{r+1}^2 + \cdots + y_n^2$. Now, in both Procedure 3.3.15 and Procedure 3.5.4 the variables $y_1, y_2, \ldots, y_r$ are fixed but $y_{r+1}, \ldots, y_n$ are free to vary. Hence the smallest $|y|^2$ is obtained by setting $y_{r+1} = \cdots = y_n = 0$. Then this gives the particular solution $x = Vy$ of smallest $|x|$.

Example 3.5.14 In the table tennis ratings of Example 3.5.3 the procedure found that the ratings were any of

$$x = \begin{bmatrix} 1 \\ \frac{1}{3} \\ -\frac{4}{3} \end{bmatrix} + \frac{y_3}{\sqrt{3}} \begin{bmatrix} 1 \\ 1 \\ 1 \end{bmatrix},$$

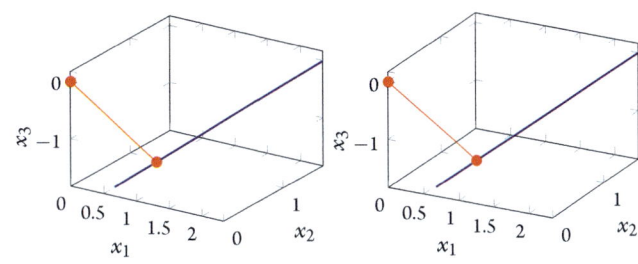

as illustrated in stereo at the right (blue). Verify that $|x|$ is a minimum only when the free variable $y_3 = 0$ (a red disc in the plot).

Solution:

$$|x|^2 = x \cdot x$$

$$= \left(\begin{bmatrix} 1 \\ \frac{1}{3} \\ -\frac{4}{3} \end{bmatrix} + \frac{y_3}{\sqrt{3}} \begin{bmatrix} 1 \\ 1 \\ 1 \end{bmatrix} \right) \cdot \left(\begin{bmatrix} 1 \\ \frac{1}{3} \\ -\frac{4}{3} \end{bmatrix} + \frac{y_3}{\sqrt{3}} \begin{bmatrix} 1 \\ 1 \\ 1 \end{bmatrix} \right)$$

$$= \begin{bmatrix} 1 \\ \frac{1}{3} \\ -\frac{4}{3} \end{bmatrix} \cdot \begin{bmatrix} 1 \\ \frac{1}{3} \\ -\frac{4}{3} \end{bmatrix} + \frac{2y_3}{\sqrt{3}} \begin{bmatrix} 1 \\ 1 \\ 1 \end{bmatrix} \cdot \begin{bmatrix} 1 \\ \frac{1}{3} \\ -\frac{4}{3} \end{bmatrix} + \frac{y_3^2}{3} \begin{bmatrix} 1 \\ 1 \\ 1 \end{bmatrix} \cdot \begin{bmatrix} 1 \\ 1 \\ 1 \end{bmatrix}$$

$$= \frac{26}{9} + 0y_3 + y_3^2 = \frac{26}{9} + y_3^2$$

This quadratic is minimized for $y_3 = 0$. Hence the length $|x|$ is minimized by the free variable $y_3 = 0$. $\square$

Example 3.5.15 *(closest point to the origin)* What is the point on the line $3x_1 + 4x_2 = 25$ that is closest to the origin? I am sure you could think of several methods, perhaps inspired by the right-hand graph, but here use an SVD and Theorem 3.5.13. Confirm the Octave computation A\b gives this same closest point, but MATLAB gives a different 'answer' (one that is not relevant here).

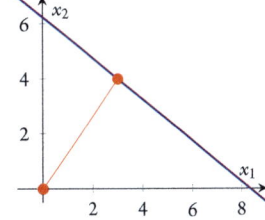

Solution: The point on the line $3x_1 + 4x_2 = 25$ closest to the origin is the smallest solution of $3x_1 + 4x_2 = 25$. Rephrase as the matrix-vector system $A\boldsymbol{x} = \boldsymbol{b}$ for matrix $A = \begin{bmatrix} 3 & 4 \end{bmatrix}$ and $b = 25$, and apply Procedure 3.3.15.

(a) Factorize $A = USV^{\mathsf{T}}$ in MATLAB/Octave via $[\mathtt{U,S,V}] = \mathtt{svd}(\,[\,3\ \,4\,]\,)$:

```
U =    1
S =
       5    0
V =
       0.6000   -0.8000
       0.8000    0.6000
```

(b) Solve $Uz = b = 25$ which here gives $z = 25$.

(c) Solve $S\boldsymbol{y} = z = 25$ with general solution here of $\boldsymbol{y} = (5, y_2)$. Obtain the smallest solution with free variable $y_2 = 0$.

(d) Solve $V^{\mathsf{T}}\boldsymbol{x} = \boldsymbol{y}$ by $\boldsymbol{x} = V\boldsymbol{y} = V(5, 0) = (3, 4)$.

This is the smallest solution and hence the point on the line closest to the origin (as plotted).

Computing $\mathtt{x=A\backslash b}$, which here is simply $\mathtt{x= [3\ \ 4]\backslash 25}$, gives answer $\boldsymbol{x} = (3, 4)$ in Octave; as determined by the SVD, this point is the closest on the line to the origin. In MATLAB (R2017b), $\mathtt{x= [3\ \ 4]\backslash 25}$ gives $\boldsymbol{x} = (0, 6.25)$ which the above graph shows is a valid solution, but not the smallest solution. □

Activity 3.5.16 What is the closest point to the origin of the plane $2x + 3y + 6z = 98$? Use the following SVD:

$$\begin{bmatrix} 2 & 3 & 6 \end{bmatrix} = \begin{bmatrix} 1 \end{bmatrix} \begin{bmatrix} 7 & 0 & 0 \end{bmatrix} \begin{bmatrix} \frac{2}{7} & -\frac{3}{7} & -\frac{6}{7} \\ \frac{3}{7} & \frac{6}{7} & -\frac{2}{7} \\ \frac{6}{7} & -\frac{2}{7} & \frac{3}{7} \end{bmatrix}^{\mathsf{T}}.$$

(a) $(-12, -4, 6)$ (b) $(2, 3, 6)$ (c) $(-3, 6, -2)$ (d) $(4, 6, 12)$

Example 3.5.17 *(computed tomography)*

A CT-scan, also called X-ray computed tomography (X-ray CT) or computerized axial tomography scan (CAT scan), makes use of computer-processed combinations of many X-ray images taken from different angles to produce cross-sectional (tomographic) images (virtual 'slices') of specific areas of a scanned object, allowing the user to see inside the object without cutting.[24]

Wikipedia, 2015

[24] https://en.wikipedia.org/wiki/CT_scan

Table 3.7 As well as the MATLAB/Octave commands and operations listed in Tables 1.2, 2.3 and 3.1 to 3.3 we may invoke these functions for drawing images—functions which are otherwise not needed.

- reshape(A,p,q) for an $m \times n$ matrix/vector A, provided $mn = pq$, generates a $p \times q$ matrix with entries taken column-wise from A. Either p or q can be [], in which case MATLAB/Octave uses $p = mn/q$ or $q = mn/p$ respectively.
- colormap(gray) MATLAB/Octave usually draws graphs with colour, but for many images we need greyscale; this command changes the current figure to 64 shades of grey. (colormap(jet) is the default, colormap(hot) is good for both colour and greyscale reproductions, colormap('list') lists the available colormaps you can try.)
- imagesc(A) where A is an $m \times n$ matrix of values draws an $m \times n$ image in the current figure window using the values of A (scaled to fit) to determine the colour from the current colormap (e.g., greyscale).
- log(x) where x is a matrix, vector, or scalar computes the natural logarithm to the base e of each element, and returns the result(s) as a correspondingly sized matrix, vector, or scalar.
- exp(x) where x is a matrix, vector, or scalar computes the exponential of each element, and returns the result(s) as a correspondingly sized matrix, vector, or scalar.[a]

[a] In advanced linear algebra, for application to differential equations and Markov chains, we define the exponential of a matrix, denoted $\exp A$ or e^A. This mathematical function is *not* the same as MATLAB/Octave's exp(A); instead one computes expm(A) to get e^A.

Importantly for medical diagnosis and industrial purposes, the computed tomography answer must not have artificial features. Artificial features must not be generated because of deficiencies in the measurements. If there is any ambiguity about the answer, then the answer computed should be the 'greyest'—the 'greyest' corresponds to the mathematical smallest solution.

Let's analyse a toy example, as real-life examples have millions of unknowns and equations.[25] Suppose we divide a cross-section of a body into nine squares (large pixels) in a 3×3 grid. Inside each square the body's material has some unknown density represented by transmission factors, $r_1, r_2, \ldots, r_9$, as shown to the right. The CT-scan is to find these transmission factors. The factor r_j is the fraction of the incident X-ray that emerges after passing through the jth square: typically, smaller r_i corresponds to higher density in the body.

r_1	r_4	r_7
r_2	r_5	r_8
r_3	r_6	r_9

As indicated next to the right, six X-ray measurements are made through the body where $f_1, f_2, \ldots, f_6$ denote the fraction of energy in the measurements relative to the incident power of the X-ray beam. Thus we need to solve six equations for the nine unknown transmission factors:

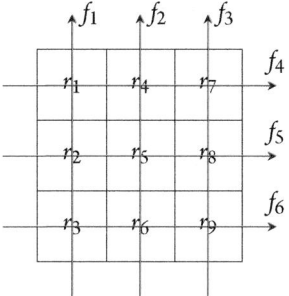

$$r_1 r_2 r_3 = f_1, \quad r_4 r_5 r_6 = f_2, \quad r_7 r_8 r_9 = f_3,$$
$$r_1 r_4 r_7 = f_4, \quad r_2 r_5 r_8 = f_5, \quad r_3 r_6 r_9 = f_6.$$

[25] For those interested in reading further, Kress (2015) [§8] introduces the advanced, highly mathematical, approach to computerized tomography.

Turn such nonlinear equations into linear equations that we can handle by taking the logarithm (to any base, but here say the natural logarithm to base e) of both sides of all equations (computers almost always use "log" to denote the natural logarithm, so we do too. Herein, unsubscripted "log" means the same as "ln"):

$$r_i r_j r_k = f_l \iff (\log r_i) + (\log r_j) + (\log r_k) = (\log f_l).$$

That is, letting new unknowns $x_i = \log r_i$ and new right-hand sides $b_i = \log f_i$, we solve six linear equations for nine unknowns:

$$x_1 + x_2 + x_3 = b_1, \quad x_4 + x_5 + x_6 = b_2, \quad x_7 + x_8 + x_9 = b_3,$$
$$x_1 + x_4 + x_7 = b_4, \quad x_2 + x_5 + x_8 = b_5, \quad x_3 + x_6 + x_9 = b_6.$$

This forms the matrix-vector system $Ax = b$ for 6×9 matrix

$$A = \begin{bmatrix} 1 & 1 & 1 & 0 & 0 & 0 & 0 & 0 & 0 \\ 0 & 0 & 0 & 1 & 1 & 1 & 0 & 0 & 0 \\ 0 & 0 & 0 & 0 & 0 & 0 & 1 & 1 & 1 \\ 1 & 0 & 0 & 1 & 0 & 0 & 1 & 0 & 0 \\ 0 & 1 & 0 & 0 & 1 & 0 & 0 & 1 & 0 \\ 0 & 0 & 1 & 0 & 0 & 1 & 0 & 0 & 1 \end{bmatrix}.$$

For example, let's find an answer for the factors when the measurements give vector $b = (-0.91, -1.04, -1.54, -1.52, -1.43, -0.53)$ (all negative as they are the logarithms of fractions f_i less than one)

```
A= [1 1 1 0 0 0 0 0 0
    0 0 0 1 1 1 0 0 0
    0 0 0 0 0 0 1 1 1
    1 0 0 1 0 0 1 0 0
    0 1 0 0 1 0 0 1 0
    0 0 1 0 0 1 0 0 1 ]
b= [-0.91 -1.04 -1.54 -1.52 -1.43 -0.53]'
x=A\b
r=reshape(exp(x),3,3)
colormap(gray),imagesc(r)
```

- The answer from Octave is (2 d.p.)

$$x = (-.42, -.39, -.09, -.47, -.44, -.14, -.63, -.60, -.30).$$

These are logarithms so to get the corresponding physical transmission factors compute the exponential of each component, denoted as $\exp(x)$,

$$r = \exp(x) = (.66, .68, .91, .63, .65, .87, .53, .55, .74),$$

although it is perhaps more appealing to put these factors into the shape of the 3×3 array of pixels as in (and as illustrated to the right)

$$\begin{bmatrix} r_1 & r_4 & r_7 \\ r_2 & r_5 & r_8 \\ r_3 & r_6 & r_9 \end{bmatrix} = \begin{bmatrix} 0.66 & 0.63 & 0.53 \\ 0.68 & 0.65 & 0.55 \\ 0.91 & 0.87 & 0.74 \end{bmatrix}.$$

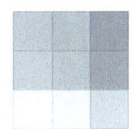

Octave's answer predicts that there is less transmitting, more absorbing, denser, material to the top-right; and more transmitting, less absorbing, less dense, material to the bottom-left.

- However, the answer from MATLAB's A\b is (2 d.p.)

$$\mathbf{x} = (-0.91, 0, 0, -0.61, -1.43, 1.01, 0, 0, -1.54),$$

as illustrated below in the leftmost picture (Table 3.7). This is quite a different answer![26]

Furthermore, MATLAB could give other 'answers', as illustrated in the other pictures above. Reordering the rows in the matrix A and right-hand side $\mathbf{b}$ does not change the system of equations. But after such reordering the answer from MATLAB's x=A\b variously predicts each of the above four pictures.

The reason for such a multiplicity of mathematically valid answers is that the problem is underdetermined. There are nine unknowns but only six equations, so in linear algebra there are typically an infinity of valid answers (as in Theorem 2.2.30): just five of these are illustrated above. *In this application to CT-scans, we add the extra information that we desire the answer that is the 'greyest', the most 'washed out', the answer with fewest features. Finding the answer $\mathbf{x}$ that minimizes $|\mathbf{x}|$ is a reasonable way to quantify this desire.*[27]

The SVD procedure guarantees that we find such a smallest answer. Procedure 3.5.4 in MATLAB/Octave gives the following process to satisfy the experimental measurements expressed in $A\mathbf{x} = \mathbf{b}$.

(a) First, find an SVD, $A = USV^{\mathsf{T}}$, via [U,S,V]=svd(A) and get (2 d.p.)

```
U =
  -0.41 -0.00   0.82 -0.00   0.00   0.41
  -0.41 -0.00  -0.41 -0.57  -0.42   0.41
```

---

[26] MATLAB does give a warning in this instance (Warning: Rank deficient, ...), but it does not always. For example, it does not warn of issues when you ask it to solve $\frac{1}{2}(x_1 + x_2) = 3$ via [0.5 0.5]\3:; it simply computes its 'answer' $\mathbf{x} = (6, 0)$.

[27] Another possibility is to increase the number of measurements in order to increase the number of equations to match the number of unknown pixels. However, measurements are often prohibitively expensive. Further, increasing the number of measurements tempts us to increase the resolution by having more smaller pixels: in which case we again have to deal with the same issue of more variables than known equations.

```
-0.41 -0.00 -0.41   0.57   0.42   0.41
-0.41  0.81 -0.00   0.07 -0.09 -0.41
-0.41 -0.31 -0.00 -0.45   0.61 -0.41
-0.41 -0.50  0.00   0.38 -0.52 -0.41
```

S =

2.45	0	0	0	0	0	0	0	0
0	1.73	0	0	0	0	0	0	0
0	0	1.73	0	0	0	0	0	0
0	0	0	1.73	0	0	0	0	0
0	0	0	0	1.73	0	0	0	0
0	0	0	0	0	0.00	0	0	0

V =

```
-0.33  0.47  0.47  0.04 -0.05  0.03 -0.58 -0.21 -0.25
-0.33 -0.18  0.47 -0.26  0.35 -0.36  0.49 -0.27 -0.07
-0.33 -0.29  0.47  0.22 -0.30  0.33  0.09  0.47  0.33
-0.33  0.47 -0.24 -0.29 -0.29 -0.48  0.11  0.37  0.26
-0.33 -0.18 -0.24 -0.59  0.11  0.41 -0.24 -0.27  0.38
-0.33 -0.29 -0.24 -0.11 -0.54  0.07  0.13 -0.10 -0.64
-0.33  0.47 -0.24  0.37  0.19  0.45  0.47 -0.16 -0.00
-0.33 -0.18 -0.24  0.07  0.59 -0.05 -0.25  0.53 -0.31
-0.33 -0.29 -0.24  0.55 -0.06 -0.40 -0.22 -0.37  0.32
```

(b) Solve $Uz = b$ by $z = U' *b$ to find

$$z = (2.85, -0.52, 0.31, 0.05, -0.67, -0.00).$$

(c) Because the sixth singular value is zero, ignore the sixth equation: because $z_6 = 0.00$ (2 d.p.), this is only a small inconsistency error. Now set $y_i = z_i/\sigma_i$ for $i = 1, \ldots, 5$ and for the smallest magnitude answer set the free variables $y_6 = y_7 = y_8 = y_9 = 0$ (Theorem 3.5.13). Obtain the nonzero values via $y=z(1:5)./diag(S(1:5,1:5))$ to find

$$y = (1.16, -0.30, 0.18, 0.03, -0.39, 0, 0, 0, 0)$$

(d) Then, via $x=V(:,1:5)*y$, solve $V^\mathsf{T}x = y$ to determine the smallest solution is $x = (-0.42, -0.39, -0.09, -0.47, -0.44, -0.14, -0.63, -0.60, -0.30)$. This is the same answer as computed by Octave's $A\backslash b$ to give the pixel image shown that has minimal artifices.

In practice, *each* slice of a real CT-scan would involve finding the absorption of tens of millions of pixels. That is, a CT-scan needs to best solve many systems of tens of millions of equations in tens of millions of unknowns! □

3.5.3 Orthogonal projection resolves vector components

Reconsider the task of making a minimal change to the right-hand side of a system of linear equations, and let's connect it to the so-called orthogonal projection. This important connection occurs because of the geometry that the closest point on a line or plane to another given point is the one that forms a right-angle; that is, it forms an orthogonal vector.

This optional section does usefully support least square approximation, and provides examples of transformations for the next Section 3.6. Such orthogonal projections are extensively used in applications.

Project onto a direction

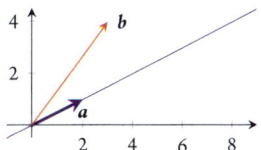

Example 3.5.18 Consider 'solving' the inconsistent system $ax = b$ where $a = (2,1)$ and $b = (3,4)$; that is, solve

$$\begin{bmatrix} 2 \\ 1 \end{bmatrix} x = \begin{bmatrix} 3 \\ 4 \end{bmatrix}.$$

As illustrated to the right, the impossible task is to find some multiple of the vector $a = (2,1)$ (all multiples plotted) that equals $b = (3,4)$. It cannot be done. Question: how may we change the right-hand side vector b so that the task is possible? A partial answer is to replace b by some vector b' which is in the column space of matrix $A = \begin{bmatrix} a \end{bmatrix}$. But we could choose any b' in the column space, so any answer for the multiple x would be possible! Surely any answer is not acceptable.

Instead, often the preferred answer is, out of all vectors in the column space of matrix $A = \begin{bmatrix} a \end{bmatrix}$, find the vector b' in the column space *which is closest to b*—as illustrated to the right here, where it looks like $b' = (4,2)$.

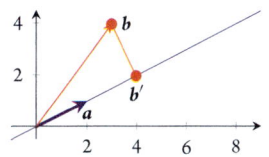

The SVD approach of Procedure 3.5.4 to find b' and x is the following.

(a) Use [U,S,V] =svd ([2;1]) to find here the SVD factorization

$$A = USV^{\mathrm{T}} = \begin{bmatrix} 0.89 & -0.45 \\ 0.45 & 0.89 \end{bmatrix} \begin{bmatrix} 2.24 \\ 0 \end{bmatrix} [1]^{\mathrm{T}} \quad (\text{2 d.p.}).$$

(b) Then $z = U^{\mathrm{T}}b = (4.47, 2.24)$.

(c) Treat the second component of $Sy = z$ as an error—it is the magnitude $|b - b'|$—to deduce $y = 4.47/2.24 = 2.00$ (2 d.p.) from the first component.

(d) Then $x = Vy = 1y = 2$ solves the changed problem.

From this solution, the vector $b' = ax = (2,1)2 = (4,2)$, as is recognizable in the graphs. □

Now let's derive the same result but with two differences: firstly, use more elementary arguments, not the SVD, and secondly, derive the result for general vectors a and b (although continuing to use the same illustration). Start with the crucial observation that the closest point/vector b' in the column space of $A = \begin{bmatrix} a \end{bmatrix}$ is such that $b - b'$ is

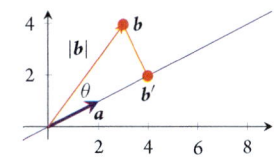

at right-angles, orthogonal, to a. (If $b - b'$ were not orthogonal, then we would be able to slide b' along the line span$\{a\}$ to reduce the length of $b - b'$.) Thus we form a right-angle triangle with hypotenuse of length $|b|$ and angle θ as shown to the right. Trigonometry then gives the adjacent length $|b'| = |b| \cos\theta$. But the angle θ is that between the given vectors a and b, so the dot product gives the cosine as $\cos\theta = a \cdot b/(|a||b|)$ (Theorem 1.3.5). Hence the adjacent length

$$|b'| = |b|\cos\theta = |b|\frac{a \cdot b}{|a||b|} = \frac{a \cdot b}{|a|}.$$

To approximately solve $ax = b$, replace the inconsistent $ax = b$ by the consistent $ax = b'$. Then as x is a scalar, we solve this consistent equation via the ratio of lengths, $x = |b'|/|a| = a \cdot b/|a|^2$. For Example 3.5.18, this gives the 'solution' $x = (2,1) \cdot (3,4)/(2^2 + 1^2) = 10/5 = 2$ as before.

A crucial part of such solutions is the general formula for $b' = ax = a(a \cdot b)/|a|^2$. Geometrically the formula gives the 'shadow' b' of vector b when projected by a 'sun' high above the line of the vector a, as illustrated schematically to the right. As such, the formula is called an orthogonal projection.

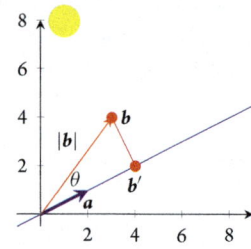

Definition 3.5.19 (*orthogonal projection onto 1D*) Let $u, v \in \mathbb{R}^n$ and vector $u \neq 0$, then the **orthogonal projection** of v onto u is

$$\text{proj}_u(v) := u\frac{u \cdot v}{|u|^2}. \tag{3.5a}$$

In the special, but common, case when u is a unit vector,

$$\text{proj}_u(v) := u(u \cdot v). \tag{3.5b}$$

Example 3.5.20 For the following pairs of vectors, draw the named orthogonal projection, and for the given inconsistent system, determine whether the 'best' approximate solution is in the range $x < -1$, $-1 < x < 0$, $0 < x < 1$, or $1 < x$.

(a) $\text{proj}_u(v)$ and $ux = v$

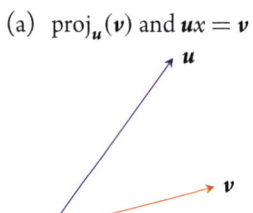

Solution:

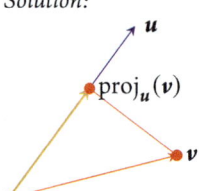

Draw a line perpendicular to u that passes through the tip of v. Then $\text{proj}_u(v)$ is the shown brown vector. To 'best solve' $ux = v$, approximate the equation $ux = v$ by $ux = \text{proj}_u(v)$. Since $\text{proj}_u(v)$ is smaller than u and the same direction, $0 < x < 1$. □

(b) $\mathrm{proj}_q(p)$ and $qx = p$

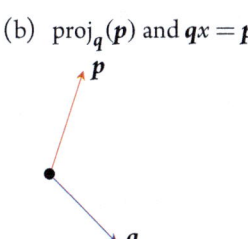

Solution:

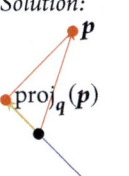

Vector q in $\mathrm{proj}_q(p)$ gives the direction of a line, so we can and do project onto the negative direction of q. To 'best solve' $qx = p$, approximate the equation $qx = p$ by $qx = \mathrm{proj}_q(p)$. Since the brown vector $\mathrm{proj}_q(p)$ is smaller than q and in the opposite direction, $-1 < x < 0$. ☐

Example 3.5.21 For the following pairs of vectors, compute the given orthogonal projection, and hence find the 'best' approximate solution to the given inconsistent system.

(a) Find $\mathrm{proj}_u(v)$ for vectors $u = (3,4)$ and $v = (4,1)$, and hence best solve $ux = v$.

Solution:

$$\mathrm{proj}_u(v) = (3,4)\frac{(3,4) \cdot (4,1)}{|(3,4)|^2} = (3,4)\frac{16}{25} = (\tfrac{48}{25}, \tfrac{64}{25}).$$

Approximate equation $ux = v$ by $ux = \mathrm{proj}_u(v)$, that is, $(3,4)x = (\tfrac{48}{25}, \tfrac{64}{25})$ with solution $x = \tfrac{16}{25}$ (from either component, or the ratio of lengths). ☐

(b) Find $\mathrm{proj}_p(q)$ for vectors $p = (\tfrac{1}{3}, \tfrac{2}{3}, \tfrac{2}{3})$ and $q = (3,2,1)$, and best solve $px = q$.

Solution: Vector p is a unit vector, so we use the simpler formula that

$$\begin{aligned}
\mathrm{proj}_p(q) &= (\tfrac{1}{3}, \tfrac{2}{3}, \tfrac{2}{3})[(\tfrac{1}{3}, \tfrac{2}{3}, \tfrac{2}{3}) \cdot (3,2,1)] \\
&= (\tfrac{1}{3}, \tfrac{2}{3}, \tfrac{2}{3})[1 + \tfrac{4}{3} + \tfrac{2}{3}] \\
&= (\tfrac{1}{3}, \tfrac{2}{3}, \tfrac{2}{3})3 = (1,2,2).
\end{aligned}$$

Then 'best solve' equation $px = q$ by the approximation $px = \mathrm{proj}_p(q)$, that is, $(\tfrac{1}{3}, \tfrac{2}{3}, \tfrac{2}{3})x = (1,2,2)$ with solution $x = 3$ (from any component, or the ratio of lengths). ☐

Activity 3.5.22 Use projection to best solve the inconsistent equation $(1, 4, 8)x = (4, 4, 2)$. The best answer is which of the following?

(a) $x = 21/4$ (b) $x = 10/13$ (c) $x = 4$ (d) $x = 4/9$

Project onto a subspace

The previous subsection develops a geometric view of the 'best' solution to the inconsistent system $ax = b$. The discussion introduced that the conventional 'best' solution—that determined by Procedure 3.5.4—is to replace b by its projection $\text{proj}_a(b)$, namely to solve $ax = \text{proj}_a(b)$. The rationale is that this is the *smallest* change to the right-hand side b that enables the equation to be solved. This subsection introduces that solving inconsistent equations in more variables may be viewed as an analogous projection onto a subspace.

Definition 3.5.23 *(project onto a subspace) Let $\mathbb{W}$ be a k-dimensional subspace of $\mathbb{R}^n$ with an orthonormal basis $\{w_1, w_2, \ldots, w_k\}$. For every vector $v \in \mathbb{R}^n$, the **orthogonal projection** of vector v onto subspace $\mathbb{W}$ is*

$$\text{proj}_{\mathbb{W}}(v) = w_1(w_1 \cdot v) + w_2(w_2 \cdot v) + \cdots + w_k(w_k \cdot v).$$

Example 3.5.24

(a) Let $\mathbb{X}$ be the xy-plane in xyz-space, find $\text{proj}_{\mathbb{X}}(3, -4, 2)$.

Solution: An orthonormal basis for the xy-plane (blue plane in the stereo picture to the right) are the two unit 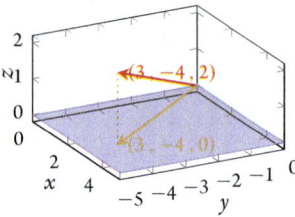 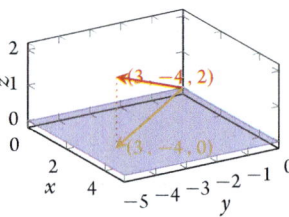 vectors $i = (1, 0, 0)$ and $j = (0, 1, 0)$. Hence

$$\begin{aligned}
\text{proj}_{\mathbb{W}}(3, -4, 2) &= i(i \cdot (3, -4, 2)) + j(j \cdot (3, -4, 2)) \\
&= i(3 + 0 + 0) + j(0 - 4 + 0) \\
&= (3, -4, 0) \qquad \text{(shown in brown)}.
\end{aligned}$$

That is, just set the third component of $(3, -4, 2)$ to zero.   ☐

(b) For the subspace $\mathbb{W} = \text{span}\{(2, -2, 1), (2, 1, -2)\}$, determine the $\text{proj}_{\mathbb{W}}(3, 2, 1)$ (illustrated to the right).

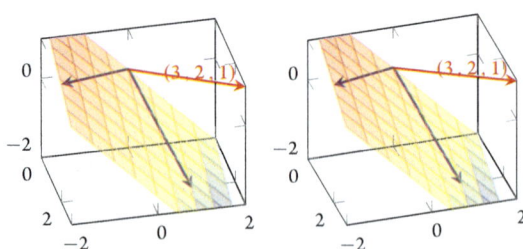

Solution: Although the two vectors in the span are orthogonal (blue in the stereo pictures), they are not unit vectors. Normalize the vectors by dividing by their length

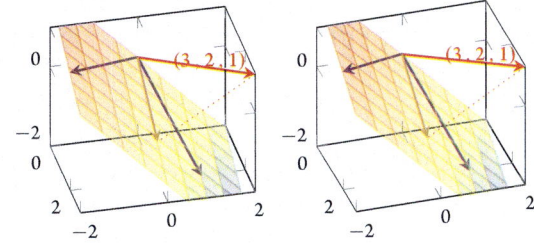

$\sqrt{2^2 + (-2)^2 + 1^2} = \sqrt{2^2 + 1^2 + (-2)^2} = 3$ to find that the vectors $\boldsymbol{w}_1 = (\frac{2}{3}, -\frac{2}{3}, \frac{1}{3})$ and $\boldsymbol{w}_2 = (\frac{2}{3}, \frac{1}{3}, -\frac{2}{3})$ are an orthonormal basis for $\mathbb{W}$ (a plane). Hence

$$\text{proj}_{\mathbb{W}}(3, 2, 1) = \boldsymbol{w}_1(\boldsymbol{w}_1 \cdot (3, 2, 1)) + \boldsymbol{w}_2(\boldsymbol{w}_2 \cdot (3, 2, 1))$$
$$= \boldsymbol{w}_1(2 - \tfrac{4}{3} + \tfrac{1}{3}) + \boldsymbol{w}_2(2 + \tfrac{2}{3} - \tfrac{2}{3})$$
$$= \boldsymbol{w}_1 + 2\boldsymbol{w}_2$$
$$= (\tfrac{2}{3}, -\tfrac{2}{3}, \tfrac{1}{3}) + 2(\tfrac{2}{3}, \tfrac{1}{3}, -\tfrac{2}{3})$$
$$= (2, 0, -1) \qquad \text{(shown in brown)}.$$

$\square$

(c) Recall the table tennis ranking Examples 3.5.3 and 3.3.13. To rank the players we seek to solve the matrix-vector system, $A\boldsymbol{x} = \boldsymbol{b}$,

$$\begin{bmatrix} 1 & -1 & 0 \\ 1 & 0 & -1 \\ 0 & 1 & -1 \end{bmatrix} \boldsymbol{x} = \begin{bmatrix} 1 \\ 2 \\ 2 \end{bmatrix}.$$

Letting $\mathbb{A}$ denote the column space of matrix A, determine $\text{proj}_{\mathbb{A}}(\boldsymbol{b})$.

Solution:

We need to find an orthonormal basis for the column space (the illustrated plane spanned by the three shown column vectors)—an SVD gives it to us. Example 3.3.12 found an SVD $A = USV^{\text{T}}$, in MATLAB/Octave via [U, S, V] =svd(A), to be

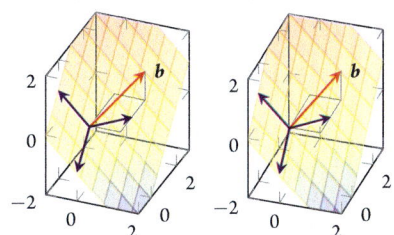

U =
```
     0.4082    -0.7071     0.5774
    -0.4082    -0.7071    -0.5774
    -0.8165    -0.0000     0.5774
```
S =
```
     1.7321          0          0
          0     1.7321          0
          0          0     0.0000
```

$$V = $$
$$
\begin{array}{ccc}
0.0000 & -0.8165 & 0.5774 \\
-0.7071 & 0.4082 & 0.5774 \\
0.7071 & 0.4082 & 0.5774
\end{array}
$$

Since there are only two nonzero singular values, the column space $\mathbb{A}$ is 2D and spanned by the first two orthonormal columns of matrix U: that is, an orthonormal basis for $\mathbb{A}$ is the two vectors (as illustrated to the right)

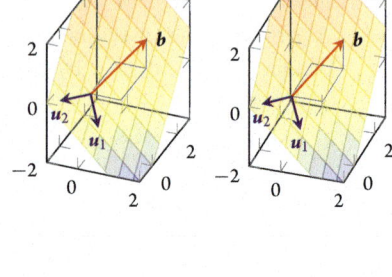

$$
u_1 = \begin{bmatrix} 0.4082 \\ -0.4082 \\ -0.8165 \end{bmatrix} = \frac{1}{\sqrt{6}} \begin{bmatrix} 1 \\ -1 \\ -2 \end{bmatrix},
$$

$$
u_2 = \begin{bmatrix} -0.7071 \\ -0.7071 \\ -0.0000 \end{bmatrix} = \frac{1}{\sqrt{2}} \begin{bmatrix} -1 \\ -1 \\ 0 \end{bmatrix}.
$$

Hence the projection of the right-hand side b onto the column space $\mathbb{A}$ is

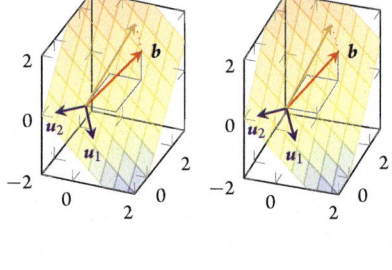

$\text{proj}_{\mathbb{A}}(1, 2, 2)$

$= u_1(u_1 \cdot (1, 2, 2)) + u_2(u_2 \cdot (1, 2, 2))$

$= u_1(1 - 2 - 4)/\sqrt{6} + u_2(-1 - 2 + 0)/\sqrt{2}$

$= -\frac{5}{\sqrt{6}} u_1 - \frac{3}{\sqrt{2}} u_2$

$= \frac{1}{6}(-5, 5, 10) + \frac{1}{2}(3, 3, 0)$

$= \frac{1}{3}(2, 7, 5)$ (shown in brown).

□

(d) Find the projection of the vector $(1,2,2)$ onto the plane $2x - \frac{1}{2}y + 4z = 6$.

Solution: This plane is not a subspace as it does not pass through the origin. Definition 3.5.23 only defines projection onto a subspace, so we cannot answer this problem (as yet).   □

(e) Use an SVD to find the projection of the vector $(1,2,2)$ onto the plane $2x - \frac{1}{2}y + 4z = 0$ (illustrated to the right).

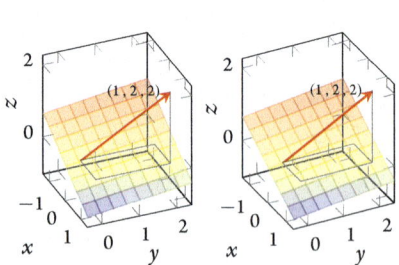

Solution: This plane does pass through the origin so it forms a subspace, call it $\mathbb{P}$ (illustrated above). To project, we need two orthonormal basis vectors. Recall that a normal to the plane is its vectors of coefficients, here $(2, -\frac{1}{2}, 4)$, so we need to find two orthonormal vectors which are orthogonal to $(2, -\frac{1}{2}, 4)$. Further, recall that the columns of an orthogonal matrix are orthonormal (Theorem 3.2.48(b)), so use an SVD to find orthonormal vectors to $(2, -\frac{1}{2}, 4)$. In MATLAB/Octave, compute an SVD with [U,S,V] =svd([2;-1/2;4]) to find

```
U  =
   -0.4444     0.1111    -0.8889
    0.1111     0.9914     0.0684
   -0.8889     0.0684     0.4530
S  =
    4.5000
         0
         0
V  =  -1
```

The first column $u_1 = (-4, 1, -8)/9$ of orthogonal matrix U is in the direction of a normal to the plane as it must, since it must be in the span of $(2, -\frac{1}{2}, 4)$. Since matrix U is orthogonal, the last two columns (say u_2 and u_3, drawn in blue to the right) are not only orthonormal, but also orthogonal to u_1 and hence an orthonormal basis for the plane $\mathbb{P}$.

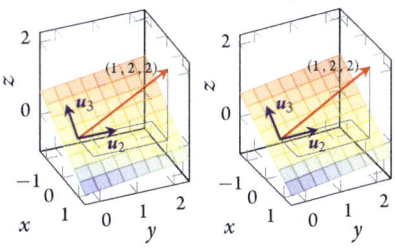

Hence the projection of the given vector onto the plane is

$$\text{proj}_{\mathbb{P}}(1,2,2) = u_2(u_2 \cdot (1,2,2)) + u_3(u_3 \cdot (1,2,2))$$
$$= 2.2308\,u_2 + 0.1539\,u_3$$
$$= 2.2308(0.1111, 0.9914, 0.0684)$$
$$\quad + 0.1539(-0.8889, 0.0684, 0.4530)$$
$$= (0.1111, 2.2222, 0.2222)$$
$$= \tfrac{1}{9}(1, 10, 2) \qquad \text{(shown in brown)}.$$

This answer may be computed in MATLAB/Octave via the two dot products cs=U(:,2:3)'*[1;2;2], giving the two coefficients 2.2308 and 0.1539, and then the linear combination proj=U(:,2:3)*cs.

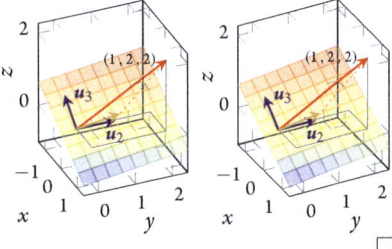

Activity 3.5.25 Determine which of the following is $\text{proj}_{\mathbb{W}}(1,1,-2)$ for the subspace $\mathbb{W} = \text{span}\{(2,3,6),(-3,6,-2)\}$.

(a) $(-\frac{1}{7},\frac{9}{7},\frac{4}{7})$ (b) $(\frac{5}{7},-\frac{3}{7},\frac{8}{7})$ (c) $(-\frac{5}{7},\frac{3}{7},-\frac{8}{7})$ (d) $(\frac{1}{7},-\frac{9}{7},-\frac{4}{7})$

Example 3.5.24(c) determines that the orthogonal projection of the given table tennis results $b = (1,2,2)$ onto the column space of matrix A is the vector $b' = \frac{1}{3}(2,7,5)$. Recall that Example 3.5.3 invokes Procedure 3.5.4 to find the 'approximate' solution of the impossible $Ax = b$ to be $x = (1,\frac{1}{3},-\frac{4}{3})$. Now see that $Ax = (1 - \frac{1}{3}, 1 - (-\frac{4}{3}), \frac{1}{3} - (-\frac{4}{3})) = (\frac{2}{3},\frac{7}{3},\frac{5}{3}) = b'$. That is, the approximate solution method of Procedure 3.5.4 solved the problem $Ax = \text{proj}_{\mathbb{A}}(b)$. The following Theorem 3.5.26 confirms that this is no accident: orthogonally projecting the right-hand side onto the column space of the matrix in a system of linear equations is equivalent to solving the system with the smallest change to the right-hand side that makes it consistent.

Theorem 3.5.26 *The 'least square' solution/s of the system $Ax = b$ determined by Procedure 3.5.4 is/are the solution/s of $Ax = \text{proj}_{\mathbb{A}}(b)$ where $\mathbb{A}$ denotes the column space of A.*

Proof. For any $m \times n$ matrix A, Procedure 3.5.4 first finds an SVD $A = USV^{\mathsf{T}}$ and sets $r = \text{rank}\, A$. Second, it computes $z = U^{\mathsf{T}}b$ but disregards z_i for $i = r+1,\ldots,m$ as errors. That is, instead of using $z = U^{\mathsf{T}}b$, Procedure 3.5.4 solves the equations with $z' = (z_1, z_2, \ldots, z_r, 0, \ldots, 0)$. This vector z' corresponds to a modified right-hand side b' satisfying $z' = U^{\mathsf{T}}b'$; that is, $b' = Uz'$ as matrix U is orthogonal. Recalling that u_i denotes the ith column of U and that components $z_i = u_i \cdot b$ from $z = U^{\mathsf{T}}b$, the matrix-vector product $b' = Uz'$ is the linear combination (Example 2.3.6)

$$\begin{aligned} b' &= u_1 z_1' + u_2 z_2' + \cdots + u_r z_r' + u_{r+1}0 + \cdots + u_m 0 \\ &= u_1(u_1 \cdot b) + u_2(u_2 \cdot b) + \cdots + u_r(u_r \cdot b) \\ &= \text{proj}_{\text{span}\{u_1,u_2,\ldots,u_r\}}(b), \end{aligned}$$

by Definition 3.5.23 since the columns u_i of U are orthonormal (Theorem 3.2.48). Theorem 3.4.32 establishes that this span is the column space $\mathbb{A}$ of matrix A. Hence, $b' = \text{proj}_{\mathbb{A}}(b)$ and so Procedure 3.5.4 solves the system $Ax = \text{proj}_{\mathbb{A}}(b)$. $\qquad\square$

Example 3.5.27 Recall that Example 3.5.1 rationalizes four apparently contradictory weighings: in kg the weights are $84.8, 84.1, 84.7,$ and 84.4. Denoting the 'uncertain' weight by x, we write these weighings as the inconsistent matrix-vector system

$$Ax = b, \quad \text{namely} \quad \begin{bmatrix} 1 \\ 1 \\ 1 \\ 1 \end{bmatrix} x = \begin{bmatrix} 84.8 \\ 84.1 \\ 84.7 \\ 84.4 \end{bmatrix}.$$

Let's see that the orthogonal projection of the right-hand side onto the column space of A is the same as the minimal change of Example 3.5.1, which in turn is the well known average.

To find the orthogonal projection, observe that matrix A has one column $\boldsymbol{a}_1 = (1,1,1,1)$ so by Definition 3.5.19 the orthogonal projection

$$\text{proj}_{\text{span}\{\boldsymbol{a}_1\}}(84.8, 84.1, 84.7, 84.4)$$
$$= \boldsymbol{a}_1 \frac{\boldsymbol{a}_1 \cdot (84.8, 84.1, 84.7, 84.4)}{|\boldsymbol{a}_1|^2}$$
$$= \boldsymbol{a}_1 \frac{84.8 + 84.1 + 84.7 + 84.4}{1+1+1+1}$$
$$= \boldsymbol{a}_1 \cdot 84.5$$
$$= (84.5, 84.5, 84.5, 84.5).$$

The projected system $A\boldsymbol{x} = (84.5, 84.5, 84.5, 84.5)$ is now consistent. Its solution is $x = 84.5$ kg. As in Example 3.5.1, this solution is the well-known averaging of the four weights. □

Example 3.5.28 Recall the round robin tournament among four players of Example 3.5.6. To estimate the player ratings of the four players from the results of six matches we want to solve the inconsistent system $A\boldsymbol{x} = \boldsymbol{b}$ where

$$A = \begin{bmatrix} 1 & -1 & 0 & 0 \\ 1 & 0 & -1 & 0 \\ 0 & 1 & -1 & 0 \\ 1 & 0 & 0 & -1 \\ 0 & 1 & 0 & -1 \\ 0 & 0 & 1 & -1 \end{bmatrix}, \quad \boldsymbol{b} = \begin{bmatrix} 1 \\ 3 \\ 1 \\ -2 \\ 4 \\ -1 \end{bmatrix}.$$

Let's see that the orthogonal projection of $\boldsymbol{b}$ onto the column space of A is the same as the minimal change of Example 3.5.6.

An SVD finds an orthonormal basis for the column space $\mathbb{A}$ of matrix A: Example 3.5.6 uses the SVD (2 d.p.)

```
U =
   0.31  -0.26  -0.58  -0.26   0.64  -0.15
   0.07   0.40  -0.58   0.06  -0.49  -0.51
  -0.24   0.67   0.00  -0.64   0.19   0.24
  -0.38  -0.14  -0.58   0.21  -0.15   0.66
  -0.70   0.13   0.00   0.37   0.45  -0.40
  -0.46  -0.54  -0.00  -0.58  -0.30  -0.26
S =
   2.00      0      0      0
      0   2.00      0      0
      0      0   2.00      0
```

$$
\begin{array}{cccc}
0 & 0 & 0 & 0.00 \\
0 & 0 & 0 & 0 \\
0 & 0 & 0 & 0 \\
\end{array}
$$

$V = \ldots$

As there are three nonzero singular values in S, the first three columns of U are an orthonormal basis for the column space $\mathbb{A}$. Letting u_j denote the columns of U, Definition 3.5.23 gives the orthogonal projection (2 d.p.)

$$
\begin{aligned}
\mathrm{proj}_{\mathbb{A}}(b) &= u_1(u_1 \cdot b) + u_2(u_2 \cdot b) + u_3(u_3 \cdot b) \\
&= -1.27\, u_1 + 2.92\, u_2 - 1.15\, u_3 \\
&= (-0.50, 1.75, 2.25, 0.75, 1.25, -1.00).
\end{aligned}
$$

Compute these three dot products in MATLAB/Octave via cs=U(:,1:3)'*b, and then compute the linear combination with projb=U(:,1:3)*cs. To confirm that Procedure 3.5.4 solves $Ax = \mathrm{proj}_{\mathbb{A}}(b)$ we check that the ratings that were found by Example 3.5.6, $x = (\frac{1}{2}, 1, -\frac{5}{4}, -\frac{1}{4})$, satisfy $Ax = \mathrm{proj}_{\mathbb{A}}(b)$: in MATLAB/Octave compute A*[0.50; 1.00;-1.25;-0.25] and see the product is $\mathrm{proj}_{\mathbb{A}}(b)$. □

Section 3.6 uses orthogonal projection as an example of a linear transformation. The section shows that a linear transformation always corresponds to multiplying by a specific matrix, which for orthogonal projection is here WW^{T}.

There is a useful feature of Example 3.5.24(e) and Example 3.5.28. In both, we use MATLAB/Octave to compute the projection in two steps: letting matrix W denote the matrix of appropriate columns of orthogonal U (respectively $W = U(:,2:3)$ and $W = U(:,1:3)$), first the examples compute cs=W'*b, that is, the vector $c = W^{\mathrm{T}}b$; and second the examples compute proj=W*cs, that is, $\mathrm{proj}_{W}(b) = Wc$. Combining these two steps into one (using associativity) gives

$$
\mathrm{proj}_{W}(b) = Wc = W(W^{\mathrm{T}})b = (WW^{\mathrm{T}})b.
$$

The interesting feature is that the orthogonal projection formula of Definition 3.5.23 is equivalent to the multiplication by matrix (WW^{T}) for an appropriate matrix W.[28]

Theorem 3.5.29 (*orthogonal projection matrix*) *Let* W *be a k-dimensional subspace of* $\mathbb{R}^n$ *with an orthonormal basis* $\{w_1, w_2, \ldots, w_k\}$, *then for every vector* $v \in \mathbb{R}^n$, *the orthogonal projection*

$$
\mathrm{proj}_{W}(v) = (WW^{\mathrm{T}})v \tag{3.6}
$$

for the n × k matrix $W = \begin{bmatrix} w_1 & w_2 & \cdots & w_k \end{bmatrix}$.

---

[28] However, to minimize computation time compute $\mathrm{proj}_{W}(v)$ via the two matrix-vector products in $W(W^{\mathrm{T}}v)$ because computing the projection matrix WW^{T} and then the product $(WW^{\mathrm{T}})v$ involves many more computations. Like the inverse A^{-1}, a projection matrix WW^{T} is crucial theoretically rather than practically.

Proof. Directly from Definition 3.5.23,

$$\text{proj}_{\mathbb{W}}(v) = w_1(w_1 \cdot v) + w_2(w_2 \cdot v) + \cdots + w_k(w_k \cdot v)$$

$$\text{(then using that } w \cdot v = w^T v, \text{ Example 3.1.19)}$$

$$= w_1 w_1^T v + w_2 w_2^T v + \cdots + w_k w_k^T v$$

$$= \left(w_1 w_1^T + w_2 w_2^T + \cdots + w_k w_k^T\right) v.$$

Let the components of the vector $w_j = (w_{1j}, w_{2j}, \ldots, w_{nj})$, then from the matrix product Definition 3.1.12, the k products in the sum

$$w_1 w_1^T + w_2 w_2^T + \cdots + w_k w_k^T$$

$$= \begin{bmatrix} w_{11}w_{11} & w_{11}w_{21} & \cdots & w_{11}w_{n1} \\ w_{21}w_{11} & w_{21}w_{21} & \cdots & w_{21}w_{n1} \\ \vdots & \vdots & & \vdots \\ w_{n1}w_{11} & w_{n1}w_{21} & \cdots & w_{n1}w_{n1} \end{bmatrix} + \begin{bmatrix} w_{12}w_{12} & w_{12}w_{22} & \cdots & w_{12}w_{n2} \\ w_{22}w_{12} & w_{22}w_{22} & \cdots & w_{22}w_{n2} \\ \vdots & \vdots & & \vdots \\ w_{n2}w_{12} & w_{n2}w_{22} & \cdots & w_{n2}w_{n2} \end{bmatrix}$$

$$+ \cdots + \begin{bmatrix} w_{1k}w_{1k} & w_{1k}w_{2k} & \cdots & w_{1k}w_{nk} \\ w_{2k}w_{1k} & w_{2k}w_{2k} & \cdots & w_{2k}w_{nk} \\ \vdots & \vdots & & \vdots \\ w_{nk}w_{1k} & w_{nk}w_{2k} & \cdots & w_{nk}w_{nk} \end{bmatrix}.$$

So the (i, j)th entry of this sum is

$$w_{i1}w_{j1} + w_{i2}w_{j2} + \cdots + w_{ik}w_{jk}$$

$$= w_{i1}(W^T)_{1j} + w_{i2}(W^T)_{2j} + \cdots + w_{ik}(W^T)_{kj},$$

which, from Definition 3.1.12 again, is the (i, j)th entry of the product WW^T. Hence $\text{proj}_{\mathbb{W}}(v) = (WW^T)v$.

Example 3.5.30 Find the matrices of the following orthogonal projections (from Example 3.5.21), and use the matrix to find the given projection.

(a) $\text{proj}_u(v)$ for vector $u = (3, 4)$ and $v = (4, 1)$.

Solution: First, normalize u to the unit vector $w = u/|u| = (3, 4)/5$. Second, the matrix is

$$WW^T = ww^T = \begin{bmatrix} \frac{3}{5} \\ \frac{4}{5} \end{bmatrix} \begin{bmatrix} \frac{3}{5} & \frac{4}{5} \end{bmatrix} = \begin{bmatrix} \frac{9}{25} & \frac{12}{25} \\ \frac{12}{25} & \frac{16}{25} \end{bmatrix}.$$

Then the projection

$$\text{proj}_u(v) = (WW^T)v = \begin{bmatrix} \frac{9}{25} & \frac{12}{25} \\ \frac{12}{25} & \frac{16}{25} \end{bmatrix} \begin{bmatrix} 4 \\ 1 \end{bmatrix} = \begin{bmatrix} 48/25 \\ 64/25 \end{bmatrix}$$

□

(b) $\text{proj}_p(q)$ for vector $p = (\frac{1}{3}, \frac{2}{3}, \frac{2}{3})$ and $q = (3,3,0)$.

Solution: Vector p is already a unit vector so the matrix is

$$WW^T = pp^T = \begin{bmatrix} \frac{1}{3} \\ \frac{2}{3} \\ \frac{2}{3} \end{bmatrix} \begin{bmatrix} \frac{1}{3} & \frac{2}{3} & \frac{2}{3} \end{bmatrix} = \begin{bmatrix} \frac{1}{9} & \frac{2}{9} & \frac{2}{9} \\ \frac{2}{9} & \frac{4}{9} & \frac{4}{9} \\ \frac{2}{9} & \frac{4}{9} & \frac{4}{9} \end{bmatrix}.$$

Then the projection

$$\text{proj}_p(q) = (WW^T)q = \begin{bmatrix} \frac{1}{9} & \frac{2}{9} & \frac{2}{9} \\ \frac{2}{9} & \frac{4}{9} & \frac{4}{9} \\ \frac{2}{9} & \frac{4}{9} & \frac{4}{9} \end{bmatrix} \begin{bmatrix} 3 \\ 3 \\ 0 \end{bmatrix} = \begin{bmatrix} 1 \\ 2 \\ 2 \end{bmatrix}.$$

□

Activity 3.5.31 The projection $\text{proj}_u(v)$ for vectors $u = (2,6,3)$ and $v = (1,4,8)$ could be done by premultiplying by which of the following matrices?

(a) $\begin{bmatrix} \frac{1}{81} & \frac{4}{81} & \frac{8}{81} \\ \frac{4}{81} & \frac{16}{81} & \frac{32}{81} \\ \frac{8}{81} & \frac{32}{81} & \frac{64}{81} \end{bmatrix}$
(b) $\begin{bmatrix} \frac{4}{49} & \frac{12}{49} & \frac{6}{49} \\ \frac{12}{49} & \frac{36}{49} & \frac{18}{49} \\ \frac{6}{49} & \frac{18}{49} & \frac{9}{49} \end{bmatrix}$
(c) $\begin{bmatrix} \frac{2}{63} & \frac{2}{21} & \frac{1}{21} \\ \frac{8}{63} & \frac{8}{21} & \frac{4}{21} \\ \frac{16}{63} & \frac{16}{21} & \frac{8}{21} \end{bmatrix}$
(d) $\begin{bmatrix} \frac{2}{63} & \frac{8}{63} & \frac{16}{63} \\ \frac{2}{21} & \frac{8}{21} & \frac{16}{21} \\ \frac{1}{21} & \frac{4}{21} & \frac{8}{21} \end{bmatrix}$

Example 3.5.32 Find the matrices of the following orthogonal projections.

(a) $\text{proj}_{\mathbb{X}}(v)$ where $\mathbb{X}$ is the xy-plane in xyz-space.

Solution: The two unit vectors $i = (1,0,0)$ and $j = (0,1,0)$ form an orthonormal basis, so matrix

$$W = \begin{bmatrix} i & j \end{bmatrix} = \begin{bmatrix} 1 & 0 \\ 0 & 1 \\ 0 & 0 \end{bmatrix},$$

hence the matrix of the projection is

$$WW^T = \begin{bmatrix} 1 & 0 \\ 0 & 1 \\ 0 & 0 \end{bmatrix} \begin{bmatrix} 1 & 0 & 0 \\ 0 & 1 & 0 \end{bmatrix} = \begin{bmatrix} 1 & 0 & 0 \\ 0 & 1 & 0 \\ 0 & 0 & 0 \end{bmatrix}.$$

☐

(b) $\text{proj}_{\mathbb{W}}(v)$ for the subspace $\mathbb{W} = \text{span}\{(2, -2, 1), (2, 1, -2)\}$.

Solution: The two given vectors are orthogonal, so $w_1 = (\frac{2}{3}, -\frac{2}{3}, \frac{1}{3})$ and $w_2 = (\frac{2}{3}, \frac{1}{3}, -\frac{2}{3})$ form an orthonormal basis for $\mathbb{W}$. Then let matrix

$$W = \begin{bmatrix} w_1 & w_2 \end{bmatrix} = \frac{1}{3} \begin{bmatrix} 2 & 2 \\ -2 & 1 \\ 1 & -2 \end{bmatrix}.$$

Hence the matrix of the projection is

$$WW^T = \frac{1}{3} \begin{bmatrix} 2 & 2 \\ -2 & 1 \\ 1 & -2 \end{bmatrix} \frac{1}{3} \begin{bmatrix} 2 & -2 & 1 \\ 2 & 1 & -2 \end{bmatrix} = \frac{1}{9} \begin{bmatrix} 8 & -2 & -2 \\ -2 & 5 & -4 \\ -2 & -4 & 5 \end{bmatrix}.$$

☐

(c) The orthogonal projection onto the column space of matrix

$$A = \begin{bmatrix} 1 & -1 & 0 \\ 1 & 0 & -1 \\ 0 & 1 & -1 \end{bmatrix}.$$

Solution: The SVD of Example 3.5.24(c) determines an orthonormal basis of the column space is $u_1 = (1, -1, -2)/\sqrt{6}$ and $u_2 = (-1, -1, 0)/\sqrt{2}$. Hence the matrix of the projection is

$$WW^T = \begin{bmatrix} \frac{1}{\sqrt{6}} & -\frac{1}{\sqrt{2}} \\ -\frac{1}{\sqrt{6}} & -\frac{1}{\sqrt{2}} \\ -\frac{2}{\sqrt{6}} & 0 \end{bmatrix} \begin{bmatrix} \frac{1}{\sqrt{6}} & -\frac{1}{\sqrt{6}} & -\frac{2}{\sqrt{6}} \\ -\frac{1}{\sqrt{2}} & -\frac{1}{\sqrt{2}} & 0 \end{bmatrix} = \begin{bmatrix} \frac{2}{3} & \frac{1}{3} & -\frac{1}{3} \\ \frac{1}{3} & \frac{2}{3} & \frac{1}{3} \\ -\frac{1}{3} & \frac{1}{3} & \frac{2}{3} \end{bmatrix}.$$

Alternatively, recall the SVD of matrix A from Example 3.3.12, and recall that the first two columns of U are the orthonormal basis vectors. Hence matrix $W = U (: , 1:2)$ and so MATLAB/Octave computes the matrix of the projection, WW^T, via WWT=U (: , 1 : 2) * U (: , 1 : 2) ' to give the answer

```
WWT  =
        0.6667       0.3333     -0.3333
        0.3333       0.6667      0.3333
       -0.3333       0.3333      0.6667
```

□

Orthogonal decomposition separates

Because orthogonal projection has such a close connection to the geometry underlying important tasks such as 'least square' approximation (Theorem 3.5.26), this section develops further some orthogonal properties.

For any subspace $\mathbb{W}$ of interest, it is often useful to be able to discuss the set of vectors orthogonal to all those in $\mathbb{W}$, called the orthogonal complement. Such a set forms a subspace, called $\mathbb{W}^\perp$, read as "W perp", as illustrated below and defined by Definition 3.5.34.

(a) Given the blue subspace $\mathbb{W}$ in $\mathbb{R}^2$ (the origin is a black dot), consider the set of all vectors at right-angles to $\mathbb{W}$ (drawn arrows). Move the base of these vectors to the origin, and then they all lie in the red subspace $\mathbb{W}^\perp$.

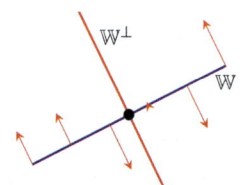

(b) Given the blue plane subspace $\mathbb{W}$ in $\mathbb{R}^3$ (the origin is a black dot), the red line subspace $\mathbb{W}^\perp$ contains all vectors orthogonal to $\mathbb{W}$ (when drawn with their base at the origin).

(c) Conversely, given the blue line subspace $\mathbb{W}$ in $\mathbb{R}^3$ (the origin is a black dot), the red plane subspace $\mathbb{W}^\perp$ contains all vectors orthogonal to $\mathbb{W}$ (when drawn with their base at the origin).

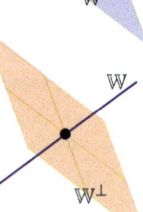

Activity 3.5.33 Given the above qualitative description of an orthogonal complement, which of the following red lines is the orthogonal complement to the shown (blue) subspace $\mathbb{W}$?

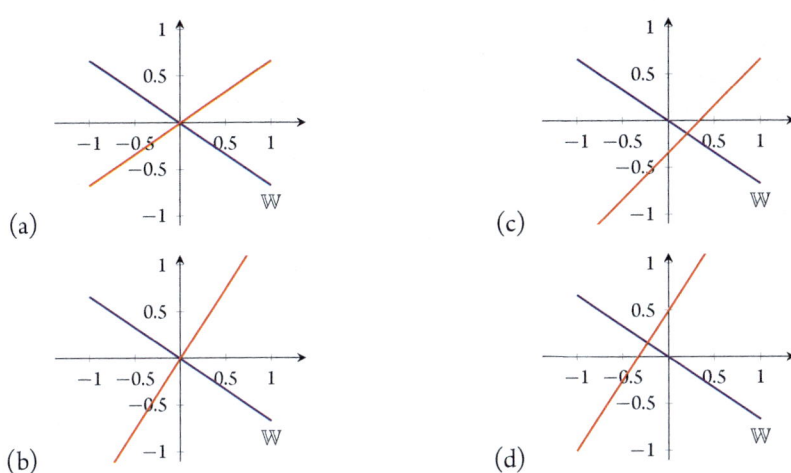

(a)

(c)

(b)

(d)

Definition 3.5.34 *(orthogonal complement)* Let $\mathbb{W}$ be a k-dimensional subspace of $\mathbb{R}^n$. The set of all vectors $\mathbf{u} \in \mathbb{R}^n$ (together with $\mathbf{0}$) that are each orthogonal to all vectors in $\mathbb{W}$ is called the **orthogonal complement** $\mathbb{W}^\perp$ ("W-perp"); that is,

$$\mathbb{W}^\perp = \{\mathbf{u} \in \mathbb{R}^n : \mathbf{u} \cdot \mathbf{w} = 0 \text{ for all } \mathbf{w} \in \mathbb{W}\}.$$

Example 3.5.35 *(orthogonal complement)*

(a) Given the subspace $\mathbb{W} = \text{span}\{(3,4)\}$, find its orthogonal complement $\mathbb{W}^\perp$.

Solution: Every vector in $\mathbb{W}$ is of the form $\mathbf{w} = (3c, 4c)$. For any vector $\mathbf{v} = (u, v) \in \mathbb{R}^2$ the dot product

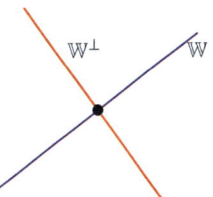

$$\mathbf{w} \cdot \mathbf{v} = (3c, 4c) \cdot (u, v) = c(3u + 4v).$$

This dot product is zero for all c if and only if $3u + 4v = 0$. That is, when $u = -4v/3$. Hence $\mathbf{v} = (-\frac{4}{3}v, v) = (-\frac{4}{3}, 1)v$, for every v, and so $\mathbb{W}^\perp = \text{span}\{(-\frac{4}{3}, 1)\}$. □

(b) Describe the orthogonal complement $\mathbb{X}^{\perp}$ to the subspace $\mathbb{X} = \text{span}\{(4, -4, 7)\}$.

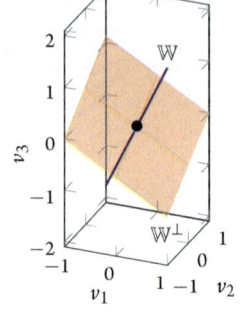 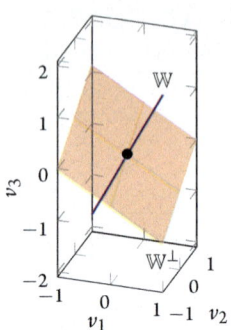

Solution: Every vector in $\mathbb{W}$ is of the form $w = (4, -4, 7)c$. Seek all vectors v such that $w \cdot v = 0$. For vectors $v = (v_1, v_2, v_3)$ the dot product

$$w \cdot v = c(4, -4, 7) \cdot (v_1, v_2, v_3)$$
$$= c(4v_1 - 4v_2 + 7v_3)$$

is zero for all c if and only if $4v_1 - 4v_2 + 7v_3 = 0$. That is, the orthogonal complement is all vectors v in the plane $4v_1 - 4v_2 + 7v_3 = 0$ (illustrated in stereo). □

(c) Describe the orthogonal complement of the set $\mathbb{W} = \{(t, t^2) : t \in \mathbb{R}\}$.

Solution: It does not exist, as an orthogonal complement is only defined for a subspace, and the parabola (t, t^2) is not a subspace. □

(d) Given the subspace $\mathbb{W} = \text{span}\{(2, -2, 1), (2, 1, -2)\}$, determine the orthogonal complement of $\mathbb{W}$.

Solution: Let $w_1 = (2, -2, 1)$ and $w_2 = (2, 1, -2)$ then all vectors $w \in \mathbb{W}$ are of the form $w = c_1 w_1 + c_2 w_2$ for all c_1 and c_2. Every vector $v \in \mathbb{W}^{\perp}$ must satisfy, for all c_1 and c_2,

$$w \cdot v = (c_1 w_1 + c_2 w_2) \cdot v = c_1 w_1 \cdot v + c_2 w_2 \cdot v = 0.$$

The only way to be zero for all c_1 and c_2 is for *both* $w_1 \cdot v = 0$ and $w_2 \cdot v = 0$. For vectors $v = (v_1, v_2, v_3)$ these two equations become the pair

$$2v_1 - 2v_2 + v_3 = 0 \quad \text{and} \quad 2v_1 + v_2 - 2v_3 = 0.$$

Adding twice the second to the first, and subtracting the first from the second gives the equivalent pair

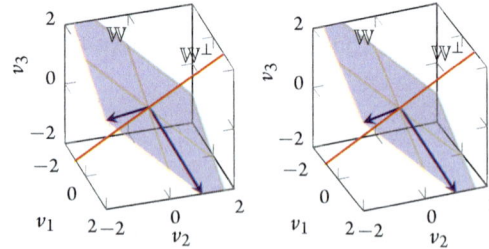

$$6v_1 - 3v_3 = 0 \quad \text{and}$$
$$3v_2 - 3v_3 = 0.$$

Both are satisfied for all $v_3 = t$ with $v_1 = t/2$ and $v_2 = t$. Therefore all possible v in the complement $\mathbb{W}^{\perp}$ are those in the form of the line $v = (\frac{1}{2}t, t, t)$. That is, $\mathbb{W}^{\perp} = \text{span}\{(\frac{1}{2}, 1, 1)\}$ (as illustrated in stereo). □

Activity 3.5.36 Which of the following vectors are in the orthogonal complement of the vector space spanned by $(3, -1, 1)$?

(a) $(6, -2, 2)$ (b) $(-1, -1, 1)$ (c) $(1, 3, -1)$ (d) $(3, 5, -4)$

Example 3.5.37 Prove $\{\mathbf{0}\}^{\perp} = \mathbb{R}^n$ and $(\mathbb{R}^n)^{\perp} = \{\mathbf{0}\}$.

Solution:

- The only vector in $\{\mathbf{0}\}$ is $\mathbf{w} = \mathbf{0}$. Since every vector $\mathbf{v} \in \mathbb{R}^n$ satisfies $\mathbf{w} \cdot \mathbf{v} = \mathbf{0} \cdot \mathbf{v} = 0$, by Definition 3.5.34 $\{\mathbf{0}\}^{\perp} = \mathbb{R}^n$.

- Certainly, $\mathbf{0} \in (\mathbb{R}^n)^{\perp}$ as $\mathbf{w} \cdot \mathbf{0} = 0$ for every vector $\mathbf{w} \in \mathbb{R}^n$. Establish that there are no others by contradiction. Assume a nonzero vector $\mathbf{v} \in (\mathbb{R}^n)^{\perp}$. Now set $\mathbf{w} = \mathbf{v} \in \mathbb{R}^n$, then $\mathbf{w} \cdot \mathbf{v} = \mathbf{v} \cdot \mathbf{v} = |\mathbf{v}|^2 \neq 0$ as $\mathbf{v}$ is nonzero. Consequently, a nonzero $\mathbf{v}$ cannot be in the complement. Thus $(\mathbb{R}^n)^{\perp} = \{\mathbf{0}\}$. $\quad\square$

These examples find that orthogonal complements are lines, planes, or the entire space. These indicate that an orthogonal complement is generally a subspace as proved next.

Theorem 3.5.38 *(orthogonal complement is a subspace)* *For every subspace W of $\mathbb{R}^n$, the orthogonal complement $W^{\perp}$ is a subspace of $\mathbb{R}^n$. Further, the intersection $W \cap W^{\perp} = \{\mathbf{0}\}$; that is, the zero vector is the only vector in both W and $W^{\perp}$.*

Proof. Recall the Definition 3.4.3 of a subspace: we need to establish that $W^{\perp}$ has the zero vector, and is closed under addition and scalar multiplication. We invoke its Definition 3.5.34.

- For every $\mathbf{w} \in W$, $\mathbf{0} \cdot \mathbf{w} = 0$ and so $\mathbf{0} \in W^{\perp}$.
- Let $\mathbf{v}_1, \mathbf{v}_2 \in W^{\perp}$, then for every $\mathbf{w} \in W$ the dot product $(\mathbf{v}_1 + \mathbf{v}_2) \cdot \mathbf{w} = \mathbf{v}_1 \cdot \mathbf{w} + \mathbf{v}_2 \cdot \mathbf{w} = 0 + 0 = 0$ and so $\mathbf{v}_1 + \mathbf{v}_2 \in W^{\perp}$.
- Let scalar $c \in \mathbb{R}$ and $\mathbf{v} \in W^{\perp}$, then for every $\mathbf{w} \in W$ the dot product $(c\mathbf{v}) \cdot \mathbf{w} = c(\mathbf{v} \cdot \mathbf{w}) = c0 = 0$ and so $c\mathbf{v} \in W^{\perp}$.

Hence, by Definition 3.4.3, $W^{\perp}$ is a subspace.

Further, as they are both subspaces, the zero vector is in both W and $W^{\perp}$. Let vector $\mathbf{u}$ be any vector in both W and $W^{\perp}$. As $\mathbf{u} \in W^{\perp}$, by Definition 3.5.34 $\mathbf{u} \cdot \mathbf{w} = 0$ for every $\mathbf{w} \in W$. But $\mathbf{u} \in W$ also, so using this for $\mathbf{w}$ in the previous equation gives $\mathbf{u} \cdot \mathbf{u} = 0$; that is, $|\mathbf{u}|^2 = 0$. Hence vector $\mathbf{u}$ has to be the zero vector (Theorem 1.1.13). That is, $W \cap W^{\perp} = \{\mathbf{0}\}$. $\quad\square$

Activity 3.5.39 Vectors in which of the following (red) sets form the orthogonal complement to the shown (blue) subspace W?

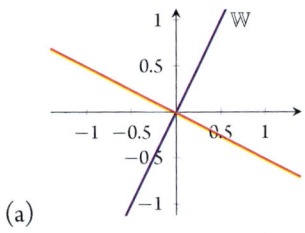

(a)

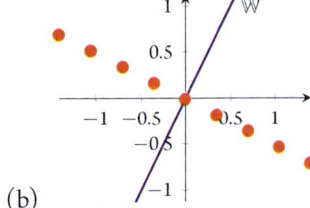

(b)

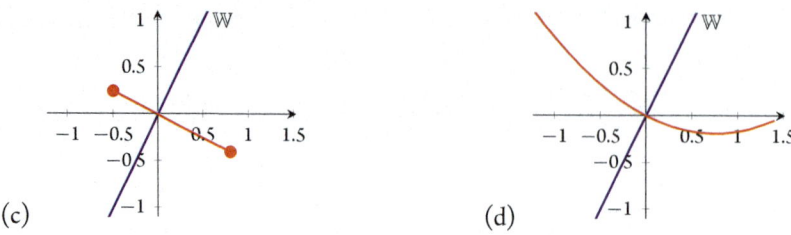

(c) (d)

When orthogonal complements arise, they are often usefully written as the nullspace of a matrix.

Theorem 3.5.40 (*nullspace complementarity*) *For every* $m \times n$ *matrix* A, *the column space of* A *has* $\text{null}(A^\mathsf{T})$ *as its orthogonal complement in* $\mathbb{R}^m$. *That is, identifying the columns of matrix* $A = \begin{bmatrix} a_1 & a_2 & \cdots & a_n \end{bmatrix}$, *and denoting the column space by* $\mathbb{A} = \text{span}\{a_1, a_2, \ldots, a_n\}$, *then the orthogonal complement* $\mathbb{A}^{\perp} = \text{null}(A^\mathsf{T})$. *Further,* $\text{null}(A)$ *in* $\mathbb{R}^n$ *is the orthogonal complement of the row space of* A.

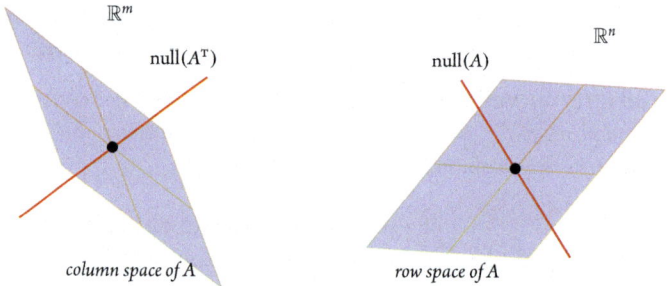

Proof. First, by Definition 3.5.34, any vector $v \in \mathbb{A}^{\perp}$ is orthogonal to all vectors in the column space of A, in particular it is orthogonal to the columns of A:

$$a_1 \cdot v = 0, \; a_2 \cdot v = 0, \; \ldots, \; a_k \cdot v = 0$$
$$\iff a_1^\mathsf{T} v = 0, \; a_2^\mathsf{T} v = 0, \; \ldots, \; a_k^\mathsf{T} v = 0$$
$$\iff \begin{bmatrix} a_1^\mathsf{T} \\ a_2^\mathsf{T} \\ \vdots \\ a_k^\mathsf{T} \end{bmatrix} v = 0$$
$$\iff A^\mathsf{T} v = 0$$
$$\iff v \in \text{null}(A^\mathsf{T}).$$

That is, $\mathbb{A}^{\perp} \subseteq \text{null}(A^\mathsf{T})$. Second, for any $v \in \text{null}(A^\mathsf{T})$, recall that by Definition 3.4.10 for any vector w in the column space of A, there exists a linear combination $w = c_1 a_1 + c_2 a_2 + \cdots + c_n a_n$. Then

$$w \cdot v = (c_1 a_1 + c_2 a_2 + \cdots + c_n a_n) \cdot v$$
$$= c_1 (a_1 \cdot v) + c_2 (a_2 \cdot v) + \cdots + c_n (a_n \cdot v)$$
$$= c_1 0 + c_2 0 + \cdots + c_n 0 \quad \text{(from above } \Longleftrightarrow \text{)}$$
$$= 0,$$

and so by Definition 3.5.34 vector $v \in \mathbb{A}^\perp$; that is, null$(A^T) \subseteq \mathbb{A}^\perp$. Putting these two together, null$(A^T) = \mathbb{A}^\perp$.

Lastly, that the null(A) in $\mathbb{R}^n$ is the orthogonal complement of the row space of A follows from applying the above result to the matrix A^T. □

Example 3.5.41

(a) Let the subspace $\mathbb{W} = \text{span}\{(2, -1)\}$. Find the orthogonal complement $\mathbb{W}^\perp$.

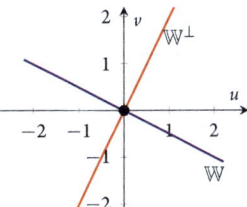

Solution: Here the subspace $\mathbb{W}$ is the column space of the matrix $W = \begin{bmatrix} 2 \\ -1 \end{bmatrix}$. To find $\mathbb{W}^\perp = \text{null}(W^T)$, solve $W^T v = 0$, that is, for vectors $v = (u, v)$

$$\begin{bmatrix} 2 & -1 \end{bmatrix} v = 2u - v = 0.$$

All solutions are $v = 2u$ (as illustrated). Hence $v = (u, 2u) = (1, 2)u$, and so $\mathbb{W}^\perp = \text{span}\{(1, 2)\}$. □

(b) Describe the subspace of $\mathbb{R}^3$ whose orthogonal complement is the plane $-\frac{1}{2}x - y + 2z = 0$.

Solution:
The equation of the plane in $\mathbb{R}^3$ may be written

$$\begin{bmatrix} -\frac{1}{2} & -1 & 2 \end{bmatrix} \begin{bmatrix} x \\ y \\ z \end{bmatrix} = 0,$$

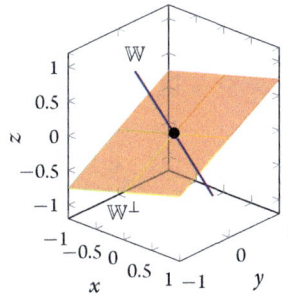
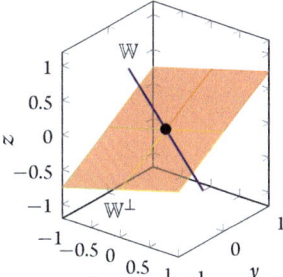

that is $W^T v = 0$ for matrix $W = \begin{bmatrix} w_1 \end{bmatrix}$ and vectors $w_1 = (-\frac{1}{2}, -1, 2)$ and $v = (x, y, z)$. Since the plane is the nullspace of matrix W^T, the plane must be the orthogonal complement of the line $\mathbb{W} = \text{span}\{w_1\}$ (as illustrated above). □

(c) Find the orthogonal complement to the column space of matrix

$$A = \begin{bmatrix} 1 & -1 & 0 \\ 1 & 0 & -1 \\ 0 & 1 & -1 \end{bmatrix}.$$

Solution: The required orthogonal complement is the nullspace of A^T. Recall from Section 2.1 that for such small problems we find all solutions of $A^T v = 0$ by algebraic elimination; that is,

$$\begin{bmatrix} 1 & 1 & 0 \\ -1 & 0 & 1 \\ 0 & -1 & -1 \end{bmatrix} v = 0 \iff \begin{cases} v_1 + v_2 = 0, \\ -v_1 + v_3 = 0, \\ -v_2 - v_3 = 0, \end{cases}$$

$$\iff \begin{cases} v_2 = -v_1, \\ v_3 = v_1, \\ -v_2 - v_3 = v_1 - v_1 = 0. \end{cases}$$

Therefore all solutions of $A^T v = 0$ are of the form $v_1 = t, v_2 = -v_1 = -t$ and $v_3 = v_1 = t$; that is, $v = (1, -1, 1)t$. Hence the orthogonal complement is $\text{span}\{(1, -1, 1)\}$. ☐

(d) Describe the orthogonal complement of the subspace spanned by the four vectors $(1, 1, 0, 1, 0, 0)$, $(-1, 0, 1, 0, 1, 0)$, $(0, -1, -1, 0, 0, 1)$ and $(0, 0, 0, -1, -1, -1)$.

Solution: Arrange these vectors as the four columns of a matrix, say

$$A = \begin{bmatrix} 1 & -1 & 0 & 0 \\ 1 & 0 & -1 & 0 \\ 0 & 1 & -1 & 0 \\ 1 & 0 & 0 & -1 \\ 0 & 1 & 0 & -1 \\ 0 & 0 & 1 & -1 \end{bmatrix},$$

then seek null(A^T), the solutions of $A^T x = 0$. Adapt Procedure 3.3.15 to solve $A^T x = 0$:

i. Example 3.5.6 computed an SVD $A = USV^T$ for this matrix A, which gives the SVD $A^T = VS^T U^T$ for the transpose where (2 d.p.)

```
U =
     0.31  -0.26  -0.58  -0.26   0.64  -0.15
     0.07   0.40  -0.58   0.06  -0.49  -0.51
    -0.24   0.67   0.00  -0.64   0.19   0.24
    -0.38  -0.14  -0.58   0.21  -0.15   0.66
    -0.70   0.13   0.00   0.37   0.45  -0.40
    -0.46  -0.54  -0.00  -0.58  -0.30  -0.26
```

$$S =$$

$$
\begin{array}{cccc}
2.00 & 0 & 0 & 0 \\
0 & 2.00 & 0 & 0 \\
0 & 0 & 2.00 & 0 \\
0 & 0 & 0 & 0.00 \\
0 & 0 & 0 & 0 \\
0 & 0 & 0 & 0
\end{array}
$$

$$V = \ldots$$

ii. $Vz = 0$ determines $z = 0$.

iii. $S^Ty = z = 0$ determines $y_1 = y_2 = y_3 = 0$ as there are three nonzero singular values, and y_4, y_5 and y_6 are free variables; that is, $y = (0,0,0,y_4,y_5,y_6)$.

iv. Denoting the columns of U by u_1, u_2, ..., u_6, the solutions of $U^Tx = y$ are $x = Uy = u_4y_4 + u_5y_5 + u_6y_6$.

That is, the orthogonal complement is the three-dimensional subspace span$\{u_4, u_5, u_6\}$ in $\mathbb{R}^6$, where (2 d.p.)

$$u_4 = (-0.26, 0.06, -0.64, 0.21, 0.37, -0.58),$$
$$u_5 = (0.64, -0.49, 0.19, -0.15, 0.45, -0.30),$$
$$u_6 = (-0.15, -0.51, 0.24, 0.66, -0.40, -0.26).$$

□

In the previous Example 3.5.41(d) there are three nonzero singular values in the first three rows of S. These three nonzero singular values determine that the first three columns of U form a basis for the column space of A. The example argues that the remaining three columns of U form a basis for the orthogonal complement of the column space. That is, all six of the columns of the orthogonal U are used in either the column space or its complement. This is generally true.

Activity 3.5.42 A given matrix A has column space W such that dim $W = 4$ and dim $W^\perp = 3$. What size could the matrix be?

(a) 7×3 (b) 3×4 (c) 7×5 (d) 4×3

Example 3.5.43 Recall the cases of Example 3.5.41.

3.5.41(a) dim W + dim $W^\perp = 1 + 1 = 2 = $ dim $\mathbb{R}^2$.

3.5.41(b) dim W + dim $W^\perp = 1 + 2 = 3 = $ dim $\mathbb{R}^3$.

3.5.41(c) dim W + dim $W^\perp = 2 + 1 = 3 = $ dim $\mathbb{R}^3$.

3.5.41(d) dim W + dim $W^\perp = 3 + 3 = 6 = $ dim $\mathbb{R}^6$.   □

Recall that the Rank Theorem 3.4.39 connects the dimension of a space with the dimensions of a nullspace and column space of a matrix. Since a subspace is closely connected to matrices, and its orthogonal complement is connected to nullspaces, then the Rank Theorem should say something general here.

Theorem 3.5.44 *Let* W *be a subspace of* $\mathbb{R}^n$, *then* $\dim W + \dim W^\perp = n$; *equivalently,* $\dim W^\perp = n - \dim W$.

Proof. Let the columns of a matrix W form an orthonormal basis for the subspace W (Theorem 3.4.29 asserts that a basis exists). Theorem 3.5.40 establishes that $W^\perp = \text{null}(W^\mathsf{T})$. Equating dimensions of both sides,

$$
\begin{aligned}
\dim W^\perp &= \text{nullity}(W^\mathsf{T}) \quad \text{(from Definition 3.4.36)} \\
&= n - \text{rank}(W^\mathsf{T}) \quad \text{(from Rank Theorem 3.4.39)} \\
&= n - \text{rank}(W) \quad \text{(from Theorem 3.3.24)} \\
&= n - \dim W \quad \text{(from Procedure 3.4.23),}
\end{aligned}
$$

as required.

Since the dimension of the whole space is the sum of the dimension of a subspace plus the dimension of its orthogonal complement, surely we must be able to separate vectors into two corresponding components.

Example 3.5.45 Recall from Example 3.5.35(a) that subspace $W = \text{span}\{(3,4)\}$ has orthogonal complement $W^\perp = \text{span}\{(-4,3)\}$, as illustrated.

As shown, for example, write the brown vector $(2,4) = (3.2, 2.4) + (-1.2, 1.6) = \text{proj}_W(2,4) + \text{perp}$, where here the vector $\text{perp} = (-1.2, 1.6) \in W^\perp$. Indeed, any vector can be written as a component in subspace W and a component in the orthogonal complement $W^\perp$ (Theorem 3.5.51). For another example, write the green vector $(-5,1) = (-2.72, -2.04) + (-2.28, 3.04) = \text{proj}_W(-5,1) + \text{perp}$, where in this case the vector $\text{perp} = (-2.28, 3.04) \in W^\perp$.

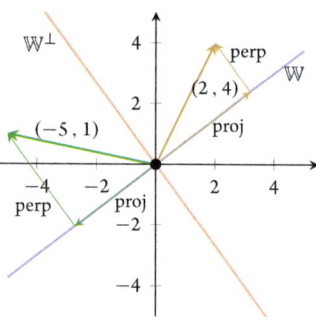

□

Activity 3.5.46 Let subspace $W = \text{span}\{(1,1)\}$ and its orthogonal complement $W^\perp = \text{span}\{(1,-1)\}$. Which of the following writes vector $(5,-9)$ as a sum of two vectors, one from each of W and $W^\perp$?

(a) $(-2,-2) + (7,-7)$ (c) $(5,5) + (0,-14)$

(b) $(7,7) + (-2,2)$ (d) $(9,-9) + (-4,0)$

Further, such a separation can be done for any pair of complementary subspaces W and $W^\perp$ within any space $\mathbb{R}^n$. To proceed, let's define what is meant by "perp" in such a context.

Definition 3.5.47 *(perpendicular component)* Let W be a subspace of $\mathbb{R}^n$. For every vector $v \in \mathbb{R}^n$, the **perpendicular component** of v to W is the vector $\text{perp}_W(v) := v - \text{proj}_W(v)$.

Example 3.5.48

(a) Let the subspace $\mathbb{W}$ be the span of $(-2,-3,6)$. Find the perpendicular component to $\mathbb{W}$ of the vector $(4,1,3)$. Verify that the perpendicular component lies in the plane $-2x - 3y + 6z = 0$.

Solution: Projection is easiest with a unit vector. Obtain a unit vector to span $\mathbb{W}$ by normalizing the basis vector to $\boldsymbol{w}_1 = (-2,-3,6)/\sqrt{2^2 + 3^2 + 6^2} = (-2,-3,6)/7$. Then

$$\text{perp}_{\mathbb{W}}(4,1,3) = (4,1,3) - \boldsymbol{w}_1(\boldsymbol{w}_1 \cdot (4,1,3))$$
$$= (4,1,3) - \boldsymbol{w}_1(-8 - 3 + 18)/7$$
$$= (4,1,3) - \boldsymbol{w}_1 = (30,10,15)/7.$$

For $(x,y,z) = (30, 10, 15)/7$ we find

$$-2x - 3y + 6z$$
$$= \tfrac{1}{7}(-60 - 30 + 90)$$
$$= \tfrac{1}{7}0 = 0.$$

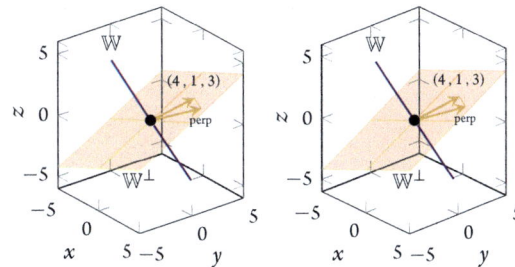

Hence $\text{perp}_{\mathbb{W}}(4, 1, 3)$ lies in the plane $-2x - 3y + 6z = 0$ (which is the orthogonal complement $\mathbb{W}^{\perp}$, as illustrated in stereo to the right). ☐

(b) For the vector $(-5,-1,6)$ find its perpendicular component to the subspace $\mathbb{W}$ spanned by $(-2,-3,6)$. Verify that the perpendicular component lies in the plane $-2x - 3y + 6z = 0$.

Solution: As in the previous case, use the basis vector $\boldsymbol{w}_1 = (-2,-3,6)/7$. Then

$$\text{perp}_{\mathbb{W}}(-5,-1,6) = (-5,-1,6) - \boldsymbol{w}_1(\boldsymbol{w}_1 \cdot (-5,-1,6))$$
$$= (-5,-1,6) - \boldsymbol{w}_1(10 + 3 + 36)/7$$
$$= (-5,-1,6) - \boldsymbol{w}_1 7 = (-3,2,0).$$

For $(x,y,z) = (-3,2,0)$ we find $-2x - 3y + 6z = 6 - 6 + 0 = 0$. Hence $\text{perp}_{\mathbb{W}}(-5, -1, 6)$ lies in the plane $-2x - 3y + 6z = 0$ (which is the orthogonal complement $\mathbb{W}^{\perp}$, as illustrated to the right in stereo). ☐

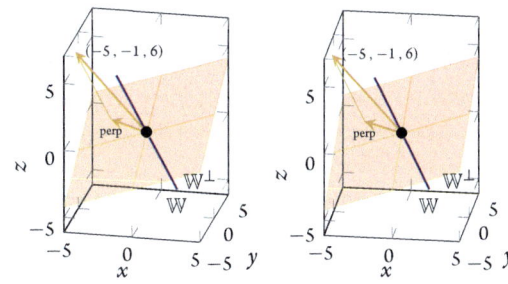

(c) Let the subspace $\mathbb{X} = \text{span}\{(2,-2,1), (2,1,-2)\}$. Determine the perpendicular component of each of the two vectors $y = (3,2,1)$ and $z = (3,-3,-3)$.

Solution: Computing $\text{proj}_{\mathbb{X}}$ needs an orthonormal basis for $\mathbb{X}$ (Definition 3.5.23). The two vectors in the span are orthogonal, so normalize them to $w_1 = (2,-2,1)/3$ and $w_2 = (2,1,-2)/3$.

- Then for the first vector $y = (3,2,1)$ (illustrated below in brown),

$$\begin{aligned}
\text{perp}_{\mathbb{X}}(y) &= y - \text{proj}_{\mathbb{X}}(y) \\
&= y - w_1(w_1 \cdot y) - w_2(w_2 \cdot y) \\
&= y - w_1(6-4+1)/3 - w_2(6+2-2)/3 \\
&= y - w_1 - 2w_2 \\
&= (3,2,1) - (2,-2,1)/3 - (4,2,-4)/3 \\
&= (1,2,2)
\end{aligned}$$

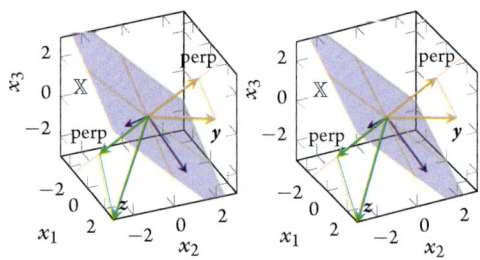

- For the second vector $z = (3,-3,-3)$ (green in the picture above),

$$\begin{aligned}
\text{perp}_{\mathbb{X}}(z) &= z - \text{proj}_{\mathbb{X}}(z) \\
&= z - w_1(w_1 \cdot z) - w_2(w_2 \cdot z) \\
&= z - w_1(6+6-3)/3 - w_2(6-3+6)/3 \\
&= z - 3w_1 - 3w_2 \\
&= (3,-3,-3) - (2,-2,1) - (2,1,-2) \\
&= (-1,-2,-2).
\end{aligned}$$

$\square$

As seen in all these examples, the perpendicular component of a vector always lies in the orthogonal complement to the subspace (as suggested by the naming).

Theorem 3.5.49 (*perpendicular component is orthogonal*) *Let $\mathbb{W}$ be a subspace of $\mathbb{R}^n$ and let v be any vector in $\mathbb{R}^n$, then the perpendicular component $\text{perp}_{\mathbb{W}}(v) \in \mathbb{W}^{\perp}$.*

Proof. Let vectors $w_1, w_2, \ldots, w_k$ form an orthonormal basis for the subspace $\mathbb{W}$ (a basis exists by Theorem 3.4.29). Let the $n \times k$ matrix $W = \begin{bmatrix} w_1 & w_2 & \cdots & w_k \end{bmatrix}$ so subspace $\mathbb{W}$ is the column space of matrix W, then Theorem 3.5.40 asserts that we just need to check that $W^{\mathsf{T}} \text{perp}_{\mathbb{W}}(v) = 0$. Consider

$$W^{\mathrm{T}} \mathrm{perp}_{\mathbb{W}}(v) = W^{\mathrm{T}}\left[v - \mathrm{proj}_{\mathbb{W}}(v)\right] \quad \text{(from Definition 3.5.47)}$$
$$= W^{\mathrm{T}}\left[v - (WW^{\mathrm{T}})v\right] \quad \text{(from Theorem 3.5.29)}$$
$$= W^{\mathrm{T}}v - W^{\mathrm{T}}(WW^{\mathrm{T}})v \quad \text{(by distributivity)}$$
$$= W^{\mathrm{T}}v - (W^{\mathrm{T}}W)W^{\mathrm{T}}v \quad \text{(by associativity)}$$
$$= W^{\mathrm{T}}v - I_k W^{\mathrm{T}}v \quad \text{(only if } W^{\mathrm{T}}W = I_k)$$
$$= W^{\mathrm{T}}v - W^{\mathrm{T}}v = 0.$$

Hence $\mathrm{perp}_{\mathbb{W}}(v) \in \mathrm{null}(W^{\mathrm{T}})$ and so is in $\mathbb{W}^{\perp}$ (by Theorem 3.5.40).

But this proof only holds if $W^{\mathrm{T}}W = I_k$. To establish this identity, use the same argument as in the proof of Theorem 3.2.48(a) $\Longleftrightarrow$ 3.2.48(b):

$$W^{\mathrm{T}}W = \begin{bmatrix} w_1^{\mathrm{T}} \\ w_2^{\mathrm{T}} \\ \vdots \\ w_k^{\mathrm{T}} \end{bmatrix} \begin{bmatrix} w_1 & w_2 & \cdots & w_k \end{bmatrix} = \begin{bmatrix} w_1^{\mathrm{T}}w_1 & w_1^{\mathrm{T}}w_2 & \cdots & w_1^{\mathrm{T}}w_k \\ w_2^{\mathrm{T}}w_1 & w_2^{\mathrm{T}}w_2 & \cdots & w_2^{\mathrm{T}}w_k \\ \vdots & \vdots & \ddots & \vdots \\ w_k^{\mathrm{T}}w_1 & w_k^{\mathrm{T}}w_2 & \cdots & w_k^{\mathrm{T}}w_k \end{bmatrix}$$

$$= \begin{bmatrix} w_1 \cdot w_1 & w_1 \cdot w_2 & \cdots & w_1 \cdot w_k \\ w_2 \cdot w_1 & w_2 \cdot w_2 & \cdots & w_2 \cdot w_k \\ \vdots & \vdots & \ddots & \vdots \\ w_k \cdot w_1 & w_k \cdot w_2 & \cdots & w_k \cdot w_k \end{bmatrix} = I_k$$

as vectors $w_1, w_2, \ldots, w_k$ are an orthonormal set (from Definition 3.2.38, the dot product $w_i \cdot w_j = 0$ for $i \neq j$ and $|w_i|^2 = w_i \cdot w_i = 1$).

Example 3.5.50 The previous examples' calculation of the perpendicular component confirm that $v = \mathrm{proj}_{\mathbb{W}}(v) + \mathrm{perp}_{\mathbb{W}}(v)$, where we now know that $\mathrm{perp}_{\mathbb{W}}$ is orthogonal to $\mathbb{W}$:

Example 3.5.45 $(2,4) = (3.2, 2.4) + (-1.2, 1.6)$ and
$(-5, 1) = (-2.72, -2.04) + (-2.28, 3.04)$;

Example 3.5.48(b) $(-5, -1, 6) = (-2, -3, 6) + (-3, 2, 0)$;

Example 3.5.48(c) $(3, 2, 1) = (2, 0, -1) + (1, 2, 2)$ and
$(3, -3, -3) = (4, -1, -1) + (-1, -2, -2)$. □

Given any subspace $\mathbb{W}$, this theorem indicates that every vector can be written as a sum of two vectors: one in the subspace $\mathbb{W}$; and one in its orthogonal complement $\mathbb{W}^{\perp}$.

Theorem 3.5.51 (orthogonal decomposition) *Let $\mathbb{W}$ be a subspace of $\mathbb{R}^n$ and vector $v \in \mathbb{R}^n$, then there exist unique vectors $w \in \mathbb{W}$ and $n \in \mathbb{W}^{\perp}$ such that vector $v = w + n$; this particular sum is called an* **orthogonal decomposition** *of v.*

Proof. First, establish existence. By Definition 3.5.47, $\mathrm{perp}_{\mathbb{W}}(v) = v - \mathrm{proj}_{\mathbb{W}}(v)$, so it follows that $v = \mathrm{proj}_{\mathbb{W}}(v) + \mathrm{perp}_{\mathbb{W}}(v) = w + n$ when we set $w = \mathrm{proj}_{\mathbb{W}}(v) \in \mathbb{W}$ and $n = \mathrm{perp}_{\mathbb{W}}(v) \in \mathbb{W}^{\perp}$.

Second, establish uniqueness by contradiction. Suppose there is another decomposition $v = w' + n'$ where $w' \in \mathbb{W}$ and $n' \in \mathbb{W}^{\perp}$. Then $w + n = v = w' + n'$. Rearranging gives

$w - w' = n' - n$. By closure of a subspace under vector addition (Definition 3.4.3), the left-hand side is in $\mathbb{W}$ and the right-hand side is in $\mathbb{W}^\perp$, so the two sides must be both in $\mathbb{W}$ and $\mathbb{W}^\perp$. The zero vector is the only common vector to the two subspaces (Theorem 3.5.38), so $w - w' = n' - n = 0$, and hence both $w = w'$ and $n = n'$. That is, the decomposition must be unique.

Example 3.5.52 For each pair of the shown subspaces $\mathbb{X} = \text{span}\{x\}$ and vectors v, draw the decomposition of vector v into the sum of vectors in $\mathbb{X}$ and $\mathbb{X}^\perp$.

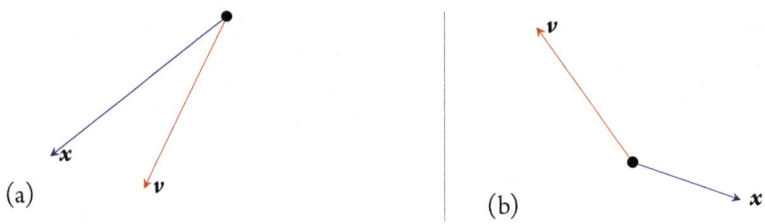

(a) (b)

Solution: In each case, the two brown vectors shown are the decomposition, with proj $\in \mathbb{X}$ and perp $\in \mathbb{X}^\perp$.

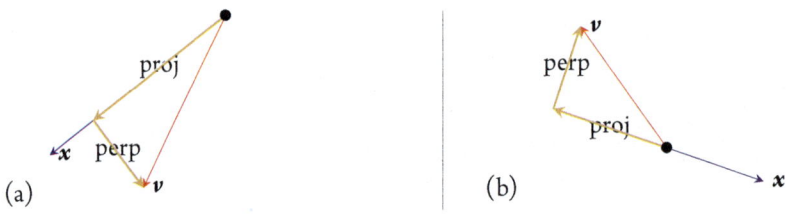

(a) (b)

In two or even three dimensions, that a decomposition has such a nice physical picture is appealing. What is powerful is that the same decomposition works in any number of dimensions: it works no matter how complicated the scenario, no matter how much data. In particular, the next Theorem 3.5.53 gives a geometric view of the 'least square' solution of Procedure 3.5.4: in that procedure the minimal change of the right-hand side b to make the linear equation $Ax = b$ consistent (Theorem 3.5.8) is also to be viewed as the projection of the right-hand side b to the *closest* point in the column space of the matrix. That is, the 'least square' procedure solves $Ax = \text{proj}_A(b)$.

Theorem 3.5.53 (*best approximation*) For every vector v in $\mathbb{R}^n$, and every subspace $\mathbb{W}$ in $\mathbb{R}^n$, $\text{proj}_\mathbb{W}(v)$ is the closest vector in $\mathbb{W}$ to v; that is, $|v - \text{proj}_\mathbb{W}(v)| \leq |v - w|$ for every $w \in \mathbb{W}$.

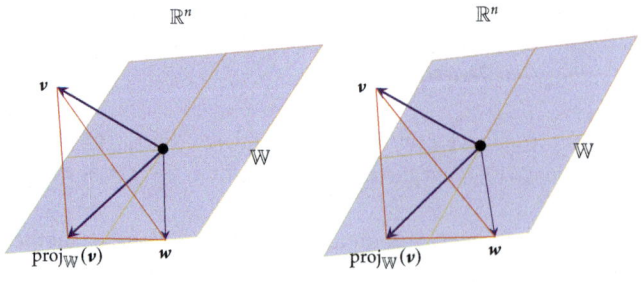

Proof. For any vector $w \in \mathbb{W}$, consider the triangle formed by the three vectors $v - \mathrm{proj}_{\mathbb{W}}(v)$, $v - w$, and $w - \mathrm{proj}_{\mathbb{W}}(v)$ (the stereo illustration above schematically plots this triangle in red). This is a right-angle triangle as $w - \mathrm{proj}_{\mathbb{W}}(v) \in \mathbb{W}$ by closure of the subspace $\mathbb{W}$, and as $v - \mathrm{proj}_{\mathbb{W}}(v) = \mathrm{perp}_{\mathbb{W}}(v) \in \mathbb{W}^{\perp}$. Then Pythagoras tells us

$$|v - w|^2 = |v - \mathrm{proj}_{\mathbb{W}}(v)|^2 + |w - \mathrm{proj}_{\mathbb{W}}(v)|^2$$
$$\geq |v - \mathrm{proj}_{\mathbb{W}}(v)|^2.$$

Hence $|v - w| \geq |v - \mathrm{proj}_{\mathbb{W}}(v)|$ for every $w \in \mathbb{W}$.

3.5.4 Exercises

Exercise 3.5.1 During an experiment on the strength of beams, you and your partner measure the length of a crack in the beam. With vernier callipers you measure, in millimetres, the crack as 17.8 mm long, whereas your partner measures it as 18.4 mm long.

- Write this information as a simple matrix-vector equation for the as yet to be decided length x, and involving the matrix $A = \begin{bmatrix} 1 \\ 1 \end{bmatrix}$.
- Confirm that an SVD of this 2×1 matrix is

$$A = \begin{bmatrix} \frac{1}{\sqrt{2}} & -\frac{1}{\sqrt{2}} \\ \frac{1}{\sqrt{2}} & \frac{1}{\sqrt{2}} \end{bmatrix} \begin{bmatrix} \sqrt{2} \\ 0 \end{bmatrix} [1]^{\mathrm{T}}.$$

- Use the SVD to 'best' solve the inconsistent equations and estimate the length of the crack is $x \approx 18.1$ mm—the average of the two measurements.

Exercise 3.5.2 In measuring the amount of butter to use in cooking a recipe you weigh a container to have 207 g (grams), then a bit later weigh it at 211 g. Wanting to be more accurate you weigh the butter container a third time and find 206 g.

- Write this information as a simple matrix-vector equation for the as yet to be decided weight x, and involving the matrix $B = \begin{bmatrix} 1 \\ 1 \\ 1 \end{bmatrix}$.
- Confirm that an SVD of this 3×1 matrix is

$$B = \begin{bmatrix} \frac{1}{\sqrt{3}} & -\frac{1}{\sqrt{2}} & \frac{1}{\sqrt{6}} \\ \frac{1}{\sqrt{3}} & 0 & -\frac{2}{\sqrt{6}} \\ \frac{1}{\sqrt{3}} & \frac{1}{\sqrt{2}} & \frac{1}{\sqrt{6}} \end{bmatrix} \begin{bmatrix} \sqrt{3} \\ 0 \\ 0 \end{bmatrix} [1]^{\mathrm{T}}.$$

- Use the SVD to 'best' solve the inconsistent equations and estimate that the butter container weighs $x \approx 208$ g—the average of the three measurements.

Exercise 3.5.3 An astro-geologist wants to measure the mass of a space rock. The lander accelerates the rock by applying different forces, and the astro-geologist measures the resulting acceleration. For the three forces of 1, 2, and 3 N (newtons) the measured accelerations are 0.0027,

Table 3.8 Stock prices (in $) of three banks, each a week apart (for Exercise 3.5.6).

week	ANZ	WBC	CBA
1	29.86	32.22	81.05
2	30.88	32.86	82.95
3	31.32	33.37	83.99
4	31.16	33.45	85.34

0.0062, and 0.0086 m/s^2, respectively. Using Newton's law that $F = ma$, formulate a system of three equations for the unknown mass m, and solve using Procedure 3.5.4 to best estimate the mass of the space rock.

Exercise 3.5.4 A school experiment aims to measure the acceleration of gravity g. Dropping a ball from a height, a camera takes a burst of photographs of the falling ball, one every 0.2 seconds. From the photographs the ball falls 0.21, 0.79, and 1.77 m after times 0.2, 0.4, and 0.6 s, respectively. Physical laws say that the distance fallen $s = \frac{1}{2}gt^2$ at time t. Use this law to formulate a system of three equations for gravity g, and solve using Procedure 3.5.4 to best estimate g.

Exercise 3.5.5 A spring under different loads stretches to different lengths according to Hooke's law that the length $L = a + bF$ where F is the applied load (force), a is the unknown rest length of the spring, and b is the unknown stiffness of the spring. An experiment applies the load forces 15, 30, and 40 N, and measures that the resultant spring length is 35, 48, and 61 mm. Formulate this as a system of three equations, and solve using Procedure 3.5.4 to best estimate the spring parameters a and b.

Exercise 3.5.6 Table 3.8 lists the share price of three banks. The prices fluctuate from week to week, as shown. Suspecting that these three prices *tend* to move up and down together according to the rule CBA $\approx a \cdot$ WBC $+ b \cdot$ ANZ, use the share prices to formulate a system of four equations, and solve using Procedure 3.5.4 to best estimate the coefficients a and b.

Exercise 3.5.7 Consider three sporting teams that play each other in a round robin event: Newark, Yonkers, and Edison: Yonkers beat Newark, 2 to 0; Edison beat Newark 5 to 2; and Edison beat Yonkers 3 to 2. Assuming the teams can be rated, and based upon the scores, write three equations that ideally relate the team ratings. Use Procedure 3.5.4 to estimate the ratings.

Recall that to 'rate' teams we seek to assign a (real) number to their ability in the competition. These unknown numbers are to be determined from equations formed from the results of each of the games.

Exercise 3.5.8 Consider four sporting teams that play each other in a round robin event: Acton, Barbican, Clapham, and Dalston. Table 3.9 summarizes the results of the six matches played. Assuming the teams can be rated, and based upon the scores, write six equations that ideally relate the team ratings. Use Procedure 3.5.4 to estimate the ratings.

Exercise 3.5.9 Consider five sporting teams that play each other in a round robin event: Atlanta, Boston, Concord, Denver, and Frankfort. Table 3.10 summarizes the results of the ten matches played. Assuming the teams can be rated, and based upon the scores, write ten equations that ideally relate the team ratings. Use Procedure 3.5.4 to estimate the ratings.

Table 3.9 The results of six matches played in a round robin: the scores are games/goals/points scored by each when playing the others. For example, Clapham beat Acton 4 to 2. Exercise 3.5.8 rates these teams.

	Acton	Barbican	Clapham	Dalston
Acton	—	2	2	6
Barbican	2	—	2	6
Clapham	4	4	—	5
Dalston	3	1	0	—

Table 3.10 The results of ten matches played in a round robin: the scores are games/goals/points scored by each when playing the others. For example, Atlanta beat Concord 3 to 2. Exercise 3.5.9 rates these teams.

	Atlanta	Boston	Concord	Denver	Frankfort
Atlanta	—	3	3	2	5
Boston	2	—	2	3	8
Concord	2	7	—	6	1
Denver	2	2	1	—	5
Frankfort	2	3	6	7	—

Table 3.11 The body weight and heat production of various mammals (Kleiber, 1947). Recall that numbers written as xEn denote the number $x \cdot 10^n$.

animal	body weight (kg)	heat prod. (kcal/day)
mouse	1.95E−2	3.06E+0
rat	2.70E−1	2.61E+1
cat	3.62E+0	1.56E+2
dog	1.28E+1	4.35E+2
goat	2.58E+1	7.50E+2
sheep	5.20E+1	1.14E+3
cow	5.34E+2	7.74E+3
elephant	3.56E+3	4.79E+4

Exercise 3.5.10 Consider six sporting teams in a weekly competition: Algeria, Botswana, Chad, Djibouti, Ethiopia, and Gabon. In the first week of competition Algeria beat Botswana 3 to 0, Chad and Djibouti drew 3 all, and Ethiopia beat Gabon 4 to 2. In the second week of competition Chad beat Algeria 4 to 2, Botswana beat Ethiopia 4 to 2, Djibouti beat Gabon 4 to 3. In the third week of competition Algeria beat Ethiopia 4 to 1, Botswana beat Djibouti 3 to 1, Chad drew with Gabon 2 all. Assuming the teams can be rated, and based upon the scores after the first three weeks, write nine equations that ideally relate the ratings of the six teams. Use Procedure 3.5.4 to estimate the ratings.

Discover power laws Exercises 3.5.11 to 3.5.14 use log-log plots as examples of the scientific inference of some surprising patterns in nature. These are simple examples of what, in modern parlance, might be termed 'data mining', 'knowledge discovery', or 'artificial intelligence'.

Table 3.12 River length and basin area for some Russian rivers (Arnold, 2014, p.154).

river	basin area (km^2)	length (km)
Moscow	17640	502
Protva	4640	275
Vorya	1160	99
Dubna	5474	165
Istra	2120	112
Nara	2170	156
Pakhra	2720	129
Skhodnya	259	47
Volgusha	265	40
Pekhorka	513	42
Setun	187	38
Yauza	452	41

Exercise 3.5.11 Table 3.11 lists data on the body weight and heat production of various mammals. As in Example 3.5.11, use this data to discover Kleiber's power law that (heat) $\propto$ (weight)$^{3/4}$. Graph the data on a log-log plot, fit a best straight line, check the correspondence between neglected parts of the right-hand side and the quality of the graphical fit, and describe the power law.

Exercise 3.5.12 Table 3.12 lists data on river lengths and basin areas of some Russian rivers. As in Example 3.5.11, use this data to discover Hack's exponent in the power law that (length) $\propto$ (area)$^{0.58}$. Graph the data on a log-log plot, fit a best straight line, check the correspondence between neglected parts of the right-hand side and the quality of the graphical fit, and describe the power law.

Exercise 3.5.13 Find, for another country, some river length and basin area data akin to that of Exercise 3.5.12. Confirm, or otherwise, Hack's exponent for your data. Write a short report.

Exercise 3.5.14 The basin area to river length relationship of a river is expected to be (length) $\propto$ (area)$^{1/2}$, so it is a puzzle as to why one consistently finds Hack's exponent (e.g., Exercise 3.5.12). The puzzle may be answered by the surprising notion that rivers do not have a well-defined length! L. F. Richardson first established this remarkable notion for coastlines.

Table 3.13 lists data on the length of the west coast of Britain computed by using measuring sticks of various lengths: as one uses a smaller and smaller measuring stick, more and more bays and inlets are resolved and measured, which increases the computed coast length. As in Example 3.5.11, use this data to discover the power law that the coast-length $\propto$ (stick-length)$^{-1/4}$. Hence as the measuring stick length goes to 'zero', the coast length goes to 'infinity'! Graph the data on a log-log plot, fit a best straight line, check the correspondence between neglected parts of the right-hand side and the quality of the graphical fit, and describe the power law.

Exercise 3.5.15 Table 3.14 lists nine of the US universities ranked by some organization

Table 3.13 Given a measuring stick of some length, compute the length of the west coast of Britain (Mandelbrot, 1982, Plate 33).

stick length (km)	coast length (km)
10.4	2845
30.2	2008
99.6	1463
202.	1138
532.	929
933.	914

Table 3.14 A selection of nine of the US universities ranked in 2013 by *The Center for Measuring University Performance* [http://mup.asu.edu/research_data.html]. Among others, these particular nine universities are listed by the Center in the following order. The other three columns give just three of the attributes used to create their ranked list.

Institution	Research fund(M$)	Faculty awards	Median SAT U/G
Stanford University	868	45	1455
Yale University	654	45	1500
University of California, San Diego	1004	35	1270
University of Pittsburgh, Pittsburgh	880	22	1270
Vanderbilt University	535	19	1440
Pennsylvania State University, University Park	677	20	1195
Purdue University, West Lafayette	520	22	1170
University of Utah	410	12	1110
University of California, Santa Barbara	218	11	1205

in 2013, in the order they list.[29] The table also lists three of the attributes used to generate the ranked list. Find a formula that approximately reproduces the listed ranking from the three given attributes.

(a) Pose that the rank of the ith institution is a linear function of the attributes and a constant, say the rank $i = x_1 f_i + x_2 a_i + x_3 s_i + x_4$, where f_i denotes the funding, a_i denotes the awards, and s_i denotes the SAT.

(b) Form a system of nine equations that we would ideally solve to find the coefficients $\boldsymbol{x} = (x_1, x_2, x_3, x_4)$.

(c) Enter the data into MATLAB/Octave and find a best approximate solution (you should find that the formula is roughly that rank $\approx 97 - 0.01 f_i - 0.07 a_i - 0.01 s_i$).

(d) Discuss briefly how well the approximation reproduces the ranking of the list.

---

[29] I neither condone nor endorse such naive one-dimensional ranking of complex multi-faceted institutions. This exercise simply illustrates a technique that deconstructs such a credulous endeavour.

Exercise 3.5.16 For each of the following lines and planes, use an SVD to find the point closest to the origin in the line or plane. For the lines in 2D, draw a graph to show the answer is correct.

(a) $5x_1 - 12x_2 = 169$

(b) $x_1 - 2x_2 = 5$

(c) $2x_1 - 3x_2 + 6x_3 = 7$

(d) $q_1 + q_2 - 5q_3 = 2$

Exercise 3.5.17 Following the computed tomography Example 3.5.17, predict the densities in the body if the fraction of X-ray energy measured in the six paths is $f = (0.9, 0.2, 0.8, 0.9, 0.8, 0.2)$ respectively. Draw an image of your predictions (see Table 3.7). Which region is the most absorbing (least transmitting)?

Exercise 3.5.18 In an effort to remove the need for requiring the 'smallest', most washed out, CT-scan, you make three more measurements, as illustrated to the right, so that you obtain nine equations for the nine unknowns.

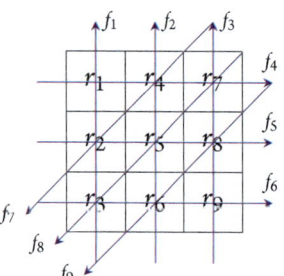

(a) Write down the nine equations for the transmission factors in terms of the fraction of X-ray energy measured after passing through the body. Take logarithms to form a system of linear equations.

(b) Encode the matrix A of the system and check $\texttt{rcond(A)}$: curses, $\texttt{rcond}$ is terrible, so we must still use an SVD.

(c) Suppose the measured fractions of X-ray energy are $f = (0.05, 0.35, 0.33, 0.31, 0.05, 0.36, 0.07, 0.32, 0.51)$. Use an SVD to find the 'greyest' transmission factors consistent with the measurements.

(d) Which part of the body is predicted to be the most absorbing?

Exercise 3.5.19 Use a little higher resolution in computed tomography: suppose the two-dimensional 'body' is notionally divided into 16 regions as illustrated to the right. Suppose a CT-scan takes 13 measurements of the intensity of an X-ray after passing through the shown paths, and that the fraction of the X-ray energy that is measured is $f = (0.29, 0.33, 0.07, 0.35, 0.36, 0.07, 0.31, 0.32, 0.62, 0.40, 0.06, 0.47, 0.58)$.

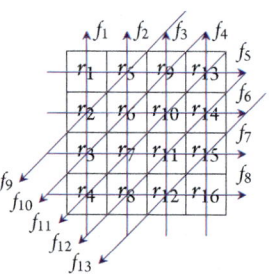

(a) Write down the 13 equations for the 16 transmission factors in terms of the fraction of X-ray energy measured after passing through the body. Take logarithms to form a system of linear equations.

(b) Encode the matrix A of the system and find it has rank 12.

(c) Use an SVD to find the 'greyest' transmission factors consistent with the measurements.

(d) In which square pixel is the 'lump' of dense material?

Exercise 3.5.20 This exercise is for those who, in calculus courses, have studied constrained optimization with Lagrange multipliers. In some applications, we would like to solve as best we can $Ax = b$ but only for unknowns x of limited magnitude. So the aim of this exercise is to derive how to use the SVD to find the vector x that minimizes $|Ax - b|$ such that the length $|x| \le \alpha$ for some given prescribed largest allowable magnitude α.

(a) As a first simpler problem, you are given vector $z \in \mathbb{R}^n$ and $n \times n$ diagonal matrix $S = \operatorname{diag}(\sigma_1, \sigma_2, \ldots, \sigma_n)$, with real $\sigma_1, \sigma_2, \ldots, \sigma_n > 0$. Minimize $|Sy - z|^2$ such that $|y|^2 \le \alpha^2$ for some given magnitude α. Consider the two following possible cases.

  • Solve $Sy^* - z = 0$: if $|y^*| \le \alpha$, then this solution is the desired minimum.

  • Otherwise, when $|y^*| > \alpha$, use a Lagrange multiplier λ to find the components of vector y (as a function of λ and z) that minimizes $|Sy - z|^2$ such that $|y|^2 = \alpha^2$: show that the multiplier λ satisfies a polynomial equation of degree $2n$.

(b) What can be further deduced if one or more $\sigma_j = 0$?

(c) Hence use an SVD of $n \times n$ real matrix A to find the vector $x \in \mathbb{R}^n$ that minimizes $|Ax - b|$ such that the length $|x| \le \alpha$ for some given magnitude α. Use that multiplication by orthogonal matrices preserves lengths. Report on all cases.

Exercise 3.5.21 For each pair of vectors, draw the orthogonal projection $\operatorname{proj}_u(v)$.

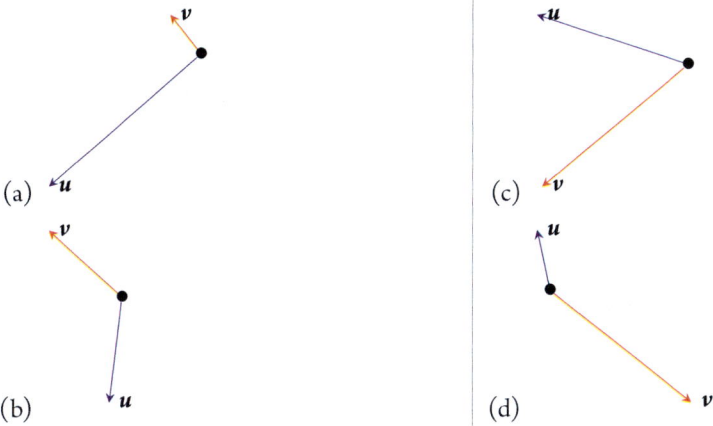

(a)

(b)

(c)

(d)

Exercise 3.5.22 For the following pairs of vectors: compute the orthogonal projection $\operatorname{proj}_u(v)$; and hence find the 'best' approximate solution to the inconsistent system $ux = v$.

(a) $u = (2,1)$, $v = (2,0)$

(b) $u = (4,-1)$, $v = (-1,1)$

(c) $u = (4,5,-1)$, $v = (-1,2,-1)$

(d) $u = (-3,2,2)$, $v = (0,1,-1)$

(e) $u = (0,2,0)$, $v = (-2,1,1)$

(f) $u = (-1,-7,5)$, $v = (1,1,-1)$

Exercise 3.5.23 For each of the following subspaces $\mathbb{W}$ (given as the span of orthogonal vectors), and the given vectors v, find the orthogonal projection $\operatorname{proj}_{\mathbb{W}}(v)$.

(a) $\mathbb{W} = \operatorname{span}\{(-6,-6,7), (2,-9,-6)\}$,
  $v = (0,1,-2)$

(b) $\mathbb{W} = \operatorname{span}\{(1,8,-4), (-8,-1,-4)\}$,
  $v = (-2,2,0)$

(c) $\mathbb{W} = \operatorname{span}\{(-1,2,-2), (-2,1,2), (2,2,1)\}$,
  $v = (3,-1,1)$

(d) $W =$
span$\{(-1,3,1,5),(-3,-1,-5,1)\}$,
$v = (-3,2,3,-2)$

(e) $W =$
span$\{(-1,5,3,-1),(-1,-1,1,-1)\}$,
$v = (0,-2,-5,-5)$

(f) $W = $ span$\{(1,4,-2,-2),$
$(-4,1,4,-4),(-2,4,2,5)\}$,
$v = (-2,-4,3,-1)$

Exercise 3.5.24 For each of the following matrices, compute an SVD in MATLAB/Octave to find an orthonormal basis for the column space of the matrix, and then compute the matrix of the orthogonal projection onto the column space.

(a) $A = \begin{bmatrix} 0 & -2 & 4 \\ 4 & -1 & -14 \\ 1 & -1 & -2 \end{bmatrix}$

(b) $B = \begin{bmatrix} -3 & 4 \\ -1 & 5 \\ -3 & -1 \end{bmatrix}$

(c) $C = \begin{bmatrix} 12 & 0 & 10 & 5 \\ -26 & -5 & 5 & 0 \\ -1 & -2 & -16 & 1 \\ -29 & -9 & 29 & 8 \end{bmatrix}$

(d) $D = \begin{bmatrix} -12 & 4 & 8 & 16 & 8 \\ 15 & -5 & -10 & -20 & -10 \end{bmatrix}$

(e) $E = \begin{bmatrix} 1 & 26 & -13 & 10 \\ -13 & 2 & 9 & 10 \\ -4 & -2 & 4 & 2 \\ -21 & 32 & 1 & 28 \\ -1 & -9 & 5 & -3 \end{bmatrix}$

(f) $F = \begin{bmatrix} 51 & -15 & -19 & -35 & 11 \\ -7 & 2 & 5 & 6 & -5 \\ 14 & -17 & -2 & -8 & -4 \\ 10 & -12 & -2 & -6 & -2 \\ -40 & 30 & 14 & 27 & -4 \end{bmatrix}$

Exercise 3.5.25 Generally, each of the following systems of equations are inconsistent. Use your answers to the previous Exercise 3.5.24 to find the right-hand side vector b' that is the closest vector to the given right-hand side among all the vectors in the column space of the matrix. What is the magnitude of the difference between b' and the given right-hand side? Hence write down a system of *consistent* equations that best approximates the original system.

(a) $\begin{bmatrix} 0 & -2 & 4 \\ 4 & -1 & -14 \\ 1 & -1 & -2 \end{bmatrix} x = \begin{bmatrix} 6 \\ -19 \\ -3 \end{bmatrix}$

(b) $\begin{bmatrix} 0 & -2 & 4 \\ 4 & -1 & -14 \\ 1 & -1 & -2 \end{bmatrix} x = \begin{bmatrix} 2 \\ -8 \\ -1 \end{bmatrix}$

(c) $\begin{bmatrix} -3 & 4 \\ -1 & 5 \\ -3 & -1 \end{bmatrix} x = \begin{bmatrix} 9 \\ 11 \\ -1 \end{bmatrix}$

(d) $\begin{bmatrix} -3 & 4 \\ -1 & 5 \\ -3 & -1 \end{bmatrix} x = \begin{bmatrix} -1 \\ 2 \\ -3 \end{bmatrix}$

(e) $\begin{bmatrix} 12 & 0 & 10 & 5 \\ -26 & -5 & 5 & 0 \\ -1 & -2 & -16 & 1 \\ -29 & -9 & 29 & 8 \end{bmatrix} x = \begin{bmatrix} 4 \\ -45 \\ 27 \\ -98 \end{bmatrix}$

(f) $\begin{bmatrix} 12 & 0 & 10 & 5 \\ -26 & -5 & 5 & 0 \\ -1 & -2 & -16 & 1 \\ -29 & -9 & 29 & 8 \end{bmatrix} x = \begin{bmatrix} -11 \\ -4 \\ 18 \\ -37 \end{bmatrix}$

Exercise 3.5.26 Theorems 3.5.8 and 3.5.26, and some examples and exercises, solve an inconsistent system of equations by some specific 'best approximation' that forms a consistent system of equations to solve. Describe briefly the key idea of this 'best approximation'. Discuss other possibilities for a 'best approximation' that might be developed.

Exercise 3.5.27 For any matrix A, suppose you know an orthonormal basis for the column space of A. Form the matrix W from all the vectors of the orthonormal basis. What is the result of the product $(WW^T)A$? Explain why.

Exercise 3.5.28 For each of the following subspaces, draw its orthogonal complement on the plot.

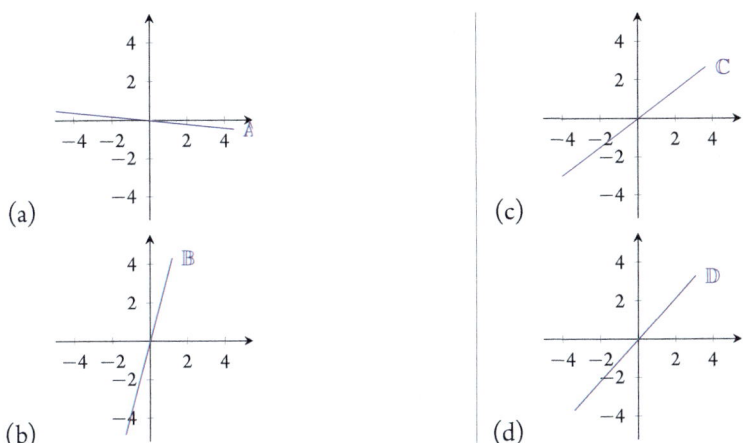

(a)

(b)

(c)

(d)

Exercise 3.5.29 Describe the orthogonal complement of each of the sets given below, if the set has one.

(a) $A = \text{span}\{(-1,2)\}$
(b) $B = \text{span}\{(5,-1)\}$
(c) C is the plane $5x + 2y + 3z = 3$
(d) $D = \text{span}\{(-5,5,-3), (-2,1,1)\}$
(e) $E = \text{span}\{(-2,2,8), (5,3,5)\}$
(f) $F = \text{span}\{(6,5,1,-3)\}$

Exercise 3.5.30 Compute, using MATLAB/Octave when necessary, an orthonormal basis for the orthogonal complement, if it exists, to each of the following sets. Use that the orthogonal complement is the nullspace of the transpose of a matrix of column vectors.

(a) The $\mathbb{R}^3$ vectors in the plane $-6x + 2y - 3z = 0$.
(b) The $\mathbb{R}^3$ vectors in the plane $x + 4y + 8z = 0$.
(c) The span of vectors $(3, -2, 1), (-3, 2, -1), (-9, 6, -3), (-6, 4, -2)$.
(d) The span of vectors $(26, -2, -4, 20), (23, -3, 2, 6), (2, -2, 8, -16), (21, -5, 12, -16)$.
(e) The span of vectors $(7, -5, 1, -6, -4), (6, -4, -2, -8, -4), (-5, 5, -15, -10, 0), (8, -6, 4, -4, -4)$.
(f) The intersection in $\mathbb{R}^4$ of the two hyper-planes $-3x_1 + x_2 + 4x_3 - 7x_4 = 0$ and $-6x_2 - x_3 - 2x_4 = 0$.

Exercise 3.5.31 For the subspace $\mathbb{X} = \text{span}\{x\}$ and the vector v, draw the decomposition of v into the sum of vectors in $\mathbb{X}$ and $\mathbb{X}^{\perp}$.

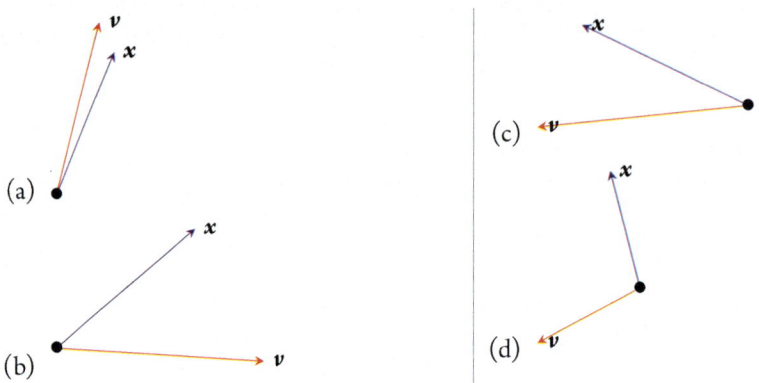

Exercise 3.5.32 For each of the following vectors, find the perpendicular component to the subspace $\mathbb{W} = \text{span}\{(4,-4,7)\}$. Verify that the perpendicular component lies in the plane $4x - 4y + 7z = 0$.

(a) $(4,2,4)$

(b) $(0,1,-2)$

(c) $(5,1,5)$

(d) (p,q,r)

Exercise 3.5.33 For each of the following vectors, find the perpendicular component to the subspace $\mathbb{W} = \text{span}\{(1,5,5,7),(-5,1,-7,5)\}$.

(a) $(1,2,-1,-1)$

(b) $(-2,4,5,0)$

(c) $(2,-6,1,-3)$

(d) (p,q,r,s)

Exercise 3.5.34 Let $\mathbb{W}$ be a subspace of $\mathbb{R}^n$ and let v be any vector in $\mathbb{R}^n$. Prove that $\text{perp}_{\mathbb{W}}(v) = (I_n - WW^{\mathsf{T}})v$ where the columns of the matrix W are an orthonormal basis for $\mathbb{W}$.

Exercise 3.5.35 For each of the following vectors in $\mathbb{R}^2$, write the vector as the orthogonal decomposition with respect to the subspace $\mathbb{W} = \text{span}\{(3,4)\}$.

(a) $(-2,4)$

(b) $(-3,3)$

(c) $(0,0)$

(d) $(3,1)$

Exercise 3.5.36 For each of the following vectors in $\mathbb{R}^3$, write the vector as the orthogonal decomposition with respect to the subspace $\mathbb{W} = \text{span}\{(3,-6,2)\}$.

(a) $(-5,4,-5)$

(b) $(0,5,-1)$

(c) $(1,-1,-2)$

(d) $(-3,1,-1)$

Exercise 3.5.37 For each of the following vectors in $\mathbb{R}^4$, write the vector as the orthogonal decomposition with respect to the subspace $\mathbb{W} = \text{span}\{(3, -1, 9, 3), (-9, 3, 3, 1)\}$.

(a) $(5, -5, 1, -3)$

(b) $(-4, -2, 5, 5)$

(c) $(2, -1, -4, -3)$

(d) $(5, 4, 0, 3)$

Exercise 3.5.38 The vector $(-3, 4)$ has an orthogonal decomposition $(1, 2) + (-4, 2)$. Draw in $\mathbb{R}^2$ the possibilities for the subspace $\mathbb{W}$ and its orthogonal complement.

Exercise 3.5.39 The vector $(2, 0, -3)$ in $\mathbb{R}^3$ has an orthogonal decomposition $(2, 0, 0) + (0, 0, -3)$. Describe the possibilities for the subspace $\mathbb{W}$ and its orthogonal complement.

Exercise 3.5.40 The vector $(0, -2, 5, 0)$ in $\mathbb{R}^4$ has an orthogonal decomposition $(0, -2, 0, 0) + (0, 0, 5, 0)$. Describe the possibilities for the subspace $\mathbb{W}$ and its orthogonal complement.

Exercise 3.5.41 In a few sentences, answer/discuss each of the following.

(a) How does rating sports teams often lead to an inconsistent system of linear equations?

(b) For an inconsistent system of equations, $A\boldsymbol{x} = \boldsymbol{b}$, why does solving $A\boldsymbol{x} = \boldsymbol{b}'$ for a slightly different right-hand side $\boldsymbol{b}'$ give a reasonable approximate solution?

(c) How does Procedure 3.5.4 ensure that we approximate an inconsistent system, $A\boldsymbol{x} = \boldsymbol{b}$, by making the smallest change to the right-hand side $\boldsymbol{b}$?

(d) Why does attempting to solve an inconsistent system of equations have so many applications in science and engineering?

(e) In solving systems of equations, $A\boldsymbol{x} = \boldsymbol{b}$, that have many possible solutions, the command A\b in MATLAB/Octave computes one answer for you; how is it that MATLAB and Octave often give different answers? Search for information about what 'answers' they each compute.

(f) What causes Procedures 3.3.15 and 3.5.4 to give the smallest solution when all free variables are set to zero?

(g) Why should the smallest solution be the 'best' answer for computed tomography?

(h) What causes an orthogonal projection to be relevant to approximately solving an inconsistent system of equations?

(i) Why is the concept of orthogonal complement relevant to matrices? and to linear equations?

3.6 Introducing linear transformations

This optional section unifies the transformation examples seen so far, and forms a foundation for more advanced algebra.

Recall function notation. For example, $f(x) = x^2$ means that for every $x \in \mathbb{R}$ the function $f(x)$ gives a result in $\mathbb{R}$, namely the value x^2, as plotted to the right. We often write $f : \mathbb{R} \to \mathbb{R}$ to denote this functionality: that is, $f : \mathbb{R} \to \mathbb{R}$ means that function f transforms any given real number into another real number by some specific rule.

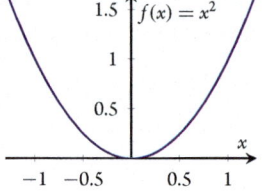

There is analogous functionality in multiple dimensions with vectors: given any vector,

- multiplication by a diagonal matrix stretches and/or shrinks the vector (Section 3.2.2);
- multiplication by an orthogonal matrix rotates and/or reflects the vector (Section 3.2.3); and
- projection finds a vector's components in a subspace (Section 3.5.3).

Correspondingly, we use the notation $f : \mathbb{R}^n \to \mathbb{R}^m$ to mean that the function f transforms a given vector with n components (in $\mathbb{R}^n$) into another vector with m components (in $\mathbb{R}^m$) according to some rule. For example, suppose the function $f(x)$ is to denote multiplication by the matrix

$$
A = \begin{bmatrix} 1 & -\frac{1}{3} \\ \frac{1}{2} & -1 \\ -1 & -\frac{1}{2} \end{bmatrix}.
$$

Then the function

$$
f(x) = Ax = \begin{bmatrix} 1 & -\frac{1}{3} \\ \frac{1}{2} & -1 \\ -1 & -\frac{1}{2} \end{bmatrix} \begin{bmatrix} x_1 \\ x_2 \end{bmatrix} = \begin{bmatrix} x_1 - x_2/3 \\ x_1/2 - x_2 \\ -x_1 - x_2/2 \end{bmatrix}.
$$

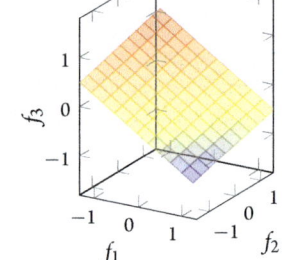

That is, here $f : \mathbb{R}^2 \to \mathbb{R}^3$. Given any vector in the 2D-plane, the function f, also called a transformation, returns a vector in 3D-space. Such a function can be evaluated for every vector $x \in \mathbb{R}^2$, so we ask what is the shape, the structure, of all the possible results of the function? The plot to the above-right illustrates the subspace formed by this $f(x)$ for all 2D vectors x.

There is a major difference between 'curvaceous' functions like the parabola above, and matrix multiplication functions such as rotation and projection. The difference is that linear algebra empowers many practical results in the latter case.

Definition 3.6.1 *A transformation/function $T : \mathbb{R}^n \to \mathbb{R}^m$ is called a **linear transformation** if*

(a) $T(u + v) = T(u) + T(v)$ *for every $u, v \in \mathbb{R}^n$, and*

(b) $T(cv) = cT(v)$ *for every $v \in \mathbb{R}^n$ and every scalar c.*

Example 3.6.2 *(1D cases)*

(a) Show that the parabolic function $f : \mathbb{R} \to \mathbb{R}$ where $f(x) = x^2$ is not a linear transformation.

Solution: To test Property 3.6.1(a), for any real x and y consider $f(x+y) = (x+y)^2 = x^2 + 2xy + y^2 = f(x) + 2xy + f(y) \neq f(x) + f(y)$ in general (it is equal if either are zero, but the test requires equality to hold for every x and y). Alternatively one could test Property 3.6.1(b) and consider $f(cx) = (cx)^2 = c^2x^2 = c^2f(x) \neq cf(x)$ for every c. Either of these proves that f is not a linear transformation. □

(b) Is the function $T(x) = |x|$, $T : \mathbb{R} \to \mathbb{R}$, a linear transformation?

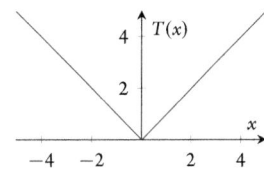

Solution: To prove not, it is sufficient to find just one instance when Definition 3.6.1 fails. Let $u = -1$ and $v = 2$, then $T(u+v) = |-1+2| = |1| = 1$ whereas $T(u) + T(v) = |-1| + |2| = 1 + 2 = 3 \neq T(u+v)$ so the function T fails the additivity property Definition 3.6.1(a) and so it is not a linear transformation. □

(c) Is the function $g : \mathbb{R} \to \mathbb{R}$ such that $g(x) = -x/2$ a linear transformation?

Solution:

Because the graph of g is a straight line (as in the right-hand picture) we suspect that it is a linear transformation. Thus check the properties in full generality:

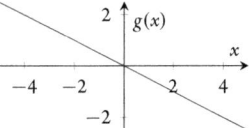

Definition 3.6.1(a) for every $u, v \in \mathbb{R}$, $g(u+v) =$
$-(u+v)/2 = -u/2 - v/2 = (-u/2) + (-v/2) = g(u) + g(v)$;

Definition 3.6.1(b) for every $u, c \in \mathbb{R}$, $g(cu) = -(cu)/2 = c(-u/2) = cg(u)$.

Hence g is a linear transformation. □

(d) Show that the function $h(y) = 2y - 3$, $h : \mathbb{R} \to \mathbb{R}$, is not a linear transformation.

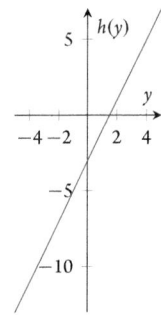

Solution: Because the graph of $h(y)$ is a straight line (as shown to the right) we suspect that it may be a linear transformation. To prove not, it is enough to find one instance when Definition 3.6.1 fails. Let $u = 0$ and $c = 2$, then $h(cu) = h(2 \cdot 0) = h(0) = -3$ whereas $ch(u) = 2h(0) = 2 \cdot (-3) = -6 \neq h(cu)$ so the function g fails the multiplication property 3.6.1(b) and hence it is not a linear transformation. This function fails because every linear transformation has to pass through the origin: that is, here we need $h(0) = 0$ but this does not hold, so h is not a linear transform. □

(e) Is the function $S : \mathbb{Z} \to \mathbb{Z}$ given by $S(n) = -n/2$ a linear transformation? Here $\mathbb{Z}$ denotes the set of integers $\ldots, -2,$ $-1, 0, 1, 2, \ldots$.

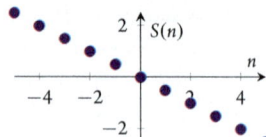

Solution: No, because the function S is here only defined for integers $\mathbb{Z}$ (as plotted to the right) whereas Definition 3.6.1 requires the function to be defined for all reals.[30] ☐

Activity 3.6.3 Which of the following is the graph of a linear transformation?

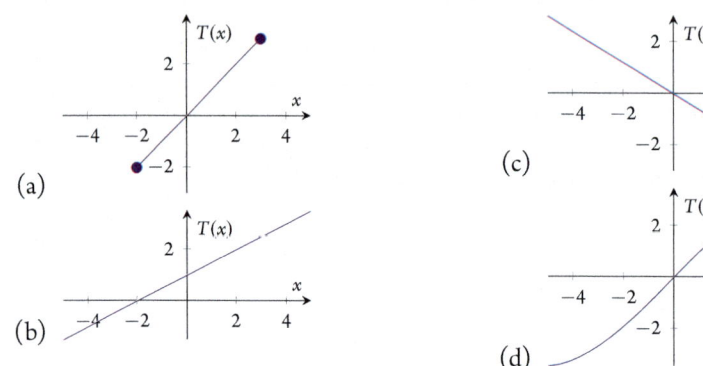

Example 3.6.4 *(higher-D cases)*

(a) Let function $T : \mathbb{R}^3 \to \mathbb{R}^3$ be $T(x,y,z) = (y,z,x)$. Is T a linear transformation?

Solution: Yes, because it satisfies the two properties of Definition 3.6.1:

3.6.1(a) for every $u = (x,y,z)$ and $v = (x',y',z')$ in $\mathbb{R}^3$ consider $T(u+v) = T(x+x', y+y', z+z') = (y+y', z+z', x+x') = (y,z,x) + (y',z',x') = T(x,y,z) + T(x',y',z') = T(u) + T(v)$;

3.6.1(b) for every $u = (x,y,z)$ and every scalar c consider $T(cu) = T(cx, cy, cz) = (cy, cz, cx) = c(y,z,x) = cT(x,y,z) = cT(u)$.

Hence, T is a linear transformation. ☐

(b) Consider the function $f(x,y,z) = x+y+1$, $f : \mathbb{R}^3 \to \mathbb{R}$. Is f a linear transformation?

Solution: No. For example, choose $u = 0$ and scalar $c = 2$ then $f(cu) = f(2 \cdot 0) = f(0) = 1$ whereas $cf(u) = 2f(0) = 2 \cdot 1 = 2$. Hence f fails the scalar multiplication property 3.6.1(b). ☐

---

[30] More advanced linear algebra generalizes the definition of a linear transformation to non-reals, but not here.

(c) Consider the following illustrated transformations of the plane, $T : \mathbb{R}^2 \rightarrow \mathbb{R}^2$. Which *cannot* be that of a linear transformation? In each illustration of a transformation T, the four corners of the blue unit square $((0,0), (1,0), (1,1)$ and $(0,1))$, are transformed to the four corners of the red figure $(T(0,0), T(1,0), T(1,1)$ and $T(0,1)$—the 'roof' of the unit square clarifies which side goes where).

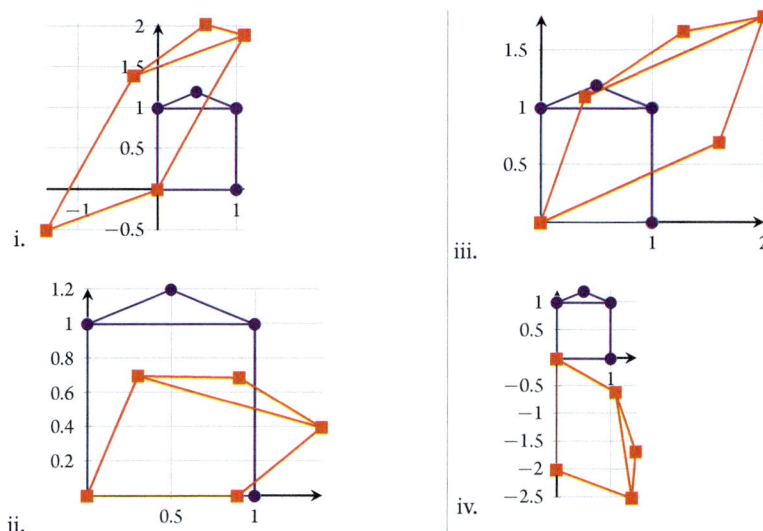

i.

ii.

iii.

iv.

Solution: To *test*, we check the addition property 3.6.1(a). First, with $\boldsymbol{u} = \boldsymbol{v} = \boldsymbol{0}$ property 3.6.1(a) requires $T(\boldsymbol{0} + \boldsymbol{0}) = T(\boldsymbol{0}) + T(\boldsymbol{0})$, but the left-hand side is just $T(\boldsymbol{0})$ which cancels with one on the right-hand side to leave that a linear transformation has to satisfy $T(\boldsymbol{0}) = \boldsymbol{0}$: all the shown transformations satisfy $T(\boldsymbol{0}) = \boldsymbol{0}$ as the (blue) origin point is transformed to the (red) origin point. Second, with $\boldsymbol{u} = (1,0)$, $\boldsymbol{v} = (0,1)$ and $\boldsymbol{u} + \boldsymbol{v} = (1,1)$ property 3.6.1(a) requires $T(1,1) = T(1,0) + T(0,1)$; let's see which do not pass this test.

i. Here $T(1,1) \approx (-0.3, 1.4)$, whereas $T(1,0) + T(0,1) \approx (-1.4, -0.5) + (1.1, 1.9) = (-0.3, 1.4) \approx T(1,1)$ so this may be a linear transformation.

ii. Here $T(1,1) \approx (1.4, 0.4)$, whereas $T(1,0) + T(0,1) \approx (0.9, 0) + (0.3, 0.7) = (1.2, 0.7) \napprox T(1,1)$ so this *cannot* be a linear transformation.

iii. Here $T(1,1) \approx (2.0, 1.8)$, whereas $T(1,0) + T(0,1) \approx (1.6, 0.7) + (0.4, 1.1) = (2.0, 1.8) \approx T(1,1)$ so this may be a linear transformation.

iv. Here $T(1,1) \approx (1.4, -2.5)$, whereas $T(1,0) + T(0,1) \approx (0, -2) + (1.1, -0.6) = (1.1, -2.6) \napprox T(1,1)$ so this *cannot* be a linear transformation.

The ones that pass this test may fail other tests: all we are sure of is that those that fail such tests *cannot* be linear transformations. □

(d) The previous Example 3.6.4(c) illustrated that a linear transformation of the square seems to transform the unit square to a parallelogram: if a function transforms the unit square to something that is not a parallelogram, then the function cannot be a linear transformation. Analogously in higher dimensions: for example, if a function transforms the unit cube to something that is not a parallelepiped, then the function is not a linear transformation. Using this information, which of the following illustrated functions, $f : \mathbb{R}^3 \to \mathbb{R}^3$, cannot be a linear transformation? Each of these stereo illustrations, plot the unit cube in blue (with a 'roof' and 'door' to help orientate), and the transform of the unit cube in red (with its transformed 'roof' and 'door').

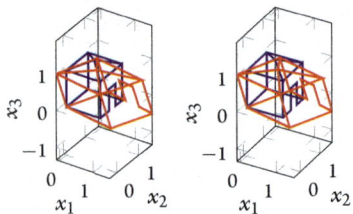

i.   This *may* be a linear transformation as the transform of the unit cube looks like a parallelepiped, with the origin fixed.

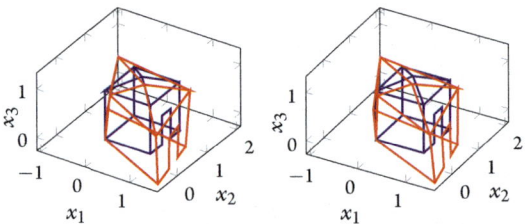

ii.   This *cannot* be a linear transformation as the unit cube transforms to something not a parallelepiped.

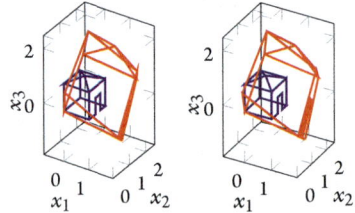

iii.   This *cannot* be a linear transformation as the unit cube transforms to something not a parallelepiped.

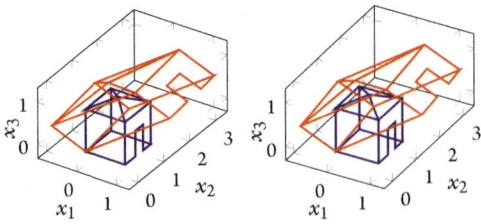

iv.   This *may* be a linear transformation as the transform of the unit cube looks like a parallelepiped, with the origin fixed.

Activity 3.6.5 Which of the following functions $f : \mathbb{R}^3 \to \mathbb{R}^2$ is *not* a linear transformation?

(a) $f(x,y,z) = (0,0)$

(b) $f(x,y,z) = (2.7x + 3y, 1 - 2z)$

(c) $f(x,y,z) = (y, x + z)$

(d) $f(x,y,z) = (0, 13x + \pi y)$

Example 3.6.6 For any given nonzero vector $w \in \mathbb{R}^n$, prove that the projection $P : \mathbb{R}^n \to \mathbb{R}^n$ defined by $P(u) := \mathrm{proj}_w(u)$ is a linear transformation (as a function of u). But, for any given nonzero vector $u \in \mathbb{R}^n$, prove that the projection $Q : \mathbb{R}^n \to \mathbb{R}^n$ defined by $Q(w) := \mathrm{proj}_w(u)$ is not a linear transformation (as a function of w).

Solution: For the function P consider the two properties of Definition 3.6.1.

3.6.1(a) for every $u, v \in \mathbb{R}^n$, from Definition 3.5.19 for a projection, $P(u + v) = \mathrm{proj}_w(u + v) = w[w \cdot (u + v)]/|w|^2 = w[(w \cdot u) + (w \cdot v)]/|w|^2 = w(w \cdot u)/|w|^2 + w(w \cdot v)/|w|^2 = \mathrm{proj}_w(u) + \mathrm{proj}_w(v) = P(u) + P(v)$;

3.6.1(b) for every $u \in \mathbb{R}^n$ and every scalar c, $P(cu) = \mathrm{proj}_w(cu) = w[w \cdot (cu)]/|w|^2 = w[c(w \cdot u)]/|w|^2 = c[w(w \cdot u)/|w|^2] = c\,\mathrm{proj}_w(u) = cP(u)$.

Hence, the projection P is a linear transformation.

Now consider $Q(w) = \mathrm{proj}_w(u)$. For any $u, w \in \mathbb{R}^n$ let's check $Q(2w) = \mathrm{proj}_{(2w)}(u) = (2w)[(2w) \cdot u]/|2w|^2 = 4w(w \cdot u)/(4|u|^2) = w(w \cdot u)/|u|^2 = \mathrm{proj}_w(u) = Q(w) \neq 2Q(w)$ and so the projection is not a linear transformation when considered as a function of the direction of the transform w for some given u. $\square$

3.6.1 Matrices correspond to linear transformations

One important class of linear transformations are the transformations that can be written as matrix multiplications. One reason for the importance is that Theorem 3.6.10 establishes that all linear transformations may be written as matrix multiplications! This in turn justifies why we define matrix multiplication to be as it is (Section 3.1.2): *matrix multiplication is defined just so that all linear transformations are encompassed.*

Example 3.6.7 But first, the following Theorem 3.6.8 proves, among many other possibilities, that the following transformations we have already met are linear transformations:

- stretching/shrinking along coordinate axes, because these are multiplication by a diagonal matrix (Section 3.2.2);

- rotations and/or reflections, because they arise as multiplications by an orthogonal matrix (Section 3.2.3);

- orthogonal projection onto a subspace, because all such projections may be expressed as multiplication by a matrix (the matrix WW^{T} in Theorem 3.5.29). $\square$

Theorem 3.6.8 *Let A be any given $m \times n$ matrix and define the transformation $T_A : \mathbb{R}^n \to \mathbb{R}^m$ by the matrix multiplication $T_A(\mathbf{x}) := A\mathbf{x}$ for every $\mathbf{x} \in \mathbb{R}^n$. Then T_A is a linear transformation.*

Proof. For every pair of vectors $\mathbf{u}, \mathbf{v} \in \mathbb{R}^n$ and every scalar $c \in \mathbb{R}$, consider the two properties of Definition 3.6.1.

3.6.1(a). By the distributivity of matrix-vector multiplication (Theorem 3.1.24), $T_A(\mathbf{u} + \mathbf{v}) = A(\mathbf{u} + \mathbf{v}) = A\mathbf{u} + A\mathbf{v} = T_A(\mathbf{u}) + T_A(\mathbf{v})$.

3.6.1(b). By commutativity of scalar multiplication (Theorem 3.1.26), $T_A(c\mathbf{u}) = A(c\mathbf{u}) = c(A\mathbf{u}) = cT_A(\mathbf{u})$.

Hence T_A is a linear transformation. $\qquad\square$

Example 3.6.9 Prove that a matrix multiplication with a nonzero shift $\mathbf{b}$, $S : \mathbb{R}^n \to \mathbb{R}^m$ where $S(\mathbf{x}) = A\mathbf{x} + \mathbf{b}$ for some given vector $\mathbf{b} \neq \mathbf{0}$, is not a linear transformation.

Solution: Just consider the addition property 3.6.1(a) for the zero vectors $\mathbf{u} = \mathbf{v} = \mathbf{0}$: on the one hand, $S(\mathbf{u} + \mathbf{v}) = S(\mathbf{0} + \mathbf{0}) = S(\mathbf{0}) = A\mathbf{0} + \mathbf{b} = \mathbf{b}$; on the other hand $S(\mathbf{u}) + S(\mathbf{v}) - S(\mathbf{0}) + S(\mathbf{0}) = A\mathbf{0} + \mathbf{b} + A\mathbf{0} + \mathbf{b} = 2\mathbf{b}$. Hence when the shift $\mathbf{b}$ is nonzero, there are vectors for which $S(\mathbf{u} + \mathbf{v}) \neq S(\mathbf{u}) + S(\mathbf{v})$ and so S is not a linear transformation. $\qquad\square$

Now let's establish the important converse to Theorem 3.6.8: that every linear transformation can be written as a matrix multiplication.

Theorem 3.6.10 *Let $T : \mathbb{R}^n \to \mathbb{R}^m$ be any linear transformation. Then T is the transformation corresponding to the $m \times n$ matrix*

$$A = \begin{bmatrix} T(\mathbf{e}_1) & T(\mathbf{e}_2) & \cdots & T(\mathbf{e}_n) \end{bmatrix}$$

*where $\mathbf{e}_j$ are the standard unit vectors in $\mathbb{R}^n$. This matrix A, often denoted $[T]$, is called the **standard matrix** of the linear transformation T.*

The matrix $[T]$ is called the *standard matrix* because it is defined in terms of the *standard* unit vectors $\mathbf{e}_1$, $\mathbf{e}_2$, ..., $\mathbf{e}_n$. Later, Theorem 7.2.35 begins to show how to use non-standard vectors to construct and interpret matrices corresponding a given linear transformation.

Proof. For every vector $\mathbf{x}$ in $\mathbb{R}^n$ we write $\mathbf{x} = (x_1, x_2, \ldots, x_n) = x_1\mathbf{e}_1 + x_2\mathbf{e}_2 + \cdots + x_n\mathbf{e}_n$ for standard unit vectors $\mathbf{e}_1$, $\mathbf{e}_2$, ..., $\mathbf{e}_n$. Then

$$
\begin{aligned}
T(\mathbf{x}) &= T(x_1\mathbf{e}_1 + x_2\mathbf{e}_2 + \cdots + x_n\mathbf{e}_n) \\
&= x_1 T(\mathbf{e}_1) + x_2 T(\mathbf{e}_2) + \cdots + x_n T(\mathbf{e}_n) \quad \text{(using identity of Exercise 3.6.6)} \\
&= \begin{bmatrix} T(\mathbf{e}_1) & T(\mathbf{e}_2) & \cdots & T(\mathbf{e}_n) \end{bmatrix} \begin{bmatrix} x_1 \\ x_2 \\ \vdots \\ x_n \end{bmatrix} = A\mathbf{x}
\end{aligned}
$$

for matrix A of the theorem. Since $T(e_1), T(e_2), \dots, T(e_n)$ are n (column) vectors in $\mathbb{R}^m$, the matrix A is $m \times n$.

Example 3.6.11

(a) Find the standard matrix of the linear transformation $T : \mathbb{R}^3 \to \mathbb{R}^4$ where $T(x, y, z) = (y, z, x, 3x - 2y + z)$.

Solution: We need to find the transform of the three standard unit vectors in $\mathbb{R}^3$:

$$T(e_1) = T(1,0,0) = (0,0,1,3);$$
$$T(e_2) = T(0,1,0) = (1,0,0,-2);$$
$$T(e_3) = T(0,0,1) = (0,1,0,1).$$

Form the standard matrix with these as its three columns, in order,

$$[T] = \begin{bmatrix} T(e_1) & T(e_2) & T(e_3) \end{bmatrix} = \begin{bmatrix} 0 & 1 & 0 \\ 0 & 0 & 1 \\ 1 & 0 & 0 \\ 3 & -2 & 1 \end{bmatrix}.$$

□

(b) Find the standard matrix of the rotation of the plane by $60°$ anticlockwise about the origin.

Solution: Denote the rotation of the plane by the function $R : \mathbb{R}^2 \to \mathbb{R}^2$. Since $60° = \frac{\pi}{3}$ then, as illustrated to the right,

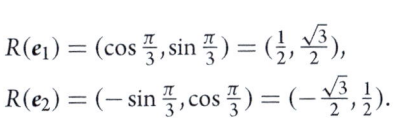

$$R(e_1) = (\cos \tfrac{\pi}{3}, \sin \tfrac{\pi}{3}) = (\tfrac{1}{2}, \tfrac{\sqrt{3}}{2}),$$

$$R(e_2) = (-\sin \tfrac{\pi}{3}, \cos \tfrac{\pi}{3}) = (-\tfrac{\sqrt{3}}{2}, \tfrac{1}{2}).$$

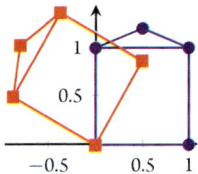

Form the standard matrix with these as its columns, in order,

$$[R] = \begin{bmatrix} R(e_1) & R(e_2) \end{bmatrix} = \begin{bmatrix} \tfrac{1}{2} & -\tfrac{\sqrt{3}}{2} \\ \tfrac{\sqrt{3}}{2} & \tfrac{1}{2} \end{bmatrix}.$$

□

(c) Find the standard matrix of the rotation about the point $(1,0)$ of the plane by $45°$ anticlockwise.

Solution: Since the origin $(0,0)$ is transformed by the rotation to $(1,-1)$ which is nonzero, this transformation cannot be of the form Ax, so cannot have a standard matrix, and hence is not a linear transformation. □

(d) Estimate the standard matrix for each of the illustrated transformations given that they transform the unit square as shown.

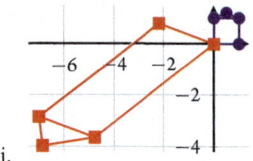

i.

Solution: Here $T(1,0) \approx (-2.2,0.8)$ and $T(0,1) \approx (-4.8,-3.6)$ so the approximate standard matrix is

$$\begin{bmatrix} -2.2 & -4.8 \\ 0.8 & -3.6 \end{bmatrix}.$$

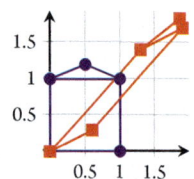

ii.

Solution: Here $T(1,0) \approx (0.6,0.3)$ and $T(0,1) \approx (1.3,1.4)$ so the approximate standard matrix is $\begin{bmatrix} 0.6 & 1.3 \\ 0.3 & 1.4 \end{bmatrix}$.

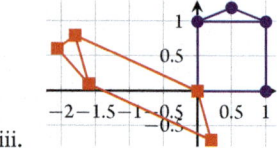

iii.

Solution: Here $T(1,0) \approx (0.2,-0.7)$ and $T(0,1) \approx (-1.8,0.8)$ so the approximate standard matrix is

$$\begin{bmatrix} 0.2 & -1.8 \\ -0.7 & 0.8 \end{bmatrix}.$$

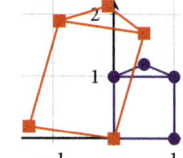

iv.

Solution: Here $T(1,0) \approx (-1.4,0.2)$ and $T(0,1) \approx (0.5,1.7)$ so the approximate standard matrix is

$$\begin{bmatrix} -1.4 & 0.5 \\ 0.2 & 1.7 \end{bmatrix}.$$

Activity 3.6.12 Which of the following is the standard matrix for the transformation $T(x,y,z) = (4.5y - 1.6z, 1.9x - 2z)$?

(a) $\begin{bmatrix} 0 & 1.9 \\ 4.5 & 0 \\ -1.6 & -2 \end{bmatrix}$
(b) $\begin{bmatrix} 4.5 & -1.6 \\ 1.9 & -2 \end{bmatrix}$
(c) $\begin{bmatrix} 0 & 4.5 & -1.6 \\ 1.9 & 0 & -2 \end{bmatrix}$
(d) $\begin{bmatrix} 4.5 & 1.9 \\ -1.6 & -2 \end{bmatrix}$

Example 3.6.13 For a fixed scalar a, let the function $H : \mathbb{R}^n \to \mathbb{R}^n$ be $H(u) = au$. Show that H is a linear transformation, and then find its standard matrix.

Solution: Let $u, v \in \mathbb{R}^n$ and c be any scalar. Function H is a linear transformation because

- $H(u+v) = a(u+v) = au + av = H(u) + H(v)$, and

- $H(cu) = a(cu) = (ac)u = (ca)u = c(au) = cH(u)$.

To find the standard matrix consider

$$H(e_1) = ae_1 = (a,0,0,\ldots,0),$$
$$H(e_2) = ae_2 = (0,a,0,\ldots,0),$$
$$\vdots$$
$$H(e_n) = ae_n = (0,0,\ldots,0,a).$$

Hence the standard matrix

$$[H] = \begin{bmatrix} H(e_1) & H(e_2) & \cdots & H(e_n) \end{bmatrix} = \begin{bmatrix} a & 0 & \cdots & 0 \\ 0 & a & & 0 \\ \vdots & & \ddots & \vdots \\ 0 & 0 & \cdots & a \end{bmatrix} = \mathrm{diag}(a,a,\ldots,a) = aI_n.$$

□

Consider this last Example 3.6.13 in the case $a = 1$: then $H(u) = u$ is the identity and so the example shows that the standard matrix of the identity transformation is I_n.

3.6.2 The pseudo-inverse of a matrix

This subsection is an optional extension.

In attempting to solve inconsistent linear equations, $Ax = b$ for some given A, Procedure 3.5.4 constructs a result x, a result that is meant to be a 'solution' of sorts, and that depends upon the right-hand side b. That is, any given b is transformed by the procedure to some result x: the result is a function of the given b. This section establishes that the result given by the procedure is a linear transformation of b, and hence there must be a matrix corresponding to the procedure. Let's denote this matrix by A^+. This matrix gives the result $x = A^+b$. We call the matrix A^+ the pseudo-inverse of A.

Example 3.6.14 Find the pseudo-inverse of the matrix $A = \begin{bmatrix} 3 \\ 4 \end{bmatrix}$.

Solution: Apply Procedure 3.5.4 to 'best' solve $Ax = b$ for every right-hand side $b = (b_1,b_2)$. (I use symbol x rather than $\boldsymbol{x}$ because here the unknown has only one component and hence is a scalar x.)

(a) This matrix has an SVD

$$A = \begin{bmatrix} 3 \\ 4 \end{bmatrix} = \begin{bmatrix} \frac{3}{5} & -\frac{4}{5} \\ \frac{4}{5} & \frac{3}{5} \end{bmatrix} \begin{bmatrix} 5 \\ 0 \end{bmatrix} [1]^{\mathrm{T}} = USV^{\mathrm{T}}.$$

(b) Hence $z = U^T b = \begin{bmatrix} \frac{3}{5}b_1 + \frac{4}{5}b_2 \\ -\frac{4}{5}b_1 + \frac{3}{5}b_2 \end{bmatrix}$.

(c) Then the diagonal system $Sy = z$ is $\begin{bmatrix} 5 \\ 0 \end{bmatrix} y = \begin{bmatrix} \frac{3}{5}b_1 + \frac{4}{5}b_2 \\ -\frac{4}{5}b_1 + \frac{3}{5}b_2 \end{bmatrix}$. Approximately solve this system by neglecting the second component in the equations, and so from the first component just set $y = \frac{3}{25}b_1 + \frac{4}{25}b_2$.

(d) Then the procedure's result is $x = Vy = 1(\frac{3}{25}b_1 + \frac{4}{25}b_2) = \frac{3}{25}b_1 + \frac{4}{25}b_2$.

That is, for all right-hand side vectors b, this result, the least square approximate solution, is

$$x = A^+ b \quad \text{for pseudo-inverse } A^+ = \begin{bmatrix} \frac{3}{25} & \frac{4}{25} \end{bmatrix}.$$

□

Activity 3.6.15 By finding the smallest magnitude, least square, solution to $Dx = b$ for matrix $D = \begin{bmatrix} 5 & 0 \\ 0 & 0 \end{bmatrix}$ and arbitrary b, determine that the pseudo-inverse of the diagonal matrix D is which of the following?

(a) $\begin{bmatrix} 0 & 0.2 \\ 0 & 0 \end{bmatrix}$ (b) $\begin{bmatrix} 0 & 0 \\ 0.2 & 0 \end{bmatrix}$ (c) $\begin{bmatrix} 0 & 0 \\ 0 & 0.2 \end{bmatrix}$ (d) $\begin{bmatrix} 0.2 & 0 \\ 0 & 0 \end{bmatrix}$

A pseudo-inverse A^+ of a non-invertible matrix A is only an 'inverse' because the pseudo-inverse builds in extra information that *you may sometimes choose to be desirable*. This extra information rationalizes all the contradictions encountered in trying to construct an inverse of a non-invertible matrix. Namely, for some applications *we choose* to desire that the pseudo-inverse solves the *nearest* consistent system to the one specified, and *we choose* the smallest of all possibilities then allowed. However, although there are many situations where these choices are useful, beware that there are also many situations where such choices are not appropriate. That is, although sometimes the pseudo-inverse is useful, beware that many times the pseudo-inverse is not appropriate.

Theorem 3.6.16 (*pseudo-inverse*) *In the context of a system of linear equations $Ax = b$ with $m \times n$ matrix A and vector $b \in \mathbb{R}^m$, recall that Procedure 3.5.4 finds the smallest solution $x \in \mathbb{R}^n$ (Theorem 3.5.13) to the closest consistent system $Ax = b'$ (Theorem 3.5.8). Procedure 3.5.4 forms a linear transformation $T : \mathbb{R}^m \to \mathbb{R}^n$, $x = T(b)$. This linear transformation has an $n \times m$ standard matrix A^+ called the **pseudo-inverse**, or **Moore–Penrose inverse**, of matrix A.*

Proof. First, for each right-hand side vector b in $\mathbb{R}^m$, the procedure gives a result x in $\mathbb{R}^n$ and so is some function $T : \mathbb{R}^m \to \mathbb{R}^n$. We proceed to confirm that Procedure 3.5.4 satisfies the two defining properties of a linear transformation (Definition 3.6.1). For each of any two right-hand side vectors $b^\dagger, b^\ddagger \in \mathbb{R}^m$ (b dagger and double-dagger, respectively) let Procedure 3.5.4 generate similarly daggered intermediaries $(z^\dagger, z^\ddagger, y^\dagger, y^\ddagger)$ through to two corresponding least square solutions $x^\dagger, x^\ddagger \in \mathbb{R}^n$. That is, $x^\dagger = T(b^\dagger)$ and $x^\ddagger = T(b^\ddagger)$. Throughout, let the matrix A have SVD $A = USV^\mathsf{T}$ and set $r = \operatorname{rank} A$.

3.6.1(a) To check $T(b^\dagger + b^\ddagger) = T(b^\dagger) + T(b^\ddagger)$, apply the procedure when the right-hand side is $b = b^\dagger + b^\ddagger$:

2. solve $Uz = b$ by $z = U^\mathsf{T} b = U^\mathsf{T}(b^\dagger + b^\ddagger) = U^\mathsf{T} b^\dagger + U^\mathsf{T} b^\ddagger = z^\dagger + z^\ddagger$;

3.
  - for $i = 1, \ldots, r$ set $y_i = z_i/\sigma_i = (z_i^\dagger + z_i^\ddagger)/\sigma_i = z_i^\dagger/\sigma_i + z_i^\ddagger/\sigma_i = y_i^\dagger + y_i^\ddagger$, and
  - for $i = r+1, \ldots, n$ set the free variables $y_i = 0 = 0 + 0 = y_i^\dagger + y_i^\ddagger$ to obtain the smallest solution (Theorem 3.5.13),

  and hence $y = y^\dagger + y^\ddagger$;

4. solve $V^\mathsf{T} x = y$ with $x = Vy = V(y^\dagger + y^\ddagger) = Vy^\dagger + Vy^\ddagger = x^\dagger + x^\ddagger$.

Since the result $x = x^\dagger + x^\ddagger$, thus $T(b^\dagger + b^\ddagger) = T(b^\dagger) + T(b^\ddagger)$.

3.6.1(b) To check $T(cb^\dagger) = cT(b^\dagger)$ for any scalar c, apply the procedure when the right-hand side is $b = cb^\dagger$:

2. solve $Uz = b$ by $z = U^\mathsf{T} b = U^\mathsf{T}(cb^\dagger) = cU^\mathsf{T} b^\dagger = cz^\dagger$;

3.
  - for $i = 1, \ldots, r$ set $y_i = z_i/\sigma_i = (cz_i^\dagger)/\sigma_i = c(z_i^\dagger/\sigma_i) = cy_i^\dagger$, and
  - for $i = r+1, \ldots, n$ set the free variables $y_i = 0 = c0 = cy_i^\dagger$ to obtain the smallest solution (Theorem 3.5.13),

  and hence $y = cy^\dagger$;

4. solve $V^\mathsf{T} x = y$ with $x = Vy = V(cy^\dagger) = cVy^\dagger = cx^\dagger$.

Since the result $x = cx^\dagger$, consequently $T(cb^\dagger) = cT(b^\dagger)$.

Since Procedure 3.5.4, denoted by T, is a linear transformation $T : \mathbb{R}^m \to \mathbb{R}^n$, Theorem 3.6.10 assures us that T has a corresponding $n \times m$ standard matrix which we denote A^+ and which we call the pseudo-inverse. □

Example 3.6.17 Find the pseudo-inverse of the matrix $A = \begin{bmatrix} 5 & 12 \end{bmatrix}$.

Solution: Apply Procedure 3.5.4 to solve $Ax = b$ for any right-hand side b.

(a) This matrix has an SVD

$$A = \begin{bmatrix} 5 & 12 \end{bmatrix} = \begin{bmatrix} 1 \end{bmatrix} \begin{bmatrix} 13 & 0 \end{bmatrix} \begin{bmatrix} \frac{5}{13} & -\frac{12}{13} \\ \frac{12}{13} & \frac{5}{13} \end{bmatrix}^{\mathsf{T}} = USV^\mathsf{T}.$$

(b) Hence $z = U^\mathsf{T} b = 1b = b$.

(c) The diagonal system $Sy = z$ becomes $\begin{bmatrix} 13 & 0 \end{bmatrix} y = b$ with general solution $y = (b/13, y_2)$. The smallest of these solutions is $y = (b/13, 0)$.

(d) Then the procedure's result is

$$x = Vy = \begin{bmatrix} \frac{5}{13} & -\frac{12}{13} \\ \frac{12}{13} & \frac{5}{13} \end{bmatrix} \begin{bmatrix} b/13 \\ 0 \end{bmatrix} = \begin{bmatrix} \frac{5}{169}b \\ -\frac{12}{169}b \end{bmatrix}.$$

That is, for every right-hand side b, this procedure's result is

$$x = A^+ b \quad \text{for pseudo-inverse } A^+ = \begin{bmatrix} \frac{5}{169} \\ -\frac{12}{169} \end{bmatrix}.$$

☐

Activity 3.6.18 Following the steps of Procedure 3.5.4, find the pseudo-inverse of the matrix $\begin{bmatrix} 1 & 1 \\ 2 & 2 \end{bmatrix}$ given that this matrix has the SVD

$$\begin{bmatrix} 1 & 1 \\ 2 & 2 \end{bmatrix} = \begin{bmatrix} \frac{1}{\sqrt{5}} & -\frac{2}{\sqrt{5}} \\ \frac{2}{\sqrt{5}} & \frac{1}{\sqrt{5}} \end{bmatrix} \begin{bmatrix} 10 & 0 \\ 0 & 0 \end{bmatrix} \begin{bmatrix} \frac{1}{\sqrt{2}} & -\frac{1}{\sqrt{2}} \\ \frac{1}{\sqrt{2}} & \frac{1}{\sqrt{2}} \end{bmatrix}^{\mathrm{T}}.$$

The pseudo-inverse is which of these?

(a) $\begin{bmatrix} 0.1 & 0.2 \\ 0.1 & 0.2 \end{bmatrix}$ (b) $\begin{bmatrix} 0.1 & 0.1 \\ 0.2 & 0.2 \end{bmatrix}$ (c) $\begin{bmatrix} 0.1 & -0.1 \\ -0.2 & 0.2 \end{bmatrix}$ (d) $\begin{bmatrix} 0.1 & -0.2 \\ -0.1 & 0.2 \end{bmatrix}$

Example 3.6.19 Recall that Example 3.5.1 explored how to best determine a weight from four apparently contradictory measurements. The exploration showed that Procedure 3.5.4 agrees with the traditional method of simple averaging. Let's see that the pseudo-inverse implements the simple average of the four measurements.

Recall that Example 3.5.1 sought to solve an inconsistent system $Ax = b$, specifically

$$\begin{bmatrix} 1 \\ 1 \\ 1 \\ 1 \end{bmatrix} x = \begin{bmatrix} 84.8 \\ 84.1 \\ 84.7 \\ 84.4 \end{bmatrix}.$$

To find the pseudo-inverse of the left-hand side matrix A, seek to solve the system for arbitrary right-hand side b.

(a) As used previously, this matrix A of ones has an SVD of

$$A = \begin{bmatrix} 1 \\ 1 \\ 1 \\ 1 \end{bmatrix} = \begin{bmatrix} \frac{1}{2} & \frac{1}{2} & \frac{1}{2} & \frac{1}{2} \\ \frac{1}{2} & \frac{1}{2} & -\frac{1}{2} & -\frac{1}{2} \\ \frac{1}{2} & -\frac{1}{2} & -\frac{1}{2} & \frac{1}{2} \\ \frac{1}{2} & -\frac{1}{2} & \frac{1}{2} & -\frac{1}{2} \end{bmatrix} \begin{bmatrix} 2 \\ 0 \\ 0 \\ 0 \end{bmatrix} [1]^{\mathrm{T}} = USV^{\mathrm{T}}.$$

(b) Solve $Uz = b$ by computing

$$z = U^{\mathrm{T}}b = \begin{bmatrix} \frac{1}{2} & \frac{1}{2} & \frac{1}{2} & \frac{1}{2} \\ \frac{1}{2} & \frac{1}{2} & -\frac{1}{2} & -\frac{1}{2} \\ \frac{1}{2} & -\frac{1}{2} & -\frac{1}{2} & \frac{1}{2} \\ \frac{1}{2} & -\frac{1}{2} & \frac{1}{2} & -\frac{1}{2} \end{bmatrix} b = \begin{bmatrix} \frac{1}{2}b_1 + \frac{1}{2}b_2 + \frac{1}{2}b_3 + \frac{1}{2}b_4 \\ \frac{1}{2}b_1 + \frac{1}{2}b_2 - \frac{1}{2}b_3 - \frac{1}{2}b_4 \\ \frac{1}{2}b_1 - \frac{1}{2}b_2 - \frac{1}{2}b_3 + \frac{1}{2}b_4 \\ \frac{1}{2}b_1 - \frac{1}{2}b_2 + \frac{1}{2}b_3 - \frac{1}{2}b_4 \end{bmatrix}.$$

(c) Now try to solve $Sy = z$, that is,

$$\begin{bmatrix} 2 \\ 0 \\ 0 \\ 0 \end{bmatrix} y = \begin{bmatrix} \frac{1}{2}b_1 + \frac{1}{2}b_2 + \frac{1}{2}b_3 + \frac{1}{2}b_4 \\ \frac{1}{2}b_1 + \frac{1}{2}b_2 - \frac{1}{2}b_3 - \frac{1}{2}b_4 \\ \frac{1}{2}b_1 - \frac{1}{2}b_2 - \frac{1}{2}b_3 + \frac{1}{2}b_4 \\ \frac{1}{2}b_1 - \frac{1}{2}b_2 + \frac{1}{2}b_3 - \frac{1}{2}b_4 \end{bmatrix}.$$

Instead of seeking an *exact* solution, we *must* adjust the last three components to zero. Hence we find a solution to a slightly different problem by solving

$$\begin{bmatrix} 2 \\ 0 \\ 0 \\ 0 \end{bmatrix} y = \begin{bmatrix} \frac{1}{2}b_1 + \frac{1}{2}b_2 + \frac{1}{2}b_3 + \frac{1}{2}b_4 \\ 0 \\ 0 \\ 0 \end{bmatrix},$$

with solution $y = \frac{1}{4}b_1 + \frac{1}{4}b_2 + \frac{1}{4}b_3 + \frac{1}{4}b_4$.

(d) Lastly, solve $V^{\mathrm{T}}x = y$ by computing

$$x = Vy = 1y = \frac{1}{4}b_1 + \frac{1}{4}b_2 + \frac{1}{4}b_3 + \frac{1}{4}b_4 = \begin{bmatrix} \frac{1}{4} & \frac{1}{4} & \frac{1}{4} & \frac{1}{4} \end{bmatrix} b.$$

Hence the pseudo-inverse of matrix A is $A^+ = \begin{bmatrix} \frac{1}{4} & \frac{1}{4} & \frac{1}{4} & \frac{1}{4} \end{bmatrix}$. Multiplication by this pseudo-inverse implements the traditional answer of averaging the four measurements. $\square$

Example 3.6.20 Recall that Example 3.5.3 rates three table tennis players, Anne, Bob, and Chris. The rating involved solving the inconsistent system $Ax = b$ for the particular matrix and vector

$$\begin{bmatrix} 1 & -1 & 0 \\ 1 & 0 & -1 \\ 0 & 1 & -1 \end{bmatrix} x = \begin{bmatrix} 1 \\ 2 \\ 2 \end{bmatrix}.$$

Find the pseudo-inverse of this matrix A. Use the pseudo-inverse to rate the players in the cases of Examples 3.5.3 and 3.3.12.

Solution: To find the pseudo-inverse, follow Procedure 3.5.4 with a general right-hand side vector b.

(a) Compute an SVD $A = USV^T$ in MATLAB/Octave via $[U,S,V] = svd(A)$:

```
U =
      0.4082   -0.7071    0.5774
     -0.4082   -0.7071   -0.5774
     -0.8165   -0.0000    0.5774
S =
      1.7321         0         0
           0    1.7321         0
           0         0    0.0000
V =
      0.0000   -0.8165    0.5774
     -0.7071    0.4082    0.5774
      0.7071    0.4082    0.5774
```

Upon recognizing various square-roots, these matrices are

$$U = \begin{bmatrix} \frac{1}{\sqrt{6}} & -\frac{1}{\sqrt{2}} & \frac{1}{\sqrt{3}} \\ -\frac{1}{\sqrt{6}} & -\frac{1}{\sqrt{2}} & -\frac{1}{\sqrt{3}} \\ -\frac{2}{\sqrt{6}} & 0 & \frac{1}{\sqrt{3}} \end{bmatrix}, \quad S = \begin{bmatrix} \sqrt{3} & 0 & 0 \\ 0 & \sqrt{3} & 0 \\ 0 & 0 & 0 \end{bmatrix}, \quad V = \begin{bmatrix} 0 & -\frac{2}{\sqrt{6}} & \frac{1}{\sqrt{3}} \\ -\frac{1}{\sqrt{2}} & \frac{1}{\sqrt{6}} & \frac{1}{\sqrt{3}} \\ \frac{1}{\sqrt{2}} & \frac{1}{\sqrt{6}} & \frac{1}{\sqrt{3}} \end{bmatrix}.$$

The system of equations for the ratings becomes

$$Ax = U \underbrace{S \overbrace{V^T x}^{=y}}_{=z} = b.$$

(b) As U is orthogonal, $Uz = b$ has unique solution

$$z = U^T b = \begin{bmatrix} \frac{1}{\sqrt{6}} & -\frac{1}{\sqrt{6}} & -\frac{2}{\sqrt{6}} \\ -\frac{1}{\sqrt{2}} & -\frac{1}{\sqrt{2}} & 0 \\ \frac{1}{\sqrt{3}} & -\frac{1}{\sqrt{3}} & \frac{1}{\sqrt{3}} \end{bmatrix} b.$$

(c) Now solve $Sy = z$. But S has a troublesome zero on the diagonal. So interpret the equation $Sy = z$ in detail as

$$\begin{bmatrix} \sqrt{3} & 0 & 0 \\ 0 & \sqrt{3} & 0 \\ 0 & 0 & 0 \end{bmatrix} y = \begin{bmatrix} \frac{1}{\sqrt{6}} & -\frac{1}{\sqrt{6}} & -\frac{2}{\sqrt{6}} \\ -\frac{1}{\sqrt{2}} & -\frac{1}{\sqrt{2}} & 0 \\ \frac{1}{\sqrt{3}} & -\frac{1}{\sqrt{3}} & \frac{1}{\sqrt{3}} \end{bmatrix} b :$$

i. the first line requires $y_1 = \frac{1}{\sqrt{3}} \begin{bmatrix} \frac{1}{\sqrt{6}} & -\frac{1}{\sqrt{6}} & -\frac{2}{\sqrt{6}} \end{bmatrix} b = \frac{1}{3} \begin{bmatrix} \frac{1}{\sqrt{2}} & -\frac{1}{\sqrt{2}} & -\sqrt{2} \end{bmatrix} b;$

ii. the second line requires $y_2 = \frac{1}{\sqrt{3}} \begin{bmatrix} -\frac{1}{\sqrt{2}} & -\frac{1}{\sqrt{2}} & 0 \end{bmatrix} b = \begin{bmatrix} -\frac{1}{\sqrt{6}} & -\frac{1}{\sqrt{6}} & 0 \end{bmatrix} b;$

iii. the third line requires $0 y_3 = \begin{bmatrix} \frac{1}{\sqrt{3}} & -\frac{1}{\sqrt{3}} & \frac{1}{\sqrt{3}} \end{bmatrix} b$ which generally cannot be satisfied, so we set $y_3 = 0$ to get the *smallest solution of the system after projecting b onto the column space of A*.

(d) Finally, as V is orthogonal, $V^T x = y$ has the solution $x = Vy$ (unique for each valid y):

$$x = Vy = \begin{bmatrix} 0 & -\frac{2}{\sqrt{6}} & \frac{1}{\sqrt{3}} \\ -\frac{1}{\sqrt{2}} & \frac{1}{\sqrt{6}} & \frac{1}{\sqrt{3}} \\ \frac{1}{\sqrt{2}} & \frac{1}{\sqrt{6}} & \frac{1}{\sqrt{3}} \end{bmatrix} \begin{bmatrix} \frac{1}{3\sqrt{2}} & -\frac{1}{3\sqrt{2}} & -\frac{\sqrt{2}}{3} \\ -\frac{1}{\sqrt{6}} & -\frac{1}{\sqrt{6}} & 0 \\ 0 & 0 & 0 \end{bmatrix} b = \frac{1}{3} \begin{bmatrix} 1 & 1 & 0 \\ -1 & 0 & 1 \\ 0 & -1 & -1 \end{bmatrix} b$$

Hence the pseudo-inverse of A is

$$A^+ = \frac{1}{3} \begin{bmatrix} 1 & 1 & 0 \\ -1 & 0 & 1 \\ 0 & -1 & -1 \end{bmatrix}.$$

- In Example 3.3.12, Anne beat Bob 3-2 games; Anne beat Chris 3-1; Bob beat Chris 3-2 so the right-hand side vector is $b = (1, 2, 1)$. The procedure's ratings are then, as before,

$$x = A^+ b = \frac{1}{3} \begin{bmatrix} 1 & 1 & 0 \\ -1 & 0 & 1 \\ 0 & -1 & -1 \end{bmatrix} \begin{bmatrix} 1 \\ 2 \\ 1 \end{bmatrix} = \begin{bmatrix} 1 \\ 0 \\ -1 \end{bmatrix}.$$

- In Example 3.5.3, Bob instead beat Chris 3-1 so the right-hand side vector is $\boldsymbol{b} = (1,2,2)$. The procedure's ratings are then, as before,

$$\boldsymbol{x} = A^+\boldsymbol{b} = \frac{1}{3}\begin{bmatrix} 1 & 1 & 0 \\ -1 & 0 & 1 \\ 0 & -1 & -1 \end{bmatrix}\begin{bmatrix} 1 \\ 2 \\ 2 \end{bmatrix} = \begin{bmatrix} 1 \\ \frac{1}{3} \\ -\frac{4}{3} \end{bmatrix}.$$

□

In some common special cases there are alternative formulas for the pseudo-inverse: specifically, the cases are when the rank of the matrix is the same as the number of rows and/or columns.

Example 3.6.21 For the 2×1 matrix $A = \begin{bmatrix} 3 \\ 4 \end{bmatrix}$, confirm that $(A^{\mathsf{T}}A)^{-1}A^{\mathsf{T}}$ is the pseudo-inverse that was found in Example 3.6.14.

Solution: Here $A^{\mathsf{T}}A = 3^2 + 4^2 = 25$, so its inverse is $(A^{\mathsf{T}}A)^{-1} = 1/25$. Then, $(A^{\mathsf{T}}A)^{-1}A^{\mathsf{T}} = \frac{1}{25}\begin{bmatrix} 3 & 4 \end{bmatrix} = \begin{bmatrix} \frac{3}{25} & \frac{4}{25} \end{bmatrix}$ as found in Example 3.6.14 for the pseudo-inverse. □

Theorem 3.6.22 *For every $m \times n$ matrix A with $\operatorname{rank}A = n$ (so $m \geq n$), the pseudo-inverse $A^+ = (A^{\mathsf{T}}A)^{-1}A^{\mathsf{T}}$.*

Proof. Apply Procedure 3.5.4 to the system $A\boldsymbol{x} = \boldsymbol{b}$ for an arbitrary $\boldsymbol{b} \in \mathbb{R}^m$.

1. Let $m \times n$ matrix A have svd $A = USV^{\mathsf{T}}$. Since $\operatorname{rank}A = n$ there are n nonzero singular values on the diagonal of $m \times n$ matrix S, and so $m \geq n$.

2. Solve $U\boldsymbol{z} = \boldsymbol{b}$ with $\boldsymbol{z} = U^{\mathsf{T}}\boldsymbol{b} \in \mathbb{R}^m$.

3. Approximately solve $S\boldsymbol{y} = \boldsymbol{z}$ by setting $y_i = z_i/\sigma_i$ for $i = 1, \ldots, n$ and neglecting the last $(m - n)$ equations. This is identical to setting $\boldsymbol{y} = S^+\boldsymbol{z} = S^+U^{\mathsf{T}}\boldsymbol{b}$ after defining the $n \times m$ matrix

$$S^+ := \begin{bmatrix} 1/\sigma_1 & 0 & \cdots & 0 & 0 & \cdots & 0 \\ 0 & 1/\sigma_2 & & 0 & 0 & \cdots & 0 \\ \vdots & & \ddots & \vdots & \vdots & \cdots & \vdots \\ 0 & 0 & \cdots & 1/\sigma_n & 0 & \cdots & 0 \end{bmatrix}.$$

4. Solve $V^{\mathsf{T}}\boldsymbol{x} = \boldsymbol{y}$ with $\boldsymbol{x} = V\boldsymbol{y} = VS^+U^{\mathsf{T}}\boldsymbol{b}$.

Hence the pseudo-inverse is $A^+ = VS^+U^{\mathsf{T}}$.
Let's find that $(A^{\mathsf{T}}A)^{-1}A^{\mathsf{T}}$ is the same expression. First, since $A^{\mathsf{T}} = (USV^{\mathsf{T}})^{\mathsf{T}} = VS^{\mathsf{T}}U^{\mathsf{T}}$,

$$A^{\mathsf{T}}A = VS^{\mathsf{T}}U^{\mathsf{T}}USV^{\mathsf{T}} = VS^{\mathsf{T}}SV^{\mathsf{T}} = V(S^{\mathsf{T}}S)V^{\mathsf{T}}$$

where the $n \times n$ matrix $(S^{\mathsf{T}}S) = \operatorname{diag}(\sigma_1^2, \sigma_2^2, \ldots, \sigma_n^2)$. Since $\operatorname{rank}A = n$, the singular values $\sigma_1, \sigma_2, \ldots, \sigma_n > 0$ and so matrix $(S^{\mathsf{T}}S)$ is invertible as it is square and diagonal with all nonzero elements in the diagonal, and the inverse is $(S^T S)^{-1} = \text{diag}(1/\sigma_1^2, 1/\sigma_2^2, \ldots, 1/\sigma_n^2)$ (Theorem 3.2.27). Second, the $n \times n$ matrix $(A^T A)$ is invertible as it has SVD $V(S^T S)V^T$ with n nonzero singular values (Theorem 3.3.27(d)), and so

$$
\begin{aligned}
(A^T A)^{-1} A^T &= (V(S^T S)V^T)^{-1} V S^T U^T \\
&= (V^T)^{-1}(S^T S)^{-1} V^{-1} V S^T U^T \\
&= V(S^T S)^{-1} V^T V S^T U^T \\
&= V(S^T S)^{-1} S^T U^T \\
&= V S^+ U^T,
\end{aligned}
$$

where the last equality follows because

$$
\begin{aligned}
(S^T S)^{-1} S^T &= \text{diag}(1/\sigma_1^2, 1/\sigma_2^2, \ldots, 1/\sigma_n^2) \, \text{diag}_{n \times m}(\sigma_1, \sigma_2, \ldots, \sigma_n) \\
&= \text{diag}_{n \times m}(1/\sigma_1, 1/\sigma_2, \ldots, 1/\sigma_n) = S^+.
\end{aligned}
$$

Hence the pseudo-inverse $A^+ = V S^+ U^T = (A^T A)^{-1} A^T$.

Theorem 3.6.23 *For every invertible matrix A, the pseudo-inverse $A^+ = A^{-1}$, the inverse.*

Proof. If A is invertible it must be square, say $n \times n$, and of rank $A = n$ (Theorem 3.3.27(f)). Further, A^T is invertible with inverse $(A^{-1})^T$ (Theorem 3.2.13(d)). Then the expression from Theorem 3.6.22 for the pseudo-inverse gives

$$
A^+ = (A^T A)^{-1} A^T = A^{-1}(A^T)^{-1} A^T = A^{-1} I_n = A^{-1}.
$$

Computer considerations Except for easy cases, we (almost) never explicitly compute the pseudo-inverse of a matrix.[31] In practical computation, forming $A^T A$ and then manipulating it is both expensive and error enhancing: for example, $\text{cond}(A^T A) = (\text{cond} A)^2$ so matrix $A^T A$ typically has a much worse condition number than matrix A. Computationally there are (almost) always better ways to proceed, such as Procedure 3.5.4. As with an inverse, a pseudo-inverse is a theoretical device, rarely a practical tool.

A major point of this subsection is to illustrate how a complicated procedure is conceptually expressible as a linear transformation, and so has associated matrix properties such as being equivalent to multiplication by some matrix—here the pseudo-inverse.

3.6.3 Function composition connects to matrix inverse

To achieve a complex goal we typically decompose the task of attaining the complex goal into a set of smaller simpler tasks and achieve those tasks one after another. The analogy in linear algebra is that we often apply linear transformations one after another to build up or solve a complex

---

[31] However, to check pseudo-inverses of small matrices in examples or exercises one may use the MATLAB/Octave function `pinv()`.

problem. This section certifies how applying a sequence of linear transformations is equivalent to one grand overall linear transformation.

Example 3.6.24 *(simple rotation)* Recall Example 3.6.11(b) on rotation by $60°$ (anticlockwise as positive, as illustrated to the right) with its standard matrix

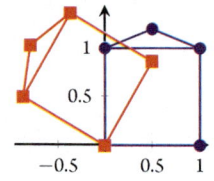

$$[R] = \begin{bmatrix} \frac{1}{2} & -\frac{\sqrt{3}}{2} \\ \frac{\sqrt{3}}{2} & \frac{1}{2} \end{bmatrix}.$$

Consider two successive rotations by $60°$: show that the standard matrix of the resultant rotation by $120°$ is the same as the matrix product $[R][R]$.

Solution: On the one hand, rotation by $120° = 2\pi/3$, call it S, transforms the unit vectors as (illustrated to the right)

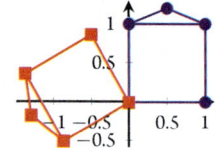

$$S(i) = (\cos\tfrac{2\pi}{3}, \sin\tfrac{2\pi}{3}) = (-\tfrac{1}{2}, \tfrac{\sqrt{3}}{2}),$$

$$S(j) = (-\sin\tfrac{2\pi}{3}, \cos\tfrac{2\pi}{3}) = (-\tfrac{\sqrt{3}}{2}, -\tfrac{1}{2}).$$

Form the standard matrix with these as its columns, in order,

$$[S] = \begin{bmatrix} S(i) & S(j) \end{bmatrix} = \begin{bmatrix} -\frac{1}{2} & -\frac{\sqrt{3}}{2} \\ \frac{\sqrt{3}}{2} & -\frac{1}{2} \end{bmatrix}.$$

On the other hand, the matrix multiplication

$$[R][R] = \begin{bmatrix} \frac{1}{2} & -\frac{\sqrt{3}}{2} \\ \frac{\sqrt{3}}{2} & \frac{1}{2} \end{bmatrix}\begin{bmatrix} \frac{1}{2} & -\frac{\sqrt{3}}{2} \\ \frac{\sqrt{3}}{2} & \frac{1}{2} \end{bmatrix} = \begin{bmatrix} \frac{1}{4} - \frac{3}{4} & -\frac{\sqrt{3}}{4} - \frac{\sqrt{3}}{4} \\ \frac{\sqrt{3}}{4} + \frac{\sqrt{3}}{4} & -\frac{3}{4} + \frac{1}{4} \end{bmatrix} = \begin{bmatrix} -\frac{1}{2} & -\frac{\sqrt{3}}{2} \\ \frac{\sqrt{3}}{2} & -\frac{1}{2} \end{bmatrix} = [S].$$

That is, multiplying the two matrices is equivalent to performing the two rotations in succession: the next theorem confirms this is generally true. ☐

Theorem 3.6.25 Let $T : \mathbb{R}^n \to \mathbb{R}^m$ and $S : \mathbb{R}^m \to \mathbb{R}^p$ be any linear transformations. Recall that the **composition** of functions is $(S \circ T)(v) = S(T(v))$. Then $S \circ T : \mathbb{R}^n \to \mathbb{R}^p$ is a linear transformation with standard matrix $[S \circ T] = [S][T]$.

Proof. For given transformations S and T, let matrix $A = [S]$ ($p \times m$) and $B = [T]$ ($m \times n$). Then for every vector u in $\mathbb{R}^n$, $(S \circ T)(u) = S(T(u)) = S(Bu) = A(Bu) = (AB)u$ (using associativity, Theorem 3.1.26(c)). Hence the effect of $S \circ T$ is identical to multiplication by the $p \times n$ matrix (AB). It is thus a matrix transformation, which is consequently linear (Theorem 3.6.8), and its standard matrix $[S \circ T] = AB = [S][T]$.

Example 3.6.26 Consider two linear transformations: first, $T : \mathbb{R}^3 \to \mathbb{R}^2$ defined by $T(x_1, x_2, x_3) := (3x_1 + x_2, -x_2 - 7x_3)$; and second, the linear transformation $S : \mathbb{R}^2 \to \mathbb{R}^4$ defined by $S(y_1, y_2) = (-y_1, -3y_1 + 2y_2, 2y_1 - y_2, 2y_2)$. Find the standard matrix of the linear transformation $S \circ T$, and also that of $T \circ S$.

Solution: From the given formulas for the two given linear transformations we write down the standard matrices

$$[T] = \begin{bmatrix} 3 & 1 & 0 \\ 0 & -1 & -7 \end{bmatrix} \quad \text{and} \quad [S] = \begin{bmatrix} -1 & 0 \\ -3 & 2 \\ 2 & -1 \\ 0 & 2 \end{bmatrix}.$$

First, Theorem 3.6.25 assures us that the standard matrix of the composition

$$[S \circ T] = [S][T] = \begin{bmatrix} -1 & 0 \\ -3 & 2 \\ 2 & -1 \\ 0 & 2 \end{bmatrix} \begin{bmatrix} 3 & 1 & 0 \\ 0 & -1 & -7 \end{bmatrix} = \begin{bmatrix} -3 & -1 & 0 \\ -9 & -5 & -14 \\ 6 & 3 & 7 \\ 0 & -2 & -14 \end{bmatrix}.$$

However, second, the standard matrix of $T \circ S$ does not exist because it would require the multiplication of a 2×3 matrix by a 4×2 matrix, and such a multiplication is not defined. The failure is rooted earlier in the question because $S : \mathbb{R}^2 \to \mathbb{R}^4$ and $T : \mathbb{R}^3 \to \mathbb{R}^2$ so a result of S, which is in $\mathbb{R}^4$, cannot be used as an argument to T, which must be in $\mathbb{R}^3$; the lack of a defined multiplication is a direct reflection of this incompatibility in '$T \circ S$' which means $T \circ S$ cannot exist. □

Example 3.6.27 Find the standard matrix of the transformation of the plane that first rotates by $45°$ about the origin (anticlockwise as positive), and then second reflects in the vertical axis.

Solution: Two possible solutions are the following.

- Let R denote the rotation about the origin of the plane by $45°$, $R :$ $\mathbb{R}^2 \to \mathbb{R}^2$ (illustrated to the right). Its standard matrix is

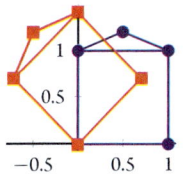

$$[R] = \begin{bmatrix} R(i) & R(j) \end{bmatrix} = \begin{bmatrix} \frac{1}{\sqrt{2}} & -\frac{1}{\sqrt{2}} \\ \frac{1}{\sqrt{2}} & \frac{1}{\sqrt{2}} \end{bmatrix}.$$

Let F denote the reflection in the vertical axis of the plane, $F :$ $\mathbb{R}^2 \to \mathbb{R}^2$ (illustrated to the right). Its standard matrix is

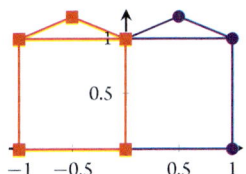

$$[F] = \begin{bmatrix} F(i) & F(j) \end{bmatrix} = \begin{bmatrix} -1 & 0 \\ 0 & 1 \end{bmatrix}.$$

Then the standard matrix of the composition

$$[F \circ R] = [F][R] = \begin{bmatrix} -1 & 0 \\ 0 & 1 \end{bmatrix} \begin{bmatrix} \frac{1}{\sqrt{2}} & -\frac{1}{\sqrt{2}} \\ \frac{1}{\sqrt{2}} & \frac{1}{\sqrt{2}} \end{bmatrix} = \begin{bmatrix} -\frac{1}{\sqrt{2}} & \frac{1}{\sqrt{2}} \\ \frac{1}{\sqrt{2}} & \frac{1}{\sqrt{2}} \end{bmatrix}$$

- Alternatively, just consider the action of the two component transformations on the standard unit vectors.

  - $(F \circ R)(i) = F(R(i))$ which first rotates $i = (1,0)$ to point to the top-right, then reflects in the vertical axis to point to the top-left and thus $(F \circ R)(i) = (-\frac{1}{\sqrt{2}}, \frac{1}{\sqrt{2}})$.

  - $(F \circ R)(j) = F(R(j))$ which first rotates $j = (0,1)$ to point to the top-left, then reflects in the vertical axis to point to the top-right and thus $(F \circ R)(j) = (\frac{1}{\sqrt{2}}, \frac{1}{\sqrt{2}})$.

  Then the standard matrix of the composition is (as illustrated)

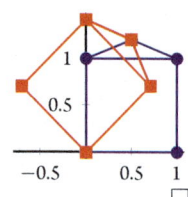

$$[F \circ R] = \left[(F \circ R)(i) \quad (F \circ R)(j) \right] = \begin{bmatrix} -\frac{1}{\sqrt{2}} & \frac{1}{\sqrt{2}} \\ \frac{1}{\sqrt{2}} & \frac{1}{\sqrt{2}} \end{bmatrix}$$

As an extension, check that although $R \circ F$ is defined, it is different to $F \circ R$; the difference is due to the non-commutativity of matrix multiplication (Section 3.1.3).

Activity 3.6.28 Given the stretching transformation S with standard matrix $[S] = \text{diag}(2, 1/2)$, and the anticlockwise rotation R by $90°$ with standard matrix $[R] = \begin{bmatrix} 0 & -1 \\ 1 & 0 \end{bmatrix}$, what is the standard matrix of the transformation composed of first the stretching and then the rotation?

(a) $\begin{bmatrix} 0 & -2 \\ \frac{1}{2} & 0 \end{bmatrix}$
(b) $\begin{bmatrix} 0 & 2 \\ -\frac{1}{2} & 0 \end{bmatrix}$
(c) $\begin{bmatrix} 0 & -\frac{1}{2} \\ 2 & 0 \end{bmatrix}$
(d) $\begin{bmatrix} 0 & \frac{1}{2} \\ -2 & 0 \end{bmatrix}$

Invert transformations Having introduced and characterized the composition of linear transformations, we now discuss when two transformations composed together end up 'cancelling' each other out.

Example 3.6.29 *(inverse transformations)*

(a) Let S be rotation of the plane by $60°$ (anticlockwise as positive), and T be rotation of the plane by $-60°$ (clockwise as negative). Then $S \circ T$ is first the $-60°$ rotation by T, and second the $60°$ rotation by S: the overall result is no change. Because $S \circ T$ is effectively the identity transformation, we call the rotations S and T the inverse transformation of each other.

(b) Let R be reflection of the plane in the line at $30°$ to the horizontal (illustrated to the right). Then $R \circ R$ is first reflection in the line at $30°$ by R, and second another reflection in the line at $30°$ by R: the overall result is no change. Because $R \circ R$ is effectively the identity transformation, the reflection R is its own inverse.

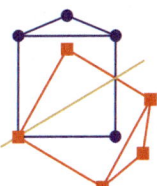

Definition 3.6.30   *Let S and T be linear transformations from $\mathbb{R}^n$ to $\mathbb{R}^n$ (the same dimension). If $S \circ T = T \circ S = I$, the identity transformation, then S and T are **inverse transformations** of each other. Further, we say S and T are **invertible**.*

Example 3.6.31   Let $S : \mathbb{R}^3 \to \mathbb{R}^3$ be rotation about the vertical axis by $120°$ (as illustrated in stereo above-right), and let $T : \mathbb{R}^3 \to \mathbb{R}^3$ be rotation about the vertical axis by $240°$ (below-right). Argue that $S \circ T = T \circ S = I$ is the identity and so S and T are inverse transformations of each other.

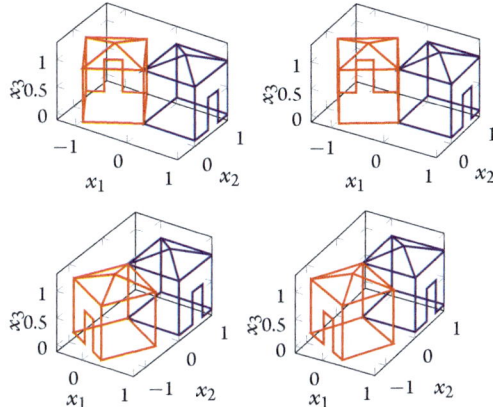

Solution: A basic argument is that rotation by $120°$ together with a rotation of $240°$ about the same axis, in either order, is the same as a rotation by $360°$ about the axis. But a $360°$ rotation leaves everything unchanged and so must be the identity.

Alternatively one could dress up the argument with some algebra as in the following. First consider $T \circ S$:

- the vertical unit vector e_3 is unchanged by both S and T so $(T \circ S)(e_3) = T(S(e_3)) = e_3$;

- the unit vector e_1 is rotated $120°$ by S, and then by $240°$ by T which is a total of $360°$, that is, it is rotated back to itself so $(T \circ S)(e_1) = T(S(e_1)) = e_1$; and

- the unit vector e_2 is rotated $120°$ by S, and then by $240°$ by T which is a total of $360°$, that is, it is rotated back to itself so $(T \circ S)(e_2) = T(S(e_2)) = e_2$.

Form these results into a matrix to deduce the standard matrix

$$[T \circ S] = \begin{bmatrix} e_1 & e_2 & e_3 \end{bmatrix} = I_3$$

which is the standard matrix of the identity transformation. Hence $T \circ S = I$, the identity.

Second, an exactly corresponding argument gives $S \circ T = I$. By Definition 3.6.30, S and T are inverse transformations of each other. $\square$

Example 3.6.32 In some violent weather, a storm passes and the strong winds lean a house sideways as in the shear transformation illustrated to the right. Esti- mate the standard matrix of the shear transformation

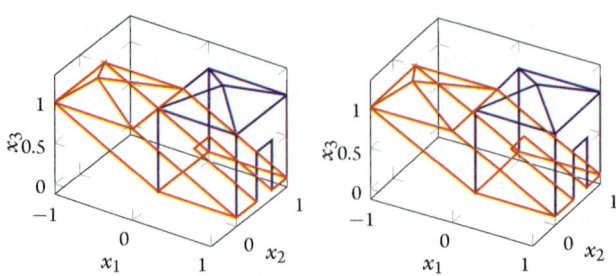

shown. To restore the house back upright, we need to shear it an equal amount in the opposite direction: hence write down the standard matrix of the inverse shear. Confirm that the product of the two standard matrices is the standard matrix of the identity.

Solution: As shown, the unit vectors e_1 and e_2 are unchanged by the storm S. However, the vertical unit vector e_3 is sheared by the storm to $S(e_3) = (-1, -0.5, 1)$. Hence the standard matrix of the storm S is

$$[S] = \begin{bmatrix} 1 & 0 & -1 \\ 0 & 1 & -\frac{1}{2} \\ 0 & 0 & 1 \end{bmatrix}.$$

To restore, R, the house upright we need to shear in the opposite direction, so the restoration shear has $R(e_3) = (1, \frac{1}{2}, 1)$; that is, the standard matrix of the inverse is

$$[R] = \begin{bmatrix} 1 & 0 & 1 \\ 0 & 1 & \frac{1}{2} \\ 0 & 0 & 1 \end{bmatrix}.$$

Multiplying these matrices together gives

$$[R \circ S] = [R][S] = \begin{bmatrix} 1 & 0 & 1 \\ 0 & 1 & \frac{1}{2} \\ 0 & 0 & 1 \end{bmatrix} \begin{bmatrix} 1 & 0 & -1 \\ 0 & 1 & -\frac{1}{2} \\ 0 & 0 & 1 \end{bmatrix}$$

$$= \begin{bmatrix} 1+0+0 & 0+0+0 & -1+0+1 \\ 0+0+0 & 0+1+0 & 0-\frac{1}{2}+\frac{1}{2} \\ 0+0+0 & 0+0+0 & 0+0+1 \end{bmatrix} = I_3.$$

This is the standard matrix of the identity transformation. ☐

Because of the exact correspondence between linear transformations and matrix multiplication, the inverse of a transformation exactly corresponds to the inverse of a matrix. In the last Example 3.6.32, because $[R][S] = I_3$ we know that the matrices $[R]$ and $[S]$ are inverses of each other. Correspondingly, the transformations R and S are inverses of each other.

Theorem 3.6.33 Let $T : \mathbb{R}^n \to \mathbb{R}^n$ be an invertible linear transformation. Then its standard matrix $[T]$ is invertible, and $[T^{-1}] = [T]^{-1}$.

Proof. As T is invertible, let the symbol T^{-1} denote its inverse. Since both are linear transformations, they both have standard matrices, $[T]$ and $[T^{-1}]$. Then $[T \circ T^{-1}] = [T][T^{-1}]$; but also $[T \circ T^{-1}] = [I] = I_n$; so $[T][T^{-1}] = I_n$. Similarly, $[T^{-1}][T] = [T^{-1} \circ T] = [I] = I_n$. Consequently the matrices $[T]$ and $[T^{-1}]$ are the inverses of each other.

Example 3.6.34 Estimate the standard matrix of the linear transformation T illustrated to the right. Then use Theorem 3.2.7 to determine the standard matrix of its inverse transformation T^{-1}. Hence sketch how the inverse transforms the unit square and write a sentence or two about how the sketch confirms that it is a reasonable inverse.

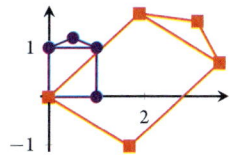

Solution: The illustrated transformation shows $T(i) \approx (1.7, -1)$ and $T(j) \approx (1.9, 1.7)$ hence its standard matrix is

$$[T] \approx \begin{bmatrix} 1.7 & 1.9 \\ -1 & 1.7 \end{bmatrix}.$$

Using Theorem 3.2.7 the inverse of this matrix is, since its determinant $= 1.7 \cdot 1.7 - (-1) \cdot 1.9 = 4.79$,

$$[T^{-1}] = [T]^{-1} = \frac{1}{4.79} \begin{bmatrix} 1.7 & -1.9 \\ 1 & 1.7 \end{bmatrix} \approx \begin{bmatrix} 0.35 & -0.40 \\ 0.21 & 0.35 \end{bmatrix}$$

This matrix determines that the inverse transforms the corners of the unit square as $T^{-1}(1,0) = (0.35, 0.21)$, $T^{-1}(0,1) = (-0.40, 0.35)$ and $T^{-1}(1,1) = (-0.05, 0.56)$. Hence the unit square is transformed as shown to the right. The original transformation, roughly, rotated the unit square clockwise and stretched it: the sketch shows that the inverse roughly rotates the unit square anticlockwise and shrinks it. Thus the inverse does appear to undo the action of the original transformation.

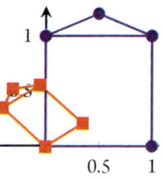

Example 3.6.35 Determine if the orthogonal projection of the plane onto the line at $30°$ to the horizontal (illustrated to the right) is an invertible transformation; if it is, find its inverse.

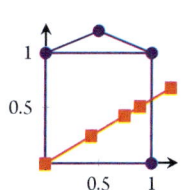

Solution: Recall that Theorem 3.5.29 gives the matrix of an orthogonal projection as WW^T where columns of W are an orthonormal basis for the projected space. Here the projected space is the line at $30°$ to the horizontal (illustrated to the right) which has orthonormal basis of the one vector $w = (\cos 30°, \sin 30°) = (\frac{\sqrt{3}}{2}, \frac{1}{2})$. Hence the standard matrix of the projection is

$$\boldsymbol{w}\boldsymbol{w}^T = \begin{bmatrix} \frac{\sqrt{3}}{2} \\ \frac{1}{2} \end{bmatrix} \begin{bmatrix} \frac{\sqrt{3}}{2} & \frac{1}{2} \end{bmatrix} = \begin{bmatrix} \frac{3}{4} & \frac{\sqrt{3}}{4} \\ \frac{\sqrt{3}}{4} & \frac{1}{4} \end{bmatrix}.$$

From Theorem 3.2.7 this matrix is invertible only if the determinant $(\det = ad - bc)$ is nonzero, but here

$$\det(\boldsymbol{w}\boldsymbol{w}^T) = \frac{3}{4} \cdot \frac{1}{4} - \frac{\sqrt{3}}{4} \cdot \frac{\sqrt{3}}{4} = \frac{3}{16} - \frac{3}{16} = 0.$$

Since its standard matrix is not invertible, the given orthogonal projection is also not invertible (as the illustration shows, the projection 'squashes' the plane onto the line, which cannot be uniquely undone, and hence is not invertible). □

3.6.4 Exercises

Exercise 3.6.1 Which of the following illustrated transformations of the plane *cannot* be that of a linear transformation? In each illustration of a transformation T, the four corners of the blue unit square $((0,0), (1,0), (1,1),$ and $(0,1))$, are transformed to the four corners of the red figure $(T(0,0), T(1,0), T(1,1),$ and $T(0,1))$—the 'roof' of the unit square clarifies which side goes where).

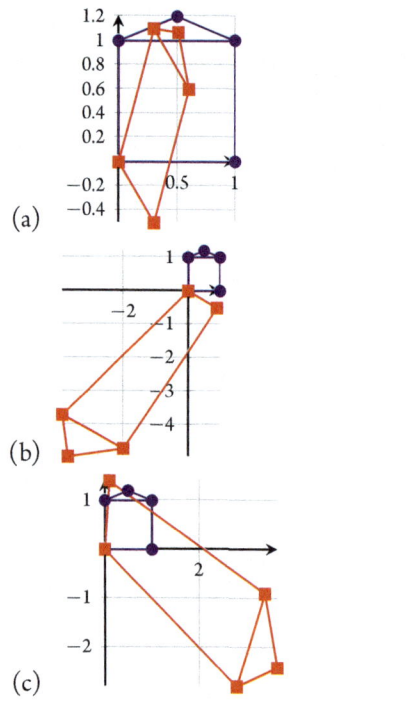

(a)

(b)

(c)

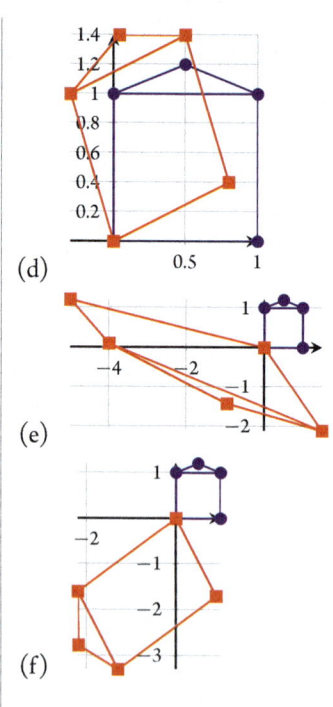

(d)

(e)

(f)

Exercise 3.6.2 Consider the transformations of Exercise 3.6.1: for those transformations that *may* be linear transformations, assume they are, and so estimate roughly the standard matrix of each such linear transformation.

Exercise 3.6.3 Consider the following illustrated transformations of $\mathbb{R}^3$. Which of these *cannot* be that of a linear transformation?

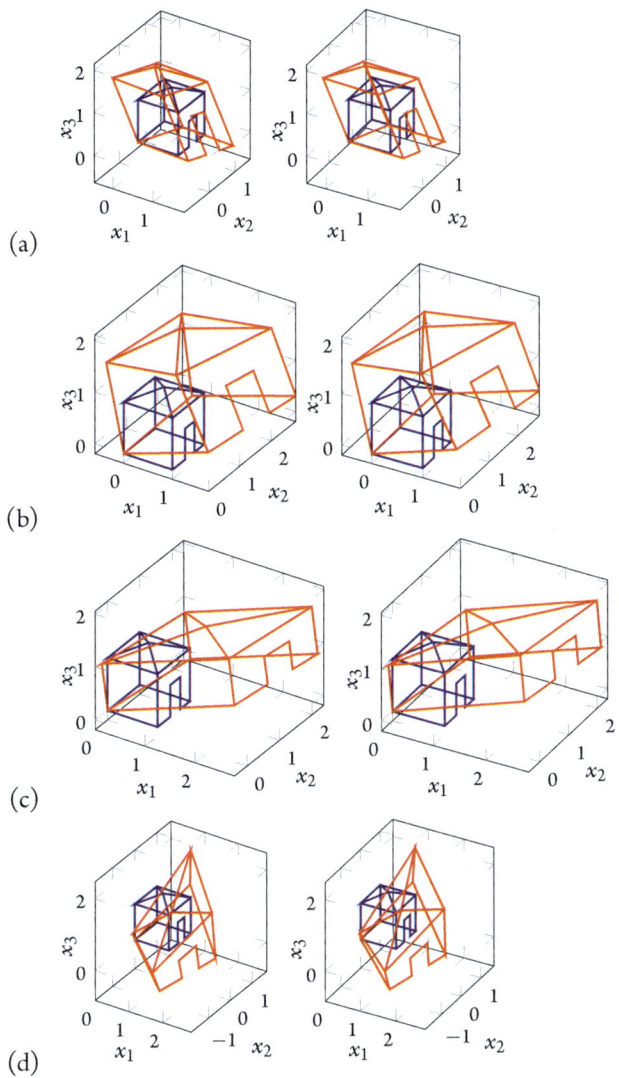

(a)

(b)

(c)

(d)

Exercise 3.6.4 Let $T : \mathbb{R}^n \to \mathbb{R}^m$ be a linear transformation. Prove from the Definition 3.6.1 that $T(\mathbf{0}) = \mathbf{0}$ and $T(\mathbf{u} - \mathbf{v}) = T(\mathbf{u}) - T(\mathbf{v})$ for every $\mathbf{u}, \mathbf{v} \in \mathbb{R}^n$.

Exercise 3.6.5 *(equivalent definition)* Consider a function $T : \mathbb{R}^n \to \mathbb{R}^m$. Prove that T is a linear transformation (Definition 3.6.1) if and only if $T(c_1 \mathbf{v}_1 + c_2 \mathbf{v}_2) = c_1 T(\mathbf{v}_1) + c_2 T(\mathbf{v}_2)$ for every $\mathbf{v}_1, \mathbf{v}_2 \in \mathbb{R}^n$ and every scalars c_1, c_2.

Exercise 3.6.6 Given $T : \mathbb{R}^n \to \mathbb{R}^m$ is a linear transformation, use induction to prove that for every k

$$T(c_1 u_1 + c_2 u_2 + \cdots + c_k u_k) = c_1 T(u_1) + c_2 T(u_2) + \cdots + c_k T(u_k)$$

for every scalar c_1, c_2, ..., c_k and every vector u_1, u_2, ..., u_k in $\mathbb{R}^n$.

Exercise 3.6.7 Consider each of the following vector transformations: if it is a linear transformation, then write down its standard matrix.

(a) $A(x,y,z) = (-3x + 2y, -z)$

(b) $B(x,y) = (0, 3x + 7y, -2y, -3y)$

(c) $C(x_1, x_2, \ldots, x_5) = (2x_1 + x_2 - 2x_3, 7x_1 + 7x_4)$

(d) $D(x,y) = (2x + 3, -5y + 3, 2x - 4y + 3, 0, -6x)$

(e) $E(p,q,r,s) = (-3p - 4r, -s, 0, p + r + 6s, 5p + 6q - s)$

(f) $F(x,y) = (5x + 4y, x^2, 2y, -4x, 0)$

Exercise 3.6.8 Use Procedure 3.5.4 to derive that the pseudo-inverse of the general 2×1 matrix $A = \begin{bmatrix} a \\ b \end{bmatrix}$ is the 1×2 matrix $A^+ = \begin{bmatrix} \frac{a}{a^2+b^2} & \frac{b}{a^2+b^2} \end{bmatrix}$. Further, what is the pseudo-inverse of the general 1×2 matrix $\begin{bmatrix} a & b \end{bmatrix}$?

Exercise 3.6.9 Consider the general $m \times n$ diagonal matrix of rank r,

$$S = \begin{bmatrix} \sigma_1 & \cdots & 0 & \\ \vdots & \ddots & \vdots & O_{r \times (n-r)} \\ 0 & \cdots & \sigma_r & \\ & O_{(m-r) \times r} & & O_{(m-r) \times (n-r)} \end{bmatrix},$$

equivalently $S = \mathrm{diag}_{m \times n}(\sigma_1, \sigma_2, \ldots, \sigma_r, 0, \ldots, 0)$. Derive that, based upon Procedure 3.5.4, the pseudo-inverse of S is the $n \times m$ diagonal matrix of rank r,

$$S^+ = \begin{bmatrix} 1/\sigma_1 & \cdots & 0 & \\ \vdots & \ddots & \vdots & O_{r \times (m-r)} \\ 0 & \cdots & 1/\sigma_r & \\ & O_{(n-r) \times r} & & O_{(n-r) \times (m-r)} \end{bmatrix},$$

equivalently $S^+ = \mathrm{diag}_{n \times m}(1/\sigma_1, 1/\sigma_2, \ldots, 1/\sigma_r, 0, \ldots, 0)$.

Exercise 3.6.10 For every $m \times n$ matrix A, let $A = USV^{\mathsf{T}}$ be its SVD, and use the result of Exercise 3.6.9 to establish that the pseudo-inverse $A^+ = VS^+U^{\mathsf{T}}$.

Exercise 3.6.11 Use the results of Exercises 3.6.9 and 3.6.10 to prove the following properties of the pseudo-inverse hold for every matrix A:

(a) $AA^+A = A$;

(b) $A^+AA^+ = A^+$;

(c) AA^+ is symmetric;

(d) A^+A is symmetric.

Exercise 3.6.12 Use MATLAB/Octave and the identity of Exercise 3.6.10 to compute the pseudo-inverse of each of the following matrices.

(a) $A = \begin{bmatrix} 0.2 & 0.3 \\ 0.2 & 1.5 \\ 0.1 & -0.3 \end{bmatrix}$

(b) $B = \begin{bmatrix} 2.5 & 0.6 & 0.3 \\ 0.5 & 0.4 & 0.2 \end{bmatrix}$

(c) $C = \begin{bmatrix} 0.1 & -0.5 \\ 0.6 & -3.0 \\ 0.4 & -2.0 \end{bmatrix}$

(d) $D = \begin{bmatrix} 0.3 & -0.3 & -1.2 \\ 0.1 & 0.3 & 1.4 \\ -3.1 & -0.5 & -3.8 \\ 1.5 & -0.3 & -0.6 \end{bmatrix}$

(e) $E = \begin{bmatrix} 4.1 & 1.8 & -0.4 & 0.0 & -0.1 & -1.4 \\ -3.3 & -3.9 & 0.6 & -2.2 & 0.5 & 0.1 \\ -0.9 & -1.9 & 0.6 & -2.2 & 0.1 & 0.5 \\ -4.3 & -3.6 & 0.8 & -2.0 & 0.3 & 1.2 \end{bmatrix}$

(f) $F = \begin{bmatrix} -0.6 & -1.3 & -1.2 & -1.9 & 1.6 & 1.6 \\ -0.7 & 0.6 & -0.2 & 0.9 & -0.6 & -0.7 \\ 0.0 & 0.2 & 0.7 & 1.1 & -0.4 & -0.6 \\ 0.8 & 0.1 & -1.6 & -0.9 & -0.5 & -0.8 \\ -0.5 & 0.9 & 1.3 & 1.7 & -0.3 & -0.6 \end{bmatrix}$

Exercise 3.6.13 Invent matrices A and B such that $(AB)^+ \neq B^+A^+$.

Exercise 3.6.14 Prove that in the case of an $m \times n$ matrix A with rank $A = m$ (so $m \le n$), the pseudo-inverse is the $n \times m$ matrix $A^+ = A^{\mathsf{T}}(AA^{\mathsf{T}})^{-1}$.

Exercise 3.6.15 Use Theorem 3.6.22 and the identity in Exercise 3.6.14 to prove that $(A^+)^+ = A$ in the case when $m \times n$ matrix A has rank $A = n$. (Be careful as many plausible looking steps are incorrect.)

Exercise 3.6.16 Confirm that each of the following composition of two linear transformations in $\mathbb{R}^2$ has a standard matrix that is the same as multiplying the two standard matrices of the specified linear transformations.

(a) Rotation by $30°$ followed by rotation by $60°$.

(b) Rotation by $120°$ followed by rotation by $-60°$ (clockwise $60°$).

(c) Reflection in the x-axis followed by reflection in the line $y = x$.

(d) Reflection in the line $y = x$ followed by reflection in the x-axis.

(e) Reflection in the line $y = x$ followed by rotation by $90°$.

(f) Reflection in the line $y = \sqrt{3}x$ followed by rotation by $-30°$ (clockwise $30°$).

Exercise 3.6.17 For each of the following pairs of linear transformations S and T, if possible determine the standard matrices of the compositions $S \circ T$ and $T \circ S$.

(a) $S(x) = (-5x, 2x, -x, -x)$ and $T(y_1, y_2, y_3, y_4) = -4y_1 - 3y_2 - 4y_3 + 5y_4$

(b) $S(x_1, x_2, x_3, x_4) = (-2x_2 - 3x_3 + 6x_4, -3x_1 + 2x_2 - 4x_3 + 3x_4)$ and $T(y) = (-4y, 0, 0, -y)$

(c) $S(x, y) = (-5x - 3y, -5x + 5y)$ and $T(z_1, z_2, z_3, z_4) = (3z_1 + 3z_2 - 2z_3 - 2z_4, 4z_1 + 7z_2 + 4z_3 + 3z_4)$

(d) $S(x, y, z) = 5x - y + 4z$ and $T(p) = (6p, -3p, 3p, p)$

Exercise 3.6.18 For each of the illustrated transformations, estimate the standard matrix of the linear transformation. Then use Theorem 3.2.7 to determine the standard matrix of its inverse transformation. Hence sketch how the inverse transforms the unit square and write a sentence or two about how the sketch confirms it is a reasonable inverse.

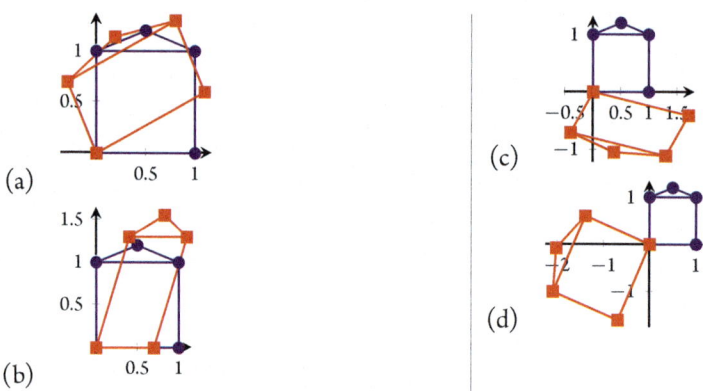

Exercise 3.6.19 In a few sentences, answer/discuss each of the following.

(a) How does Definition 3.6.1 compare with the definition of a subspace?

(b) Why should the linear transformation of a square/cube be a parallelogram/parallelepiped?

(c) What causes multiplication by a matrix to be a linear transformation?

(d) What causes the transform of the standard unit vectors to form the matrix of the linear transformation?

(e) How does the concept of a pseudo-inverse of a matrix compare with that of the inverse of a matrix?

(f) For an $m \times n$ matrix A with rank $A < n \leq m$, what is it that fails in the formula for the pseudo-inverse $A^+ = (A^T A)^{-1} A^T$?

(g) How does the composition of linear transformations connect to the inverse of a matrix?

3.7 Summary of matrices

Matrix operations and algebra

- First, some basic terminology (Section 3.1.1) corresponds to commands in MATLAB/Octave.
  - ⋆⋆ A **matrix** is a rectangular array of real numbers, written inside **brackets** $[\cdots]$—create in MATLAB/Octave via [. . . ; . . . ; . . .].
  - ⋆⋆ The **size** of a matrix is written $m \times n$ where m is the number of rows and n is the number of columns—compute with size(A) for matrix A. If $m = n$, then it is called a **square matrix**.
    - – A **column vector** means a matrix of size $m \times 1$ for some m. We often write a column vector horizontally within **parentheses** $(\cdots)$.
    - – The numbers appearing in a matrix are called the **entries**, **elements** or **components** of the matrix. For a matrix A, the entry in row i and column j is denoted by a_{ij} — compute with A(i,j).
  - ⋆ $O_{m \times n}$ denotes the $m \times n$ zero matrix, O_n denotes the square zero matrix of size $n \times n$—compute with zeros(m,n) and zeros(n), respectively—whereas O denotes a zero matrix whose size is apparent from the context.
  - ⋆ The **identity matrix** I_n denotes a $n \times n$ square matrix which has zero entries except for the diagonal from the top-left to the bottom-right which are all ones—compute with eye(n). Non-square 'identity' matrices are denoted $I_{m \times n}$—compute with eye(m,n). The symbol I denotes an identity matrix whose size is apparent from the context.
    - – In MATLAB/Octave, randn(m,n) computes a $m \times n$ matrix with random entries (distributed Normally, mean zero, standard deviation one).
    - – Two matrices are **equal** $(=)$ if they both have the same size *and* their corresponding entries are equal. Otherwise the two matrices are not equal.
- Basic matrix operations include the following (Section 3.1.2).
  - – When A and B are both $m \times n$ matrices, then their **sum** or **addition**, $A + B$, is the $m \times n$ matrix whose (i,j)th entry is $a_{ij} + b_{ij}$ —compute by A+B. Similarly, the **difference** or **subtraction** $A - B$ is the $m \times n$ matrix whose (i,j)th entry is $a_{ij} - b_{ij}$ — compute by A-B.
  - – For an $m \times n$ matrix A, the **scalar product** by c, denoted either cA or Ac—and compute by c*A or A*c—is the $m \times n$ matrix whose (i,j)th entry is ca_{ij}.
  - ⋆⋆ For $m \times n$ matrix A and vector x in $\mathbb{R}^n$, the **matrix-vector product** Ax —compute by A*x—is the following vector in $\mathbb{R}^m$,

$$
Ax := \begin{bmatrix} a_{11}x_1 + a_{12}x_2 + \cdots + a_{1n}x_n \\ a_{21}x_1 + a_{22}x_2 + \cdots + a_{2n}x_n \\ \vdots \\ a_{m1}x_1 + a_{m2}x_2 + \cdots + a_{mn}x_n \end{bmatrix}.
$$

Multiplication of a vector by a square matrix transforms the vector into another vector in the same space.

★ In modelling the age structure of populations, the so-called Leslie matrix encodes the birth, ageing, and death in the population (usually of females) to empower predictions of the future population. Such prediction often involves repeated matrix-matrix multiplication, that is, computing the powers of a matrix.

★★ For $m \times n$ matrix A, and $n \times p$ matrix B, the **matrix product** $C = AB$ —compute by C=A*B—is the $m \times p$ matrix whose (i,j)th entry is

$$c_{ij} = a_{i1}b_{1j} + a_{i2}b_{2j} + \cdots + a_{in}b_{nj}.$$

★ Matrix addition and scalar multiplication satisfy familiar properties (Theorem 3.1.24): $A + B = B + A$ (commutativity); $(A + B) + C = A + (B + C)$ (associativity); $A \pm O = A = O + A$; $c(A \pm B) = cA \pm cB$ (distributivity); $(c \pm d)A = cA \pm dA$ (distributivity); $c(dA) = (cd)A$ (associativity); $1A = A$; and $0A = O$.

★ Matrix multiplication also satisfies familiar properties (Theorem 3.1.26): $A(B \pm C) = AB \pm AC$ (distributivity); $(A \pm B)C = AC \pm BC$ (distributivity); $A(BC) = (AB)C$ (associativity); $c(AB) = (cA)B = A(cB)$; $I_m A = A = AI_n$ for $m \times n$ matrix A (multiplicative identity); $O_m A = O_{m \times n} = AO_n$ for $m \times n$ matrix A; $A^p A^q = A^{p+q}$, $(A^p)^q = A^{pq}$ and $(cA)^p = c^p A^p$ for square A and integer p, q.
But matrix multiplication is *not* commutative: generally $AB \neq BA$.

★★ The **transpose** of an $m \times n$ matrix A is the $n \times m$ matrix $B = A^T$ with entries $b_{ij} = a_{ji}$ (Definition 3.1.17)—compute by A'.
A (real) matrix A is a **symmetric matrix** if $A^T = A$ (Definition 3.1.20). A symmetric matrix must be a square matrix

★ The matrix transpose satisfies (Theorem 3.1.29): $(A^T)^T = A$; $(A \pm B)^T = A^T \pm B^T$; $(cA)^T = c(A^T)$; $(AB)^T = B^T A^T$ (remember to reverse the order); $(A^p)^T = (A^T)^p$; and $A + A^T, A^T A$ and AA^T are symmetric matrices.

The inverse of a matrix

★★ An **inverse** of a square matrix A is a matrix B such that both $AB = I$ and $BA = I$ (Definition 3.2.2). If such a matrix B exists, then matrix A is called **invertible**.

★ If A is an invertible matrix, then its inverse is unique, and denoted by A^{-1} (Theorem 3.2.6).

★★ For every 2×2 matrix $A = \begin{bmatrix} a & b \\ c & d \end{bmatrix}$, the matrix A is invertible if and only if the **determinant** $ad - bc \neq 0$ (Theorem 3.2.7), in which case

$$A^{-1} = \frac{1}{ad - bc} \begin{bmatrix} d & -b \\ -c & a \end{bmatrix}.$$

★ If a matrix A is invertible, then $Ax = b$ has the unique solution $x = A^{-1}b$ for every b (Theorem 3.2.10).

★ For all invertible matrices A and B, the inverse has the properties (Theorem 3.2.13):
  - matrix A^{-1} is invertible and $(A^{-1})^{-1} = A$;
  - if scalar $c \neq 0$, then matrix cA is invertible and $(cA)^{-1} = \frac{1}{c}A^{-1}$;
  - matrix AB is invertible and $(AB)^{-1} = B^{-1}A^{-1}$ (remember the reversed order);
  - matrix A^T is invertible and $(A^T)^{-1} = (A^{-1})^T$;
  - matrices A^p are invertible for all $p = 1, 2, 3, \ldots$ and $(A^p)^{-1} = (A^{-1})^p$.

- For every invertible matrix A, define $A^0 = I$ and for every positive integer p define $A^{-p} := (A^{-1})^p = (A^p)^{-1}$ (Definition 3.2.15).
- ★★ The **diagonal entries** of an $m \times n$ matrix A are a_{11}, a_{22}, ..., a_{pp} where $p = \min(m,n)$. A matrix whose off-diagonal entries are all zero is called a **diagonal matrix**: $\text{diag}(v_1, v_2, ..., v_n)$ denotes the $n \times n$ square matrix with diagonal entries $v_1, v_2, ..., v_n$; whereas $\text{diag}_{m \times n}(v_1, v_2, ..., v_p)$ denotes an $m \times n$ matrix with diagonal entries $v_1, v_2, ..., v_p$ (Definition 3.2.21).
- For every $n \times n$ diagonal matrix $D = \text{diag}(d_1, d_2, ..., d_n)$, if $d_i \neq 0$ for $i = 1, 2, ..., n$, then D is invertible and the inverse $D^{-1} = \text{diag}(1/d_1, 1/d_2, ..., 1/d_n)$ (Theorem 3.2.27).
- Multiplication by a diagonal matrix just stretches or squashes and/or reflects in the directions of the coordinate axes. Consequently, in applications we often choose coordinate systems such that the matrices which appear are diagonal.
- In MATLAB/Octave:
  - `diag(v)` where v is a row/column vector of size p generates the $p \times p$ matrix

$$\text{diag}(v_1, v_2, ..., v_p) = \begin{bmatrix} v_1 & 0 & \cdots & 0 \\ 0 & v_2 & & \vdots \\ \vdots & & \ddots & \\ 0 & \cdots & & v_p \end{bmatrix}.$$

  - ★★ In MATLAB/Octave (but not in algebra), `diag` also does the opposite: for an $m \times n$ matrix A such that both $m, n \geq 2$, `diag(A)` returns the (column) vector $(a_{11}, a_{22}, ..., a_{pp})$ of diagonal entries where the result vector size $p = \min(m,n)$.
  - ★★ The dot operators `./` and `.*` perform element-by-element division and multiplication of two matrices/vectors of the same size (but see Table 5.1).
    - `log10(v)` finds the logarithm to base 10 of each component of v and returns the results in a vector of the same size; `log(v)` does the same but for the natural logarithm (not `ln(v)`).
- ★★ A set of nonzero vectors $\{q_1, q_2, ..., q_k\}$ is called an **orthogonal set** if all pairs of distinct vectors in the set are orthogonal: that is, $q_i \cdot q_j = 0$ whenever $i \neq j$ for $i, j = 1, 2, ..., k$ (Definition 3.2.38). A set of vectors is called an **orthonormal set** if it is an orthogonal set of unit vectors.
- ★★ A square matrix Q is called an **orthogonal matrix** if $Q^T Q = I$ (Definition 3.2.43). Multiplication by an orthogonal matrix is called a **rotation and/or reflection**.
- ★ For every square matrix Q, the following statements are equivalent (Theorem 3.2.48):
  - Q is an orthogonal matrix;
  - the column vectors of Q form an orthonormal set;
  - Q is invertible and $Q^{-1} = Q^T$;
  - Q^T is an orthogonal matrix;
  - the row vectors of Q form an orthonormal set;
  - multiplication by Q preserves all lengths and angles (and hence corresponds to our intuition of a rotation and/or reflection).

Factorize to the singular value decomposition

★★ Every $m \times n$ real matrix A can be factored into a product of three matrices $A = USV^{\mathsf{T}}$ (Theorem 3.3.6), called a **singular value decomposition** (SVD), where

- $m \times m$ matrix $U = \begin{bmatrix} u_1 & u_2 & \cdots & u_m \end{bmatrix}$ is orthogonal,
- $n \times n$ matrix $V = \begin{bmatrix} v_1 & v_2 & \cdots & v_n \end{bmatrix}$ is orthogonal, and
- $m \times n$ diagonal matrix S is zero except for the unique non-negative diagonal elements called **singular values** $\sigma_1 \geq \sigma_2 \geq \cdots \geq \sigma_{\min(m,n)} \geq 0$.

The orthonormal sets of vectors $\{u_j\}$ and $\{v_j\}$ are called **singular vectors**.

Almost always use MATLAB/Octave to find an SVD of a given matrix.

★★ Procedure 3.3.15 derives a general solution of the system $Ax = b$ using an SVD.

1. Obtain an SVD factorization $A = USV^{\mathsf{T}}$.

2. Solve $Uz = b$ by $z = U^{\mathsf{T}}b$ (unique given U).

3. To solve $Sy = z$, identify the nonzero and the zero singular values: suppose $\sigma_1 \geq \sigma_2 \geq \cdots \geq \sigma_r > 0$ and $\sigma_{r+1} = \cdots = \sigma_{\min(m,n)} = 0$:
   - if $z_i \neq 0$ for any $i = r+1, \ldots, m$, then there is no solution (the equations are **inconsistent**);
   - otherwise determine the ith component of y by $y_i = z_i / \sigma_i$ for $i = 1, \ldots, r$, and let y_i be a free variable for $i = r+1, \ldots, n$.

4. Solve $V^{\mathsf{T}}x = y$ (unique for each y given V) to derive that a general solution is $x = Vy$.

★★ In MATLAB/Octave:
- [U,S,V] =svd(A) computes the three matrices U, S and V in a singular value decomposition (SVD) of the $m \times n$ matrix: $A = USV^{\mathsf{T}}$.
  svd(A) just reports the singular values in a vector.
- To extract and compute with a subset of rows/columns of a matrix, specify the vector of indices.

★ The **condition number** of a matrix A is the ratio of the largest to smallest of its singular values: $\operatorname{cond} A := \sigma_1 / \sigma_{\min(m,n)}$ (Definition 3.3.16); if $\sigma_{\min(m,n)} = 0$, then $\operatorname{cond} A := \infty$; also, $\operatorname{cond} O_{m \times n} := \infty$.

★★ The **rank** of a matrix A is the number of *nonzero* singular values in an SVD, $A = USV^{\mathsf{T}}$ (Definition 3.3.19).

★ For every matrix A, let an SVD of A be USV^{T}, then the transpose A^{T} has an SVD of $V(S^{\mathsf{T}})U^{\mathsf{T}}$ (Theorem 3.3.24). Further, $\operatorname{rank}(A^{\mathsf{T}}) = \operatorname{rank} A$.

• For every $n \times n$ square matrix A, the following statements are equivalent:
  - A is invertible;
  - $Ax = b$ has a unique solution for every b in $\mathbb{R}^n$;
  - $Ax = 0$ has *only* the zero solution;
  - all n singular values of A are nonzero;
  - the condition number of A is finite ($\text{rcond} > 0$);
  - $\operatorname{rank} A = n$.

• The condition number determines the reliability of solutions to linear equations. Consider solving $Ax = b$ for $n \times n$ matrix A with full rank $A = n$. When the right-hand side b has relative error ϵ, then the solution $\boldsymbol{x}$ has relative error $\leq \epsilon \operatorname{cond} A$, with equality in the worst case (Theorem 3.3.30).

Subspaces, basis and dimension

★★ A **subspace** $\mathbb{W}$ of $\mathbb{R}^n$, is a set of vectors, including $\mathbf{0} \in \mathbb{W}$, such that $\mathbb{W}$ is **closed** under addition and scalar multiplication: that is, for all $c \in \mathbb{R}$ and $\boldsymbol{u}, \boldsymbol{v} \in \mathbb{W}$, then both $\boldsymbol{u} + \boldsymbol{v} \in \mathbb{W}$ and $c\boldsymbol{u} \in \mathbb{W}$ (Definition 3.4.3).

- Let $\boldsymbol{v}_1, \boldsymbol{v}_2, \ldots, \boldsymbol{v}_k$ be k vectors in $\mathbb{R}^n$, then span$\{\boldsymbol{v}_1, \boldsymbol{v}_2, \ldots, \boldsymbol{v}_k\}$ is a subspace of $\mathbb{R}^n$ (Theorem 3.4.6).
- The **column space** of any $m \times n$ matrix A is the subspace of $\mathbb{R}^m$ spanned by the n column vectors of A (Definition 3.4.10).

  The **row space** of any $m \times n$ matrix A is the subspace of $\mathbb{R}^n$ spanned by the m row vectors of A.
- ★ For any $m \times n$ matrix A, define null(A) to be the set of all solutions $\boldsymbol{x}$ to the homogeneous system $A\boldsymbol{x} = \mathbf{0}$. The set null$(A)$ is a subspace of $\mathbb{R}^n$ called the **nullspace** of A (Theorem 3.4.14).
- ★ An **orthonormal basis** for a subspace $\mathbb{W}$ of $\mathbb{R}^n$ is an orthonormal set of vectors that span $\mathbb{W}$ (Definition 3.4.18).
- ★★ Procedure 3.4.23 finds an orthonormal basis for the subspace span$\{\boldsymbol{a}_1, \boldsymbol{a}_2, \ldots, \boldsymbol{a}_n\}$, where $\{\boldsymbol{a}_1, \boldsymbol{a}_2, \ldots, \boldsymbol{a}_n\}$ is a set of n vectors in $\mathbb{R}^m$.

  1. Form matrix $A := \begin{bmatrix} \boldsymbol{a}_1 & \boldsymbol{a}_2 & \cdots & \boldsymbol{a}_n \end{bmatrix}$.
  2. Factorize A into an SVD, $A = USV^{\mathsf{T}}$, let $\boldsymbol{u}_j$ denote the columns of U (singular vectors), and let $r = \operatorname{rank} A$ be the number of nonzero singular values.
  3. Then $\{\boldsymbol{u}_1, \boldsymbol{u}_2, \ldots, \boldsymbol{u}_r\}$ is an orthonormal basis for the subspace span$\{\boldsymbol{a}_1, \boldsymbol{a}_2, \ldots, \boldsymbol{a}_n\}$.

- Singular Spectrum Analysis seeks patterns over time by using an SVD to find orthonormal bases for 'sliding windows' in time of the data (Example 3.4.27).
- Any two orthonormal bases for a given subspace have the same number of vectors (Theorem 3.4.28).
- Let $\mathbb{W}$ be a subspace of $\mathbb{R}^n$, then there exists an orthonormal basis for $\mathbb{W}$ (Theorem 3.4.29).
- ★★ For every subspace $\mathbb{W}$, the number of vectors in an orthonormal basis for $\mathbb{W}$ is called the **dimension** of $\mathbb{W}$, denoted dim $\mathbb{W}$ (Definition 3.4.30). By convention, dim$\{\mathbf{0}\} = 0$.
- ★ The row space and column space of a matrix A have the same dimension (Theorem 3.4.32). Further, given an SVD of the matrix, say $A = USV^{\mathsf{T}}$ and setting $r = \operatorname{rank} A$, an orthonormal basis for the column space is the first r columns of U, and that for the row space is the first r columns of V.
- The **nullity** of a matrix A is the dimension of its nullspace, and is denoted by nullity(A) (Definition 3.4.36).
- ★★ For every $m \times n$ matrix A, rank A + nullity $A = n$, the number of columns of A (Theorem 3.4.39).
- ★ For every $n \times n$ square matrix A, and extending Theorem 3.3.27, the following statements are equivalent (Theorem 3.4.43):
  - A is invertible;

- $Ax = b$ has a unique solution for every $b \in \mathbb{R}^n$;
- $Ax = 0$ has only the zero solution;
- all n singular values of A are nonzero;
- the condition number of A is finite ($\texttt{rcond} > 0$);
- $\operatorname{rank} A = n$;
- nullity $A = 0$;
- the column vectors of A span $\mathbb{R}^n$;
- the row vectors of A span $\mathbb{R}^n$.

Project to solve inconsistent equations

★★ Procedure 3.5.4 computes the 'least square' approximate solution(s) of inconsistent equations $Ax = b$:

1. factorize $A = USV^\mathsf{T}$ and set $r = \operatorname{rank} A$ (relatively small singular values are effectively zero);

2. solve $Uz = b$ by $z = U^\mathsf{T} b$;

3. set $y_i = z_i / \sigma_i$ for $i = 1, \ldots, r$, with y_i free for $i = r+1, \ldots, n$, and consider z_i for $i = r+1, \ldots, n$ as errors;

4. solve $V^\mathsf{T} x = y$ to obtain a general approximate solution as $x = Vy$.

- A robust way to use the results of pairwise competitions to rate a group of players or teams is to approximately solve the set of equations $x_i - x_j = \text{result}_{ij}$ where x_i and x_j denote the unknown ratings of the players/teams to be determined, and result_{ij} is the result or score (to i over j) when the two compete against each other.
  Beware of Arrow's Impossibility Theorem that all 1D ranking systems are flawed!

★ All approximate solutions obtained by Procedure 3.5.4 solve the linear system $Ax = b'$ for the unique consistent right-hand side vector b' that minimizes the distance $|b' - b|$ (Theorem 3.5.8).

- To fit the best straight line through some data, express as a linear algebra approximation problem. Say the task is to find the linear relation between v ('vertical' values) and h ('horizontal' values), $v = x_1 + x_2 h$, for as yet unknown coefficients x_1 and x_2. Form the system of linear equations from all data points (h, v) by $x_1 + x_2 h = v$, and solve the system via Procedure 3.5.4.
  In science and engineering one often seeks power laws $v = c_1 h^{c_2}$ in which case take logarithms and use the data to find $x_1 = \log c_1$ and $x_2 = c_2$ via Procedure 3.5.4. Taking logarithms of equations also empowers forming linear equations to be approximately solved in the computed tomography of medical and industrial CT-scans.

★★ Obtain the smallest solution, whether exact or as an approximation, to a system of linear equations by Procedures 3.3.15 and 3.5.4, as appropriate, and setting to zero the free variables, $y_{r+1} = \cdots = y_n = 0$ (Theorem 3.5.13).

- For application to image analysis, in MATLAB/Octave:
  - $\texttt{reshape (A,p,q)}$ for a $m \times n$ matrix/vector A, provided $mn = pq$, generates a $p \times q$ matrix with entries taken column-wise from A. Either p or q can be $[\,]$ in which case MATLAB/Octave uses $p = mn/q$ or $q = mn/p$ respectively.

- colormap(gray) draws the current figure with 64 shades of grey (colormap('list') lists the available colormaps).
- imagesc(A) where A is a $m \times n$ matrix of values draws an $m \times n$ image using the values of A to determine the colour.
- log(x) where x is a matrix, vector or scalar computes the natural logarithm to the base e of each element, and returns the result(s) as a correspondingly sized matrix, vector or scalar.
- exp(x) where x is a matrix, vector or scalar computes the exponential of each element, and returns the result(s) as a correspondingly sized matrix, vector or scalar.

- Let $u, v \in \mathbb{R}^n$ and vector $u \neq 0$, then the **orthogonal projection** of v onto u is $\text{proj}_u(v) := u(u \cdot v)/|u|^2$ (Definition 3.5.19). When u is a unit vector, $\text{proj}_u(v) := u(u \cdot v)$.

★★ Let $\mathbb{W}$ be a k-dimensional subspace of $\mathbb{R}^n$ with an orthonormal basis $\{w_1, w_2, \ldots, w_k\}$. For every vector $v \in \mathbb{R}^n$, the **orthogonal projection** of vector v onto subspace $\mathbb{W}$ is (Definition 3.5.23)

$$\text{proj}_{\mathbb{W}}(v) = w_1(w_1 \cdot v) + w_2(w_2 \cdot v) + \cdots + w_k(w_k \cdot v).$$

★ The 'least square' solution(s) of the system $Ax = b$ determined by Procedure 3.5.4 is(are) the solution(s) of $Ax = \text{proj}_{\mathbb{A}}(b)$ where $\mathbb{A}$ is the column space of A (Theorem 3.5.26).

★ Let $\mathbb{W}$ be a k-dimensional subspace of $\mathbb{R}^n$ with an orthonormal basis $\{w_1, w_2, \ldots, w_k\}$, then for every vector $v \in \mathbb{R}^n$, the orthogonal projection $\text{proj}_{\mathbb{W}}(v) = (WW^T)v$ for the $n \times k$ matrix $W = \begin{bmatrix} w_1 & w_2 & \cdots & w_k \end{bmatrix}$ (Theorem 3.5.29).

- Let $\mathbb{W}$ be a k-dimensional subspace of $\mathbb{R}^n$. The set of all vectors $u \in \mathbb{R}^n$ (together with 0) that are each orthogonal to all vectors in $\mathbb{W}$ is called the **orthogonal complement** $\mathbb{W}^\perp$ (Definition 3.5.34); that is,

$$\mathbb{W}^\perp = \{u \in \mathbb{R}^n : u \cdot w = 0 \text{ for all } w \in \mathbb{W}\}.$$

- For every subspace $\mathbb{W}$ of $\mathbb{R}^n$, the orthogonal complement $\mathbb{W}^\perp$ is a subspace of $\mathbb{R}^n$ (Theorem 3.5.38). Further, the intersection $\mathbb{W} \cap \mathbb{W}^\perp = \{0\}$.

★ For every $m \times n$ matrix A, and denoting the column space of A by $\mathbb{A} = \text{span}\{a_1, a_2, \ldots, a_n\}$, the orthogonal complement $\mathbb{A}^\perp = \text{null}(A^T)$ (Theorem 3.5.40). Further, $\text{null}(A)$ is the orthogonal complement of the row space of A.

- For every subspace $\mathbb{W}$ of $\mathbb{R}^n$, $\dim \mathbb{W} + \dim \mathbb{W}^\perp = n$ (Theorem 3.5.44).

- Let $\mathbb{W}$ be a subspace of $\mathbb{R}^n$. For every vector $v \in \mathbb{R}^n$, the **perpendicular component** of v to $\mathbb{W}$ is the vector $\text{perp}_{\mathbb{W}}(v) := v - \text{proj}_{\mathbb{W}}(v)$ (Definition 3.5.47).

- Let $\mathbb{W}$ be a subspace of $\mathbb{R}^n$, then for every vector $v \in \mathbb{R}^n$ the perpendicular component $\text{perp}_{\mathbb{W}}(v) \in \mathbb{W}^\perp$ (Theorem 3.5.49).

★ Let $\mathbb{W}$ be a subspace of $\mathbb{R}^n$, then for every vector $v \in \mathbb{R}^n$ there exist unique vectors $w \in \mathbb{W}$ and $n \in \mathbb{W}^\perp$ such that vector $v = w + n$ (Theorem 3.5.51); this sum is called an **orthogonal decomposition** of v.

★ For every vector v in $\mathbb{R}^n$, and every subspace $\mathbb{W}$ in $\mathbb{R}^n$, $\text{proj}_{\mathbb{W}}(v)$ is the closest vector in $\mathbb{W}$ to v (Theorem 3.5.53).

Introducing linear transformations

★ A transformation/function $T : \mathbb{R}^n \to \mathbb{R}^m$ is called a **linear transformation** if (Definition 3.6.1)
  - $T(\boldsymbol{u} + \boldsymbol{v}) = T(\boldsymbol{u}) + T(\boldsymbol{v})$ for every $\boldsymbol{u}, \boldsymbol{v} \in \mathbb{R}^n$, and
  - $T(c\boldsymbol{v}) = cT(\boldsymbol{v})$ for every $\boldsymbol{v} \in \mathbb{R}^n$ and every scalar c.

  A linear transform maps the unit square to a parallelogram, a unit cube to a parallelepiped, and so on.

- Let A be any given $m \times n$ matrix and define the transformation $T_A : \mathbb{R}^n \to \mathbb{R}^m$ by the matrix multiplication $T_A(\boldsymbol{x}) := A\boldsymbol{x}$ for all $\boldsymbol{x} \in \mathbb{R}^n$. Then T_A is a linear transformation (Theorem 3.6.8).

★ For every linear transformation $T : \mathbb{R}^n \to \mathbb{R}^m$, T is the transformation corresponding to the $m \times n$ matrix (Theorem 3.6.10)

$$A = \begin{bmatrix} T(e_1) & T(e_2) & \cdots & T(e_n) \end{bmatrix}.$$

  This matrix A, often denoted $[T]$, is called the **standard matrix** of the linear transformation T.

★ Recall that in the context of a system of linear equations $A\boldsymbol{x} = \boldsymbol{b}$ with $m \times n$ matrix A, for every $\boldsymbol{b} \in \mathbb{R}^m$ Procedure 3.5.4 finds the smallest solution $\boldsymbol{x} \in \mathbb{R}^n$ (Theorem 3.5.13) to the closest consistent system $A\boldsymbol{x} = \boldsymbol{b}'$ (Theorem 3.5.8). Procedure 3.5.4 forms a linear transformation $T : \mathbb{R}^m \to \mathbb{R}^n$, $\boldsymbol{x} = T(\boldsymbol{b})$. This linear transformation has an $n \times m$ standard matrix A^+ called the **pseudo-inverse**, or **Moore–Penrose inverse**, of matrix A (Theorem 3.6.16).

- For every $m \times n$ matrix A with $\operatorname{rank} A = n$ (so $m \geq n$), the pseudo-inverse $A^+ = (A^\mathsf{T} A)^{-1} A^\mathsf{T}$ (Theorem 3.6.22).

- For every invertible matrix A, the pseudo-inverse $A^+ = A^{-1}$, the inverse (Theorem 3.6.23).

- Let $T : \mathbb{R}^n \to \mathbb{R}^m$ and $S : \mathbb{R}^m \to \mathbb{R}^p$ be linear transformations. Then the composition $S \circ T : \mathbb{R}^n \to \mathbb{R}^p$ is a linear transformation with standard matrix $[S \circ T] = [S][T]$ (Theorem 3.6.25).

- Let S and T be linear transformations from $\mathbb{R}^n$ to $\mathbb{R}^n$ (the same dimension). If $S \circ T = T \circ S = I$, the identity transformation, then S and T are **inverse transformation**s of each other (Definition 3.6.30). Further, we say S and T are **invertible**.

★ Let $T : \mathbb{R}^n \to \mathbb{R}^n$ be an invertible linear transformation. Then its standard matrix $[T]$ is invertible, and $[T^{-1}] = [T]^{-1}$ (Theorem 3.6.33).

Answers to selected activities

3.1.1b, 3.1.3c, 3.1.6c, 3.1.13a, 3.1.18b, 3.1.22c, 3.2.4b, 3.2.9b, 3.2.14c, 3.2.17a, 3.2.23a, 3.2.26c, 3.2.33b, 3.2.41d, 3.2.44a, 3.2.51b, 3.3.3d, 3.3.10a, 3.3.18b, 3.3.21b, 3.3.26d, 3.3.32d, 3.4.2b, 3.4.5d, 3.4.12d, 3.4.17a, 3.4.20d, 3.4.26c, 3.4.34a, 3.4.41d, 3.5.2b, 3.5.7c, 3.5.10b, 3.5.16d, 3.5.22d, 3.5.25c, 3.5.31b, 3.5.33b, 3.5.36d, 3.5.39a, 3.5.42c, 3.5.46a, 3.6.3c, 3.6.5b, 3.6.12c, 3.6.15d, 3.6.18a, 3.6.28c,

Answers to selected exercises

3.1.1b Only B and D.

3.1.2 $a_1 = (-0.3, -5.9), a_2 = (2.1, 3.6),$
$a_3 = (-4.8, -1.3); a_{13} = -4.8,$
$a_{21} = -5.9.$

3.1.4 The columns are the standard unit
vectors $e_1, e_2, e_3,$ and $e_4.$

3.1.5b $C + D = [2\ 0\ -9],$
$C - D = [-6\ -4\ -5]$

3.1.6a $-2A = \begin{bmatrix} 6 & 4 \\ -8 & 4 \\ -4 & 8 \end{bmatrix}, 2A = \begin{bmatrix} -6 & -4 \\ 8 & -4 \\ 4 & -8 \end{bmatrix},$
$3A = \begin{bmatrix} -9 & -6 \\ 12 & -6 \\ 6 & -12 \end{bmatrix}.$

3.1.6c $1.3V = \begin{bmatrix} -3.38 & -4.16 \\ 4.29 & -1.04 \\ -0.39 & 0.39 \end{bmatrix},$
$-3.7V = \begin{bmatrix} 9.62 & 11.84 \\ -12.21 & 2.96 \\ 1.11 & -1.11 \end{bmatrix},$
$2.5V = \begin{bmatrix} -6.5 & -8. \\ 8.25 & -2. \\ -0.75 & 0.75 \end{bmatrix}.$

3.1.9b $Bp = \begin{bmatrix} -21 \\ 3 \end{bmatrix}, Bq = \begin{bmatrix} 8 \\ 3 \end{bmatrix}, Br = \begin{bmatrix} 7 \\ -30 \end{bmatrix}.$

3.1.10(a) $Au = (7, -5), Av = (-6, 3),$
$Aw = (9, -6).$

3.1.10(c) $Cx_1 = (-4.41, 9.66),$
$Cx_2 = (1.42, -1.56),$
$Cx_3 = (-0.47, -0.38).$

3.1.11b $Qp = (4.2, 3.8), Qq = (5.6, -2.2),$
$Qr = (1.6, 6.6).$ Stretches by a factor of
two.

3.1.12a $\begin{bmatrix} -9 & 4 \\ -13 & -16 \end{bmatrix}, \begin{bmatrix} -9 & -15 \\ -13 & 11 \end{bmatrix},$
$\begin{bmatrix} 4 & -15 \\ -16 & 11 \end{bmatrix}, \begin{bmatrix} -9 & 4 & -15 \\ -13 & -16 & 11 \end{bmatrix}.$

3.1.12c $\begin{bmatrix} 6 & 3 \\ 3 & 4 \end{bmatrix}, \begin{bmatrix} 24 & 3 \\ -5 & 4 \end{bmatrix}, \begin{bmatrix} 24 & 6 \\ -5 & 3 \end{bmatrix}, \begin{bmatrix} 24 & 3 & 6 \\ -5 & 4 & 3 \end{bmatrix}.$

3.1.17a $\begin{bmatrix} -2 & 3 & -8 & -2 \\ 3 & -0 & 2 & -4 \end{bmatrix}$

3.1.17c $\begin{bmatrix} 14 & 5 & 3 & 2 \\ 5 & 0 & -1 & 1 \\ 3 & -1 & -6 & -4 \\ 2 & 1 & -4 & 4 \end{bmatrix},$ symmetric

3.1.17e $\begin{bmatrix} -1.5 & -1 \\ -0.6 & -0.4 \\ -1.7 & -5.6 \end{bmatrix}$

3.1.18 yes, no, yes, no.

3.2.1b Inverse.

3.2.1d Not inverse.

3.2.1f Inverse.

3.2.2b No inverse.

3.2.2d $\begin{bmatrix} -1/4 & -1/4 \\ 1/8 & -3/8 \end{bmatrix}$

3.2.3b $x = 0, m = 1$

3.2.3d $x = (-1, 1, 1, 4)$

3.2.4a Theorem 3.2.10 cannot determine
(but Theorem 3.3.27 establishes it must
be invertible).

3.2.4c Not invertible.

3.2.4e Theorem 3.2.10 cannot determine
(but Theorem 3.3.27 establishes it must
be invertible).

3.2.8a $\begin{bmatrix} 0 & 1/16 \\ -1/16 & -1/16 \end{bmatrix}$

3.2.8c $\begin{bmatrix} 10 & -6 & -6 \\ -2 & 6 & 0 \\ -2 & 2 & 4 \end{bmatrix}$

3.2.9a $\operatorname{diag}(9, -5, 4)$

3.2.9c $\operatorname{diag}(-5, 1, 9, 1, 0)$

3.2.9e $\operatorname{diag}_{5 \times 3}(0, -1, -5)$

3.2.10a $x = (-3/2, -1, 2, -2, -4/3)$

3.2.10c $(p, q, r, s) = (-2, 2, -8/3, t)$ for
all t

3.2.11a $\operatorname{diag}(2, 1)$

3.2.11c Not diagonal.

3.2.11e $\operatorname{diag}(0.4, 1.5)$

3.2.12a $\operatorname{diag}(2, 1.5, 1)$

3.2.12c Not diagonal.

3.2.15a Not orthogonal.

3.2.15c Orthogonal.

3.2.15e Orthonormal.

3.2.16a Not orthogonal.

3.2.16c Orthonormal.

3.2.17a Orthogonal set, divide each by seven.

3.2.17c Orthogonal set, divide each by two.

3.2.18a Orthogonal matrix.

3.2.18c Orthogonal matrix.

3.2.18e Orthogonal matrix.

3.2.18g Not orthogonal matrix.

3.2.19a $\theta = 21.04°$

3.2.19c $\theta = 60°$

3.2.20a $(x, y) = (1.4, 5.2)$

3.2.20c $(a, b, c) = (8, 32, -1)/11$

3.2.24a No—the square is deformed.

3.2.24c No—the square is squashed (and rotated/reflected).

3.2.25a Yes—the cube appears rotated.

3.2.25c Yes—the cube appears rotated.

3.3.2b $x = (-2, 2)$

3.3.2d No solution.

3.3.2f $x = (2, -\frac{7}{4}, -\frac{9}{2})$

3.3.3b $x =$
$(27, -12, -24, 9) + (\frac{1}{2}, \frac{1}{2}, \frac{1}{2}, \frac{1}{2})t$

3.3.4a $x = (-8, -6, 1, 2)$

3.3.4c No solution.

3.3.4e No solution.

3.3.9 The theorem applies to the square matrix systems of Exercises 3.3.2 and 3.3.3. The cases with no zero singular value, full rank, have a unique solution. The cases with a zero singular value, rank less than n, either have no solution or an infinite number.

3.3.14b $v_1 \approx (0.1, 1.0), \sigma_1 \approx 1.7$,
$v_2 \approx (1.0, -0.1), \sigma_2 \approx 0.3$.

3.3.14d $v_1 \approx (0.9, -0.4), \sigma_1 \approx 2.3$,
$v_2 \approx (0.4, 0.9), \sigma_2 \approx 0.3$.

3.4.1b Not a subspace.

3.4.1d Not a subspace.

3.4.1f Subspace.

3.4.1h Not a subspace.

3.4.2b Not a subspace.

3.4.2d Subspace.

3.4.2f Not a subspace.

3.4.2h Subspace.

3.4.4a b_1 is in column space; r_1 is in row space.

3.4.4c b_3 is in column space; r_3 is not in row space.

3.4.4e b_5 is in column space; r_5 is in row space.

3.4.5b no

3.4.5d yes

3.4.6b $(2, -8, -2, -3)/9, (8, 2, -1, 2)/9$

3.4.8b (2 d.p.) column space
$\{(0.30, 0.80, 0.07, -0.52),$
$(0.78, 0.07, 0.22, 0.58),$
$(0.14, 0.13, -0.97, 0.16)\}$; row space
$\{(-0.20, 0.90, -0.09, 0.17, -0.34),$
$(-0.65, -0.08, 0.67, -0.31, -0.16),$
$(0.08, 0.31, -0.18, -0.84, 0.41)\}$;
nullspace
$\{(0.73, 0.18, 0.63, -0.10, -0.20),$
$(-0.07, 0.25, 0.33, 0.41, 0.81)\}$.

3.4.8d (2 d.p.) column space
$\{(0.94, -0.24, 0.23, -0.01),$
$(-0.32, -0.43, 0.83, -0.14),$
$(-0.08, -0.50, -0.14, 0.85),$
$(0.07, 0.71, 0.48, 0.50)\}$; row space
$\{(-0.86, -0.03, 0.46, -0.24, 0.01),$
$(0.41, -0.62, 0.54, -0.37, 0.15),$
$(0.24, 0.44, 0.70, 0.42, -0.30),$
$(-0.20, -0.63, -0.02, 0.75,$
$-0.09)\}$; nullspace
$\{(0.00, 0.18, 0.13, 0.27, 0.94)\}$.

3.4.10a 3,4,5

3.4.10c 0,1,2

3.4.10e 0,1,2,3,4

3.4.11a no.

3.4.11c no.

3.5.4 $9.847 \, \text{m/s}^2$

3.5.6 CBA ≈ 2.9WBC $- 0.4$ANZ

3.5.8 To within an arbitrary constant: Acton, 0.25; Barbican, 0.75; Clapham, 2.25; Dalston, -3.25.

3.5.10 To within an arbitrary constant: Algeria, 1.4; Botswana, 0.4; Chad, 0.6; Djibouti, -0.4; Ethiopia, -0.8; Gabon, -1.2.

3.5.16b $x = (1, -2)$

3.5.16d $q = (0.0741, 0.0741, -0.3704)$

3.5.18 $r = (0.70, 0.14, 0.51, 0.50, 0.70, 1.00, 0.90, 0.51, 0.71)$ so the middle left is the most absorbing.

3.5.22a $\operatorname{proj}_u(v) = (1.6, 0.8), x = 0.8$

3.5.22c (2 d.p.)
$\operatorname{proj}_u(v) = (0.67, 0.83, -0.17),$
$x = 0.17$

3.5.22e $\operatorname{proj}_u(v) = e_2, x = 0.5$

3.5.23a (2 d.p.)
$\operatorname{proj}_\mathbb{W}(v) = (1.04, 0.77, -1.31)$

3.5.23c $\operatorname{proj}_\mathbb{W}(v) = v$

3.5.23e (2 d.p.) $\operatorname{proj}_\mathbb{W}(v) =$
$(0.06, -3.28, -1.17, 0.06)$

3.5.24a (2 d.p.) $\begin{bmatrix} 0.88 & -0.08 & 0.31 \\ -0.08 & 0.95 & 0.21 \\ 0.31 & 0.21 & 0.17 \end{bmatrix}$

3.5.24c (2 d.p.) $\begin{bmatrix} 0.63 & -0.42 & 0.05 & 0.23 \\ -0.42 & 0.51 & 0.05 & 0.26 \\ 0.05 & 0.05 & 0.99 & -0.03 \\ 0.23 & 0.26 & -0.03 & 0.86 \end{bmatrix}$

3.5.24e (2 d.p.)
$\begin{bmatrix} 0.66 & -0.30 & -0.16 & 0.21 & -0.25 \\ -0.30 & 0.39 & 0.15 & 0.33 & 0.13 \\ -0.16 & 0.15 & 0.06 & 0.08 & 0.06 \\ 0.21 & 0.33 & 0.08 & 0.79 & -0.06 \\ -0.25 & 0.13 & 0.06 & -0.06 & 0.09 \end{bmatrix}$

3.5.25a (2 d.p.) $b' = (5.84, -19.10, -2.58)$, difference 0.46

3.5.25c (2 d.p.) $b' = (9.27, 10.75, -1.18)$, difference 0.41

3.5.25e (2 d.p.)
$b' = (0.71, -48.77, 27.40, -96.00)$, difference 5.40

3.5.29a The line $x = 2y$

3.5.29c It is not a subspace as it does not include $\mathbf{0}$, and so does not have an orthogonal complement.

3.5.29e The line span$\{(7,-25,8)\}$

3.5.30a $\{(-\frac{6}{7}, \frac{2}{7}, -\frac{3}{7})\}$ is one possibility.

3.5.30c $\{(-0.53, -0.43, 0.73), (0.28, 0.73, 0.63)\}$ is one possibility (2 d.p.).

3.5.30e $\{(-0.73, -0.30, 0.29, -0.22, -0.50), (-0.18, -0.33, 0.24, -0.43, 0.79), (-0.06, -0.74, -0.51, 0.43, 0.06)\}$ is one possibility (2 d.p.).

3.5.32a $\frac{1}{9}(20, 34, 8)$

3.5.32c $\frac{1}{27}(67, 95, 16)$

3.5.33a $(0.96, 2.06, -1.02, -0.88)$

3.5.33c $(0.54, -3.42, 0.54, 1.98)$

3.5.35a $(-2, 4) = (\frac{6}{5}, \frac{8}{5}) + (-\frac{16}{5}, \frac{12}{5})$

3.5.35c $(0, 0) = (0, 0) + (0, 0)$

3.5.36a $(-3, 6, -2) + (-2, -2, -3)$

3.5.36c $(0.31, -0.61, 0.20) + (0.69, -0.39, -2.20)$ (2 d.p.)

3.5.37a $(6, -2, 0, 0) + (-1, -3, 1, -3)$

3.5.37c $(2.1, -0.7, -4.5, -1.5) + (-0.1, -0.3, 0.5, -1.5)$

3.5.38 Either $\mathbb{W} = \operatorname{span}\{(1, 2)\}$ and $\mathbb{W}^\perp = \operatorname{span}\{(-2, 1)\}$, or vice-versa.

3.5.40 Use $x_1x_2x_3x_4$-space. Either $\mathbb{W}$ is x_2-axis, or $x_1x_2x_4$-space, or any plane in $x_1x_2x_4$-space that contains the x_2-axis, and $\mathbb{W}^\perp$ corresponding complement, or vice-versa.

3.6.1b Not a LT.

3.6.1d Maybe a LT, with standard matrix $\begin{bmatrix} 0.8 & -0.3 \\ 0.4 & 1.0 \end{bmatrix}$

3.6.1f Maybe a LT, with standard matrix $\begin{bmatrix} 0.9 & -2.2 \\ -1.7 & -1.6 \end{bmatrix}$

3.6.3b Not a LT.

3.6.3d Not a LT.

3.6.7b $[B] = \begin{bmatrix} 0 & 0 \\ 3 & 7 \\ 0 & -2 \\ 0 & -3 \end{bmatrix}$

3.6.7d Not a LT.

3.6.7f Not a LT.

3.6.12b $B^+ = \begin{bmatrix} 0.57 & -0.86 \\ -0.57 & 2.86 \\ -0.29 & 1.43 \end{bmatrix}$ (2 d.p.)

3.6.12d $D^+ = \begin{bmatrix} 0.17 & -0.12 & -0.15 & 0.31 \\ -0.05 & 0.04 & -0.01 & -0.08 \\ -0.16 & 0.14 & -0.12 & -0.20 \end{bmatrix}$

(2 d.p.)

3.6.12f $F^+ =$
$\begin{bmatrix} -0.07 & -0.89 & 0.28 & 0.45 & 0.33 \\ -0.24 & -0.25 & -2.02 & 0.36 & 1.32 \\ -0.33 & -0.35 & -0.27 & -0.40 & -0.07 \\ 0.36 & 0.07 & 1.10 & 0.15 & 0.13 \\ 1.16 & -1.07 & 0.61 & 1.24 & 2.11 \\ -0.57 & 0.36 & -1.05 & -0.89 & -0.82 \end{bmatrix}$

(2 d.p.)

3.6.16b Equivalent to rotation by $60°$ with

matrix $\begin{bmatrix} 1/2 & -\sqrt{3}/2 \\ \sqrt{3}/2 & 1/2 \end{bmatrix}$.

3.6.16d Equivalent to rotation by $-90°$ (clockwise $90°$) with matrix $\begin{bmatrix} 0 & 1 \\ -1 & 0 \end{bmatrix}$.

3.6.16f Equivalent to reflection in the line $y = x$ with matrix $\begin{bmatrix} 0 & 1 \\ 1 & 0 \end{bmatrix}$.

3.6.17b $[S \circ T] = \begin{bmatrix} -6 \\ 9 \end{bmatrix}$ and $[T \circ S]$ does not exist.

3.6.17d $[S \circ T]$ does not exist, and

$[T \circ S] = \begin{bmatrix} 30 & -6 & 24 \\ -15 & 3 & -12 \\ 15 & -3 & 12 \\ 5 & -1 & 4 \end{bmatrix}$

3.6.18b Inverse $\approx \begin{bmatrix} 1.43 & -0.44 \\ -0.00 & 0.77 \end{bmatrix}$

3.6.18d Inverse $\approx \begin{bmatrix} -0.23 & -0.53 \\ -0.60 & 0.26 \end{bmatrix}$

4 Eigenvalues and eigenvectors of symmetric matrices

Recall (Section 3.1.2) that a symmetric matrix A is a square matrix such that $A^{\mathrm{T}} = A$, that is, $a_{ij} = a_{ji}$. For example, of the following two matrices, the first is symmetric, but the second is not:

$$\begin{bmatrix} -2 & 4 & 0 \\ 4 & 2 & -3 \\ 0 & -3 & 1 \end{bmatrix}; \qquad \begin{bmatrix} -1 & 3 & 0 \\ 1 & 1 & 0 \\ 0 & -3 & 1 \end{bmatrix}.$$

Example 4.0.1 Compute some SVDs of random symmetric matrices, $A = USV^{\mathrm{T}}$, observe in the SVDs that the columns of U are always $\pm$ the columns of V (well, almost always).

Solution: Repeat as often us you like for any size of square matrix that you like (one example is recorded here to two decimal places).

(a) Generate in MATLAB/Octave some random symmetric matrix by adding a random matrix to its transpose with A=randn(5); A=A+A' (Table 3.1):

Linear Algebra for the 21st Century. A. J. Roberts, Oxford University Press (2020). © A. J. Roberts.
DOI: 10.1093/oso/9780198856399.003.0004

A =

-0.45	-0.18	1.59	-0.96	-0.54
-0.18	-0.24	-1.04	0.14	0.80
1.59	-1.04	-2.87	-0.40	1.11
-0.96	0.14	-0.40	-0.26	-1.90
-0.54	0.80	1.11	-1.90	1.64

This matrix is symmetric as $a_{ij} = a_{ji}$.

(b) Find an SVD via [U, S, V] =svd (A) (2 d.p.)

U =

-0.41	-0.09	-0.28	-0.67	0.55
0.25	-0.11	-0.05	0.53	0.80
0.82	-0.19	-0.40	-0.36	-0.07
-0.15	0.51	-0.80	0.27	-0.11
-0.27	-0.83	-0.36	0.25	-0.22

S =

4.28	0	0	0	0
0	3.12	0	0	0
0	0	1.65	0	0
0	0	0	1.14	0
0	0	0	0	0.51

V =

0.41	-0.09	0.28	-0.67	-0.55
-0.25	-0.11	0.05	0.53	-0.80
-0.82	-0.19	0.40	-0.36	0.07
0.15	0.51	0.80	0.27	0.11
0.27	-0.83	0.36	0.25	0.22

Observe that the second and fourth columns of U and V are identical, and the other pairs of columns of U and V have opposite signs.

Repeat for different random symmetric matrices and observe that $u_j = \pm v_j$ for every column j (almost always). ☐

Why, for symmetric matrices, are the columns of U (almost) always ± the columns of V? The answer is connected to the following rearrangement of an SVD. Because $A = USV^T$, post-multiplying by V gives $AV = USV^TV = US$, and then the jth column of the two sides of $AV = US$ determines $Av_j = \sigma_j u_j$. Example 4.0.1 indicates that for symmetric matrices A we find $u_j = \pm v_j$ (almost always) so this last equation becomes $Av_j = (\pm\sigma_j)v_j$. This equation is of the important form $Av = \lambda v$. (The symbol λ is the Greek letter lambda, and denotes eigenvalues.) This form is important because it is the mathematical expression of the following geometric question: for what vectors v does multiplication by A just stretch/shrink v by some scalar λ?

Solid modelling Lean with a hand on a table/wall: the force changes depending upon the orientation of the surface. Similarly inside any solid: the internal forces $= A\mathbf{v}$ where $\mathbf{v}$ is the orthogonal unit vector to the internal 'surface'. In this solid-force scenario, the matrix A is always symmetric. To know whether a material breaks apart under pulling, or crumbles under compression, we need to know where the extreme forces are. The extreme forces are found as solutions to $A\mathbf{v} = \lambda\mathbf{v}$ where $\mathbf{v}$ gives the direction and λ the strength of the extreme force. To understand the potential failure of the material we thus need to solve equations in the form $A\mathbf{v} = \lambda\mathbf{v}$.

4.1 Introduction to eigenvalues and eigenvectors

This chapter focuses on some marvellous properties of symmetric matrices. Nonetheless it defines some basic concepts which also apply to general matrices. Chapter 7 explores analogous properties for such general matrices. The marvellously useful properties developed here result from asking: for which vectors does multiplication by a given matrix simply stretch or shrink the vector, without changing direction?

Definition 4.1.1   *Let A be a square matrix. A scalar λ (lambda) is called an **eigenvalue** of A if there is a nonzero vector $\mathbf{x}$ such that $A\mathbf{x} = \lambda\mathbf{x}$.[1] Such a vector $\mathbf{x}$ is called an **eigenvector** of A corresponding to the eigenvalue λ.*

Example 4.1.2   Consider the symmetric matrix

$$A = \begin{bmatrix} 1 & -1 & 0 \\ -1 & 2 & -1 \\ 0 & -1 & 1 \end{bmatrix}.$$

(a) Verify an eigenvector is $(1,0,-1)$. What is the corresponding eigenvalue?

    Solution: The simplest approach is to multiply the matrix by the given vector and see what happens. For vector $\mathbf{x} = (1,0,-1)$,

$$A\mathbf{x} = \begin{bmatrix} 1 & -1 & 0 \\ -1 & 2 & -1 \\ 0 & -1 & 1 \end{bmatrix} \begin{bmatrix} 1 \\ 0 \\ -1 \end{bmatrix} = \begin{bmatrix} 1 \\ 0 \\ -1 \end{bmatrix} = 1 \cdot \mathbf{x}.$$

    Hence $(1,0,-1)$ is an eigenvector of A corresponding to the eigenvalue $\lambda = 1$.   □

(b) Verify that $(2,-4,2)$ is an eigenvector. What is its corresponding eigenvalue.

---

[1] The prefix "eigen-" comes from the German word for 'owning' or 'belonging to' in relation to the originating matrix.

Solution: For vector $x = (2, -4, 2)$,

$$Ax = \begin{bmatrix} 1 & -1 & 0 \\ -1 & 2 & -1 \\ 0 & -1 & 1 \end{bmatrix} \begin{bmatrix} 2 \\ -4 \\ 2 \end{bmatrix} = \begin{bmatrix} 6 \\ -12 \\ 6 \end{bmatrix} = 3 \cdot x.$$

Hence $(2, -4, 2)$ is an eigenvector of A corresponding to the eigenvalue $\lambda = 3$. ☐

(c) Verify that $(1, 2, 1)$ is not an eigenvector.

Solution: For vector $x = (1, 2, 1)$,

$$Ax = \begin{bmatrix} 1 & -1 & 0 \\ -1 & 2 & -1 \\ 0 & -1 & 1 \end{bmatrix} \begin{bmatrix} 1 \\ 2 \\ 1 \end{bmatrix} = \begin{bmatrix} -1 \\ 2 \\ -1 \end{bmatrix} \not\propto \begin{bmatrix} 1 \\ 2 \\ 1 \end{bmatrix}.$$

If there was a constant of proportionality (an eigenvalue), then the first component would require the constant $\lambda = -1$ but the second component would require $\lambda = +1$ which is a contradiction. Hence $(1, 2, 1)$ is not an eigenvector of A. ☐

(d) Use inspection to guess and verify another eigenvector (not proportional to either of the above two). What is its eigenvalue?

Solution: Inspection is useful if it is quick: here one might quickly spot that the elements in each row of A sum to the same thing, namely zero, so try vector $x = (1, 1, 1)$:

$$Ax = \begin{bmatrix} 1 & -1 & 0 \\ -1 & 2 & -1 \\ 0 & -1 & 1 \end{bmatrix} \begin{bmatrix} 1 \\ 1 \\ 1 \end{bmatrix} = \begin{bmatrix} 0 \\ 0 \\ 0 \end{bmatrix} = 0 \cdot x.$$

Hence $(1, 1, 1)$ is an eigenvector of A corresponding to the eigenvalue $\lambda = 0$. ☐

Activity 4.1.3 Which of the following vectors is an eigenvector of the symmetric matrix $\begin{bmatrix} -1 & 12 \\ 12 & 6 \end{bmatrix}$?

(a) $\begin{bmatrix} -3 \\ 1 \end{bmatrix}$ (b) $\begin{bmatrix} 4 \\ -3 \end{bmatrix}$ (c) $\begin{bmatrix} -1 \\ 2 \end{bmatrix}$ (d) $\begin{bmatrix} 2 \\ 1 \end{bmatrix}$

Importantly, eigenvectors tell us the key directions of a given matrix: the directions in which the multiplication by a matrix is to simply stretch, shrink, or reverse by a factor: the factor being the corresponding eigenvalue. In two-dimensional plots we can graphically estimate eigenvectors and eigenvalues. For some examples and exercises, we plot a given vector x and join onto its head the vector Ax:

- if both x and Ax are aligned in the same direction, or the opposite direction, then x is an eigenvector;
- if they form some other angle, then x is not an eigenvector.

Example 4.1.4 Let the matrix $A = \begin{bmatrix} 1 & -1/2 \\ -1/2 & 1 \end{bmatrix}$. The plot below-left shows the vector $x = (1, \frac{1}{2})$, and adjoined to its head the matrix-vector product $Ax = (\frac{3}{4}, 0)$: because the two are at an angle, $(1, \frac{1}{2})$ is not an eigenvector.

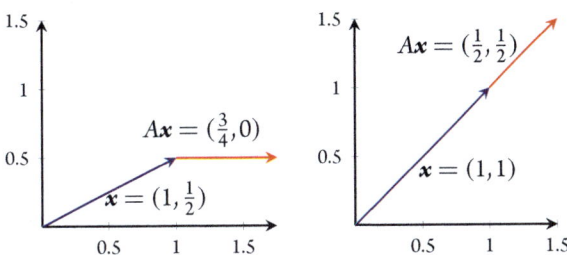

However, as plotted above-right, for the vector $x = (1, 1)$ the matrix-vector product $Ax = (\frac{1}{2}, \frac{1}{2})$ and the plot of these vectors head-to-tail illustrates that they are aligned in the same direction. Because of the alignment, $(1, 1)$ is an eigenvector of this matrix. The constant of proportionality is the corresponding eigenvalue: here $Ax = (\frac{1}{2}, \frac{1}{2}) = \frac{1}{2}(1, 1) = \frac{1}{2}x$ so the eigenvalue is $\lambda = \frac{1}{2}$. This eigenvalue of $\frac{1}{2}$ is seen graphically by the (red) vector Ax being half the length of the (blue) vector x. □

Activity 4.1.5 For some matrix A, the following pictures plot a vector x and the corresponding product Ax, head-to-tail. Which picture indicates that x is an eigenvector of the matrix?

(a)   (b)   (c)   (d)

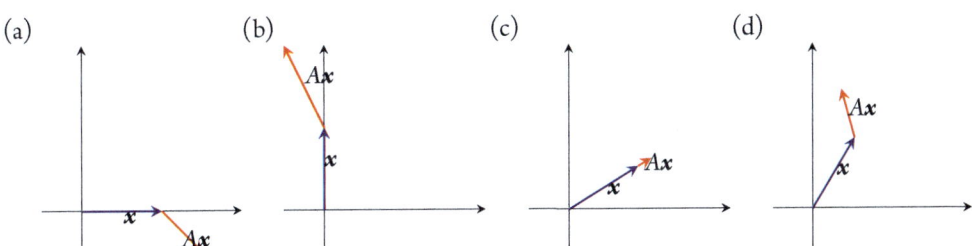

Activity 4.1.6 Further, for the picture in Activity 4.1.5 that indicates x is an eigenvector, is the corresponding eigenvalue λ:

(a) $0 > \lambda$ (b) $\lambda > 1$ (c) $1 > \lambda > 0.5$ (d) $0.5 > \lambda > 0$

As in the next example, we sometimes plot for many directions x a diagram of vector Ax adjoined head-to-tail to vector x. Then inspection estimates the eigenvectors and corresponding eigenvalues (Schonefeld, 1995).

Example 4.1.7 *(graphical eigenvectors one)* The plot on the right shows many unit vectors x (blue), and for some symmetric matrix A the corresponding vectors Ax (red) adjoined. Estimate which directions x are eigenvectors, and for each eigenvector estimate the corresponding eigenvalue.

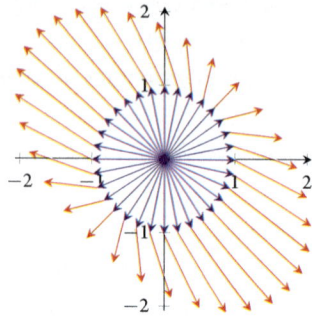

(The MATLAB function eigshow(A) provides an interactive alternative to this static view. Download from the internet if using a recent version of MATLAB.)

Solution: We seek vectors x, such that Ax is in the same direction (to graphical accuracy). It appears that vectors at $45°$ to the axes are the only ones for which Ax is in the same direction as x:

- the two (blue) vectors $\pm(0.7, 0.7)$ appear to be shrunk to length a half (red) so we estimate that two eigenvectors are $x \approx \pm(0.7, 0.7)$ and the corresponding eigenvalue $\lambda \approx 0.5$;
- the two (blue) vectors $\pm(0.7, -0.7)$ appear to be stretched by a factor about 1.5 (red) so we estimate that two eigenvectors are $x \approx \pm(0.7, -0.7)$ and the corresponding eigenvalue is $\lambda \approx 1.5$;
- and for no other (unit) vector x is Ax aligned with x.

Any multiple of these eigenvectors are also eigenvectors so we may report the directions more simply, perhaps $(1, 1)$ and $(1, -1)$ respectively. □

Example 4.1.8 *(graphical eigenvectors two)*
The plot on the right shows many unit vectors x (blue), and for some symmetric matrix A the corresponding vectors Ax (red) adjoined. Estimate which directions x are eigenvectors, and for each eigenvector estimate the corresponding eigenvalue.

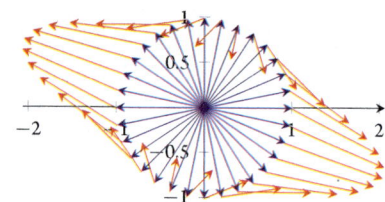

Solution: We seek vectors x such that Ax is in the same direction (to graphical accuracy):

- the two (blue) vectors $\pm(0.9, -0.3)$ appear stretched a little by a factor about 1.2 (red) so we estimate that eigenvectors are $x \propto (0.9, -0.3)$ and the corresponding eigenvalue is $\lambda \approx 1.2$;
- the two (blue) vectors $\pm(0.3, 0.9)$ appear shrunk and *reversed* by a factor about 0.4 (red) so we estimate that eigenvectors are $x \propto (0.3, 0.9)$ and the corresponding eigenvalue is $\lambda \approx -0.4$ —negative because the direction is reversed;
- and for no other (unit) vector x is Ax aligned with x.

If this matrix arose in the description of forces inside a solid, then the forces would be compressive in directions $\pm(0.3, 0.9)$, and the forces would be (tension) 'ripping apart' the solid in directions $\pm(0.9, -0.3)$. □

Example 4.1.9 *(diagonal matrix)* The eigenvalues of a (square) diagonal matrix are the entries on the diagonal. Consider an $n \times n$ diagonal matrix

$$D = \begin{bmatrix} d_1 & 0 & \cdots & 0 \\ 0 & d_2 & & 0 \\ \vdots & & \ddots & \vdots \\ 0 & 0 & \cdots & d_n \end{bmatrix}.$$

Multiply by the standard unit vectors $e_1, e_2, \ldots, e_n$ in turn:

$$D e_1 = \begin{bmatrix} d_1 & 0 & \cdots & 0 \\ 0 & d_2 & & 0 \\ \vdots & & \ddots & \vdots \\ 0 & 0 & \cdots & d_n \end{bmatrix} \begin{bmatrix} 1 \\ 0 \\ \vdots \\ 0 \end{bmatrix} = \begin{bmatrix} d_1 \\ 0 \\ \vdots \\ 0 \end{bmatrix} = d_1 e_1 ;$$

$$D e_2 = \begin{bmatrix} d_1 & 0 & \cdots & 0 \\ 0 & d_2 & & 0 \\ \vdots & & \ddots & \vdots \\ 0 & 0 & \cdots & d_n \end{bmatrix} \begin{bmatrix} 0 \\ 1 \\ \vdots \\ 0 \end{bmatrix} = \begin{bmatrix} 0 \\ d_2 \\ \vdots \\ 0 \end{bmatrix} = d_2 e_2 ;$$

$$\vdots$$

$$D e_n = \begin{bmatrix} d_1 & 0 & \cdots & 0 \\ 0 & d_2 & & 0 \\ \vdots & & \ddots & \vdots \\ 0 & 0 & \cdots & d_n \end{bmatrix} \begin{bmatrix} 0 \\ \vdots \\ 0 \\ 1 \end{bmatrix} = \begin{bmatrix} 0 \\ \vdots \\ 0 \\ d_n \end{bmatrix} = d_n e_n .$$

Thus, by Definition 4.1.1, each diagonal element d_j is an eigenvalue of the diagonal matrix, and the standard unit vector e_j is a corresponding eigenvector. □

Eigenvalues The 3×3 matrix of Example 4.1.2 has three eigenvalues. The 2×2 matrices underlying Examples 4.1.7 and 4.1.8 both have two eigenvalues. Example 4.1.9 shows an $n \times n$ diagonal matrix has n eigenvalues. The next section establishes the general pattern that an $n \times n$ *symmetric matrix* generally has n real eigenvalues.

However, the eigenvalues of non-symmetric matrices are more complex (in both senses of the word) as explored by Chapter 7.

Eigenvectors It is the direction of eigenvectors that is important. In Example 4.1.2 any nonzero multiple of $(1, -2, 1)$, positive or negative, is also an eigenvector corresponding to eigenvalue $\lambda = 3$. In the diagonal matrices of Example 4.1.9, a straightforward extension of the working shows that any nonzero multiple of the standard unit vector e_j is an eigenvector corresponding to the eigenvalue d_j. Let's collect all possible eigenvectors into a subspace.

Hereafter, "iff" is short for "if and only if".

Theorem 4.1.10 *Let A be a square matrix. A scalar λ is an eigenvalue of A iff the homogeneous linear system $(A - \lambda I)x = 0$ has nonzero solutions x. The set of all eigenvectors corresponding to any* one eigenvalue λ, together with the zero vector, is a subspace; the subspace is called the **eigenspace** of λ and is denoted by $\mathbb{E}_\lambda$.

Proof. From Definition 4.1.1, $Ax = \lambda x$ rearranged is $Ax - \lambda x = 0$, that is, $Ax - \lambda Ix = 0$, which upon factoring becomes $(A - \lambda I)x = 0$, and vice versa. Also, eigenvectors x must be nonzero, so the homogeneous system $(A - \lambda I)x = 0$ must have nonzero solutions. Theorem 3.4.14 assures us that the set of solutions to a homogeneous system, here $(A - \lambda I)x = 0$ for any given λ, is a subspace. Hence the set of eigenvectors for any given eigenvalue λ, nonzero solutions, together with 0, form a subspace. $\qquad\square$

Example 4.1.11 Reconsider the symmetric matrix

$$A = \begin{bmatrix} 1 & -1 & 0 \\ -1 & 2 & -1 \\ 0 & -1 & 1 \end{bmatrix}$$

of Example 4.1.2. Find the eigenspaces $\mathbb{E}_1$, $\mathbb{E}_3$, and $\mathbb{E}_0$.

Solution:

- The eigenspace $\mathbb{E}_1$ is the set of solutions of

$$(A - 1I)x = \begin{bmatrix} 0 & -1 & 0 \\ -1 & 1 & -1 \\ 0 & -1 & 0 \end{bmatrix} x = 0.$$

That is, $-x_2 = 0$, $-x_1 + x_2 - x_3 = 0$, and $-x_2 = 0$. Hence, $x_2 = 0$ and $x_1 = -x_3$. A general solution is $x = (-t, 0, t)$ so the eigenspace $\mathbb{E}_1 = \{(-t, 0, t) : t \in \mathbb{R}\} =$ span$\{(-1, 0, 1)\}$.

- The eigenspace $\mathbb{E}_3$ is the set of solutions of

$$(A - 3I)x = \begin{bmatrix} -2 & -1 & 0 \\ -1 & -1 & -1 \\ 0 & -1 & -2 \end{bmatrix} x = 0.$$

That is, $-2x_1 - x_2 = 0$, $-x_1 - x_2 - x_3 = 0$, and $-x_2 - 2x_3 = 0$. From the first $x_2 = -2x_1$ which substituted into the third gives $2x_3 = -x_2 = 2x_1$. This suggests that we try $x_1 = t$, $x_2 = -2t$ and $x_3 = t$; that is, $x = (t, -2t, t)$. This also satisfies the second equation and so is a general solution. So the eigenspace $\mathbb{E}_3 = \{(t, -2t, t) : t \in \mathbb{R}\} =$ span$\{(1, -2, 1)\}$.

- The eigenspace $\mathbb{E}_0$ is the set of solutions of

$$(A - 0I)x = \begin{bmatrix} 1 & -1 & 0 \\ -1 & 2 & -1 \\ 0 & -1 & 1 \end{bmatrix} x = 0.$$

That is, $x_1 - x_2 = 0$, $-x_1 + 2x_2 - x_3 = 0$, and $-x_2 + x_3 = 0$. The first and third of these require $x_1 = x_2 = x_3$ which also satisfies the second. Thus a general solution is $\mathbf{x} = (t, t, t)$ so the eigenspace $\mathbb{E}_0 = \{(t, t, t) : t \in \mathbb{R}\} = \text{span}\{(1, 1, 1)\}$.

☐

Activity 4.1.12 Which line, in the xy-plane, is the eigenspace corresponding to the eigenvalue -5 of the matrix $\begin{bmatrix} 3 & 4 \\ 4 & -3 \end{bmatrix}$?

(a) $x + 2y = 0$ (c) $2x + y = 0$

(b) $y = 2x$ (d) $x = 2y$

Example 4.1.13 *(graphical eigenspaces)* The plot to the right shows unit vectors $\mathbf{x}$ (blue), and for the matrix A of Example 4.1.8 the corresponding vectors $A\mathbf{x}$ (red) adjoined. Estimate and draw the eigenspaces of matrix A.

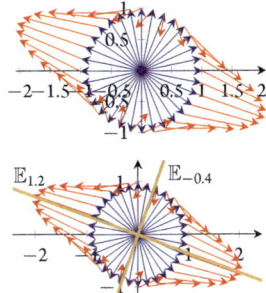

Solution: Example 4.1.8 found directions in which $A\mathbf{x}$ is aligned with $\mathbf{x}$. Then the corresponding eigenspace (brown to the right) is every vector in the line aligned with that direction, including the opposite direction.

☐

Example 4.1.14 Eigenspaces may be multi-dimensional. Find the eigenspaces of the diagonal matrix

$$D = \begin{bmatrix} -\frac{1}{3} & 0 & 0 \\ 0 & \frac{3}{2} & 0 \\ 0 & 0 & \frac{3}{2} \end{bmatrix}.$$

Solution: Example 4.1.9 shows that this diagonal matrix has two distinct eigenvalues $\lambda = -\frac{1}{3}$ and $\lambda = \frac{3}{2}$.

- Eigenvectors corresponding to eigenvalue $\lambda = -\frac{1}{3}$ satisfy

$$(D + \tfrac{1}{3}I)\mathbf{x} = \begin{bmatrix} 0 & 0 & 0 \\ 0 & \frac{11}{6} & 0 \\ 0 & 0 & \frac{11}{6} \end{bmatrix} \mathbf{x} = \mathbf{0}.$$

Hence $\mathbf{x} = t\mathbf{e}_1$ are eigenvectors, for every nonzero t. The eigenspace $\mathbb{E}_{-1/3} = \{t\mathbf{e}_1 : t \in \mathbb{R}\} = \text{span}\{\mathbf{e}_1\}$.

- Eigenvectors corresponding to eigenvalue $\lambda = \frac{3}{2}$ satisfy

$$(D - \tfrac{3}{2}I)\mathbf{x} = \begin{bmatrix} -\frac{11}{6} & 0 & 0 \\ 0 & 0 & 0 \\ 0 & 0 & 0 \end{bmatrix} \mathbf{x} = \mathbf{0}.$$

Hence $x = te_2 + se_3$ are eigenvectors, for every nonzero t and s. Then the eigenspace $\mathbb{E}_{3/2} = \{te_2 + se_3 : t,s \in \mathbb{R}\} = \text{span}\{e_2, e_3\}$ is two-dimensional. ☐

Definition 4.1.15 *For every real symmetric matrix A, the **multiplicity** of an eigenvalue λ of A is the dimension of the corresponding eigenspace $\mathbb{E}_\lambda$.*

Be aware that for general matrices (non-symmetric) we subsequently define the multiplicity of an eigenvalue differently (Definition 7.1.7). But for real symmetric matrices the two definitions are the same.[2]

Example 4.1.16 The multiplicity of the various eigenvalues in earlier examples are the following.

Example 4.1.11 Recall that in this example:

- the eigenspace $\mathbb{E}_1 = \text{span}\{(1,0,-1)\}$ has dimension one, so the multiplicity of eigenvalue $\lambda = 1$ is one;
- the eigenspace $\mathbb{E}_3 = \text{span}\{(1,-2,1)\}$ has dimension one, so the multiplicity of eigenvalue $\lambda = 3$ is one; and
- the eigenspace $\mathbb{E}_0 = \text{span}\{(1,1,1)\}$ has dimension one, so the multiplicity of eigenvalue $\lambda = 0$ is one.

Example 4.1.14 Recall that in this example:

- the eigenspace $\mathbb{E}_{-1/3} = \text{span}\{e_1\}$ has dimension one, so the multiplicity of eigenvalue $\lambda = -1/3$ is one; and
- the eigenspace $\mathbb{E}_{3/2} = \text{span}\{e_2, e_3\}$ has dimension two, so the multiplicity of eigenvalue $\lambda = 3/2$ is two. ☐

4.1.1 Systematically find eigenvalues and eigenvectors

Computer packages easily compute eigenvalues and eigenvectors for us. Sometimes we need to explicitly see dependence upon a parameter, so this subsection also develops how to find by hand the eigenvalues and eigenvectors of small matrices. We start with computation.

Compute eigenvalues and eigenvectors

Compute in MATLAB/Octave `[V,D]=eig(A)` computes eigenvalues and eigenvectors (Table 4.1). The function `eig()` places eigenvalues in the diagonal of $D = \text{diag}(\lambda_1, \lambda_2, \ldots, \lambda_n)$. Then the jth column of V is a unit eigenvector corresponding to the jth eigenvalue λ_j: $V = \begin{bmatrix} v_1 & v_2 & \cdots & v_n \end{bmatrix}$. If the matrix A is real and symmetric, then V is an orthogonal matrix (Theorem 4.2.19).

[2] Section 7.3 discusses that for non-symmetric matrices the dimension of an eigenspace may be less than the multiplicity of an eigenvalue (Theorem 7.3.14).

Table 4.1 As well as the MATLAB/Octave commands and operations listed in Tables 1.2, 2.3 and 3.1 to 3.3 we need the eigenvector function eig().

- [V,D] =eig(A) computes eigenvectors and the eigenvalues of the $n \times n$ square matrix A.
  - The n eigenvalues of A (repeated according to their multiplicity, Definition 4.1.15) form the diagonal of $n \times n$ square matrix $D = \text{diag}(\lambda_1, \lambda_2, \ldots, \lambda_n)$ (in no specific order).
  - Corresponding to the jth eigenvalue λ_j, the jth column of $n \times n$ square matrix V is an eigenvector (of unit length).
- eig(A) by itself just reports, in a vector, the eigenvalues of square matrix A (repeated according to their multiplicity, Definitions 4.1.15 and 7.1.7).
- If the matrix A is a real symmetric matrix, then the eigenvalues and eigenvectors are all real (Theorem 4.2.9), and the eigenvector matrix V is orthogonal (Theorem 4.2.11). If the matrix A is either not symmetric, or is complex valued, then the eigenvalues and eigenvectors may be non-real complex valued.

Example 4.1.17 Reconsider the symmetric matrix of Example 4.1.2:

$$A = \begin{bmatrix} 1 & -1 & 0 \\ -1 & 2 & -1 \\ 0 & -1 & 1 \end{bmatrix}.$$

Use MATLAB/Octave to find its eigenvalues and corresponding eigenvectors. Confirm that $AV = VD$ for matrices $V = \begin{bmatrix} v_1 & v_2 & \cdots & v_n \end{bmatrix}$ and $D = \text{diag}(\lambda_1, \lambda_2, \ldots, \lambda_n)$, and confirm that the computed V is an orthogonal matrix.

Solution: Enter the matrix into MATLAB/Octave and execute eig():

```
A= [1 -1 0;-1 2 -1;0 -1 1]
[V,D] =eig(A)
```

The output is

```
A =
      1    -1     0
     -1     2    -1
      0    -1     1
V =
  -0.5774   -0.7071    0.4082
  -0.5774    0.0000   -0.8165
  -0.5774    0.7071    0.4082
D =
   0.0000        0         0
        0   1.0000         0
        0        0    3.0000
```

- The first diagonal element of D is zero[3] so eigenvalue $\lambda_1 = 0$. A corresponding eigenvector is the first column of V, namely $v_1 = -0.5774(1,1,1)$; since eigenvectors can be scaled by a constant, we could more simply say an eigenvector is $v_1 = (1,1,1)$.

- The second diagonal element of D is one so eigenvalue $\lambda_2 = 1$. A corresponding eigenvector is the second column of V, namely $v_2 = 0.7071(-1,0,1)$; we could more simply say an eigenvector is $v_2 = (-1,0,1)$.

- The third diagonal element of D is three so eigenvalue $\lambda_3 = 3$. A corresponding eigenvector is the third column of V, namely $v_3 = 0.4082(1,-2,1)$; we could also say an eigenvector is $v_3 = (1,-2,1)$.

Confirm $AV = VD$ simply by computing $A*V-V*D$ and seeing it is zero (to numerical error of circa 10^{-16} shown by e-16 in the computer's output):

```
ans =
    5.7715e-17   -1.1102e-16    4.4409e-16
    1.6874e-16    1.2490e-16    0.0000e+00
   -5.3307e-17   -1.1102e-16   -2.2204e-16
```

To verify the computed matrix V is orthogonal (Definition 3.2.43), check that $V' *V$ gives the identity:

```
ans =
    1.0000   -0.0000    0.0000
   -0.0000    1.0000    0.0000
    0.0000    0.0000    1.0000
```

□

Activity 4.1.18 The statement [V,D]=eig(A) returns the following result (2 d.p.)

```
V =
    0.50    0.50   -0.10   -0.70
    0.10   -0.70    0.50   -0.50
   -0.70   -0.10   -0.50   -0.50
    0.50   -0.50   -0.70    0.10
D =
   -0.10       0       0       0
       0    0.10       0       0
       0       0    0.30       0
       0       0       0    0.50
```

Which of the following is *not* an eigenvalue of the matrix A?

(a) −0.1   (b) 0.1   (c) −0.5   (d) 0.5

---

[3] The function eig() actually computes the eigenvalue to be 10^{-16}, which is effectively zero, as 10^{-15} is the typical level of relative error in computer calculations.

Example 4.1.19 *(application to vibrations)* Consider three masses in a row connected by two springs: on a tiny scale this could represent a molecule of carbon dioxide (CO_2). For simplicity suppose the three masses are equal, and the spring strengths are equal. Define $y_i(t)$ to be the distance from equilibrium of the ith mass. Newton's law for bodies says the acceleration of the mass, d^2y_i/dt^2, is proportional to the forces due to the springs. Hooke's law for springs says the force is proportional to the stretching/compression of the springs, here $y_2 - y_1$ and $y_3 - y_2$. For algebraic simplicity, suppose the constants of proportionality are all one.

- The left mass (y_1) is accelerated by the spring connecting it to the middle mass (y_2); that is, $d^2y_1/dt^2 = y_2 - y_1$.
- The middle mass (y_2) is accelerated by the springs connecting it to the left mass (y_1) and to the right mass (y_3); that is, $d^2y_2/dt^2 = (y_1 - y_2) + (y_3 - y_2) = y_1 - 2y_2 + y_3$.
- The right mass (y_3) is accelerated by the spring connecting it to the middle mass (y_2); that is, $d^2y_3/dt^2 = y_2 - y_3$.

Guess that there are solutions oscillating in time, so let's see if we can find solutions $y_i(t) = x_i \cos(ft)$ for some as yet unknown frequency f. Substitute and the three differential equations become

$$-f^2 x_1 \cos(ft) = x_2 \cos(ft) - x_1 \cos(ft),$$
$$-f^2 x_2 \cos(ft) = x_1 \cos(ft) - 2x_2 \cos(ft) + x_3 \cos(ft),$$
$$-f^2 x_3 \cos(ft) = x_2 \cos(ft) - x_3 \cos(ft).$$

These are satisfied for all time t only if the coefficients of the cosines are equal on each side of each equation:

$$-f^2 x_1 = x_2 - x_1,$$
$$-f^2 x_2 = x_1 - 2x_2 + x_3,$$
$$-f^2 x_3 = x_2 - x_3.$$

Moving the terms on the left to the right, and all terms on the right to the left, this becomes the eigen-problem $Ax = \lambda x$ for symmetric matrix A of Example 4.1.17 and for eigenvalue $\lambda = f^2$, the square of the as yet unknown frequency. The symmetry of matrix A reflects Newton's law that every action has an equal and opposite reaction: symmetric matrices arise commonly in applications.

Example 4.1.17 tells us that there are three possible eigenvalue and eigenvector solutions for us to interpret.

- Eigenvalue $\lambda = 1$ and corresponding eigenvector $x \propto (-1, 0, 1)$ corresponds to oscillations of frequency $f = \sqrt{\lambda} = \sqrt{1} = 1$. The eigenvector $(-1, 0, 1)$ shows that the middle mass is stationary while the outer two masses oscillate in and out, in opposition to each other.
- Eigenvalue $\lambda = 3$ and corresponding eigenvector $x \propto (1, -2, 1)$ corresponds to oscillations of higher frequency $f = \sqrt{\lambda} = \sqrt{3}$. The eigenvector $(1, -2, 1)$ shows the outer two masses oscillate together, and the middle mass moves opposite to them.

- Eigenvalue $\lambda = 0$ and corresponding eigenvector $x \propto (1,1,1)$ appears as oscillations of zero frequency $f = \sqrt{\lambda} = \sqrt{0} = 0$ which is a static displacement. The eigenvector $(1,1,1)$ shows that the static displacement is that of all three masses moved all together as a unit.

That these three solutions combine together to form a general solution of the system of differential equations is a topic for a course on differential equations. □

Example 4.1.20 (Sierpinski network) Consider three triangles formed into a single triangle (as shown to the right)—perhaps because triangles make strong structures, or perhaps because of a hierarchical computer/social network. Such a picture is called a *network* because it is formed from a set of nodes (the discs) connected by links (the lines). Let's analyse the matrix used to represent such a network.

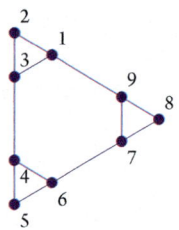

Form a matrix $A = \begin{bmatrix} a_{ij} \end{bmatrix}$ of ones if node i is connected to node j; set the diagonal a_{ii} to be minus the number of other nodes to which node i is connected; and all other components of A are zero. The symmetry of the matrix A follows from the symmetry of the connections: construct the matrix, check it is symmetric, and find the eigenvalues and eigenspaces with MATLAB/Octave, and their multiplicity. For the computed matrices V and D, check that $AV = VD$ and also that V is an orthogonal matrix.

Solution: In MATLAB/Octave use (Table 4.1)

```
A= [-3  1  1  0  0  0  0  0  1
     1 -2  1  0  0  0  0  0  0
     1  1 -3  1  0  0  0  0  0
     0  0  1 -3  1  1  0  0  0
     0  0  0  1 -2  1  0  0  0
     0  0  0  1  1 -3  1  0  0
     0  0  0  0  0  1 -3  1  1
     0  0  0  0  0  0  1 -2  1
     1  0  0  0  0  0  1  1 -3 ]
A-A'
[V,D] =eig(A)
```

To two decimal places so that the results fit the page, the computation may give

```
V =
 -0.41   0.51 -0.16 -0.21 -0.45   0.18 -0.40   0.06   0.33
  0.00 -0.13   0.28   0.63   0.13 -0.18 -0.58 -0.08   0.33
  0.41 -0.20 -0.49 -0.42   0.32   0.01 -0.36 -0.17   0.33
 -0.41 -0.11   0.52 -0.42   0.32   0.01   0.14 -0.37   0.33
 -0.00 -0.18 -0.26   0.37 -0.22   0.51   0.36 -0.46   0.33
  0.41   0.53   0.07   0.05 -0.10 -0.51   0.33 -0.23   0.33
 -0.41 -0.39 -0.36   0.05 -0.10 -0.51   0.25   0.31   0.33
  0.00   0.31 -0.03   0.16   0.55   0.34   0.22   0.55   0.33
  0.41 -0.33   0.42 -0.21 -0.45   0.18   0.03   0.40   0.33
```

D =

-5.00	0	0	0	0	0	0	0	0
0	-4.30	0	0	0	0	0	0	0
0	0	-4.30	0	0	0	0	0	0
0	0	0	-3.00	0	0	0	0	0
0	0	0	0	-3.00	0	0	0	0
0	0	0	0	0	-3.00	0	0	0
0	0	0	0	0	0	-0.70	0	0
0	0	0	0	0	0	0	-0.70	0
0	0	0	0	0	0	0	0	-0.00

The five eigenvalues are -5.00, -4.30, -3.00, -0.70, and 0.00 (to two decimal places). Three of the eigenvalues are repeated as a consequence of the geometric symmetry in the network (different from the symmetry in the matrix). The following are the eigenspaces.

- Corresponding to eigenvalue $\lambda = -5$ are eigenvectors $\boldsymbol{x} \propto (-0.41, 0, 0.41, -0.41, 0, 0.41, -0.41, 0, 0.41)$; that is, the eigenspace $\mathbb{E}_{-5} = \text{span}\{(-1, 0, 1, -1, 0, 1, -1, 0, 1)\}$. From Definition 4.1.15 the multiplicity of eigenvalue $\lambda = -5$ is one.

- Corresponding to eigenvalue $\lambda = -4.30$ there are two eigenvectors computed by MAT-LAB/Octave. These two eigenvectors are orthogonal (you should check). Because these arise as the solutions of the homogeneous system $(A - \lambda I)\boldsymbol{x} = \boldsymbol{0}$, any (nonzero) linear combination of these is also an eigenvector corresponding to the same eigenvalue. That is, the eigenspace

$$\mathbb{E}_{-4.30} = \text{span}\left\{ \begin{bmatrix} 0.51 \\ -0.13 \\ -0.20 \\ -0.11 \\ -0.18 \\ 0.53 \\ -0.39 \\ 0.31 \\ -0.33 \end{bmatrix}, \begin{bmatrix} -0.16 \\ 0.28 \\ -0.49 \\ 0.52 \\ -0.26 \\ 0.07 \\ -0.36 \\ -0.03 \\ 0.42 \end{bmatrix} \right\}.$$

Hence the eigenvalue $\lambda = -4.30$ has multiplicity two.

- Corresponding to eigenvalue $\lambda = -3$ there are three eigenvectors computed by MATLAB/Octave. These three eigenvectors are orthogonal (you should check). Thus the eigenspace

$$\mathbb{E}_{-3} = \text{span}\left\{ \begin{bmatrix} -0.21 \\ 0.63 \\ -0.42 \\ -0.42 \\ 0.37 \\ 0.05 \\ 0.05 \\ 0.16 \\ -0.21 \end{bmatrix}, \begin{bmatrix} -0.45 \\ 0.13 \\ 0.32 \\ 0.32 \\ -0.22 \\ -0.10 \\ -0.10 \\ 0.55 \\ -0.45 \end{bmatrix}, \begin{bmatrix} 0.18 \\ -0.18 \\ 0.01 \\ 0.01 \\ 0.51 \\ -0.51 \\ -0.51 \\ 0.34 \\ 0.18 \end{bmatrix} \right\},$$

and so eigenvalue $\lambda = -3$ has multiplicity three.

- Corresponding to eigenvalue $\lambda = -0.70$ there are two eigenvectors computed by MATLAB/Octave. These two eigenvectors are orthogonal (you should check). Thus the eigenspace

In 1966 Mark Kac asked "Can one hear the shape of the drum?" That is, from just knowing the eigenvalues of a network such as the one in Example 4.1.20, can one infer the connectivity of the network? The question for 2D drums was answered "no" in 1992 by Gordon, Webb, and Wolpert who constructed two different-shaped 2D drums which have the same set of frequencies of oscillation: that is, the same set of eigenvalues. A challenge for an advanced learner is to find the two smallest connected networks that have different connectivity and yet the same eigenvalues (unit strength connections).

$$\mathbb{E}_{-0.70} = \text{span} \left\{ \begin{bmatrix} -0.40 \\ -0.58 \\ -0.36 \\ 0.14 \\ 0.36 \\ 0.33 \\ 0.25 \\ 0.22 \\ 0.03 \end{bmatrix}, \begin{bmatrix} 0.06 \\ -0.08 \\ -0.17 \\ -0.37 \\ -0.46 \\ -0.23 \\ 0.31 \\ 0.55 \\ 0.40 \end{bmatrix} \right\},$$

and so eigenvalue $\lambda = -0.70$ has multiplicity two.

- Lastly, corresponding to eigenvalue $\lambda = 0$ are eigenvectors $x \propto (0.33, 0.33, 0.33, 0.33, 0.33, 0.33, 0.33, 0.33, 0.33)$; that is, the eigenspace $\mathbb{E}_0 = \text{span}\{(1,1,1,1,1,1,1,1,1)\}$, and so eigenvalue $\lambda = 0$ has multiplicity one.

Then check that A*V-V*D is zero (2 d.p.),

```
ans =
    0.00    0.00    0.00    0.00    0.00  -0.00  -0.00    0.00  -0.00
    0.00    0.00  -0.00    0.00  -0.00    0.00  -0.00  -0.00    0.00
    0.00    0.00    0.00    0.00  -0.00  -0.00  -0.00    0.00  -0.00
   -0.00    0.00  -0.00  -0.00    0.00  -0.00    0.00  -0.00    0.00
   -0.00  -0.00    0.00  -0.00    0.00  -0.00  -0.00    0.00  -0.00
    0.00    0.00    0.00  -0.00    0.00    0.00  -0.00    0.00    0.00
   -0.00  -0.00    0.00  -0.00  -0.00    0.00  -0.00  -0.00  -0.00
    0.00    0.00    0.00  -0.00    0.00  -0.00    0.00  -0.00    0.00
    0.00    0.00    0.00    0.00  -0.00  -0.00    0.00  -0.00  -0.00
```

and confirm V is orthogonal by checking that V' *V is the identity (2 d.p.)

```
ans =
    1.00  -0.00  -0.00    0.00    0.00    0.00  -0.00    0.00    0.00
   -0.00    1.00  -0.00  -0.00    0.00    0.00  -0.00  -0.00  -0.00
   -0.00  -0.00    1.00  -0.00    0.00    0.00  -0.00    0.00    0.00
    0.00  -0.00  -0.00    1.00    0.00  -0.00  -0.00    0.00  -0.00
    0.00    0.00    0.00    0.00    1.00    0.00  -0.00    0.00  -0.00
    0.00    0.00    0.00  -0.00    0.00    1.00  -0.00    0.00  -0.00
   -0.00  -0.00  -0.00  -0.00  -0.00  -0.00    1.00  -0.00    0.00
    0.00  -0.00    0.00    0.00    0.00    0.00  -0.00    1.00    0.00
    0.00  -0.00    0.00  -0.00  -0.00  -0.00    0.00    0.00    1.00
```

□

Why write "the computation may give" in Example 4.1.20? The reason is associated with the repeated eigenvalues. What is important is the eigenspace. When an eigenvalue of a symmetric matrix is duplicated in the diagonal D (or triplicated), then there are many choices of eigenvectors that form an orthonormal basis (Definition 3.4.18) of the eigenspace (the same holds for singular vectors of a repeated singular value). Different algorithms may report different orthonormal bases of the same eigenspace. The bases given in Example 4.1.20 are just one possibility for each eigenspace.

Theorem 4.1.21 *For every $n \times n$ square matrix A (not just symmetric), $\lambda_1, \lambda_2, \ldots, \lambda_m$ are eigenvalues of A with corresponding eigenvectors $v_1, v_2, \ldots, v_m$, for some m (commonly $m = n$), iff $AV = VD$ for diagonal matrix $D = \text{diag}(\lambda_1, \lambda_2, \ldots, \lambda_m)$ and $n \times m$ matrix $V = \begin{bmatrix} v_1 & v_2 & \cdots & v_m \end{bmatrix}$ for nonzero $v_1, v_2, \ldots, v_m$.*

Proof. From Definition 4.1.1, λ_j are eigenvalues and nonzero v_j are eigenvectors iff $Av_j = \lambda_j v_j$. Form all the cases $j = 1, 2, \ldots, m$ into the one matrix equation

$$\begin{bmatrix} Av_1 & Av_2 & \cdots & Av_m \end{bmatrix} = \begin{bmatrix} \lambda_1 v_1 & \lambda_2 v_2 & \cdots & \lambda_m v_m \end{bmatrix}$$

By the definition of matrix products this matrix equation is identical to

$$A\begin{bmatrix} v_1 & v_2 & \cdots & v_m \end{bmatrix} = \begin{bmatrix} v_1 & v_2 & \cdots & v_m \end{bmatrix} \begin{bmatrix} \lambda_1 & 0 & \cdots & 0 \\ 0 & \lambda_2 & & 0 \\ \vdots & & \ddots & \vdots \\ 0 & 0 & \cdots & \lambda_m \end{bmatrix}.$$

Since the matrix of eigenvectors is called V, and the diagonal matrix of eigenvalues is called D, this equation is the same as $AV = VD$. $\square$

Example 4.1.22 Use MATLAB/Octave to compute eigenvectors and the eigenvalues of the (symmetric) matrix

$$A = \begin{bmatrix} 2 & 2 & -2 & 0 \\ 2 & -1 & -2 & -3 \\ -2 & -2 & 4 & 0 \\ 0 & -3 & 0 & 1 \end{bmatrix}.$$

Confirm $AV = VD$ for the computed matrices.

Solution: First compute (Table 4.1)

```
A= [2  2  -2  0
    2 -1  -2 -3
   -2 -2   4  0
    0 -3   0  1]
[V,D] =eig(A)
```

The output is (2 d.p.)

```
V =
    -0.23    0.83    0.08    0.50
     0.82    0.01   -0.40    0.42
     0.15    0.52   -0.42   -0.72
     0.51    0.20    0.81   -0.23
D =
    -3.80       0       0       0
        0    0.77       0       0
        0       0    2.50       0
        0       0       0    6.53
```

Hence the eigenvalues are (2 d.p.) $\lambda_1 = -3.80$, $\lambda_2 = 0.77$, $\lambda_3 = 2.50$, and $\lambda_4 = 6.53$. From the columns of V, corresponding eigenvectors are (2 d.p.) $v_1 \propto (-0.23, 0.82, 0.15, 0.51)$, $v_2 \propto (0.83, 0.01, 0.52, 0.20)$, $v_3 \propto (0.08, -0.40, -0.42, 0.81)$, and $v_4 \propto (0.50, 0.42, -0.72, -0.23)$. Then confirm A*V-V*D is zero:

```
ans =
     0.00    0.00    0.00    0.00
     0.00    0.00   -0.00    0.00
    -0.00    0.00   -0.00   -0.00
    -0.00   -0.00   -0.00    0.00
```

□

Find eigenvalues and eigenvectors by hand

- Recall from previous study (Theorem 3.2.7) that a 2×2 matrix $A = \begin{bmatrix} a & b \\ c & d \end{bmatrix}$ has determinant $\det A = |A| = ad - bc$, and that A is not invertible iff $\det A = 0$.

- Similarly, although not justified until Chapter 6, a 3×3 matrix $A = \begin{bmatrix} a & b & c \\ d & e & f \\ g & h & i \end{bmatrix}$ has determinant $\det A = |A| = aei + bfg + cdh - ceg - afh - bdi$, and A is not invertible iff $\det A = 0$.

This section shows that these two formulas for a determinant are useful for hand calculations on small problems. The formulas are best remembered via the following diagrams where products along the red lines are subtracted from the sum of products along the blue lines, respectively:

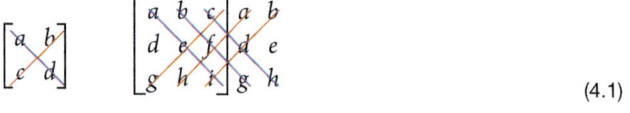

(4.1)

Chapter 6 extends such determinants to any size of matrix, and explores more useful properties, but for now this is all the information we need on determinants.

For hand calculation on small matrices (2×2 or 3×3) the key property is the following. By Definition 4.1.1 eigenvalues and eigenvectors are determined from $A x = \lambda x$. Rearranging, this equation is equivalent to $(A - \lambda I)x = 0$. Both Theorem 3.2.7 (2×2 matrices) and Theorem 6.1.29 (general matrices) establish that $(A - \lambda I)x = 0$ has *nonzero* solutions x iff the determinant $\det(A - \lambda I) = 0$. Since eigenvectors must be nonzero, the eigenvalues of a square matrix are precisely the solutions of $\det(A - \lambda I) = 0$. This reasoning leads to the following procedure.

Procedure 4.1.23 (eigenvalues and eigenvectors) *To find by hand eigenvalues and eigenvectors of any (small) square matrix A:*

1. *find all eigenvalues by solving the **characteristic equation** of A, $\det(A - \lambda I) = 0$;*

2. *for each eigenvalue λ, solve the homogeneous matrix-vector equation $(A - \lambda I)x = 0$ to find the corresponding eigenspace $\mathbb{E}_\lambda$;*

3. *write each eigenspace as the span of a few chosen eigenvectors.*

This procedure applies to general matrices A, as fully established in Section 7.1, but this chapter uses it only for small symmetric matrices. Further, this chapter uses it only as a convenient method to illustrate some properties by hand calculation. None of the beautiful theorems of the next Section 4.2 for symmetric matrices are based upon this 'by-hand' procedure.

Example 4.1.24 Use Procedure 4.1.23 to find the eigenvalues and eigenvectors of the matrix

$$A = \begin{bmatrix} 1 & -\frac{1}{2} \\ -\frac{1}{2} & 1 \end{bmatrix}$$

(this is the matrix illustrated in Examples 4.1.4 and 4.1.7).

Solution: Follow the first two steps of Procedure 4.1.23.

(a) Solve $\det(A - \lambda I) = 0$ for the eigenvalues λ. Using (4.1),

$$\det(A - \lambda I) = \begin{vmatrix} 1 - \lambda & -\frac{1}{2} \\ -\frac{1}{2} & 1 - \lambda \end{vmatrix} = (1 - \lambda)^2 - \frac{1}{4} = 0.$$

That is, $(\lambda - 1)^2 = \frac{1}{4}$. Taking account of both square-roots, this quadratic gives $\lambda - 1 = \pm\frac{1}{2}$; that is, $\lambda = 1 \pm \frac{1}{2} = \frac{1}{2}, \frac{3}{2}$ are the only two eigenvalues.

(b) For eigenvectors, consider the two eigenvalues in turn.

i. For eigenvalue $\lambda = \frac{1}{2}$ solve $(A - \lambda I)x = 0$. That is,

$$(A - \tfrac{1}{2}I)x = \begin{bmatrix} 1 - \frac{1}{2} & -\frac{1}{2} \\ -\frac{1}{2} & 1 - \frac{1}{2} \end{bmatrix} x = \begin{bmatrix} \frac{1}{2} & -\frac{1}{2} \\ -\frac{1}{2} & \frac{1}{2} \end{bmatrix} x = 0.$$

The first component of this system says $x_1 - x_2 = 0$; that is, $x_2 = x_1$. The second component of this system says $-x_1 + x_2 = 0$; that is, $x_2 = x_1$ (the same). So a general solution for a corresponding eigenvector is $x = (1, 1)t$ for any nonzero t. That is, the eigenspace $\mathbb{E}_{1/2} = \text{span}\{(1, 1)\}$.

ii. For eigenvalue $\lambda = \frac{3}{2}$ solve $(A - \lambda I)x = 0$. That is,

$$(A - \tfrac{3}{2}I)x = \begin{bmatrix} 1 - \frac{3}{2} & -\frac{1}{2} \\ -\frac{1}{2} & 1 - \frac{3}{2} \end{bmatrix} x = \begin{bmatrix} -\frac{1}{2} & -\frac{1}{2} \\ -\frac{1}{2} & -\frac{1}{2} \end{bmatrix} x = 0.$$

The first component of this system says $x_1 + x_2 = 0$, as does the second component; that is, $x_2 = -x_1$. So a general solution for a corresponding eigenvector is $x = (1, -1)t$ for any nonzero t. That is, the eigenspace $\mathbb{E}_{3/2} = \text{span}\{(1, -1)\}$.

□

Activity 4.1.25 Consider the matrix $A = \begin{bmatrix} 3 & 2 \\ 2 & 0 \end{bmatrix}$. Use its characteristic equation to determine that all its eigenvalues are which of the following?

(a) $-4, 1$ (b) $3, 4$ (c) $0, 3$ (d) $-1, 4$

Example 4.1.26 Use the determinant (4.1) to confirm that $\lambda = 0, 1, 3$ are the *only* eigenvalues of the matrix

$$A = \begin{bmatrix} 1 & -1 & 0 \\ -1 & 2 & -1 \\ 0 & -1 & 1 \end{bmatrix}.$$

(Example 4.1.11 already found the eigenspaces corresponding to these three eigenvalues.)

Solution: To find all eigenvalues, find all solutions of the characteristic equation $\det(A - \lambda I) = 0$. Using (4.1),

$$\det(A - \lambda I) = \begin{vmatrix} 1 - \lambda & -1 & 0 \\ -1 & 2 - \lambda & -1 \\ 0 & -1 & 1 - \lambda \end{vmatrix}$$

$$= (1 - \lambda)^2 (2 - \lambda) + 0 + 0 - 0 - (1 - \lambda) - (1 - \lambda)$$
$$= (1 - \lambda)[(1 - \lambda)(2 - \lambda) - 2]$$
$$= (1 - \lambda)[2 - 3\lambda + \lambda^2 - 2]$$
$$= (1 - \lambda)[-3\lambda + \lambda^2]$$
$$= (1 - \lambda)(-3 + \lambda)\lambda.$$

So the characteristic equation is $(1 - \lambda)(-3 + \lambda)\lambda = 0$. In this factored form we see the *only* solutions are the three eigenvalues $\lambda = 0, 1, 3$ as previously identified.

□

Example 4.1.27 Use Procedure 4.1.23 to find all eigenvalues and the corresponding eigenspaces of the symmetric matrix

$$A = \begin{bmatrix} -2 & 0 & -6 \\ 0 & 4 & 6 \\ -6 & 6 & -9 \end{bmatrix}.$$

Solution: Follow the steps of Procedure 4.1.23.

(a) Solve $\det(A - \lambda I) = 0$ for the eigenvalues. Using (4.1),

$$
\begin{aligned}
\det(A - \lambda I) &= \begin{vmatrix} -2 - \lambda & 0 & -6 \\ 0 & 4 - \lambda & 6 \\ -6 & 6 & -9 - \lambda \end{vmatrix} \\
&= (-2 - \lambda)(4 - \lambda)(-9 - \lambda) + 0 \cdot 6 \cdot (-6) + (-6) \cdot 0 \cdot 6 \\
&\quad - (-6)(4 - \lambda)(-6) - (-2 - \lambda) \cdot 6 \cdot 6 - 0 \cdot 0 \cdot (-9 - \lambda) \\
&= (2 + \lambda)(4 - \lambda)(9 + \lambda) + 36(-4 + \lambda) + 36(2 + \lambda) \\
&= -\lambda^3 - 7\lambda^2 + 98\lambda \\
&= -\lambda(\lambda^2 - 7\lambda + 98) \\
&= -\lambda(\lambda - 14)(\lambda + 7).
\end{aligned}
$$

This determinant is zero only for the three eigenvalues $\lambda = 0, -7, 14$.

(b) For eigenvectors, consider the three eigenvalues in turn.

i. For eigenvalue $\lambda = 0$ solve $(A - \lambda I)v = \mathbf{0}$. That is,

$$(A - 0I)v = \begin{bmatrix} -2 & 0 & -6 \\ 0 & 4 & 6 \\ -6 & 6 & -9 \end{bmatrix} v = \begin{bmatrix} -2v_1 - 6v_3 \\ 4v_2 + 6v_3 \\ -6v_1 + 6v_2 - 9v_3 \end{bmatrix} = \mathbf{0}.$$

The first row says $v_1 = -3v_3$, the second row says $v_2 = -\frac{3}{2}v_3$. Substituting these into the left-hand side of the third row gives $-6v_1 + 6v_2 - 9v_3 = 18v_3 - 9v_3 - 9v_3 = 0$ for every v_3, which confirms that there are nonzero solutions to form eigenvectors. Eigenvectors may be written in the form $v = (-3v_3, -\frac{3}{2}v_3, v_3)$; that is, the eigenspace $\mathbb{E}_0 = \operatorname{span}\{(-6, -3, 2)\}$.

ii. For eigenvalue $\lambda = 7$ solve $(A - \lambda I)v = \mathbf{0}$. That is,

$$(A - 7I)v = \begin{bmatrix} -9 & 0 & -6 \\ 0 & -3 & 6 \\ -6 & 6 & -16 \end{bmatrix} v = \begin{bmatrix} -9v_1 - 6v_3 \\ -3v_2 + 6v_3 \\ -6v_1 + 6v_2 - 16v_3 \end{bmatrix} = \mathbf{0}.$$

The first row says $v_1 = -\frac{2}{3}v_3$, the second row says $v_2 = 2v_3$. Substituting these into the left-hand side of the third row gives $-6v_1 + 6v_2 - 16v_3 = 4v_3 + 12v_3 - 16v_3 = 0$ for every v_3, which confirms that there are nonzero solutions to form eigenvectors. Eigenvectors may be written in the form $v = (-\frac{2}{3}v_3, 2v_3, v_3)$; that is, the eigenspace $\mathbb{E}_7 = \operatorname{span}\{(-2, 6, 3)\}$.

iii. For eigenvalue $\lambda = -14$ solve $(A - \lambda I)v = 0$. That is,

$$(A + 14I)v = \begin{bmatrix} 12 & 0 & -6 \\ 0 & 18 & 6 \\ -6 & 6 & 5 \end{bmatrix} v = \begin{bmatrix} 12v_1 - 6v_3 \\ 18v_2 + 6v_3 \\ -6v_1 + 6v_2 + 5v_3 \end{bmatrix} = 0.$$

The first row says $v_1 = \frac{1}{2}v_3$, the second row says $v_2 = -\frac{1}{3}v_3$. Substituting these into the left-hand side of the third row gives $-6v_1 + 6v_2 + 5v_3 = -3v_3 - 2v_3 + 5v_3 = 0$ for every v_3, which confirms that there are nonzero solutions to form eigenvectors. Eigenvectors may be written in the form $v = (\frac{1}{2}v_3, -\frac{1}{3}v_3, v_3)$; that is, the eigenspace $\mathbb{E}_{-14} = \text{span}\{(3, -2, 6)\}$.   □

General matrices may have non-real complex valued eigenvalues and eigenvectors, as seen in the next example, and for good reasons in some applications. One of the key results of the next Section 4.2 is to prove that real symmetric matrices always have real eigenvalues and eigenvectors. There are many applications where this reality is crucial.

Example 4.1.28 Find the eigenvalues and a corresponding eigenvector for the non-symmetric matrix $A = \begin{bmatrix} 0 & 1 \\ -1 & 0 \end{bmatrix}$.

(This example aims to recall basic properties of complex numbers as a prelude to the proof of the reality of eigenvalues for every symmetric matrix.)

Solution:

(a) Solve the characteristic equation $\det(A - \lambda I) = 0$ for the eigenvalues λ. Using (4.1),

$$\det(A - \lambda I) = \begin{vmatrix} -\lambda & 1 \\ -1 & -\lambda \end{vmatrix} = \lambda^2 + 1 = 0.$$

That is, $\lambda^2 = -1$. Taking square-roots we find that there are two complex eigenvalues $\lambda = \pm\sqrt{-1} = \pm i$ (the upright roman character $i = \sqrt{-1}$ distinguishes it from the integer index i). Despite the appearance of complex numbers, all our arithmetic, algebra, and properties continue to hold. Thus we proceed to find complex valued eigenvectors.

There is a Fundamental Theorem of Algebra (proved in more advanced courses) that establishes that when real numbers fail us, then complex numbers are always sufficient for all eigenvalues.

(b) Consider the two eigenvalues in turn.

i. For eigenvalue $\lambda = i$ solve $(A - \lambda I)x = 0$. That is,

$$(A - iI)x = \begin{bmatrix} -i & 1 \\ -1 & -i \end{bmatrix} x = 0.$$

The first component of this system says $-ix_1 + x_2 = 0$; that is, $x_2 = ix_1$. The second component of this system says $-x_1 - ix_2 = 0$; that is, $x_2 = ix_1$ (the same). So a general corresponding eigenvector is the complex $\pmb{x} = (1, i)t$ for any nonzero t.

ii. For eigenvalue $\lambda = -i$ solve $(A - \lambda I)\pmb{x} = \pmb{0}$. That is,

$$(A + iI)\pmb{x} = \begin{bmatrix} i & 1 \\ -1 & i \end{bmatrix} \pmb{x} = \pmb{0}.$$

The first component of this system says $ix_1 + x_2 = 0$; that is, $x_2 = -ix_1$. The second component of this system says $-x_1 + ix_2 = 0$; that is, $x_2 = -ix_1$ (the same). So a general corresponding eigenvector is the complex $\pmb{x} = (1, -i)t$ for any nonzero t.

□

Example 4.1.28 is a problem that might arise using calculus to describe the dynamics of a mass on a spring. Let the displacement of the mass be $y_1(t)$ then Newton's law says the acceleration $d^2 y_1 / dt^2 \propto -y_1$, the negative of the displacement; for simplicity, let the constant of proportionality be one. Introduce $y_2(t) = dy_1/dt$, then Newton's law becomes $dy_2/dt = -y_1$. Seek solutions of these two first-order differential equations in the form $y_j(t) = x_j e^{\lambda t}$ and the differential equations become $x_2 = \lambda x_1$ and $\lambda x_2 = -x_1$ respectively. Forming into a matrix-vector problem these are

$$\begin{bmatrix} x_2 \\ -x_1 \end{bmatrix} = \lambda \begin{bmatrix} x_1 \\ x_2 \end{bmatrix} \iff \begin{bmatrix} 0 & 1 \\ -1 & 0 \end{bmatrix} \pmb{x} = \lambda \pmb{x}.$$

We need to find the eigenvalues and eigenvectors of the matrix: we derive that eigenvalues are $\lambda = \pm\sqrt{-1} = \pm i$. Physically, such complex eigenvalues represent oscillations in time t since, for example, $e^{\lambda t} = e^{it} = \cos t + i \sin t$ by Euler's formula.

4.1.2 Exercises

Exercise 4.1.1   Each plot below shows (unit) vectors $\pmb{x}$ (blue), and for some matrix the corresponding vectors $A\pmb{x}$ (red) adjoined. Estimate which directions $\pmb{x}$ are eigenvectors of matrix A, and for each eigenvector estimate the corresponding eigenvalue.

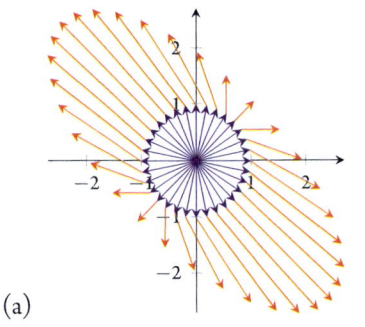

(a)

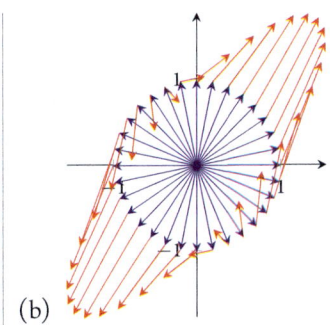

(b)

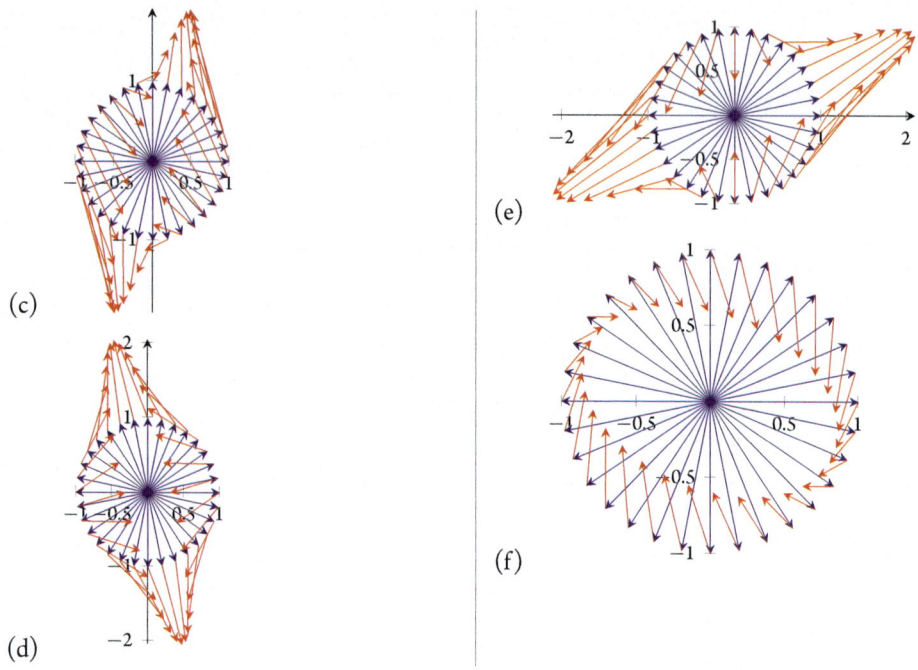

(c)

(e)

(d)

(f)

Exercise 4.1.2 In each case, use the matrix-vector product to determine which of the given vectors are eigenvectors of the given matrix. For each such eigenvector what is the corresponding eigenvalue?

(a) $\begin{bmatrix} 6 & 2 \\ 3 & 1 \end{bmatrix}$, $\begin{bmatrix} 2 \\ 1 \end{bmatrix}$ $\begin{bmatrix} 1 \\ -3 \end{bmatrix}$, $\begin{bmatrix} 3 \\ -2 \end{bmatrix}$, $\begin{bmatrix} -\frac{1}{2} \\ -\frac{1}{4} \end{bmatrix}$ $\begin{bmatrix} 0 \\ 0 \end{bmatrix}$, $\begin{bmatrix} 2 \\ -3 \end{bmatrix}$

(b) $\begin{bmatrix} -2 & 1 \\ 3 & 0 \end{bmatrix}$, $\begin{bmatrix} 1 \\ 1 \end{bmatrix}$ $\begin{bmatrix} -2 \\ 2 \end{bmatrix}$, $\begin{bmatrix} 1 \\ -1 \end{bmatrix}$ $\begin{bmatrix} 1 \\ 3 \end{bmatrix}$, $\begin{bmatrix} 0 \\ 0 \end{bmatrix}$, $\begin{bmatrix} -\frac{1}{3} \\ -1 \end{bmatrix}$ $\begin{bmatrix} \frac{1}{2} \\ 1 \end{bmatrix}$

(c) $\begin{bmatrix} 3 & 3 & 4 \\ 0 & -4 & 0 \\ -2 & -1 & -6 \end{bmatrix}$, $\begin{bmatrix} 4 \\ 0 \\ -1 \end{bmatrix}$, $\begin{bmatrix} -1 \\ 0 \\ 2 \end{bmatrix}$, $\begin{bmatrix} -2 \\ 6 \\ -1 \end{bmatrix}$, $\begin{bmatrix} 1 \\ 1 \\ 1 \end{bmatrix}$, $\begin{bmatrix} \frac{1}{3} \\ -1 \\ \frac{1}{6} \end{bmatrix}$ $\begin{bmatrix} 1 \\ 0 \\ 2 \end{bmatrix}$

(d) $\begin{bmatrix} 3 & 0 & 0 \\ 2 & -5 & -3 \\ -4 & -2 & 0 \end{bmatrix}$, $\begin{bmatrix} 2 \\ 1 \\ -1 \end{bmatrix}$, $\begin{bmatrix} 0 \\ -1 \\ 2 \end{bmatrix}$, $\begin{bmatrix} 0 \\ 1 \\ 2 \end{bmatrix}$, $\begin{bmatrix} 0 \\ 3 \\ 1 \end{bmatrix}$, $\begin{bmatrix} \frac{1}{2} \\ \frac{1}{2} \\ -1 \end{bmatrix}$ $\begin{bmatrix} 0 \\ 0 \\ 0 \end{bmatrix}$, $\begin{bmatrix} 0 \\ -1 \\ -\frac{1}{3} \end{bmatrix}$

Exercise 4.1.3 Use the MATLAB/Octave function eig() to determine the eigenvalues and corresponding eigenspaces of the following symmetric matrices.

(a) $\begin{bmatrix} -1 & -\frac{4}{5} & \frac{12}{5} & -2 \\ -\frac{4}{5} & \frac{1}{5} & -\frac{8}{5} & -\frac{12}{5} \\ \frac{12}{5} & -\frac{8}{5} & -\frac{11}{5} & -\frac{14}{5} \\ -2 & -\frac{12}{5} & -\frac{14}{5} & 2 \end{bmatrix}$

(b) $\begin{bmatrix} \frac{7}{5} & -\frac{2}{5} & 0 & -\frac{2}{5} \\ -\frac{2}{5} & 2 & \frac{2}{5} & 0 \\ 0 & \frac{2}{5} & \frac{13}{5} & \frac{2}{5} \\ -\frac{2}{5} & 0 & \frac{2}{5} & 2 \end{bmatrix}$

$$\text{(c)} \quad \begin{bmatrix} 1.4 & -7.1 & -0.7 & 6.2 \\ -7.1 & -1.0 & -2.2 & -2.5 \\ -0.7 & -2.2 & -3.4 & -4.1 \\ 6.2 & -2.5 & -4.1 & -1.0 \end{bmatrix}$$

$$\text{(d)} \quad \begin{bmatrix} -1 & 1 & 4 & -3 & 1 \\ 1 & 0 & -2 & 1 & 0 \\ 4 & -2 & 1 & -3 & 0 \\ -3 & 1 & -3 & 2 & -1 \\ 1 & 0 & 0 & -1 & -1 \end{bmatrix}$$

Exercise 4.1.4 For each of the given symmetric matrices, determine all eigenvalues by finding and solving the characteristic equation of the matrix.

(a) $\begin{bmatrix} 2 & 3 \\ 3 & 2 \end{bmatrix}$

(b) $\begin{bmatrix} 6 & \frac{11}{2} \\ \frac{11}{2} & 6 \end{bmatrix}$

(c) $\begin{bmatrix} -5 & 1 \\ 2 & -2 \end{bmatrix}$

(d) $\begin{bmatrix} -5 & 5 \\ 5 & -5 \end{bmatrix}$

(e) $\begin{bmatrix} 2 & -3 & -3 \\ -3 & 2 & -3 \\ -3 & -3 & 2 \end{bmatrix}$

(f) $\begin{bmatrix} 4 & -4 & 3 \\ -4 & -2 & 6 \\ 3 & 6 & -8 \end{bmatrix}$

(g) $\begin{bmatrix} 8 & 4 & 2 \\ 4 & 0 & 0 \\ 2 & 0 & 0 \end{bmatrix}$

(h) $\begin{bmatrix} 0 & 0 & -3 \\ 0 & 2 & 0 \\ -3 & 0 & 0 \end{bmatrix}$

Exercise 4.1.5 For each symmetric matrix, find the eigenspace of the given 'eigenvalues' by hand solution of linear equations, or determine from your solution that the given value cannot be an eigenvalue.

(a) $\begin{bmatrix} 1 & 3 \\ 3 & 1 \end{bmatrix}, 4, -2$

(b) $\begin{bmatrix} 4 & -2 \\ -2 & 7 \end{bmatrix}, 3, 6$

(c) $\begin{bmatrix} -7 & 0 & 2 \\ 0 & -7 & -2 \\ 2 & -2 & -0 \end{bmatrix}, -8, -7, 1$

(d) $\begin{bmatrix} 0 & 6 & -3 \\ 6 & 0 & 7 \\ -3 & 7 & 3 \end{bmatrix}, -6, 4, 9$

Exercise 4.1.6 For each symmetric matrix, find by hand all eigenvalues and an orthonormal basis for all the corresponding eigenspaces. What is the multiplicity of each eigenvalue?

(a) $\begin{bmatrix} -8 & 3 \\ 3 & 0 \end{bmatrix}$

(b) $\begin{bmatrix} 6 & -5 \\ -5 & 6 \end{bmatrix}$

(c) $\begin{bmatrix} 11 & 4 & -2 \\ 4 & 5 & 4 \\ -2 & 4 & 11 \end{bmatrix}$

(d) $\begin{bmatrix} -7 & 2 & 2 \\ 2 & -6 & 0 \\ 2 & 0 & -8 \end{bmatrix}$

(e) $\begin{bmatrix} 6 & 10 & -5 \\ 10 & 13 & -2 \\ -5 & -2 & -2 \end{bmatrix}$
(f) $\begin{bmatrix} 4 & 3 & 1 \\ 3 & -4 & -3 \\ 1 & -3 & 4 \end{bmatrix}$

Exercise 4.1.7 *(Seven Bridges of Königsberg)* The picture to the right shows the abstract network of the seven bridges of Königsberg during the time of Euler: the small disc nodes represent the islands of Königsberg; the lines between represent the seven different bridges that connect the islands. This abstract network is famous for its role in founding the theory of such networks, but this exercise addresses an aspect relevant to well-used web search software. Number the nodes from 1 to 4. Form the  4×4 symmetric matrix of the number of lines from each node to the other nodes (and zero for the number of lines from a node to itself)—called the adjacency matrix. Use MATLAB/Octave function eig() to find the eigenvalues and eigenvectors for this matrix. Analogous to well-known web search software, identify the largest eigenvalue and a corresponding eigenvector; then we choose to rank the importance of each node in order of the magnitude of the component in the corresponding eigenvector.

Exercise 4.1.8 For each of the following networks:[4]

- label the nodes;
- construct the symmetric adjacency matrix A such that a_{ij} is one if node i is linked to node j, and a_{ij} is zero otherwise (and zero on the diagonal);
- in MATLAB/Octave use eig() to find all eigenvalues and eigenvectors;
- rank the 'importance' of the nodes from the magnitude of their component in the eigenvector corresponding to the largest (most positive) eigenvalue.

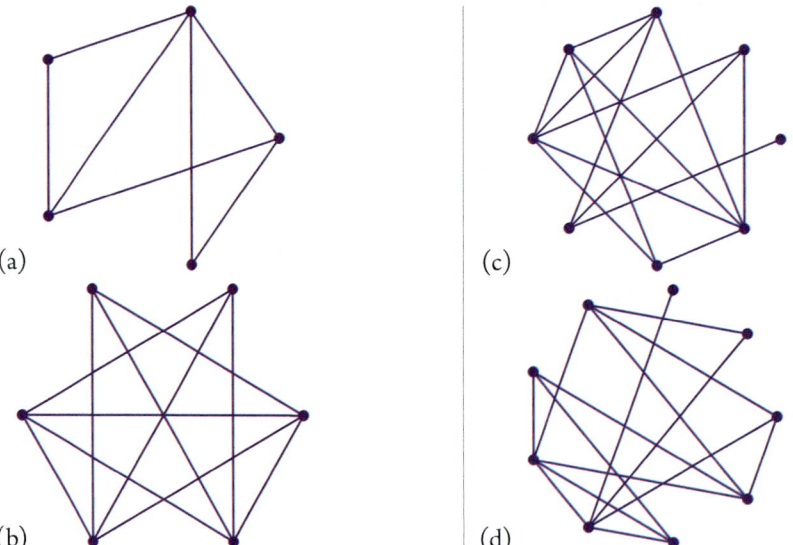

(a)

(b)

(c)

(d)

---

[4] Although a well-known web search engine computes eigenvectors for all the web pages on the internet, it uses an approximate iterative algorithm more suited to the mind-bogglingly vast size of the internet.

Exercise 4.1.9 In a few sentences, answer/discuss each of the following.

(a) In an SVD, USV^{T}, what is important about singular vectors for which $u_j = \pm v_j$?

(b) What fundamental geometric question corresponds to seeking eigenvalues and eigenvectors?

(c) Why do we require eigenvectors to be nonzero?

(d) For a symmetric matrix, how do its singular values compare to its eigenvalues?

(e) What geometric reason underlies the simplicity of the eigenvalues and eigenvectors of diagonal matrices?

(f) How does the concept of an eigenspace follow from that of eigenvalues and eigenvectors?

(g) How did the concept of multiplicity of eigenvalues for symmetric matrices arise?

(h) How is it that complex eigenvalues can arise for real matrices?

4.2 Beautiful properties for symmetric matrices

This section starts by exploring two properties for eigenvalues of general matrices, and then proceeds to the special case of real symmetric matrices. Symmetric matrices have the beautifully useful properties of always having real eigenvalues and always having orthogonal eigenvectors.

4.2.1 Matrix powers maintain eigenvectors

Recall that Section 3.2 introduced the inverse of a matrix (Definition 3.2.2). This first theorem links an eigenvalue of zero to the non-existence of an inverse, and hence links a zero eigenvalue to problematic linear equations.

Theorem 4.2.1 *A square matrix is invertible iff zero is not an eigenvalue of the matrix.*

Proof. From Definition 4.1.1, zero is an eigenvalue ($\lambda = 0$) iff $Ax = 0x$ has nonzero solutions x; that is, iff the homogeneous system $Ax = 0$ has nonzero solutions x. But the Unique Solution Theorems 3.3.27 and 3.4.43 assure us that this occurs iff matrix A is not invertible. Consequently a matrix is invertible iff zero is not an eigenvalue. $\quad\square$

Example 4.2.2

- The 3×3 matrix of Examples 4.1.2, 4.1.11, 4.1.17, and 4.1.26 is not invertible as among its eigenvalues of 0, 1, and 3 it has zero as an eigenvalue.

- The plot to the right shows (unit) vectors x (blue), and for some matrix A the corresponding vectors Ax (red) adjoined. There are no directions x for which $Ax = 0 = 0x$. Hence zero cannot be an eigenvalue, and the matrix A must be invertible.

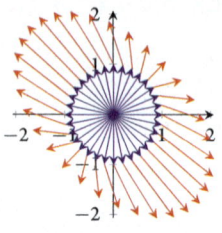

  Similarly for Example 4.1.8.

- The 3×3 diagonal matrix of Example 4.1.14 has eigenvalues of only $-\frac{1}{3}$ and $\frac{3}{2}$. Since zero is not an eigenvalue, the matrix is invertible.

- The 9×9 matrix of the Sierpinski network in Example 4.1.20 is not invertible as it has zero among its five eigenvalues.

- The 2×2 matrix of Example 4.1.24 is invertible as its eigenvalues are $\lambda = \frac{1}{2}, \frac{3}{2}$, neither of which are zero. Indeed, matrix multiplication confirms that the matrix

$$A = \begin{bmatrix} 1 & -\frac{1}{2} \\ -\frac{1}{2} & 1 \end{bmatrix}, \quad \text{has inverse } A^{-1} = \begin{bmatrix} \frac{4}{3} & \frac{2}{3} \\ \frac{2}{3} & \frac{4}{3} \end{bmatrix}.$$

- The 2×2 non-symmetric matrix of Example 4.1.28 is invertible because zero is not among its eigenvalues of $\lambda = \pm i$. Indeed, matrix multiplication confirms that the matrix

$$A = \begin{bmatrix} 0 & 1 \\ -1 & 0 \end{bmatrix}, \quad \text{has inverse } A^{-1} = \begin{bmatrix} 0 & -1 \\ 1 & 0 \end{bmatrix}.$$ □

Example 4.2.3 The next theorem considers eigenvalues and eigenvectors of powers of a matrix. Two examples are the following.

- Recall the matrix $A = \begin{bmatrix} 0 & 1 \\ -1 & 0 \end{bmatrix}$ has eigenvalues $\lambda = \pm i$. The square of this matrix

$$A^2 = \begin{bmatrix} 0 & 1 \\ -1 & 0 \end{bmatrix}\begin{bmatrix} 0 & 1 \\ -1 & 0 \end{bmatrix} = \begin{bmatrix} -1 & 0 \\ 0 & -1 \end{bmatrix}$$

  is diagonal so its eigenvalues are the diagonal elements (Example 4.1.9), namely the only eigenvalue is -1. Observe that the eigenvalue of A^2, $-1 = (\pm i)^2$, is the square of the eigenvalues of A. That the eigenvalues of A^2 are the square of those of A holds generally.

- Also recall that matrix

$$A = \begin{bmatrix} 1 & -\frac{1}{2} \\ -\frac{1}{2} & 1 \end{bmatrix}, \quad \text{has inverse } A^{-1} = \begin{bmatrix} \frac{4}{3} & \frac{2}{3} \\ \frac{2}{3} & \frac{4}{3} \end{bmatrix}.$$

  Let's determine the eigenvalues of this inverse. Its characteristic equation (defined in Procedure 4.1.23) is

$$\det(A^{-1} - \lambda I) = \begin{vmatrix} \frac{4}{3} - \lambda & \frac{2}{3} \\ \frac{2}{3} & \frac{4}{3} - \lambda \end{vmatrix} = (\tfrac{4}{3} - \lambda)^2 - \tfrac{4}{9} = 0.$$

That is, $(\lambda - \frac{4}{3})^2 = \frac{4}{9}$. Taking the square-root of both sides gives $\lambda - \frac{4}{3} = \pm\frac{2}{3}$; that is, the two eigenvalues of the inverse A^{-1} are $\lambda = \frac{4}{3} \pm \frac{2}{3} = 2, \frac{2}{3}$. Observe that these eigenvalues of the inverse are the reciprocals of the eigenvalues $\frac{1}{2}, \frac{3}{2}$ of A. This reciprocal relation also holds generally.

The right-hand pictures illustrates the reciprocal relation graphically: the left picture shows Ax for various x, the right picture shows $A^{-1}x$. The eigenvector directions are the same for both matrix and inverse. But in those eigenvector directions where the matrix stretches, the inverse shrinks, and where the matrix shrinks, the inverse stretches. In contrast, in directions that are not eigenvectors, the relationship between Ax and $A^{-1}x$ is somewhat obscure. ☐

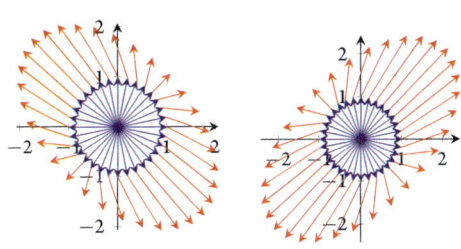

Theorem 4.2.4 *Let A be a square matrix with eigenvalue λ and corresponding eigenvector x.*

(a) *For every positive integer n, λ^n is an eigenvalue of A^n with corresponding eigenvector x.*

(b) *If A is invertible, then $1/\lambda$ is an eigenvalue of A^{-1} with corresponding eigenvector x.*

(c) *If A is invertible, then for every integer n (including negative n), λ^n is an eigenvalue of A^n with corresponding eigenvector x.*

Proof. Consider each property in turn.

4.2.4(a) First, the result holds for power $n = 1$ by Definition 4.1.1, that $Ax = \lambda x$. Second, for the case of power $n = 2$ consider $A^2x = (AA)x = A(Ax) = A(\lambda x) = \lambda(Ax) = \lambda(\lambda x) = (\lambda^2)x$. Hence by Definition 4.1.1 λ^2 is an eigenvalue of A^2 corresponding to eigenvector x.

Third, use induction to extend to any power: assume the result for $n = k$ (and proceed to prove it for $n = k + 1$). Consider $A^{k+1}x = (A^kA)x = A^k(Ax) = A^k(\lambda x) = \lambda(A^kx) = \lambda(\lambda^kx) = \lambda^{k+1}x$. Hence by Definition 4.1.1 λ^{k+1} is an eigenvalue of A^{k+1} corresponding to eigenvector x. By induction, the property 4.2.4(a) holds for all integer $n \geq 1$.

4.2.4(b) For invertible A, we know that none of the eigenvalues are zero: thus $1/\lambda$ exists. Pre-multiply $Ax = \lambda x$ by $\frac{1}{\lambda}A^{-1}$ to deduce $\frac{1}{\lambda}A^{-1}Ax = \frac{1}{\lambda}A^{-1}\lambda x$, which gives $\frac{1}{\lambda}Ix = \frac{1}{\lambda}\lambda A^{-1}x$, that is, $\frac{1}{\lambda}x = A^{-1}x$. Consequently, $1/\lambda$ is an eigenvalue of A^{-1} with corresponding eigenvector x. ☐

4.2.4(c) Proved by Exercise 4.2.11.

Example 4.2.5 Recall from Example 4.1.24 that matrix

$$A = \begin{bmatrix} 1 & -\frac{1}{2} \\ -\frac{1}{2} & 1 \end{bmatrix}$$

has eigenvalues $1/2$ and $3/2$ with corresponding eigenvectors $(1,1)$ and $(1,-1)$ respectively. Confirm that matrix A^2 has eigenvalues which are these squared, and corresponding to the same eigenvectors.

Solution: Compute

$$A^2 = \begin{bmatrix} 1 & -\frac{1}{2} \\ -\frac{1}{2} & 1 \end{bmatrix} \begin{bmatrix} 1 & -\frac{1}{2} \\ -\frac{1}{2} & 1 \end{bmatrix} = \begin{bmatrix} \frac{5}{4} & -1 \\ -1 & \frac{5}{4} \end{bmatrix}.$$

Then

$$A^2 \begin{bmatrix} 1 \\ 1 \end{bmatrix} = \begin{bmatrix} \frac{1}{4} \\ \frac{1}{4} \end{bmatrix} = \frac{1}{4} \begin{bmatrix} 1 \\ 1 \end{bmatrix}, \qquad A^2 \begin{bmatrix} 1 \\ -1 \end{bmatrix} = \begin{bmatrix} \frac{9}{4} \\ -\frac{9}{4} \end{bmatrix} = \frac{9}{4} \begin{bmatrix} 1 \\ -1 \end{bmatrix},$$

and so A^2 has eigenvalues $1/4 = (1/2)^2$ and $9/4 = (3/2)^2$ with the same corresponding eigenvectors $(1,1)$ and $(1,-1)$ respectively. □

Activity 4.2.6 You are given that -3 and 2 are eigenvalues of the matrix $A = \begin{bmatrix} 1 & 2 \\ 2 & -2 \end{bmatrix}$.

• Which of the following matrices has an eigenvalue of 8?

(a) A^{-1} (b) A^{-2} (c) A^2 (d) A^3

• Further, which of the above matrices has eigenvalue $1/9$?

Example 4.2.7 Consider the matrix

$$A = \begin{bmatrix} 1 & 1 & 0 \\ 1 & 0 & 1 \\ 0 & 1 & 1 \end{bmatrix}.$$

You are given that this matrix has eigenvalues 2, 1, and -1 with corresponding eigenvectors $(1,1,1)$, $(-1,0,1)$, and $(1,-2,1)$ respectively. Confirm that matrix A^2 has eigenvalues which are these squared, and corresponding to the same eigenvectors. Given the inverse

$$A^{-1} = \begin{bmatrix} \frac{1}{2} & \frac{1}{2} & -\frac{1}{2} \\ \frac{1}{2} & -\frac{1}{2} & \frac{1}{2} \\ -\frac{1}{2} & \frac{1}{2} & \frac{1}{2} \end{bmatrix}$$

confirm that its eigenvalues are the reciprocals of those of A, and for corresponding eigenvectors.

Solution:

• Compute

$$A^2 = \begin{bmatrix} 1 & 1 & 0 \\ 1 & 0 & 1 \\ 0 & 1 & 1 \end{bmatrix} \begin{bmatrix} 1 & 1 & 0 \\ 1 & 0 & 1 \\ 0 & 1 & 1 \end{bmatrix} = \begin{bmatrix} 2 & 1 & 1 \\ 1 & 2 & 1 \\ 1 & 1 & 2 \end{bmatrix}.$$

Then

$$A^2 \begin{bmatrix} 1 \\ 1 \\ 1 \end{bmatrix} = \begin{bmatrix} 4 \\ 4 \\ 4 \end{bmatrix} = 4 \begin{bmatrix} 1 \\ 1 \\ 1 \end{bmatrix},$$

$$A^2 \begin{bmatrix} -1 \\ 0 \\ 1 \end{bmatrix} = \begin{bmatrix} -1 \\ 0 \\ 1 \end{bmatrix} = 1 \begin{bmatrix} -1 \\ 0 \\ 1 \end{bmatrix},$$

$$A^2 \begin{bmatrix} 1 \\ -2 \\ 1 \end{bmatrix} = \begin{bmatrix} 1 \\ -2 \\ 1 \end{bmatrix} = 1 \begin{bmatrix} 1 \\ -2 \\ 1 \end{bmatrix},$$

has eigenvalues $4 = 2^2$, and $1 = (\pm 1)^2$ with corresponding eigenvectors $(1,1,1)$, and the pair $(-1,0,1)$, and $(1,-2,1)$. Thus here $\text{span}\{(-1,0,1),(1,-2,1)\}$ is the eigenspace of A^2 corresponding to eigenvalue one.

- For the inverse

$$A^{-1} \begin{bmatrix} 1 \\ 1 \\ 1 \end{bmatrix} = \begin{bmatrix} \frac{1}{2} \\ \frac{1}{2} \\ \frac{1}{2} \end{bmatrix} = \frac{1}{2} \begin{bmatrix} 1 \\ 1 \\ 1 \end{bmatrix},$$

$$A^{-1} \begin{bmatrix} -1 \\ 0 \\ 1 \end{bmatrix} = \begin{bmatrix} -1 \\ 0 \\ 1 \end{bmatrix} = 1 \begin{bmatrix} -1 \\ 0 \\ 1 \end{bmatrix},$$

$$A^{-1} \begin{bmatrix} 1 \\ -2 \\ 1 \end{bmatrix} = \begin{bmatrix} -1 \\ 2 \\ -1 \end{bmatrix} = (-1) \begin{bmatrix} 1 \\ -2 \\ 1 \end{bmatrix},$$

has eigenvalues $1/2$, $1 = 1/1$ and $-1 = 1/(-1)$ with corresponding eigenvectors $(1,1,1)$, $(-1,0,1)$ and $(1,-2,1)$. □

Example 4.2.8 (*long-term age structure*) Recall that Example 3.1.9 introduced how to use a Leslie matrix to predict the future population of an animal. In the example, letting $x = (x_1,x_2,x_3)$ be the current number of pups, juveniles, and mature females respectively, then for the Leslie matrix

$$L = \begin{bmatrix} 0 & 0 & 4 \\ \frac{1}{2} & 0 & 0 \\ 0 & \frac{1}{3} & \frac{1}{3} \end{bmatrix}$$

the predicted population number after a year is $x' = Lx$, after two years is $x'' = Lx' = L^2 x$, and so on. Predict what happens after many generations: does the population die out? grow? oscillate?

Solution: Consider what happens after n generations for large n, say $n = 10$ or 100. The predicted population is $x^{(n)} = L^n x$; that is, the matrix L^n transforms the current population to that after n generations. The stretching and/or shrinking of matrix L^n is summarized by its eigenvectors and eigenvalues (Section 4.1). By Theorem 4.2.4 the eigenvalues of L^n are λ^n in terms of the eigenvalues λ of L. By hand (Procedure 4.1.23), the characteristic equation of L is

$$\det(L - \lambda I) = \begin{vmatrix} -\lambda & 0 & 4 \\ \frac{1}{2} & -\lambda & 0 \\ 0 & \frac{1}{3} & \frac{1}{3} - \lambda \end{vmatrix}$$

$$= \lambda^2(\tfrac{1}{3} - \lambda) + 0 + \tfrac{2}{3} - 0 - 0 - 0$$

$$= (1 - \lambda)(\lambda^2 + \tfrac{2}{3}\lambda + \tfrac{2}{3})$$

$$= (1 - \lambda)[(\lambda + \tfrac{1}{3})^2 + \tfrac{5}{9}] = 0$$

$$\implies \quad \lambda = 1, (-1 \pm i\sqrt{5})/3.$$

Such complex-valued eigenvalues may arise in real applications when the matrix is not symmetric, as here—the next Theorem 4.2.9 proves that such complexities do not arise for symmetric matrices.

But the algebra still works with complex eigenvalues (Chapter 7). Here, the eigenvalues of L^n are λ^n (Theorem 4.2.4) namely $1^n = 1$ for all n and $[(-1 \pm i\sqrt{5})/3]^n$. Because the absolute value $|(-1 \pm i\sqrt{5})/3| = |-1 \pm i\sqrt{5}|/3 = \sqrt{1+5}/3 = \sqrt{6}/3 = \sqrt{2/3} = 0.81650$ (recall that we take five significant digits to be effectively exact in most practice, Section 1.5), then the absolute value of $[(-1 \pm i\sqrt{5})/3]^n$ is 0.81650^n which becomes negligibly small for large n; for example, $0.81650^{34} \approx 0.001$. Since the eigenvectors of L^n are the same as those of L (Theorem 4.2.4), these negligibly small eigenvalues of L^n imply that any component in the initial population in the direction of the corresponding eigenvectors is shrunk to zero by L^n. For large n, it is only the component in the eigenvector corresponding to eigenvalue $\lambda = 1$ that remains. Find the eigenvector by solving $(L - I)x = 0$, namely

$$\begin{bmatrix} -1 & 0 & 4 \\ \frac{1}{2} & -1 & 0 \\ 0 & \frac{1}{3} & -\frac{2}{3} \end{bmatrix} x = \begin{bmatrix} -x_1 + 4x_3 \\ \frac{1}{2}x_1 - x_2 \\ \frac{1}{3}x_2 - \frac{2}{3}x_3 \end{bmatrix} = 0.$$

The first row gives that $x_1 = 4x_3$, the third row that $x_2 = 2x_3$, and the second row confirms that these are correct as $\frac{1}{2}x_1 - x_2 = \frac{1}{2}4x_3 - 2x_3 = 0$. Eigenvectors corresponding to $\lambda = 1$ are then of the form $(4x_3, 2x_3, x_3) = (4, 2, 1)x_3$. Because the corresponding eigenvalue of $L^n = 1^n = 1$ the component of x in this direction remains in $L^n x$ whereas all other components decay to zero. Thus the model predicts that after many generations the population reaches a steady state of the pups, juveniles, and mature females being in the ratio of $4 : 2 : 1$. □

4.2.2 Symmetric matrices are orthogonally diagonalizable

General real matrices may have non-real complex valued eigenvalues (as in Examples 4.2.8 and 4.1.28). That real symmetric matrices always have real eigenvalues (such as in all matrices of

Examples 4.2.5 and 4.2.7) is a special property that marvellously often reflects the physical reality of many applications.

To establish the reality of eigenvalues (Theorem 4.2.9), we have to eliminate the possibility that they are complex valued with a non zero imaginary part. Consequently, the proof of the next Theorem 4.2.9 needs to use some complex numbers and some properties of complex numbers. Recall that any complex number $z = a + b\,i$ has a complex conjugate $\bar{z} = a - b\,i$ (denoted by the overbar, and similarly for vectors), and that a complex number equals its conjugate only if it is real valued (the imaginary part is zero). Such properties of complex numbers and operations also hold for complex valued vectors, complex valued matrices, and arithmetic operations with complex valued matrices and vectors.

Theorem 4.2.9 *For every real symmetric matrix A, every eigenvalue of A is real valued.*

Proof. Let λ be any eigenvalue of real matrix A with corresponding eigenvector x; that is, $Ax = \lambda x$ for $x \neq 0$. First, involving the complex conjugate $\bar{x}$, we establish $x^T\bar{x} > 0$, which is used in the second part of the proof. In general, the nonzero eigenvector x is non-real complex valued, say

$$x = (a_1 + b_1\,i, a_2 + b_2\,i, \ldots, a_n + b_n\,i)$$
$$\implies \bar{x} = (a_1 - b_1\,i, a_2 - b_2\,i, \ldots, a_n - b_n\,i).$$

Then the product

$$x^T\bar{x} = \begin{bmatrix} a_1 + b_1\,i & a_2 + b_2\,i & \cdots & a_n + b_n\,i \end{bmatrix} \begin{bmatrix} a_1 - b_1\,i \\ a_2 - b_2\,i \\ \vdots \\ a_n - b_n\,i \end{bmatrix}$$
$$= (a_1 + b_1\,i)(a_1 - b_1\,i) + (a_2 + b_2\,i)(a_2 - b_2\,i)$$
$$+ \cdots + (a_n + b_n\,i)(a_n - b_n\,i)$$
$$= (a_1^2 + b_1^2) + (a_2^2 + b_2^2) + \cdots + (a_n^2 + b_n^2)$$
$$> 0$$

since x is an eigenvector which necessarily is nonzero and so at least one term in the sum is positive, as required.

Second, consider $x^T A\bar{x}$ in two different ways:

$$x^T A\bar{x} = x^T (A\bar{x}) = x^T (\bar{A}\bar{x}) \quad \text{(as } A \text{ is real)}$$
$$= x^T (\overline{Ax}) = x^T (\overline{\lambda x}) = x^T (\bar{\lambda}\bar{x}) = \bar{\lambda} x^T\bar{x};$$
$$x^T A\bar{x} = (x^T A)\bar{x} = (A^T x)^T\bar{x} = (Ax)^T\bar{x} \quad \text{(as } A \text{ is symmetric)}$$
$$= (\lambda x)^T\bar{x} = \lambda x^T\bar{x}.$$

Equating the two ends of this identity gives $\bar{\lambda} x^T\bar{x} = \lambda x^T\bar{x}$. Rearrange to $\bar{\lambda} x^T\bar{x} - \lambda x^T\bar{x} = 0$, which factors to $(\bar{\lambda} - \lambda)x^T\bar{x} = 0$. Because this product is zero, and $x^T\bar{x} > 0$, consequently

$\bar{\lambda} - \lambda = 0$. Hence $\bar{\lambda} = \lambda$ and so λ cannot have any imaginary part. That is, the eigenvalue λ must be real. ∎

The other property that we have seen graphically for 2D matrices is that the eigenvectors of symmetric matrices are orthogonal. For Example 4.2.3, both the matrices A and A^{-1} in the second part are symmetric and from the illustration their eigenvectors are proportional to $(1,1)$ and $(-1,1)$ which are orthogonal directions—they are at right-angles in the illustration.

Example 4.2.10 Recall that Example 4.1.27 found the 3×3 symmetric matrix

$$\begin{bmatrix} -2 & 0 & -6 \\ 0 & 4 & 6 \\ -6 & 6 & -9 \end{bmatrix}$$

has eigenspaces $\mathbb{E}_0 = \text{span}\{(-6,-3,2)\}$, $\mathbb{E}_7 = \text{span}\{(-2,6,3)\}$ and $\mathbb{E}_{-14} = \text{span}\{(3,-2,6)\}$. These eigenspaces are orthogonal as evidenced by the dot products of the basis vectors in each span:

$$\begin{aligned} \mathbb{E}_0, \mathbb{E}_7, \quad & (-6,-3,2) \cdot (-2,6,3) = 12 - 18 + 6 = 0; \\ \mathbb{E}_7, \mathbb{E}_{-14}, \quad & (-2,6,3) \cdot (3,-2,6) = -6 - 12 + 18 = 0; \\ \mathbb{E}_{-14}, \mathbb{E}_0, \quad & (3,-2,6) \cdot (-6,-3,2) = -18 + 6 + 12 = 0. \end{aligned}$$

☐

Theorem 4.2.11 *Let A be a real symmetric matrix, then for every two distinct eigenvalues of A, any corresponding two eigenvectors are orthogonal.*

Proof. Let eigenvalues $\lambda_1 \neq \lambda_2$, and let $\boldsymbol{x}_1$ and $\boldsymbol{x}_2$ be any corresponding eigenvectors, respectively; that is, $A\boldsymbol{x}_1 = \lambda_1\boldsymbol{x}_1$ and $A\boldsymbol{x}_2 = \lambda_2\boldsymbol{x}_2$. Consider $\boldsymbol{x}_1^T A\boldsymbol{x}_2$ in two different ways:

$$\begin{aligned} \boldsymbol{x}_1^T A\boldsymbol{x}_2 &= \boldsymbol{x}_1^T (A\boldsymbol{x}_2) = \boldsymbol{x}_1^T (\lambda_2\boldsymbol{x}_2) = \lambda_2\boldsymbol{x}_1^T\boldsymbol{x}_2 = \lambda_2\boldsymbol{x}_1 \cdot \boldsymbol{x}_2; \\ \boldsymbol{x}_1^T A\boldsymbol{x}_2 &= \boldsymbol{x}_1^T A^T\boldsymbol{x}_2 \quad \text{(as A is symmetric)} \\ &= (\boldsymbol{x}_1^T A^T)\boldsymbol{x}_2 = (A\boldsymbol{x}_1)^T\boldsymbol{x}_2 \\ &= (\lambda_1\boldsymbol{x}_1)^T\boldsymbol{x}_2 = \lambda_1\boldsymbol{x}_1^T\boldsymbol{x}_2 = \lambda_1\boldsymbol{x}_1 \cdot \boldsymbol{x}_2. \end{aligned}$$

Equating the two ends of this identity gives $\lambda_2\boldsymbol{x}_1 \cdot \boldsymbol{x}_2 = \lambda_1\boldsymbol{x}_1 \cdot \boldsymbol{x}_2$. Rearrange to $\lambda_2\boldsymbol{x}_1 \cdot \boldsymbol{x}_2 - \lambda_1\boldsymbol{x}_1 \cdot \boldsymbol{x}_2 = (\lambda_2 - \lambda_1)(\boldsymbol{x}_1 \cdot \boldsymbol{x}_2) = 0$. Since $\lambda_1 \neq \lambda_2$, the factor $\lambda_2 - \lambda_1 \neq 0$, and so it follows that the dot product $\boldsymbol{x}_1 \cdot \boldsymbol{x}_2 = 0$. Hence (Definition 1.3.19) the two eigenvectors are orthogonal. ∎

Example 4.2.12 The plots below show (unit) vectors $\boldsymbol{x}$ (blue), and for some matrix A (different for different plots) the corresponding vectors $A\boldsymbol{x}$ (red) adjoined. By estimating eigenvectors, determine which cases *cannot* be the plot of a real symmetric matrix.

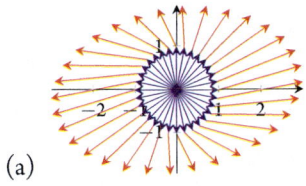

(a)

Solution: Estimate eigenvectors $(0.8, 0.6)$ and $(0.5, 0.9)$ which are not orthogonal, so cannot be from a symmetric matrix.   ☐

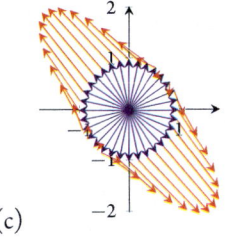

(c)

Solution: Estimate eigenvectors $(0.8, 0.7)$ and $(-0.7, 0.8)$ which are orthogonal, so may be a symmetric matrix.   ☐

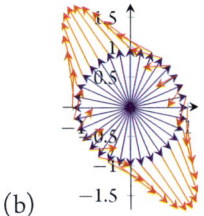

(b)

Solution: Estimate eigenvectors $(0.8, 0.5)$ and $(-0.5, 0.8)$ which are orthogonal, so may be a symmetric matrix.   ☐

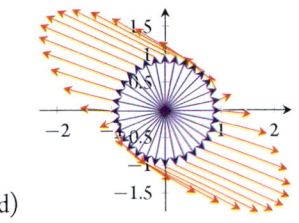

(d)

Solution: Estimate eigenvectors $(1, 0.1)$ and $(1, -0.3)$ which are not orthogonal, so cannot be from a symmetric matrix.   ☐

Example 4.2.13 By hand, find eigenvectors corresponding to the two distinct eigenvalues of the following matrices. Confirm that symmetric matrix A has orthogonal eigenvectors, and that non-symmetric matrix B does not:

$$A = \begin{bmatrix} 1 & \frac{3}{2} \\ \frac{3}{2} & -3 \end{bmatrix}; \quad B = \begin{bmatrix} 0 & -3 \\ -2 & 1 \end{bmatrix}.$$

Solution:

- For matrix A, the eigenvalues come from the characteristic equation

$$\det(A - \lambda I) = (1 - \lambda)(-3 - \lambda) - \tfrac{9}{4}$$
$$= \lambda^2 + 2\lambda - \tfrac{21}{4} = (\lambda + 1)^2 - \tfrac{25}{4} = 0,$$

so eigenvalues are $\lambda = -1 \pm \tfrac{5}{2} = -\tfrac{7}{2}, \tfrac{3}{2}$.

  – Corresponding to eigenvalue $\lambda = -7/2$, eigenvectors $\boldsymbol{x}$ satisfy $(A + \tfrac{7}{2}I)\boldsymbol{x} = \boldsymbol{0}$, that is

$$\begin{bmatrix} \frac{9}{2} & \frac{3}{2} \\ \frac{3}{2} & \frac{1}{2} \end{bmatrix} x = \frac{1}{2}\begin{bmatrix} 9x_1 + 3x_2 \\ 3x_1 + x_2 \end{bmatrix} = \begin{bmatrix} 0 \\ 0 \end{bmatrix},$$

giving $x_2 = -3x_1$. Eigenvectors must be $x \propto (1, -3)$.

- Corresponding to eigenvalue $\lambda = 3/2$, eigenvectors x satisfy $(A - \frac{3}{2}I)x = 0$, that is

$$\begin{bmatrix} -\frac{1}{2} & \frac{3}{2} \\ \frac{3}{2} & -\frac{9}{2} \end{bmatrix} x = \frac{1}{2}\begin{bmatrix} -x_1 + 3x_2 \\ 3x_1 - 9x_2 \end{bmatrix} = \begin{bmatrix} 0 \\ 0 \end{bmatrix},$$

giving $x_1 = 3x_2$. Eigenvectors must be $x \propto (3, 1)$.

The dot product of the two basis eigenvectors is $(1, -3) \cdot (3, 1) = 3 - 3 = 0$ and hence eigenvectors of $\lambda = -\frac{7}{2}$ are orthogonal to eigenvectors of $\lambda = \frac{3}{2}$.

- For matrix B, the eigenvalues come from the characteristic equation

$$\det(B - \lambda I) = (-\lambda)(1 - \lambda) - 6 = \lambda^2 - \lambda - 6 = (\lambda - 3)(\lambda + 2) = 0,$$

so eigenvalues are $\lambda = 3, -2$.

- Corresponding to eigenvalue $\lambda = -2$, eigenvectors x satisfy $(B + 2I)x = 0$, that is

$$\begin{bmatrix} 2 & -3 \\ -2 & 3 \end{bmatrix} x = \begin{bmatrix} 2x_1 - 3x_2 \\ -2x_1 + 3x_2 \end{bmatrix} = \begin{bmatrix} 0 \\ 0 \end{bmatrix},$$

giving $x_2 = \frac{2}{3}x_1$. Eigenvectors must be $x \propto (3, 2)$.

- Corresponding to eigenvalue $\lambda = 3$, eigenvectors x satisfy $(B - 3I)x = 0$, that is

$$\begin{bmatrix} -3 & -3 \\ -2 & -2 \end{bmatrix} x = \begin{bmatrix} -3x_1 - 3x_2 \\ -2x_1 - 2x_2 \end{bmatrix} = \begin{bmatrix} 0 \\ 0 \end{bmatrix},$$

giving $x_1 = -x_2$. Eigenvectors must be $x \propto (-1, 1)$.

The dot product of the two basis eigenvectors is $(3, 2) \cdot (-1, 1) = -3 + 2 = -1 \neq 0$ and hence eigenvectors of $\lambda = 3$ are *not* orthogonal to eigenvectors of $\lambda = -2$. □

Example 4.2.14 Use MATLAB/Octave to compute eigenvectors of the following matrices. Confirm that the eigenvectors are orthogonal for the symmetric matrix.

(a)
$$\begin{bmatrix} 0 & 3 & 2 & -1 \\ 0 & 3 & 0 & 0 \\ 3 & 0 & -1 & -1 \\ -3 & 1 & 3 & 0 \end{bmatrix}$$

(b)
$$\begin{bmatrix} -6 & 0 & 1 & 1 \\ 0 & 0 & 2 & 2 \\ 1 & 2 & 2 & -1 \\ 1 & 2 & -1 & -1 \end{bmatrix}$$

Solution: For each matrix, enter the matrix as say A, then execute [V, D] = eig (A) to give eigenvectors as the columns of V. Then confirm orthogonality of all pairs of eigenvectors by computing V' *V and confirm that the off-diagonal dot products are zero, or confirm the lack of orthogonality if nonzero. (In the case of repeated eigenvalues, MATLAB/Octave generates an orthonormal basis for the corresponding eigenspace so the returned matrix V of eigenvectors is still orthogonal for symmetric A.)

(a) The MATLAB/Octave code

```
A= [0 3 2 -1
  0 3 0 0
  3 0 -1 -1
  -3 1 3 0]
[V,D] =eig (A)
V' *V
```

gives the following (2 d.p.)

```
V =
  -0.49   -0.71    0.41    0.74
   0.00    0.00    0.00    0.34
   0.32   -0.71    0.41    0.57
  -0.81   -0.00    0.82   -0.06
D =
  -3.00       0       0       0
      0    2.00       0       0
      0       0    0.00       0
      0       0       0    3.00
```

so eigenvectors corresponding to the four distinct eigenvalues are $(-0.49, 0, 0.32, -0.81)$, $(-0.71, 0, -0.71, 0)$, $(0.41, 0, 0.41, 0.82)$, and $(0.74, 0.34, 0.57, -0.6)$. Then V' *V is (2 d.p.)

```
   1.00    0.11   -0.73   -0.13
   0.11    1.00   -0.58   -0.93
  -0.73   -0.58    1.00    0.49
  -0.13   -0.93    0.49    1.00
```

As the off-diagonal elements are nonzero, the pairs of dot products $v_i \cdot v_j = v_i^T v_j$ for $i \neq j$ are nonzero, indicating that the column vectors are not orthogonal. Hence the matrix A cannot be symmetric.

(b) The MATLAB/Octave code

```
B= [-6 0 1 1
  0 0 2 2
  1 2 2 -1
  1 2 -1 -1]
[V,D] =eig (B)
V' *V
```

gives the following (2 d.p.)

```
V =
    0.94     0.32    -0.04    -0.10
    0.13    -0.63    -0.53    -0.55
   -0.17     0.32     0.43    -0.83
   -0.25     0.63    -0.73    -0.08
D =
   -6.45        0        0        0
       0    -3.00        0        0
       0        0     1.11        0
       0        0        0     3.34
```

so eigenvectors corresponding to the four distinct eigenvalues are $(0.94, 0.13, -0.17, -0.25)$, $(0.32, -0.63, 0.32, 0.63)$, $(-0.04, -0.53, 0.43, -0.73)$, and $(-0.10, -0.55, -0.83, -0.08)$. Then V' *V is (2 d.p.)

```
    1.00    -0.00    -0.00    -0.00
   -0.00     1.00     0.00    -0.00
   -0.00     0.00     1.00    -0.00
   -0.00    -0.00    -0.00     1.00
```

As the off-diagonal elements are zero, the pairs of dot products are zero, indicating that the column vectors are orthogonal. The symmetry of this matrix B requires such orthogonality. □

Recall that to find eigenvalues by hand for 2×2 or 3×3 matrices we solve a quadratic or cubic characteristic equation, respectively. Thus we find at most two or three eigenvalues, respectively. Further, when we ask MATLAB/Octave to compute eigenvalues of an $n \times n$ matrix, it always returns n eigenvalues in an $n \times n$ diagonal matrix.

Theorem 4.2.15 *Every $n \times n$ real symmetric matrix A has at most n distinct eigenvalues.*

Proof. Let's use contradiction. Assume that there are more than n distinct eigenvalues. Then there would be more than n eigenvectors corresponding to distinct eigenvalues. Theorem 4.2.11 asserts that all such eigenvectors are orthogonal. But there cannot be more than n vectors in an orthogonal set in $\mathbb{R}^n$ (Theorem 1.3.25). Hence the assumption is wrong: there cannot be any more than n distinct eigenvalues. □

The previous theorem establishes that there are at most n distinct eigenvalues (here for symmetric matrices, but Theorem 7.1.1 establishes that it is true for general matrices). Now we establish that typically there exist n distinct eigenvalues of an $n \times n$ matrix—here symmetric.

Example 4.0.1 started this chapter by observing that in an SVD of a *symmetric* matrix, $A = USV^T$, the columns of U appear to be (almost) always plus/minus the corresponding columns of V. Exceptions possibly arise in the degenerate cases when two or more singular values are identical. We now prove this close relation between U and V in all non-degenerate cases.

Theorem 4.2.16 *Let A be an $n \times n$ real symmetric matrix with SVD $A = USV^T$. If all the singular values are distinct or zero, $\sigma_1 > \cdots > \sigma_r > \sigma_{r+1} = \cdots = \sigma_n = 0$, then v_j is an eigenvector of A corresponding to an eigenvalue of either $\lambda_j = +\sigma_j$ or $\lambda_j = -\sigma_j$ (not both, except for the trivial case of $\sigma_j = 0$).*

If nonzero singular values are repeated, then one can always choose an SVD such that the result of this theorem still holds. However, the proof is too involved to give here.

This proof modifies parts of the proof of the SVD Theorem 3.3.6 to the specific case of a symmetric matrix.

Proof. First, for any zero singular value, $\sigma_j = 0$, then the result is immediate as from the SVD, $AV = US$ the jth column gives $Av_j = 0u_j = \mathbf{0} = 0v_j$ for the nonzero v_j.

Second, for the singular values $\sigma_j > 0$ we use a form of induction allied with a contradiction to prove $u_j = \pm v_j$. The induction starts with the case of u_1 and v_1. We seek a contradiction by assuming $u_1 \neq \pm v_1$. Since u_1 and v_1 are unit vectors, we can write $u_1 = v_1 \cos t + v' \sin t$ for some unit vector v' orthogonal to v_1 ($v' \propto \text{perp}_{v_1} u_1$) and for angle $0 < t < \pi$ (as illustrated to the right). Multiply by A giving 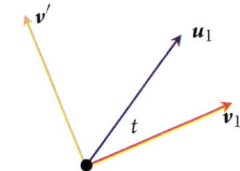 the identity $Au_1 = Av_1 \cos t + Av' \sin t$. Now the first column of $AV = US$ gives us that $Av_1 = \sigma_1 u_1$. Also for the symmetric matrix $A = A^T = (USV^T)^T = VS^T U^T = VSU^T$ is an alternative SVD of A, so $AU = VS$ giving us in its first column that $Au_1 = \sigma_1 v_1$. That is, the identity becomes $\sigma_1 v_1 = \sigma_1 u_1 \cos t + (Av') \sin t$. Further, since v' is orthogonal to v_1, the proof of the SVD Theorem 3.3.6 establishes that Av' is orthogonal to u_1. Equate the lengths of both sides: $\sigma_1^2 = \sigma_1^2 \cos^2 t + |Av'|^2 \sin^2 t$ which rearranging implies $(\sigma_1^2 - |Av'|^2) \sin^2 t = 0$. For angles $0 < t < \pi$ this implies $|Av'| = \sigma_1$; that is, $Av' = \sigma_1 u'$ for the unit vector v' orthogonal to v_1, and for some unit vector u'. Hence v' is a singular vector corresponding to singular value σ_1, and so v' must also be orthogonal to $v_2, v_3, \ldots, v_n$. Thus $\{v', v_1, v_2, \ldots, v_n\}$ is an orthogonal set of unit vectors. This is a set of $n + 1$ orthogonal vectors in $\mathbb{R}^n$ which Theorem 1.3.25 asserts is a contradiction. Hence the assumption that $u_1 \neq \pm v_1$ must be wrong. Consequently $u_1 = \pm v_1$ (for one of the signs, not both).

Recall the induction proof for the SVD Theorem 3.3.6 (Section 3.3.3). Here, since $u_1 = \pm v_1$, we can and do choose $\bar{U} = \bar{V}$. Hence $B = \bar{U}^T A \bar{V} = \bar{V}^T A \bar{V}$ is symmetric as $B^T = (\bar{V}^T A \bar{V})^T = \bar{V}^T A^T \bar{V} = \bar{V}^T A \bar{V} = B$. Consequently, the same argument applies at all steps in the induction for the proof of an SVD and hence establishes $u_j = \pm v_j$ (each for one of the signs, not both).

Third and last, from the jth column of $AV = US$, $Av_j = \sigma_j u_j = \sigma_j(\pm v_j) = \lambda_j v_j$ for eigenvalue λ_j one of $\pm \sigma_j$ but not both. □

Recall that for every real matrix A an SVD is $A = USV^T$. But specifically for symmetric A, the proof of the previous Theorem 4.2.16 identified that the columns of US, $\sigma_j u_j$, are generally the same as $\lambda_j v_j$ and hence are the columns of VD where $D = \text{diag}(\lambda_1, \lambda_2, \ldots, \lambda_n)$. In which case the SVD becomes $A = VDV^T$. This form of an SVD is intimately connected to the following definition.

Definition 4.2.17 *A real square matrix A is **orthogonally diagonalizable** if there exists an orthogonal matrix V and a diagonal matrix D such that $V^T A V = D$, equivalently $AV = VD$, equivalently $A = VDV^T$ is a factorization of A.*

The equivalences in this definition arise immediately from the orthogonality of matrix V (Definition 3.2.43): pre-multiplying $V^T AV = D$ by V gives $VV^T AV = AV = VD$; and so on.

Example 4.2.18

(a) Recall from Example 4.2.13 that the symmetric matrix $A = \begin{bmatrix} 1 & 3/2 \\ 3/2 & -3 \end{bmatrix}$ has eigenvalues $\lambda = -\frac{7}{2}, \frac{3}{2}$ with corresponding orthogonal eigenvectors $(1, -3)$ and $(3, 1)$. Normalize these eigenvectors to unit length as the columns of the orthogonal matrix

$$V = \begin{bmatrix} \frac{1}{\sqrt{10}} & \frac{3}{\sqrt{10}} \\ -\frac{3}{\sqrt{10}} & \frac{1}{\sqrt{10}} \end{bmatrix} = \frac{1}{\sqrt{10}} \begin{bmatrix} 1 & 3 \\ -3 & 1 \end{bmatrix} \quad \text{then}$$

$$V^T AV = \frac{1}{\sqrt{10}} \begin{bmatrix} 1 & -3 \\ 3 & 1 \end{bmatrix} \begin{bmatrix} 1 & \frac{3}{2} \\ \frac{3}{2} & -3 \end{bmatrix} \frac{1}{\sqrt{10}} \begin{bmatrix} 1 & 3 \\ -3 & 1 \end{bmatrix}$$

$$= \frac{1}{10} \begin{bmatrix} -\frac{7}{2} & \frac{21}{2} \\ \frac{9}{2} & \frac{3}{2} \end{bmatrix} \begin{bmatrix} 1 & 3 \\ -3 & 1 \end{bmatrix} = \frac{1}{10} \begin{bmatrix} -35 & 0 \\ 0 & 15 \end{bmatrix} = \begin{bmatrix} -\frac{7}{2} & 0 \\ 0 & \frac{3}{2} \end{bmatrix}.$$

Hence this matrix is orthogonally diagonalizable.

(b) Recall from Example 4.2.14 that the symmetric matrix

$$B = \begin{bmatrix} -6 & 0 & 1 & 1 \\ 0 & 0 & 2 & 2 \\ 1 & 2 & 2 & -1 \\ 1 & 2 & -1 & -1 \end{bmatrix}$$

has orthogonal eigenvectors computed by MATLAB/Octave into the orthogonal matrix V. By additionally computing V′ *B*V we get the following diagonal result (2 d.p.)

```
ans =
    -6.45      0.00      0.00      0.00
     0.00     -3.00      0.00     -0.00
     0.00      0.00      1.11     -0.00
    -0.00     -0.00     -0.00      3.34
```

and see that this matrix B is orthogonally diagonalizable. □

These examples of orthogonal diagonalization invoke symmetric matrices. Also, the connection between an SVD and orthogonal matrices was previously discussed only for symmetric matrices. The next theorem establishes that all real symmetric matrices are orthogonally diagonalizable, and vice versa. That is, eigenvectors of a matrix form an orthogonal set if and only if the matrix is symmetric.

Theorem 4.2.19 (spectral) *For every real square matrix A, matrix A is symmetric iff it is orthogonally diagonalizable.*

Proof. The "if" and the "only if" lead to two parts in the proof.

- If matrix A is orthogonally diagonalizable, then $A = VDV^{\mathrm{T}}$ for orthogonal V and diagonal D (and recall that for a diagonal matrix, $D^{\mathrm{T}} = D$). Consider

$$A^{\mathrm{T}} = (VDV^{\mathrm{T}})^{\mathrm{T}} = (V^{\mathrm{T}})^{\mathrm{T}}D^{\mathrm{T}}V^{\mathrm{T}} = VDV^{\mathrm{T}} = A.$$

Consequently the matrix A is symmetric.

- Theorem 4.2.16 establishes the converse for the generic case of distinct singular values. If matrix A is symmetric, then Theorem 4.2.16 asserts that an svd $A = USV^{\mathrm{T}}$ has matrix U such that columns $u_j = \pm v_j$. That is, we can write $U = VR$ for diagonal matrix $R = \mathrm{diag}(\pm 1, \pm 1, \ldots, \pm 1)$ for appropriately chosen signs. Then by the svd $A = USV^{\mathrm{T}} = VRSV^{\mathrm{T}} = VDV^{\mathrm{T}}$ for diagonal matrix $D = RS = \mathrm{diag}(\pm \sigma_1, \pm \sigma_2, \ldots, \pm \sigma_n)$ for the same pattern of signs. Hence matrix A is orthogonally diagonalizable.

We omit proving the degenerate case when nonzero singular values are repeated. □

4.2.3 Change orthonormal basis to classify quadratics

The following preliminary example illustrates the important principle, applicable throughout mathematics, that we often either choose or change to a coordinate system in which the mathematical algebra is simplest.

This optional subsection has many uses—although it is not an application itself, as it does not involve real data.

Example 4.2.20 (*choose useful coordinates*) Consider the following two quadratic curves. For each curve draw a coordinate system in which the algebraic description of the curve would be most straightforward.

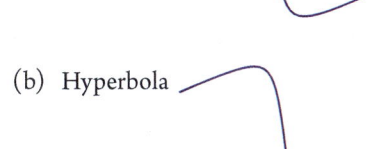

(b) Hyperbola

(a) Ellipse

Solution: Among several possibilities is the following.

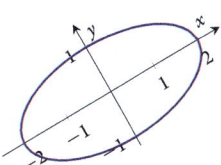

In this coordinate system the ellipse is algebraically $(x/2)^2 + y^2 = 1$. □

Solution: Possibly

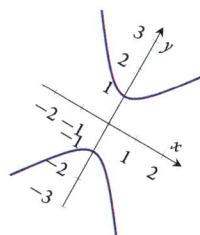

In this coordinate system the hyperbola is algebraically $y^2 = 1 + 2x^2$. □

Now let's proceed to see how to implement in algebra this geometric idea of choosing good coordinates to fit a given physical curve.

Graph quadratic equations

Example 4.2.20 illustrated an ellipse and a hyperbola. These curves are examples of the so-called **conic sections**, which arise as solutions of the quadratic equation in two variables, say x and y,

$$ax^2 + bxy + cy^2 + dx + ey + f = 0 \tag{4.2}$$

(where a, b, c cannot all be zero). As invoked in the example, the canonical simplest algebraic form of such curves are the following. The challenge of this subsection is to choose good new coordinates so that a given quadratic equation (4.2) becomes one of the following recognized canonical forms.

Ellipse or circle $\frac{x^2}{a^2} + \frac{y^2}{b^2} = 1$

- ellipse $a > b$

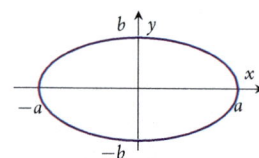

- the circle $a = b$

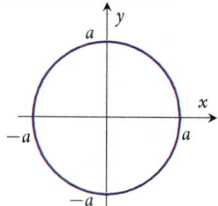

- ellipse $a < b$

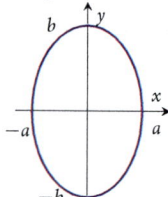

Hyperbola $\frac{x^2}{a^2} - \frac{y^2}{b^2} = 1$ or $-\frac{x^2}{a^2} + \frac{y^2}{b^2} = 1$

- $\frac{x^2}{a^2} - \frac{y^2}{b^2} = 1$

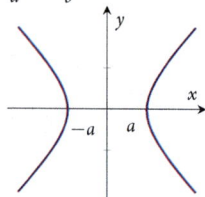

- $-\frac{x^2}{a^2} + \frac{y^2}{b^2} = 1$

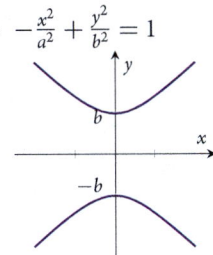

Parabola $y = ax^2$ or $x = ay^2$

- $y = ax^2$

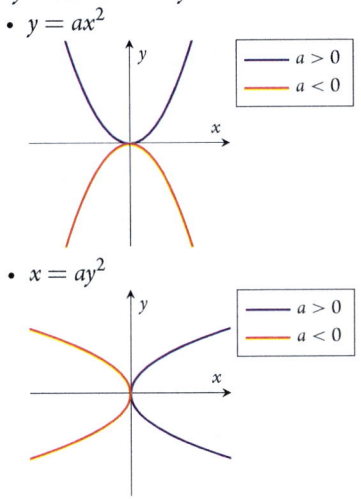

- $x = ay^2$

Example 4.2.20 implicitly has two steps: first, we decide upon an orientation for the coordinate axes; second, we decide that the coordinate system should be 'centred' in the picture. Algebra follows the same two steps.

Example 4.2.21 *(centre coordinates)* By shifting coordinates, identify the conic section whose equation is

$$2x^2 + y^2 - 4x + 4y + 2 = 0.$$

Solution: Group the linear terms with corresponding quadratic powers and seek to rewrite each as a perfect square: the equation is

$$(2x^2 - 4x) + (y^2 + 4y) + 2 = 0$$
$$\Longleftrightarrow \quad 2(x^2 - 2x) + (y^2 + 4y) = -2$$
$$\Longleftrightarrow \quad 2(x^2 - 2x + 1) + (y^2 + 4y + 4) = -2 + 2 + 4$$
$$\Longleftrightarrow \quad 2(x - 1)^2 + (y + 2)^2 = 4.$$

Thus changing to a new (dashed) coordinate system $x' = x - 1$ and $y' = y + 2$, that is, choosing the origin of the dashed coordinate system at $(x, y) = (1, -2)$, the quadratic equation becomes

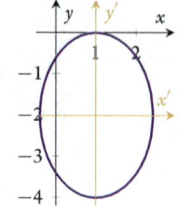

$$2x'^2 + y'^2 = 4, \quad \text{that is,} \quad \frac{x'^2}{2} + \frac{y'^2}{4} = 1.$$

In this new coordinate system the equation is that of an ellipse with horizontal axis of half-length $\sqrt{2}$ and vertical axis of half-length $\sqrt{4} = 2$ (as illustrated to the right). □

Example 4.2.22 *(rotate coordinates)* By rotating the coordinate system, identify the conic section whose equation is

$$x^2 + 3xy - 3y^2 - \tfrac{1}{2} = 0.$$

(There are no terms linear in x and y so we do not shift coordinates.)

Solution: The equation contains the product xy. To identify the conic we must eliminate the xy term. To use matrix algebra, and in terms of the vector $\boldsymbol{x} = (x, y)$, recognize that the quadratic terms may be written as $\boldsymbol{x}^T A \boldsymbol{x}$ for symmetric matrix $A = \begin{bmatrix} 1 & 3/2 \\ 3/2 & -3 \end{bmatrix}$ as then

$$\boldsymbol{x}^T A \boldsymbol{x} = \boldsymbol{x}^T \begin{bmatrix} 1 & \frac{3}{2} \\ \frac{3}{2} & -3 \end{bmatrix} \boldsymbol{x} = \begin{bmatrix} x & y \end{bmatrix} \begin{bmatrix} x + \frac{3}{2}y \\ \frac{3}{2}x - 3y \end{bmatrix}$$

$$= x(x + \tfrac{3}{2}y) + y(\tfrac{3}{2}x - 3y) = x^2 + 3xy - 3y^2.$$

(The matrix form $\boldsymbol{x}^T A \boldsymbol{x}$ splits the cross product term $3xy$ into two equal halves represented by the two off-diagonal elements $\frac{3}{2}$ in matrix A.) Suppose we change to some new (dashed) coordinate system with its standard unit vectors $\boldsymbol{v}_1$ and $\boldsymbol{v}_2$ as illustrated to the right. The vectors in the plane may then be written as the linear combination $\boldsymbol{x} = \boldsymbol{v}_1 x' + \boldsymbol{v}_2 y'$. That is, $\boldsymbol{x} = V \boldsymbol{x}'$ for new coordinate vector $\boldsymbol{x}' = (x', y')$ and matrix $V = \begin{bmatrix} \boldsymbol{v}_1 & \boldsymbol{v}_2 \end{bmatrix}$.

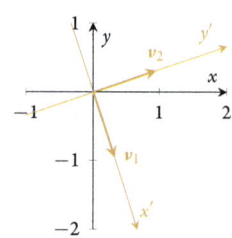

Since the new coordinate system is related to the old by $\boldsymbol{x} = V\boldsymbol{x}'$, then the quadratic terms transform as follows:

$$\boldsymbol{x}^T A \boldsymbol{x} = (V\boldsymbol{x}')^T A (V\boldsymbol{x}') = \boldsymbol{x}'^T V^T A V \boldsymbol{x}' = \boldsymbol{x}'^T (V^T A V) \boldsymbol{x}'.$$

Thus choose V to simplify $V^T A V$. Because matrix A is symmetric, Theorem 4.2.19 asserts that it is orthogonally diagonalizable (using eigenvectors). Indeed, Example 4.2.18(a) orthogonally diagonalized this particular matrix A, via its eigenvalues and eigenvectors, using the orthogonal matrix

$$V = \begin{bmatrix} \boldsymbol{v}_1 & \boldsymbol{v}_2 \end{bmatrix} = \begin{bmatrix} \frac{1}{\sqrt{10}} & \frac{3}{\sqrt{10}} \\ -\frac{3}{\sqrt{10}} & \frac{1}{\sqrt{10}} \end{bmatrix}.$$

Using this V, in the new dashed coordinate system (illustrated) the quadratic terms in the equation become

$$\boldsymbol{x}'^T(V^T AV)\boldsymbol{x}' = \boldsymbol{x}'^T D\boldsymbol{x}' = \boldsymbol{x}'^T \begin{bmatrix} -\frac{7}{2} & 0 \\ 0 & \frac{3}{2} \end{bmatrix} \boldsymbol{x}' = -\frac{7}{2}x'^2 + \frac{3}{2}y'^2$$

Hence the quadratic equation becomes

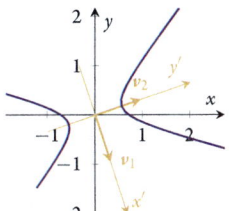

$$-\tfrac{7}{2}x'^2 + \tfrac{3}{2}y'^2 - \tfrac{1}{2} = 0 \iff -7x'^2 + 3y'^2 = 1$$

which is the equation of a hyperbola intersecting the y'-axis at $y' = \pm 1/\sqrt{3}$, as illustrated to the right.

□

Example 4.2.23   Identify the conic section whose equation is

$$x^2 - xy + y^2 + \tfrac{5}{2\sqrt{2}}x - \tfrac{7}{2\sqrt{2}}y + \tfrac{1}{8} = 0.$$

Solution: When there are both the cross product xy and linear terms, it is easier to first rotate coordinates, and second shift coordinates.

(a) Rewrite the quadratic terms using vector $\boldsymbol{x} = (x, y)$, splitting the cross product into two equal halves:

$$x^2 - xy + y^2 = x(x - \tfrac{1}{2}y) + y(-\tfrac{1}{2}x + y) = \begin{bmatrix} x & y \end{bmatrix} \begin{bmatrix} x - \tfrac{1}{2}y \\ -\tfrac{1}{2}x + y \end{bmatrix}$$

$$= \boldsymbol{x}^T \begin{bmatrix} 1 & -\tfrac{1}{2} \\ -\tfrac{1}{2} & 1 \end{bmatrix} \begin{bmatrix} x \\ y \end{bmatrix} = \boldsymbol{x}^T A \boldsymbol{x} \quad \text{for matrix } A = \begin{bmatrix} 1 & -1/2 \\ -1/2 & 1 \end{bmatrix}.$$

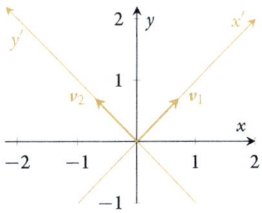

Recall that Example 4.1.24 found that the eigenvalues of this matrix are $\lambda = \tfrac{1}{2}, \tfrac{3}{2}$ with corresponding orthonormal eigenvectors $\boldsymbol{v}_1 = (1, 1)/\sqrt{2}$ and $\boldsymbol{v}_2 = (-1, 1)/\sqrt{2}$, respectively. Let's change to a new (dashed) coordinate system (x', y') with $\boldsymbol{v}_1$ and $\boldsymbol{v}_2$ as its standard unit vectors (as illustrated to the right). Then throughout the 2D-plane every vector/position

$$\boldsymbol{x} = \boldsymbol{v}_1 x' + \boldsymbol{v}_2 y' = \begin{bmatrix} \boldsymbol{v}_1 & \boldsymbol{v}_2 \end{bmatrix} \begin{bmatrix} x' \\ y' \end{bmatrix} = V\boldsymbol{x}'$$

for orthogonal matrix

$$V = \begin{bmatrix} \boldsymbol{v}_1 & \boldsymbol{v}_2 \end{bmatrix} = \frac{1}{\sqrt{2}} \begin{bmatrix} 1 & -1 \\ 1 & 1 \end{bmatrix}.$$

In the new coordinates:

- the quadratic terms

$$x^2 - xy + y^2 = \mathbf{x}^T A \mathbf{x}$$
$$= (V\mathbf{x}')^T A (V\mathbf{x}')$$
$$= \mathbf{x}'^T V^T A V \mathbf{x}'$$
$$= \mathbf{x}'^T \begin{bmatrix} \frac{1}{2} & 0 \\ 0 & \frac{3}{2} \end{bmatrix} \mathbf{x}' \quad (\text{as } V^T A V = D)$$
$$= \tfrac{1}{2} x'^2 + \tfrac{3}{2} y'^2;$$

- whereas the linear terms

$$\tfrac{5}{2\sqrt{2}} x - \tfrac{7}{2\sqrt{2}} y = \begin{bmatrix} \tfrac{5}{2\sqrt{2}} & -\tfrac{7}{2\sqrt{2}} \end{bmatrix} \mathbf{x}$$
$$= \begin{bmatrix} \tfrac{5}{2\sqrt{2}} & -\tfrac{7}{2\sqrt{2}} \end{bmatrix} V\mathbf{x}'$$
$$= \begin{bmatrix} -\tfrac{1}{2} & -3 \end{bmatrix} \mathbf{x}'$$
$$= -\tfrac{1}{2} x' - 3y';$$

- so the quadratic equation transforms to

$$\tfrac{1}{2} x'^2 + \tfrac{3}{2} y'^2 - \tfrac{1}{2} x' - 3y' + \tfrac{1}{8} = 0.$$

(b) The second step is to shift coordinates via completing the squares:

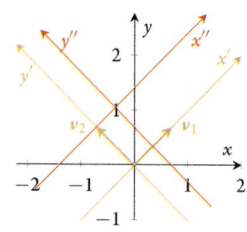

$$\tfrac{1}{2} x'^2 + \tfrac{3}{2} y'^2 - \tfrac{1}{2} x' - 3y' + \tfrac{1}{8} = 0$$
$$\Longleftrightarrow \tfrac{1}{2}(x'^2 - x') + \tfrac{3}{2}(y'^2 - 2y') = -\tfrac{1}{8}$$
$$\Longleftrightarrow \tfrac{1}{2}(x'^2 - x' + \tfrac{1}{4}) + \tfrac{3}{2}(y'^2 - 2y' + 1) = -\tfrac{1}{8} + \tfrac{1}{8} + \tfrac{3}{2}$$
$$\Longleftrightarrow \tfrac{1}{2}(x' - \tfrac{1}{2})^2 + \tfrac{3}{2}(y' - 1)^2 = \tfrac{3}{2}$$

Thus let's change to a new (double dashed) coordinate system $x'' = x' - \tfrac{1}{2}$ and $y'' = y' - 1$ (equivalently, choose the origin of a new coordinate system to be at $(x', y') = (\tfrac{1}{2}, 1)$ as illustrated above-right). In this new coordinate system the quadratic equation becomes

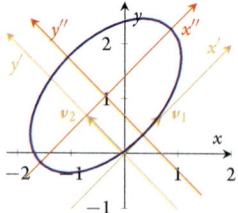

$$\tfrac{1}{2} x''^2 + \tfrac{3}{2} y''^2 = \tfrac{3}{2}, \quad \text{that is} \quad \frac{x''^2}{3} + \frac{y''^2}{1} = 1.$$

In this new coordinate system the equation is that of an ellipse with x''-axis of half-length $\sqrt{3}$ and y''-axis of half-length 1 (as illustrated to the right). □

Simplify quadratic forms

To understand the response and strength of built structures like bridges, buildings, and cars, engineers need to analyse the dynamics of energy distribution in the structure. The potential energy in such structures is expressed and analysed as the following quadratic form. Such quadratic forms are also important in distinguishing maxima from minima in economic optimization.

Definition 4.2.24 *A **quadratic form** in variables $x \in \mathbb{R}^n$ is a function $q : \mathbb{R}^n \to \mathbb{R}$ that may be written as $q(x) = x^{\mathsf{T}} A x$ for some real symmetric $n \times n$ matrix A.*

Example 4.2.25

(a) The dot product of a vector with itself is a quadratic form. For all $x \in \mathbb{R}^n$ consider

$$x \cdot x = x^{\mathsf{T}} x = x^{\mathsf{T}} I_n x,$$

which is the quadratic form associated with the identity matrix I_n.

(b) Example 4.2.22 found the hyperbola satisfying equation $x^2 + 3xy - 3y^2 - \frac{1}{2} = 0$. This equation may be written in terms of a quadratic form as $x^{\mathsf{T}} A x - \frac{1}{2} = 0$ for vector $x = (x, y)$ and symmetric matrix $A = \begin{bmatrix} 1 & 3/2 \\ 3/2 & -3 \end{bmatrix}$.

(c) Example 4.2.23 found the ellipse satisfying the equation $x^2 - xy + y^2 + \frac{5}{2\sqrt{2}} x - \frac{7}{2\sqrt{2}} y + \frac{1}{8} = 0$ via writing the quadratic part of the equation as $x^{\mathsf{T}} A x$ for vector $x = (x, y)$ and symmetric matrix $A = \begin{bmatrix} 1 & -1/2 \\ -1/2 & 1 \end{bmatrix}$. $\qquad\square$

Theorem 4.2.26 (*principal axes*) *For every quadratic form, there exists an orthogonal coordinate system that diagonalizes the quadratic form. Specifically, for the quadratic form $x^{\mathsf{T}} A x$ find the eigenvalues λ_1, λ_2, ..., λ_n and an orthonormal set of eigenvectors v_1, v_2, ..., v_n of symmetric A, and then in the new coordinate system $(y_1, y_2, ..., y_n)$ with unit vectors $\{v_1, v_2, ..., v_n\}$ the quadratic form has the **canonical form** $x^{\mathsf{T}} A x = \lambda_1 y_1^2 + \lambda_2 y_2^2 + \cdots + \lambda_n y_n^2$.*

Proof. In the new coordinate system $(y_1, y_2, ..., y_n)$ the orthonormal vectors v_1, v_2, ..., v_n (called the **principal axes**) act as the standard unit vectors. Hence any vector $x \in \mathbb{R}^n$ may be written as a linear combination

$$x = y_1 v_1 + y_2 v_2 + \cdots + y_n v_n = \begin{bmatrix} v_1 & v_2 & \cdots & v_n \end{bmatrix} \begin{bmatrix} y_1 \\ y_2 \\ \vdots \\ y_n \end{bmatrix} = Vy$$

for orthogonal matrix $V = \begin{bmatrix} v_1 & v_2 & \cdots & v_n \end{bmatrix}$ and vector $y = (y_1, y_2, ..., y_n)$. Then the quadratic form

$$x^{\mathsf{T}} A x = (Vy)^{\mathsf{T}} A (Vy) = y^{\mathsf{T}} V^{\mathsf{T}} A V y = y^{\mathsf{T}} D y,$$

since $V^{\mathsf{T}}AV = D = \operatorname{diag}(\lambda_1, \lambda_2, \ldots, \lambda_n)$ by Theorem 4.2.19. Consequently,

$$x^{\mathsf{T}}Ax = y^{\mathsf{T}}Dy = \lambda_1 y_1^2 + \lambda_2 y_2^2 + \cdots + \lambda_n y_n^2.$$

Example 4.2.27 Consider the quadratic form $f(x,y) = x^2 + 3xy - 3y^2$. That is, consider $f(x) = x^{\mathsf{T}}Ax$ for $x = (x,y)$ and matrix $A = \begin{bmatrix} 1 & 3/2 \\ 3/2 & -3 \end{bmatrix}$. The stereo illustration to the right shows the surface $f(x,y)$. Also plotted in

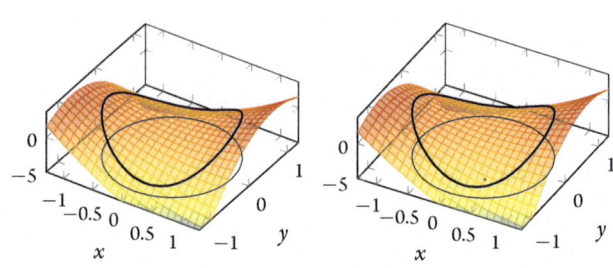

black is the curve of values of $f(x,y)$ on the unit circle $x^2 + y^2 = 1$ (also shown); that is, $f(x)$ for unit vectors x. Find the maxima and minima of f on this unit circle (for unit vectors x). Relate to the eigenvalues of Example 4.2.13.

Solution: Let's express the unit vectors as $x = (\cos t, \sin t)$ and consider f as a function of t. Then

$$\begin{aligned} f(t) &= x^2 + 3xy - 3y^2 \quad (\text{for } x = \cos t \text{ and } y = \sin t) \\ &= \cos^2 t + 3\cos t \sin t - 3\sin^2 t \\ &= \tfrac{1}{2} + \tfrac{1}{2}\cos 2t + \tfrac{3}{2}\sin 2t - \tfrac{3}{2} + \tfrac{3}{2}\cos 2t \\ &= -1 + 2\cos 2t + \tfrac{3}{2}\sin 2t. \end{aligned}$$

From calculus, maxima and minima occur when the derivative is zero, $df/dt = 0$. Here $df/dt = -4\sin 2t + 3\cos 2t$ which to be zero requires $4\sin 2t = 3\cos 2t$, that is, $\tan 2t = \tfrac{3}{4}$. From the classic Pythagorean triplet of the $3:4:5$ triangle, this tangent being $3/4$ requires either

- $\cos 2t = \tfrac{4}{5}$ and $\sin 2t = \tfrac{3}{5}$ giving the function value $f(t) = -1 + 2 \cdot \tfrac{4}{5} + \tfrac{3}{2} \cdot \tfrac{3}{5} = \tfrac{3}{2} = 1.5$,

- or the negative $\cos 2t = -\tfrac{4}{5}$ and $\sin 2t = -\tfrac{3}{5}$ giving the function value $f(t) = -1 - 2 \cdot \tfrac{4}{5} - \tfrac{3}{2} \cdot \tfrac{3}{5} = -\tfrac{7}{2} = -3.5$.

These maximum and minimum values of f seem reasonable from the plot. Observe that these extreme values are precisely the two eigenvalues of the matrix A—the next theorem shows this connection is no accident. □

Theorem 4.2.28   *Let A be an $n \times n$ symmetric matrix with eigenvalues $\lambda_1 \leq \lambda_2 \leq \cdots \leq \lambda_n$ (sorted). Then for every unit vector $x \in \mathbb{R}^n$ (that is, $|x| = 1$), the quadratic form $x^{\mathsf{T}}Ax$ has the following properties:*

(a) $\lambda_1 \leq x^{\mathsf{T}}Ax \leq \lambda_n$;

(b) the minimum of $x^{\mathsf{T}}Ax$ is λ_1, and occurs when x is a unit eigenvector corresponding to λ_1;

(c) the maximum of $x^{\mathsf{T}}Ax$ is λ_n, and occurs when x is a unit eigenvector corresponding to λ_n.

Proof. Change to an orthogonal coordinate system y that diagonalizes the matrix A (Theorem 4.2.26): say coordinates $x = Vy$ for orthogonal matrix V whose columns are orthogonal eigenvectors of A in order so that $D = V^T A V = \text{diag}(\lambda_1, \lambda_2, \ldots, \lambda_n)$. Then the quadratic form

$$x^T A x = (Vy)^T A (Vy) = y^T V^T A V y = y^T D y.$$

Since V is orthogonal it preserves lengths (Theorem 3.2.48(f)) so the unit vector condition $|x| = 1$ is the same as $|y| = 1$.

(a) To prove the lower bound, consider

$$
\begin{aligned}
x^T A x &= y^T D y \\
&= \lambda_1 y_1^2 + \lambda_2 y_2^2 + \cdots + \lambda_n y_n^2 \\
&= \lambda_1 y_1^2 + \lambda_1 y_2^2 + \cdots + \lambda_1 y_n^2 \\
&\quad + \underbrace{(\lambda_2 - \lambda_1) y_2^2}_{\geq 0} + \cdots + \underbrace{(\lambda_n - \lambda_1) y_n^2}_{\geq 0} \\
&\geq \lambda_1 y_1^2 + \lambda_1 y_2^2 + \cdots + \lambda_1 y_n^2 \\
&= \lambda_1 (y_1^2 + y_2^2 + \cdots + y_n^2) = \lambda_1 |y|^2 = \lambda_1.
\end{aligned}
$$

Similarly for the upper bound (Exercise 4.2.22). Thus $\lambda_1 \leq x^T A x \leq \lambda_n$ for all unit vectors x.

(b) Let v_1 be a unit eigenvector of A corresponding to the minimal eigenvalue λ_1; that is, $A v_1 = \lambda_1 v_1$ and $|v_1| = 1$. Then, setting $x = v_1$, the quadratic form

$$x^T A x = v_1^T A v_1 = v_1^T \lambda_1 v_1 = \lambda_1 (v_1^T v_1) = \lambda_1 |v_1|^2 = \lambda_1.$$

Thus the quadratic form $x^T A x$ takes on the minimum value λ_1 and it occurs when $x = v_1$ (at least). □

(c) Exercise 4.2.22 proves the maximum value occurs.

Activity 4.2.29 Recall Example 4.1.27 found that the 3×3 symmetric matrix

$$A = \begin{bmatrix} -2 & 0 & -6 \\ 0 & 4 & 6 \\ -6 & 6 & -9 \end{bmatrix}$$

has eigenvalues $7, 0$, and -14.

- What is the maximum of the quadratic form $x^T A x$ over unit vectors x?

(a) 14   (b) 7   (c) 0   (d) −14

- Further, what is the minimum of the quadratic form $x^T A x$ over unit vectors x?

4.2.4 Exercises

Exercise 4.2.1 Each plot below shows (unit) vectors x (blue), and for some 2×2 matrix A the corresponding vectors Ax (red) adjoined. By assessing whether there are any zero eigenvalues, estimate if the matrix A is invertible or not.

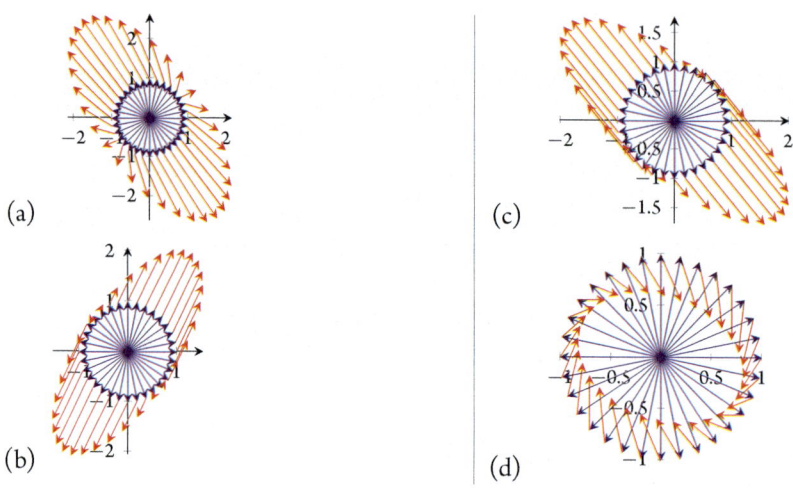

(a)

(b)

(c)

(d)

Exercise 4.2.2 For each of the following symmetric matrices, from a hand derivation of the characteristic equation (defined in Procedure 4.1.23), determine whether each matrix has a zero eigenvalue or not, and hence determine whether it is invertible or not.

(a) $\begin{bmatrix} -1/2 & -3/4 \\ -3/4 & -1/2 \end{bmatrix}$

(b) $\begin{bmatrix} 4 & -2 \\ -2 & 1 \end{bmatrix}$

(c) $\begin{bmatrix} 0 & -2/5 \\ -2/5 & 3/5 \end{bmatrix}$

(d) $\begin{bmatrix} 2 & 1 & -2 \\ 1 & 3 & -1 \\ -2 & -1 & 2 \end{bmatrix}$

(e) $\begin{bmatrix} -2 & -1 & 1 \\ -1 & -0 & -1 \\ 1 & -1 & -2 \end{bmatrix}$

(f) $\begin{bmatrix} 2 & 1 & 1 \\ 1 & 2 & 1 \\ 1 & 1 & 2 \end{bmatrix}$

(g) $\begin{bmatrix} -1/2 & 3/2 & 1 \\ 3/2 & -3 & -3/2 \\ 1 & -3/2 & -1/2 \end{bmatrix}$

(h) $\begin{bmatrix} 1 & -1 & -1 \\ -1 & -1/2 & 1/2 \\ -1 & 1/2 & -1/2 \end{bmatrix}$

Exercise 4.2.3 For each of the following (symmetric) matrices, find by hand the eigenvalues and eigenvectors. Using these eigenvectors, confirm that the eigenvalues of the matrix squared are the square of its eigenvalues. If the matrix has an inverse, what are the eigenvalues of the inverse?

(a) $A = \begin{bmatrix} 0 & -2 \\ -2 & 3 \end{bmatrix}$

(b) $B = \begin{bmatrix} 5/2 & -2 \\ -2 & 5/2 \end{bmatrix}$

(c) $C = \begin{bmatrix} 0 & -1 & -1 \\ -1 & 1 & 0 \\ -1 & 0 & 1 \end{bmatrix}$

(d) $D = \begin{bmatrix} -1 & 3/2 & 3/2 \\ 3/2 & -3 & -1/2 \\ 3/2 & -1/2 & 3 \end{bmatrix}$

(e) $E = \begin{bmatrix} -1 & -2 & 0 \\ -2 & 0 & 2 \\ 0 & 2 & 1 \end{bmatrix}$

(f) $F = \begin{bmatrix} 2 & 1 & 3 \\ 1 & 0 & -1 \\ 3 & -1 & 2 \end{bmatrix}$

(g) $G = \begin{bmatrix} 3 & 8 \\ 8 & -9 \end{bmatrix}$

(h) $H = \begin{bmatrix} -2 & 1 \\ 1 & 14/5 \end{bmatrix}$

Exercise 4.2.4 Each plot below shows (unit) vectors x (blue), and for some 2×2 matrix A the corresponding vectors Ax (red) adjoined. For each plot of a matrix A there is a companion plot for the inverse matrix A^{-1}. By roughly estimating eigenvalues and eigenvectors by eye, identify the pairs of plots corresponding to each matrix and its inverse.

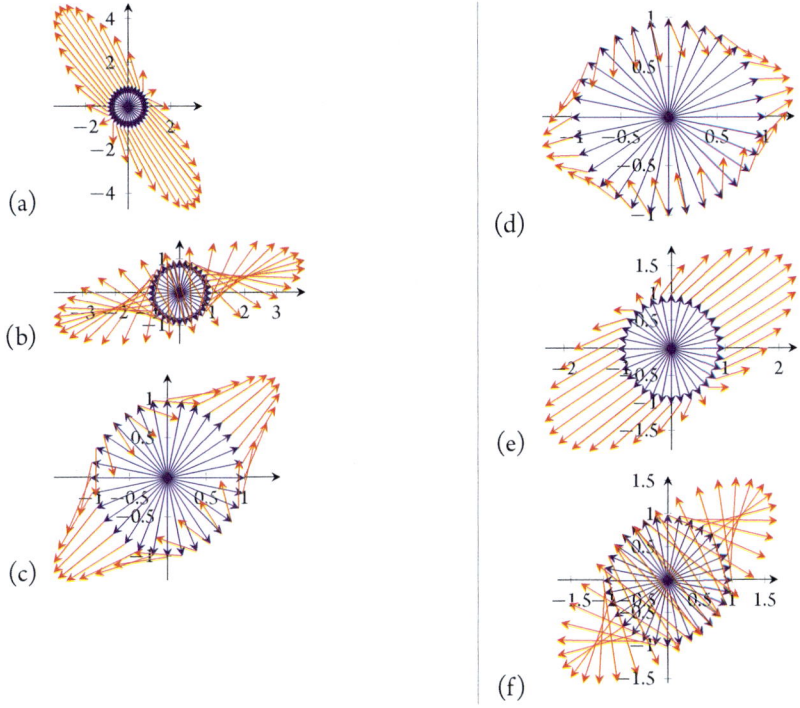

(a)

(b)

(c)

(d)

(e)

(f)

Exercise 4.2.5 For the symmetric matrices of Exercise 4.2.3, confirm that eigenvectors corresponding to distinct eigenvalues are orthogonal (Theorem 4.2.11). Show your working.

Exercise 4.2.6 For every $n \times n$ symmetric matrix,

- eigenvectors corresponding to distinct eigenvalues are orthogonal (Theorem 4.2.11), and
- there are generally n eigenvalues.

Which of the illustrated 2D examples of Exercise 4.1.1 appear to come from symmetric matrices, and which appear to come from non-symmetric matrices?

Exercise 4.2.7 For each of the following *non-symmetric* matrices, confirm that eigenvectors corresponding to distinct eigenvalues are *not* orthogonal. Show and comment on your working. Find eigenvectors by hand for 2×2 and 3×3 matrices, and compute with MATLAB/Octave for 4×4 matrices and larger (using `eig()` and V'*V).

(a) $A = \begin{bmatrix} -2 & 2 \\ 3 & -3 \end{bmatrix}$

(c) $C = \begin{bmatrix} 2 & 1 & -2 & 1 \\ 1 & -3 & 4 & -4 \\ 3 & 2 & 4 & -5 \\ -3 & -1 & -3 & 0 \end{bmatrix}$

(b) $B = \begin{bmatrix} -1 & -2 & 9 \\ 3 & -6 & 3 \\ 0 & 0 & 3 \end{bmatrix}$

(d) $D = \begin{bmatrix} 2 & 0 & 2 & -1 & -1 \\ 2 & 1 & -1 & 2 & -0 \\ 5 & -2 & 6 & 2 & -1 \\ 4 & 0 & -2 & 6 & -5 \\ 2 & 0 & -5 & -3 & -5 \end{bmatrix}$

Exercise 4.2.8 For the symmetric matrices of Exercise 4.2.3, use MATLAB/Octave to compute an SVD (USV^{T}) of each matrix. Confirm that each column of V is an eigenvector of the matrix (that is, proportional to what the exercise found) and the corresponding singular value is the magnitude of the corresponding eigenvalue (Theorem 4.2.16). Show and discuss your working.

Exercise 4.2.9 To complement the previous exercise, for each of the *non-symmetric* matrices of Exercise 4.2.7, use MATLAB/Octave to compute an SVD (USV^{T}) of each matrix. Confirm that usually each column of V is *not* an eigenvector of the matrix, and the singular values do *not* appear closely related to the eigenvalues. Show and discuss your working.

Exercise 4.2.10 Let A be an $m \times n$ matrix with SVD $A = USV^{\mathsf{T}}$. Prove that for any $j = 1, 2, \ldots, n$, the jth column of V, v_j, is an eigenvector of the $n \times n$ symmetric matrix $A^{\mathsf{T}}A$ corresponding to the eigenvalue $\lambda_j = \sigma_j^2$ (or $\lambda_j = 0$ if $m < j \le n$).

Exercise 4.2.11 Prove Theorem 4.2.4(c) using parts 4.2.4(a) and 4.2.4(b): that if matrix A is invertible, then for every integer n (including negative n), λ^n is an eigenvalue of A^n with corresponding eigenvector x.

Exercise 4.2.12 For each of the following matrices, give reasons as to whether the matrix is orthogonally diagonalizable, and if it is, then find an orthogonal matrix V that does so and the corresponding diagonal matrix D. Use MATLAB/Octave for the larger matrices.

(a) $A = \begin{bmatrix} 2 & 0 & 1 \\ 0 & 2 & -1 \\ 1 & -1 & 3 \end{bmatrix}$

(b) $B = \begin{bmatrix} 3 & 1 & 1 & 1 & -1 \\ -1 & 0 & 0 & 0 & 1 \\ 0 & 0 & 0 & 0 & 1 \\ 0 & 1 & -1 & 1 & -1 \\ 0 & 0 & 0 & 0 & 1 \end{bmatrix}$

(c) $C = \begin{bmatrix} 2 & 0 & 2 \\ 1 & -1 & 0 \\ -1 & 0 & -1 \end{bmatrix}$

(d) $D = \begin{bmatrix} 3 & 0 & 2 & 0 & 0 \\ 0 & 3 & 1 & 0 & -1 \\ 2 & 1 & 1 & 0 & -1 \\ 0 & 0 & 0 & 3 & 2 \\ 0 & -1 & -1 & 2 & 1 \end{bmatrix}$

Exercise 4.2.13 Let matrix A be invertible and orthogonally diagonalizable. Show that the inverse A^{-1} is orthogonally diagonalizable.

Exercise 4.2.14 Suppose matrices A and B are orthogonally diagonalizable by the same orthogonal matrix V. Show that $AB = BA$ and that the product AB is orthogonally diagonalizable.

Exercise 4.2.15 For each of the given symmetric matrices, say A, find a symmetric matrix X such that $X^2 = A$. That is, find a square-root of the matrix.

(a) $A = \begin{bmatrix} 5/2 & 3/2 \\ 3/2 & 5/2 \end{bmatrix}$

(b) $B = \begin{bmatrix} 6 & -5 & -5 \\ -5 & 10 & 1 \\ -5 & 1 & 10 \end{bmatrix}$

(c) $C = \begin{bmatrix} 2 & 1 & 3 \\ 1 & 2 & 3 \\ 3 & 3 & 6 \end{bmatrix}$

(d) How many possible answers are there for each of the given matrices? Why?

(e) For each symmetric matrix, say A, show that if every eigenvalue of A is non-negative, then there exists a symmetric matrix X such that $A = X^2$.

(f) Continuing the previous part, how many such matrices X exist? Justify your answer.

Exercise 4.2.16 For each of the following conic sections, draw a pair of coordinate axes for a coordinate system in which the algebraic description of the curves should be simplest.

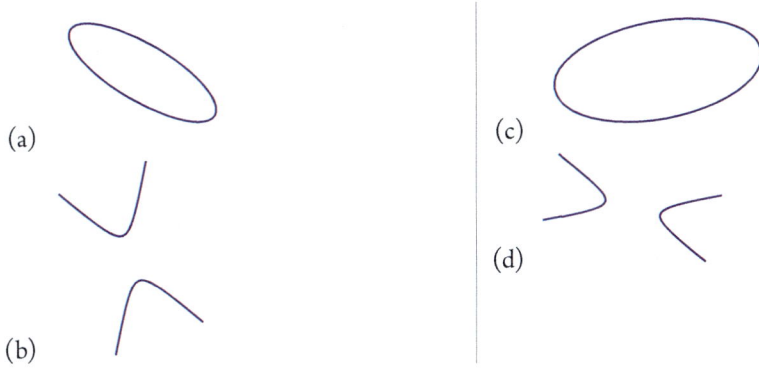

(a)

(b)

(c)

(d)

Exercise 4.2.17 By shifting to new coordinate axes, find the canonical form of each of the following quadratic equations, and hence describe each curve.

(a) $-4x^2 + 5y^2 + 4x + 4y - 1 = 0$

(b) $-4y^2 - 6x - 4y + 2 = 0$

(c) $5x^2 - y^2 - 2y + 4 = 0$

(d) $3x^2 + 5y^2 + 6x + 4y + 2 = 0$

(e) $-2x^2 - 4y^2 + 7x - 2y + 1 = 0$

Exercise 4.2.18 By rotating to new coordinate axes, identify each of the following conic sections. Write the quadratic terms in the form $x^\mathsf{T} A x$, and use the eigenvalues and eigenvectors of matrix A.

(a) $4xy + 3y^2 - 3 = 0$

(b) $3x^2 + 8xy - 3y^2 = 0$

(c) $2x^2 - 3xy + 6y^2 - 5 = 0$

(d) $-4x^2 + 3xy - 4y^2 - 2 = 0$

Exercise 4.2.19 By rotating and shifting to new coordinate axes, identify each of the following conic sections from its equation.

(a) $-2x^2 - 5xy - 2y^2 - \frac{33}{2}x - 15y - 32 = 0$

(b) $-7x^2 + 3xy - 3y^2 - 52x + \frac{33}{2}y - \frac{381}{4} = 0$

(c) $-4xy - 3y^2 - 18x - 11y + \frac{37}{4} = 0$

(d) $6x^2 + 6y^2 - 12x - 42y + \frac{155}{2} = 0$

(e) $2x^2 - 4xy + 5y^2 + 34x - 52y + \frac{335}{2} = 0$

Exercise 4.2.20 For each of the following matrices, say A, consider the corresponding quadratic form $q(x) = x^\mathsf{T} A x$. Find coordinate axes, the principal axes, such that the quadratic has the canonical form in the new coordinates y_1, y_2, ..., y_n. Use eigenvalues and eigenvectors, and use MATLAB/Octave for the larger matrices. Over all unit vectors, what is the maximum value of $q(x)$? and what is the minimum value?

(a) $A = \begin{bmatrix} 0 & -5 \\ -5 & 0 \end{bmatrix}$

(b) $B = \begin{bmatrix} 2 & 1 & -1 \\ 1 & 3 & 2 \\ -1 & 2 & 3 \end{bmatrix}$

(c) $C = \begin{bmatrix} 6 & 0 & -3 & 1 \\ 0 & -1 & 5 & -3 \\ -3 & 5 & -4 & -7 \\ 1 & -3 & -7 & 0 \end{bmatrix}$

(d) $D = \begin{bmatrix} 12 & -3 & -3 & -4 & -6 \\ -3 & 0 & 0 & 2 & -5 \\ -3 & 0 & -4 & 1 & -3 \\ -4 & 2 & 1 & -5 & 2 \\ -6 & -5 & -3 & 2 & 0 \end{bmatrix}$

Exercise 4.2.21 For any given $n \times n$ symmetric matrix A consider the quadratic form $q(x) = x^\mathsf{T} A x$. For general vectors x, not necessarily unit vectors, what is the maximum value of $q(x)$ in terms of $|x|$? and what is the minimum value of $q(x)$? Justify your answer.

Exercise 4.2.22 Complete the proof of Theorem 4.2.28 by detailing the proof of the upper bound, and that the upper bound is achieved for an appropriate unit eigenvector.

Exercise 4.2.23 For a symmetric matrix, discuss the similarities and differences between the SVD and the diagonalization factorization, the singular values and the eigenvalues, and the singular vectors and the eigenvectors.

Exercise 4.2.24 In a few sentences, answer/discuss each of the following.

(a) What is the geometric reason for a matrix with a zero eigenvalue being not invertible?

(b) Recall that if λ is an eigenvalue of a matrix A, then λ^2 is an eigenvalue of A^2. Why is it that we cannot generally say the equivalent for singular values of a matrix?

(c) What is the key to establishing that all eigenvalues of a symmetric matrix are real?

(d) How is it that we now know that every $n \times n$ symmetric matrix has at most n eigenvalues?

(e) How does orthogonal diagonalization of a matrix compare with singular value decomposition?

(f) Why is it important to be flexible about the choice of coordinate system in a given application problem?

(g) How do symmetric matrices lie at the heart of multivariable quadratic functions?

(h) Why are the extreme values of the quadratic form $x^{\mathsf{T}} A x$ determined by the extreme eigenvalues of A? and how is this connected to the proof of existence of a singular value decomposition?

4.3 Summary of symmetric eigen-problems

- Eigenvalues and eigenvectors arise from the following important geometric question: for what vectors v does multiplication by A just stretch/shrink v by some scalar factor λ?

Introduction to eigenvalues and eigenvectors

⋆⋆ For every square matrix A, a scalar λ is called an **eigenvalue** of A if there is a nonzero vector x such that $Ax = \lambda x$ (Definition 4.1.1). Such a vector x is called an **eigenvector** of A corresponding to the eigenvalue λ.
It is the direction of an eigenvector that is important, not its length.

- The eigenvalues of a diagonal matrix are the entries on the diagonal, and the unit vectors e_1, e_2, ..., e_n are corresponding eigenvectors (Example 4.1.9).

- For every square matrix A, a scalar λ is an eigenvalue of A iff the homogeneous linear system $(A - \lambda I)x = 0$ has nonzero solutions x (Theorem 4.1.10). The set of all eigenvectors corresponding to any one eigenvalue λ, together with the zero vector, is a subspace; the subspace is called the **eigenspace** of λ and is denoted by $\mathbb{E}_\lambda$.

⋆ For every real symmetric matrix A, the **multiplicity** of an eigenvalue λ of A is the dimension of the corresponding eigenspace $\mathbb{E}_\lambda$ (Definition 4.1.15).

- Symmetric matrices arise in many mechanical and physical problems due to Newton's third law that every action has an equal and opposite reaction. Symmetric matrices arise in many networking problems in the cases when every network connection is two-way.

⋆ For every $n \times n$ square matrix A (not just symmetric), λ_1, λ_2, ..., λ_m are eigenvalues of A with corresponding eigenvectors v_1, v_2, ..., v_m, for some m (commonly $m = n$), iff $AV = VD$ for diagonal matrix $D = \operatorname{diag}(\lambda_1, \lambda_2, \ldots, \lambda_m)$ and $n \times m$ matrix $V = \begin{bmatrix} v_1 & v_2 & \cdots & v_m \end{bmatrix}$ for nonzero v_1, v_2, ..., v_m (Theorem 4.1.21).

- In MATLAB/Octave:

⋆⋆ [V,D]=eig(A) computes eigenvectors and the eigenvalues of the $n \times n$ square matrix A.

⋆ The n eigenvalues of A (repeated according to their multiplicity, Definition 4.1.15) form the diagonal of $n \times n$ square matrix $D = \operatorname{diag}(\lambda_1, \lambda_2, \ldots, \lambda_n)$.

★ Corresponding to the jth eigenvalue λ_j, the jth column of $n \times n$ square matrix V is an eigenvector (of unit length).

- `eig(A)` reports a vector of the eigenvalues of square matrix A (repeated according to their multiplicity, Definition 4.1.15).
- If the matrix A is a real symmetric matrix, then the computed eigenvalues and eigenvectors are all real, and the eigenvector matrix V is orthogonal. If the matrix A is either not symmetric, or is complex valued, then the eigenvalues and eigenvectors may be non-real complex valued.

★ Procedure 4.1.23 finds by hand eigenvalues and eigenvectors of a (small) square matrix A:

1. find all eigenvalues by solving the **characteristic equation** of A, $\det(A - \lambda I) = 0$ (using (4.1));

2. for each eigenvalue λ, solve the homogeneous $(A - \lambda I)x = \mathbf{0}$ to find the eigenspace $\mathbb{E}_\lambda$;

3. write each eigenspace as the span of a few chosen eigenvectors.

Beautiful properties for symmetric matrices

- A square matrix is invertible iff zero is *not* an eigenvalue of the matrix (Theorem 4.2.1).
- ★ Let A be a square matrix with eigenvalue λ and corresponding eigenvector x (Theorem 4.2.4).
  - For every positive integer k, λ^k is an eigenvalue of A^k with corresponding eigenvector x.
  - If A is invertible, then $1/\lambda$ is an eigenvalue of A^{-1} with corresponding eigenvector x.
  - If A is invertible, then for every integer k (including negative k), λ^k is an eigenvalue of A^k with corresponding eigenvector x.
- ★★ For every real symmetric matrix A, every eigenvalue of A is real valued (Theorem 4.2.9). This marvellous reality often reflects the physical reality of many applications.
- ★ Let A be a real symmetric matrix, then for every two distinct eigenvalues of A, any corresponding two eigenvectors are orthogonal (Theorem 4.2.11).
- Every $n \times n$ real symmetric matrix A has at most n distinct eigenvalues (Theorem 4.2.15).
- ★ Let A be an $n \times n$ real symmetric matrix with SVD $A = USV^{\mathsf{T}}$. If all the singular values are distinct or zero, $\sigma_1 > \cdots > \sigma_r > \sigma_{r+1} = \cdots = \sigma_n = 0$, then v_j is an eigenvector of A, corresponding to an eigenvalue of either $\lambda_j = +\sigma_j$ or $\lambda_j = -\sigma_j$ (not both, except the zero case) (Theorem 4.2.16).
- A real square matrix A is **orthogonally diagonalizable** if there exists an orthogonal matrix V and a diagonal matrix D such that $V^{\mathsf{T}}AV = D$, equivalently $AV = VD$, equivalently $A = VDV^{\mathsf{T}}$ is a factorization of A (Theorem 4.2.17).
- ★★ For every real square matrix A, matrix A is symmetric iff it is orthogonally diagonalizable (Theorem 4.2.19).
- The conic sections of ellipses, hyperbolas, and parabolas arise as solutions of quadratic equations (Theorem 4.2.3). Changes in the coordinate system discover their shape, location, and orientation.
- A **quadratic form** in variables $x \in \mathbb{R}^n$ is a function $q : \mathbb{R}^n \to \mathbb{R}$ that may be written as $q(x) = x^{\mathsf{T}}Ax$ for some real symmetric $n \times n$ matrix A (Definition 4.2.24).
- For every quadratic form, there exists an orthogonal coordinate system that diagonalizes the quadratic form (Theorem 4.2.26). Specifically, for the quadratic form $x^{\mathsf{T}}Ax$ find

the eigenvalues λ_1, λ_2, ..., λ_n and orthonormal set of eigenvectors v_1, v_2, ..., v_n of symmetric A, and then in the new coordinate system $(y_1, y_2, \ldots, y_n)$ with unit vectors $\{v_1, v_2, \ldots, v_n\}$ the quadratic form has the **canonical form** $x^T A x = \lambda_1 y_1^2 + \lambda_2 y_2^2 + \cdots + \lambda_n y_n^2$.

- Let A be an $n \times n$ symmetric matrix with eigenvalues $\lambda_1 \leq \lambda_2 \leq \cdots \leq \lambda_n$ (sorted). Then for all unit vectors $x \in \mathbb{R}^n$, the quadratic form $x^T A x$ has the following properties (Theorem 4.2.28):

  - $\lambda_1 \leq x^T A x \leq \lambda_n$;
  - the minimum of $x^T A x$ is λ_1, and occurs when x is a (unit) eigenvector corresponding to λ_1;
  - the maximum of $x^T A x$ is λ_n, and occurs when x is a (unit) eigenvector corresponding to λ_n.

Answers to selected activities

4.1.3b, 4.1.5c, 4.1.6d, 4.1.12c, 4.1.18c, 4.1.25d, 4.2.6d, 4.2.29b,

Answers to selected exercises

4.1.1b $v_1 \propto \pm(-0.5, 0.9)$, $\lambda_1 \approx -0.3$; and $v_2 \propto \pm(0.6, 0.8)$, $\lambda_2 \approx 1.1$.

4.1.1d $v_1 \propto \pm(1, -0.2)$, $\lambda_1 \approx -0.7$; and $v_2 \propto \pm(-0.2, 1)$, $\lambda_2 \approx 1.1$.

4.1.1f There appear to be none.

4.1.2b Corresponding eigenvalues are: n/a, $-3, -3, 1,$ n/a, $1,$ n/a

4.1.2d Corresponding eigenvalues are: n/a, 1, n/a, $-6, 3,$ n/a, -6

4.1.3b Eigenvalues 2, 3, 1, eigenspace $\mathbb{E}_2$ is 2D.

4.1.3d Eigenvalues $-5.1461, -1.6639,$ $-0.7427, 0.7676, 7.7851$.

4.1.4b $\frac{1}{2}$, $\frac{23}{2}$

4.1.4d $-10, 0$

4.1.4f $-13, 1, 6$

4.1.4h $-3, 2, 3$

4.1.5b $\mathbb{E}_3 = \text{span}\{(-1, 2)\}$, 6 is not an eigenvalue.

4.1.5d -6 is not an eigenvalue, $\mathbb{E}_4 = \text{span}\{(3, 1, -2)\}$, $\mathbb{E}_9 = \text{span}\{(1, 3, 3)\}$.

4.1.6b $\mathbb{E}_1 = \text{span}\{(1, 1)\}$, $\mathbb{E}_{11} = \text{span}\{(-1, 1)\}$. Both eigenvalues have multiplicity one.

4.1.6d $\mathbb{E}_{-10} = \text{span}\{(-2, 1, 2)\}$, $\mathbb{E}_{-7} = \text{span}\{(-1, 2, -2)\}$, $\mathbb{E}_{-4} = \text{span}\{(2, 2, 1)\}$. All eigenvalues have multiplicity one.

4.1.6f $\mathbb{E}_{-6} = \text{span}\{(-1, 3, 1)\}$, $\mathbb{E}_5 = \text{span}\{(3, 1, 0), (-1, 3, -10)\}$ Eigenvalue $\lambda = -6$ has multiplicity one, whereas $\lambda = 5$ has multiplicity two.

4.2.1b not invertible

4.2.1d invertible

4.2.2b eigenvalues 0, 5 so not invertible.

4.2.2d eigenvalues 0, 2, 5 so not invertible.

4.2.2f eigenvalues 1, 4 so invertible.

4.2.2h eigenvalues $-1, 2$ so invertible.

4.2.3b Eigenvalues $1/2, 9/2$, and corresponding eigenvectors proportional to $(1, 1), (-1, 1)$. The inverse has eigenvalues 2, $2/9$.

4.2.3d Eigenvalues $-4, -1/2, 7/2$, and corresponding eigenvectors proportional to $(-3, 5, 1)$,

$(3, 2, -1)$, $(1, 0, 3)$. The inverse has eigenvalues $-1/4$, -2, $2/7$.

4.2.3f Eigenvalues -2, 1, 5, and corresponding eigenvectors proportional to $(-1, 1, 1)$, $(1, 2, -1)$, $(1, 0, 1)$. The inverse has eigenvalues $-1/2$, 1, $1/5$.

4.2.3h Eigenvalues $-11/5$, 3, and corresponding eigenvectors proportional to $(-5, 1)$, $(1, 5)$. The inverse has eigenvalues $-5/11$, $1/3$.

4.2.7a Eigenvectors proportional to $(1, 1)$, $(-2, 3)$.

4.2.7c Eigenvectors proportional to
$(-.39, .44, .76, -.28)$,
$(.58, -.41, -.68, .18)$,
$(.22, -.94, .23, .11)$,
$(.21, -.53, .53, .62)$ (2 d.p.).

4.2.12a $V = \begin{bmatrix} -\frac{1}{\sqrt{3}} & \frac{1}{\sqrt{2}} & \frac{1}{\sqrt{6}} \\ \frac{1}{\sqrt{3}} & \frac{1}{\sqrt{2}} & -\frac{1}{\sqrt{6}} \\ \frac{1}{\sqrt{3}} & 0 & \frac{2}{\sqrt{6}} \end{bmatrix}$, and
$D = \text{diag}(1, 2, 4)$.

4.2.12c Not symmetric, so not orthogonally diagonalizable.

4.2.15a $X = \begin{bmatrix} 3/2 & 1/2 \\ 1/2 & 3/2 \end{bmatrix}$

4.2.15c $X = \begin{bmatrix} 1 & 0 & 1 \\ 0 & 1 & 1 \\ 1 & 1 & 2 \end{bmatrix}$

4.2.17a hyperbola, centred $(1/2, -2/5)$,
$$-\frac{x'^2}{1/5} + \frac{y'^2}{4/25} = 1$$

4.2.17c hyperbola, centred $(0, -1)$,
$$-\frac{x'^2}{1} + \frac{y'^2}{5} = 1$$

4.2.17e ellipse, centred $(7/4, -1/4)$,
$$\frac{x'^2}{59/16} + \frac{y'^2}{59/32} = 1$$

4.2.18b With axes $i' = (\frac{1}{\sqrt{5}}, -\frac{2}{\sqrt{5}})$ and $j' = (\frac{2}{\sqrt{5}}, \frac{1}{\sqrt{5}})$, $-5x'^2 + 5y'^2 = 0$ is pair of straight lines.

4.2.18d No solution points exist.

4.2.19b ellipse centred $(-7/2, 1)$ at angle $78°$

4.2.19d circle centred $(1, 7/2)$

4.2.20a $q = -5y_1^2 + 5y_2^2$, max= 5, min=-5

4.2.20c (2 d.p.) $q = -10.20y_1^2 - 3.56y_2^2 + 4.81y_3^2 + 9.95y_4^2$, max= 9.95, min=$-10.20$

5 Approximate matrices

This chapter develops how concepts associated with length and distance not only apply to vectors but also apply to matrices. More advanced courses on linear algebra place these in a unifying framework that also encompasses much you see both in solving differential equations (and integral equations) and in problems involving complex numbers (such as those in electrical engineering or quantum physics).

This chapter could be studied any time after Chapter 3 to help the transition to more abstract linear algebra. It also is good revision of the SVD, rank, orthogonality, and so on.

5.1 Measure changes to matrices

5.1.1 Compress images optimally

Photographs and other images require a lot of storage. Reducing the amount of storage for an image is essential, both for storage and for transmission. The well-known jpeg format for compressing photographs is incredibly useful: the SVD provides a related effective method of compression.

Linear Algebra for the 21st Century. A. J. Roberts, Oxford University Press (2020). © A. J. Roberts.
DOI: 10.1093/oso/9780198856399.003.0005

These svd methods approximate images by matrices of various ranks. Recall that a matrix of rank k (Definition 3.3.19) means that matrix has precisely k nonzero singular values, that is, an $m \times n$ matrix

$$A = USV^{\mathsf{T}}$$

$$= \begin{bmatrix} u_1 & \cdots & u_k & \cdots & u_m \end{bmatrix} \begin{bmatrix} \sigma_1 & \cdots & 0 & \\ \vdots & \ddots & \vdots & O_{k \times (n-k)} \\ 0 & \cdots & \sigma_k & \\ O_{(m-k) \times k} & & O_{(m-k) \times (n-k)} \end{bmatrix} V^{\mathsf{T}}$$

(then multiplying the form of the first two matrices)

$$= \begin{bmatrix} \sigma_1 u_1 & \cdots & \sigma_k u_k & O_{m \times (n-k)} \end{bmatrix} \begin{bmatrix} v_1^{\mathsf{T}} \\ \vdots \\ v_k^{\mathsf{T}} \\ \vdots \\ v_n^{\mathsf{T}} \end{bmatrix}$$

(then multiplying the form of these two matrices)

$$= \sigma_1 u_1 v_1^{\mathsf{T}} + \sigma_2 u_2 v_2^{\mathsf{T}} + \cdots + \sigma_k u_k v_k^{\mathsf{T}}.$$

This last sum precisely constructs matrix A. Further, when the rank k is low compared to size m and n, this last sum *has relatively few components*.

Example 5.1.1 Invent and write down a rank three representation of the following 5×5 'bulls eye' matrix (illustrated to the right)

$$A = \begin{bmatrix} 0 & 1 & 1 & 1 & 0 \\ 1 & 0 & 0 & 0 & 1 \\ 1 & 0 & 1 & 0 & 1 \\ 1 & 0 & 0 & 0 & 1 \\ 0 & 1 & 1 & 1 & 0 \end{bmatrix}.$$

Solution: Here we set all coefficients $\sigma_1 = \sigma_2 = \sigma_3 = 1$ and let the vectors set the magnitude (for the moment, here these are not singular vectors and σ_j are not singular values—but subsequently such a meaning is restored).

(a) Arbitrarily start by addressing together the first and last rows of the image (as illustrated): they are computed by choosing $u_1 = (1,0,0,0,1)$ and $v_1 = (0,1,1,1,0)$, and using the rank one matrix

$$u_1 v_1^{\mathsf{T}} = \begin{bmatrix} 1 \\ 0 \\ 0 \\ 0 \\ 1 \end{bmatrix} \begin{bmatrix} 0 & 1 & 1 & 1 & 0 \end{bmatrix} = \begin{bmatrix} 0 & 1 & 1 & 1 & 0 \\ 0 & 0 & 0 & 0 & 0 \\ 0 & 0 & 0 & 0 & 0 \\ 0 & 0 & 0 & 0 & 0 \\ 0 & 1 & 1 & 1 & 0 \end{bmatrix}.$$

(b) Next choose to add in the first and last columns of the image (as illustrated): they are computed by choosing $u_2 = (0,1,1,1,0)$ and $v_2 = (1,0,0,0,1)$, and using the rank two matrix

$$u_1 v_1^T + u_2 v_2^T = u_1 v_1^T + \begin{bmatrix} 0 \\ 1 \\ 1 \\ 1 \\ 0 \end{bmatrix} [1\ 0\ 0\ 0\ 1]$$

$$= u_1 v_1^T + \begin{bmatrix} 0&0&0&0&0 \\ 1&0&0&0&1 \\ 1&0&0&0&1 \\ 1&0&0&0&1 \\ 0&0&0&0&0 \end{bmatrix} = \begin{bmatrix} 0&1&1&1&0 \\ 1&0&0&0&1 \\ 1&0&0&0&1 \\ 1&0&0&0&1 \\ 0&1&1&1&0 \end{bmatrix}.$$

(c) Lastly put the dot in the middle of the image (to form the original image): choose $u_3 = v_3 = (0,0,1,0,0)$, and compute the rank three matrix

$$u_1 v_1^T + u_2 v_2^T + u_3 v_3^T = u_1 v_1^T + u_2 v_2^T + \begin{bmatrix} 0 \\ 0 \\ 1 \\ 0 \\ 0 \end{bmatrix} [0\ 0\ 1\ 0\ 0]$$

$$= u_1 v_1^T + u_2 v_2^T + \begin{bmatrix} 0&0&0&0&0 \\ 0&0&0&0&0 \\ 0&0&1&0&0 \\ 0&0&0&0&0 \\ 0&0&0&0&0 \end{bmatrix} = \begin{bmatrix} 0&1&1&1&0 \\ 1&0&0&0&1 \\ 1&0&1&0&1 \\ 1&0&0&0&1 \\ 0&1&1&1&0 \end{bmatrix}.$$

□

Activity 5.1.2 Which pair of vectors gives a rank one representation, uv^T of the matrix

$$\begin{bmatrix} 0&1&1&0 \\ 0&0&0&0 \\ 0&1&1&0 \\ 0&1&1&0 \end{bmatrix} ?$$

(a) $u = (1,1,0,1)$, $v = (0,1,1,0)$

(b) $u = (0,1,1,0)$, $v = (1,1,0,1)$

(c) $u = (0,1,1,0)$, $v = (1,0,1,1)$

(d) $u = (1,0,1,1)$, $v = (0,1,1,0)$

Procedure 5.1.3 (approximate images) *Consider any image stored as scalars in an $m \times n$ matrix A.*[1]

1. *Compute an SVD $A = USV^T$ with $[U,S,V] = \text{svd}(A)$.*

2. *Choose a desired rank k based upon the singular values (Theorem 5.1.16): typically there are k 'large' singular values and all the rest are 'small'.*

---

[1] Some of you may wonder about compressing three-dimensional (3D) images such as the details in 3D space found by CT-scans of your body, or such as a movie (2D in space and 1D in time). As yet there is no one outstanding, clear, and efficient generalization of the SVD to represent a 3D array of numbers, nor for nD with $n \geq 3$ (search the internet for tensor rank decomposition).

3. *Then the 'best' rank k approximation to the image matrix A is (using the subscript k on the matrix name to denote the rank k approximation)*

$$A_k := \sigma_1 u_1^T v_1 + \sigma_2 u_2^T v_2 + \cdots + \sigma_k u_k^T v_k$$
$$= U(:,1:k)*S(1:k,1:k)*V(:,1:k)'$$

Example 5.1.4 Use Procedure 5.1.3 to find the 'best' rank two matrix, and also the 'best' rank three matrix, to approximate the 'bulls eye' image matrix (illustrated to the right)

$$A = \begin{bmatrix} 0 & 1 & 1 & 1 & 0 \\ 1 & 0 & 0 & 0 & 1 \\ 1 & 0 & 1 & 0 & 1 \\ 1 & 0 & 0 & 0 & 1 \\ 0 & 1 & 1 & 1 & 0 \end{bmatrix}.$$

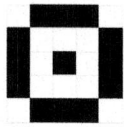

Solution: Enter the matrix into MATLAB/Octave and compute an SVD, $A = USV^T$, with [U,S,V]=svd(A) to find (2 d.p.)

```
U =
    -0.47     0.51     0.14     0.71     0.05
    -0.35    -0.44     0.43    -0.05     0.71
    -0.56    -0.31    -0.77     0.00     0.00
    -0.35    -0.44     0.43     0.05    -0.71
    -0.47     0.51     0.14    -0.71    -0.05
S =
     2.68        0        0        0        0
        0     2.32        0        0        0
        0        0     0.64        0        0
        0        0        0     0.00        0
        0        0        0        0     0.00
V =
    -0.47    -0.51     0.14    -0.68    -0.18
    -0.35     0.44     0.43    -0.18     0.68
    -0.56     0.31    -0.77     0.00    -0.00
    -0.35     0.44     0.43     0.18    -0.68
    -0.47    -0.51     0.14     0.68     0.18
```

- For this matrix there are three 'large' singular values of 2.68, 2.32, and 0.64, and two 'small' singular values of 0.00 (they are precisely zero), thus construct a rank three approximation to the image matrix as

$$A_3 = \sigma_1 u_1 v_1^T + \sigma_2 u_2 v_2^T + \sigma_3 u_3 v_3^T,$$

computed with A3=U(:,1:3)*S(1:3,1:3)*V(:,1:3)', giving (2 d.p.)

A3 =

0.00	1.00	1.00	1.00	0.00
1.00	0.00	0.00	0.00	1.00
1.00	0.00	1.00	0.00	1.00
1.00	0.00	0.00	0.00	1.00
-0.00	1.00	1.00	1.00	-0.00

The rank three matrix A_3 exactly reproduces the image matrix A. This exactness is due to the fourth and fifth singular values being precisely zero (to numerical error).

- Alternatively, in the context of some application, we could subjectively decide that there are two 'large' singular values of 2.68 and 2.32, and three 'small' singular values of 0.64 and 0.00. In such a case, construct a rank two approximation to the image matrix as (illustrated to the right)

$$A_2 = \sigma_1 u_1 v_1^T + \sigma_2 u_2 v_2^T,$$

computed with A2=U(:,1:2)*S(1:2,1:2)*V(:,1:2)', giving (2 d.p.)

A2 =

-0.01	0.96	1.07	0.96	-0.01
0.96	-0.12	0.21	-0.12	0.96
1.07	0.21	0.62	0.21	1.07
0.96	-0.12	0.21	-0.12	0.96
-0.01	0.96	1.07	0.96	-0.01

This rank two approximation A_2 is indeed roughly the same as the image matrix A, albeit with errors of 20% or so. Subsequent theory confirms that the relative error is characterised by $\sigma_3/\sigma_1 = 0.24$ here. ☐

Activity 5.1.5 Consider the image shown to the right. The image has matrix with svd given below (2 d.p.). What rank representation exactly reproduces the matrix/image?

(a) 4   (b) 3   (c) 2   (d) 1

U =

-0.72	0.48	0.50	-0.00
-0.22	-0.84	0.50	-0.00
-0.47	-0.18	-0.50	-0.71
-0.47	-0.18	-0.50	0.71

S =

2.45	0	0	0
0	0.37	0	0
0	0	0.00	0
0	0	0	0.00

V =

-0.43	-0.07	0.87	-0.24
-0.43	-0.07	-0.44	-0.78
-0.48	0.83	-0.11	0.26
-0.62	-0.55	-0.21	0.51

Example 5.1.6 To the left is a 292×277 image of Blaise Pascal (French mathematician, physicist, inventor, and writer, 1623–62). The image is coded as 80 884 numbers. Let's find a good approximation to the image that uses much fewer numbers, and hence takes less storage. That is, we effectively compress the image for storage or transmission.

Solution: We use an SVD to approximate the image of Pascal to controllable levels of approximation. For example, Figure 5.1 shows four approximations to the image of Pascal, ranging from the hopeless (labelled "rank 3") to the original image (labelled "rank 277").

Procedure 5.1.3 is as follows.

- First obtain the image, say as a png file (e.g., download from a website or export a photograph). Then read the image into MATLAB/Octave using

```
rgb=imread('pascal.png');
A=mean(rgb,3);
```

The imread command sets the $292 \times 277 \times 3$ array rgb to the red-green-blue values of the image data. Then convert into a greyscale image matrix A by averaging the red-green-

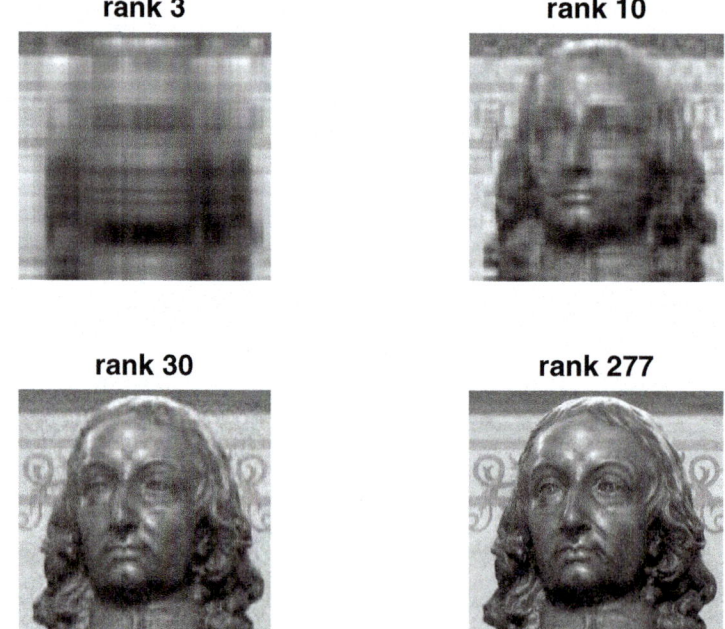

Figure 5.1 Four approximate images of Pascal ranging from the hopeless rank 3, via the rough rank 10, the good rank 30, to the original rank 277.

blue values via the function mean(rgb,3) which computes the mean over the third dimension of the array (over the three colours).

- Compute an SVD, $A = USV^\mathsf{T}$, of the matrix A with the usual [U,S,V]=svd(A). Here orthogonal U is 292×292, diagonal S $= \mathrm{diag}(36822, 7073, \ldots, 1.2859, 1.1291)$ is 292×277, and orthogonal V is 277×277. Such matrices are far too big to record in this text.

- Figure 5.2 plots the nonzero singular values from largest to smallest: they cover a range of five orders of magnitude (the vertical axis is logarithmic). Choose some number k of singular vectors to use: k is the rank of the approximate images in Figure 5.1. Guide your choice by the decrease of the singular values: as discussed later (Theorem 5.1.16), for a say 1% error, choose k such that $\sigma_k \approx 0.01\sigma_1$ which from the index j axis of Figure 5.2 is around $k \approx 30$.

- Construct the approximate rank k image

$$A_k = \sigma_1 u_1 v_1^\mathsf{T} + \sigma_2 u_2 v_2^\mathsf{T} + \cdots + \sigma_k u_k v_k^\mathsf{T}$$
$$= \mathtt{U(:,1:k)*S(1:k,1:k)*V(:,1:k)'}$$

- Let's say the rank 30 image of Figure 5.1 is the desired good approximation. To reconstruct it we need 30 singular values $\sigma_1, \sigma_2, \ldots, \sigma_{30}$, 30 columns $u_1, u_2, \ldots, u_{30}$ of U, and 30 columns $v_1, v_2, \ldots, v_{30}$ of V making a total of

$$30 + 30 \times 292 + 30 \times 277 = 17\,100 \text{ numbers.}$$

These 17 100 numbers are much fewer than (one fifth of) the $292 \times 277 = 80\,884$ numbers of the original image.

The SVD provides an effective flexible data compression. □

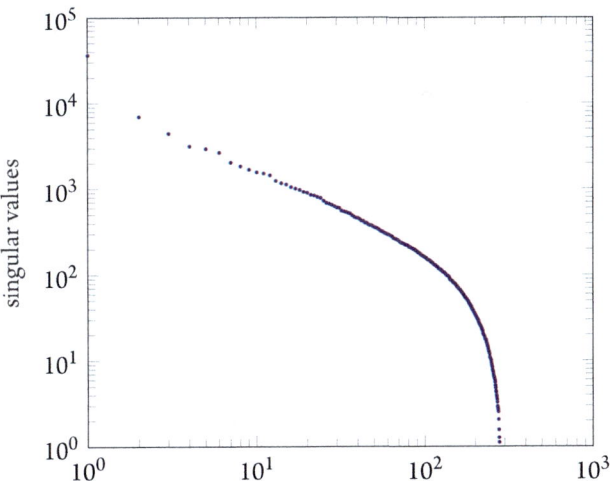

Figure 5.2 Singular values of the image of Pascal.

Table 5.1 As well as the MATLAB/Octave commands and operations listed in Tables 1.2, 2.3, 3.1 to 3.3, and 3.7 we may invoke these functions.

- norm (A) computes the matrix norm of Definition 5.1.7, namely the largest singular value of the matrix A.

  Also recall that (Table 1.2) norm (v) for a vector v computes the length, or magnitude, $\sqrt{v_1^2 + v_2^2 + \cdots + v_n^2}$.

- scatter (x, y, [], c) draws a 2D scatter plot of points with coordinates in vectors x and y, each point with a colour determined by the corresponding entry of vector c. Similarly for scatter3 (x, y, z, [], c) but in 3D.

- [U, S, V] =svds (A, k) computes the k largest singular values of the matrix A in the diagonal of $k \times k$ matrix S, and the k columns of U and V are the corresponding singular vectors.

- imread ('filename') typically reads an image from a file into an $m \times n \times 3$ array of red-green-blue values. The values are all integers in the range $[0, 255]$.

- csvread ('filename') reads data from a file into a matrix. When each of the m lines in the file is n numbers separated by commas, then the result is an $m \times n$ matrix.

- mean (A) of an $m \times n$ array computes the n elements in the row vector of averages (the arithmetic mean) over each column of A.

  Whereas mean (A, p) for an ℓ-dimensional array A of dimension $m_1 \times m_2 \times \cdots \times m_\ell$, computes the mean over the pth index to give an array of size $m_1 \times \cdots \times m_{p-1} \times 1 \times m_{p+1} \times \cdots \times m_\ell$.

- std (A) of an $m \times n$ array computes the n elements in the row vector of the standard deviation over each column of A. That is, for each column of A, the corresponding elements of std (A) are proportional to the *spread* in values of that column.

- For the element-by-element operations of +, -, .*, ./, .^, and so on, Matlab/Octave auto-*replicates scalars and vectors as appropriate*. Where this auto-replication becomes extremely useful is that if either of the matrices is $m \times 1$ or $1 \times n$, or one of each, then first its column/row is replicated to a matrix of size $m \times n$, and second the operation is performed.

  The general rule is that if all the non-1 sizes of the operands agree, then the dimensions of size 1 in each operand are replicated as needed to match the other operand.

- axis sets some properties of a drawn figure:
  - axis equal ensures that horizontal and vertical directions are scaled the same—so there is no distortion of the image;
  - axis off means that the horizontal and vertical axes are not drawn—so the image is unadorned.

5.1.2 Relate matrix changes to the SVD

We need to define what 'best' means in the approximation Procedure 5.1.3 and then show the procedure achieves this best. We need a measure of the magnitude of matrices and of distances between matrices.

In linear algebra we use a pair of double vertical bars, $\| \cdot \|$, to denote the magnitude of a matrix. The double vertical bars avoid a notational clash with the well-established use of $| \cdot |$ for the determinant of a matrix (Chapter 6).

Definition 5.1.7 *Let A be an m × n matrix. Define the **matrix norm** (or called the spectral norm)*

$$\|A\| := \max_{|x|=1} |Ax|, \qquad\qquad \text{equivalently } \|A\| = \sigma_1 \qquad\qquad (5.1)$$

the largest singular value of the matrix A.[2]

The equivalence, that $\max_{|x|=1} |Ax| = \sigma_1$, is due to the definition of the largest singular value in the proof of the existence of an SVD (Section 3.3.3).

Example 5.1.8 The two following 2 × 2 matrices have the product Ax plotted (red), adjoined to x (blue), for a complete range of unit vectors x (as in Section 4.1 for eigenvectors). From Definition 5.1.7, the norm of the matrix A is then the length of the longest such plotted Ax (this norm is a measure of the magnitude of the matrix). (The MATLAB function eigshow(A)[3] provides an interactive alternative to such static views.)

For each matrix, use the plot to roughly estimate their norm.

(a) $A = \begin{bmatrix} 0.5 & 0.5 \\ -0.6 & 1.2 \end{bmatrix}$

    Solution: The longest vectors Ax appear to be near the top and bottom of the plot, and these vectors appear to be a little longer than one, so we estimate the norm $\|A\| \approx 1.3$. □

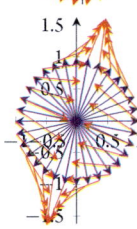

(b) $B = \begin{bmatrix} -0.7 & 0.4 \\ 0.6 & 0.5 \end{bmatrix}$

    Solution: Near the top and bottom of the plot, the vectors Bx appear to be of length 0.6. But the vectors pointing inwards from the right and left appear longer, at about 0.9. So we estimate the norm $\|B\| \approx 0.9$. □

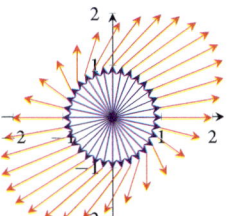

Example 5.1.9 Consider the 2 × 2 matrix $A = \begin{bmatrix} 1 & 1 \\ 0 & 1 \end{bmatrix}$. Algebraically explore products Ax for unit vectors x, as illustrated to the right, and then find the matrix norm $\|A\|$.

- The standard unit vector $e_2 = (0,1)$ has $|e_2| = 1$ and $Ae_2 = (1,1)$ has length $|Ae_2| = \sqrt{2}$. Since the matrix norm is the maximum of all possible $|Ax|$, so $\|A\| \geq |Ae_2| = \sqrt{2} \approx 1.41$.

- Another unit vector is $x = (\frac{3}{5}, \frac{4}{5})$. Here $Ax = (\frac{7}{5}, \frac{4}{5})$ has length $\sqrt{49+16}/5 = \sqrt{65}/5 \approx 1.61$. Hence the matrix norm $\|A\| \geq |Ax| \approx 1.61$.

- To systematically find the norm, recall that all unit vectors in 2D are of the form $x = (\cos t, \sin t)$. Then

---

[2] Sometimes this matrix norm is more specifically called a 2-norm and correspondingly denoted by $\|A\|_2$: but not in this book because at other times and places $\|A\|_2$ denotes something slightly different.

[3] Download for R2017b etc.

$$|Ax|^2 = |(\cos t + \sin t, \sin t)|^2$$
$$= (\cos t + \sin t)^2 + \sin^2 t$$
$$= \cos^2 t + 2 \cos t \sin t + \sin^2 t + \sin^2 t$$
$$= \tfrac{3}{2} + \sin 2t - \tfrac{1}{2} \cos 2t.$$

This length (squared) is maximized (and minimized) for some t determined by calculus. Differentiating with respect to t leads to

$$\frac{d|Ax|^2}{dt} = 2 \cos 2t + \sin 2t = 0 \quad \text{for stationary points.}$$

Rearranging this equation determines that we require $\tan 2t = -2$. The right-angle triangles, drawn to the right, illustrate that these stationary points of $|Ax|^2$ occur for $\sin 2t = \mp 2/\sqrt{5}$ and correspondingly $\cos 2t = \pm 1/\sqrt{5}$ (one gives a minimum and one gives the desired maximum). Substituting these two cases gives

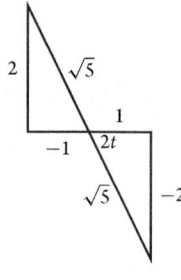

$$|Ax|^2 = \tfrac{3}{2} + \sin 2t - \tfrac{1}{2} \cos 2t = \tfrac{3}{2} \mp \tfrac{2}{\sqrt{5}} \mp \tfrac{1}{2}\tfrac{1}{\sqrt{5}}$$

$$= \tfrac{1}{2}(3 \mp \sqrt{5}) = \left(\frac{1 \mp \sqrt{5}}{2} \right)^2.$$

The plus alternative is the larger, so gives the maximum, hence

$$\|A\| = \max_{|x|=1} |Ax| = \frac{1 + \sqrt{5}}{2} = 1.6180.$$

- Confirm with MATLAB/Octave via svd([1 1;0 1]), which gives the singular values $\sigma_1 = 1.6180$ and $\sigma_2 = 0.6180$. Hence confirming the norm $\|A\| = \sigma_1 = 1.6180$.

Alternatively, see Table 5.1, execute norm([1 1;0 1]) to compute the norm $\|A\| = 1.6180$. □

Activity 5.1.10 A given 3×3 matrix A has the following products

$$A \begin{bmatrix} 1 \\ 0 \\ 0 \end{bmatrix} = \begin{bmatrix} 1 \\ 1 \\ 1 \end{bmatrix}, \quad A \begin{bmatrix} -1/3 \\ -2/3 \\ 2/3 \end{bmatrix} = \begin{bmatrix} 1/3 \\ 1/3 \\ -7/3 \end{bmatrix}, \quad A \begin{bmatrix} 1 \\ -2 \\ -2 \end{bmatrix} = \begin{bmatrix} 11 \\ 3 \\ 3 \end{bmatrix}.$$

Which of the following is the 'best' statement about the norm of matrix A (best in the sense of giving the largest valid lower bound)?

(a) $\|A\| \geq 3.9$ (b) $\|A\| \geq 11.7$ (c) $\|A\| \geq 1.7$ (d) $\|A\| \geq 2.3$

Example 5.1.11 MATLAB/Octave readily computes the matrix norm either via an SVD or using the norm() function directly (Table 5.1). Compute the norm of the 4×6 matrix

$$B = \begin{bmatrix} 0 & -2 & -1 & -4 & -5 & 0 \\ 2 & 0 & 1 & -2 & -6 & -2 \\ -2 & 0 & 4 & 2 & 3 & -3 \\ 1 & 2 & -4 & 2 & 1 & 3 \end{bmatrix}$$

Solution: Enter the matrix into MATLAB/Octave then executing svd(B) returns the vector of singular values $(10.1086, 7.6641, 3.2219, 0.8352)$, so $\|B\| = \sigma_1 = 10.1086$. Alternatively, executing norm(B) directly gives $\|B\| = 10.1086$. □

The Definition 5.1.7 of the magnitude/norm of a matrix may appear a little strange. But, in addition to some other marvellously useful properties, the norm nonetheless has all the familiar properties of a magnitude/length. Recall from Chapter 1 that for vectors:

- $|v| = 0$ if and only if $v = 0$ (Theorem 1.1.13);
- $|u \pm v| \leq |u| + |v|$ (the triangle inequality of Theorem 1.3.17);
- $|tv| = |t| \cdot |v|$ (Theorem 1.3.17).

Analogous 'magnitude' properties hold for the matrix norm as established in the next theorem.

Theorem 5.1.12 (*matrix norm properties.*) *For every $m \times n$ real matrix A:*

(a) $\|A\| = 0$ if and only if $A = O_{m \times n}$;

(b) $\|I_n\| = 1$;

(c) $\|A \pm B\| \leq \|A\| + \|B\|$, for every $m \times n$ matrix B, is like a triangle inequality (Theorem 1.3.17(c));

(d) $\|tA\| = |t| \cdot \|A\|$ for every scalar t;

(e) $\|A\| = \|A^T\|$;

(f) $\|Q_m A\| = \|A\| = \|AQ_n\|$ for every $m \times m$ orthogonal matrix Q_m and every $n \times n$ orthogonal matrix Q_n;

(g) $|Ax| \leq \|A\| \cdot |x|$ for all $x \in \mathbb{R}^n$, is like a Cauchy–Schwarz inequality (Theorem 1.3.17(b)), as is the next;

(h) $\|AB\| \leq \|A\| \|B\|$ for every $n \times p$ matrix B.

Proof. Alternative proofs to the following may be invoked (Exercise 5.1.3). Where necessary in the following, let matrix A have the SVD $A = USV^T$.

5.1.12(a) If $A = O_{m \times n}$, then from Definition 5.1.7

$$\|A\| = \max_{|x|=1} |Ox| = \max_{|x|=1} |0| = \max_{|x|=1} 0 = 0.$$

Conversely, if $\|A\| = 0$, then the largest singular value $\sigma_1 = 0$ (Definition 5.1.7), which implies that all singular values are zero, so the matrix A has an SVD of the form $A = UO_{m \times n}V^T$, which reduces to $A = O_{m \times n}$.

5.1.12(b) From Definition 5.1.7,

$$\|I_n\| = \max_{|x|=1} |I_n x| = \max_{|x|=1} |x| = \max_{|x|=1} 1 = 1.$$

5.1.12(c) Using Definition 5.1.7 at the first and last steps:

$$\begin{aligned} \|A \pm B\| &= \max_{|x|=1} |(A \pm B)x| = \max_{|x|=1} |Ax \pm Bx| \quad \text{(by distributivity)} \\ &\leq \max_{|x|=1} (|Ax| + |Bx|) \quad \text{(by triangle inequality)} \\ &\leq \max_{|x|=1} |Ax| + \max_{|x|=1} |Bx| = \|A\| + \|B\|. \end{aligned}$$

5.1.12(d) Using Definition 5.1.7,

$$\begin{aligned} \|tA\| &= \max_{|x|=1} |(tA)x| = \max_{|x|=1} |t(Ax)| \quad \text{(by associativity)} \\ &= \max_{|x|=1} |t||Ax| \quad \text{(by Theorem 1.3.17)} \\ &= |t| \max_{|x|=1} |Ax| = |t| \cdot \|A\|. \end{aligned}$$

5.1.12(e) Recall that the transpose A^T has an SVD $A^T = (USV^T)^T = VS^T U^T$, and so has the same singular values as A. So the largest singular value of A^T is the same as that of A. Hence $\|A\| = \|A^T\|$.

5.1.12(f) Recall that multiplication by an orthogonal matrix is a rotation/reflection and so does not change lengths (Theorem 3.2.48(f)); correspondingly, it also does not change the norm of a matrix as established here.

Now $Q_m A = Q_m(USV^T) = (Q_m U)SV^T$. But $Q_m U$ is an orthogonal matrix (Exercise 3.2.21), so $(Q_m U)SV^T$ is an SVD for $Q_m A$ and so has the same singular values as those in S. Hence, $\|Q_m A\| = \sigma_1 = \|A\|$.

Also, using 5.1.12(e) twice: $\|AQ_n\| = \|(AQ_n)^T\| = \|Q_n^T A^T\| = \|A^T\| = \|A\|$.

5.1.12(g) Split into two cases. In the case $x = 0$, then $|Ax| = |A0| = |0| = 0$ whereas $\|A\||x| = \|A\||0| = \|A\|0 = 0$, so $|Ax| \leq \|A\||x|$. Alternatively, in the case $x \neq 0$, then we write $x = \hat{x}|x|$ for unit vector $\hat{x} = x/|x|$ so that

$$\begin{aligned} |Ax| &= |A\hat{x}|x|| = |A\hat{x}||x| \quad \text{(as } |x| \text{ is a scalar)} \\ &\leq \left(\max_{|\hat{x}|=1} |A\hat{x}| \right)|x| \quad \text{(as } \hat{x} \text{ is a unit vector)} \\ &= \|A\||x| \quad \text{(by Definition 5.1.7)} \end{aligned}$$

5.1.12(h) See Exercise 5.1.3(e). □

Since the matrix norm has these familiar properties of a measure of magnitude, we use the matrix norm to measure the 'distance' between matrices.

Example 5.1.13

(a) Use the matrix norm to estimate the 'distance' between matrices

$$B = \begin{bmatrix} -0.7 & 0.4 \\ 0.6 & 0.5 \end{bmatrix} \quad \text{and} \quad C = \begin{bmatrix} -0.2 & 0.9 \\ 0 & 1.7 \end{bmatrix}.$$

Solution: The 'distance' between matrices B and C is, via the matrix A of Example 5.1.8(a) and its estimated norm,

$$\|C - B\| = \left\| \begin{bmatrix} 0.5 & 0.5 \\ -0.6 & 1.2 \end{bmatrix} \right\| = \|A\| \approx 1.3.$$

$\square$

(b) Find the distance between the matrices

$$A = \begin{bmatrix} 10 & 2 \\ 5 & 11 \end{bmatrix} \quad \text{and } B = \begin{bmatrix} 4 & -4 \\ -3 & 3 \end{bmatrix}, \quad \text{and between } A \text{ and } A_1 = \begin{bmatrix} 6 & 6 \\ 8 & 8 \end{bmatrix}.$$

Recall from Example 3.3.2 that the matrix A has an SVD of

$$USV^{\mathsf{T}} = \begin{bmatrix} \frac{3}{5} & -\frac{4}{5} \\ \frac{4}{5} & \frac{3}{5} \end{bmatrix} \begin{bmatrix} 10\sqrt{2} & 0 \\ 0 & 5\sqrt{2} \end{bmatrix} \begin{bmatrix} \frac{1}{\sqrt{2}} & -\frac{1}{\sqrt{2}} \\ \frac{1}{\sqrt{2}} & \frac{1}{\sqrt{2}} \end{bmatrix}^{\mathsf{T}}.$$

i. Find $\|A - B\|$ for the rank one matrix

$$B = \sigma_2 u_2 v_2^{\mathsf{T}} = 5\sqrt{2} \begin{bmatrix} -\frac{4}{5} \\ \frac{3}{5} \end{bmatrix} \begin{bmatrix} -\frac{1}{\sqrt{2}} & \frac{1}{\sqrt{2}} \end{bmatrix} = \begin{bmatrix} 4 & -4 \\ -3 & 3 \end{bmatrix}.$$

Solution: Let's write matrix

$$B = \begin{bmatrix} \frac{3}{5} & -\frac{4}{5} \\ \frac{4}{5} & \frac{3}{5} \end{bmatrix} \begin{bmatrix} 0 & 0 \\ 0 & 5\sqrt{2} \end{bmatrix} \begin{bmatrix} \frac{1}{\sqrt{2}} & -\frac{1}{\sqrt{2}} \\ \frac{1}{\sqrt{2}} & \frac{1}{\sqrt{2}} \end{bmatrix}^{\mathsf{T}} = U \begin{bmatrix} 0 & 0 \\ 0 & 5\sqrt{2} \end{bmatrix} V^{\mathsf{T}}.$$

Then the difference is

$$A - B = U \begin{bmatrix} 10\sqrt{2} & 0 \\ 0 & 5\sqrt{2} \end{bmatrix} V^{\mathsf{T}} - U \begin{bmatrix} 0 & 0 \\ 0 & 5\sqrt{2} \end{bmatrix} V^{\mathsf{T}}$$

$$= U \left(\begin{bmatrix} 10\sqrt{2} & 0 \\ 0 & 5\sqrt{2} \end{bmatrix} - \begin{bmatrix} 0 & 0 \\ 0 & 5\sqrt{2} \end{bmatrix} \right) V^{\mathsf{T}} = U \begin{bmatrix} 10\sqrt{2} & 0 \\ 0 & 0 \end{bmatrix} V^{\mathsf{T}}.$$

This last expression is an SVD for $A - B$ with singular values $10\sqrt{2}$ and 0, so by Definition 5.1.7 its norm $\|A - B\| = \sigma_1 = 10\sqrt{2}$. $\square$

ii. Find $\|A - A_1\|$ for the rank one matrix

$$A_1 = \sigma_1 u_1 v_1^T = 10\sqrt{2} \begin{bmatrix} \frac{3}{5} \\ \frac{4}{5} \end{bmatrix} \begin{bmatrix} \frac{1}{\sqrt{2}} & \frac{1}{\sqrt{2}} \end{bmatrix} = \begin{bmatrix} 6 & 6 \\ 8 & 8 \end{bmatrix}.$$

Solution: Let's write matrix

$$A_1 = \begin{bmatrix} \frac{3}{5} & -\frac{4}{5} \\ \frac{4}{5} & \frac{3}{5} \end{bmatrix} \begin{bmatrix} 10\sqrt{2} & 0 \\ 0 & 0 \end{bmatrix} \begin{bmatrix} \frac{1}{\sqrt{2}} & -\frac{1}{\sqrt{2}} \\ \frac{1}{\sqrt{2}} & \frac{1}{\sqrt{2}} \end{bmatrix}^T = U \begin{bmatrix} 10\sqrt{2} & 0 \\ 0 & 0 \end{bmatrix} V^T.$$

Then the difference is

$$A - A_1 = U \begin{bmatrix} 10\sqrt{2} & 0 \\ 0 & 5\sqrt{2} \end{bmatrix} V^T - U \begin{bmatrix} 10\sqrt{2} & 0 \\ 0 & 0 \end{bmatrix} V^T$$

$$= U \left(\begin{bmatrix} 10\sqrt{2} & 0 \\ 0 & 5\sqrt{2} \end{bmatrix} - \begin{bmatrix} 10\sqrt{2} & 0 \\ 0 & 0 \end{bmatrix} \right) V^T = U \begin{bmatrix} 0 & 0 \\ 0 & 5\sqrt{2} \end{bmatrix} V^T.$$

This last expression is an SVD for $A - A_1$ with singular values $5\sqrt{2}$ and 0, albeit out of order, so by Definition 5.1.7 the norm $\|A - A_1\|$ is the largest singular value which here is $5\sqrt{2}$. □

Out of these two matrices, A_1 and B, the matrix A_1 is 'closer' to A as $\|A - A_1\| = 5\sqrt{2} < 10\sqrt{2} = \|A - B\|$.

Activity 5.1.14 Which of the following matrices is *not* a distance one from the matrix
$F = \begin{bmatrix} 9 & -1 \\ 1 & 5 \end{bmatrix}$?

(a) $\begin{bmatrix} 10 & -1 \\ 1 & 6 \end{bmatrix}$ (b) $\begin{bmatrix} 8 & -2 \\ 2 & 4 \end{bmatrix}$ (c) $\begin{bmatrix} 9 & -1 \\ 1 & 6 \end{bmatrix}$ (d) $\begin{bmatrix} 8 & -1 \\ 1 & 5 \end{bmatrix}$

Example 5.1.15 From Example 5.1.4, recall the 'bulls eye' matrix

$$A = \begin{bmatrix} 0 & 1 & 1 & 1 & 0 \\ 1 & 0 & 0 & 0 & 1 \\ 1 & 0 & 1 & 0 & 1 \\ 1 & 0 & 0 & 0 & 1 \\ 0 & 1 & 1 & 1 & 0 \end{bmatrix},$$

and its rank two and three approximations A_2 and A_3. Find $\|A - A_2\|$ and $\|A - A_3\|$.

Solution: Example 5.1.4 found $A_3 = A$ hence $\|A - A_3\| = \|O_5\| = 0$.

Although $\|A - A_2\|$ is nontrivial, finding it is straightforward using SVDs. Recall that, from the given SVD $A = USV^{\mathsf{T}}$,

$$A_2 = \sigma_1 u_1 v_1^{\mathsf{T}} + \sigma_2 u_2 v_2^{\mathsf{T}} = \sigma_1 u_1 v_1^{\mathsf{T}} + \sigma_2 u_2 v_2^{\mathsf{T}} + 0 u_3 v_3^{\mathsf{T}} + 0 u_4 v_4^{\mathsf{T}} + 0 u_5 v_5^{\mathsf{T}}$$

$$= U \begin{bmatrix} \sigma_1 & 0 & 0 & 0 & 0 \\ 0 & \sigma_2 & 0 & 0 & 0 \\ 0 & 0 & 0 & 0 & 0 \\ 0 & 0 & 0 & 0 & 0 \\ 0 & 0 & 0 & 0 & 0 \end{bmatrix} V^{\mathsf{T}}.$$

Hence the difference

$$A - A_2 = U \begin{bmatrix} \sigma_1 & 0 & 0 & 0 & 0 \\ 0 & \sigma_2 & 0 & 0 & 0 \\ 0 & 0 & \sigma_3 & 0 & 0 \\ 0 & 0 & 0 & \sigma_4 & 0 \\ 0 & 0 & 0 & 0 & \sigma_5 \end{bmatrix} V^{\mathsf{T}} - U \begin{bmatrix} \sigma_1 & 0 & 0 & 0 & 0 \\ 0 & \sigma_2 & 0 & 0 & 0 \\ 0 & 0 & 0 & 0 & 0 \\ 0 & 0 & 0 & 0 & 0 \\ 0 & 0 & 0 & 0 & 0 \end{bmatrix} V^{\mathsf{T}}$$

$$= U \begin{bmatrix} 0 & 0 & 0 & 0 & 0 \\ 0 & 0 & 0 & 0 & 0 \\ 0 & 0 & \sigma_3 & 0 & 0 \\ 0 & 0 & 0 & \sigma_4 & 0 \\ 0 & 0 & 0 & 0 & \sigma_5 \end{bmatrix} V^{\mathsf{T}}.$$

This last expression is an SVD for $A - A_2$, albeit with the singular values out of order, with singular values of $0, 0, \sigma_3 = 0.64$, and $\sigma_4 = \sigma_5 = 0$. The largest of these singular values, σ_3, gives the norm $\|A - A_2\| = 0.64$ (2 d.p.).

To be discussed later, the relative error in the approximate A_2 is $\|A - A_2\|/\|A\| = \sigma_3/\sigma_1 = 0.64/2.68 = 0.24 = 24\%$ (2 d.p.).   □

Theorem 5.1.16 (*Eckart–Young.*) *Let A be any $m \times n$ matrix of rank r with SVD $A = USV^{\mathsf{T}}$. Then for every $k < r$ the matrix*

$$A_k := US_k V^{\mathsf{T}} = \sigma_1 u_1 v_1^{\mathsf{T}} + \sigma_2 u_2 v_2^{\mathsf{T}} + \cdots + \sigma_k u_k v_k^{\mathsf{T}} \tag{5.2}$$

where $S_k := \mathrm{diag}(\sigma_1, \sigma_2, \ldots, \sigma_k, 0, \ldots, 0)$, is a closest rank k matrix approximating A, in the matrix norm. The distance between A and A_k is $\|A - A_k\| = \sigma_{k+1}$.

That is, obtain a closest rank k matrix A_k by 'setting' the singular values $\sigma_{k+1} = \cdots = \sigma_r = 0$ from an SVD for A.

Proof. As a prelude to this difficult proof, let's establish the distance between A and A_k. Using their SVDs,

$$\begin{aligned}
A - A_k &= USV^{\mathsf{T}} - US_kV^{\mathsf{T}} = U(S - S_k)V^{\mathsf{T}} \\
&= U\operatorname{diag}(0,\ldots,0,\sigma_{k+1},\ldots,\sigma_r,0,\ldots,0)V^{\mathsf{T}},
\end{aligned}$$

and so $A - A_k$ has largest singular value σ_{k+1}. Then from Definition 5.1.7, $\|A - A_k\| = \sigma_{k+1}$.

Now let's use contradiction to prove there is no matrix of rank k closer to A when using $\|\cdot\|$ to measure matrix distances (Trefethen and Bau, 1997, p.36). Assume there is some $m \times n$ matrix B with rank $B \le k$ and B is closer to A than is A_k, that is, $\|A - B\| < \|A - A_k\|$. First, the Rank Theorem 3.4.39 asserts that the nullspace of B has dimension nullity $B = n - \operatorname{rank} B \ge n - k$ as rank $B \le k$. For every $w \in \operatorname{null} B$, as $Bw = 0$, $Aw = Aw - Bw = (A - B)w$. Then

$$\begin{aligned}
|Aw| &= |(A - B)w| \\
&\le \|A - B\||w| \quad \text{(by Theorem 5.1.12(g))} \\
&< \|A - A_k\||w| \quad \text{(by assumption)} \\
&= \sigma_{k+1}|w|
\end{aligned}$$

That is, under the assumption there exists an (at least) $(n - k)$-dimensional subspace in which $|Aw| < \sigma_{k+1}|w|$.

Second, consider any vector v in the $(k + 1)$-dimensional subspace $\operatorname{span}\{v_1, v_2, \ldots, v_{k+1}\}$. Say $v = c_1v_1 + c_2v_2 + \cdots + c_{k+1}v_{k+1} = Vc$ for some vector of coefficients $c = (c_1, c_2, \ldots, c_{k+1}, 0, \ldots, 0) \in \mathbb{R}^n$. Then

$$\begin{aligned}
|Av| &= |USV^{\mathsf{T}}Vc| = |USc| \quad \text{(as } V^{\mathsf{T}}V = I) \\
&= |Sc| \quad \text{(as } U \text{ is orthogonal)} \\
&= |(\sigma_1 c_1, \sigma_2 c_2, \ldots, \sigma_{k+1} c_{k+1}, 0, \ldots, 0)| \\
&= \sqrt{\sigma_1^2 c_1^2 + \sigma_2^2 c_2^2 + \cdots + \sigma_{k+1}^2 c_{k+1}^2} \\
&\ge \sqrt{\sigma_{k+1}^2 c_1^2 + \sigma_{k+1}^2 c_2^2 + \cdots + \sigma_{k+1}^2 c_{k+1}^2} \\
&= \sigma_{k+1}\sqrt{c_1^2 + c_2^2 + \cdots + c_{k+1}^2} \\
&= \sigma_{k+1}|c| = \sigma_{k+1}|Vc| \quad \text{(as } V \text{ is orthogonal)} \\
&= \sigma_{k+1}|v|.
\end{aligned}$$

That is, there exists a $(k + 1)$-dimensional subspace in which $|Av| \ge \sigma_{k+1}|v|$.

Lastly, since the sum of the dimensions of these two subspaces of $\mathbb{R}^n$ is at least $(n - k) + (k + 1) > n$, there must be a nonzero vector, say u, lying in both. So for this u, simultaneously $|Au| < \sigma_{k+1}|u|$ and $|Au| \ge \sigma_{k+1}|u|$. These two statements contradict each other. Hence the assumption is wrong: there is no rank k matrix more closely approximating A than A_k. □

Example 5.1.17 *(the letter R)* In digital displays with low resolution, letters and numbers are displayed with noticeable pixel patterns: for example, the letter R is pixellated to the right. Let's see how such pixel patterns are best approximated by matrices of different ranks. (This example is illustrative: it is not a practical image compression since the required singular vectors are more complicated than a small-sized pattern of pixels.)

Solution: Use Procedure 5.1.3. First, form and enter into MATLAB/Octave the 7×5 matrix of the pixel pattern as illustrated above-right

$$R = \begin{bmatrix} 1 & 1 & 1 & 1 & 0 \\ 1 & 0 & 0 & 0 & 1 \\ 1 & 0 & 0 & 0 & 1 \\ 1 & 1 & 1 & 1 & 0 \\ 1 & 0 & 1 & 0 & 0 \\ 1 & 0 & 0 & 1 & 0 \\ 1 & 0 & 0 & 0 & 1 \end{bmatrix}.$$

Second, compute an SVD via $[U, S, V] = svd(R)$ to find (2 d.p.)

```
U =
   -0.53    0.38   -0.00   -0.29   -0.70   -0.06   -0.07
   -0.28   -0.49    0.00   -0.13    0.10   -0.69   -0.42
   -0.28   -0.49   -0.00   -0.13   -0.02    0.72   -0.39
   -0.53    0.38   -0.00   -0.29    0.70    0.06    0.07
   -0.32    0.03   -0.71    0.63   -0.00   -0.00   -0.00
   -0.32    0.03    0.71    0.63   -0.00    0.00    0.00
   -0.28   -0.49   -0.00   -0.13   -0.08   -0.02    0.81
S =
    3.47       0       0       0       0
       0    2.09       0       0       0
       0       0    1.00       0       0
       0       0       0    0.75       0
       0       0       0       0    0.00
       0       0       0       0       0
       0       0       0       0       0
V =
   -0.73   -0.32    0.00    0.40   -0.45
   -0.30    0.36   -0.00   -0.76   -0.45
   -0.40    0.37   -0.71    0.07    0.45
   -0.40    0.37    0.71    0.07    0.45
   -0.24   -0.70   -0.00   -0.50    0.45
```

The singular values are $\sigma_1 = 3.47, \sigma_2 = 2.09, \sigma_3 = 1.00, \sigma_4 = 0.75$ and $\sigma_5 = 0$. Four successively better approximations to the image are the following.

- The coarsest approximation is $R_1 = \sigma_1 u_1 v_1^T$, that is

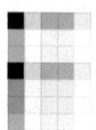

$$R_1 = 3.47 \begin{bmatrix} -0.53 \\ -0.28 \\ -0.28 \\ -0.53 \\ -0.32 \\ -0.32 \\ -0.28 \end{bmatrix} \begin{bmatrix} -0.73 & -0.30 & -0.40 & -0.40 & -0.24 \end{bmatrix}.$$

Compute with R1=U(:,1)*S(1,1)*V(:,1)' to find (2 d.p.), as illustrated above,

R1 =

1.34	0.55	0.72	0.72	0.44
0.71	0.29	0.39	0.39	0.24
0.71	0.29	0.39	0.39	0.24
1.34	0.55	0.72	0.72	0.44
0.83	0.34	0.45	0.45	0.27
0.83	0.34	0.45	0.45	0.27
0.71	0.29	0.39	0.39	0.24

This has difference $\|R - R_1\| = \sigma_2 = 2.09$ which at 60% of σ_1 is large: indeed, the letter R is not recognizable.

- Computing R2=U(:,1:2)*S(1:2,1:2)*V(:,1:2)' finds the second approximation $R_2 = \sigma_1 u_1 v_1^T + \sigma_2 u_2 v_2^T$ is (2 d.p.), as illustrated,

R2 =

1.09	0.83	1.02	1.02	-0.11
1.04	-0.07	0.01	0.01	0.95
1.04	-0.07	0.01	0.01	0.95
1.09	0.83	1.02	1.02	-0.11
0.81	0.36	0.47	0.47	0.24
0.81	0.36	0.47	0.47	0.24
1.04	-0.07	0.01	0.01	0.95

This has difference $\|R - R_2\| = \sigma_3 = 1.00$ which at 29% of σ_1 is large: but one can begin to imagine the letter R in the image.

- The third approximation is

$$R_3 = \sigma_1 u_1 v_1^T + \sigma_2 u_2 v_2^T + \sigma_3 u_3 v_3^T.$$

Compute with R3=U(:,1:3)*S(1:3,1:3)*V(:,1:3)' to find (2 d.p.), as illustrated,

R3 =

1.09	0.83	1.02	1.02	-0.11
1.04	-0.07	0.01	0.01	0.95
1.04	-0.07	0.01	0.01	0.95

```
1.09     0.83     1.02     1.02    -0.11
0.81     0.36     0.97    -0.03     0.24
0.81     0.36    -0.03     0.97     0.24
1.04    -0.07     0.01     0.01     0.95
```

This has difference $\|R - R_3\| = \sigma_4 = 0.75$ which at 22% of σ_1 is moderate and one can see the letter R emerging.

- The fourth approximation is

$$R_4 = \sigma_1 u_1 v_1^T + \sigma_2 u_2 v_2^T + \cdots + \sigma_4 u_4 v_4^T.$$

Compute with R4=U(:,1:4)*S(1:4,1:4)*V(:,1:4)' to find (2 d.p.), as illustrated,

```
R4 =
    1.00     1.00     1.00     1.00    -0.00
    1.00     0.00    -0.00     0.00     1.00
    1.00     0.00    -0.00    -0.00     1.00
    1.00     1.00     1.00     1.00    -0.00
    1.00    -0.00     1.00     0.00    -0.00
    1.00    -0.00    -0.00     1.00     0.00
    1.00     0.00    -0.00    -0.00     1.00
```

This has difference $\|R - R_4\| = \sigma_5 = 0.00$ and so R_4 exactly reproduces R. ☐

Activity 5.1.18 A given image has singular values 12.74, 8.38, 3.06, 1.96, 1.08, ... What rank approximation has a relative error of just a little less than 25%?

(a) 1   (b) 2   (c) 3   (d) 4

Example 5.1.19 Recall that Example 5.1.6 approximated the image of Blaise Pascal with various rank k approximations, and that these approximations came from an SVD of the image. Let the image be denoted by matrix A. From Figure 5.2 the largest singular value of the image is $\|A\| = \sigma_1 \approx 37\,000$.

- From Theorem 5.1.16, the rank 3 approximation in Figure 5.1 is a distance $\|A - A_3\| = \sigma_4 \approx 3\,200$ (from Figure 5.2) away from the image. That is, image A_3 has a relative error roughly $3\,200/37\,000 \approx 9\%$.

- From Theorem 5.1.16, the rank 10 approximation in Figure 5.1 is a distance $\|A - A_{10}\| = \sigma_{11} \approx 1\,500$ (from Figure 5.2) away from the image. That is, image A_{10} has a relative error roughly $1\,500/37\,000 \approx 4\%$.

- From Theorem 5.1.16, the rank 30 approximation in Figure 5.1 is a distance $\|A - A_{30}\| = \sigma_{31} \approx 600$ (from Figure 5.2) away from the image. That is, image A_{30} has a relative error roughly $600/37\,000 \approx 2\%$. ☐

5.1.3 Principal component analysis

In its 'best' approximation property, Theorem 5.1.16 establishes the effectiveness of an SVD in image compression. Scientists and engineers also use this result for so-called data reduction: often using just a rank two (or three) 'best' approximation to high-dimensional data, one then plots 2D (or 3D) graphics. Such an approach is often termed a principal component analysis (PCA).

The technique introduced here is so useful that more-or-less the same approach has been invented independently in many fields. Consequently, much the same technique has alternative names such as the Karhunen–Loève decomposition, proper orthogonal decomposition, empirical orthogonal functions, and the Hotelling transform.

Example 5.1.20 *(toy items)* Suppose you are given data about six items, three blue and three red. Suppose each item has two measured properties/attributes called h and v as in the following table:

h	v	colour
-3	-3	blue
-2	1	blue
1	-2	blue
-1	2	red
2	-1	red
3	3	red

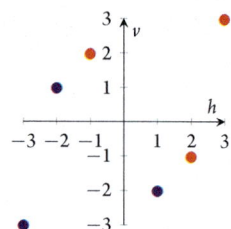

The item properties/attributes are the points (h, v) in 2D as illustrated to the above-right (h for horizontal, and v for vertical). But we humans always prefer simple one-dimensional summaries: we do it all the time when we rank sport teams, schools, web pages, political parties, and so on.

Challenge: is there a one-dimensional summary of these six items' data that clearly separates the blue from the red? Using just one of the attributes h or v on their own would not suffice:

- using h alone leads to a 1D view where the red and the blue are intermingled as shown to the right;

  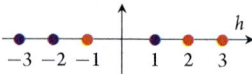

- similarly, using v alone leads to a 1D view where the red and the blue are intermingled as shown to the right.

  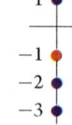

Solution: Use an SVD to automatically find the 'best' 1D view of the data.

(a) Enter the 6×2 matrix of data into MATLAB/Octave with

```
A= [-3  -3
    -2  1
     1  -2
    -1  2
     2  -1
     3  3 ]
```

(b) Then [U,S,V]=svd(A) computes an SVD, $A = USV^{\text{T}}$, of the data (2 d.p.):

U =

-0.69	0.00	-0.09	0.09	0.14	0.70
-0.11	-0.50	0.50	-0.50	0.48	-0.08
-0.11	0.50	0.82	0.18	-0.18	0.01
0.11	-0.50	0.18	0.82	0.18	-0.01
0.11	0.50	-0.14	0.14	0.83	-0.09
0.69	-0.00	0.12	-0.12	0.02	0.70

S =

6.16	0
0	4.24
0	0
0	0
0	0
0	0

V =

0.71	0.71
0.71	-0.71

(c) Now what does such an SVD tell us? Recall from the proof of the SVD (Section 3.3.3) that $A\boldsymbol{v}_1 = \sigma_1\boldsymbol{u}_1$. Further recall from the proof that $\boldsymbol{v}_1$ is the unit vector that maximizes $|A\boldsymbol{v}_1|$ so in some sense it is the direction in which the data in A is most spread out ($\boldsymbol{v}_1$ is called the principal vector). We find here (2 d.p.)

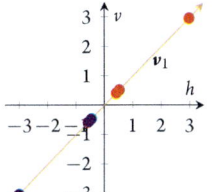

$$A\boldsymbol{v}_1 = \sigma_1\boldsymbol{u}_1 = (-4.24,\, -0.71,\, -0.71,\, 0.71,\, 0.71,\, 4.24)$$

which neatly separates the blue items (negative) from the red (positive). In essence, the product $A\boldsymbol{v}_1$ orthogonally projects (Section 3.5.3) the items' $(h,\, v)$ data onto the subspace span$\{\boldsymbol{v}_1\}$ as illustrated to the above-right. □

Although this Example 5.1.20 is just a toy to illustrate concepts, the above steps generalize straightforwardly to be immensely useful on vastly bigger and more challenging data. The next example takes the next step in complexity by introducing how to automatically find a good 2D view of some data in 4D.

Example 5.1.21 (*iris flower data set*) Table 5.2 lists part of Edgar Anderson's data on the lengths and widths of sepals and petals of iris flowers. There are three species of irises in the data (Setosa, Versicolor, Virginia). The data is 4D: each instance of thirty iris flowers is characterised by the four measurements of sepals and petals. Our challenge is to plot a 2D picture of this data in such a way that separates the flowers as best as possible. For high-D data (although 4D is not really that high), simply plotting one characteristic against another is rarely useful. For example, Figure 5.3 plots the attributes of sepal widths versus sepal lengths: the plot shows the three species

Table 5.2 Part of Edgar Anderson's iris data, lengths in centimetres (cm). The measurements come from the flowers of ten each of three different species of iris. http://archive.ics.uci.edu/ml/datasets/Iris gives the full dataset (Dua and Graff, 2019).

Sepal length	Sepal width	Petal length	Petal width	Species
4.9	3.0	1.4	0.2	
4.6	3.4	1.4	0.3	
4.8	3.4	1.6	0.2	
5.4	3.9	1.3	0.4	
5.1	3.7	1.5	0.4	Setosa
5.0	3.4	1.6	0.4	
5.4	3.4	1.5	0.4	
5.5	3.5	1.3	0.2	
4.5	2.3	1.3	0.3	
5.1	3.8	1.6	0.2	
6.4	3.2	4.5	1.5	
6.3	3.3	4.7	1.6	
5.9	3.0	4.2	1.5	
5.6	3.0	4.5	1.5	
6.1	2.8	4.0	1.3	Versicolor
6.8	2.8	4.8	1.4	
5.5	2.4	3.7	1.0	
6.7	3.1	4.7	1.5	
6.1	3.0	4.6	1.4	
5.7	2.9	4.2	1.3	
5.8	2.7	5.1	1.9	
4.9	2.5	4.5	1.7	
6.4	2.7	5.3	1.9	
6.5	3.0	5.5	1.8	
5.6	2.8	4.9	2.0	Virginia
6.2	2.8	4.8	1.8	
7.9	3.8	6.4	2.0	
6.3	3.4	5.6	2.4	
6.9	3.1	5.1	2.3	
6.3	2.5	5.0	1.9	

being intermingled together rather than reasonably separated. Our aim is to instead plot Figure 5.4 which successfully separates the three species.

Solution: Use an SVD to find a best low-rank view of the data.

(a) Enter the 30×5 matrix of iris data (Table 5.2) into MATLAB/Octave with a complete version of

```
iris=[
4.9 3.0 1.4 0.2 1
4.6 3.4 1.4 0.3 1
...
6.3 2.5 5.0 1.9 3
]
```

where the fifth column of $1,2,3$ corresponds to the species Setosa, Versicolor or Virginia, respectively. Then a scatter plot such as Figure 5.3 may be drawn with the command

```
scatter(iris(:,1),iris(:,2),[],iris(:,5))
```

The above command `scatter(x,y,[],s)` plots a scatter plot of points with colour depending upon s, which here corresponds to each different species.

(b) If we were on a walk to a scenic lookout to get a view of the countryside, then the scenic lookout would be in the countryside: it is no good going to a lookout a long way away from the scene we wish to view. Correspondingly, to best view a dataset we typically look it at from the very centre of the data, namely from its mean. That is, here we use an SVD of the data matrix only after subtracting the mean of each attribute. Then the SVD analyses the variations from the mean.

Here the mean iris sepal length and width is 5.81 cm and 3.09 cm (the black "+" in Figure 5.3), and the mean petal length and width is 3.69 cm and 1.22 cm. The row vector of these means are computed in MATLAB/Octave by mean(iris(:,1:4)), giving $\begin{bmatrix} 5.81 & 3.09 & 3.69 & 1.22 \end{bmatrix}$. Thus we subtract 5.81 cm from the first column of data, 3.09 cm from the second column, and so on. The auto-replication of MATLAB/Octave (Table 5.1) provides a convenient method to do this subtraction as it replicates the

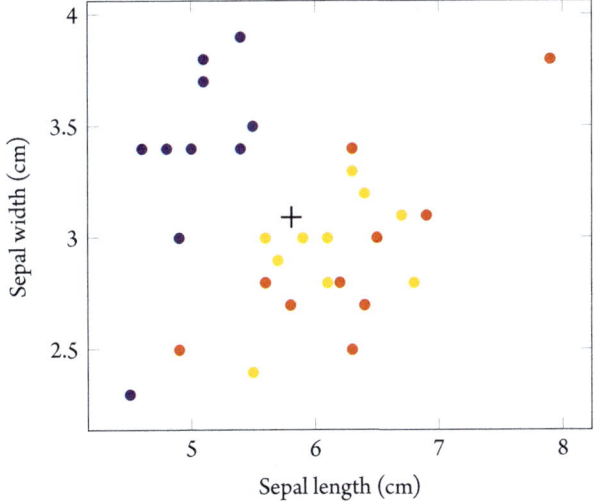

Figure 5.3 Scatter plot of sepal widths versus lengths for Edgar Anderson's iris data of Table 5.2: blue, Setosa; brown, Versicolor; red, Virginia. The black "+" marks the mean sepal width and length.

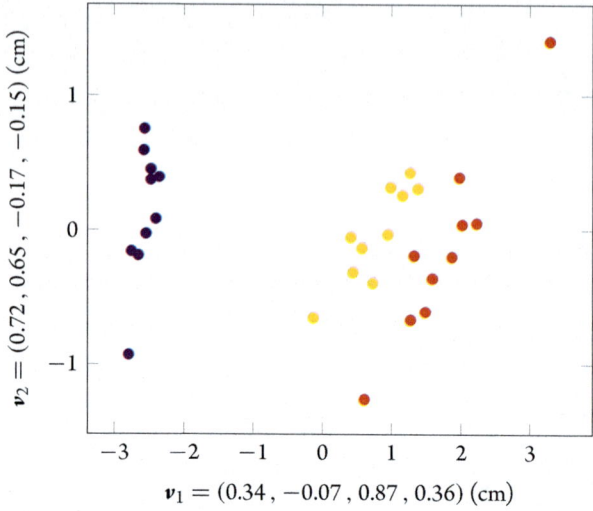

$v_1 = (0.34, -0.07, 0.87, 0.36)$ (cm)

Figure 5.4 Best 2D scatter plot of Edgar Anderson's iris data: blue, Setosa; brown, Versicolor; red, Virginia.

row vector of means to suit the number of rows in the 2D data array (Table 5.2). In MATLAB/Octave execute the following to form a matrix A of the variations from the mean, and then compute an SVD:

```
A=iris(:,1:4)-mean(iris(:,1:4))
[U,S,V]=svd(A)
```

The resulting SVD is (2 d.p.)

```
U = ...
S =
    10.46        0        0        0
        0     2.86        0        0
        0        0     1.47        0
        0        0        0     0.85
      ...      ...      ...      ...
V =
    0.34     0.72    -0.56    -0.20
   -0.07     0.65     0.74     0.14
    0.87    -0.17     0.14     0.45
    0.36    -0.15     0.33    -0.86
```

where a ... indicates information that is not directly of interest.

(c) As justified shortly, the two most important components of a flower's shape are those in the directions of v_1 and v_2 (called the two principal vectors). Because v_1 and v_2 are orthonormal vectors, the first component for each iris flower is $x = Av_1$ and the second component for each is $y = Av_2$. The beautiful Figure 5.4 is a scatter plot of the components of y versus the components of x that untangles the three species. Obtain Figure 5.4 in MATLAB/Octave with the command

```
scatter(A*V(:,1),A*V(:,2),[],iris(:,5))
```

Figure 5.4 shows that our SVD based analysis largely separates the three species using these two different combinations of the flowers' attributes.

□

Transpose the usual mathematical convention   Perhaps you noticed that the previous Example 5.1.21 flips our usual mathematical convention that vectors are column vectors. The example uses row vectors of the four attributes of each flower: Table 5.2 lists that the first iris Setosa flower has a row vector of attributes $\begin{bmatrix} 4.9 & 3.0 & 1.4 & 0.2 \end{bmatrix}$ (cm) corresponding to the sepal length and width, and the petal length and width, respectively. Similarly, the last Virginia iris flower has row vector of attributes of $\begin{bmatrix} 46.3 & 2.5 & 5.0 & 1.9 \end{bmatrix}$ (cm), and the mean vector is the row vector $\begin{bmatrix} 5.81 & 3.09 & 3.69 & 1.22 \end{bmatrix}$ (cm). The reason for this mathematical transposition is that throughout science and engineering, data results are most often presented as rows of different instances of flowers, animals, clients, or experiments: each row contains the list of characteristic measured or derived properties/attributes. Table 5.2 has this most common structure. Thus in this sort of application, the mathematics we do needs to reflect this most common structure. Hence many vectors in this subsection appear as row vectors. When they do appear, they are called row vectors: the term vector on its own still means a column vector.

Definition 5.1.22 *(principal components) Given an $m \times n$ data matrix A (usually with zero mean when averaged over all rows) with SVD $A = USV^\mathsf{T}$, then the jth column v_j of V is called the jth **principal vector** and the vector $x_j := Av_j$ is called the jth **principal components** of the data matrix A.*

Now what does an SVD tell us for 2D plots of data? We know A_2 is the best rank two approximation to the data matrix A (Theorem 5.1.16). That is, if we are only to plot two components, those two components are best to come from A_2. Recall from (5.2) that

$$A_2 = US_2V^\mathsf{T} = \sigma_1 u_1 v_1^\mathsf{T} + \sigma_2 u_2 v_2^\mathsf{T} = (\sigma_1 u_1) v_1^\mathsf{T} + (\sigma_2 u_2) v_2^\mathsf{T}.$$

That is, in this best rank two approximation of the data, the row vector of attributes of the ith iris are the linear combination of row vectors $(\sigma_1 u_{i1}) v_1^\mathsf{T} + (\sigma_2 u_{i2}) v_2^\mathsf{T}$. The vectors v_1 and v_2 are orthonormal vectors so we treat them as the horizontal and vertical unit vectors of a scatter plot. That is, $x_i = \sigma_1 u_{i1}$ and $y_i = \sigma_2 u_{i2}$ are horizontal and vertical coordinates of the ith iris in the best 2D plot. Consequently, in MATLAB/Octave we draw a scatter plot of the components of vectors $x = \sigma_1 u_1$ and $y = \sigma_2 u_2$ (Figure 5.4).

Theorem 5.1.23   *Using the matrix norm to measure 'best' (Definition 5.1.7), the best k-dimensional summary of the $m \times n$ data matrix A (usually of zero mean) are the first k principal components in the directions of the first k principal vectors.*

Proof. Let $A = USV^T$ be an SVD of matrix A. For every $k < \text{rank} A$, Theorem 5.1.16 establishes that

$$A_k := US_k V^T = \sigma_1 u_1 v_1^T + \sigma_2 u_2 v_2^T + \cdots + \sigma_k u_k v_k^T$$

is the best rank k approximation to A in the matrix norm. Letting matrix $U = [u_{ij}]$, write the ith row of A_k as $(\sigma_1 u_{i1}) v_1^T + (\sigma_2 u_{i2}) v_2^T + \cdots + (\sigma_k u_{ik}) v_k^T$ and hence the transpose of each row of A_k lies in the kD subspace $\text{span}\{v_1, v_2, \ldots, v_k\}$. This establishes that these are principal vectors.

Since $\{v_1, v_2, \ldots, v_k\}$ is an orthonormal set, we now use them as standard unit vectors of a coordinate system for the kD subspace. From the above linear combination, the components of the ith data point approximation in this subspace coordinate system are $\sigma_1 u_{i1}, \sigma_2 u_{i2}, \ldots, \sigma_k u_{ik}$. That is, the jth coordinate for all data points, the principal components, is $\sigma_j u_j$. By postmultiplying the SVD $A = USV^T$ by orthogonal V, recall that $AV = US$ which is written in terms of the first $r = \text{rank} A$ columns is

$$\begin{bmatrix} Av_1 & Av_2 & \cdots & Av_r \end{bmatrix} = \begin{bmatrix} \sigma_1 u_1 & \sigma_2 u_2 & \cdots & \sigma_r u_r \end{bmatrix}.$$

Consequently, the vector $\sigma_j u_j$ of jth coordinates in the subspace are equal to Av_j, the principal components. $\qquad \square$

Activity 5.1.24 A given data matrix from some experiment has singular values 12.76, 10.95, 7.62, 0.95, 0.48, ... How many dimensions should you expect to be needed for a good view of the data?

(a) 4D   (b) 3D   (c) 1D   (d) 2D

Example 5.1.25 *(wine recognition)* From the Dua and Graff (2019) repository[4] download the data file `wine.data` and its description file `wine.names`. The wine data has 178 rows of different wine samples, and 14 columns of attributes of which the first column is the cultivar class number and the remaining 13 columns are the amounts of different chemicals measured in the wine. Question: is there a two-dimensional view of these chemical measurements that largely separates the cultivars?

Solution: Use an SVD to find the best two-dimensional, rank two, view of the data.

(a) Read in the 178×14 matrix of data into MATLAB/Octave with the commands

```
wine=csvread('wine.data')
[m,n]=size(wine)
scatter(wine(:,2),wine(:,3),[],wine(:,1))
```

---

[4] http://archive.ics.uci.edu/ml/datasets/Wine (Dua and Graff, 2019)

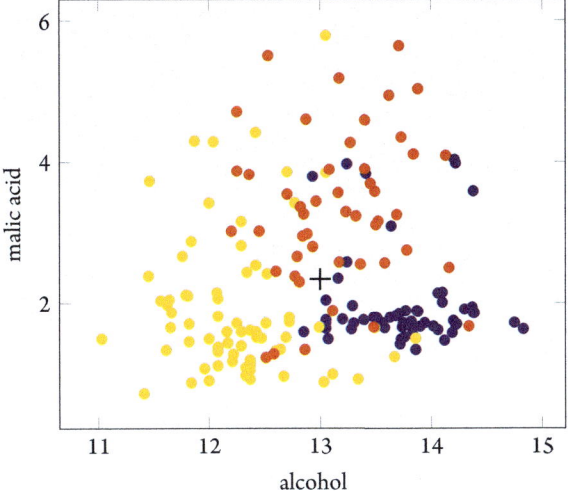

Figure 5.5 For the wine data of Example 5.1.25, a plot of the measured malic acid versus measured alcohol, and coloured depending upon the cultivar, shows that these measurements alone cannot effectively discriminate between the cultivars.

The scatter plot, Figure 5.5, shows that if we just plot the first two chemicals, alcohol and malic acid, then the three cultivars are inextricably intermingled. Our aim is to automatically draw Figure 5.6 in which the three cultivars are largely separated.

(b) To find the principal components of the wine chemicals it is best to subtract the mean. In MATLAB/Octave, recall that `mean(X)` computes the row vector of the mean/average of each column of X (Table 5.1). Then auto-replication (Table 5.1) provides a convenient method to do this subtraction as it replicates the row vector of means, to suit the number of rows in the data; thus invoke

```
A=wine(:,2:14)-mean(wine(:,2:14));
```

But now a further issue arises: the values in the columns are of widely different magnitudes; moreover, each column has different physical units (in contrast, the iris flower measurements were all cm). In practice we *must not* mix together quantities with different physical units. The general rule, after making each column zero mean, is to scale each column by dividing by its standard deviation, equivalently by its root-mean-square. This scaling does two practically useful things:

- since the standard deviation measures the spread of data in a column, it has the same physical units as the column of data, so dividing by it renders the results dimensionless, and so suitable for mixing with other scaled columns;
- also the spread of data in each column is now comparable to each other, namely around about magnitude one, instead of some columns being of the order of one-tenths and other columns being in the hundreds.

In MATLAB/Octave (Table 5.1), std(X) computes the row vector of the standard deviation of each column of X. Then auto-replication provides a convenient method to do the division by these standard deviations as it replicates the row vector to suit the number of rows in the data. Consequently, form the 178×13 matrix to analyse by the commands

```
A=wine(:,2:14)-mean(wine(:,2:14));
A=A./std(A);
```

Or both in one command

```
A=(wine(:,2:14)-mean(wine(:,2:14)))./std(wine(:,2:14)));
```

(c) Now compute and use an SVD $A = USV^{\mathsf{T}}$. But for low rank approximations we only ever use the first few singular values and the first few singular vectors. Thus it is pointless computing a full SVD which here has 178×178 matrix U and 13×13 matrix V.[5] Consequently, use [U,S,V]=svds(A,4) to economically compute only the first four singular values and singular vectors (change the number four to suit your purpose) to find (2 d.p.)

```
U = ...
S =
    28.86        0           0           0
        0      21.02         0           0
        0          0       16.00         0
        0          0           0       12.75
V =
   -0.14      0.48     -0.21     -0.02
    0.25      0.22      0.09      0.54
    0.00      0.32      0.63     -0.21
    0.24     -0.01      0.61      0.06
   -0.14      0.30      0.13     -0.35
   -0.39      0.07      0.15      0.20
   -0.42     -0.00      0.15      0.15
    0.30      0.03      0.17     -0.20
   -0.31      0.04      0.15      0.40
    0.09      0.53     -0.14      0.07
   -0.30     -0.28      0.09     -0.43
   -0.38     -0.16      0.17      0.18
   -0.29      0.36     -0.13     -0.23
```

where the ... indicates that we do not here need to know U.

(d) Recall that the orthonormal columns of this V are the principal vectors v_1, v_2, ..., v_4, and the jth principal components of the data are $x_j = Av_j$. We form a 2D plotted view of the data, Figure 5.6, by drawing a scatter plot of the first two principal components with

---

[5] Yes, on modern computers this is here done within a millisecond. But for modern datasets with thousands to billions of rows a full SVD is infeasible, so let's see how to analyse such modern large datasets.

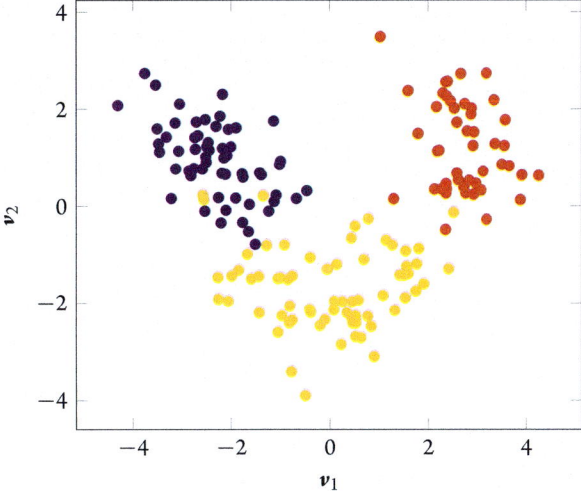

Figure 5.6 For the wine data of Example 5.1.25, a plot of the first two principal components almost entirely separates the three cultivars.

```
scatter(A*V(:,1),A*V(:,2),[],wine(:,1))
```

Figure 5.6 shows these two principal components do an amazingly good job of almost completely disentangling the three wine cultivars (use `scatter3()` to explore the first three principal components). ☐

The previous three examples develop the following procedure for 'best' viewing data in low dimensions. However, any additional information about the data or about preferred results may modify this procedure.

Procedure 5.1.26 (principal component analysis) *Consider the case when you have data values consisting of n attributes for each of m instances, and the aim is to find a good k-dimensional summary/ view of the data.*

1. *Form/enter the $m \times n$ data matrix B.*

2. *Scale the data matrix B to form $m \times n$ matrix A:*

   (a) *usually make each column have zero mean by subtracting its mean $\bar{b}_j$, algebraically*
   $$a_j = b_j - \bar{b}_j \, ;$$

   (b) *but ensure each column has the same 'physical dimensions', usually by dividing by the standard deviation s_j of each column, algebraically $a_j = (b_j - \bar{b}_j)/s_j$.*

   Compute in Matlab/Octave *using auto-replication:*

   `A=(B-mean(B))./std(B)`

3. *Economically compute an* SVD *for the best rank k approximation to the scaled data matrix with* [U,S,V]=svds(A,k).

4. *Then the jth column of* V *is the jth principal vector, and the principal components are the entries of the* $m \times k$ *matrix* A*V.

Courses on multivariate statistics prove that, for every (usually zero mean) data matrix A, the first k principal vectors v_1, v_2, ..., v_k are orthonormal vectors that *maximize the total variance* in the principal components $x_j = Av_j$; that is, that maximize $|x_1|^2 + |x_2|^2 + \cdots + |x_k|^2$. Indeed, this maximization of the variance corresponds closely to the constructive proof of the existence of SVDs (Section 3.3.3) which successively maximizes $|Av|$ subject to v being orthonormal to the singular/principal vectors already determined. Consequently, when data is approximated in the space of the first k principal vectors, then the data is the most spread out it can be in k dimensions. When the data is most spread out in kD, then (roughly) it retains the most information possible in kD.

Application to latent semantic indexing

> This ability to retrieve relevant information based upon meaning rather than literal term usage is the main motivation for using LSI [latent semantic indexing].   *Berry et al. (1995)*

Information searches based upon word matching results in surprisingly poor retrieval of relevant documents (Berry et al., 1995, §5.5). Instead, the so-called method of latent semantic indexing improves retrieval by replacing individual words with nearness of word vectors. The word vectors being derived via the singular value decomposition. This section introduces such latent semantic indexing via a very small example.

The Society for Industrial and Applied Mathematics (SIAM) reviews many mathematical books. In 2015, six of those books had the following titles:

1 Introduction to Finite and Spectral Element Methods using MATLAB

2 Iterative Methods for Linear Systems: Theory and Applications

3 Singular Perturbations: Introduction to System Order Reduction Methods with Applications

4 Risk and Portfolio Analysis: Principles and Methods

5 Stochastic Chemical Kinetics: Theory and Mostly Systems Biology Applications

6 Quantum Theory for Mathematicians

Consider the capitalized words. For those words that appear in more than one title, let's form a word vector (Example 1.1.7) for each title, then use principal components to summarize these six books on a 2D plane. This task is part of what is called latent semantic indexing (Berry et al., 1995). (We should also count words that are used only once, but for simplicity this example omits such once-used words.)

Follow the principal component analysis Procedure 5.1.26.

1. First find the set of words that are used more than once. Ignoring pluralization, they are, in alphabetical order, Application, Introduction, Method, System, Theory. The corresponding word vector for each book title is then the following:

- $w_1 = (0,1,1,0,0)$ *Introduction* to Finite and Spectral Element *Methods* using MAT-LAB
- $w_2 = (1,0,1,1,1)$ Iterative *Methods* for Linear *Systems: Theory* and *Applications*
- $w_3 = (1,1,1,1,0)$ Singular Perturbations: *Introduction* to *System* Order Reduction *Methods* with *Applications*
- $w_4 = (0,0,1,0,0)$ Risk and Portfolio Analysis: Principles and *Methods*
- $w_5 = (1,0,0,1,1)$ Stochastic Chemical Kinetics: *Theory* and Mostly *Systems* Biology *Applications*
- $w_6 = (0,0,0,0,1)$ Quantum *Theory* for Mathematicians

2. Second, form the data matrix with w_1, w_2, ..., w_6 as rows (not columns). We could remove the mean word vector, but choose not to: here the position of each book title relative to an empty title (the origin) is interesting. There is no need to scale each column as each column has the same 'physical' dimensions, namely a word count. The data matrix of word vectors is then

$$A = \begin{bmatrix} 0 & 1 & 1 & 0 & 0 \\ 1 & 0 & 1 & 1 & 1 \\ 1 & 1 & 1 & 1 & 0 \\ 0 & 0 & 1 & 0 & 0 \\ 1 & 0 & 0 & 1 & 1 \\ 0 & 0 & 0 & 0 & 1 \end{bmatrix}.$$

3. Third, to compute a representation in the 2D plane, principal components uses, as an orthonormal basis, the singular vectors corresponding to the two largest singular values. So compute the economical SVD with [U, S, V] =svds (A, 2) giving (2 d.p.)

```
U = ...
S =
     3.14        0
        0     1.85
V =
   +0.52    -0.20
   +0.26    +0.52
   +0.50    +0.57
   +0.52    -0.20
   +0.37    -0.57
```

4. Columns of V are word vectors in the 5D space of counts of Application, Introduction, Method, System, and Theory. The two given columns of $V = \begin{bmatrix} v_1 & v_2 \end{bmatrix}$ are the two orthonormal principal vectors:

- the first v_1, from its largest components, mainly identifies the overall direction of Application, Method, and System;
- whereas the second v_2, from its largest positive and negative components, mainly distinguishes Introduction and Method from Theory.

The corresponding principal components are the entries of the 6×2 matrix

$$AV = \begin{bmatrix} 0.76 & 1.09 \\ 1.92 & -0.40 \\ 1.80 & 0.69 \\ 0.50 & 0.57 \\ 1.41 & -0.97 \\ 0.37 & -0.57 \end{bmatrix} :$$

for each of the six books, the book title has components in the two principal directions given by the corresponding row in this product. We plot the six books on a 2D plane with the MATLAB/Octave command

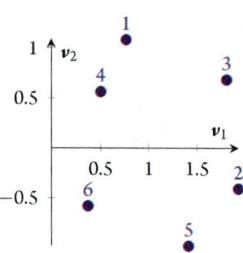

```
scatter(A*V(:,1),A*V(:,2),[],1:6)
```

to produce a picture like that to the right. The SVD analysis nicely distributes the six books in this plane.

The above procedure would approximate the original word vector data, formed into a matrix, by the following rank two matrix (2 d.p.)

$$A_2 = US_2 V^{\mathsf{T}} = \begin{bmatrix} 0.18 & 0.77 & 1.01 & 0.18 & -0.33 \\ 1.08 & 0.29 & 0.74 & 1.08 & 0.95 \\ 0.80 & 0.82 & 1.30 & 0.80 & 0.28 \\ 0.15 & 0.43 & 0.58 & 0.15 & -0.14 \\ 0.93 & -0.14 & 0.16 & 0.93 & 1.08 \\ 0.31 & -0.20 & -0.14 & 0.31 & 0.46 \end{bmatrix}.$$

The largest components in each row do correspond to the ones in the original word vector matrix A. However, in this application we work with the representation in the low-dimensional, 2D, subspace spanned by the first two principal vectors v_1 and v_2.

Angles measure similarity Recall that Example 1.3.9 introduced using the dot product to measure the similarity between word vectors. We could use the dot product in the 5D space of the word vectors to find the 'angles' between the book titles. However, we know that the 2D view just plotted is the 'best' 2D summary of the book titles, so we could more economically estimate the angle between book titles using just the 2D summary.

Example 5.1.27 What is the 'angle' between the first two listed books?

- Introduction to Finite and Spectral Element Methods using MATLAB
- Iterative Methods for Linear Systems: Theory and Applications

Solution: Find the angle in two ways.

(a) First, the corresponding 5D word vectors are $w_1 = (0,1,1,0,0)$ and $w_2 = (1,0,1,1,1)$, with lengths $|w_1| = \sqrt{2}$ and $|w_2| = \sqrt{4} = 2$. The dot product then determines

$$\cos\theta = \frac{w_1 \cdot w_2}{|w_1||w_2|} = \frac{0 + 0 + 1 + 0 + 0}{2\sqrt{2}} = 0.3536.$$

Hence the angle $\theta = 69.30°$.

(b) Secondly, estimate the angle using the 2D view. For these two books the principal component vectors are $(0.76, 1.09)$ and $(1.92, -0.40)$, respectively, with lengths 1.33 and 1.96 (2 d.p.). The dot product gives

$$\cos\theta \approx \frac{(0.76, 1.09) \cdot (1.92, -0.40)}{1.33 \cdot 1.96} = \frac{1.02}{2.61} = 0.39.$$

Hence the angle $\theta \approx 67°$ which is effectively the same as the first exact calculation.

Because of the relatively large 'angle' between these two book titles, we deduce that the two books are quite dissimilar. □

We can also use the 2D plane to economically measure similarity between the book titles and any other title or words of interest.

Example 5.1.28 Let's ask which of the six books is 'closest' to a book about Applications.

Solution: The word Application has word vector $w = (1,0,0,0,0)$. So we could do some computations in the original 5D space of word vectors finding precise angles between this word vector and the word vectors of all titles. Alternatively, let's draw a picture in 2D. The Application word vector w projects onto the 2D plane of principal components by computing $w \cdot v_1 = w^T v_1$ and $w \cdot v_2 = w^T v_2$, that is, $w^T V$. Here the Application word vector $w = (1,0,0,0,0)$, so $w^T V = \begin{bmatrix} 0.52 & -0.20 \end{bmatrix}$, as plotted to the right.

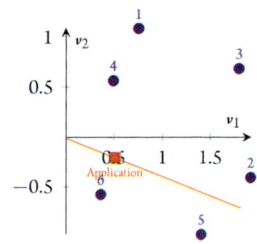

Which of the six books makes the smallest angle with the line through $(0.52, -0.20)$? Visually, books 2 and 5 are closest, and book 2 appears to have slightly smaller angle to the line than book 5. On this data, we deduce that closest to "Application" is book 2: "Iterative Methods for Linear Systems: Theory and Applications" □

Search for information from more books Berry et al. (1995) reviewed the application of the SVD to the problem of searching for information. Let's explore this further with more data, albeit still very restricted. Berry et al. (1995) listed some mathematical books including the following fourteen titles.

1. A Course on Integral Equations
2. Automatic Differentiation of Algorithms: Theory, Implementation, and Application
3. Geometrical Aspects of Partial Differential Equations
4. Introduction to Hamiltonian Dynamical Systems and the n-Body Problem
5. Knapsack Problems: Algorithms and Computer Implementations
6. Methods of Solving Singular Systems of Ordinary Differential Equations

7. Nonlinear Systems

8. Ordinary Differential Equations

9. Oscillation Theory of Delay Differential Equations

10. Pseudodifferential Operators and Nonlinear Partial Differential Equations

11. Sinc Methods for Quadrature and Differential Equations

12. Stability of Stochastic Differential Equations with Respect to Semi-Martingales

13. The Boundary Integral Approach to Static and Dynamic Contact Problems

14. The Double Mellin–Barnes Type Integrals and their Applications to Convolution Theory

Principal component analysis summarizes and relates these titles. Follow Procedure 5.1.26.

1. The significant (capitalized) words which appear more than once in these titles (ignoring pluralization) are the fourteen words, in alphabetical order,

$$\text{Algorithm, Application, Differential/tion, Dynamic/al, Equation,} \qquad (5.3)$$
Implementation, Integral, Method, Nonlinear, Ordinary, Partial, Problem, System, and Theory.

With this dictionary of significant words, the titles have the following word vectors.

- $w_1 = (0,0,0,0,1,0,1,0,0,0,0,0,0,0)$ a Course on Integral Equations
- $w_2 = (1,1,1,0,0,1,0,0,0,0,0,0,0,1)$ Automatic Differentiation of Algorithms: Theory, Implementation, and Application
- ...
- $w_{14} = (0,1,0,0,0,0,1,0,0,0,0,0,0,1)$ the Double Mellin–Barnes Type Integrals and their Applications to Convolution Theory

2. Form the 14×14 data matrix with the word count for each title in rows

$$A = \begin{bmatrix} 0 & 0 & 0 & 0 & 1 & 0 & 1 & 0 & 0 & 0 & 0 & 0 & 0 & 0 \\ 1 & 1 & 1 & 0 & 0 & 1 & 0 & 0 & 0 & 0 & 0 & 0 & 0 & 1 \\ 0 & 0 & 1 & 0 & 1 & 0 & 0 & 0 & 0 & 0 & 1 & 0 & 0 & 0 \\ 0 & 0 & 0 & 1 & 0 & 0 & 0 & 0 & 0 & 0 & 0 & 1 & 1 & 0 \\ 1 & 0 & 0 & 0 & 0 & 1 & 0 & 0 & 0 & 0 & 0 & 1 & 0 & 0 \\ 0 & 0 & 1 & 0 & 1 & 0 & 0 & 1 & 0 & 1 & 0 & 0 & 1 & 0 \\ 0 & 0 & 0 & 0 & 0 & 0 & 0 & 0 & 1 & 0 & 0 & 0 & 1 & 0 \\ 0 & 0 & 1 & 0 & 1 & 0 & 0 & 0 & 0 & 1 & 0 & 0 & 0 & 0 \\ 0 & 0 & 1 & 0 & 1 & 0 & 0 & 0 & 0 & 0 & 0 & 0 & 0 & 1 \\ 0 & 0 & 1 & 0 & 1 & 0 & 0 & 0 & 1 & 0 & 1 & 0 & 0 & 0 \\ 0 & 0 & 1 & 0 & 1 & 0 & 0 & 1 & 0 & 0 & 0 & 0 & 0 & 0 \\ 0 & 0 & 1 & 0 & 1 & 0 & 0 & 0 & 0 & 0 & 0 & 0 & 0 & 0 \\ 0 & 0 & 0 & 1 & 0 & 0 & 1 & 0 & 0 & 0 & 0 & 1 & 0 & 0 \\ 0 & 1 & 0 & 0 & 0 & 0 & 1 & 0 & 0 & 0 & 0 & 0 & 0 & 1 \end{bmatrix}.$$

Each row corresponds to a book title, and each column corresponds to a word.

3. To compute a representation of the titles in 3D space, principal components uses, as an orthonormal basis, the singular vectors corresponding to the three largest singular values. So in MATLAB/Octave compute the economical SVD with [U,S,V]=svds(A,3) giving (2 d.p.)

```
U = ...
S =
     4.20        0        0
        0     2.65        0
        0        0     2.36
V =
     0.07     0.40     0.14
     0.07     0.38     0.25
     0.65     0.00     0.15
     0.01     0.23    -0.46
     0.64    -0.21    -0.07
     0.07     0.40     0.14
     0.06     0.30    -0.18
     0.19    -0.09    -0.12
     0.10    -0.05    -0.11
     0.19    -0.09    -0.12
     0.17    -0.09     0.02
     0.02     0.40    -0.50
     0.12     0.05    -0.48
     0.16     0.41     0.32
```

4. The three orthonormal columns of V are word vectors in the 14D space of counts of the dictionary words (5.3) Algorithm, Application, Differential, Dynamic, Equation, Implementation, Integral, Method, Nonlinear, Ordinary, Partial, Problem, System, and Theory.

   - The first column v_1 of V, from its largest components, mainly identifies the two most common words of Differential and Equation.
   - The second column v_2 of V, from its largest components, identifies books with Algorithms, Applications, Implementations, Problems, and Theory.
   - The third column v_3 of V, from its largest components, largely distinguishes Dynamics, Problems, and Systems from Differential and Theory.

The corresponding principal components are the entries of the 14×3 matrix (2 d.p.)

$$AV = \begin{bmatrix} 0.70 & 0.09 & -0.25 \\ 1.02 & 1.59 & 1.00 \\ 1.46 & -0.29 & 0.10 \\ 0.16 & 0.67 & -1.44 \\ 0.16 & 1.19 & -0.22 \\ 1.78 & -0.34 & -0.64 \\ 0.22 & -0.00 & -0.58 \\ 1.48 & -0.29 & -0.04 \\ 1.45 & 0.21 & 0.40 \\ 1.56 & -0.34 & -0.01 \\ 1.48 & -0.29 & -0.04 \\ 1.29 & -0.20 & 0.08 \\ 0.10 & 0.92 & -1.14 \\ 0.29 & 1.09 & 0.39 \end{bmatrix}.$$

Each of the fourteen books is represented in 3D space by the corresponding row of these coordinates. Plot these books in MATLAB/Octave with

```
scatter3 (A*V(:,1),A*V(:,2),A*V(:,3), [],1:14)
```

as shown below in stereo.

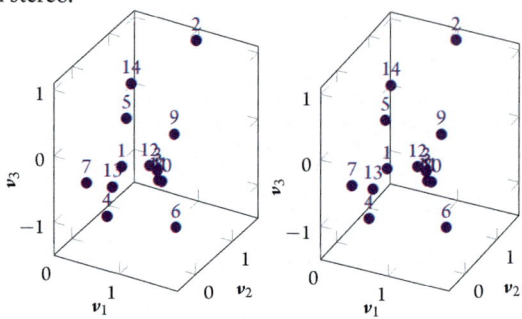

There is a cluster of five books near the front along the v_1-axis (numbered 3, 8, 10, 11, and 12, their focus is Differential Equations), the other nine are spread out.

Queries Suppose we search for books on *Application and Theory*. In our dictionary (5.3), the corresponding word vector for this search is $w = (1,0,0,0,0,0,0,0,0,0,0,0,0,1)$. Project this query into the 3D space of principal components with the product $w^T V$ which evaluates to the query vector $q = (0.22, 0.81, 0.46)$ whose direction is added to the picture as shown to the right.

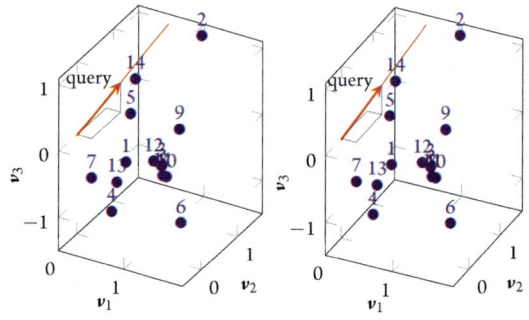

Books 2 and 14 appear close to the direction of the query vector and so should be returned as a match: these books are no surprise as each has both *Application* and *Theory* in its title. But the above plot also suggests that Book 5 is near to the direction of the query vector, and so is also worth considering despite not having either of the search words in its title! The power of this latent semantic indexing is that it extracts additional titles that are relevant to the query yet share no common words with the query—as commented at the start of this section.

The angles, in 3D, between this query vector and the book title vectors confirm the graphical appearance claimed above. Recall that the dot product determines the angle between vectors (Theorem 1.3.5).

- From the second row of the above product AV, Book 2 has the principal component vector $(1.02, 1.59, 1.00)$ which has length 2.14. Consequently, it is at small angle 15° to the 3D query vector $q = (0.22, 0.81, 0.46)$, of length $|q| = 0.96$, because its cosine

$$\cos \theta = \frac{(1.02, 1.59, 1.00) \cdot q}{2.14 \cdot 0.96} = 0.97.$$

- Similarly, Book 14 has the principal component vector $(0.29, 1.09, 0.39)$ which has length 1.20. Consequently, it is at small angle $10°$ to the 3D query vector $q = (0.22, 0.81, 0.46)$ because its cosine

$$\cos\theta = \frac{(0.29, 1.09, 0.39) \cdot q}{1.20 \cdot 0.96} = 0.99.$$

- Whereas Book 5 has the principal component vector $(0.16, 1.19, -0.22)$ which has length 1.22. Consequently, it is at moderate angle $40°$ to the 3D query vector $q = (0.22, 0.81, 0.46)$ because its cosine

$$\cos\theta = \frac{(0.16, 1.19, -0.22) \cdot q}{1.20 \cdot 0.96} = 0.76.$$

Such a significant cosine suggests that Book 5 is also of interest.

If we were to compute the angles in the original 14D space of the full dictionary (5.3), then the title of Book 5 would be orthogonal to the query, because it has no words in common, and so Book 5 would not be flagged as of interest. The principal component analysis reduces the dimensionality to those relatively few directions that are important, and it is in these important directions that the title of Book 5 appears promising for the query.

- All the other book titles have angles greater than $62°$ and so are significantly less related to the query.

Latent semantic indexing in practice This application of principal components to analysing a few book titles is purely indicative. In practice, one would analyse the many thousands of words used throughout hundreds or thousands of documents. Moreover, one would be interested in not just plotting the documents in a 2D plane or 3D space, but in representing the documents in say a 70D space of 70 principal components. Berry et al. (1995) reviews how such statistically derived principal word vectors are a more robust indicator of meaning than individual terms. Hence this SVD analysis of documents becomes an effective way of retrieving information from a search without requiring the results actually match any of the words in the search request—the results just need to match cognate words.

5.1.4 Exercises

Exercise 5.1.1 For some distinct 2×2 matrices A the following plots adjoin the product Ax to x for a complete range of unit vectors x. Use each plot to roughly estimate the norm of the underlying matrix A for that plot.

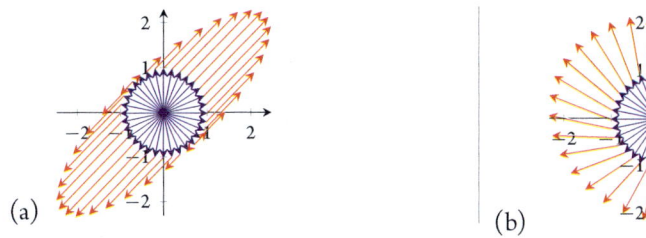

(a)

(b)

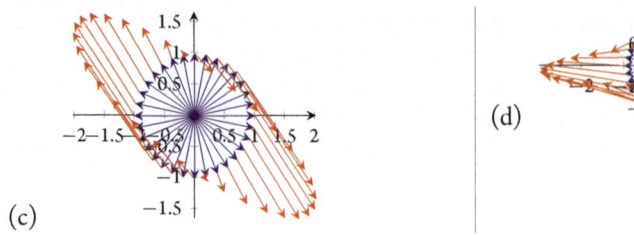

(c)

(d)

Exercise 5.1.2 For the following matrices, use a few unit vectors x to illustrate how the matrix-vector product varies with x. Using calculus to find the appropriate maximum, find from definition the norm of the matrices (hint: all norms here are integers).

(a) $\begin{bmatrix} 2 & -3 \\ 0 & 2 \end{bmatrix}$

(b) $\begin{bmatrix} -4 & -1 \\ -1 & -4 \end{bmatrix}$

(c) $\begin{bmatrix} 2 & -1 \\ 1 & -2 \end{bmatrix}$

(d) $\begin{bmatrix} 2 & -2 \\ 2 & 1 \end{bmatrix}$

Exercise 5.1.3 Many properties of a norm may be proved in other ways than those given for Theorem 5.1.12.

(a) Use an SVD factorization of I_n to prove Theorem 5.1.12(b).

(b) Use the existence of an SVD factorization of A to prove Theorem 5.1.12(d).

(c) Use the definition of a norm as a maximum to prove Theorem 5.1.12(f).

(d) Use the existence of an SVD factorization of A to prove Theorem 5.1.12(g).

(e) Prove Theorem 5.1.12(h) using 5.1.12(g) and the definition of a norm as a maximum.

Exercise 5.1.4 Let $m \times n$ matrix A have in each row and column at most one nonzero element. Argue that there exists an SVD which establishes that the norm $\|A\| = \max_{i,j} |a_{ij}|$.

Exercise 5.1.5 The left-hand picture to the right shows a 7×5 pixel image of the letter G. Compute an SVD of the pixel image. By inspecting various rank approximations from this SVD, determine the rank of the approximation to G shown in the right-hand picture to the right.

Exercise 5.1.6 Write down two different rank two representations of the pixel image of the letter L, as shown to the right. Compute SVD representations of the letter L. Compare and comment on the various representations.

Exercise 5.1.7 *(Sierpinski triangle)* Whereas mostly we deal with the smooth geometry of lines, planes, and curves, the subject of fractal geometry recognizes that there is much in the world around us that has a rich fractured structure: from clouds, and rocks, to the cardiovascular system within each of us. The Sierpinski triangle, illustrated to the right, is a simple fractal. Generate such fractals using recursion as in the following MATLAB/Octave code:[6] the recursion is that the next generation image A is computed from the previous generation A.

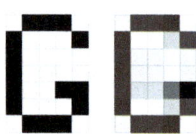

---

[6] Many of you may know that a for-loop would more concisely compute the recursion; if so, then do so.

```
A=1
A=[A 0*A;A A]
A=[A 0*A;A A]
A=[A 0*A;A A]
A=[A 0*A;A A]
imagesc(1-A)
colormap('gray')
axis equal,axis off
```

(a) Add code to this recursion to compute and print the singular values of each generation of the Sierpinski triangle image. What do you conjecture about the number of distinct singular values as a function of generation number k? Test your conjecture for more iterations in the recursion.

(b) Returning to the 16×16 Sierpinski triangle formed after four iterations, use an SVD to form the best rank five approximation of the Sierpinski triangle (as illustrated to the right, it has a beautiful structure). Comment on why such a rank five approximation may be a reasonable one to draw. What is the next rank for an approximation that is reasonable to draw? Justify your answer.

(c) Modify the code to compute and draw the 256×256 image of the Sierpinski triangle. Use an SVD to generate and draw the best rank nine approximation to the image.

Exercise 5.1.8 *(Sierpinski carpet)* The Sierpinski carpet is another fractal which is easily generated by recursion (as illustrated to the right after three iterations in the recursion).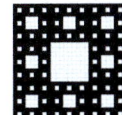

(a) Invent a modification to the recursive MATLAB/Octave code of the previous Exercise 5.1.7 to generate such an image.

(b) For a range of generations of the image, compute the singular values and comment on the apparent patterns in the singular values.

Exercise 5.1.9 *(another fractal)* Illustrated to the right is another fractal generated by recursion.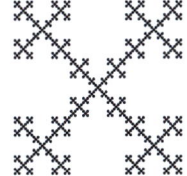

(a) Invent a modification to the recursive MATLAB/Octave code of Exercise 5.1.7 to generate such an image.

(b) For a range of generations of the image, compute the singular values and comment on the apparent patterns in the singular values.

Exercise 5.1.10 Countess Ada Lovelace (1815–52) is recognized as the first computer programmer. She invented, developed and wrote programs for Charles Babbage's analytical engine. Download the 249×178 image of Ada Lovelace at the web address below.[7] Using an SVD, draw various rank approximations to the image. Using the matrix norm to measure errors, what is the smallest rank to reproduce the image of Ada Lovelace to an error of 5%? and of 1%?

---

[7] https://www.maa.org/press/periodicals/convergence/portrait-gallery [Oct 2019] provides many other portraits to be downloaded for corresponding exercises.

http://www.maa.org/sites/default/files/images/upload_library/46/Portraits/Lovelace_Ada.jpg [Oct 2019]

Exercise 5.1.11 As in Example 5.1.28, consider the 2D plot of the six books. Add to the plot the word vector corresponding to a query for books relevant to "Introduction". Using angle to measure closeness, which book is closest to "Introduction"? Confirm by computing the angles, in the 2D plane, between "Introduction" and all six books.

Exercise 5.1.12 As in Example 5.1.28, consider the 2D plot of the six books. Add to the plot the word vector corresponding to a query for books relevant to "Application and Method". Using angle to measure closeness, which book is closest to "Application and Method"? Confirm by computing, in the 2D plane, the angles between "Application and Method" and all six books.

Exercise 5.1.13 Reconsider the word vectors of the group of 14 mathematical books listed in the text. Instead of computing a representation of these books in 3D space, here use principal component analysis to compute a representation in a 2D plane.

(a) Plot the titles of the 14 books in the 2D plane of the two principal vectors.
(b) Add the vector corresponding to the query of *Differential and Equation*: which books appear to have a small angle to this query in this 2D representation?
(c) Add the vector corresponding to the query of *Application and Theory*: which books appear to have a small angle to this query in this 2D representation?

Exercise 5.1.14 The discussion on latent semantic indexing only considered queries which were "and", such as a search for books relevant to *Application and Theory*. What if we wanted to search for books relevant to *either Application or Theory*? Discuss how such an "or" query might be phrased in terms of the angle between vectors and a multi-D subspace.

Exercise 5.1.15 Table 5.3 lists 20 short reviews about bathrooms of a major chain of hotels. There are about 17 meaningful words common to more than one review: create a list of such words. Then form corresponding word vectors for each review. Use an SVD to best plot these reviews in 2D. Discuss any patterns in the results.

Exercise 5.1.16 In a few sentences, answer/discuss each of the following.

(a) What is it about an SVD that makes it useful in compressing images?
(b) Why does the matrix norm behave like a measure of the magnitude of a matrix? What else might be used as a measure?
(c) Why is principal component analysis important?
(d) In a principal component analysis, via the SVD $A = USV^T$, why are the principal vectors the columns of V and not the columns of U?
(e) What causes the first few principal components of some data to often form the basis for a good view of the data?
(f) How is it that we can discuss and analyse the "angle between books"?
(g) When searching for books with a specified keyword, what is it that enables latent semantic indexing to return book titles that do not have the keyword?

Table 5.3 Twenty user reviews of bathrooms in a major chain of hotels. The data is part of the Opinosis Opinion/Review (Ganesan et al., 2010) in the UCI Machine Learning Repository (Dua and Graff, 2019) [https://archive.ics.uci.edu/ml/datasets/Opinosis+Opinion+%26frasl%3B+Review].

- The room was not overly big, but clean and very comfortable beds, a great shower and very clean bathrooms
- The second room was smaller, with a very inconvenient bathroom layout, but at least it was quieter and we were able to sleep
- Large comfortable room, wonderful bathroom
- The rooms were nice, very comfy bed and very clean bathroom
- Bathroom was spacious too and very clean
- The bathroom only had a single sink, but it was very large
- The room was a standard but nice motel room like any other, bathroom seemed upgraded if I remember
- The room was quite small but perfectly formed with a super bathroom
- You could eat off the bathroom floor it was so clean
- The bathroom door does the same thing, making the bathroom seem slightly larger
- bathroom spotless and nicely appointed
- The rooms are exceptionally clean and also the bathrooms
- The bathroom was clean and the bed was comfy
- They provide you with great aveda products in the bathroom
- Also, the bathroom was a bit dirty , brown water came out of the bath tub faucet initially and the sink wall by the toilet was dirty
- If your dog tends to be a little disruptive or on the noisy side, there is a bathroom fan that you can keep on to make noise
- The bathroom was big and clean as well
- Also, the bathrooms were quite well set up, with a separate toilet shower to basin, so whilst one guest is showering another can use the basin
- The bathroom was marble and we had luxurious bathrobes and really, every detail attended to
- It was very clean, had a beautiful bathroom, and was comfortable

5.2 Regularize linear equations

Singularity is almost invariably a clue.

Sherlock Holmes, in The Boscombe Valley Mystery, by Sir Arthur Conan Doyle, 1892

Often we need to approximate the matrix in a linear equation. Such approximation is especially likely when the matrix itself comes from experimental measurements and so has errors. We do not want such errors to affect results. By avoiding division with small singular values, the procedure developed in this section avoids unwarranted magnification of errors. Sometimes such error magnification is disastrous, so avoiding it is essential.

Example 5.2.1 Suppose from measurements in some experiment we want to solve the two linear equations

$$0.5x + 0.3y = 1 \quad \text{and} \quad 1.1x + 0.7y = 2,$$

where all the coefficients on *both* of the left-hand sides and the right-hand sides are determined from experimental measurements. Suppose they are measured to experimental errors ± 0.05. Solve the equations.

Solution: Using Procedure 2.2.5 in MATLAB/Octave, form the matrix and the right-hand side

```
A= [0.5 0.3;1.1 0.7]
b= [1.0;2.0]
```

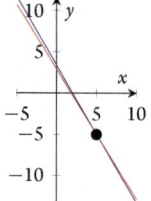

Then check the condition number with `rcond(A)` to find that it is 0.007 which previously we would call only just outside the 'good' range (Procedure 2.2.5). So proceed warily to compute the solution with A\b to find $(x, y) = (5, -5)$ (as illustrated by the intersection of the two lines to the above-right).

Is this solution reasonable? No. Not when the matrix itself has errors. Let's perturb the matrix A by amounts consistent with its experimental error of ± 0.05 and explore the predicted solutions (the first two perturbations illustrated):

$$A = \begin{bmatrix} 0.47 & 0.29 \\ 1.06 & 0.68 \end{bmatrix} \implies x = A\backslash b = \begin{bmatrix} 8.2 \\ -9.8 \end{bmatrix};$$

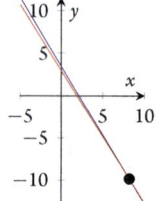

$$A = \begin{bmatrix} 0.45 & 0.32 \\ 1.05 & 0.67 \end{bmatrix} \implies x = A\backslash b = \begin{bmatrix} -0.9 \\ 4.3 \end{bmatrix};$$

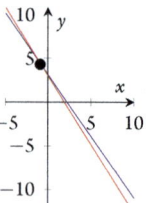

$$A = \begin{bmatrix} 0.46 & 0.31 \\ 1.06 & 0.73 \end{bmatrix} \implies x = A\backslash b = \begin{bmatrix} 15.3 \\ -19.4 \end{bmatrix}.$$

For equally valid matrices A, to within experimental error, the predicted solutions are all over the place!

If the matrix itself has errors, then we *must* reconsider Procedure 2.2.5. The SVD empowers a sensible resolution. Compute an SVD of the matrix, $A = USV^T$, with $[U,S,V] = \text{svd}(A)$ to find (2 d.p.)

$$A = \begin{bmatrix} -0.41 & -0.91 \\ -0.91 & 0.41 \end{bmatrix} \begin{bmatrix} 1.43 & 0 \\ 0 & 0.01 \end{bmatrix} \begin{bmatrix} -0.85 & -0.53 \\ -0.53 & 0.85 \end{bmatrix}^T$$

Because the matrix A has errors ± 0.05, the small singular value of 0.01 might as well be zero: it is zero to experimental error. That is, because the matrix A_1 is a distance $\|A - A_1\| = 0.01$ away from A, and this distance is less than the experimental error of 0.05, then it is better to solve the equation with the singular A_1 instead of the original A. The appropriate solution algorithm is then Procedure 3.5.4 for inconsistent equations, not Procedure 2.2.5. Thus, using the above SVD, (2 d.p.)

(a) $z = U^T b = (-2.23, -0.10)$;

(b) due to the effectively zero singular value, neglect $z_2 = -0.10$ as an error, and solve $\begin{bmatrix} 1.43 & 0 \\ 0 & 0 \end{bmatrix} y = \begin{bmatrix} -2.23 \\ 0 \end{bmatrix}$ to deduce $y = (-1.56, y_2)$;

(c) consequently, we find that reasonable solutions are $x = Vy = (1.32, 0.83) + y_2(-0.53, 0.85)$ —the parametric equation of a line (Section 1.2.3).

The four different 'answers' computed by A\b above are just four different points (nearly) on this line. Other different 'answers' computed by A\b would be other points (nearly) on this line.

To choose between this infinitude of solutions on the line, extra information must be provided by the context/application/modelling. For example, often one prefers the solution of smallest length/magnitude, obtained by setting $y_2 = 0$ (Theorem 3.5.13); that is, often one prefers $x_{\text{smallest}} = (1.32, 0.83)$. $\qquad\square$

Activity 5.2.2 The coefficients in the following pair of linear equations are obtained from an experiment and so the coefficients have errors of roughly ± 0.05:

$$0.8x + 1.1y = 4, \quad 0.6x + 0.8y = 3.$$

By checking how well the equations are satisfied, which of the following *cannot* be a plausible solution (x, y) of the pair of equations?

(a) $(6.6, -1.2)$ (b) $(5, 0)$ (c) $(5.6, 0.8)$ (d) $(3.6, 1)$

5.2.1 The SVD illuminates regularization

I think it is much more interesting to live with uncertainty than to live with answers that might be wrong.
Richard Feynman

Procedure 5.2.3 (approximate linear equations) *Suppose the system of linear equations $Ax = b$ arises from an experiment where both the $m \times n$ matrix A and the right-hand side vector b are subject to experimental error. Suppose the expected error in the matrix and vector entries are of magnitude ϵ. (Recall that (Theorem 3.3.30) the symbol ϵ is the Greek letter epsilon, and often denotes errors.)*

1. *When forming the matrix A and vector b, scale the data so that*
   - *all $m \times n$ components in A have the same physical units, and they are of roughly the same magnitude; and*
   - *similarly for the m components of b.*
   Estimate the error ϵ corresponding to this matrix A.

2. *Compute an SVD $A = USV^\mathsf{T}$.*

3. *Choose 'rank' k to be the number of singular values bigger than the error ϵ; that is, $\sigma_1 \geq \sigma_2 \geq \cdots \geq \sigma_k > \epsilon > \sigma_{k+1} \geq \cdots \geq 0$. Then the rank k approximation to A is*

$$A_k := US_kV^\mathsf{T}$$
$$= \sigma_1 u_1 v_1^\mathsf{T} + \sigma_2 u_2 v_2^\mathsf{T} + \cdots + \sigma_k u_k v_k^\mathsf{T}$$
$$= \text{U}(:,1:k)*\text{S}(1:k,1:k)*\text{V}(:,1:k)'.$$

   But do not construct A_k, as we only need its SVD to solve the system.

4. *Solve the approximating linear equation $A_k x = b$ as in Theorems 3.5.8 and 3.5.13 (often as an inconsistent set of equations). Usually use the SVD $A_k = US_kV^\mathsf{T}$.*

5. *Among all the solutions allowed, choose the 'best', according to some explicit additional need of the application—often the smallest solution overall, or just as often a solution with the most zero components.*

That is, the procedure is to treat as zero all singular values smaller than the expected error in the matrix entries. For example, modern computers have nearly 16 significant decimal digits accuracy, so even in 'exact' computation there is a background relative error of about 10^{-15}. Consequently, in computation on modern computers, every singular value smaller than $10^{-15}\sigma_1$ must be treated as zero. For safety, even in 'exact' computation, every singular value smaller than say $10^{-8}\sigma_1$ should be treated as zero.

Activity 5.2.4 In some system of linear equations the five singular values of the matrix are

$$1.5665, \quad 0.2222, \quad 0.0394, \quad 0.0107, \quad 0.0014.$$

Given the matrix components have errors of about 0.02, what is the effective rank of the matrix?

(a) 4   (b) 3   (c) 1   (d) 2

The final step in Procedure 5.2.3 arises because in many cases an infinite number of possible solutions are derived. The linear algebra cannot presume which is best for your application. Consequently, in future applications you will have to be aware of the freedom, and make a choice based on extra information. For two examples:

- in a CT-scan such as Example 3.5.17 one would usually prefer the greyest result in order to avoid diagnosing artifices;

- in the data mining task of fitting curves or surfaces to data, one would instead usually prefer a curve or surface with the fewest nonzero coefficients.

Such extra information from the application is essential.

Example 5.2.5 For the following matrix A and right-hand side vector b, solve $Ax = b$. But suppose the matrix entries come from experiments and are only known to within errors ± 0.05, solve $A'x' = b$ for some chosen matrices A' which approximate A to this error. Finally, use an SVD to find a general solution consistent with the error in matrix A. Report to two decimal places.

$$A = \begin{bmatrix} -0.2 & -0.6 & 1.8 \\ 0.0 & 0.2 & -0.4 \\ -0.3 & 0.7 & 0.3 \end{bmatrix}, \quad b = \begin{bmatrix} -0.5 \\ 0.1 \\ -0.2 \end{bmatrix}.$$

Solution: Enter the matrix and vector into MATLAB/Octave, note the poor `rcond`, and solve with `x=A\b` to determine $x = (0.06, -0.13, -0.31)$.

To within the experimental error of ± 0.05 the following two matrices approximate A: that is, they might have been what was measured for A. Then MATLAB/Octave `x=A\b` gives the corresponding equally valid solutions.

- $A' = \begin{bmatrix} -0.16 & -0.58 & 1.83 \\ 0.01 & 0.16 & -0.45 \\ -0.28 & 0.74 & 0.30 \end{bmatrix}$ gives $x' = \begin{bmatrix} 0.85 \\ 0.12 \\ -0.16 \end{bmatrix}$ as compared to $x = \begin{bmatrix} 0.06 \\ -0.13 \\ -0.31 \end{bmatrix}$.

- $A'' = \begin{bmatrix} -0.22 & -0.62 & 1.77 \\ 0.01 & 0.17 & -0.42 \\ -0.26 & 0.66 & 0.26 \end{bmatrix}$ gives $x'' = \begin{bmatrix} 0.42 \\ -0.04 \\ -0.24 \end{bmatrix}$ as compared to x and x'.

There are major differences between these equally valid solutions x, x', and x''. The problem is that, relative to the experimental error, there is a small singular value in the matrix A. We must use an SVD to find all solutions consistent with the experimental error. Consequently, compute `[U,S,V]=svd(A)` to find (2 d.p.)

```
U =
   -0.97    0.03   -0.23
    0.22   -0.10   -0.97
   -0.05   -0.99    0.09
S =
    1.96       0       0
       0    0.82       0
       0       0    0.02
V =
    0.11    0.36    0.93
    0.30   -0.90    0.31
   -0.95   -0.25    0.20
```

The singular value 0.02 is less than the error ± 0.05 so is effectively zero. Hence we solve the system as if this singular value is zero; that is, as if matrix A has rank two. Compute the smallest consistent solution with the three steps z=U(:,1:2)'*b, y=z./diag(S(1:2,1:2)), and x=V(:,1:2)*y. Then add an arbitrary multiple of the last column of V to determine a general solution

$$x = (0.10, -0.11, -0.30) + t(0.93, 0.31, 0.20). \qquad \square$$

This example gives infinitely many solutions which are equally valid as far as the linear algebra is concerned. In such an example, more information from an application is needed to choose which to *prefer* among the infinity of solutions.

Most often the singular values are spread over a wide range of orders of magnitude. In such cases, an assessment of the errors in the matrix is crucial in what one reports as a solution. The following artificial example illustrates the range.

Example 5.2.6 *(various errors)* The matrix

$$A = \begin{bmatrix} 1 & \frac{1}{2} & \frac{1}{3} & \frac{1}{4} & \frac{1}{5} \\ \frac{1}{2} & \frac{1}{3} & \frac{1}{4} & \frac{1}{5} & \frac{1}{6} \\ \frac{1}{3} & \frac{1}{4} & \frac{1}{5} & \frac{1}{6} & \frac{1}{7} \\ \frac{1}{4} & \frac{1}{5} & \frac{1}{6} & \frac{1}{7} & \frac{1}{8} \\ \frac{1}{5} & \frac{1}{6} & \frac{1}{7} & \frac{1}{8} & \frac{1}{9} \end{bmatrix}$$

is an example of a so-called Hilbert matrix. Explore the effects of various assumptions about possible errors in A upon the solution to $Ax = \mathbf{1}$ where $\mathbf{1} := (1,1,1,1,1)$.

Solution: Enter the matrix A into MATLAB/Octave with A=hilb(5) for the above 5×5 Hilbert matrix, and enter the right-hand side with b=ones(5,1).

- First assume that there is insignificant error in A (there is always the base error of 10^{-15} in computation). Then Procedure 2.2.5 finds that although the reciprocal of the condition number rcond(A) $\approx 10^{-6}$ is bad, the unique solution to $Ax = \mathbf{1}$, obtained via x=A\b, is

$$x = (5, -120, 630, -1120, 630).$$

- Second suppose the errors in A are roughly 10^{-5}. This level of error is a concern as rcond $\approx 10^{-6}$ so errors would be magnified by 10^6 in a direct solution of $Ax = \mathbf{1}$ (Theorem 3.3.30). Here we explore when all errors are in A and none in the right-hand side vector $\mathbf{1}$. To explore, adopt Procedure 5.2.3.

  (a) Find an SVD $A = USV^{\mathsf{T}}$ via [U,S,V]=svd(A) (2 d.p.)

  U =

     -0.77   0.60   -0.21   0.05   0.01

$$
\begin{array}{rrrrr}
-0.45 & -0.28 & 0.72 & -0.43 & -0.12 \\
-0.32 & -0.42 & 0.12 & 0.67 & 0.51 \\
-0.25 & -0.44 & -0.31 & 0.23 & -0.77 \\
-0.21 & -0.43 & -0.57 & -0.56 & 0.38
\end{array}
$$

$S =$

$$
\begin{array}{rrrrr}
1.57 & 0 & 0 & 0 & 0 \\
0 & 0.21 & 0 & 0 & 0 \\
0 & 0 & 0.01 & 0 & 0 \\
0 & 0 & 0 & 0.00 & 0 \\
0 & 0 & 0 & 0 & 0.00
\end{array}
$$

$V =$

$$
\begin{array}{rrrrr}
-0.77 & 0.60 & -0.21 & 0.05 & 0.01 \\
-0.45 & -0.28 & 0.72 & -0.43 & -0.12 \\
-0.32 & -0.42 & 0.12 & 0.67 & 0.51 \\
-0.25 & -0.44 & -0.31 & 0.23 & -0.77 \\
-0.21 & -0.43 & -0.57 & -0.56 & 0.38
\end{array}
$$

More informatively, the singular values have the following wide range of magnitudes,

$$
\sigma_1 = 1.57, \quad \sigma_2 = 0.21, \quad \sigma_3 = 1.14 \cdot 10^{-2},
$$
$$
\sigma_4 = 3.06 \cdot 10^{-4}, \quad \sigma_5 = 3.29 \cdot 10^{-6}.
$$

(b) Because the assumed error 10^{-5} lies between σ_4 and σ_5, $\sigma_4 > 10^{-5} > \sigma_5$, the matrix A is effectively of rank four, $k = 4$.

(c) Solving the system $Ax = USV^{\mathsf{T}}x = 1$ as rank four, in the least square sense, Procedure 3.5.4 gives (2 d.p.)

i. $z = U^{\mathsf{T}}1 = (-2.00, -0.97, -0.24, -0.04, 0.00)$,

ii. neglect the fifth component of z as an error and obtain the first four components of y via y=z (1:4) ./diag (S (1:4, 1:4)) so that

$$
y = (-1.28, -4.66, -21.43, -139.69, y_5),
$$

iii. then the smallest, least square, solution determined with x=V (: , 1:4) *y is

$$
x = (-3.82, 46.78, -93.41, -23.53, 92.27),
$$

and a general solution includes the arbitrary multiple y_5 of the last column of V to be

$$
x = (-3.82, 46.78, -93.41, -23.53, 92.27)
$$
$$
+ y_5(0.01, -0.12, 0.51, -0.77, 0.38).
$$

- Third suppose that the errors in A are roughly 10^{-3}. Re-adopt Procedure 5.2.3.

  (a) Use the same SVD, $A = USV^T$.

  (b) Because the assumed error 10^{-3} satisfies $\sigma_3 > 10^{-3} > \sigma_4$ the matrix A is effectively of rank three, $k = 3$.

  (c) Solving the system $Ax = USV^Tx = 1$ as rank three, in the least square sense, Procedure 3.5.4 gives (2 d.p.) the same z, and the same first three components in

  $$y = (-1.28, -4.66, -21.43, y_4, y_5),$$

  then x=V (: , 1 : 3) *y determines the smallest, least square, solution

  $$x = (2.76, -13.66, -0.19, 9.03, 14.38),$$

  and a general solution includes the arbitrary multiples of the last columns of V to be

  $$x = (2.76, -13.66, -0.19, 9.03, 14.38)$$
  $$+ y_4(0.05, -0.43, 0.67, 0.23, -0.56)$$
  $$+ y_5(0.01, -0.12, 0.51, -0.77, 0.38).$$

- Lastly suppose the errors in A are roughly 0.05. Re-adopt Procedure 5.2.3.

  (a) Use the same SVD, $A = USV^T$.

  (b) Because the assumed error 0.05 satisfies $\sigma_2 > 0.05 > \sigma_3$ the matrix A is effectively of rank two, $k = 2$.

  (c) Solving the system $Ax = USV^Tx = 1$ as rank two, in the least square sense, Procedure 3.5.4 gives (2 d.p.) the same z, and the same first two components in

  $$y = (-1.28, -4.66, y_3, y_4, y_5),$$

  then x=V (: , 1 : 2) *y determines the smallest, least square, solution

  $$x = (-1.83, 1.85, 2.39, 2.39, 2.27),$$

  and a general solution includes the arbitrary multiples of the last columns of V to be

  $$x = (-1.83, 1.85, 2.39, 2.39, 2.27)$$
  $$+ y_3(-0.21, 0.72, 0.12, -0.31, -0.57)$$
  $$+ y_4(0.05, -0.43, 0.67, 0.23, -0.56)$$
  $$+ y_5(0.01, -0.12, 0.51, -0.77, 0.38).$$

The level of error makes a major difference in the qualitative nature of allowable solutions: here from a unique solution through to a three parameter family of equally valid solutions. To appropriately solve systems of linear equations we must know the level of error. □

Example 5.2.7 *(translating temperatures)* Recall that Example 2.2.11 attempts to fit a quartic polynomial to observations (plotted to the right) of the relation between Celsius and Fahrenheit temperature. The attempt failed because rcond is too small. Let's try again now that we can cater for matrices with errors. Recall that the data between temperatures reported by a European and an American are the following:

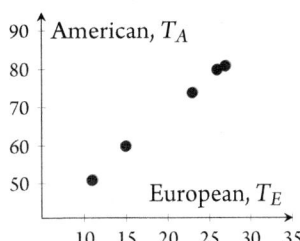

$$\begin{array}{c|ccccc} T_E & 15 & 26 & 11 & 23 & 27 \\ \hline T_A & 60 & 80 & 51 & 74 & 81 \end{array}$$

Example 2.2.11 attempts to fit the data with the quartic polynomial

$$T_A = c_1 + c_2 T_E + c_3 T_E^2 + c_4 T_E^3 + c_5 T_E^4,$$

and deduced the following system of equations for the coefficients

$$\begin{bmatrix} 1 & 15 & 225 & 3375 & 50625 \\ 1 & 26 & 676 & 17576 & 456976 \\ 1 & 11 & 121 & 1331 & 14641 \\ 1 & 23 & 529 & 12167 & 279841 \\ 1 & 27 & 729 & 19683 & 531441 \end{bmatrix} \begin{bmatrix} c_1 \\ c_2 \\ c_3 \\ c_3 \\ c_4 \\ c_5 \end{bmatrix} = \begin{bmatrix} 60 \\ 80 \\ 51 \\ 74 \\ 81 \end{bmatrix}.$$

In order to find a robust solution, here let's approximate both the matrix and the right-hand side vector because both the matrix and the vector come from real data with errors of about up to $\pm 0.5°$.

Solution: Now invoke Procedure 5.2.3 to approximate the system of linear equations, and solve the approximate problem.

(a) There is a problem in approximating the matrix: the columns are of wildly different magnitudes. In contrast, our mathematical analysis treats all columns the same. The problem is that each column comes from different powers of temperatures. To avoid the problem we must scale the temperature data for the matrix. The simplest scaling is to divide by a typical temperature. That is, instead of seeking a fit in terms of powers of T_E, we seek a fit in powers of $T_E/20°$ because 20 degrees is a typical temperature in the data. Hence let's here fit the data with the quartic polynomial

$$T_A = c_1 + c_2 \frac{T_E}{20} + c_3 \left(\frac{T_E}{20}\right)^2 + c_4 \left(\frac{T_E}{20}\right)^3 + c_5 \left(\frac{T_E}{20}\right)^4,$$

which gives the following system for the coefficients (2 d.p.)

$$
\begin{bmatrix}
1 & 0.75 & 0.56 & 0.42 & 0.32 \\
1 & 1.30 & 1.69 & 2.20 & 2.86 \\
1 & 0.55 & 0.30 & 0.17 & 0.09 \\
1 & 1.15 & 1.32 & 1.52 & 1.75 \\
1 & 1.35 & 1.82 & 2.46 & 3.32
\end{bmatrix}
\begin{bmatrix}
c_1 \\ c_2 \\ c_3 \\ c_3 \\ c_4 \\ c_5
\end{bmatrix}
=
\begin{bmatrix}
60 \\ 80 \\ 51 \\ 74 \\ 81
\end{bmatrix}.
$$

Now all the components of the matrix are roughly the same magnitude, as required.

There is no need to scale the right-hand side vector as all components are roughly the same magnitude; they are all simply 'American temperatures'.

In script, construct the scaled matrix and right-hand side vector with

```
te= [15;26;11;23;27]
ta= [60;80;51;74;81]
tes=te/20
A= [ones(5,1) tes tes.^2 tes.^3 tes.^4]
```

(b) Compute an SVD, $A = USV^\mathsf{T}$, with [U, S, V] = svd (A) to get (2 d.p.)

```
U =
   -0.16    0.64    0.20    0.72   -0.12
   -0.59   -0.13   -0.00   -0.15   -0.78
   -0.10    0.67   -0.59   -0.45    0.05
   -0.42    0.23    0.68   -0.42    0.36
   -0.66   -0.28   -0.39    0.29    0.49
S =
    7.26       0       0       0       0
       0    1.44       0       0       0
       0       0    0.21       0       0
       0       0       0    0.02       0
       0       0       0       0    0.00
V =
   -0.27    0.78   -0.49   -0.27    0.09
   -0.32    0.39    0.36    0.66   -0.42
   -0.40    0.09    0.55   -0.09    0.72
   -0.50   -0.17    0.25   -0.62   -0.52
   -0.65   -0.44   -0.51    0.32    0.14
```

(c) Now choose the effective rank of the matrix to be the number of singular values bigger than the error. Here recall that the temperatures in the matrix have been divided by $20°$. Hence the errors of roughly $\pm0.5°$ in each temperature becomes roughly $\pm0.5/20 = \pm0.025$ in the scaled components in the matrix.[8] There are three singular values larger than the

---

[8] Discerning people may comment that raising to a power also amplifies errors so that the jth column has errors more like $j \cdot 0.5/20$. However, let's ignore this extra complication here.

error 0.025, so the matrix effectively has rank three. The two singular values less than the error 0.025 are effectively zero. That is, although it is not necessary to construct, we approximate the matrix A by (2 d.p.)

$$A_3 = US_3V^{\mathsf{T}} = \begin{bmatrix} 1 & 0.74 & 0.56 & 0.43 & 0.31 \\ 1 & 1.30 & 1.69 & 2.20 & 2.86 \\ 1 & 0.56 & 0.30 & 0.16 & 0.09 \\ 1 & 1.16 & 1.32 & 1.52 & 1.75 \\ 1 & 1.35 & 1.82 & 2.46 & 3.32 \end{bmatrix} :$$

the differences between this approximate A_3 and the original A are only ± 0.01, so matrix A_3 is indeed close to A.

(d) Solve the equations as if matrix A has rank three.

 i. Find $z = U'b$ via z=U' *ta to find

```
z  =
  -146.53
    56.26
     0.41
     1.06
    -0.69
```

 As matrix A has effective rank of three, we approximate the right-hand side data by neglecting the last two components in this z. That the last two components in z are small compared to the others indicates that this neglect is a reasonable approximation.

 ii. Via the command y=z(1:3)./diag(S(1:3,1:3)), find y by solving $Sy = z$ as a rank three system to find (2 d.p.)

$$y = (-20.19, 39.15, 1.95, y_4, y_5).$$

 The smallest solution would be obtained by setting $y_4 = y_5 = 0$.

 iii. Finally determine the coefficients $c = Vy$ with command c=V(:,1:3)*y and then add arbitrary multiples of the remaining columns of V to obtain the general solution (2 d.p.)

$$c = \begin{bmatrix} 35.09 \\ 22.50 \\ 12.77 \\ 3.99 \\ -5.28 \end{bmatrix} + y_4 \begin{bmatrix} -0.27 \\ 0.66 \\ -0.09 \\ -0.62 \\ 0.32 \end{bmatrix} + y_5 \begin{bmatrix} 0.09 \\ -0.42 \\ 0.72 \\ -0.52 \\ 0.14 \end{bmatrix}.$$

(e) Obtain the solution with smallest coefficients by setting $y_4 = y_5 = 0$. This would fit the data with the quartic polynomial

$$T_A = 35.09 + 22.50\frac{T_E}{20} + 12.77\left(\frac{T_E}{20}\right)^2$$

$$+ 3.99\left(\frac{T_E}{20}\right)^3 - 5.28\left(\frac{T_E}{20}\right)^4.$$

But choosing the polynomial with smallest coefficients has little meaning in this application. Surely we prefer a polynomial with fewer terms, fewer nonzero coefficients. Surely we would prefer, say, the quadratic (plotted to the right)

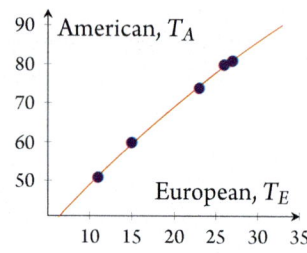

$$T_A = 25.93 + 49.63\frac{T_E}{20} - 6.45\left(\frac{T_E}{20}\right)^2.$$

In this application, let's use the freedom in y_4 and y_5 to set two of the coefficients to zero in the quartic. Since $c_4 = 3.99$ and $c_5 = -5.28$ are the smallest coefficients, and because they correspond to the highest powers in the quartic, it is natural to choose to make them both zero. Let's redo the last step in the procedure.[9]

The last step in the procedure is to solve $V^T c = y$ for c. With the last two components of c set to zero, from the computed SVD this is the system of equations (2 d.p.)

$$\begin{bmatrix} -0.27 & -0.32 & -0.40 & -0.50 & -0.65 \\ 0.78 & 0.39 & 0.09 & -0.17 & -0.44 \\ -0.49 & 0.36 & 0.55 & 0.25 & -0.51 \\ -0.27 & 0.66 & -0.09 & -0.62 & 0.32 \\ 0.09 & -0.42 & 0.72 & -0.52 & 0.14 \end{bmatrix} \begin{bmatrix} c_1 \\ c_2 \\ c_3 \\ 0 \\ 0 \end{bmatrix} = \begin{bmatrix} -20.19 \\ 39.15 \\ 1.95 \\ y_4 \\ y_5 \end{bmatrix},$$

where y_4 and y_5 can be anything for equally good solutions. Considering only the first three rows of this system, and using the zeros in c, this system becomes

$$\begin{bmatrix} -0.27 & -0.32 & -0.40 \\ 0.78 & 0.39 & 0.09 \\ -0.49 & 0.36 & 0.55 \end{bmatrix} \begin{bmatrix} c_1 \\ c_2 \\ c_3 \end{bmatrix} = \begin{bmatrix} -20.19 \\ 39.15 \\ 1.95 \end{bmatrix}.$$

This is a basic system of three equations for three unknowns. Since the matrix is the first three rows and columns of V^T and the right-hand side is the three components of y already computed, we solve the equation by

[9] Alternatively, one could redo the linear algebra to seek a quadratic from the outset rather than a quartic. The two alternative answers for a quadratic are not the same, but they are nearly the same. The small differences in the answers are because one approach modifies the matrix by recognizing its errors, and the other approach only modifies the right-hand side vector.

- checking the condition number, rcond(V(1:3,1:3)) is 0.05 which is good, and

- then c=V(1:3,1:3)'\y determines the coefficients $(c_1, c_2, c_3) = (25.93, 49.63, -6.45)$ (2 d.p.).

That is, a just as good polynomial fit, consistent with errors in the data, is the simpler quadratic polynomial (as plotted previously)

$$T_A = 25.93 + 49.63 \frac{T_E}{20} - 6.45 \left(\frac{T_E}{20} \right)^2.$$

☐

Occam's razor: Non sunt multiplicanda entia sine necessitate [Entities must not be multiplied beyond necessity]   John Punch (1639)

Example 5.2.8 Recall that Exercise 3.5.18 introduced extra 'diagonal' measurements into a 2D CT-scan. As shown to the right, the 2D region is divided into a 3×3 grid of nine blocks. Then measurements are taken of the X-rays not absorbed along the shown nine paths: three horizontal, three vertical, and three diagonal. Suppose the measured fractions of X-ray energy are $f = (0.048, 0.081, 0.042, 0.020, 0.106, 0.075, 0.177, 0.181, 0.105)$. Use an SVD to find the 'greyest' transmission factors consistent with the measurements and likely errors.

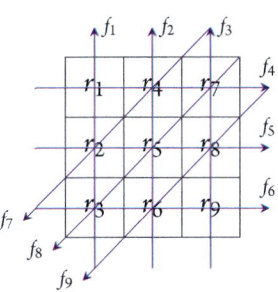

Solution: Nine X-ray measurements are made through the body where $f_1, f_2, \ldots, f_9$ denote the fraction of energy in the measurements relative to the power of the X-ray beam. Thus we need to solve nine equations for the nine unknown transmission factors:

$$r_1 r_2 r_3 = f_1, \quad r_4 r_5 r_6 = f_2, \quad r_7 r_8 r_9 = f_3,$$
$$r_1 r_4 r_7 = f_4, \quad r_2 r_5 r_8 = f_5, \quad r_3 r_6 r_9 = f_6,$$
$$r_2 r_4 = f_7, \quad r_3 r_5 r_7 = f_8, \quad r_6 r_8 = f_9.$$

Turn such nonlinear equations into linear equations by taking the logarithm (to any base, but here say the natural logarithm to base e) of both sides of all equations:

$$r_i r_j r_k = f_l \iff (\log r_i) + (\log r_j) + (\log r_k) = (\log f_l).$$

That is, letting new unknowns $x_i = \log r_i$ and new right-hand sides $b_i = \log f_i$, we aim to solve a system of nine linear equations for nine unknowns:

$$x_1 + x_2 + x_3 = b_1, \quad x_4 + x_5 + x_6 = b_2, \quad x_7 + x_8 + x_9 = b_3,$$
$$x_1 + x_4 + x_7 = b_4, \quad x_2 + x_5 + x_8 = b_5, \quad x_3 + x_6 + x_9 = b_6,$$
$$x_2 + x_4 \quad\quad = b_7, \quad x_3 + x_5 + x_7 = b_8, \quad x_6 + x_8 \quad\quad = b_9.$$

These form the matrix-vector system $Ax = b$ for 9×9 matrix

$$A = \begin{bmatrix} 1 & 1 & 1 & 0 & 0 & 0 & 0 & 0 & 0 \\ 0 & 0 & 0 & 1 & 1 & 1 & 0 & 0 & 0 \\ 0 & 0 & 0 & 0 & 0 & 0 & 1 & 1 & 1 \\ 1 & 0 & 0 & 1 & 0 & 0 & 1 & 0 & 0 \\ 0 & 1 & 0 & 0 & 1 & 0 & 0 & 1 & 0 \\ 0 & 0 & 1 & 0 & 0 & 1 & 0 & 0 & 1 \\ 0 & 1 & 0 & 1 & 0 & 0 & 0 & 0 & 0 \\ 0 & 0 & 1 & 0 & 1 & 0 & 1 & 0 & 0 \\ 0 & 0 & 0 & 0 & 0 & 1 & 0 & 1 & 0 \end{bmatrix}, \quad b = \log \begin{bmatrix} .048 \\ .081 \\ .042 \\ .020 \\ .106 \\ .075 \\ .177 \\ .181 \\ .105 \end{bmatrix} = \begin{bmatrix} -3.04 \\ -2.51 \\ -3.17 \\ -3.91 \\ -2.24 \\ -2.59 \\ -1.73 \\ -1.71 \\ -2.25 \end{bmatrix}.$$

Implement Procedure 5.2.3.

(a) Here there is no need to scale the vector b as all entries are roughly the same. There is no need to scale the matrix A as all entries mean the same, namely simply zero or one, depending upon whether the beam passes through the pixel square. However, the entries of A are in error in two ways.

- A diagonal beam has a path through a pixel that is up to 41% longer than the horizontal or vertical beam—which is not accounted for. Further, a beam has finite width so it also passes through part of some off-diagonal pixels—which is not represented.

- Similarly, a horizontal or vertical beam has finite width and may underrepresent the sides of the pixels it goes through, and/or involve parts of neighbouring pixels— neither effect is represented.

Consequently the entries in the matrix A could easily have error of ± 0.5. Let's use knowledge of this error to ensure that the predictions by the scan are reliable.

(b) Compute an SVD, $A = USV^{\mathsf{T}}$, via [U, S, V] =svd (A) (2 d.p.)

```
U =
   0.33 -0.41   0.27   0.21   0.54   0.23 -0.29   0.13 -0.41
   0.38 -0.00 -0.47   0.36 -0.45 -0.19 -0.00   0.32 -0.41
   0.33   0.41   0.27 -0.57 -0.08   0.23   0.29   0.13 -0.41
   0.33 -0.41   0.27 -0.21 -0.54   0.23 -0.29   0.13   0.41
   0.38 -0.00 -0.47 -0.36   0.45 -0.19 -0.00   0.32   0.41
   0.33   0.41   0.27   0.57   0.08   0.23   0.29   0.13   0.41
   0.23 -0.41 -0.29 -0.00   0.00   0.30   0.58 -0.52 -0.00
   0.41   0.00   0.33   0.00   0.00 -0.74 -0.00 -0.43   0.00
   0.23   0.41 -0.29 -0.00 -0.00   0.30 -0.58 -0.52   0.00
S =
   2.84      0      0      0      0      0      0      0      0
      0   2.00      0      0      0      0      0      0      0
      0      0   1.84      0      0      0      0      0      0
      0      0      0   1.73      0      0      0      0      0
      0      0      0      0   1.73      0      0      0      0
      0      0      0      0      0   1.51      0      0      0
      0      0      0      0      0      0   1.00      0      0
      0      0      0      0      0      0      0   0.51      0
      0      0      0      0      0      0      0      0      0
```

$$V =$$

0	0	0	0	0	0	0	0	0.00
0.23	-0.41	0.29	0.00	0.00	0.30	-0.58	0.52	-0.00
0.33	-0.41	-0.27	-0.08	0.57	0.23	0.29	-0.13	-0.41
0.38	0.00	0.47	0.45	0.36	-0.19	-0.00	-0.32	0.41
0.33	-0.41	-0.27	0.08	-0.57	0.23	0.29	-0.13	0.41
0.41	-0.00	-0.33	0.00	-0.00	-0.74	-0.00	0.43	-0.00
0.33	0.41	-0.27	0.54	-0.21	0.23	-0.29	-0.13	-0.41
0.38	0.00	0.47	-0.45	-0.36	-0.19	0.00	-0.32	-0.41
0.33	0.41	-0.27	-0.54	0.21	0.23	-0.29	-0.13	0.41
0.23	0.41	0.29	0.00	0.00	0.30	0.58	0.52	0.00

(c) Here choose the rank of the matrix to be effectively seven because two of the nine singular values, namely 0.51 and 0.00, are less than or about the magnitude of the expected error, roughly 0.5.

(d) Use the rank seven SVD to solve the approximate system as in Procedure 3.5.4.

    i. Find $z = U^\mathsf{T}b$ via z=U' *b to find

```
z =
    -7.63
     0.27
    -0.57
     0.42
     0.64
    -1.95
     0.64
    -0.40
    -0.01
```

    ii. Neglect the last two rows in solving $S_7y = z$ to find via y=z(1:7)./diag (S(1:7,1:7)) that the first seven components of y are

```
y =
    -2.69
     0.14
    -0.31
     0.24
     0.37
    -1.29
     0.64
```

The last two components of y, y_8 and y_9, are free variables.

    iii. Obtain a particular solution to $V^\mathsf{T}x = y$, the one of smallest magnitude, by setting $y_8 = y_9 = 0$ and determining x from x=V(:,1:7)*y to get the smallest solution

```
x =
   -1.53
   -0.78
   -0.67
   -1.16
   -0.05
   -1.18
   -1.16
   -1.28
   -0.68
```

Obtain other equally valid solutions, in the context of the identified error in matrix A, by adding arbitrary multiples of the last two columns of V.

(e) Here we aim to make predictions from the CT-scan. The 'best' solution in this application is the one with least artificial features. The smallest magnitude x seems to reasonably implement this criterion. Thus use the above particular x to determine the transmission factors, $r_i = \exp(x_i)$. Here use `r=reshape(exp(x),3,3)` to compute and form into the 3×3 array of pixels

0.22	0.31	0.31
0.46	0.95	0.28
0.51	0.31	0.51

$\mapsto$

when illustrated with `colormap(gray),imagesc(r)`

The CT-scan and linear algebra robustly identifies that there is a significant 'hole' in the middle of the body being scanned. ☐

5.2.2 Tikhonov regularization explained

Regularization of poorly posed linear equations is a widely invoked practical necessity. Many people have invented alternative techniques. Many have independently re-invented techniques. Perhaps the most common is the so-called Tikhonov regularization. This section introduces and discusses Tikhonov regularization.

This optional subsection connects to much established practice that graduates may encounter.

In statistics, the method is known as ridge regression, and with multiple independent discoveries, it is also variously known as the Tikhonov–Miller method, the Phillips–Twomey method, the constrained linear inversion method, and the method of linear regularization.[10]

Wikipedia (2015)

---

[10] https://en.wikipedia.org/wiki/Tikhonov_regularization

Definition 5.2.9 *In seeking to solve the poorly posed system $Ax = b$ for $m \times n$ matrix A, a **Tikhonov regularization** is the system $(A^T A + \alpha^2 I_n)x = A^T b$ for some chosen regularization parameter value $\alpha > 0$.[11] (The greek letter α is 'alpha', and is different to the 'proportional to' symbol $\propto$.)*

Example 5.2.10 Use Tikhonov regularization to solve the system of Example 5.2.1:

$$0.5x + 0.3y = 1 \quad \text{and} \quad 1.1x + 0.7y = 2,$$

Solution:　　Here the matrix and right-hand side vector are

$$A = \begin{bmatrix} 0.5 & 0.3 \\ 1.1 & 0.7 \end{bmatrix}, \quad b = \begin{bmatrix} 1 \\ 2 \end{bmatrix}.$$

Evaluating $A^T A$ and $A^T b$, a Tikhonov regularization, $(A^T A + \alpha^2 I_n)x = A^T b$, is then the system

$$\begin{bmatrix} 1.46 + \alpha^2 & 0.92 \\ 0.92 & 0.58 + \alpha^2 \end{bmatrix} x = \begin{bmatrix} 2.7 \\ 1.7 \end{bmatrix}.$$

Choose the regularization parameter α to be roughly the error; here the error is ± 0.05, so let's choose $\alpha = 0.1$ ($\alpha^2 = 0.01$). Enter into MATLAB/Octave with

```
A= [0.5 0.3;1.1 0.7]
b= [1.0;2.0]
AtA=A'*A+0.01*eye(2)
rcondAtA=rcond(AtA)
x=AtA\(A'*b)
```

to find that the Tikhonov regularized solution is $x = (1.39, 0.72)$.[12] This solution is reasonably close to the smallest solution found by the SVD which is $(1.32, 0.83)$. However, Tikhonov regularization gives no hint of the reasonable general solutions found by the SVD approach of Example 5.2.1.

Change the regularization parameter to $\alpha = 0.01$ and $\alpha = 1$ and see that both of these choices degrade the Tikhonov solution. □

Activity 5.2.11 In the linear system for $x = (x, y)$,

$$4x - y = -4 \quad \text{and} \quad -2x + y = 3,$$

---

[11] Some may notice that a Tikhonov regularization is closely connected to the so-called normal equation $(A^T A)x = A^T b$. Tikhonov regularization shares with the normal equation some practical limitations as well as some strengths.

[12] Interestingly, $rcond = 0.003$ for the Tikhonov system which is worse than $rcond(A)$. The regularization only works at all because pre-multiplying by A^T pushes both sides into the row space of A (except for numerical error and the small $\alpha^2 I$ factor).

the coefficients on the left-hand side of each equation are in error by about ± 0.3. Tikhonov regularization should solve which one of the following systems?

(a) $\begin{bmatrix} 20.1 & -6 \\ -6 & 2.1 \end{bmatrix} x = \begin{bmatrix} -22 \\ 7 \end{bmatrix}$

(c) $\begin{bmatrix} 18.1 & -5 \\ -10 & 3.1 \end{bmatrix} x = \begin{bmatrix} -19 \\ 11 \end{bmatrix}$

(b) $\begin{bmatrix} 18.3 & -5 \\ -10 & 3.3 \end{bmatrix} x = \begin{bmatrix} -19 \\ 11 \end{bmatrix}$

(d) $\begin{bmatrix} 20.3 & -6 \\ -6 & 2.3 \end{bmatrix} x = \begin{bmatrix} -22 \\ 7 \end{bmatrix}$

Do not apply Tikhonov regularization blindly as it does introduce biases. The following example illustrates the bias.

Example 5.2.12 Recall Example 3.5.1 at the start of Section 3.5.1 where scales variously reported my weight in kg as 84.8, 84.1, 84.7, and 84.4. To best estimate my weight x we rewrote the problem in matrix-vector form

$$Ax = b, \quad \text{namely} \quad \begin{bmatrix} 1 \\ 1 \\ 1 \\ 1 \end{bmatrix} x = \begin{bmatrix} 84.8 \\ 84.1 \\ 84.7 \\ 84.4 \end{bmatrix}.$$

A Tikhonov regularization of this inconsistent system is

$$\left(\begin{bmatrix} 1 & 1 & 1 & 1 \end{bmatrix} \begin{bmatrix} 1 \\ 1 \\ 1 \\ 1 \end{bmatrix} + \alpha^2 \right) x = \begin{bmatrix} 1 & 1 & 1 & 1 \end{bmatrix} \begin{bmatrix} 84.8 \\ 84.1 \\ 84.7 \\ 84.4 \end{bmatrix}.$$

That is, $(4 + \alpha^2)x = 338$ kg with solution $x = 338/(4 + \alpha^2) = 84.5/(1 + \alpha^2/4)$ kg. Because of the division by $1 + \alpha^2/4$, this Tikhonov answer is biased as it is systematically below the average 84.5 kg. For small Tikhonov parameter α the bias is small, but even so, such a bias in the methodology is unpleasant.

Example 5.2.13 Use Tikhonov regularization to solve $Ax = b$ for the matrix and vector of Example 5.2.5.

Solution: Here the matrix and right-hand side vector are

$$A = \begin{bmatrix} -0.2 & -0.6 & 1.8 \\ 0.0 & 0.2 & -0.4 \\ -0.3 & 0.7 & 0.3 \end{bmatrix}, \quad b = \begin{bmatrix} -0.5 \\ 0.1 \\ -0.2 \end{bmatrix}.$$

A Tikhonov regularization, $(A^T A + \alpha^2 I_n)x = A^T b$, is then the system

$$\begin{bmatrix} 0.13 + \alpha^2 & -0.09 & -0.45 \\ -0.09 & 0.89 + \alpha^2 & -0.95 \\ -0.45 & -0.95 & 3.49 + \alpha^2 \end{bmatrix} x = \begin{bmatrix} 0.16 \\ 0.18 \\ -1.00 \end{bmatrix}.$$

Choose regularization parameter α to be roughly the error; here the error is ± 0.05, so let's choose $\alpha = 0.1$ ($\alpha^2 = 0.01$). Enter into and solve with MATLAB/Octave via

```
A=[-0.2 -0.6 1.8
    0.0 0.2 -0.4
   -0.3 0.7 0.3 ]
b=[-0.5;0.1;-0.2]
AtA=A'*A+0.01*eye(3)
rcondAtA=rcond(AtA)
x=AtA\(A'*b)
```

which finds the Tikhonov regularized solution is $x = (0.10, -0.11, -0.30)$. To two decimal places, this is the same as the smallest solution found by an SVD. However, Tikhonov regularization gives no hint of the reasonable general solutions found by the SVD approach of Example 5.2.5. $\square$

Although Definition 5.2.9 does not look like it, Tikhonov regularization relates directly to the SVD regularization of Section 5.2.1. The next theorem establishes the connection.

Theorem 5.2.14 (*Tikhonov regularization*) *Solving $Ax = b$ by Tikhonov regularization, with parameter $\alpha > 0$, is equivalent to finding the smallest, least square solution of the system $A'x = b$, where the matrix A' is obtained from A by replacing each of its nonzero singular values σ_i by $\sigma_i' :=$ $\sigma_i + \alpha^2/\sigma_i$.*

Proof. Let's use an SVD to understand Tikhonov regularization. Suppose $m \times n$ matrix A has SVD $A = USV^\mathsf{T}$. First, the left-hand side matrix in a Tikhonov regularization is

$$
\begin{aligned}
A^\mathsf{T}A + \alpha^2 I_n &= (USV^\mathsf{T})^\mathsf{T}USV^\mathsf{T} + \alpha^2 I_n VV^\mathsf{T} \\
&= VS^\mathsf{T}U^\mathsf{T}USV^\mathsf{T} + \alpha^2 VI_n V^\mathsf{T} \\
&= VS^\mathsf{T}SV^\mathsf{T} + V(\alpha^2 I_n)V^\mathsf{T} \\
&= V(S^\mathsf{T}S + \alpha^2 I_n)V^\mathsf{T},
\end{aligned}
$$

whereas the right-hand side is

$$
A^\mathsf{T}b = (USV^\mathsf{T})^\mathsf{T}b = VS^\mathsf{T}U^\mathsf{T}b.
$$

Corresponding to the variables used in previous procedures, let $z = U^\mathsf{T}b \in \mathbb{R}^m$ and as yet unknown $y = V^\mathsf{T}x \in \mathbb{R}^n$. Then equating the above two sides, and premultiplying by the orthogonal V^T, means that the Tikhonov regularization is equivalent to solving $(S^\mathsf{T}S + \alpha^2 I_n)y = S^\mathsf{T}z$ for the as yet unknown y. (Beautifully, this equation could be interpreted as the Tikhonov regularization of the equation $Sy = z$.)

Second, let $r = \text{rank}\, A$ so that the singular value matrix

$$S = \begin{bmatrix} \sigma_1 & \cdots & 0 & \\ \vdots & \ddots & \vdots & O_{r\times(n-r)} \\ 0 & \cdots & \sigma_r & \\ & & & \\ O_{(m-r)\times r} & & O_{(m-r)\times(n-r)} \end{bmatrix}$$

(where the bottom-right zero block contains all the zero singular values). Consequently, the equivalent Tikhonov regularization, $(S^{\mathsf{T}}S + \alpha^2 I_n)y = S^{\mathsf{T}}z$, becomes

$$\begin{bmatrix} \sigma_1^2 + \alpha^2 & \cdots & 0 & \\ \vdots & \ddots & \vdots & O_{r\times(n-r)} \\ 0 & \cdots & \sigma_r^2 + \alpha^2 & \\ & & & \\ O_{(n-r)\times r} & & \alpha^2 I_{n-r} \end{bmatrix} y = \begin{bmatrix} \sigma_1 z_1 \\ \vdots \\ \sigma_r z_r \\ \\ 0_{n-r} \end{bmatrix}.$$

Dividing each of the first r rows by the corresponding nonzero singular value, $\sigma_1, \sigma_2, \ldots, \sigma_r$, the equivalent system is

$$\begin{bmatrix} \sigma_1 + \alpha^2/\sigma_1 & \cdots & 0 & \\ \vdots & \ddots & \vdots & O_{r\times(n-r)} \\ 0 & \cdots & \sigma_r + \alpha^2/\sigma_r & \\ & & & \\ O_{(n-r)\times r} & & \alpha^2 I_{n-r} \end{bmatrix} y = \begin{bmatrix} z_1 \\ \vdots \\ z_r \\ \\ 0_{n-r} \end{bmatrix},$$

with solution

- $y_i = z_i/(\sigma_i + \alpha^2/\sigma_i)$ for $i = 1, \ldots, r$, and
- $y_i = 0$ for $i = r+1, \ldots, n$ (since $\alpha^2 > 0$).

This establishes that solving the Tikhonov system is equivalent to performing the SVD Procedure 3.5.4 for the least square solution to $Ax = b$ but with two changes in Step 3:

- for $i = 1, \ldots, r$ divide by $\sigma_i' := \sigma_i + \alpha^2/\sigma_i$ instead of the true singular value σ_i (the upcoming right-hand plot shows σ' versus σ), and
- for $i = r+1, \ldots, n$ set $y_i = 0$ to obtain the smallest possible solution (Theorem 3.5.13).

Thus Tikhonov regularization of $Ax = b$ is equivalent to finding the smallest, least square, solution of the system $A'x = b$. ☐

There is another reason to be careful when using Tikhonov regularization. Yes, it gives a nice, neat, unique solution, but it does not hint that there may be an infinite number of equally good nearby solutions (as found through Procedure 5.2.3). Among those equally good nearby solutions may be ones that you prefer in your application.

Choose a good regularization parameter

- One strategy to choose the regularization parameter α is that the effective change in the matrix, from A to A', should be about the magnitude of errors expected in A. (This strategic choice is sometimes called the discrepancy principle (Kress, 2015, §7).) Since changes in the matrix are largely measured by the singular values, we need to consider the relation between $\sigma' = \sigma + \alpha^2/\sigma$ and σ. From the graph to the right, the small singular values are changed by a lot, but these are the ones for which we want σ' large, in order to give a 'least square' approximation. Significantly, the above right graph also shows that singular values larger than α change by less than α. Thus the parameter α should not be much larger than the expected error in the elements of the matrix A.

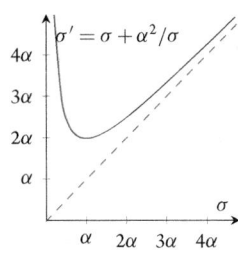

- Another consideration is the effect of regularization upon errors in the right-hand side vector. The condition number of A may be very bad. However, as the marginal graph shows the smallest $\sigma' \geq 2\alpha$. Thus, in the regularized system the condition number of the effective matrix A' is approximately $\sigma_1/(2\alpha)$. We need to choose the regularization parameter α large enough so that $\frac{\sigma_1}{2\alpha} \times$ (relative error in $\boldsymbol{b}$) is an acceptable relative error in the solution $\boldsymbol{x}$ (Theorem 3.3.30). It is only when the regularization parameter α is big enough that the regularization is effective in finding a least square approximation.

5.2.3 Exercises

Exercise 5.2.1   For each of the following matrices, say A, and right-hand side vectors, say $\boldsymbol{b}_1$, solve $A\boldsymbol{x} = \boldsymbol{b}_1$. But suppose the matrix entries come from experiments and are only known to within errors ± 0.05. Thus within experimental error the given matrices A' and A'' may be the 'true' matrix A. Solve $A'\boldsymbol{x}' = \boldsymbol{b}_1$ and $A''\boldsymbol{x}'' = \boldsymbol{b}_1$ and comment on the results. Finally, use an SVD to find a general solution consistent with the error in the matrix.

(a) $A = \begin{bmatrix} -1.3 & -0.4 \\ 0.7 & 0.2 \end{bmatrix}$, $\boldsymbol{b}_1 = \begin{bmatrix} 2.4 \\ -1.3 \end{bmatrix}$, $A' = \begin{bmatrix} -1.27 & -0.43 \\ 0.71 & 0.19 \end{bmatrix}$, $A'' = \begin{bmatrix} -1.27 & -0.38 \\ 0.66 & 0.22 \end{bmatrix}$.

(b) $B = \begin{bmatrix} -1.8 & -1.1 \\ -0.2 & -0.1 \end{bmatrix}$, $\boldsymbol{b}_2 = \begin{bmatrix} -0.7 \\ -0.1 \end{bmatrix}$, $B' = \begin{bmatrix} -1.81 & -1.13 \\ -0.24 & -0.12 \end{bmatrix}$, $B'' = \begin{bmatrix} -1.81 & -1.13 \\ -0.18 & -0.1 \end{bmatrix}$.

(c) $C = \begin{bmatrix} 0.6 & -0.8 & -0.2 \\ -0.9 & 1.0 & 1.2 \\ -0.9 & 0.9 & 1.4 \end{bmatrix}$, $\boldsymbol{b}_3 = \begin{bmatrix} 1.1 \\ -3.7 \\ -4.1 \end{bmatrix}$, $C' = \begin{bmatrix} 0.57 & -0.78 & -0.23 \\ -0.91 & 0.99 & 1.22 \\ -0.93 & 0.9 & 1.39 \end{bmatrix}$,

$C'' = \begin{bmatrix} 0.56 & -0.77 & -0.21 \\ -0.87 & 1.01 & 1.22 \\ -0.87 & 0.9 & 1.39 \end{bmatrix}$.

(d) $D = \begin{bmatrix} -0.9 & -0.5 & -0.3 & -0.4 \\ -0.1 & 0.1 & -0.2 & 0.8 \\ -1.0 & 0.4 & -1.1 & 0.6 \\ 1.0 & 2.2 & -1.0 & -0.1 \end{bmatrix}$, $b_4 = \begin{bmatrix} 0.4 \\ 0.3 \\ 0.2 \\ -2.0 \end{bmatrix}$,

$D' = \begin{bmatrix} -0.88 & -0.52 & -0.33 & -0.41 \\ -0.11 & 0.13 & -0.17 & 0.78 \\ -0.96 & 0.44 & -1.12 & 0.61 \\ 0.98 & 2.19 & -0.99 & -0.13 \end{bmatrix}$, $D'' = \begin{bmatrix} -0.86 & -0.49 & -0.29 & -0.37 \\ -0.06 & 0.14 & -0.18 & 0.83 \\ -0.96 & 0.38 & -1.11 & 0.58 \\ 1.01 & 2.21 & -1.04 & -0.13 \end{bmatrix}$.

Exercise 5.2.2 Recall that Example 5.2.6 explores the effective rank of the 5×5 Hilbert matrix depending upon a supposed level of error. Similarly, explore the effective rank of the 7×7 Hilbert matrix (hilb(7) in MATLAB/Octave) depending upon supposed levels of error in the matrix. What levels of error in the components would give what effective rank of the matrix?

Exercise 5.2.3 Recall that Exercise 2.2.11 considered the inner four planets in the solar system. The exercise fitted a quadratic polynomial to the orbital period $T = c_1 + c_2 R + c_3 R^2$ as a function of distance R using the data of Table 2.4. In view of the bad condition number, $\text{rcond} = 6 \cdot 10^{-6}$, revisit the task with the more powerful techniques of this section. Use the data for Mercury, Venus, and Earth to fit the quadratic and predict the period for Mars. Discuss how the bad condition number is due to the failure in Exercise 2.2.11 of scaling the data in the matrix.

Exercise 5.2.4 Recall Exercise 3.5.19 used a 4×4 grid of pixels in the computed tomography of a CT-scan. Redo this exercise recognizing that the entries in matrix A have errors up to roughly 0.5. Discuss any change in the prediction.

Exercise 5.2.5 Reconsider each of the matrix-vector systems you explored in Exercise 5.2.1. Also solve each system using Tikhonov regularization; for example, in the first system solve $Ax = b_1$, $A'x' = b_1$, and $A''x'' = b_1$. Discuss why x, x', and x'' are all reasonably close to the smallest solution of those obtained via an SVD.

Exercise 5.2.6 Recall that Example 5.2.6 explores the effective rank of the 5×5 Hilbert matrix, say A, depending upon a supposed level of error. Here do the alternative and solve the system $Ax = 1$ via Tikhonov regularization using a wide range of various regularization parameters α. Comment on the relation between the solutions obtained for various α and those obtained in the example for the various presumed error—perhaps plot the components of x versus parameter α (on a log-log plot).

Exercise 5.2.7 Recall that Example 5.2.8 used a 3×3 grid of pixels in the computed tomography of a CT-scan. Redo this example with Tikhonov regularization, recognizing that the entries in matrix A have errors up to roughly 0.5. Discuss the relation between the solution of Example 5.2.8 and that of Tikhonov regularization.

Exercise 5.2.8 Recall that Exercise 3.5.19 used a 4×4 grid of pixels in the computed tomography of a CT-scan. Redo this exercise with Tikhonov regularization, recognizing that the entries in matrix A have errors up to roughly 0.5. Discuss any change in the prediction, and compare to the answer for Exercise 5.2.4.

Exercise 5.2.9 In a few sentences, answer/discuss each of the following.

(a) How can errors in the components of a matrix get magnified to badly affect the solution of a system of linear equations?

(b) Recall that in some example linear equations, we reported on the variety of solutions found upon changing the matrix by a typical error in the components. What is the relation between the variety of solutions found and the solutions predicted by the SVD regularization Procedure 5.2.3?

(c) Why does the effective rank of a matrix typically depend upon the expected errors in the matrix?

(d) What are the advantages and disadvantages of Tikhonov regularization compared to using an SVD?

(e) How is dividing by $\sigma_i + \alpha^2/\sigma_i$ in Tikhonov regularization roughly like neglecting equations in the SVD regularization?

5.3 Summary of matrix approximation

Measure changes to matrices

★★ Procedure 5.1.3 approximates matrices. For an image stored as scalars in an $m \times n$ matrix A.

1. Compute an SVD $A = USV^\mathsf{T}$ with $[\mathtt{U},\mathtt{S},\mathtt{V}]=\mathtt{svd(A)}$.

2. Choose a desired rank k based upon the singular values (Theorem 5.1.16); typically there are k 'large' singular values and the rest are 'small'.

3. Then the 'best' rank k approximation to the image matrix A is

$$A_k := \sigma_1 u_1 v_1^\mathsf{T} + \sigma_2 u_2 v_2^\mathsf{T} + \cdots + \sigma_k u_k v_k^\mathsf{T}$$
$$= \mathtt{U(:,1:k)*S(1:k,1:k)*V(:,1:k)'}$$

- In MATLAB/Octave:
  - ★ `norm(A)` computes the matrix norm, Definition 5.1.7, namely the largest singular value of the matrix A. Also, `norm(v)` for vector v computes the length, or magnitude, $\sqrt{v_1^2 + v_2^2 + \cdots + v_n^2}$.
  - – `scatter(x,y,[],c)` draws a 2D scatter plot of points with coordinates in vectors x and y, each point with a colour determined by the corresponding entry of vector c.
    Similarly for `scatter3(x,y,z,[],c)` but in 3D.
  - ★ `[U,S,V]=svds(A,k)` computes the k largest singular values of the matrix A in the diagonal of $k \times k$ matrix S, and the k columns of U and V are the corresponding singular vectors.
  - – `imread('filename')` typically reads an image from a file into an $m \times n \times 3$ array of red-green-blue values. The values are all integers in the range $[0,255]$.

- ★ mean(A) of an $m \times n$ array computes the n elements in the row vector of averages (the arithmetic mean) over each column of A.

  Whereas mean(A,p) for an ℓ-dimensional array A of dimension $m_1 \times m_2 \times \cdots \times m_\ell$, computes the mean over the pth index to give an array of size $m_1 \times \cdots \times m_{p-1} \times 1 \times m_{p+1} \times \cdots \times m_\ell$.

- ★ std(A) of an $m \times n$ array computes the n elements in the row vector of the standard deviation over each column of A.

- − For the element-by-element operations of +, -, .*, ./, .^, and so on, MAT-LAB/*Octave automatically replicates scalars and vectors as appropriate.* Where this functionality becomes extremely useful is that if either of the matrices is $m \times 1$ or $1 \times n$ or one of each, then its column/row is replicated to a matrix of size $m \times n$ and the operation then performed.

- − csvread('filename') reads data from a file into a matrix. When each of the m lines in the file is n numbers separated by commas, then the result is an $m \times n$ matrix.

- − axis sets some properties of a drawn figure:
  - - axis equal ensures that horizontal and vertical directions are scaled the same—so here there is no distortion of the image;
  - - axis off means that the horizontal and vertical axes are not drawn—so here the image is unadorned.

★★ Define the **matrix norm** (sometimes called the spectral norm) such that for every $m \times n$ matrix A,

$$\|A\| := \max_{|x|=1} |Ax|, \quad \text{equivalently } \|A\| = \sigma_1$$

the largest singular value of the matrix A (Definition 5.1.7). This norm usefully measures the 'length' or 'magnitude' of a matrix, and hence also the 'distance' between two matrices as $\|A - B\|$.

★ For every $m \times n$ real matrix A (Theorem 5.1.12):
  - − $\|A\| = 0$ if and only if $A = O_{m \times n}$;
  - − $\|I_n\| = 1$;
  - − $\|A \pm B\| \leq \|A\| + \|B\|$, for every $m \times n$ matrix B—like a triangle inequality;
  - − $\|tA\| = |t| \cdot \|A\|$ for every scalar t;
  - − $\|A\| = \|A^{\mathsf{T}}\|$;
  - − $\|Q_m A\| = \|A\| = \|A Q_n\|$ for every $m \times m$ orthogonal matrix Q_m and every $n \times n$ orthogonal matrix Q_n;
  - − $|Ax| \leq \|A\| \cdot |x|$ for all $x \in \mathbb{R}^n$—like a Cauchy–Schwarz inequality, as is the following;
  - − $\|AB\| \leq \|A\| \|B\|$ for every $n \times p$ matrix B.

★ Let A be an $m \times n$ matrix of rank r with SVD $A = USV^{\mathsf{T}}$. Then for every $k < r$ the matrix

$$A_k := US_k V^{\mathsf{T}} = \sigma_1 u_1 v_1^{\mathsf{T}} + \sigma_2 u_2 v_2^{\mathsf{T}} + \cdots + \sigma_k u_k v_k^{\mathsf{T}}$$

where $S_k := \text{diag}(\sigma_1, \sigma_2, \ldots, \sigma_k, 0, \ldots, 0)$, is a *closest* rank k matrix approximating A (Theorem 5.1.16). The distance between A and A_k is $\|A - A_k\| = \sigma_{k+1}$.

- Given a $m \times n$ data matrix A (usually with zero mean when averaged over all rows) with SVD $A = USV^\mathsf{T}$, then the jth column v_j of V is called the jth **principal vector** and the vector $x_j := Av_j$ is called the jth **principal components** of the data matrix A (Definition 5.1.22).

- Using the matrix norm to measure 'best', the best k-dimensional summary of the $m \times n$ data matrix A (usually of zero mean) is the first k principal components in the directions of the first k principal vectors (Theorem 5.1.23).

★ Procedure 5.1.26 considers the case when you have data values consisting of n attributes for each of m instances: it finds a good k-dimensional summary/view of the data.

    1. Form/enter the $m \times n$ data matrix B.

    2. Scale the data matrix B to form $m \times n$ matrix A:

        (a) usually make each column have zero mean by subtracting its mean $\bar{b}_j$, algebraically $a_j = b_j - \bar{b}_j$;

        (b) but ensure that each column has the same 'physical dimensions', often by dividing by the standard deviation s_j of each column, algebraically $a_j = (b_j - \bar{b}_j)/s_j$.

    Compute in MATLAB/Octave via the auto-replication of

```
A=(B-mean(B))./std(B);
```

    3. Economically compute an SVD for the best rank k approximation to the scaled data matrix with `[U,S,V]=svds(A,k)`.

    4. Then the jth column of V is the jth principal vector, and the principal components are the entries of the $m \times k$ matrix `A*V`.

- Latent semantic indexing uses SVDs to form useful low-rank approximations to word data and queries.

Regularize linear equations

★★ Procedure 5.2.3 approximates linear equations. Suppose the system of linear equations $Ax = b$ arises from an experiment where both the $m \times n$ matrix A and the right-hand side vector b are subject to experimental error. Suppose the expected errors in the matrix and vector entries are of magnitude ϵ.

    1. When forming the matrix A and vector b, scale the data so that

        – all $m \times n$ components in A have the same physical units, and they are of roughly the same magnitude; and

        – similarly for the m components of b.

    Estimate the error ϵ corresponding to this matrix A.

    2. Compute an SVD $A = USV^\mathsf{T}$.

    3. Choose 'rank' k to be the number of singular values bigger than the error ϵ; that is, $\sigma_1 \geq \sigma_2 \geq \cdots \geq \sigma_k > \epsilon > \sigma_{k+1} \geq \cdots \geq 0$. Then the best rank k approximation to A has SVD

$$A_k = US_k V^{\mathsf{T}}$$
$$= \sigma_1 u_1 v_1^{\mathsf{T}} + \sigma_2 u_2 v_2^{\mathsf{T}} + \cdots + \sigma_k u_k v_k^{\mathsf{T}}$$
$$= \texttt{U(:,1:k)*S(1:k,1:k)*V(:,1:k)'}.$$

4. Solve the approximating linear equation $A_k x = b$ as in Theorems 3.5.8 and 3.5.13 (often as an inconsistent set of equations). Use the SVD $A_k = US_k V^{\mathsf{T}}$.

5. Among all the solutions allowed, choose the 'best' according to some explicit additional need of the application often the smallest solution overall, or just as often a solution with the most zero components.

- In seeking to solve the poorly posed system $Ax = b$ for $m \times n$ matrix A, a **Tikhonov regularization** is the system $(A^{\mathsf{T}}A + \alpha^2 I_n)x = A^{\mathsf{T}}b$ for some chosen regularization parameter value $\alpha > 0$ (Definition 5.2.9).
- Solving the Tikhonov regularization, with parameter α, of $Ax = b$ is equivalent to finding the smallest, least square, solution of the system $A'x = b$ where the matrix A' is obtained from A by replacing each of its nonzero singular values σ_i by $\sigma_i' := \sigma_i + \alpha^2/\sigma_i$ (Theorem 5.2.14).

Answers to selected activities

5.1.2d, 5.1.5c, 5.1.10a, 5.1.14b, 5.1.18b, 5.1.24b, 5.2.2c, 5.2.4b, 5.2.11a,

Answers to selected exercises

5.1.1b 1.4

5.1.1d 2.3

5.1.2b 5

5.1.2d 3

5.1.5 Rank three.

5.1.8a Use A= [A A A;A 0*A A;
A A A]

5.1.9a A= [A 0*A A;0*A A 0*A;
A 0*A A]

5.1.10 ranks 6, 24

5.1.12 Book 3: angles $35°, 32°, 1°, 29°, 54°,$ $76°$.

5.1.13c Books 14, 4, 5, 13 and 2 are at angles $1°, 3°, 9°, 10°$ and $16°$, respectively.

5.2.1b $x = (1.0, -1.0), x' = (0.54, -0.24),$ $x'' = (1.92, -2.46), x = (0.28,$ $0.17) + t(-0.52, 0.85)$ (2 d.p.).

5.2.1d $x = (-0.76, -0.23, 0.69, 0.48),$ $x' = (-1.36, 0.33, 1.35, 0.43),$ $x'' = (-0.09, -0.92, -0.17, 0.47),$ $x = (-0.38, -0.63, 0.2, 0.47) +$ $t(0.52, -0.54, -0.67, -0.02)$ (2 d.p.).

5.2.4 The matrix has effective rank of eleven. Pixel ten is still the most absorbing. The corner pixels are the most affected.

6 Determinants distinguish matrices

Although much of the theoretical role of determinants is usurped by the SVD, nonetheless, determinants aid in establishing forthcoming properties of eigenvalues and eigenvectors, and empower graduates to connect to much extant practice.

Recall from previous study (Section 4.1.1, e.g.)

- a 2×2 matrix $A = \begin{bmatrix} a & b \\ c & d \end{bmatrix}$ has determinant $\det A = |A| = ad - bc$, and that the matrix A is invertible if and only if $\det A \neq 0$;

- a 3×3 matrix $A = \begin{bmatrix} a & b & c \\ d & e & f \\ g & h & i \end{bmatrix}$ has determinant $\det A = |A| = aei + bfg + cdh - ceg - afh - bdi$, and that the matrix A is invertible if and only if $\det A \neq 0$.

For hand calculations, these two formulas for a determinant are best remembered via the following diagrams where products along the red lines are subtracted from the products along the blue lines, respectively:

$$
\begin{bmatrix} a & b \\ c & d \end{bmatrix}
\qquad
\begin{bmatrix} a & b & c \\ d & e & f \\ g & h & i \end{bmatrix} \begin{matrix} a & b \\ d & e \\ g & h \end{matrix}
\tag{6.1}
$$

This chapter extends these determinants to any size matrix, and explores more of the useful properties of a determinant—especially those properties useful for understanding and developing the general eigenvalue problems and applications of Chapter 7.

Linear Algebra for the 21st Century. A. J. Roberts, Oxford University Press (2020). © A. J. Roberts.
DOI: 10.1093/oso/9780198856399.003.0006

6.1 Geometry underlies determinants

Sections 3.2.2, 3.2.3 and 3.6 introduced that multiplication by a matrix transforms areas and volumes. Determinants give precisely how much a square matrix transforms such areas and volumes.

Example 6.1.1 Consider the square matrix $A = \begin{bmatrix} 1/2 & 1 \\ 0 & 1 \end{bmatrix}$. Use matrix-vector multiplication to find the image of the unit square under the transformation by A. How much is the area of the unit square scaled up/down? Compare with the determinant.

Solution: Consider the corner points of the unit square, under multiplication by A: using the 'mapsto' symbol $\mapsto$ to denote that vector $\mathbf{x}$ transforms to $A\mathbf{x}$, $(0,0) \mapsto (0,0)$, $(1,0) \mapsto (\frac{1}{2},0)$, $(0,1) \mapsto (1,1)$, and $(1,1) \mapsto (\frac{3}{2},1)$, as illustrated to the right (the 'roof' is only plotted to uniquely identify the sides). The resultant parallelogram has area of $\frac{1}{2}$ as its base is $\frac{1}{2}$ and its height is 1. This parallelogram area of $\frac{1}{2}$ is the same as the determinant since here (6.1) gives $\det A = \frac{1}{2} \cdot 1 - 0 \cdot 1 = \frac{1}{2}$. □

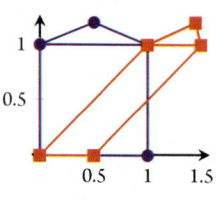

Example 6.1.2 Consider the square matrix $B = \begin{bmatrix} -1 & 1 \\ 1 & 1 \end{bmatrix}$. Use matrix-vector multiplication to find the image of the unit square under the transformation by B. How much is the unit area scaled up/down? Compare with the determinant.

Solution: Consider the corner points of the unit square, under multiplication by B: $(0,0) \mapsto (0,0)$, $(1,0) \mapsto (-1,1)$, $(0,1) \mapsto (1,1)$, and $(1,1) \mapsto (0,2)$, as shown to the right. Through multiplication by the matrix B, the unit square is expanded, rotated, and reflected. The resultant square has area of 2 as its sides are all of length $\sqrt{2}$. This area has the same magnitude as the determinant since here (6.1) gives $\det B = (-1) \cdot 1 - 1 \cdot 1 = -2$. □

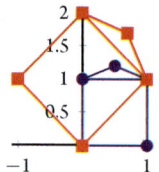

Activity 6.1.3 Upon multiplication by the matrix $\begin{bmatrix} -2 & 5 \\ -3 & -2 \end{bmatrix}$ the unit square transforms to a parallelogram. Use the determinant of the matrix to find the area of the parallelogram is which of the following.

(a) 4   (b) 11   (c) 19   (d) 16

Example 6.1.4 Let the square matrix $C = \begin{bmatrix} 2 & 0 & 0 \\ 1 & 1 & 0 \\ 0 & 0 & 3/2 \end{bmatrix}$. Use matrix-vector multiplication to find the image of the unit cube under the transformation by C. How much is the volume of the unit cube scaled up/down? Compare with the determinant.

Solution: Consider the corner points of the unit cube under multiplication by C:
$(0,0,0) \mapsto (0,0,0)$, $(1,0,0) \mapsto (2,1,0)$,
$(0,1,0) \mapsto (0,1,0)$, $(0,0,1) \mapsto (0,0,\frac{3}{2})$, and
so on, as shown right (in stereo). Through
multiplication by the matrix C, the unit
cube is deformed to a parallelepiped. The
resultant parallelepiped has volume of 3 as

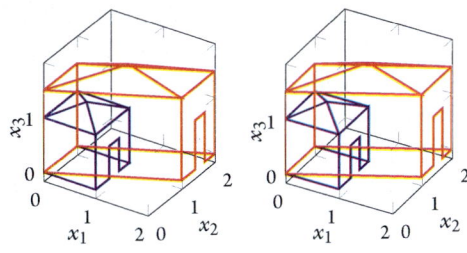

it has height $\frac{3}{2}$ and the parallelogram base has area $2 \cdot 1$. This volume is the same as the matrix determinant since (6.1) gives $\det C = 2 \cdot 1 \cdot \frac{3}{2} + 0 + 0 - 0 - 0 - 0 = 3$.   □

Determinants determine area transformation

Consider multiplication by the general 2×2 matrix
$A = \begin{bmatrix} a & b \\ c & d \end{bmatrix}$. Under multiplication by this matrix A the
unit square becomes the parallelogram, as illustrated to
the right, with four corners at $(0,0)$, (a,c), (b,d), and
$(a+b, c+d)$. Let's determine the area of the parallelogram by that of the containing rectangle (brown) less
the two small rectangles and the four triangles. The two
small rectangles have the same area, namely bc. The two
triangles on the left and the right also have the same
area, namely $\frac{1}{2} bd$. The two triangles on the top and the

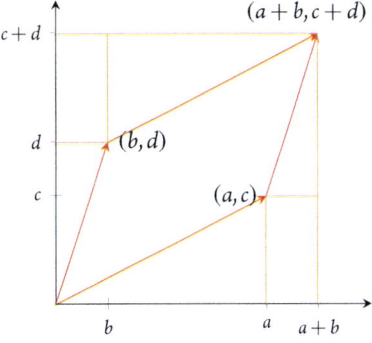

bottom have the same area, namely $\frac{1}{2} ac$. Thus, under multiplication by matrix A the image of the unit square is the parallelogram with

$$
\begin{aligned}
\text{area} &= (a+b)(c+d) - 2 \cdot bc - 2 \cdot \frac{1}{2} bd - 2 \cdot \frac{1}{2} ac \\
&= ac + ad + bc + bd - 2bc - bd - ac \\
&= ad - bc = \det A.
\end{aligned}
$$

This picture is the case when the matrix does not also reflect the image: if the matrix also reflects, as in Example 6.1.2, then the determinant is the negative of the area. In either case, the area of the unit square after transforming by the matrix A is the magnitude $|\det A|$.

Analogous geometric arguments relate determinants of 3×3 matrices
with transformations of volumes. Under multiplication by a 3×3 matrix $A = \begin{bmatrix} a_1 & a_2 & a_3 \end{bmatrix}$, the image of the unit cube is a parallelepiped with edges a_1,
a_2, and a_3 as illustrated. By computing the volumes of various rectangular
boxes, prisms, and tetrahedra, the volume of such a parallelepiped could be
expressed as the 3×3 determinant formula (6.1).

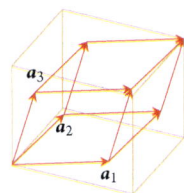

In higher dimensions we want the determinant to behave analogously and so next define the determinant to do so. We use the terms n**D-cube** to generalize a square and cube to n dimensions ($\mathbb{R}^n$), n**D-volume** to generalize the notion of area and volume to n dimensions, and so on. When the dimension of the space is unspecified, then we may say **hyper-cube**, **hyper-volume**, and so on.

Definition 6.1.5 *Let A be an $n \times n$ square matrix, and let C be the unit nD-cube in $\mathbb{R}^n$. Transform the nD-cube C by $x \mapsto Ax$ to its image C' in $\mathbb{R}^n$. Define the* **determinant** *of A, denoted either $\det A$ or sometimes $|A|$ such that:*

- *the magnitude $|\det A|$ is the nD-volume of C'; and*
- *the sign of $\det A$ to be negative iff the transformation reflects the orientation of the nD-cube.*

But what do we mean by "*orientation*" in this definition? Recall that complex-shaped objects come in two versions: humans have a left-hand and a right-hand; bolts have a left-handed thread or a right-handed thread; biological molecules are often either left-handed or right-handed (sometimes one is essential to life, and the other fatal). Such 'two-handedness' holds in every dimension (higher than 1D). Such left-hand and right-hand versions are the mirror image, the reflection, of each other. Here, a negative determinant indicates that the transformation of a geometric object changes a 'left-hand' version to a 'right-hand' version, or vice versa.

Example 6.1.6 Roughly estimate the determinant of the matrix that transforms the unit square to the parallelogram as shown to the right.

Solution: The image is a parallelogram with a vertical base of length about 0.8 and a horizontal height of about 0.7 so the area of the image is about $0.8 \times 0.7 = 0.56 \approx 0.6$. But the image has been reflected because one cannot rotate and/or stretch the unit square to get the image (remember the origin is fixed under matrix multiplication): thus the determinant must be negative. Our estimate for the matrix determinant is -0.6. ☐

Activity 6.1.7 Roughly estimate the determinant of the matrix that transforms the unit square to the rectangle as shown to the right.

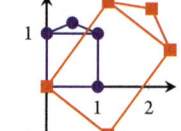

(a) 4 (b) 3 (c) 2.5 (d) 2

Basic properties of a determinant follow direct from Definition 6.1.5.

Theorem 6.1.8

(a) *For every $n \times n$ diagonal matrix D, the determinant of D is the product of the diagonal entries:* $\det D = d_{11}d_{22}\cdots d_{nn}$.

(b) *Every orthogonal matrix Q has $\det Q = \pm 1$ (only one alternative, not both). Further, $\det Q = \det(Q^{\mathsf{T}})$.*

(c) *For every $n \times n$ matrix A, $\det(kA) = k^n \det A$ for every scalar k.*

Proof. Use Definition 6.1.5.

6.1.8(a)

The unit nD-cube in $\mathbb{R}^n$ has edges e_1, e_2, ..., e_n (the unit vectors). Multiplying each of these edges by the diagonal matrix $D = \text{diag}(d_{11}, d_{22}, \ldots, d_{nn})$ maps the unit nD-cube to a nD-'rectangle' with edges $d_{11}e_1$, $d_{22}e_2$, ..., $d_{nn}e_n$ (as illustrated to the right). Being a nD-rectangle with all edges orthogonal, its nD-volume is the product of the length of the sides; that is, $|\det D| =$

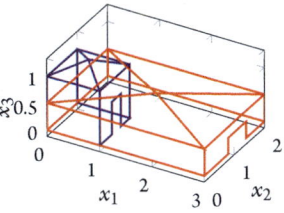

$|d_{11}| \cdot |d_{22}| \cdots |d_{nn}|$. The nD-cube is reflected only if there are an odd number of negative diagonal elements, hence the sign of the determinant is such that $\det D = d_{11}d_{22} \cdots d_{nn}$.

6.1.8(b) Multiplication by an orthogonal matrix Q is a rotation and/or reflection as it preserves all lengths and angles (Theorem 3.2.48(f)). Hence it preserves nD-volumes. Consequently the image of the unit nD-cube under multiplication by Q has the same volume of one; that is, $|\det Q| = 1$. The sign of $\det Q$ characterizes whether multiplication by Q has a reflection.

When Q is orthogonal then so is Q^{T} (Theorem 3.2.48(d)). Hence $\det Q^{\mathsf{T}} = \pm 1$. Multiplication by Q involves a reflection iff its inverse of multiplication by Q^{T} involves the reflection back again. Hence the signs of the two determinants must be the same: that is, $\det Q = \det Q^{\mathsf{T}}$.

6.1.8(c) Let the matrix A transform the unit nD-cube to a nD-parallelepiped C' that (by definition) has nD-volume $|\det A|$. Multiplication by the matrix (kA) then forms a nD-parallelepiped which is $|k|$-times bigger than C' in every direction. In $\mathbb{R}^n$ its nD-volume is then $|k|^n$-times bigger; that is, $|\det(kA)| = |k|^n |\det A| = |k^n \det A|$. If the scalar k is negative, then the orientation of the image is reversed (reflected) only in odd n dimensions; that is, the sign of the determinant is multiplied by $(-1)^n$.

For example, the unit square shown to the right is transformed through multiplication by $(-1)I_2$ and the effect is the same as rotation by $180°$, without any reflection as $(-1)^2 = 1$.

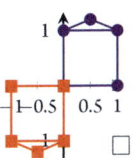

Hence for all real k, the orientation is such that $\det(kA) = k^n \det A$.

Example 6.1.9 The determinant of the $n \times n$ identity matrix is one: that is, $\det I_n = 1$. We justify this result in either of two ways.

- An identity matrix is a diagonal matrix and hence its determinant is the product of the diagonal entries (Theorem 6.1.8(a)), here all ones.
- Alternatively, multiplication by the identity does not change the unit nD-cube and so does not change its nD-volume or its orientation (Definition 6.1.5). ☐

Activity 6.1.10 What is the determinant of $-I_n$?

(a) -1

(c) $+1$

(b) $+1$ for odd n, and -1 for even n

(d) -1 for odd n, and $+1$ for even n

Example 6.1.11 Use (6.1) to compute the determinant of the orthogonal matrix

$$Q = \begin{bmatrix} \frac{1}{3} & -\frac{2}{3} & \frac{2}{3} \\ \frac{2}{3} & \frac{2}{3} & \frac{1}{3} \\ -\frac{2}{3} & \frac{1}{3} & \frac{2}{3} \end{bmatrix}.$$

Then use Theorem 6.1.8 to deduce the determinants of the following matrices:

$$\begin{bmatrix} \frac{1}{3} & \frac{2}{3} & -\frac{2}{3} \\ -\frac{2}{3} & \frac{2}{3} & \frac{1}{3} \\ \frac{2}{3} & \frac{1}{3} & \frac{2}{3} \end{bmatrix}, \quad \begin{bmatrix} -1 & 2 & -2 \\ -2 & -2 & -1 \\ 2 & -1 & -2 \end{bmatrix}, \quad \begin{bmatrix} \frac{1}{6} & -\frac{1}{3} & \frac{1}{3} \\ \frac{1}{3} & \frac{1}{3} & \frac{1}{6} \\ -\frac{1}{3} & \frac{1}{6} & \frac{1}{3} \end{bmatrix}.$$

Solution:

- Using (6.1) the determinant

$$\det Q = \frac{1}{3}\frac{2}{3}\frac{2}{3} + (-\frac{2}{3})\frac{1}{3}(-\frac{2}{3}) + \frac{2}{3}\frac{2}{3}\frac{1}{3} - \frac{2}{3}\frac{2}{3}(-\frac{2}{3}) - \frac{1}{3}\frac{1}{3}\frac{1}{3} - (-\frac{2}{3})\frac{2}{3}\frac{2}{3}$$
$$= \frac{1}{27}(4 + 4 + 4 + 8 - 1 + 8) = 1.$$

- The first matrix is Q^{T} for which $\det Q^{\mathsf{T}} = \det Q = 1$.
- The second matrix is minus three times Q so, being a 3×3 matrix, its determinant is $(-3)^3 \det Q = -27 \cdot 1 = -27$.
- The third matrix is half of Q so, being a 3×3 matrix, its determinant is $(\frac{1}{2})^3 \det Q = \frac{1}{8} \cdot 1 = \frac{1}{8}$. □

Activity 6.1.12 Given $\det \begin{bmatrix} 1 & -1 & -2 & 0 \\ 1 & 1 & 1 & 1 \\ 1 & -1 & -2 & -1 \\ -1 & 0 & 0 & -1 \end{bmatrix} = -1$, what is $\det \begin{bmatrix} -2 & 2 & 4 & 0 \\ -2 & -2 & -2 & -2 \\ -2 & 2 & 4 & 2 \\ 2 & 0 & 0 & 2 \end{bmatrix}$?

(a) -4 (b) 8   (c) 2   (d) -16

A consequence of Theorem 6.1.8(c) is that a determinant characterizes the transformation of any sized hyper-cube. Consider the transformation by a matrix A of an nD-cube of side length k ($k \geq 0$), and hence of volume k^n. The nD-cube has edges $k\boldsymbol{e}_1, k\boldsymbol{e}_2, \ldots, k\boldsymbol{e}_n$. The transformation results in an nD-parallelepiped with edges $A(k\boldsymbol{e}_1), A(k\boldsymbol{e}_2), \ldots, A(k\boldsymbol{e}_n)$, which by commutativity and associativity (Theorem 3.1.26(d)) are the same edges as $(kA)\boldsymbol{e}_1, (kA)\boldsymbol{e}_2, \ldots, (kA)\boldsymbol{e}_n$. That is, the resulting nD-parallelepiped is the same as applying matrix (kA) to the unit nD-cube, and so must have nD-volume $k^n |\det A|$. This is a factor of $|\det A|$ times the original volume. Crucially, this property that matrix multiplication multiplies all sizes of hyper-cubes by the determinant holds for all other shapes and sizes, not just hyper-cubes. Let's see a specific example before proving the general theorem.

Example 6.1.13 Multiplication by some specific matrix transforms the (blue) triangle C to the (red) triangle C' as shown to the right. By finding the ratio of the areas, estimate the magnitude of the determinant of the matrix.

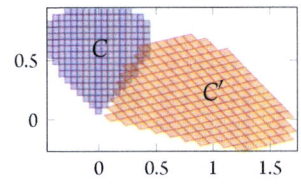

Solution: The (blue) triangle C has vertical height one and horizontal base one, so has area 0.5. The mapped (red) triangle C' has vertical base of 1.5 and horizontal height of 1.5 so its area is $\frac{1}{2} \times 1.5 \times 1.5 = 1.125$. The mapped area is thus $1.125/0.5 = 2.25$ times bigger than the initial area; hence the determinant of the transformation matrix has magnitude $|\det A| = 2.25$.

We cannot determine the sign of the determinant as we do not know about the orientation of C' relative to C. □

Theorem 6.1.14 *Consider any bounded smooth nD-volume C in $\mathbb{R}^n$ and its image C' after multiplication by $n \times n$ matrix A. Then*

$$\det A = \pm \frac{nD\text{-volume of } C'}{nD\text{-volume of } C}$$

with the negative sign when matrix A changes the orientation.

Proof. The geometric proof is analogous to integration in calculus (Hannah, 1996, p.402). In two dimensions, as drawn to the right, we divide a given region C into many small squares of side length k, each of area k^2; each of these transforms to a small parallelogram of area $k^2|\det A|$ (by Theorem 6.1.8(c)), then the sum of the transformed areas is just $|\det A|$ times the original area of C.

In n-dimensions, divide a given region C into many small nD-cubes of side length k, each of nD-volume k^n; each of these transforms to a small nD-parallelepiped of nD-volume $k^n|\det A|$ (by Theorem 6.1.8(c)), then the sum of the transformed nD-volume is just $|\det A|$ times the nD-volume of C. □

A more rigorous proof would involve upper and lower sums for the original and transformed regions, and also explicit restrictions to regions where these upper and lower sums converge to a unique nD-volume. We do not detail such a more rigorous proof here.

This property of transforming general areas and volumes also establishes the next crucial property of determinants, namely that the determinant of a matrix product is the product of the determinants: $\det(AB) = \det(A)\det(B)$ for all square matrices A and B (of the same size).

Example 6.1.15 Recall the two 2×2 matrices of Examples 6.1.1 and 6.1.2: $A = \begin{bmatrix} 1/2 & 1 \\ 0 & 1 \end{bmatrix}$ and $B = \begin{bmatrix} -1 & 1 \\ 1 & 1 \end{bmatrix}$. Check that the determinant of their product is the product of their determinants.

Solution: First, Examples 6.1.1 and 6.1.2 computed $\det A = \frac{1}{2}$ and $\det B = -2$. Thus the product of their determinants is $\det(A)\det(B) = -1$.

Secondly, calculate the matrix product

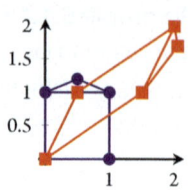

$$AB = \begin{bmatrix} \frac{1}{2} & 1 \\ 0 & 1 \end{bmatrix} \begin{bmatrix} -1 & 1 \\ 1 & 1 \end{bmatrix} = \begin{bmatrix} \frac{1}{2} & \frac{3}{2} \\ 1 & 1 \end{bmatrix},$$

whose multiplicative action upon the unit square is illustrated to the right. By (6.1), $\det(AB) = \frac{1}{2} \cdot 1 - \frac{3}{2} \cdot 1 = -1$, as required.

Thirdly, we should also check the other product (as the question does not specify the order of the product):

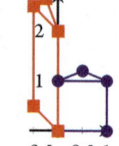

$$BA = \begin{bmatrix} -1 & 1 \\ 1 & 1 \end{bmatrix} \begin{bmatrix} \frac{1}{2} & 1 \\ 0 & 1 \end{bmatrix} = \begin{bmatrix} -\frac{1}{2} & 0 \\ \frac{1}{2} & 2 \end{bmatrix},$$

whose multiplicative action upon the unit square is illustrated to the right. By (6.1), $\det(BA) = -\frac{1}{2} \cdot 2 - 0 \cdot \frac{1}{2} = -1$, as required. ☐

Theorem 6.1.16 *For every two* $n \times n$ *matrices* A *and* B, $\det(AB) = \det(A)\det(B)$. *Also, for all* $n \times n$ *matrices* A_1, A_2, $\ldots$, A_ℓ, $\det(A_1 A_2 \cdots A_\ell) = \det(A_1)\det(A_2)\cdots\det(A_\ell)$.

Proof. Consider the unit nD-cube C, its image C' upon transforming by B, and the image C'' after transforming C' by A. That is, each edge e_j of cube C is mapped to the edge Be_j of C', which is in turn mapped to edge $A(Be_j)$ of C''. By Definition 6.1.5, C' has (signed) nD-volume $\det B$. Theorem 6.1.14 implies C'' has (signed) nD-volume $\det A$ times that of C'; that is, C'' has (signed) nD-volume $\det(A)\det(B)$.

By associativity, the jth edge $A(Be_j)$ of C'' is the same as $(AB)e_j$ and so C'' is the image of C under the transformation by matrix (AB). Consequently, the (signed) nD-volume of C'' is alternatively given by $\det(AB)$. These two expressions for the nD-volume of C'' must be equal: that is, $\det(AB) = \det(A)\det(B)$.

Exercise 6.1.13 uses induction to then prove the second statement in the theorem that $\det(A_1 A_2 \cdots A_\ell) = \det(A_1)\det(A_2)\cdots\det(A_\ell)$. ☐

Activity 6.1.17 Given that the three matrices

$$\begin{bmatrix} -1 & 0 & -1 \\ 0 & -1 & 1 \\ 0 & 1 & 1 \end{bmatrix}, \quad \begin{bmatrix} 0 & 1 & -2 \\ -1 & -1 & 1 \\ -1 & -2 & -1 \end{bmatrix}, \quad \begin{bmatrix} -1 & 0 & -1 \\ 0 & 1 & 1 \\ 1 & 2 & 0 \end{bmatrix},$$

have determinants 2, −4, and 3, respectively, what is the determinant of the product of the three matrices?

(a) 9   (b) 24   (c) 1   (d) −24

Example 6.1.18

(a) Confirm the product rule for determinants, Theorem 6.1.16, for the product

$$\begin{bmatrix} -3 & -2 \\ 3 & -3 \end{bmatrix} = \begin{bmatrix} 3 & 1 & 1 \\ 0 & -3 & 0 \end{bmatrix} \begin{bmatrix} -1 & 0 \\ -1 & 1 \\ 1 & -3 \end{bmatrix}.$$

Solution: Although the determinant of the left-hand matrix is $(-3)(-3) - 3(-2) = 9 - (-6) = 15$, we cannot confirm the product rule because it does not apply: the matrices on the right-hand side are not square matrices and so do not have determinants. □

(b) Given $\det A = 2$ and $\det B = \pi$, what is $\det(AB)$?

Solution: Strictly, there is no answer as we do not know that matrices A and B are square and of the same size. However, if we are *additionally* given that A and B are square and of the same size, then Theorem 6.1.16 gives $\det(AB) = \det(A)\det(B) = 2\pi$. □

Example 6.1.19 Use the product theorem to help find the determinant of matrix

$$C = \begin{bmatrix} 45 & -15 & 30 \\ -2\pi & \pi & 2\pi \\ \frac{1}{9} & \frac{2}{9} & -\frac{1}{3} \end{bmatrix}.$$

Solution: One route is to observe that there is a common factor in each row of the matrix so it may be factored as

$$C = \begin{bmatrix} 15 & 0 & 0 \\ 0 & \pi & 0 \\ 0 & 0 & \frac{1}{9} \end{bmatrix} \begin{bmatrix} 3 & -1 & 2 \\ -2 & 1 & 2 \\ 1 & 2 & -3 \end{bmatrix}.$$

The first matrix, being diagonal, has determinant that is the product of its diagonal elements (Theorem 6.1.8(a)) so its determinant $= 15\pi\frac{1}{9} = \frac{5}{3}\pi$. The second matrix, from (6.1), has determinant $= -9 - 2 - 8 - 2 - 12 + 6 = -27$. Theorem 6.1.16 then gives $\det C = \frac{5}{3}\pi \cdot (-27) = -45\pi$. □

We now proceed to link the determinant of a matrix to the singular values of the matrix.

Example 6.1.20 Recall that Example 3.3.4 showed that the following matrix has the given SVD:

$$A = \begin{bmatrix} -4 & -2 & 4 \\ -8 & -1 & -4 \\ 6 & 6 & 0 \end{bmatrix} = \begin{bmatrix} \frac{1}{3} & -\frac{2}{3} & \frac{2}{3} \\ \frac{2}{3} & \frac{2}{3} & \frac{1}{3} \\ -\frac{2}{3} & \frac{1}{3} & \frac{2}{3} \end{bmatrix} \begin{bmatrix} 12 & 0 & 0 \\ 0 & 6 & 0 \\ 0 & 0 & 3 \end{bmatrix} \begin{bmatrix} -\frac{8}{9} & -\frac{1}{9} & -\frac{4}{9} \\ -\frac{4}{9} & \frac{4}{9} & \frac{7}{9} \\ -\frac{1}{9} & -\frac{8}{9} & \frac{4}{9} \end{bmatrix}^T.$$

Use this SVD to find the magnitude $|\det A|$.

Solution: Given the svd $A = USV^\mathsf{T}$, the Product Theorem 6.1.16 gives $\det A = \det(USV^\mathsf{T}) = \det(U)\det(S)\det(V^\mathsf{T})$.

- $\det U = \pm 1$ by Theorem 6.1.8(b) as U is an orthogonal matrix.

- Using Theorem 6.1.8(b), $\det(V^\mathsf{T}) = \det V = \pm 1$ as V is orthogonal.

- Since $S = \operatorname{diag}(12,6,3)$ is diagonal, Theorem 6.1.8(a) asserts that its determinant is the product of the diagonal elements; that is, $\det S = 12 \cdot 6 \cdot 3 = 216$.

Consequently $\det A = (\pm 1)216(\pm 1) = \pm 216$, so $|\det A| = 216$. $\square$

Theorem 6.1.21 *For every $n \times n$ square matrix A, the magnitude of its determinant $|\det A| = \sigma_1 \sigma_2 \cdots \sigma_n$, the product of all its singular values.*

Proof. Consider an svd of the matrix $A = USV^\mathsf{T}$. Theorems 6.1.8 and 6.1.16 empower the following identities:

$$
\begin{aligned}
\det A &= \det(USV^\mathsf{T}) = \det(U)\det(S)\det(V^\mathsf{T}) \quad \text{(by Theorem 6.1.16)} \\
&= (\pm 1)\det(S)(\pm 1) \quad \text{(by Theorem 6.1.8(b))} \\
&= \pm \det S \quad \text{(and then since S is diagonal)} \\
&= \pm \sigma_1 \sigma_2 \cdots \sigma_n. \quad \text{(by Theorem 6.1.8(a))}
\end{aligned}
$$

Hence $|\det A| = \sigma_1 \sigma_2 \cdots \sigma_n$.

Example 6.1.22 Confirm Theorem 6.1.21 for the matrix $A = \begin{bmatrix} 10 & 2 \\ 5 & 11 \end{bmatrix}$ of Example 3.3.2.

Solution: Example 3.3.2 gave the svd

$$
\begin{bmatrix} 10 & 2 \\ 5 & 11 \end{bmatrix} = \begin{bmatrix} \frac{3}{5} & -\frac{4}{5} \\ \frac{4}{5} & \frac{3}{5} \end{bmatrix} \begin{bmatrix} 10\sqrt{2} & 0 \\ 0 & 5\sqrt{2} \end{bmatrix} \begin{bmatrix} \frac{1}{\sqrt{2}} & -\frac{1}{\sqrt{2}} \\ \frac{1}{\sqrt{2}} & \frac{1}{\sqrt{2}} \end{bmatrix}^\mathsf{T},
$$

so Theorem 6.1.21 asserts $|\det A| = 10\sqrt{2} \cdot 5\sqrt{2} = 100$. Using (6.1) directly, $\det A = 10 \cdot 11 - 2 \cdot 5 = 110 - 10 = 100$ which agrees with the product of the singular values. $\square$

Activity 6.1.23 Consider the following matrix and its svd:

$$
A = \begin{bmatrix} -2 & -4 & 5 \\ -6 & 0 & -6 \\ 5 & 4 & -2 \end{bmatrix} = \begin{bmatrix} \frac{1}{3} & -\frac{2}{3} & \frac{2}{3} \\ \frac{2}{3} & \frac{2}{3} & \frac{1}{3} \\ -\frac{2}{3} & \frac{1}{3} & \frac{2}{3} \end{bmatrix} \begin{bmatrix} 9 & 0 & 0 \\ 0 & 9 & 0 \\ 0 & 0 & 0 \end{bmatrix} \begin{bmatrix} -\frac{8}{9} & -\frac{1}{9} & -\frac{4}{9} \\ -\frac{4}{9} & \frac{4}{9} & \frac{7}{9} \\ -\frac{1}{9} & -\frac{8}{9} & \frac{4}{9} \end{bmatrix}^\mathsf{T}.
$$

What is the magnitude of the determinant of A, $|\det A|$?

(a) 18 (b) 0 (c) 4 (d) 81

Example 6.1.24 Use an SVD of the following matrix to find the magnitude of its determinant:

$$A = \begin{bmatrix} -2 & -1 & 4 & -5 \\ -3 & 2 & -3 & 1 \\ -3 & -1 & 0 & 3 \end{bmatrix}.$$

Solution: Although an SVD exists for this matrix and so we could form the product of its singular values, the concept of a determinant only applies to square matrices and so there is no such thing as a determinant for this 3×4 matrix A. The task is meaningless. □

Establishing this connection between determinants and singular values relied on Theorem 6.1.8(b), that transposing an orthogonal matrix does not change its determinant, $\det Q^{\mathsf{T}} = \det Q$. We now establish that this determinant-transpose property holds for the transpose of all square matrices.

Example 6.1.25 Example 6.1.18(a) determined that $\det \begin{bmatrix} -3 & -2 \\ 3 & -3 \end{bmatrix} = 15$. By (6.1), its transpose has determinant

$$\det \begin{bmatrix} -3 & 3 \\ -2 & -3 \end{bmatrix} = (-3)^2 - 3(-2) = 9 + 6 = 15.$$

The determinants are the same. □

Theorem 6.1.26 *For every square matrix A, $\det(A^{\mathsf{T}}) = \det A$.*

Proof. Use an SVD of the $n \times n$ matrix, say $A = USV^{\mathsf{T}}$. Then Theorems 6.1.8 and 6.1.16 empowers the following:

$$\begin{aligned}
\det A &= \det(USV^{\mathsf{T}}) = \det(U)\det(S)\det(V^{\mathsf{T}}) && \text{(by Theorem 6.1.16)} \\
&= \det(U^{\mathsf{T}})\det(S)\det(V) && \text{(by Theorem 6.1.8(b))} \\
&= \det(U^{\mathsf{T}})(\sigma_1\sigma_2\cdots\sigma_n)\det(V) && \text{(by Theorem 6.1.8(a))} \\
&= \det(U^{\mathsf{T}})\det(S^{\mathsf{T}})\det(V) && \text{(by Theorem 6.1.8(a))} \\
&= \det(V)\det(S^{\mathsf{T}})\det(U^{\mathsf{T}}) && \text{(by scalar commutativity)} \\
&= \det(VS^{\mathsf{T}}U^{\mathsf{T}}) && \text{(by Theorem 6.1.16)} \\
&= \det[(USV^{\mathsf{T}})^{\mathsf{T}}] = \det(A^{\mathsf{T}}).
\end{aligned}$$

□

Example 6.1.27 Every 3×3 matrix has the form $A = \begin{bmatrix} a & b & c \\ d & e & f \\ g & h & i \end{bmatrix}$ which has determinant det

$A = |A| = aei + bfg + cdh - ceg - afh - bdi$. Its transpose, $A^{\mathsf{T}} = \begin{bmatrix} a & d & g \\ b & e & h \\ c & f & i \end{bmatrix}$, from the rule (6.1)

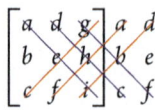

has determinant $\det A^{\mathsf{T}} = aei + dhc + gbf - gec - ahf - dbi = \det A$. □

One of the main reasons for studying determinants is to establish when solutions to linear equations may exist or not (albeit only applicable to square matrices when there are n linear equations in n unknowns). One example lies in finding eigenvalues by hand (Section 4.1.1), where we solve $\det(A - \lambda I) = 0$.

Recall that for 2×2 and 3×3 matrices we commented that a matrix is invertible only when its determinant is nonzero. Theorem 6.1.29 establishes this in general. The geometric reason for this connection between invertibility and determinants is that when a determinant is zero the action of multiplying by the matrix 'squashes' the unit nD-cube into a nD-parallelepiped of zero thickness. Such extreme squashing cannot be uniquely undone.

Example 6.1.28 Consider multiplication by the

matrix $A = \begin{bmatrix} 1 & 1/2 & 0 \\ 0 & 1/2 & 1 \\ 0 & 0 & 0 \end{bmatrix}$, whose effect on the unit

cube is illustrated to the right. As illustrated, this matrix squashes the unit cube onto the $x_1 x_2$-plane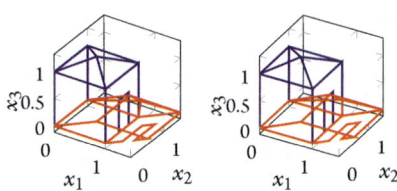
$(x_3 = 0)$. Consequently the resultant volume is zero
and so $\det A = 0$. Because many points in 3D space are squashed onto the same point in the $x_3 = 0$ plane, the action of the matrix cannot be undone. Hence the matrix is not invertible. That the matrix is not invertible and its determinant is zero is not a coincidence. □

Theorem 6.1.29 A square matrix A is invertible iff $\det A \neq 0$. If a matrix A is invertible, then $\det(A^{-1}) = 1/(\det A)$.

Proof. First, Theorem 6.1.21 establishes that $\det A \neq 0$ iff all the singular values of square matrix A are nonzero, which by Theorem 3.4.43(d) is iff matrix A is invertible.

Second, as matrix A is invertible, an inverse A^{-1} exists such that $AA^{-1} = I_n$. Then the product of determinants

$$\begin{aligned} \det(A)\det(A^{-1}) &= \det(AA^{-1}) \quad \text{(by Theorem 6.1.16)} \\ &= \det I_n \quad \text{(from } AA^{-1} = I_n) \\ &= 1 \quad \text{(by Example 6.1.9)} \end{aligned}$$

For an invertible matrix A, $\det A \neq 0$; hence dividing by $\det A$ gives $\det(A^{-1}) = 1/\det A$.

$\square$

6.1.1 Exercises

Exercise 6.1.1 For each of the given illustrations of a linear transformation of the unit square, 'guesstimate' by eye the determinant of the matrix of the transformation (estimate to an error of say 33% or so).

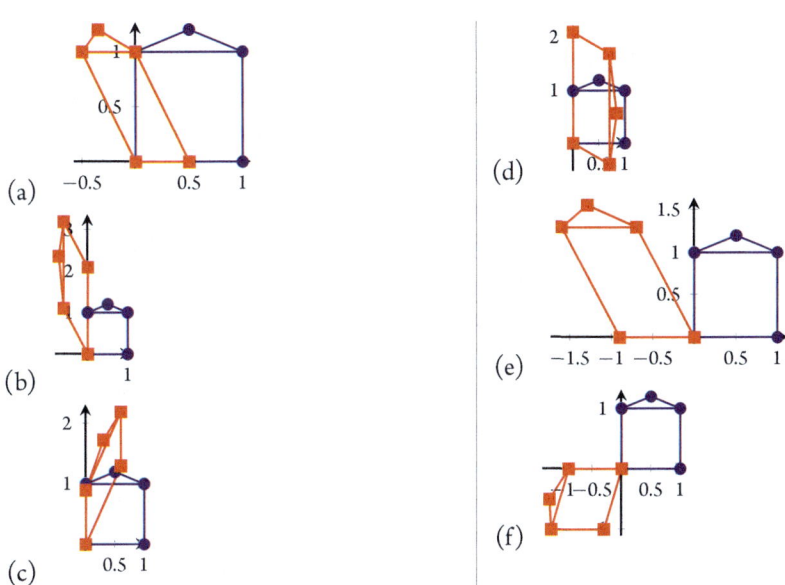

(a) (b) (c) (d) (e) (f)

Exercise 6.1.2 For each of the transformations illustrated in Exercise 6.1.1, estimate the matrix of the linear transformation (to within say 10%). Then use formula (6.1) to estimate the determinant of your matrix and confirm Exercise 6.1.1.

Exercise 6.1.3 For each of the following stereo illustrations of a linear transformation of the unit cube, estimate the matrix of the linear transformation (to within say 20%). Then use formula (6.1) to estimate the determinant of the matrix of the transformation.

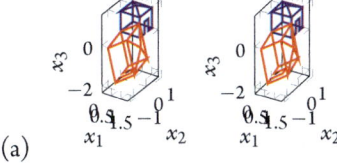

(a)

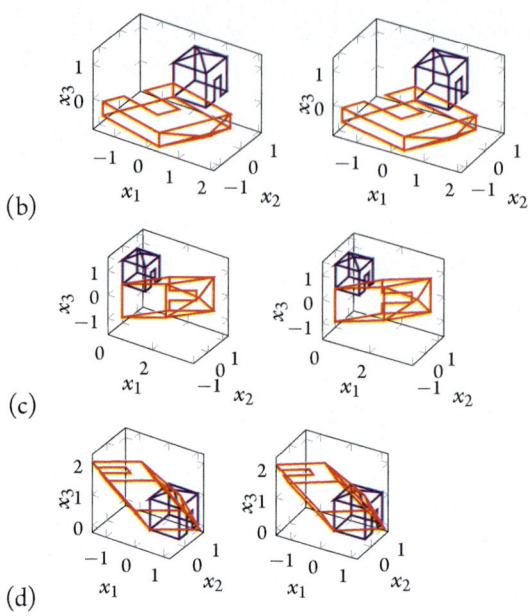

(b)

(c)

(d)

Exercise 6.1.4 For each of the following matrices, use (6.1) to find all the values of k for which the matrix is *not* invertible.

(a) $A = \begin{bmatrix} 0 & 6-2k \\ -2k & -4 \end{bmatrix}$

(b) $B = \begin{bmatrix} 3k & 4-k \\ -4 & 0 \end{bmatrix}$

(c) $C = \begin{bmatrix} 2 & 0 & -2k \\ 4k & 0 & -1 \\ 0 & k & -4+3k \end{bmatrix}$

(d) $D = \begin{bmatrix} 2 & -2-4k & -1+k \\ -1-k & 0 & -5k \\ 0 & 0 & 4+2k \end{bmatrix}$

Exercise 6.1.5 Find the determinants of each of the following matrices.

(a) $\begin{bmatrix} -3 & 0 & 0 & 0 \\ 0 & -4 & 0 & 0 \\ 0 & 0 & 1 & 0 \\ 0 & 0 & 0 & 3 \end{bmatrix}$

(b) $\begin{bmatrix} 5/6 & 0 & 0 & 0 \\ 0 & -1 & 0 & 0 \\ 0 & 0 & 7/6 & 0 \\ 0 & 0 & 0 & 2/3 \end{bmatrix}$

(c) $\begin{bmatrix} 1/3 & 0 & -4/3 & -1 \\ -1/3 & 1/3 & 5/3 & 4/3 \\ 0 & -1/3 & 5/3 & -1 \\ -1/3 & -1/3 & 8/3 & -1/3 \end{bmatrix}$

given det $\begin{vmatrix} 1 & 0 & -4 & -3 \\ -1 & 1 & 5 & 4 \\ 0 & -1 & 5 & -3 \\ -1 & -1 & 8 & -1 \end{vmatrix} = -8$

(d) $\begin{bmatrix} -1 & 3/2 & -1/2 & 1/2 \\ -2 & -1/2 & -1 & -3/2 \\ -0 & -3/2 & -5/2 & 1/2 \\ -0 & -1/2 & 3 & 3/2 \end{bmatrix}$

given det $\begin{vmatrix} 2 & -3 & 1 & -1 \\ 4 & 1 & 2 & 3 \\ 0 & 3 & 5 & -1 \\ 0 & 1 & -6 & -3 \end{vmatrix} = -524$

(e) $\begin{bmatrix} -0 & -2/3 & -2 & -4/3 \\ -0 & 1/3 & -1/3 & 1/3 \\ 1 & -1/3 & -5/3 & 2/3 \\ -7/3 & 1/3 & 4/3 & 2/3 \end{bmatrix}$

given det $\begin{bmatrix} 0 & 2 & 6 & 4 \\ 0 & -1 & 1 & -1 \\ -3 & 1 & 5 & -2 \\ 7 & -1 & -4 & -2 \end{bmatrix} = 246$

Exercise 6.1.6 Use Theorem 3.2.27 and Theorem 6.1.8(a) to prove that for every diagonal square matrix D, $\det(D^{-1}) = 1/\det D$ provided $\det D \neq 0$.

Exercise 6.1.7 For each pair of following matrices, by computing in full using (6.1) confirm $\det(AB) = \det(BA) = \det(A)\det(B)$. Show your working.

(a) $A = \begin{bmatrix} -2 & 3/2 \\ -1/2 & 0 \end{bmatrix}$, $B = \begin{bmatrix} 0 & -1/2 \\ 1/2 & -1 \end{bmatrix}$

(b) $A = \begin{bmatrix} -1 & 1 \\ -7/2 & -3/2 \end{bmatrix}$, $B = \begin{bmatrix} 1/2 & -5/2 \\ -1/2 & 3/2 \end{bmatrix}$

(c) $A = \begin{bmatrix} -2 & -1/2 & 0 \\ 0 & 0 & 2 \\ 0 & -1 & 0 \end{bmatrix}$, $B = \begin{bmatrix} -1 & -1 & 0 \\ 0 & 0 & -1 \\ 0 & -2 & 0 \end{bmatrix}$

(d) $A = \begin{bmatrix} 1 & 2 & -3/2 \\ 0 & 3/2 & 0 \\ -1 & 0 & 0 \end{bmatrix}$, $B = \begin{bmatrix} 0 & 1 & 0 \\ -4 & 2 & 0 \\ -1 & 0 & -5 \end{bmatrix}$

Exercise 6.1.8 Given that $\det(AB) = \det(A)\det(B)$ for every two square matrices of the same size, prove that $\det(AB) = \det(BA)$ (despite $AB \neq BA$ in general).

Exercise 6.1.9 Given that $n \times n$ matrices A and B have $\det A = 3$ and $\det B = -5$, determine the following determinants (if possible).

(a) $\det(AB)$

(b) $\det(B^2)$

(c) $\det(A^4)$

(d) $\det(A + B)$

(e) $\det(A^{-1}B)$

(f) $\det(2A)$

Exercise 6.1.10 Let A and P be square matrices of the same size, and let matrix P be invertible. Prove that $\det(P^{-1}AP) = \det A$.

Exercise 6.1.11 Suppose square matrix A satisfies $A^2 = A$ (called idempotent). Determine all possible values of $\det A$. Invent and verify a nontrivial example of an idempotent matrix.

Exercise 6.1.12 Suppose a square matrix A satisfies $A^p = O$ for some integer exponent $p \geq 2$ (called nilpotent). Determine all possible values of $\det A$. Invent and verify a nontrivial example of a nilpotent matrix.

Exercise 6.1.13 Recall that $\det(AB) = \det(A)\det(B)$ for every two square matrices of the same size. For $n \times n$ matrices A_1, A_2, $\ldots$, A_ℓ, use induction to prove the second part of Theorem 6.1.16, namely that $\det(A_1 A_2 \cdots A_\ell) = \det(A_1)\det(A_2)\cdots\det(A_\ell)$ for every integer $\ell \geq 2$.

Exercise 6.1.14 To complement the algebraic argument of Theorem 6.1.29, use a geometric argument based upon the transformation of nD-volumes to establish that $\det(A^{-1}) = 1/(\det A)$ for an invertible matrix A.

Exercise 6.1.15 Suppose square matrix $A = \begin{bmatrix} P & O \\ O & Q \end{bmatrix}$ for some square matrices P and Q, and appropriately sized zero matrices O. Give a geometric argument justifying that $\det A = (\det P)(\det Q)$.

Exercise 6.1.16 In a few sentences, answer/discuss each of the following.

(a) Why is it important that the determinant characterizes how a matrix transforms vectors?

(b) What causes a determinant to be only defined for square matrices?

(c) Why does every orthogonal matrix have determinant of magnitude one?

(d) What is simple about the determinant of a diagonal matrix? Why does this simplicity arise?

(e) What is the evidence for the relationship between how a matrix transforms arbitrary hyper-volumes and the determinant of the matrix?

(f) What causes the determinant of a matrix product to be the products of the determinants?

(g) What causes the determinant of a matrix to be of the same magnitude as the product of the singular values?

(h) What is the evidence for a matrix being invertible if and only if its determinant is nonzero?

6.2 Laplace expansion theorem for determinants

This section develops a so-called row/column algebraic expansion for determinants. This expansion is useful for many theoretical purposes. But there are vastly more efficient ways of computing determinants than using a row/column expansion. In MATLAB/Octave one may invoke det (A) to compute the determinant of a matrix. You may find this function useful for checking the results of some examples and exercises. However, just like computing an inverse, computing the determinant is expensive and error prone. In medium to large scale problems, avoid computing the determinant. In practice, something else is almost always better.

> The most numerically reliable way to determine whether matrices are singular [not invertible] is to test their singular values. This is far better than trying to compute determinants, which have atrocious scaling properties. *Cleve Moler, MathWorks, 2006*

Nonetheless, a row/column algebraic expansion for a determinant is useful for small matrix problems, as well as for its beautiful theoretical uses. We start with examples of row properties that underpin a row/column algebraic expansion. We use these properties to step-by-step develop determinants for matrices in a more and more general form.

Example 6.2.1 *(Theorem 6.2.5(a))* Example 6.1.28 argued geometrically that the determinant is zero for the matrix $A = \begin{bmatrix} 1 & 1/2 & 0 \\ 0 & 1/2 & 1 \\ 0 & 0 & 0 \end{bmatrix}$. Confirm this determinant algebraically.

Solution: Using (6.1), $\det A = 1 \cdot \frac{1}{2} \cdot 0 + \frac{1}{2} \cdot 1 \cdot 0 + 0 \cdot 0 \cdot 0 - 0 \cdot \frac{1}{2} \cdot 0 - 1 \cdot 1 \cdot 0 - \frac{1}{2} \cdot 0 \cdot 0 =$ 0. In every term, there is a zero from the last row of the matrix.   □

Example 6.2.2 *(Theorem 6.2.5(b))* Consider the matrix with two identical rows, $A = \begin{bmatrix} 1 & 1/2 & 1/5 \\ 1 & 1/2 & 1/5 \\ 0 & 1/2 & 1 \end{bmatrix}$. Confirm algebraically that its determinant is zero. Give a geometric reason for why its determinant has to be zero.

Solution: Using (6.1), $\det A = 1 \cdot \frac{1}{2} \cdot 1 + \frac{1}{2} \cdot \frac{1}{5} \cdot 0 + \frac{1}{5} \cdot 1 \cdot \frac{1}{2} - \frac{1}{5} \cdot \frac{1}{2} \cdot 0 - 1 \cdot \frac{1}{5} \cdot \frac{1}{2} - \frac{1}{2} \cdot 1 \cdot$ $1 = \frac{1}{2} + \frac{1}{10} - \frac{1}{10} - \frac{1}{2} = 0.$

Geometrically, consider the image of the unit cube under multiplication by this matrix A illustrated in stereo to the right. Because the first two rows of A are identical, the first two components of Ax are always identical and hence all points are mapped onto the plane $x_1 = x_2$. The image of the cube thus has zero thickness and hence zero volume. By Definition 6.1.5, $\det A = 0$.   □

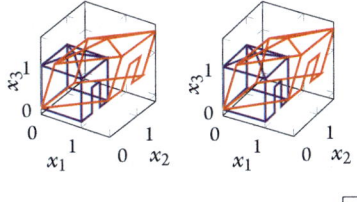

Example 6.2.3 *(Theorem 6.2.5(c))* Consider the two matrices with two rows swapped (the first two rows): $A = \begin{bmatrix} 1 & -1 & 0 \\ 0 & 1 & 1 \\ 1/5 & 1/2 & 1 \end{bmatrix}$ and $B = \begin{bmatrix} 0 & 1 & 1 \\ 1 & -1 & 0 \\ 1/5 & 1/2 & 1 \end{bmatrix}$. Confirm algebraically that their determinants are the negative of each other. Give a geometric reason why this should be so.

Solution: Using (6.1) twice:

- $\det A = 1 \cdot 1 \cdot 1 + (-1) \cdot 1 \cdot \frac{1}{5} + 0 \cdot 0 \cdot \frac{1}{2} - 0 \cdot 1 \cdot \frac{1}{5} - 1 \cdot 1 \cdot \frac{1}{2} - (-1) \cdot 0 \cdot 1 = 1 - \frac{1}{5} -$ $\frac{1}{2} = \frac{3}{10}$;

- $\det B = 0 \cdot (-1) \cdot 1 + 1 \cdot 0 \cdot \frac{1}{5} + 1 \cdot 1 \cdot \frac{1}{2} - 1 \cdot (-1) \cdot \frac{1}{5} - 0 \cdot 0 \cdot \frac{1}{2} - 1 \cdot 1 \cdot 1 = \frac{1}{2} + \frac{1}{5} -$ $1 = -\frac{3}{10} = -\det A.$

Geometrically, since the first two rows in A and B are swapped that means that multiplying by the matrix, as in Ax and Bx, has the first two components swapped. Hence Ax and Bx are always the reflection of each other in the plane $x_1 = x_2$. Consequently, the images of the unit cubes under multiplication by A and by B are the reflection of each other in the plane $x_1 = x_2$, and so the determinants must be the negative of each other.   □

Example 6.2.4 *(Theorem 6.2.5(d))* Compute the determinant of the matrix $B = \begin{bmatrix} 1 & -1 & 0 \\ 0 & 1 & 1 \\ 2 & 5 & 10 \end{bmatrix}$. Compare B with matrix A given in Example 6.2.3, and compare their determinants.

Solution: Using (6.1), $\det B = 1 \cdot 1 \cdot 10 + (-1) \cdot 1 \cdot 2 + 0 \cdot 0 \cdot 5 - 0 \cdot 1 \cdot 2 - 1 \cdot 1 \cdot 5 -$ $(-1) \cdot 0 \cdot 10 = 10 - 2 - 5 = 3$. Matrix B is the same as matrix A except that the third row is a factor of ten times bigger. Correspondingly, $\det B = 3 = 10 \times \frac{3}{10} = 10 \det A$.   □

The above four examples are specific cases of the four general properties established by the following theorem.

Theorem 6.2.5 (*row and column properties of determinants*) *For every* $n \times n$ *matrix* A *the following properties hold.*

(a) *If* A *has a zero row or column, then* $\det A = 0$.

(b) *If* A *has two identical rows or two identical columns, then* $\det A = 0$.

(c) *Let* B *be obtained by interchanging two rows or interchanging two columns of* A, *then* $\det B = -\det A$.

(d) *Let* B *be obtained by multiplying any* one *row or column of* A *by a scalar* k, *then* $\det B = k \det A$.

Proof. We establish the properties for matrix rows. Then the same properties hold for the columns because $\det(A^T) = \det(A)$ (Theorem 6.1.26).

6.2.5(d) Suppose row i of matrix A is multiplied by k to give a new matrix B. Let the diagonal matrix

$$D := \text{diag}(1, \ldots, 1, k, 1, \ldots, 1),$$

with the factor k being in the ith row and column. Then $\det D = 1 \times \cdots \times 1 \times k \times 1 \times \cdots \times 1 = k$ by Theorem 6.1.8(a). Because multiplication by D multiplies the ith row by the factor k and leaves everything else the same, $B = DA$. Equate determinants of both sides and use the product Theorem 6.1.16: $\det B = \det(DA) = \det(D) \det(A) = k \det A$.

6.2.5(a) This arises from the case $k = 0$ of property 6.2.5(d). Then $A = DA$ because multiplying the ith row of A by $k = 0$ maintains that the row is zero. Consequently, $\det A = \det(DA) = \det(D) \det(A) = 0 \det(A)$ and hence $\det A = 0$.

6.2.5(c) Suppose rows i and j of matrix A are swapped to form matrix B. Let the matrix E be the identity, except with rows i and j swapped:

$$E := \begin{bmatrix} \ddots & & & & \\ & 0 & & 1 & \\ & & \ddots & & \\ & 1 & & 0 & \\ & & & & \ddots \end{bmatrix} \begin{matrix} \\ \text{row } i \\ \\ \text{row } j \\ \\ \end{matrix}$$

where the diagonal dots $\ddots$ denote diagonals of ones, and all other unshown entries of E are zero. Then $B = EA$ as multiplication by E copies row i into row j and vice versa. Equate determinants of both sides and use Theorem 6.1.16: $\det B = \det(EA) = \det(E) \det(A)$. To find $\det E$ observe that $EE^T = E^2 = I_n$ so E is orthogonal and hence $\det E = \pm 1$ by Theorem 6.1.8(b). Geometrically, multiplication by E is a simple reflection in the $(n-1)$D-plane $x_i = x_j$ hence its determinant must be negative, so $\det E = -1$. Consequently, $\det B = \det(E) \det(A) = -\det A$.

6.2.5(b) Suppose rows i and j of matrix A are identical. Using the matrix E to swap these two identical rows results in the same matrix: that is, $A = EA$. Take determinants of both sides:

$\det A = \det(EA) = \det(E)\det(A)$. Since $\det E = -1$ it follows that $\det A = -\det A$. Zero is the only number that equals its negative: thus $\det A = 0$. ☐

Example 6.2.6 You are given that $\det A = -9$ for the matrix

$$A = \begin{bmatrix} 0 & 2 & 3 & 1 & 4 \\ -2 & 2 & -2 & 0 & -3 \\ 4 & -2 & -4 & 1 & 0 \\ 2 & -1 & -4 & 2 & 2 \\ 5 & 4 & 3 & -2 & -5 \end{bmatrix}.$$

Use Theorem 6.2.5 to find the determinant of the following matrices, giving reasons.

(a) $\begin{bmatrix} 0 & 2 & 3 & 0 & 4 \\ -2 & 2 & -2 & 0 & -3 \\ 4 & -2 & -4 & 0 & 0 \\ 2 & -1 & -4 & 0 & 2 \\ 5 & 4 & 3 & 0 & -5 \end{bmatrix}$

Solution: $\det = 0$ as the fourth column is all zeros. ☐

(c) $\begin{bmatrix} 0 & 2 & 3 & 1 & 4 \\ -2 & 2 & -2 & 0 & -3 \\ 4 & -2 & -4 & 1 & 0 \\ 2 & -1 & -4 & 2 & 2 \\ -2 & 2 & -2 & 0 & -3 \end{bmatrix}$

Solution: $\det = 0$ as the second and fifth rows are identical. ☐

(b) $\begin{bmatrix} 0 & 2 & 3 & 1 & 4 \\ -2 & 2 & -2 & 0 & -3 \\ 2 & -1 & -4 & 2 & 2 \\ 4 & -2 & -4 & 1 & 0 \\ 5 & 4 & 3 & -2 & -5 \end{bmatrix}$

Solution: $\det = -\det A = +9$ as the third and fourth rows are swapped. ☐

(d) $\begin{bmatrix} 0 & 1 & 3 & 1 & 4 \\ -2 & 1 & -2 & 0 & -3 \\ 4 & -1 & -4 & 1 & 0 \\ 2 & -\frac{1}{2} & -4 & 2 & 2 \\ 5 & 2 & 3 & -2 & -5 \end{bmatrix}$

Solution: $\det = \frac{1}{2}\det A = -\frac{9}{2}$ as the second column is half that of A. ☐

Activity 6.2.7 Now, $\det \begin{bmatrix} 2 & -3 & 1 \\ 2 & -5 & -3 \\ -4 & 1 & -3 \end{bmatrix} = -36.$

• Which of the following matrices has determinant of 18?

(a) $\begin{bmatrix} -4 & 6 & -2 \\ 2 & -5 & -3 \\ -4 & 1 & -3 \end{bmatrix}$

(c) $\begin{bmatrix} -1 & -3 & 1 \\ -1 & -5 & -3 \\ 2 & 1 & -3 \end{bmatrix}$

(b) $\begin{bmatrix} 2 & -3 & 1 \\ 2 & -5 & -3 \\ 2 & -5 & -3 \end{bmatrix}$

(d) $\begin{bmatrix} 2 & -3 & 1/3 \\ 2 & -5 & -1 \\ -4 & 1 & -1 \end{bmatrix}$

• Further, which has determinant -12? 0? 72?

Example 6.2.8 Without evaluating the determinant, use Theorem 6.2.5 to establish that the determinant equation

$$\begin{vmatrix} 1 & x & y \\ 1 & 2 & 3 \\ 1 & 4 & 5 \end{vmatrix} = 0 \tag{6.2}$$

is an equation for the straight line in the xy-plane that passes through the two points $(2,3)$ and $(4,5)$.

Solution: First, the determinant equation (6.2) is linear in variables x and y as at most one of x and y occur in each term of the determinant expansion:

$$\begin{bmatrix} 1 & x & y \\ 1 & 2 & 3 \\ 1 & 4 & 5 \end{bmatrix}\begin{matrix} 1 & x \\ 1 & 2 \\ 1 & 4 \end{matrix}$$

Since the equation is linear in x and y, any solution set of (6.2) must be a straight line. Second, equation (6.2) is satisfied when $(x,y) = (2,3)$ or when $(x,y) = (4,5)$, because then two rows in the determinant are identical, and so Theorem 6.2.5(b) assures us that the determinant is zero. Thus the solution straight line passes through the two required points.

Let's evaluate the determinant to check: equation (6.2) becomes $1 \cdot 2 \cdot 5 + x \cdot 3 \cdot 1 + y \cdot 1 \cdot 4 - y \cdot 2 \cdot 1 - 1 \cdot 3 \cdot 4 - x \cdot 1 \cdot 5 = -2 - 2x + 2y = 0$. That is, $y = x + 1$ which does indeed pass through $(2,3)$ and $(4,5)$. □

Example 6.2.9 Without evaluating the determinant, use Theorem 6.2.5 to establish that the determinant equation

$$\begin{vmatrix} x & y & z \\ -1 & -2 & 2 \\ 3 & 5 & 2 \end{vmatrix} = 0$$

is, in xyz-space, an equation of the plane that passes through the origin and the two points $(-1,-2,2)$ and $(3,5,2)$.

Solution: As in the previous example, the determinant is linear in x, y, and z, so the solutions must be those of a single linear equation, namely a plane in xyz-space (Section 1.3.4).

- The solutions include the origin, since when $x = y = z = 0$ the first row of the matrix is zero, hence the determinant is zero, and the equation satisfied.

- The solutions include the points $(-1,-2,2)$ and $(3,5,2)$ since when (x,y,z) is either of these points, then two rows in the determinant are identical, so the determinant is zero, and the equation satisfied.

Hence the solutions are the points in the plane passing through the origin and $(-1,-2,2)$ and $(3,5,2)$. □

Following Theorem 6.2.5, the next step in developing a general 'formula' for a determinant is the special class of matrices for which one column or row is zero except for one element.

Example 6.2.10 Find the determinant of $A = \begin{bmatrix} -2 & -1 & -1 \\ 1 & -3 & -2 \\ 0 & 0 & 2 \end{bmatrix}$ which has two zeros in its last row.

Solution: Let's use two different arguments, both illustrating the next theorem and proof.

- Using (6.1),

$$\begin{aligned} \det A &= (-2)(-3)2 + (-1)(-2)\cdot 0 + (-1)1\cdot 0 \\ &\quad - (-1)(-3)0 - (-2)(-2)0 - (-1)1\cdot 2 \\ &= 2[(-2)(-3) - (-1)1] = 2\cdot 7 = 14. \end{aligned}$$

Observe $\det A = 2\cdot 7 = 2\cdot \det \begin{bmatrix} -2 & -1 \\ 1 & -3 \end{bmatrix}$ which is the expression to be generalized.

- Alternatively we use the product rule for determinants. Recognize that the matrix A may be written as the product $A = FB$ where

$$F = \begin{bmatrix} 1 & 0 & -\frac{1}{2} \\ 0 & 1 & -1 \\ 0 & 0 & 1 \end{bmatrix}, \quad B = \begin{bmatrix} -2 & -1 & 0 \\ 1 & -3 & 0 \\ 0 & 0 & 2 \end{bmatrix};$$

just multiply out and see that the last column of F 'fills in' the last column of A from that of B. Consider the geometry of the two transformations arising from multiplication by F and by B.

- Multiplication by F shears the unit cube, as illustrated to the right. Thus the volume of the unit cube after multiplication by F is the square base of area one, times the height of one, which is a volume of one. Consequently, by Definition 6.1.5, $\det F = 1$.

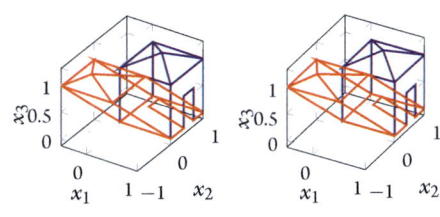

- As illustrated to the right, multiplication by B has two components. Firstly, the 2×2 top-left sub-matrix, being bordered by zeros, maps the unit square in the x_1x_2-plane into the base parallelogram in the x_1x_2-plane. The shape

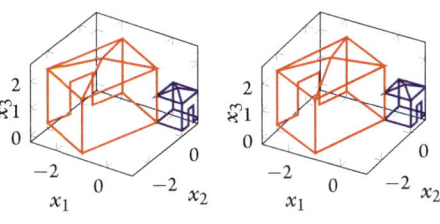

of the parallelogram is determined by the top-left sub-matrix $\begin{bmatrix} -2 & -1 \\ 1 & -3 \end{bmatrix}$ (to be called the A_{33} minor) acting on the unit square. Thus the area of the parallelogram is $\det A_{33} = (-2)(-3) - (-1)1 = 7$. Secondly, the 2 in the bottom corner of B stretches objects vertically by a factor of 2 to form a parallelepiped of height 2. Thus the volume of the parallelepiped, $\det B$ by Definition 6.1.5, is $2\cdot \det A_{33} = 2\cdot 7 = 14$.

By the product Theorem 6.1.16, $\det A = \det(FB) = \det(F)\det(B) = 1 \cdot 14 = 14$.

The key to this alternative evaluation of the determinant is the last row of matrix A which was all zero except for one element. The next Theorem 6.2.11 addresses this case in general. □

Theorem 6.2.11 *(almost zero row/column)* *For every $n \times n$ matrix A, define the (i,j)th **minor** A_{ij} to be the $(n-1) \times (n-1)$ square matrix obtained from A by omitting the ith row and jth column. If, except for the entry a_{ij}, the ith row (or jth column) of A is all zero, then*

$$\det A = (-1)^{i+j} a_{ij} \det A_{ij}. \tag{6.3}$$

The pattern of signs in this formula, $(-1)^{i+j}$, is

$$
\begin{array}{cccccc}
+ & - & + & - & + & \cdots \\
- & + & - & + & - & \cdots \\
+ & - & + & - & + & \cdots \\
- & + & - & + & - & \cdots \\
+ & - & + & - & + & \cdots \\
\vdots & \vdots & \vdots & \vdots & \vdots & \ddots
\end{array}
$$

Proof. We establish the determinant formula (6.3) for matrix rows, then the same result holds for the columns because $\det(A^{\mathsf{T}}) = \det(A)$ (Theorem 6.1.26). First, if the entry $a_{ij} = 0$, then the whole ith row (or jth column) is zero and so $\det A = 0$ by Theorem 6.2.5(a). Also, the expression $(-1)^{i+j} a_{ij} \det A_{ij} = 0$ as $a_{ij} = 0$. Consequently, the identity $\det A = (-1)^{i+j} a_{ij} \det A_{ij}$ holds. The rest of this proof addresses the case $a_{ij} \neq 0$.

Second, consider the special case when the last row *and* last column of matrix A is all zero except for $a_{nn} \neq 0$; that is,

$$
A = \begin{bmatrix} A_{nn} & \mathbf{0} \\ \mathbf{0}^{\mathsf{T}} & a_{nn} \end{bmatrix}
$$

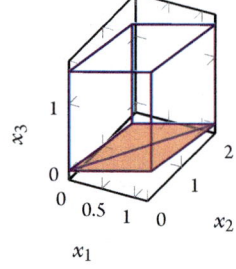

for the minor A_{nn} and $\mathbf{0} \in \mathbb{R}^{n-1}$. Recall Definition 6.1.5: the image of the nD-cube under multiplication by the matrix A is the image of the $(n-1)$D-cube under multiplication by A_{nn} extended orthogonally a length a_{nn} in the orthogonal direction $\mathbf{e}_n$ (as illustrated in 3D to the above-right). The volume of the nD-image is thus $a_{nn} \times$ (volume of the $(n-1)$D-image). Consequently, $\det A = a_{nn} \det A_{nn}$.

Third, consider the special case when the last row of matrix A is all zero except for $a_{nn} \neq 0$; that is,

$$
A = \begin{bmatrix} A_{nn} & \mathbf{a}'_n \\ \mathbf{0}^{\mathsf{T}} & a_{nn} \end{bmatrix}
$$

for the minor A_{nn}, and where $\mathbf{a}'_n = (a_{1n}, a_{2n}, \ldots, a_{n-1,n})$. Define the two $n \times n$ matrices

$$
F := \begin{bmatrix} I_{n-1} & \mathbf{a}'_n/a_{nn} \\ \mathbf{0}^{\mathsf{T}} & 1 \end{bmatrix} \quad \text{and} \quad B := \begin{bmatrix} A_{nn} & \mathbf{0} \\ \mathbf{0}^{\mathsf{T}} & a_{nn} \end{bmatrix}.
$$

Then $A = FB$ since the product

$$FB = \begin{bmatrix} I_{n-1} & a'_n/a_{nn} \\ 0^T & 1 \end{bmatrix} \begin{bmatrix} A_{nn} & 0 \\ 0^T & a_{nn} \end{bmatrix}$$

$$= \begin{bmatrix} I_{n-1}A_{nn} + a'_n/a_{nn}0^T & I_{n-1}0 + a'_n/a_{nn} \cdot a_{nn} \\ 0^T A_{nn} + 1 \cdot 0^T & 0^T 0 + 1 \cdot a_{nn} \end{bmatrix} = \begin{bmatrix} A_{nn} & a'_n \\ 0^T & a_{nn} \end{bmatrix} = A.$$

By Theorem 6.1.16, $\det A = \det(FB) = \det(F)\det(B)$. From the previous part $\det B = a_{nn}\det A_{nn}$, so we just need to determine $\det F$. As illustrated for 3D to the right, the action of matrix F on the unit nD-cube is that of a simple shear keeping the $(n-1)$D-cube base unchanged (due to the identity I_{n-1} in F). Since the height orthogonal to the $(n-1)$D-cube base is unchanged (due to the one in the bottom-right corner of F), the action of multiplying by F leaves the volume of the unit nD-cube unchanged at one. Hence $\det F = 1$. Thus $\det A = 1\det(B) = a_{nn}\det A_{nn}$ as required.

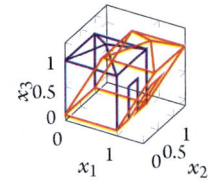

Fourth, suppose row i of matrix A is all zero except for entry a_{ij}. Swap rows i and $i+1$, then swap rows $i+1$ and $i+2$, and so on until the original row i is in the last row, and the order of all other rows are unchanged; this takes $(n-i)$ row swaps which changes the sign of the determinant $(n-i)$ times (Theorem 6.2.5(c)), that is, multiplies it by $(-1)^{n-i}$. Then swap columns j and $j+1$, then swap columns $j+1$ and $j+2$, and so on until the original column j is in the last column: this takes $(n-j)$ column swaps which changes the determinant by a factor $(-1)^{n-j}$ (Theorem 6.2.5(c)). The resulting matrix, say C, has the form

$$C = \begin{bmatrix} A_{ij} & a'_j \\ 0^T & a_{ij} \end{bmatrix}$$

for a'_j denoting the jth column of A with the ith entry omitted. Since matrix C has the form addressed in the previous part, we know $\det C = a_{ij}\det A_{ij}$. From the row and column swapping, $\det A = (-1)^{n-i}(-1)^{n-j}\det C = (-1)^{2n-i-j}\det C = (-1)^{-(i+j)}\det C = (-1)^{i+j}\det C = (-1)^{i+j}a_{ij}\det A_{ij}$. □

Example 6.2.12 Use Theorem 6.2.11 to evaluate the determinant of the following matrices.

(a) $\begin{bmatrix} -3 & -3 & -1 \\ -3 & 2 & 0 \\ 0 & 0 & 2 \end{bmatrix}$

Solution: There are two zeros in the bottom row so the determinant is
$(-1)^6 2\det\begin{bmatrix} -3 & -3 \\ -3 & 2 \end{bmatrix} = 2(-6-9) =$
-30. □

(b) $\begin{bmatrix} 2 & -1 & 7 \\ 0 & 3 & 0 \\ 2 & 2 & 5 \end{bmatrix}$

Solution: There are two zeros in the middle row so the determinant is

$(-1)^4 3\det\begin{bmatrix} 2 & 7 \\ 2 & 5 \end{bmatrix} = 3(10-14) =$
-12. □

(c) $\begin{bmatrix} 2 & 1 & 3 \\ 0 & -2 & -3 \\ 0 & 2 & 4 \end{bmatrix}$

Solution: There are two zeros in the first column so the determinant is
$(-1)^2 2\det\begin{bmatrix} -2 & -3 \\ 2 & 4 \end{bmatrix} = 2(-8+6) = -4.$ □

Activity 6.2.13 Using one of the determinants in the above Example 6.2.12, what is the determinant of the matrix

$$\begin{bmatrix} 2 & 1 & 0 & 3 \\ 5 & -2 & 15 & 2 \\ 0 & -2 & 0 & -3 \\ 0 & 2 & 0 & 4 \end{bmatrix}?$$

(a) 60 (b) 120 (c) -120 (d) -60

Example 6.2.14 Use Theorem 6.2.11 to evaluate the determinant of the so-called triangular matrix

$$A = \begin{bmatrix} 2 & -2 & 3 & 1 & 0 \\ 0 & 2 & -1 & -1 & -7 \\ 0 & 0 & 5 & -2 & -9 \\ 0 & 0 & 0 & 1 & 1 \\ 0 & 0 & 0 & 0 & 3 \end{bmatrix}$$

Solution: The last row is all zero except for the last element of 3, so

$$\det A = (-1)^{10} 3 \det \begin{bmatrix} 2 & -2 & 3 & 1 \\ 0 & 2 & -1 & -1 \\ 0 & 0 & 5 & -2 \\ 0 & 0 & 0 & 1 \end{bmatrix}$$

(then as the last row is zero except the 1)

$$= 3 \cdot (-1)^8 1 \det \begin{bmatrix} 2 & -2 & 3 \\ 0 & 2 & -1 \\ 0 & 0 & 5 \end{bmatrix}$$

(then as the last row is zero except the 5)

$$= 3 \cdot 1 \cdot (-1)^6 5 \det \begin{bmatrix} 2 & -2 \\ 0 & 2 \end{bmatrix}$$

(then as the last row is zero except the 2)

$$= 3 \cdot 1 \cdot 5 (-1)^4 2 \det [2] = 3 \cdot 1 \cdot 5 \cdot 2 \cdot 2 = 60. \qquad \square$$

The relative simplicity of finding the determinant in Example 6.2.14 indicates that there is something special and memorable about matrices with zeros in the entire lower-left 'triangle'. There is, as expressed by the following definition and theorem.

Definition 6.2.15 *A **triangular matrix** is a square matrix where all entries are zero either to the lower-left of the diagonal or to the upper-right:*[1]

- *an upper triangular matrix has the form (although any of the a_{ij} may also be zero)*

$$
\begin{bmatrix}
a_{11} & a_{12} & \cdots & a_{1\,n-1} & a_{1n} \\
0 & a_{22} & \cdots & a_{2\,n-1} & a_{2n} \\
\vdots & 0 & \ddots & \vdots & \vdots \\
0 & \vdots & \ddots & a_{n-1\,n-1} & a_{n-1\,n} \\
0 & 0 & \cdots & 0 & a_{nn}
\end{bmatrix};
$$

- *a lower triangular matrix has the form (although any of the a_{ij} may also be zero)*

$$
\begin{bmatrix}
a_{11} & 0 & \cdots & 0 & 0 \\
a_{21} & a_{22} & 0 & \cdots & 0 \\
\vdots & \vdots & \ddots & \ddots & \vdots \\
a_{n-1\,1} & a_{n-1\,2} & \cdots & a_{n-1\,n-1} & 0 \\
a_{n1} & a_{n2} & \cdots & a_{nn-1} & a_{nn}
\end{bmatrix}.
$$

Any square diagonal matrix is both an upper triangular matrix, and also a lower triangular matrix. Thus the following theorem encompasses square diagonal matrices and so generalizes Theorem 6.1.8(a).

Theorem 6.2.16 *(triangular matrix) For every $n \times n$ triangular matrix A, the determinant of A is the product of the diagonal entries, $\det A = a_{11} a_{22} \cdots a_{nn}$.*

Proof. A little induction proves that the determinant of a triangular matrix is the product of its diagonal entries: only consider upper triangular matrices as transposing the matrix caters for lower triangular matrices.

First, for 1×1 matrices the result is trivial. The result is also straightforward for 2×2 matrices since the determinant

$$
\det \begin{bmatrix} a_{11} & a_{12} \\ 0 & a_{22} \end{bmatrix} = a_{11}a_{22} - 0a_{12} = a_{11}a_{22}
$$

which is the product of the diagonal entries as required.

Second, assume the property for $(n-1) \times (n-1)$ matrices. Now, every upper triangular $n \times n$ matrix A has the form

$$
A = \begin{bmatrix} A_{nn} & a'_n \\ \mathbf{0}^{\mathsf{T}} & a_{nn} \end{bmatrix}
$$

[1] From time to time, people call an upper triangular matrix either a right triangular or an upper-right triangular matrix. Correspondingly, from time to time, people call a lower triangular matrix either a left triangular or a lower-left triangular matrix.

for $(n-1) \times (n-1)$ minor A_{nn}. Theorem 6.2.11 establishes $\det A = a_{nn} \det A_{nn}$. Since the minor A_{nn} is upper triangular and $(n-1) \times (n-1)$, by assumption $\det A_{nn} = a_{11}a_{22} \cdots a_{n-1,n-1}$. Consequently, $\det A = a_{nn} \det A_{nn} = a_{nn}a_{11}a_{22} \cdots a_{n-1,n-1}$, as required. Induction on n then establishes the theorem for every n. □

Activity 6.2.17 Which of the following matrices is *not* a triangular matrix?

(a) $\begin{bmatrix} 0 & 0 & 0 & -2 \\ 0 & 0 & -1 & -1 \\ 0 & 1 & 1 & 4 \\ -1 & -1 & 0 & 3 \end{bmatrix}$

(c) $\begin{bmatrix} 0 & 0 & 0 & 0 \\ 3 & 4 & 0 & 0 \\ 4 & -2 & -1 & 0 \\ -1 & -2 & 2 & -3 \end{bmatrix}$

(b) $\begin{bmatrix} -2 & 0 & 0 & 0 \\ 0 & -1 & 0 & 0 \\ 0 & 0 & 2 & 0 \\ 0 & 0 & 0 & 3 \end{bmatrix}$

(d) $\begin{bmatrix} -1 & -1 & 1 & 0 \\ 0 & -5 & 4 & 2 \\ 0 & 0 & 1 & -2 \\ 0 & 0 & 0 & 1 \end{bmatrix}$

Example 6.2.18 Find the determinant of those of the following matrices which are triangular.

(a) $\begin{bmatrix} -1 & -1 & -1 & -5 \\ 0 & -4 & 1 & 4 \\ 0 & 0 & 7 & 0 \\ 0 & 0 & 0 & -3 \end{bmatrix}$

Solution: This is upper triangular, and its determinant is
$$(-1) \cdot (-4) \cdot 7 \cdot (-3) = -84. \quad □$$

(b) $\begin{bmatrix} -3 & 0 & 0 & 0 \\ -4 & 2 & 0 & 0 \\ -1 & 1 & 1 & 0 \\ -2 & -3 & 7 & -1 \end{bmatrix}$

Solution: This is lower triangular, and its determinant is
$$(-3) \cdot 2 \cdot 1 \cdot (-1) = 6. \quad □$$

(c) $\begin{bmatrix} 1 & -1 & 1 & -3 \\ 0 & 0 & 0 & -5 \\ 0 & 0 & -3 & -4 \\ 0 & -2 & 1 & -2 \end{bmatrix}$

Solution: This is not triangular, so we do not have to compute its determinant. Nonetheless, if we swap the 2nd and 4th rows, then the result is the upper triangular $\begin{bmatrix} 1 & -1 & 1 & -3 \\ 0 & -2 & 1 & -2 \\ 0 & 0 & -3 & -4 \\ 0 & 0 & 0 & -5 \end{bmatrix}$

and its determinant is
$$1 \cdot (-2) \cdot (-3) \cdot (-5) = -30.$$ But the row swap changes the sign so the determinant of the original matrix is $-(-30) = 30.$ □

(d) $\begin{bmatrix} 0 & 0 & 0 & -3 \\ 0 & 0 & 2 & -4 \\ 0 & -1 & 4 & -1 \\ -6 & 1 & 5 & 1 \end{bmatrix}$

Solution: This is *not* triangular, so we do not have to compute its determinant. Nonetheless, if we swap the 1st and 4th columns, and the 2nd and 3rd columns, then the result is the lower triangular $\begin{bmatrix} -3 & 0 & 0 & 0 \\ -4 & 2 & 0 & 0 \\ -1 & 4 & -1 & 0 \\ 1 & 5 & 1 & -6 \end{bmatrix}$ and its determinant is
$$(-3) \cdot 2 \cdot (-1) \cdot (-6) = -36.$$ But each column swap changes the sign so the determinant of the original matrix is $(-1)^2(-36) = -36.$ □

The above case of triangular matrices is a short detour from the main development of this section which is to derive a formula for determinants in general. The following two examples introduce the next property that we need before establishing a general formula for determinants.

Example 6.2.19 Let's rewrite the explicit formulas (6.1) for 2×2 and 3×3 determinants explicitly as the sum of simpler determinants.

- Recall that the 2×2 determinant

$$\begin{vmatrix} a & b \\ c & d \end{vmatrix} = ad - bc = (ad - 0c) + (0d - bc) = \begin{vmatrix} a & 0 \\ c & d \end{vmatrix} + \begin{vmatrix} 0 & b \\ c & d \end{vmatrix}.$$

That is, the original determinant is the same as the sum of two determinants, each with a zero in the first row and the other row unchanged. This identity decomposes the first row as $\begin{bmatrix} a & b \end{bmatrix} = \begin{bmatrix} a & 0 \end{bmatrix} + \begin{bmatrix} 0 & b \end{bmatrix}$, while the other row is unchanged.

- Recall from (6.1) that the 3×3 determinant

$$\begin{vmatrix} a & b & c \\ d & e & f \\ g & h & i \end{vmatrix} = aei + bfg + cdh - ceg - afh - bdi$$

$$= + aei + 0fg + 0dh - 0eg - afh - 0di$$
$$+ 0ei + bfg + 0dh - 0eg - 0fh - bdi$$
$$+ 0ei + 0fg + cdh - ceg - 0fh - 0di$$

$$= \begin{vmatrix} a & 0 & 0 \\ d & e & f \\ g & h & i \end{vmatrix} + \begin{vmatrix} 0 & b & 0 \\ d & e & f \\ g & h & i \end{vmatrix} + \begin{vmatrix} 0 & 0 & c \\ d & e & f \\ g & h & i \end{vmatrix}.$$

That is, the original determinant is the same as the sum of three determinants, each with two zeros in the first row and the other rows unchanged. This identity decomposes the first row as $\begin{bmatrix} a & b & c \end{bmatrix} = \begin{bmatrix} a & 0 & 0 \end{bmatrix} + \begin{bmatrix} 0 & b & 0 \end{bmatrix} + \begin{bmatrix} 0 & 0 & c \end{bmatrix}$, whereas the other rows are unchanged.

This sort of rearrangement of a determinant makes progress because then Theorem 6.2.11 helps by finding the determinant of the resultant matrices that have an almost all zero row. □

Example 6.2.20 A 2×2 example of a more general summation property is furnished by the determinant of matrix $A = \begin{bmatrix} a_{11} & b_1 + c_1 \\ a_{21} & b_2 + c_2 \end{bmatrix}$.

$$\det A = a_{11}(b_2 + c_2) - a_{21}(b_1 + c_1) = a_{11}b_2 + a_{11}c_2 - a_{21}b_1 - a_{21}c_1$$
$$= (a_{11}b_2 - a_{21}b_1) + (a_{11}c_2 - a_{21}c_1) = \det \begin{bmatrix} a_{11} & b_1 \\ a_{21} & b_2 \end{bmatrix} + \det \begin{bmatrix} a_{11} & c_1 \\ a_{21} & c_2 \end{bmatrix}$$
$$= \det B + \det C,$$

where matrices B and C have the same first column as A, and their second columns add up to the second column of A. □

Theorem 6.2.21 (*sum formula*) *Let A, B, and C be n × n matrices. If matrices A, B and C are identical except for their ith column, and that the ith column of A is the sum of the ith columns of B and C, then* $\det A = \det B + \det C$. *Further, the same sum property holds when "column" is replaced by "row" throughout.*

Proof. We establish the theorem for matrix columns. Then the same property holds for the rows because $\det(A^{\mathsf{T}}) = \det(A)$ (Theorem 6.1.26). As a prelude to the general geometric proof, consider the 2×2 case and the second column (as also established algebraically by Example 6.2.20). Write the matrices in terms of their column vectors, and draw the determinant parallelogram areas as shown below: let the three matrices be

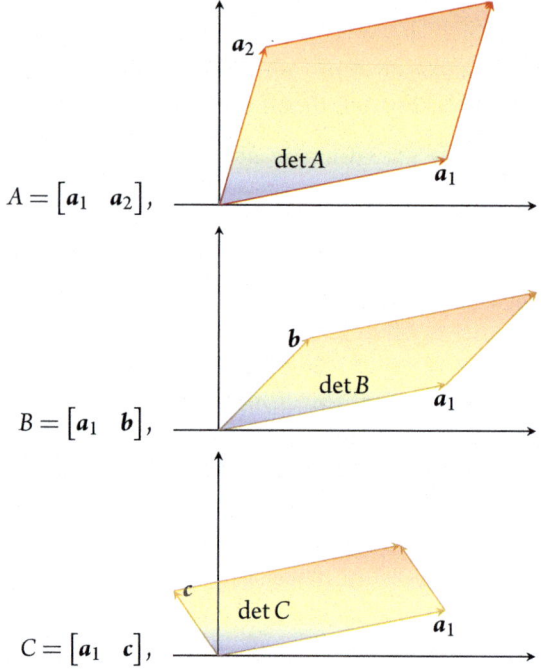

$$A = \begin{bmatrix} a_1 & a_2 \end{bmatrix},$$

$$B = \begin{bmatrix} a_1 & b \end{bmatrix},$$

$$C = \begin{bmatrix} a_1 & c \end{bmatrix},$$

The matrices A, B, and C all have the same first column a_1, whereas the second columns satisfy $a_2 = b + c$ by the condition of the theorem. Because these parallelograms have common side a_1 we can stack the area for det C on top of that for det B, and because $a_2 = b + c$ the top edge of the stack matches that for the area det A, as shown below

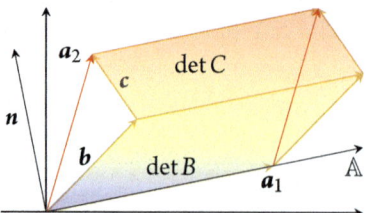

The base of the stacked shape lies on the line $\mathbb{A}$, and let vector n denote the orthogonal/normal direction (as shown). Because the shape has the same cross-section in lines parallel to $\mathbb{A}$, its

area is the length of the base times the height of the stacked shape in the direction n. But this is precisely the same height and base as the area for $\det A$, hence $\det A = \det B + \det C$.

A general proof for the last column uses the same diagrams, albeit schematically. Let matrices

$$A = \begin{bmatrix} A' & a_n \end{bmatrix}, \quad B = \begin{bmatrix} A' & b \end{bmatrix}, \quad C = \begin{bmatrix} A' & c \end{bmatrix},$$

where the $n \times (n-1)$ matrix $A' = \begin{bmatrix} a_1 & a_2 & \cdots & a_{n-1} \end{bmatrix}$ is common to all three, and where the three last columns satisfy $a_n = b + c$. Consider the nD-parallelepipeds whose nD-volumes are the three determinants, as before. Because these nD-parallelepipeds have common base of the $(n-1)$D-parallelepiped formed by the columns of A', we can and do stack the nD-volume for $\det C$ on top of that for $\det B$, and because $a_n = b + c$ the top $(n-1)$D-parallelepiped of the stack matches that for the nD-volume $\det A$, as shown schematically before. The base of the stacked shape lies on the subspace $\mathbb{A} = \mathrm{span}\{a_1, a_2, \ldots, a_{n-1}\}$, and let vector n denote the orthogonal/normal direction to $\mathbb{A}$ (as shown schematically). Because the shape has the same cross-section parallel to $\mathbb{A}$, its nD-volume is the $(n-1)$D-volume of the base $(n-1)$D-parallelepiped times the height of the stacked shape in the direction n. But this is precisely the same height and base as the nD-volume for $\det A$, hence $\det A = \det B + \det C$.

Lastly, when it is the ith column for which $a_i = b + c$ and all others columns are identical, then swap column i with column n in all matrices. Theorem 6.2.5(c) asserts that the signs of the three determinants are changed by this swapping. The above proof for the last column case then assures us that $(-\det A) = (-\det B) + (-\det C)$; that is, $\det A = \det B + \det C$, as required. □

The sum formula Theorem 6.2.21 leads to the common way to compute determinants by hand for matrices of size larger than 3×3, albeit not generally practical for matrices significantly larger in size.

Example 6.2.22 Use Theorems 6.2.11 and 6.2.21 to evaluate the determinant of matrix

$$A = \begin{bmatrix} -2 & 1 & -1 \\ 1 & -6 & -1 \\ 2 & 1 & 0 \end{bmatrix}.$$

Solution: Write the first row of A as the sum

$$\begin{bmatrix} -2 & 1 & -1 \end{bmatrix} = \begin{bmatrix} -2 & 0 & 0 \end{bmatrix} + \begin{bmatrix} 0 & 1 & -1 \end{bmatrix}$$
$$= \begin{bmatrix} -2 & 0 & 0 \end{bmatrix} + \begin{bmatrix} 0 & 1 & 0 \end{bmatrix} + \begin{bmatrix} 0 & 0 & -1 \end{bmatrix}.$$

Then using Theorem 6.2.21 twice, the determinant

$$\begin{vmatrix} -2 & 1 & -1 \\ 1 & -6 & -1 \\ 2 & 1 & 0 \end{vmatrix} = \begin{vmatrix} -2 & 0 & 0 \\ 1 & -6 & -1 \\ 2 & 1 & 0 \end{vmatrix} + \begin{vmatrix} 0 & 1 & -1 \\ 1 & -6 & -1 \\ 2 & 1 & 0 \end{vmatrix}$$

$$= \begin{vmatrix} -2 & 0 & 0 \\ 1 & -6 & -1 \\ 2 & 1 & 0 \end{vmatrix} + \begin{vmatrix} 0 & 1 & 0 \\ 1 & -6 & -1 \\ 2 & 1 & 0 \end{vmatrix} + \begin{vmatrix} 0 & 0 & -1 \\ 1 & -6 & -1 \\ 2 & 1 & 0 \end{vmatrix}$$

Each of these last three matrices has the first row zero except for one element, so Theorem 6.2.11 applies to each of the three determinants to give

$$
\begin{vmatrix} -2 & 1 & -1 \\ 1 & -6 & -1 \\ 2 & 1 & 0 \end{vmatrix} = (-1)^2(-2)\begin{vmatrix} -6 & -1 \\ 1 & 0 \end{vmatrix} + (-1)^3(1)\begin{vmatrix} 1 & -1 \\ 2 & 0 \end{vmatrix} + (-1)^4(-1)\begin{vmatrix} 1 & -6 \\ 2 & 1 \end{vmatrix}
$$

$$
= (-2)\cdot 1 - (1)\cdot 2 + (-1)\cdot 13 = -17
$$

upon using the well-known formula (6.1) for the three 2×2 determinants.

Alternatively, we could have used any row or column instead of the first row. For example, let's use the last column as it usefully already has a zero entry: write the last column of matrix A as $(-1,-1,0) = (-1,0,0) + (0,-1,0)$, then by Theorem 6.2.21 the determinant

$$
\begin{vmatrix} -2 & 1 & -1 \\ 1 & -6 & -1 \\ 2 & 1 & 0 \end{vmatrix} = \begin{vmatrix} -2 & 1 & -1 \\ 1 & -6 & 0 \\ 2 & 1 & 0 \end{vmatrix} + \begin{vmatrix} -2 & 1 & 0 \\ 1 & -6 & -1 \\ 2 & 1 & 0 \end{vmatrix} \quad \text{(so by Theorem 6.2.11)}
$$

$$
= (-1)^4(-1)\begin{vmatrix} 1 & -6 \\ 2 & 1 \end{vmatrix} + (-1)^5(-1)\begin{vmatrix} -2 & 1 \\ 2 & 1 \end{vmatrix}
$$

$$
= (-1)\cdot 13 - (-1)\cdot(-4) = -17,
$$

as before. □

Activity 6.2.23 We could compute the determinant of the matrix $\begin{bmatrix} -3 & 6 & -4 \\ 7 & 4 & 6 \\ 1 & 6 & -3 \end{bmatrix}$ as a specific sum involving three of the following four determinants. Which one of the following would not be used in the sum?

(a) $\begin{vmatrix} 6 & -4 \\ 6 & -3 \end{vmatrix}$ (b) $\begin{vmatrix} 7 & 6 \\ 1 & -3 \end{vmatrix}$ (c) $\begin{vmatrix} 4 & 6 \\ 6 & -3 \end{vmatrix}$ (d) $\begin{vmatrix} 6 & -4 \\ 4 & 6 \end{vmatrix}$

Theorem 6.2.24 (*Laplace expansion theorem*) *For every $n \times n$ matrix $A = [a_{ij}]$ $(n \geq 2)$, recall the (i,j)th minor A_{ij} to be the $(n-1) \times (n-1)$ matrix obtained from A by omitting the ith row and jth column. Then the determinant of A can be computed via expansion in any row i or any column j as, respectively,*

$$
\det A = (-1)^{i+1}a_{i1}\det A_{i1} + (-1)^{i+2}a_{i2}\det A_{i2}
$$
$$
+ \cdots + (-1)^{i+n}a_{in}\det A_{in}
$$
$$
= (-1)^{j+1}a_{1j}\det A_{1j} + (-1)^{j+2}a_{2j}\det A_{2j}
$$
$$
+ \cdots + (-1)^{j+n}a_{nj}\det A_{nj}. \tag{6.4}
$$

Proof. We establish the expansion for matrix rows: then the same property holds for the columns because $\det(A^T) = \det(A)$ (Theorem 6.1.26). First prove the expansion for a first row expansion, and then second for any row. So first use the sum Theorem 6.2.21 $(n-1)$ times to deduce

$$
\begin{vmatrix}
a_{11} & a_{12} & \cdots & a_{1n} \\
a_{21} & a_{22} & \cdots & a_{2n} \\
\vdots & \vdots & \ddots & \vdots \\
a_{n1} & a_{n2} & \cdots & a_{nn}
\end{vmatrix}
$$

$$
=
\begin{vmatrix}
a_{11} & 0 & \cdots & 0 \\
a_{21} & a_{22} & \cdots & a_{2n} \\
\vdots & \vdots & \ddots & \vdots \\
a_{n1} & a_{n2} & \cdots & a_{nn}
\end{vmatrix}
+
\begin{vmatrix}
0 & a_{12} & \cdots & a_{1n} \\
a_{21} & a_{22} & \cdots & a_{2n} \\
\vdots & \vdots & \ddots & \vdots \\
a_{n1} & a_{n2} & \cdots & a_{nn}
\end{vmatrix}
$$

$$
\vdots
$$

$$
=
\begin{vmatrix}
a_{11} & 0 & \cdots & 0 \\
a_{21} & a_{22} & \cdots & a_{2n} \\
\vdots & \vdots & \ddots & \vdots \\
a_{n1} & a_{n2} & \cdots & a_{nn}
\end{vmatrix}
+
\begin{vmatrix}
0 & a_{12} & \cdots & 0 \\
a_{21} & a_{22} & \cdots & a_{2n} \\
\vdots & \vdots & \ddots & \vdots \\
a_{n1} & a_{n2} & \cdots & a_{nn}
\end{vmatrix}
+ \cdots +
\begin{vmatrix}
0 & 0 & \cdots & a_{1n} \\
a_{21} & a_{22} & \cdots & a_{2n} \\
\vdots & \vdots & \ddots & \vdots \\
a_{n1} & a_{n2} & \cdots & a_{nn}
\end{vmatrix}
$$

As each of these n determinants has the first row zero except for one element, Theorem 6.2.11 applies to give

$$
\begin{vmatrix}
a_{11} & a_{12} & \cdots & a_{1n} \\
a_{21} & a_{22} & \cdots & a_{2n} \\
\vdots & \vdots & \ddots & \vdots \\
a_{n1} & a_{n2} & \cdots & a_{nn}
\end{vmatrix}
$$

$$
= (-1)^2 a_{11}
\begin{vmatrix}
a_{22} & \cdots & a_{2n} \\
\vdots & \ddots & \vdots \\
a_{n2} & \cdots & a_{nn}
\end{vmatrix}
+ (-1)^3 a_{12}
\begin{vmatrix}
a_{21} & \cdots & a_{2n} \\
\vdots & \ddots & \vdots \\
a_{n1} & \cdots & a_{nn}
\end{vmatrix}
$$

$$
+ \cdots + (-1)^{n+1} a_{1n}
\begin{vmatrix}
a_{21} & a_{22} & \cdots & a_{2,n-1} \\
\vdots & \vdots & \ddots & \vdots \\
a_{n1} & a_{n2} & \cdots & a_{n,n-1}
\end{vmatrix}
$$

$$
= (-1)^2 a_{11} \det A_{11} + (-1)^3 a_{12} \det A_{12} + \cdots + (-1)^{n+1} a_{1n} \det A_{1n},
$$

which is the case $i = 1$ of formula (6.4).

Second, for the general ith row expansion, let a new matrix B be obtained from A by swapping the ith row up $(i-1)$ times to form the first row of B and leaving the other rows from A in the same order. Then the elements $b_{1j} = a_{ij}$, and also the minors $B_{1j} = A_{ij}$. Apply formula (6.4) to the first row of B (just proved) to give

$$
\det B = (-1)^2 b_{11} \det B_{11} + (-1)^3 b_{12} \det B_{12} + \cdots + (-1)^{n+1} b_{1n} \det B_{1n}
$$
$$
= (-1)^2 a_{i1} \det A_{i1} + (-1)^3 a_{i2} \det A_{i2} + \cdots + (-1)^{n+1} a_{in} \det A_{in}.
$$

But by Theorem 6.2.5(c), each of the $(i-1)$ row swaps in forming B changes the sign of the determinant: hence

$$
\begin{aligned}
\det A &= (-1)^{i-1} \det B \\
&= (-1)^{i-1+2} a_{i1} \det A_{i1} + (-1)^{i-1+3} a_{i2} \det A_{i2} + \cdots + (-1)^{i-1+n+1} a_{in} \det A_{in} \\
&= (-1)^{i+1} a_{i1} \det A_{i1} + (-1)^{i+2} a_{i2} \det A_{i2} + \cdots + (-1)^{i+n} a_{in} \det A_{in},
\end{aligned}
$$

as required.

Example 6.2.25 Use the Laplace expansion (6.4) to find the determinant of the following matrices.

(a) $\begin{bmatrix} 0 & 2 & 1 & 2 \\ -1 & 2 & -1 & -2 \\ 1 & 2 & -1 & -1 \\ 0 & -1 & -1 & 1 \end{bmatrix}$

Solution: The first column has two zeros, so expand in the first column (using (6.1) for these 3×3 matrices):

$$
\det = (-1)^3(-1) \det \begin{bmatrix} 2 & 1 & 2 \\ 2 & -1 & -1 \\ -1 & -1 & 1 \end{bmatrix} + (-1)^4(1) \det \begin{bmatrix} 2 & 1 & 2 \\ 2 & -1 & -2 \\ -1 & -1 & 1 \end{bmatrix}
$$

$$
= (-2 + 1 - 4 - 2 - 2 - 2) + (-2 + 2 - 4 - 2 - 4 - 2) = -23.
$$

$\square$

(b) $\begin{bmatrix} -3 & -1 & 1 & 0 \\ -2 & 0 & -2 & 0 \\ -3 & -2 & 0 & 0 \\ 1 & -2 & 0 & 3 \end{bmatrix}$

Solution: The last column has three zeros, so expand in the last column:

$$
\det = (-1)^8(3) \det \begin{bmatrix} -3 & -1 & 1 \\ -2 & 0 & -2 \\ -3 & -2 & 0 \end{bmatrix} \qquad \text{(expand in the middle row, say, due to its zero)}
$$

$$
= 3\left\{ (-1)^3(-2) \det \begin{bmatrix} -1 & 1 \\ -2 & 0 \end{bmatrix} + (-1)^5(-2) \det \begin{bmatrix} -3 & -1 \\ -3 & -2 \end{bmatrix} \right\}
$$

$$
= 3\{2(0+2) + 2(6-3)\} = 30.
$$

$\square$

The Laplace expansion is generally too computationally expensive for all but small matrices. The reason is that computing the determinant of an $n \times n$ matrix with the Laplace expansion generally takes $n!$ operations (the next Theorem 6.2.27), and the factorial $n! = n(n-1) \cdots 3 \cdot 2 \cdot 1$ grows very quickly even for medium n. Even for just a 20×20 matrix the Laplace expansion has over two quintillion terms $(2 \cdot 10^{18})$. Exceptional matrices are those with lots of zeros, such as triangular matrices (Theorem 6.2.16). In any case, remember that except for theoretical purposes there is rarely any need to compute a medium to large determinant.

Example 6.2.26 The determinant of a 3×3 matrix has $3! = 6$ terms, each a product of three factors: diagram (6.1) gives the determinant

$$\begin{vmatrix} a & b & c \\ d & e & f \\ g & h & i \end{vmatrix} = aei + bfg + cdh - ceg - afh - bdi.$$

Further, observe that within each term the three factors come from different rows and columns. For example, a never appears in a term with the entries b, c, d, or g (the elements from either the same row or the same column). Similarly, f never appears in a term with the entries d, e, c, or i. □

Theorem 6.2.27 *The determinant of every $n \times n$ matrix expands to the sum of $n!$ terms, where each term is ± 1 times a product of n factors such that each factor comes from different rows and columns of the matrix.*

Proof. Use induction on the size of the matrix. First, the property holds for 1×1 matrices as $\det [a_{11}] = a_{11}$ is one term of one factor from the only row and column of the matrix.

Second, assume the determinant of every $(n-1) \times (n-1)$ matrix may be written the sum of $(n-1)!$ terms, where each term is (± 1) times a product of $(n-1)$ factors such that each factor comes from different rows and columns. Consider any $n \times n$ matrix A. By the Laplace Expansion Theorem 6.2.24, $\det A$ may be written as the sum of n terms of the form $\pm a_{ij} \det A_{ij}$. By induction assumption, the $(n-1) \times (n-1)$ minors A_{ij} have determinants with $(n-1)!$ terms, each of $(n-1)$ factors and so the n terms in a Laplace Expansion of $\det A$ expands to $n(n-1)! = n!$ terms, each term being of n factors through the multiplication by the entry a_{ij}. Further, recall that the minor A_{ij} is obtained from A by omitting row i and column j, and so the minor has no elements from the same row or column as a_{ij}. Consequently, each term in the determinant only has factors from different rows and columns, as required. By induction the theorem holds for all n.

6.2.1 Exercises

Exercise 6.2.1 In each of the following, the determinant of a matrix is given. Use Theorem 6.2.5 on the row and column properties of a determinant to find the determinant of each of the other four listed matrices. Give reasons for each of your answers.

(a) $\det \begin{bmatrix} -2 & 1 & -4 \\ -2 & -1 & 2 \\ -2 & 5 & -1 \end{bmatrix} = 60$

i. $\begin{bmatrix} 1 & 1 & -4 \\ 1 & -1 & 2 \\ 1 & 5 & -1 \end{bmatrix}$

iii. $\begin{bmatrix} -2 & 1 & -4 \\ -0.2 & -0.1 & 0.2 \\ -2 & 5 & -1 \end{bmatrix}$

ii. $\begin{bmatrix} -2 & 1 & -4 \\ -2 & 1 & -4 \\ -2 & 5 & -1 \end{bmatrix}$

iv. $\begin{bmatrix} -2 & 1 & -4 \\ -2 & 5 & -1 \\ -2 & -1 & 2 \end{bmatrix}$

(b) $\det \begin{bmatrix} -1 & -1 & 4 & -6 \\ 4 & -2 & -2 & -1 \\ 0 & -3 & -1 & -4 \\ 3 & 2 & 1 & 1 \end{bmatrix} = 72$

i. $\begin{bmatrix} 0 & -3 & -1 & -4 \\ 4 & -2 & -2 & -1 \\ -1 & -1 & 4 & -6 \\ 3 & 2 & 1 & 1 \end{bmatrix}$

iii. $\begin{bmatrix} -1 & -1 & 4 & -6 \\ 2 & -1 & -1 & -1/2 \\ 0 & -3 & -1 & -4 \\ 3 & 2 & 1 & 1 \end{bmatrix}$

ii. $\begin{bmatrix} -1 & -1 & 4 & -6 \\ 4 & -2 & -2 & -1 \\ 0 & 0 & 0 & 0 \\ 3 & 2 & 1 & 1 \end{bmatrix}$

iv. $\begin{bmatrix} -1 & -1 & 4 & -6 \\ -2 & 4 & -2 & -1 \\ -3 & 0 & -1 & -4 \\ 2 & 3 & 1 & 1 \end{bmatrix}$

(c) $\det \begin{bmatrix} 0.3 & -0.1 & -0.1 & 0.4 \\ 0.2 & 0.3 & 0 & 0.1 \\ 0.1 & -0.1 & -0.3 & -0.2 \\ -0.1 & -0.2 & 0.4 & 0.2 \end{bmatrix} = 0.01$

i. $\begin{bmatrix} 3 & -1 & -1 & 4 \\ 2 & 3 & 0 & 1 \\ 1 & -1 & -3 & -2 \\ -1 & -2 & 4 & 2 \end{bmatrix}$

iii. $\begin{bmatrix} 0.2 & 0.3 & 0 & 0.1 \\ 0.1 & -0.1 & -0.3 & -0.2 \\ -0.1 & -0.2 & 0.4 & 0.2 \\ 0.3 & -0.1 & -0.1 & 0.4 \end{bmatrix}$

ii. $\begin{bmatrix} 0.3 & 0.2 & 0 & 2 \\ 0.2 & -0.6 & 0 & 0.5 \\ 0.1 & 0.2 & 0 & -1 \\ -0.1 & 0.4 & 0 & 1 \end{bmatrix}$

iv. $\begin{bmatrix} 0.3 & 0.4 & -0.1 & 0.4 \\ 0.2 & 0.1 & 0 & 0.1 \\ 0.1 & -0.2 & -0.3 & -0.2 \\ -0.1 & 0.2 & 0.4 & 0.2 \end{bmatrix}$

Exercise 6.2.2 Recall Example 6.2.8. For each pair of given points, (x_1, y_1) and (x_2, y_2), evaluate the determinant in the equation

$$\det \begin{bmatrix} 1 & x & y \\ 1 & x_1 & y_1 \\ 1 & x_2 & y_2 \end{bmatrix} = 0$$

to find a linear equation for the straight line through the two given points in the xy-plane. Show your working.

(a) $(-3,-6), (2,3)$

(b) $(3,-2), (-3,0)$

(c) $(1,-4), (-3,1)$

(d) $(-1,0), (-2,1)$

Exercise 6.2.3 Using mainly the properties of Theorem 6.2.5 detail an argument that the following determinant equations each give a linear equation for the line through two given points (x_1, y_1) and (x_2, y_2) in the xy-plane.

(a) $\det \begin{bmatrix} 1 & x_1 & y_1 \\ 1 & x & y \\ 1 & x_2 & y_2 \end{bmatrix} = 0$

(b) $\det \begin{bmatrix} 1 & 1 & 1 \\ x_2 & x_1 & x \\ y_2 & y_1 & y \end{bmatrix} = 0$

Exercise 6.2.4 Recall Example 6.2.9. For each pair of given points, (x_1, y_1, z_1) and (x_2, y_2, z_2), evaluate the determinant in the equation

$$\det \begin{bmatrix} x & y & z \\ x_1 & y_1 & z_1 \\ x_2 & y_2 & z_2 \end{bmatrix} = 0$$

to find a linear equation for the plane in xyz-space that passes through the two given points and the origin. Show your working.

(a) $(-1,-1,-3), (3,-5,-1)$

(b) $(4,-2,0), (-3,-4,-1)$

(c) $(-4,-1,2), (-3,-2,2)$

(d) $(2,2,3), (2,1,4)$

Exercise 6.2.5 Using mainly the properties of Theorem 6.2.5 detail an argument that the following determinant equations each give an equation for the plane in xyz-space passing through the origin and the two given points (x_1, y_1, z_1) and (x_2, y_2, z_2).

(a) $\det \begin{bmatrix} x_1 & y_1 & z_1 \\ x_2 & y_2 & z_2 \\ x & y & z \end{bmatrix} = 0$

(b) $\det \begin{bmatrix} x_2 & x & x_1 \\ y_2 & y & y_1 \\ z_2 & z & z_1 \end{bmatrix} = 0$

Exercise 6.2.6 Prove Theorems 6.2.5(a), 6.2.5(b), and 6.2.5(d) using basic geometric arguments about the transformation of the unit nD-cube.

Exercise 6.2.7 Use Theorem 6.2.5 to prove that if a square matrix A has two nonzero rows proportional to each other, then $\det A = 0$. Why does it immediately follow that (instead of rows) if the matrix has two nonzero columns proportional to each other, then $\det A = 0$.

Exercise 6.2.8 Use Theorem 6.2.11, and then (6.1), to evaluate the following determinants. Show your working.

(a) $\det \begin{bmatrix} 6 & 1 & 1 \\ -1 & 3 & -8 \\ -6 & 0 & 0 \end{bmatrix}$

(b) $\det \begin{bmatrix} 4 & 8 & 0 \\ 3 & -2 & 0 \\ -1 & -1 & -3 \end{bmatrix}$

(c) $\det \begin{bmatrix} 2 & -4 & -2 & -2 \\ 0 & 1 & -3 & -2 \\ -2 & 0 & 0 & 0 \\ 5 & -8 & 1 & 7 \end{bmatrix}$

(e) $\det \begin{bmatrix} 0 & 2 & -4 & 3 \\ -6 & 6 & -2 & 0 \\ -1 & -8 & 4 & 0 \\ 2 & -2 & -1 & 0 \end{bmatrix}$

(d) $\det \begin{bmatrix} 0 & -5 & 0 & 0 \\ -7 & 2 & 2 & 1 \\ 1 & -2 & -2 & -5 \\ 6 & 8 & -2 & 0 \end{bmatrix}$

(f) $\det \begin{bmatrix} -3 & 4 & -1 & -1 \\ 4 & 8 & 1 & 6 \\ 0 & 7 & 0 & 0 \\ 2 & -6 & 1 & 2 \end{bmatrix}$

Exercise 6.2.9 Use the triangular matrix Theorem 6.2.16, as well as the row/column properties of Theorem 6.2.5, to find the determinants of each of the following matrices. Show your argument.

(a) $\begin{bmatrix} -6 & -4 & -7 & 2 \\ 0 & -2 & -1 & 1 \\ 0 & 0 & -4 & 1 \\ 0 & 0 & 0 & -2 \end{bmatrix}$

(d) $\begin{bmatrix} 0 & 0 & 7 & 0 \\ 0 & -3 & 7 & -4 \\ 0 & 7 & -4 & 0 \\ 1 & 2 & -1 & -3 \end{bmatrix}$

(b) $\begin{bmatrix} 2 & 0 & 0 & 0 \\ 1 & 3 & 0 & 0 \\ -5 & -1 & 2 & 0 \\ 2 & 4 & -1 & 1 \end{bmatrix}$

(e) $\begin{bmatrix} 0 & 0 & 1 & 8 & 5 \\ 6 & 1 & -5 & -8 & -1 \\ 0 & 0 & 0 & 0 & 5 \\ 0 & -1 & -6 & -5 & 4 \\ 0 & 0 & 0 & -1 & -8 \end{bmatrix}$

(c) $\begin{bmatrix} 0 & 0 & 8 & 0 \\ -5 & -6 & 6 & -1 \\ 0 & 0 & -5 & 6 \\ 0 & -6 & -4 & 3 \end{bmatrix}$

(f) $\begin{bmatrix} -6 & 0 & 0 & 0 & 0 \\ -4 & -3 & 0 & 0 & 0 \\ 0 & -4 & 0 & 4 & 0 \\ 0 & 1 & -5 & 12 & 0 \\ -2 & -1 & -5 & -2 & 5 \end{bmatrix}$

Exercise 6.2.10 Given that the determinant $\begin{vmatrix} a & b & c \\ d & e & f \\ g & h & i \end{vmatrix} = 6$, find the following determinants. Give reasons.

(a) $\begin{vmatrix} 3a & b & c \\ 3d & e & f \\ 3g & h & i \end{vmatrix}$

(d) $\begin{vmatrix} a+d & b+e & c+f \\ d & e & f \\ g & h & i \end{vmatrix}$

(b) $\begin{vmatrix} a & b & c/2 \\ -d & -e & -f/2 \\ g & h & i/2 \end{vmatrix}$

(e) $\begin{vmatrix} a & b & c-a \\ d & e & f-d \\ g & h & i-g \end{vmatrix}$

(c) $\begin{vmatrix} d & e & f \\ a & b & c \\ g & h & i \end{vmatrix}$

(f) $\begin{vmatrix} a & b & c \\ d & e & f \\ a+2g & b+2h & c+2i \end{vmatrix}$

Exercise 6.2.11 Consider a general 3×3 matrix $A = \begin{bmatrix} a & b & c \\ d & e & f \\ g & h & i \end{bmatrix}$. Derive a first column Laplace expansion (Theorem 6.2.24) of the 3×3 determinant, and rearrange to show it is the same as the determinant formula (6.1).

Exercise 6.2.12 Use the Laplace expansion (Theorem 6.2.24) to find the determinant of the following matrices. Use rows or columns with many zeros. Show your working.

(a) $\begin{bmatrix} 1 & 4 & 0 & 0 \\ 0 & 2 & -1 & 0 \\ -3 & 0 & -3 & 0 \\ 0 & 1 & 0 & 1 \end{bmatrix}$

(c) $\begin{bmatrix} 0 & 0 & -2 & 0 & -6 \\ 0 & -3 & 0 & 1 & 0 \\ -4 & -5 & 2 & 0 & 2 \\ 0 & 2 & 0 & 3 & -3 \\ 2 & 0 & 0 & -1 & 0 \end{bmatrix}$

(b) $\begin{bmatrix} -4 & -3 & 0 & -3 \\ -2 & 0 & -1 & 0 \\ -1 & 2 & 4 & 2 \\ 0 & 0 & 0 & -4 \end{bmatrix}$

(d) $\begin{bmatrix} 0 & 0 & -2 & 7 & 4 \\ 0 & 4 & -4 & 1 & 0 \\ 2 & -2 & 0 & 1 & 0 \\ 3 & 2 & 0 & 2 & 0 \\ 0 & -2 & 0 & 0 & 1 \end{bmatrix}$

Exercise 6.2.13 For each of the following matrices, use the Laplace expansion (Theorem 6.2.24) to find all the values of k for which the matrix is *not* invertible. Show your working.

(a) $\begin{bmatrix} 3 & -2k & -1 & -1-2k \\ 0 & 2 & 1 & 0 \\ 0 & 0 & 2 & 0 \\ 0 & -2 & -5 & 3+2k \end{bmatrix}$

(b) $\begin{bmatrix} -1 & -2 & 0 & 2k \\ 0 & 0 & 5 & 0 \\ 0 & 2 & -1+k & -k \\ -2+k & 1+2k & 4 & 0 \end{bmatrix}$

(c) $\begin{bmatrix} k & 1 & 0 & -5-k & -1-k \\ -4+6k & -1 & 1+3k & -3-5k & 0 \\ 0 & 2 & 0 & -2 & -5 \\ 0 & 0 & 0 & 0 & 2-k \\ -2k & 1+k & 0 & -2 & 3 \end{bmatrix}$

Exercise 6.2.14 Using Theorem 6.2.27 and the properties of Theorem 6.2.5, detail an argument that the following determinant equation generally forms an equation for the plane in xyz-space passing through the three given points (x_1, y_1, z_1), (x_2, y_2, z_2), and (x_3, y_3, z_3):

$$\det \begin{bmatrix} 1 & x & y & z \\ 1 & x_1 & y_1 & z_1 \\ 1 & x_2 & y_2 & z_2 \\ 1 & x_3 & y_3 & z_3 \end{bmatrix} = 0.$$

Exercise 6.2.15 Using Theorem 6.2.27 and the properties of Theorem 6.2.5, detail an argument that the following determinant equation generally forms an equation for the parabola passing through the three given points (x_1, y_1), (x_2, y_2), and (x_3, y_3) in the xy-plane:

$$\det \begin{bmatrix} 1 & x & x^2 & y \\ 1 & x_1 & x_1^2 & y_1 \\ 1 & x_2 & x_2^2 & y_2 \\ 1 & x_3 & x_3^2 & y_3 \end{bmatrix} = 0.$$

Exercise 6.2.16 Using Theorem 6.2.27 and the properties of Theorem 6.2.5, detail an argument that the equation

$$\det \begin{bmatrix} 1 & x & x^2 & y & y^2 & xy \\ 1 & x_1 & x_1^2 & y_1 & y_1^2 & x_1 y_1 \\ 1 & x_2 & x_2^2 & y_2 & y_2^2 & x_2 y_2 \\ 1 & x_3 & x_3^2 & y_3 & y_3^2 & x_3 y_3 \\ 1 & x_4 & x_4^2 & y_4 & y_4^2 & x_4 y_4 \\ 1 & x_5 & x_5^2 & y_5 & y_5^2 & x_5 y_5 \end{bmatrix} = 0$$

generally forms an equation for the conic section passing through the five given points (x_i, y_i), $i = 1, \ldots, 5$, in the xy-plane.

Exercise 6.2.17 Consider the task of computing the determinant of a typical $n \times n$ matrix A ('typical' in the sense that all its entries are nonzero). You are given that using the Laplace Expansion Theorem 6.2.24 to compute the determinant takes approximately $1.72\, n!$ multiplications. On a computer that does about 10^9 multiplications per second, what is the largest sized matrix for which the determinant can be computed by Laplace Expansion in less than an hour? Detail your reasons.

Exercise 6.2.18 In a few sentences, answer/discuss each of the following.

(a) What can cause a determinant to be zero?

(b) Describe the relationship between the following two statements, relevant for any $n \times n$ matrix A and any scalar k:

- $\det(kA) = k^n \det A$; and
- let matrix B be obtained by multiplying any one row or column of A by k, then $\det B = k \det A$.

(c) What properties of the determinant allow us to write the equations of lines, planes, etc, as simple determinants?

(d) What is special about a matrix with a row that is zero except for one element, that enables us to simplify the calculation of its determinant via the geometric definition of a determinant?

(e) Why is the determinant of a triangular matrix simply the product of the diagonal entries?

(f) What properties of a determinant had to be developed in order to deduce and justify the Laplace Expansion Theorem 6.2.24?

6.3 Summary of determinants

★★ The 'graphical formulas' (6.1) for 2×2 and 3×3 determinants are useful for both theory and many practical small problems.

Geometry underlies determinants

- The term **nD-cube** generalizes a square and cube to n dimensions ($\mathbb{R}^n$). The term **nD-volume** generalizes the notion of area and volume to n dimensions. When the dimension of the space is unspecified, then we may say **hyper-cube** and **hyper-volume**, respectively.

★★ Let A be an $n \times n$ square matrix, and let C be the unit nD-cube in $\mathbb{R}^n$. Transform the nD-cube C by $\boldsymbol{x} \mapsto A\boldsymbol{x}$ to its image C' in $\mathbb{R}^n$. Define the **determinant** of A, denoted either $\det A$ (or sometimes $|A|$) such that (Definition 6.1.5):
  - the magnitude $|\det A|$ is the nD-volume of C'; and
  - the sign of $\det A$ to be negative iff the transformation reflects the orientation of the nD-cube.

★★ (Theorem 6.1.8)
  - For every $n \times n$ diagonal matrix D, the determinant of D is the product of the diagonal entries: $\det D = d_{11} d_{22} \cdots d_{nn}$.
  - For every orthogonal matrix Q, $\det Q = \pm 1$ (only one alternative, not both), and $\det Q = \det(Q^{\mathsf{T}})$.
  - For every $n \times n$ matrix A and every scalar k, $\det(kA) = k^n \det A$.

- Consider any bounded smooth nD-volume C in $\mathbb{R}^n$ and its image C' after multiplication by $n \times n$ matrix A. Then (Theorem 6.1.14)

$$\det A = \pm \frac{n\text{D-volume of } C'}{n\text{D-volume of } C}$$

  with the negative sign when matrix A changes the orientation.

★★ For every two $n \times n$ matrices A and B, $\det(AB) = \det(A)\det(B)$ (Theorem 6.1.16). Further, for $n \times n$ matrices A_1, A_2, ..., A_ℓ,

$$\det(A_1 A_2 \cdots A_\ell) = \det(A_1)\det(A_2) \cdots \det(A_\ell).$$

★ For every square matrix A, $\det(A^{\mathsf{T}}) = \det A$ (Theorem 6.1.26).

- For every $n \times n$ square matrix A, the magnitude of its determinant $|\det A| = \sigma_1 \sigma_2 \cdots \sigma_n$, the product of all its singular values (Theorem 6.1.21).

★ A square matrix A is invertible iff $\det A \neq 0$ (Theorem 6.1.29). If a matrix A is invertible, then $\det(A^{-1}) = 1/(\det A)$.

Laplace expansion theorem for determinants

★★ For every $n \times n$ matrix A, the determinant has the following row and column properties (Theorem 6.2.5).
  - If A has a zero row or column, then $\det A = 0$.
  - If A has two identical rows or two identical columns, then $\det A = 0$.
  - Let B be obtained by interchanging two rows or two columns of A, then $\det B = -\det A$.

- Let B be obtained by multiplying any one row or column of A by a scalar k, then $\det B = k \det A$.
- For every $n \times n$ matrix A, define the (i,j)th **minor** A_{ij} to be the $(n-1) \times (n-1)$ square matrix obtained from A by omitting the ith row and jth column. If, except for the entry a_{ij}, the ith row (or jth column) of A is all zero, then (Theorem 6.2.11)

$$\det A = (-1)^{i+j} a_{ij} \det A_{ij}.$$

- A **triangular matrix** is a square matrix where all entries are zero either to the lower-left of the diagonal or to the upper-right (Definition 6.2.15); an upper triangular matrix has zeros in the lower-left, and a lower triangular matrix has zeros in the upper-right.
- ★ For every $n \times n$ triangular matrix A, the determinant of A is the product of the diagonal entries, $\det A = a_{11} a_{22} \cdots a_{nn}$ (Theorem 6.2.16).
- Let A, B, and C be $n \times n$ matrices. If matrices A, B, and C are identical except for their ith column, and that the ith column of A is the sum of the ith columns of B and C, then $\det A = \det B + \det C$ (Theorem 6.2.21). Further, the same sum property holds when "column" is replaced by "row" throughout.
- ★ For every $n \times n$ matrix $A = \left[a_{ij} \right]$ $(n \geq 2)$, and in terms of the (i,j)th minor A_{ij}, the determinant of A can be computed via expansion in any row i or any column j as, respectively, (Theorem 6.2.24)

$$\begin{aligned}
\det A &= (-1)^{i+1} a_{i1} \det A_{i1} + (-1)^{i+2} a_{i2} \det A_{i2} \\
&\quad + \cdots + (-1)^{i+n} a_{in} \det A_{in} \\
&= (-1)^{j+1} a_{1j} \det A_{1j} + (-1)^{j+2} a_{2j} \det A_{2j} \\
&\quad + \cdots + (-1)^{j+n} a_{nj} \det A_{nj}.
\end{aligned}$$

- ★ The determinant of every $n \times n$ matrix expands to the sum of $n!$ terms, where each term is ± 1 times a product of n factors such that each factor comes from different rows and columns of the matrix (Theorem 6.2.27).

Answers to selected activities

6.1.3c, 6.1.7b, 6.1.10d, 6.1.12d, 6.1.17d, 6.1.23b, 6.2.7c, 6.2.13a, 6.2.17a, 6.2.23b,

Answers to selected exercises

6.1.1b $\det \approx 1.26$

6.1.1d $\det \approx -1.47$

6.1.1f $\det \approx -0.9$

6.1.3b $\det \approx 1.0$

6.1.3d $\det \approx 2.3$

6.1.4b 4

6.1.4d $-\frac{1}{2}, -1, -2$

6.1.5b $-35/54$

6.1.5d $-131/4$

6.1.9a -15

6.1.9c 81

6.1.9e $-5/3$

6.2.1b $-72\,,0\,,36\,,-72$

6.2.2a $-9x+5y+3=0$

6.2.2c $-(5x+4y+11)=0$

6.2.4a $2(-7x-5y+4z)=0$

6.2.4c $2x+2y+5z=0$

6.2.8a 66

6.2.8c -208

6.2.8e 270

6.2.9a 96

6.2.9c -1440

6.2.9e 30

6.2.10a 18

6.2.10c -6

6.2.10e 6

6.2.12a 6

6.2.12c -308

6.2.13a $-3/2$

6.2.13c $-1/3\,,0\,,-9\,,2$

7 Eigenvalues and eigenvectors in general

Population modelling   Suppose two species of animals interact: how do their populations evolve in time? Let $y(t)$ and $z(t)$ be the number of female animals in each of the species at time t in years (biologists usually just count females in population models as females usually determine reproduction). Modelling might deduce that the populations interact according to the rule that one year later the population is $y(t+1) = 2y(t) - 4z(t)$ and $z(t+1) = -y(t) + 2z(t)$: that is, if it was not for the other species, then for each species the number of females would double every year (since then $y(t+1) = 2y(t)$ and $z(t+1) = 2z(t)$); but the presence of the other species causes competition for resources that decreases each of these growths via the $-4z(t)$ and $-y(t)$ terms.

Linear Algebra for the 21st Century. A. J. Roberts, Oxford University Press (2020). © A. J. Roberts.
DOI: 10.1093/oso/9780198856399.003.0007

Question: can we find special solutions in the form $(y, z) = x \lambda^t$ for some constant λ and non zero $x = (x_1, x_2)$? Let's try by substituting $y = x_1 \lambda^t$ and $z = x_2 \lambda^t$ into the equations:

$$y(t+1) = 2y(t) - 4z(t), \quad z(t+1) = -y(t) + 2z(t)$$
$$\iff x_1 \lambda^{t+1} = 2x_1 \lambda^t - 4x_2 \lambda^t, \quad x_2 \lambda^{t+1} = -x_1 \lambda^t + 2x_2 \lambda^t$$
$$\iff 2x_1 - 4x_2 = \lambda x_1, \quad -x_1 + 2x_2 = \lambda x_2$$

after dividing by the factor λ^t (assuming constant λ is nonzero). Then form these last two equations as the matrix-vector equation

$$\begin{bmatrix} 2 & -4 \\ -1 & 2 \end{bmatrix} x = \lambda x.$$

That is, this substitution $(y, z) = x \lambda^t$ shows the question about finding solutions of the population equations reduces to solving $Ax = \lambda x$, called an eigen-problem.

This chapter develops linear algebra for such eigen-problems that empowers us to construct the general solution for the population $y(t)$ and $z(t)$. Here, a general solution is, in terms of two constants c_1 and c_2, that one species has female population $y(t) = 2c_1 4^t + 2c_2$ whereas the second species has female population $z(t) = -c_1 4^t + c_2$.

The basic eigen-problem Recall from Section 4.1 that the eigen-problem equation $Ax = \lambda x$ is just asking the following: can we find directions x such that matrix A acting on x is in the same direction as x? That is, when is Ax the same as λx for some proportionality constant λ? Now $x = 0$ is always a trivial solution of the equation $Ax = \lambda x$. Consequently, we are only interested in those values of the eigenvalue λ when *nonzero* solutions for the eigenvector x exist (as it is the directions which are of interest).

Rearranging the equation $Ax = \lambda x$ as the homogeneous system $(A - \lambda I)x = 0$, let's invoke properties of linear equations to solve the eigen-problem.

- Procedure 4.1.23 establishes that one way to find the eigenvalues λ (albeit *only* suitable for matrices of small size) is to solve the characteristic equation $\det(A - \lambda I) = 0$.

- Then for each eigenvalue, solving the homogeneous system $(A - \lambda I)x = 0$ gives corresponding eigenvectors x.

- The set of eigenvectors for a given eigenvalue forms a subspace called the eigenspace $\mathbb{E}_\lambda$ (Theorem 4.1.10).

Three general difficulties in eigen-problems Recall that Section 4.1 introduced one way to visually estimate eigenvectors and eigenvalues of a given 2×2 matrix A (Schonefeld, 1995). The graphical method is to plot many unit vectors x, and at the end of each x to adjoin the vector Ax. Since eigenvectors satisfy $Ax = \lambda x$ for some scalar eigenvalue λ, we visually identify eigenvectors as those x for which Ax points in the same (or opposite) direction as x.[1] Let's use this approach to identify three general difficulties.

---

[1] Recall that the MATLAB function eigshow (A) (download for R2017b etc.) provides an interactive alternative to the static views shown herein.

1. In this first picture, for matrix $A = \begin{bmatrix} 1 & 1 \\ 1/8 & 1 \end{bmatrix}$, the eigenvectors appear to be in directions $x_1 \approx \pm(0.9, 0.3)$ and $x_2 \approx \pm(0.9, -0.3)$ corresponding to eigenvalues $\lambda_1 \approx 1.4$ and $\lambda_2 \approx 0.6$. (Recall that scalar multiples of an eigenvector are always also eigenvectors, Section 4.1, so we always see $\pm$ pairs of eigenvectors in these pictures.) The eigenvectors $\pm(0.9, 0.3)$ are not orthogonal to the other eigenvectors $\pm(0.9, -0.3)$, not at right-angles—as happens for symmetric matrices (Theorem 4.2.11). This lack of orthogonality in general means we soon generalize the concept of orthogonal sets of vectors to a new concept of linearly independent sets (Section 7.2).

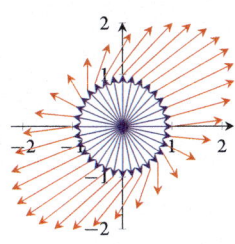

2. In this second case, for $A = \begin{bmatrix} 0 & 1 \\ -1 & 1/2 \end{bmatrix}$, there appears to be no (red) vector Ax in the same direction as the corresponding (blue) vector x. Thus there appears to be no eigenvectors at all. No eigenvectors and eigenvalues is the answer if we require real answers. However, in most applications we find it sensible to have non-real complex valued eigenvalues and eigenvectors (Section 7.1), written using $i = \sqrt{-1}$. So although we cannot see them graphically, for this matrix there are two complex eigenvalues and two families of complex eigenvectors (analogous to those found in Example 4.1.28).

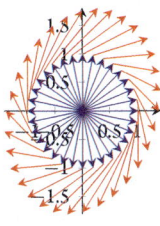

In this second case the vectors Ax all appear to be pointing clockwise. Such a consistent 'rotation' in Ax is characteristic of matrices with non-real complex valued eigenvalues and eigenvectors.

3. In this third case, for $A = \begin{bmatrix} 1 & 1 \\ 0 & 1 \end{bmatrix}$, there appear to be only the vectors $x = \pm(1, 0)$, aligned along the horizontal axis, for which $Ax = \lambda x$. Whereas for symmetric matrices there were always two pairs, here we only appear to have one pair of eigenvectors (Theorem 7.3.14). Such degeneracy occurs for matrices on the border between reality and complexity.

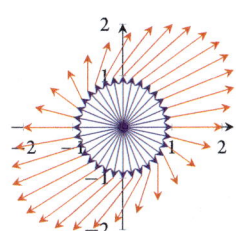

The first problem of the general lack of orthogonality of the eigenvectors is most clearly seen in the case of triangular matrices (Definition 6.2.15). The reason is linked to Theorem 6.2.16 that the determinant of a triangular matrix is simply the product of its diagonal entries.

Example 7.0.1 Find by algebra the eigenvalues and eigenvectors of the triangular matrix $A = \begin{bmatrix} 2 & 1 \\ 0 & 3 \end{bmatrix}$.

Solution: Recall Procedure 4.1.23.

(a) Find all eigenvalues by solving the characteristic equation $\det(A - \lambda I) = 0$. Here det $(A - \lambda I) = \det \begin{bmatrix} 2-\lambda & 1 \\ 0 & 3-\lambda \end{bmatrix}$ which, being a triangular matrix, has determinant that is the product of the diagonals, namely $\det(A - \lambda I) = (2 - \lambda)(3 - \lambda)$. This determinant is zero only for eigenvalues $\lambda = 2$ or 3. These are the diagonal entries in the triangular matrix.

(b) For each eigenvalue, find corresponding eigenvectors by solving the system $(A - \lambda I)x = 0$.

- For $\lambda = 2$, the system is $\begin{bmatrix} 0 & 1 \\ 0 & 1 \end{bmatrix} x = 0$ which requires $x_2 = 0$. That is, all eigenvectors are $x_1(1,0)$.

- For $\lambda = 3$, the system is $\begin{bmatrix} -1 & 1 \\ 0 & 0 \end{bmatrix} x = 0$ which requires $x_1 = x_2$. That is, all eigenvectors are $x_2(1,1)$.

The eigenvectors corresponding to the different eigenvalues are not orthogonal as their dot product $(1,0) \cdot (1,1) = 1 + 0 = 1 \neq 0$. Instead the different eigenvectors are at $45°$ to each other. □

Theorem 7.0.2 (*triangular matrices*) *The diagonal entries of a triangular matrix are the only eigenvalues of the matrix. The corresponding eigenvectors of distinct eigenvalues are generally not orthogonal.*

Proof. We detail only the case of upper triangular matrices, as the argument is similar for lower triangular matrices. First establish that the diagonal entries are eigenvalues, and second prove there are no others. Let λ be any value in the diagonal of the matrix A, and let k be the smallest index such that $a_{k,k} = \lambda$ (this 'smallest' caters for duplicated diagonal values). Let's construct an eigenvector in the form $x = (x_1, x_2, \ldots, x_{k-1}, 1, 0, \ldots, 0)$. Set $x_k = 1$ and $x_j = 0$ for $j > k$. Then set

$$x_{k-1} = -a_{k-1,k} x_k / (a_{k-1,k-1} - \lambda),$$
$$x_{k-2} = -(a_{k-2,k} x_k + a_{k-2,k-1} x_{k-1}) / (a_{k-2,k-2} - \lambda),$$
$$\vdots$$
$$x_1 = -(a_{1,k} x_k + a_{1,k-1} x_{k-1} + \cdots + a_{1,2} x_2) / (a_{1,1} - \lambda).$$

Since k is the smallest index for which $\lambda = a_{k,k}$ none of the above expressions involve divisions by zero, and so all are well-defined. Rearranging the above equations shows that this vector x satisfies, for $\lambda = a_{k,k}$,

$$\begin{aligned}
(a_{1,1} - \lambda)x_1 + a_{1,2} x_2 + \cdots + a_{1,k-1} x_{k-1} + a_{1,k} x_k &= 0, \\
(a_{2,2} - \lambda)x_2 + \cdots + a_{2,k-1} x_{k-1} + a_{2,k} x_k &= 0, \\
\ddots \qquad \vdots \qquad & \\
(a_{k-1,k-1} - \lambda)x_{k-1} + a_{k-1,k} x_k &= 0, \\
(a_{k,k} - \lambda)x_k &= 0;
\end{aligned}$$

that is, $(A - \lambda I)x = 0$. Rearranging gives $Ax = \lambda x$ for nonzero eigenvector x (nonzero since at least $x_k \neq 0$) and corresponding eigenvalue $\lambda = a_{k,k}$.

Second, there can be no other eigenvalues. Every eigenvalue has to have non-trivial solutions, nonzero eigenvectors x, to $(A - \lambda I)x = 0$, which by Theorem 6.1.29 requires $\det(A - \lambda I) = 0$. But as A is upper triangular, matrix $(A - \lambda I)$ is upper triangular and so Theorem 6.2.16 asserts the determinant

$$\det(A - \lambda I) = (a_{1,1} - \lambda)(a_{2,2} - \lambda) \cdots (a_{n,n} - \lambda).$$

This expression is zero iff the eigenvalue λ is one of the diagonal elements of A.

As an example of the non-orthogonality of eigenvectors, consider the two eigenvalues $\lambda_1 = a_{1,1}$ and $\lambda_2 = a_{2,2}$, assume distinct, with corresponding eigenvectors $\boldsymbol{x}_1 = (1,0,0,\ldots,0)$ and $\boldsymbol{x}_2 = (-a_{1,2}/(a_{1,1} - a_{2,2}), 1, 0, \ldots, 0)$. Then the dot product $\boldsymbol{x}_1 \cdot \boldsymbol{x}_2 = -a_{1,2}/(a_{1,1} - a_{2,2}) \neq 0$ in general (the dot product is zero only in the special case when $a_{1,2} = 0$). Since the dot product is generally nonzero, $\boldsymbol{x}_1$ and $\boldsymbol{x}_2$ are generally not orthogonal. Similarly for other pairs of eigenvectors corresponding to distinct eigenvalues. ☐

Example 7.0.3 Use Theorem 7.0.2 to find the eigenvalues, corresponding eigenvectors, and corresponding eigenspaces, of the following triangular matrices.

(a) $A = \begin{bmatrix} -3 & 2 & 0 \\ 0 & -4 & 2 \\ 0 & 0 & 4 \end{bmatrix}$

Solution: Matrix A is upper triangular so read off the eigenvalues from the diagonal to be -3 and ± 4.

- For $\lambda = -3$, and by inspection, all eigenvectors are proportional to $(1,0,0)$. Hence eigenspace $\mathbb{E}_{-3} = \text{span}\{(1,0,0)\}$.
- For $\lambda = -4$ we need to solve

$$(A + 4I)\boldsymbol{x} = \begin{bmatrix} 1 & 2 & 0 \\ 0 & 0 & 2 \\ 0 & 0 & 8 \end{bmatrix} \boldsymbol{x} = \boldsymbol{0}.$$

By inspection an eigenvector must be of the form $(x_1, 1, 0)$. Then the first line of the system asserts $x_1 + 2 = 0$. Hence eigenvectors are proportional to $(-2, 1, 0)$. That is, the eigenspace $\mathbb{E}_{-4} = \text{span}\{(-2, 1, 0)\}$.

- For $\lambda = +4$ we need to solve

$$(A - 4I)\boldsymbol{x} = \begin{bmatrix} -7 & 2 & 0 \\ 0 & -8 & 2 \\ 0 & 0 & 0 \end{bmatrix} \boldsymbol{x} = \boldsymbol{0}.$$

Consider eigenvectors of the form $(x_1, x_2, 1)$. The second line asserts $-8x_2 + 2 = 0$, that is $x_2 = \frac{1}{4}$. The first line asserts $-7x_1 + 2x_2 = 0$, that is $x_1 = \frac{2}{7}x_2 = \frac{1}{14}$. Hence eigenvectors are proportional to $(\frac{1}{14}, \frac{1}{4}, 1)$. That is, the eigenspace $\mathbb{E}_4 = \text{span}\{(\frac{1}{14}, \frac{1}{4}, 1)\}$. ☐

(b) $B = \begin{bmatrix} 3 & 0 & 0 & 0 \\ -2 & -4 & 0 & 0 \\ -3 & 1 & 0 & 0 \\ 0 & 0 & -3 & 1 \end{bmatrix}$

Solution: Matrix B is lower triangular so read the eigenvalues from the diagonal to be $3, -4, 0,$ and 1.

- For $\lambda = 1$, by inspection all eigenvectors are of the form $(0,0,0,1)$. Hence eigenspace $\mathbb{E}_1 = \text{span}\{(0,0,0,1)\}$.

- For $\lambda = 0$, seek an eigenvector $(0,0,1,x_4)$ then the last line of the system

$$(B - 0I)x = \begin{bmatrix} 3 & 0 & 0 & 0 \\ -2 & -4 & 0 & 0 \\ -3 & 1 & 0 & 0 \\ 0 & 0 & -3 & 1 \end{bmatrix} x = 0$$

requires $-3 + x_4 = 0$. Hence eigenvectors are proportional to $(0,0,1,3)$. That is, the eigenspace $\mathbb{E}_0 = \text{span}\{(0,0,1,3)\}$.

- For $\lambda = -4$, seek an eigenvector $(0,1,x_3,x_4)$ then the third line of the system

$$(B + 4I)x = \begin{bmatrix} 7 & 0 & 0 & 0 \\ -2 & 0 & 0 & 0 \\ -3 & 1 & 4 & 0 \\ 0 & 0 & -3 & 5 \end{bmatrix} x = 0$$

requires $1 + 4x_3 = 0$, that is $x_3 = -\frac{1}{4}$. Then the last line of the system requires $\frac{3}{4} + 5x_4 = 0$, that is $x_4 = -\frac{3}{20}$. Hence eigenvectors are proportional to $(0, 1, -\frac{1}{4}, -\frac{3}{20})$. That is, the eigenspace $\mathbb{E}_{-4} = \text{span}\{(0, 1, -\frac{1}{4}, -\frac{3}{20})\}$.

- For $\lambda = 3$, seek an eigenvector $(1, x_2, x_3, x_4)$ then the second line of the system

$$(B - 3I)x = \begin{bmatrix} 0 & 0 & 0 & 0 \\ -2 & -7 & 0 & 0 \\ -3 & 1 & -3 & 0 \\ 0 & 0 & -3 & -2 \end{bmatrix} x = 0$$

requires $-2 - 7x_2 = 0$, that is $x_2 = -\frac{2}{7}$. Then the third line of the system requires $-3 - \frac{2}{7} - 3x_3 = 0$, that is $x_3 = -\frac{23}{21}$. Lastly, the last line requires $\frac{23}{7} - 2x_4 = 0$, that is $x_4 = \frac{23}{14}$. Hence eigenvectors are proportional to $(1, -\frac{2}{7}, -\frac{23}{21}, \frac{23}{14})$. That is, the eigenspace $\mathbb{E}_3 = \text{span}\{(1, -\frac{2}{7}, -\frac{23}{21}, \frac{23}{14})\}$.

$\square$

(c) $C = \begin{bmatrix} -1 & 1 & -8 & -5 & 5 \\ -3 & 6 & 4 & -3 & 0 \\ 1 & -3 & 1 & 0 & 0 \\ -7 & 1 & 0 & 0 & 0 \\ -1 & 0 & 0 & 0 & 0 \end{bmatrix}$

Solution: Matrix C is not a triangular matrix (Definition 6.2.15), so Theorem 7.0.2 does not apply. Row or column swaps could transform C to be triangular, but we have not investigated the effect of such swaps on eigenvalues and eigenvectors. $\square$

Activity 7.0.4 What are all the eigenvalues of the matrix

$$\begin{bmatrix} 1 & 0 & 0 & 0 & 0 \\ 0 & 1 & 0 & 0 & 0 \\ 2 & 2 & 1 & 0 & 0 \\ 3 & 3 & 1 & 0 & 0 \\ 2 & 2 & 2 & 0 & 1 \end{bmatrix}?$$

(a) 1   (b) 0,1,2,3   (c) 0,1,2   (d) 0,1

One consequence of the second part of the proof of Theorem 7.0.2 is that, when counted according to multiplicity, there are precisely n eigenvalues of an $n \times n$ triangular matrix. Correspondingly, the next Section 7.1 establishes that there are precisely n eigenvalues of general $n \times n$ matrices, provided we count the eigenvalues according to multiplicity and allow complex eigenvalues.

7.0.1 Exercises

Exercise 7.0.1 Each of the following pictures applies to some specific real matrix, say called A. The pictures plot $A\boldsymbol{x}$ (red) adjoined to the end of unit vectors $\boldsymbol{x}$ (blue). By inspection decide whether the matrix, in each case, has real eigenvalues or non-real complex eigenvalues.

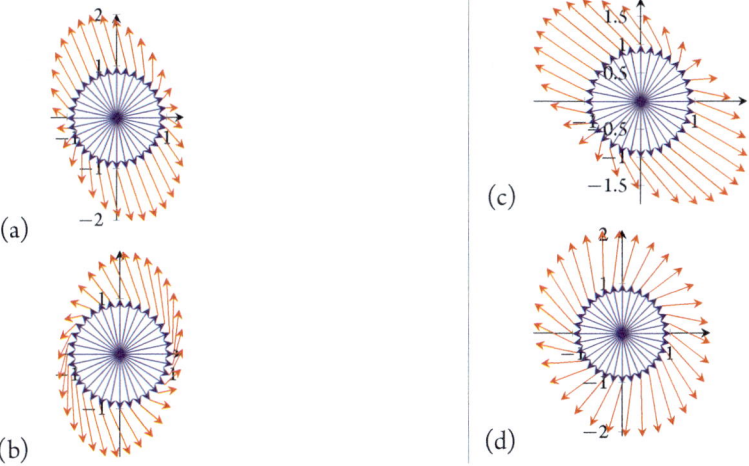

(a)

(b)

(c)

(d)

Exercise 7.0.2 For each of the following triangular matrices, write down all eigenvalues and then find the corresponding eigenspaces. Show your working.

(a) $\begin{bmatrix} 2 & 0 \\ -1 & 4 \end{bmatrix}$

(b) $\begin{bmatrix} 2 & 0 \\ -3 & 2 \end{bmatrix}$

(c) $\begin{bmatrix} -1 & 0 & 3 \\ 0 & -1 & 2 \\ 0 & 0 & -5 \end{bmatrix}$

(d) $\begin{bmatrix} 0 & 0 & 0 \\ -3 & -4 & 0 \\ 1 & 5 & -2 \end{bmatrix}$

(e) $\begin{bmatrix} 8 & -2 & 3 & 2 \\ 0 & -6 & 1 & -2 \\ 0 & 0 & 3 & -2 \\ 0 & 0 & 0 & 0 \end{bmatrix}$

(f) $\begin{bmatrix} 0 & -2 & -5 & 2 \\ 0 & 7 & -1 & 2 \\ 0 & 0 & 3 & -4 \\ 0 & 0 & 0 & 3 \end{bmatrix}$

7.1 Find eigenvalues and eigenvectors of matrices

This section begins exploring the properties and some applications of the eigen-problem $Ax = \lambda x$ for *general* matrices A. The section generalises properties established for symmetric matrices, Chapter 4, and relies on the determinant methods of Chapter 6. We establish that there are generally n eigenvalues of an $n \times n$ matrix, albeit possibly non-real complex valued, and also that repeated eigenvalues are sensitive to errors. Applications include population modelling, the computation of SVDs, and fitting exponential functions to real data.

7.1.1 A characteristic equation gives eigenvalues

The Fundamental Theorem of Algebra asserts that every polynomial equation over the complex field has a root. It is almost beneath the dignity of such a majestic theorem to mention that in fact it has precisely n roots.   *J. H. Wilkinson, 1984 (Higham, 1996, p.103)*

Recall that eigenvalues λ and nonzero eigenvectors x of a square matrix A must satisfy $(A - \lambda I)x = 0$. Theorem 6.1.29 then implies that the eigenvalues of a square matrix are precisely the solutions of the **characteristic equation** $\det(A - \lambda I) = 0$.

Theorem 7.1.1 *For every $n \times n$ square matrix A we call $\det(A - \lambda I)$ the **characteristic polynomial of A**:*[2]

---

[2] Alternatively, many call $\det(\lambda I - A)$ the characteristic polynomial, as does MATLAB/Octave. The distinction is immaterial as, for an $n \times n$ matrix A and by Theorem 6.1.8(c) with multiplicative factor $k = -1$, the only difference in the determinant is a factor of $(-1)^n$. In MATLAB/Octave, `poly (A)` computes the characteristic polynomial of the matrix, $\det(\lambda I - A)$, which might be useful for exercises, but is rarely useful in practice due to poor conditioning.

- the characteristic polynomial of A is a polynomial of nth degree in λ;
- there are at most n distinct eigenvalues of A.

Proof. Use induction on the size of the matrix. First, for 1×1 matrix $A = \begin{bmatrix} a_{11} \end{bmatrix}$, the determinant $\det(A - \lambda I) = \det \begin{bmatrix} a_{11} - \lambda \end{bmatrix} = a_{11} - \lambda$ which is of degree one in λ. Second, assume that for every $(n-1) \times (n-1)$ matrix A, $\det(A - \lambda I)$ is a polynomial of degree $n-1$ in λ. Consider any $n \times n$ matrix A. Use the Laplace Expansion Theorem 6.2.24 to give the first row expansion, in terms of minors of A and I,

$$\det(A - \lambda I) = (a_{11} - \lambda)\det(A_{11} - \lambda I_{11}) - a_{12}\det(A_{12} - \lambda I_{12})$$
$$+ \cdots - (-1)^n a_{1n}\det(A_{1n} - \lambda I_{1n}).$$

Now the minor I_{11} is precisely the $(n-1) \times (n-1)$ identity, and hence by assumption $\det(A_{11} - \lambda I_{11})$ is a polynomial of degree $(n-1)$ in λ. But differently, the minors $I_{12}, \ldots, I_{1n}$ have two of the ones removed from the $n \times n$ identity and hence $\lambda I_{12}, \ldots, \lambda I_{1n}$ each have only $(n-2)$ λs; since each term in a determinant is a product of distinct entries of the matrix (Theorem 6.2.27) it follows that for $j \geq 2$ the determinant $\det(A_{1j} - \lambda I_{1j})$ is a polynomial in λ of degree $\leq n-2$. Consequently, the first row expansion of

$$\det(A - \lambda I) = (a_{11} - \lambda)(\text{poly degree } n-1) - a_{12}(\text{poly degree } \leq n-2)$$
$$+ \cdots - (-1)^n a_{1n}(\text{poly degree } \leq n-2)$$
$$= (a_{11} - \lambda)(\text{poly degree } n-1) + (\text{poly degree } \leq n-2)$$
$$= (\text{poly degree } n)$$

as the highest power of λ cannot be cancelled by any term. Induction thus implies that for every n the characteristic polynomial of an $n \times n$ matrix A is a polynomial of nth degree in λ.

Lastly, because the characteristic polynomial of A is of nth degree in λ, the Fundamental Theorem of Algebra (established in more advanced courses) asserts that the polynomial has at most n roots (possibly complex). Hence there are at most n distinct eigenvalues (allowing non-real complex eigenvalues). □

Activity 7.1.2 A given matrix has eigenvalues of $-7, -1, 3, 4,$ and 6. The matrix must be of size $n \times n$ for n at least which of the following? (Select the smallest valid answer.)

(a) 5   (b) 6   (c) 4   (d) 7

Example 7.1.3 Find the characteristic polynomial of each of the following matrices. Where in the coefficients of the polynomial can you see the determinant? and the sum of the diagonal elements?

(a) $A = \begin{bmatrix} 1 & -1 \\ -2 & 4 \end{bmatrix}$

Solution: The characteristic polynomial is $\det(A - \lambda I) = \det \begin{bmatrix} 1-\lambda & -1 \\ -2 & 4-\lambda \end{bmatrix} = (1-\lambda)(4-\lambda) - 2 = \lambda^2 - 5\lambda + 2$. Now $\det A = 4 - 2 = 2$ which is the coefficient of the constant term in this polynomial. Whereas the sum of the diagonal of A is $1 + 4 = 5$ which is the negative of the λ coefficient in the polynomial. □

(b) $B = \begin{bmatrix} 4 & -2 & 1 \\ 1 & -2 & 0 \\ 8 & 2 & 6 \end{bmatrix}$

Solution: The characteristic polynomial is

$$\det(B - \lambda I) = \det \begin{bmatrix} 4-\lambda & -2 & 1 \\ 1 & -2-\lambda & 0 \\ 8 & 2 & 6-\lambda \end{bmatrix}$$

$$= (4-\lambda)(-2-\lambda)(6-\lambda) + 0 + 2 - (-2-\lambda)8 - 0 - (-2)(6-\lambda)$$
$$= -48 + 4\lambda + 8\lambda^2 - \lambda^3 + 2 + 16 + 8\lambda + 12 + 2\lambda$$
$$= -\lambda^3 + 8\lambda^2 + 2\lambda - 18.$$

Now $\det B = 4(-2)6 + 0 + 2 - (-2)8 - 0 - (-2)6 = -48 + 2 + 16 + 12 = -18$ which is the coefficient of the constant term in the polynomial, whereas the sum of the diagonal of B is $4 - 2 + 6 = 8$ which is the λ^2 coefficient in the polynomial. □

These observations about the coefficients in the characteristic polynomials lead to the next (optional) theorem that helps establish the nature of a characteristic polynomial.

Theorem 7.1.4 *For every $n \times n$ matrix A, the product of the eigenvalues equals $\det A$, and also equals the constant term in the characteristic polynomial. The sum of the eigenvalues equals $(-1)^{n-1}$ times the coefficient of λ^{n-1} in the characteristic polynomial, and also equals the **trace** of the matrix, defined as the sum of the diagonal elements $a_{11} + a_{22} + \cdots + a_{nn}$.*

Proof. Theorem 7.1.1, and its proof, establishes that the characteristic polynomial has the form

$$\det(A - \lambda I) = c_0 + c_1\lambda + \cdots + c_{n-1}\lambda^{n-1} + (-1)^n\lambda^n$$
$$= (\lambda_1 - \lambda)(\lambda_2 - \lambda) \cdots (\lambda_n - \lambda), \tag{7.1}$$

where the second equality follows from the Fundamental Theorem of Algebra that an nth degree polynomial factors into n linear factors (albeit possibly complex). First, substitute $\lambda = 0$, and this equation (7.1) gives $\det A = c_0 = \lambda_1\lambda_2 \cdots \lambda_n$, as required.

Second, consider the term $c_{n-1}\lambda^{n-1}$ in equation (7.1). From the factorization on the right-hand side, the λ^{n-1} term arises as

$$c_{n-1}\lambda^{n-1} = \lambda_1(-\lambda)^{n-1} + (-\lambda)\lambda_2(-\lambda)^{n-2} + \cdots + (-\lambda)^{n-1}\lambda_n$$
$$= (-1)^{n-1}(\lambda_1 + \lambda_2 + \cdots + \lambda_n)\lambda^{n-1},$$

and hence the coefficient $c_{n-1} = (-1)^{n-1}(\lambda_1 + \lambda_2 + \cdots + \lambda_n)$, as required. Now use induction to prove the trace formula. Recall that the proof of Theorem 7.1.1 establishes that

$$\det(A - \lambda I) = (a_{11} - \lambda)\det(A_{11} - \lambda I_{11})$$
$$+ (\text{poly degree } \leq n - 2).$$

Assume that the trace formula holds for $(n-1) \times (n-1)$ matrices, and so it holds for the minor A_{11}. Then the previous identity gives

$$\det(A - \lambda I) = (a_{11} - \lambda)[(-1)^{n-1}\lambda^{n-1} + (-1)^{n-2}(a_{22} + \cdots + a_{nn})\lambda^{n-2}$$
$$+ (\text{poly degree } \leq n - 3)] + (\text{poly degree } \leq n - 2)$$
$$= (-1)^n\lambda^n + (-1)^{n-1}(a_{11} + a_{22} + \cdots + a_{nn})\lambda^{n-1} + (\text{poly degree } \leq n - 2).$$

Hence the coefficient $c_{n-1} = (-1)^{n-1}(a_{11} + a_{22} + \cdots + a_{nn})$. Since the formula holds for the basic case $n = 1$, namely $c_0 = +a_{11}$, induction implies that the sum of the eigenvalues $\lambda_1 + \lambda_2 + \cdots + \lambda_n = a_{11} + a_{22} + \cdots + a_{nn}$, the trace of the matrix A.

Activity 7.1.5 What is the trace of the matrix

$$\begin{bmatrix} 4 & 5 & -4 & 3 \\ -2 & 2 & -5 & -1 \\ -1 & 2 & 2 & -6 \\ -13 & 4 & 3 & -1 \end{bmatrix}?$$

(a) -12 (b) -13 (c) 7 (d) 8

Example 7.1.6

(a) What are the two highest order terms and the constant term in the characteristic polynomial of the matrix

$$A = \begin{bmatrix} -2 & -1 & 3 & -2 \\ -1 & 3 & -2 & 2 \\ 2 & -3 & 0 & 1 \\ 0 & 1 & 0 & -3 \end{bmatrix}.$$

Solution: First compute the determinant using the Laplace expansion (Theorem 6.2.24). The two zeros in the last row suggest a last row expansion:

$$\det A = (-1)^6 1\det\begin{bmatrix} -2 & 3 & -2 \\ -1 & -2 & 2 \\ 2 & 0 & 1 \end{bmatrix} + (-1)^8(-3)\det\begin{bmatrix} -2 & -1 & 3 \\ -1 & 3 & -2 \\ 2 & -3 & 0 \end{bmatrix}$$
$$= (4 + 12 + 0 - 8 - 0 + 3) - 3(0 + 4 + 9 - 18 + 12 - 0) = -10.$$

This is the constant term in the characteristic polynomial. Second, the trace of A is $-2 + 3 + 0 - 3 = -2$ so the cubic coefficient in the characteristic polynomial is $(-1)^3(-2) = 2$. That is, the characteristic polynomial of A is of the form $\lambda^4 + 2\lambda^3 + \cdots - 10$. $\qquad\square$

(b) After laborious calculation you find the characteristic polynomial of the matrix

$$B = \begin{bmatrix} -2 & 5 & -3 & -1 & 2 \\ -2 & -5 & -1 & -1 & 3 \\ 1 & 4 & -2 & 1 & -7 \\ 1 & -5 & 1 & 4 & -5 \\ -1 & 0 & 3 & -3 & 1 \end{bmatrix}$$

is $-\lambda^5 + 2\lambda^4 - 3\lambda^3 + 234\lambda^2 + 884\lambda + 1564$. Could this polynomial be correct?

Solution: No, because the trace of B is $-2 - 5 - 2 + 4 + 1 = -4$ so the coefficient of the λ^4 term must be $(-1)^4(-4) = -4$ instead of the calculated 2. $\qquad\square$

(c) After much calculation you find the characteristic polynomial of the matrix

$$C = \begin{bmatrix} 0 & 0 & 3 & 1 & 0 & 0 \\ 0 & 0 & 0 & 0 & 0 & 1 \\ 0 & 0 & -4 & 0 & 3 & 0 \\ -5 & 0 & 3 & 0 & 0 & 0 \\ 0 & 0 & 0 & 0 & 0 & 2 \\ 0 & 0 & 0 & -6 & 0 & 0 \end{bmatrix}$$

is $\lambda^6 + 4\lambda^5 + 5\lambda^4 + 20\lambda^3 + 108\lambda^2 - 540\lambda + 668$. Could this polynomial be correct?

Solution: No. By the column of zeros in C, $\det C$ must be zero instead of the calculated 668. $\qquad\square$

Recall that an important characteristic of an eigenvalue is its multiplicity. The following definition of *multiplicity* generalizes to all matrices the somewhat different Definition 4.1.15 that applies to only symmetric matrices. For symmetric matrices the definitions are equivalent.

Definition 7.1.7 *An eigenvalue λ_0 of a matrix A is said to have* **multiplicity** *m if the characteristic polynomial factorizes to $\det(A - \lambda I) = (\lambda - \lambda_0)^m g(\lambda)$ with $g(\lambda_0) \neq 0$, and $g(\lambda)$ is a polynomial of degree $n - m$. Every eigenvalue of multiplicity $m \geq 2$ may also be called a* **repeated eigenvalue**.

Activity 7.1.8 A given matrix A has characteristic polynomial $\det(A - \lambda I) = (\lambda + 2)\lambda^2(\lambda - 2)^3(\lambda - 3)^4$. The eigenvalue 2 has what multiplicity?

(a) two   (b) four   (c) three   (d) one

Example 7.1.9 Use the characteristic polynomials for each of the following matrices to find all eigenvalues and their multiplicity.

(a) $A = \begin{bmatrix} 3 & 1 \\ 0 & 3 \end{bmatrix}$

Solution: Here, $\det(A - \lambda I) = \det \begin{bmatrix} 3-\lambda & 1 \\ 0 & 3-\lambda \end{bmatrix} = (3-\lambda)^2 - 0 \cdot 1 = (\lambda - 3)^2 = 0$ is the characteristic equation. The only eigenvalue is $\lambda = 3$ with multiplicity two. (Or more quickly: since this is an upper triangular matrix, Theorem 7.0.2 asserts that the eigenvalues are simply the diagonal elements, namely 3 twice.) □

(b) $B = \begin{bmatrix} -1 & 1 & -2 \\ -1 & 0 & -1 \\ 0 & -3 & 1 \end{bmatrix}$

Solution: The characteristic equation is

$$\det(B - \lambda I) = \det \begin{bmatrix} -1-\lambda & 1 & -2 \\ -1 & -\lambda & -1 \\ 0 & -3 & 1-\lambda \end{bmatrix}$$
$$= (1+\lambda)\lambda(1-\lambda) + 0 - 6 - 0 + 3(1+\lambda) + (1-\lambda)$$
$$= -\lambda^3 + 3\lambda - 2 = -(\lambda-1)^2(\lambda+2) = 0.$$

Eigenvalues are $\lambda = 1$ with multiplicity two, and $\lambda = -2$ with multiplicity one. □

(c) $C = \begin{bmatrix} -1 & 0 & -2 \\ 0 & -3 & 2 \\ 0 & -2 & 1 \end{bmatrix}$

Solution: The characteristic equation is

$$\det(C - \lambda I) = \det \begin{bmatrix} -1-\lambda & 0 & -2 \\ 0 & -3-\lambda & 2 \\ 0 & -2 & 1-\lambda \end{bmatrix} = (-1-\lambda)\det \begin{bmatrix} -3-\lambda & 2 \\ -2 & 1-\lambda \end{bmatrix}$$
$$= -(1+\lambda)[(-3-\lambda)(1-\lambda) + 4] = -(\lambda+1)[\lambda^2 + 2\lambda + 1]$$
$$= -(\lambda+1)^3 = 0.$$

The only eigenvalue is $\lambda = -1$ with multiplicity three. □

(d) $E = \begin{bmatrix} 0 & 1 \\ -1 & 1 \end{bmatrix}$

Solution: The characteristic equation is

$$\det(E - \lambda I) = \det\begin{bmatrix} -\lambda & 1 \\ -1 & 1 - \lambda \end{bmatrix} = -\lambda(1 - \lambda) + 1 = \lambda^2 - \lambda + 1 = 0.$$

This quadratic equation does not factor easily so use the formula

$$\lambda = \frac{-b \pm \sqrt{b^2 - 4ac}}{2a} = \frac{1 \pm \sqrt{1 - 4}}{2} = \frac{1}{2} \pm \frac{\sqrt{3}}{2}i.$$

Hence, for $i = \sqrt{-1}$, there are two eigenvalues (non-real complex valued) each of multiplicity one. □

Example 7.1.10 Use eig() in MATLAB/Octave to find the eigenvalues and their multiplicity for the following matrices. Recall (Table 4.1) that executing just eig(A) gives a column vector of eigenvalues of A, repeated according to their multiplicity.

(a) $\begin{bmatrix} 2 & 2 & -1 \\ 0 & 1 & -2 \\ 0 & -1 & 0 \end{bmatrix}$

Solution: Execute
eig([2 2 -1;0 1 -2;
0 -1 0]) to get

ans =
   2
   2
  -1

So the eigenvalue $\lambda = 2$ has multiplicity two, and the eigenvalue $\lambda = -1$ has multiplicity one. □

(b) $\begin{bmatrix} -2 & -2 & -5 & 0 \\ 0 & -2 & 2 & 1 \\ -1 & 1 & 0 & -1 \\ -2 & 1 & 4 & 0 \end{bmatrix}$

Solution: In MATLAB/Octave execute

eig([-2 -2 -5 0
  0 -2 2 1
 -1 1 0 -1
 -2 1 4 0])

to get

ans =
  -3.0000 + 0.0000i
  -3.0000 + 0.0000i
   1.0000 + 1.4142i
   1.0000 - 1.4142i

There are two complex valued eigenvalues, evidently $1 \pm \sqrt{2}i$, each of multiplicity one, and also the (real) eigenvalue $\lambda = -3$ which has multiplicity two. □

(c) $\begin{bmatrix} -1 & 0 & 0 & 0 \\ -1 & 2 & -3 & 3 \\ 3 & 1 & -1 & 0 \\ 0 & 3 & -2 & 1 \end{bmatrix}$

Solution: In MATLAB/Octave execute

eig([-1 0 0 0
 -1 2 -3 3
  3 1 -1 0
  0 3 -2 1])

to get

```
ans =
    4.0000 + 0.0000i
   -1.0000 + 0.0000i
   -1.0000 - 0.0000i
   -1.0000 + 0.0000i
```

There is one eigenvalue of multiplicity one, $\lambda = 4$. The last three components of ans appears to indicate the eigenvalue $\lambda = -1$ with multiplicity three. ☐

To find eigenvalues and eigenvectors, the following restates Procedure 4.1.23 with a little more information, and now empowered to address larger matrices upon using the determinant tools from Chapter 6.

Procedure 7.1.11 (eigenvalues and eigenvectors) *To find by hand eigenvalues and eigenvectors of a square matrix A (of small size):*

1. *find all eigenvalues (possibly non-real complex) by solving the **characteristic equation** of A, $\det(A - \lambda I) = 0$;*

2. *for each eigenvalue λ, solve the homogeneous linear equation $(A - \lambda I)x = 0$ to find the eigenspace $\mathbb{E}_\lambda$ of all eigenvectors (together with $\mathbf{0}$);*

3. *write each eigenspace as the span of a few chosen eigenvectors (Definition 7.2.20 calls such a set a basis).*

Since, for an $n \times n$ matrix, the characteristic polynomial is of nth degree in λ (Theorem 7.1.1), there are n eigenvalues (when counted according to multiplicity and allowing nonreal complex eigenvalues).

Correspondingly, the following restates the computational procedure of Section 4.1.1, but slightly more generally: the extra generality caters for non-symmetric matrices.

Compute in MATLAB/Octave   For a given square matrix A, execute [V,D]=eig(A), then the diagonal entries of D, namely diag(D), are the eigenvalues of A. Corresponding to the eigenvalue D(j,j) is an eigenvector v_j = V(:,j), the jth column of V.[3] If an eigenvalue is repeated in the diagonal of D (multiplicity more than one), then the corresponding columns of V span the eigenspace (and, as Section 7.2 discusses, when the column vectors have a property called linear independence, then they form a so-called basis for the eigenspace).

Activity 7.1.12 For the matrix $A = \begin{bmatrix} 2 & 0 & -1 \\ -5 & 3 & -5 \\ 5 & -2 & -2 \end{bmatrix}$, which one of the following vectors satisfy $(A - 3I)x = \mathbf{0}$ and hence is an eigenvector of A corresponding to eigenvalue 3?

(a) $x = (-1, 0, 1)$ (b) $x = (1, 5, 5)$ (c) $x = (1, 5, -1)$ (d) $x = (0, 1, 0)$

---

[3] Be aware that MATLAB/Octave does not use the determinant to find the eigenvalues, nor does it solve the linear equations to find eigenvectors. For any but the smallest matrices such a 'by hand' approach takes far too long. Instead, to find eigenvalues and eigenvectors, just as for the SVD, MATLAB/Octave uses an intriguing iteration based upon repeated application of what is called the QR factorization.

Example 7.1.13 Find the eigenspaces corresponding to the eigenvalues found for the first three matrices of Example 7.1.9.

7.1.9(a) $A = \begin{bmatrix} 3 & 1 \\ 0 & 3 \end{bmatrix}$

Solution: The only eigenvalue is $\lambda = 3$ with multiplicity two. Its eigenvectors x satisfy

$$(A - 3I)x = \begin{bmatrix} 0 & 1 \\ 0 & 0 \end{bmatrix} x = 0.$$

The second component of this equation is the trivial $0 = 0$. The first component of the equation is $0x_1 + 1x_2 = 0$, hence $x_2 = 0$. All eigenvectors are of the form $x = (1,0)x_1$. That is the eigenspace $\mathbb{E}_3 = \text{span}\{(1,0)\}$.

In MATLAB/Octave, executing $[V,D] = \text{eig}([3 \ 1; 0 \ 3])$ gives us

```
V =
    1.0000   -1.0000
    0.0000    0.0000
D =
    3    0
    0    3
```

The diagonal matrix D confirms that the only eigenvalue is three (multiplicity two), whereas the two columns of V confirm the eigenspace $\mathbb{E}_3 = \text{span}\{(1,0),(-1,0)\} = \text{span}\{(1,0)\}$. ∎

7.1.9(b) $B = \begin{bmatrix} -1 & 1 & -2 \\ -1 & 0 & -1 \\ 0 & -3 & 1 \end{bmatrix}$

Solution: The eigenvalues are $\lambda = -2$ (multiplicity one) and $\lambda = 1$ (multiplicity two).

- For $\lambda = -2$ solve

$$(B + 2I)x = \begin{bmatrix} 1 & 1 & -2 \\ -1 & 2 & -1 \\ 0 & -3 & 3 \end{bmatrix} x = 0.$$

The third component of this equation requires $-3x_2 + 3x_3 = 0$, that is, $x_2 = x_3$. The second component requires $-x_1 + 2x_2 - x_3 = 0$, that is, $x_1 = 2x_2 - x_3 = 2x_3 - x_3 = x_3$. The first component requires $x_1 + x_2 - 2x_3 = 0$ which is also satisfied by

$x_1 = x_2 = x_3$. All eigenvectors are of the form $x_3(1,1,1)$. That is, the eigenspace $\mathbb{E}_{-2} = \text{span}\{(1,1,1)\}$.

- For $\lambda = 1$ solve

$$(B - 1I)x = \begin{bmatrix} -2 & 1 & -2 \\ -1 & -1 & -1 \\ 0 & -3 & 0 \end{bmatrix} x = 0.$$

The third component of this equation requires $-3x_2 = 0$, that is, $x_2 = 0$. The second component requires $-x_1 - x_2 - x_3 = 0$, that is, $x_1 = -x_2 - x_3 = 0 - x_3 = -x_3$. The first component requires $-2x_1 + x_2 - 2x_3 = 0$ which is also satisfied by $x_1 = -x_3$ and $x_2 = 0$. All eigenvectors are of the form $x_3(-1,0,1)$. That is, the eigenspace $\mathbb{E}_1 = \text{span}\{(-1,0,1)\}$.

Alternatively, in MATLAB/Octave, executing

```
B= [-1  1  -2
    -1  0  -1
     0  -3  1]
[V,D] =eig (B)
```

gives us

```
V =
      -0.5774      0.7071     -0.7071
      -0.5774      0.0000      0.0000
      -0.5774     -0.7071      0.7071
D =
      -2           0           0
       0           1           0
       0           0           1
```

The diagonal matrix D confirms that the eigenvalues are $\lambda = -2$ (multiplicity one) and $\lambda = 1$ (multiplicity two). The first column of V confirms the eigenspace

$$\mathbb{E}_{-2} = \text{span}\{(-0.5774, -0.5774, -0.5774)\} = \text{span}\{(1,1,1)\}.$$

Whereas the last two columns of V confirm the eigenspace

$$\mathbb{E}_1 = \text{span}\{(0.7071, 0, -0.7071), (-0.7071, 0, 0.7071)\} = \text{span}\{(-1,0,1)\}.$$

☐

7.1.9(c) $C = \begin{bmatrix} -1 & 0 & -2 \\ 0 & -3 & 2 \\ 0 & -2 & 1 \end{bmatrix}$

Solution: The only eigenvalue is $\lambda = -1$ with multiplicity three. Its eigenvectors x satisfy

$$(C + 1I)x = \begin{bmatrix} 0 & 0 & -2 \\ 0 & -2 & 2 \\ 0 & -2 & 2 \end{bmatrix} x = 0.$$

The first component of this equation requires $x_3 = 0$. The second and third components both requires $-2x_2 + 2x_3 = 0$, hence $x_2 = x_3 = 0$. Since x_1 is unconstrained, all eigenvectors are of the form $x = x_1(1,0,0)$. That is the eigenspace $\mathbb{E}_{-1} = \text{span}\{(1,0,0)\}$.

Alternatively, in MATLAB/Octave, executing

```
C= [-1  0 -2
     0 -3  2
     0 -2  1]
[V,D] =eig(C)
```

gives us

V =

$$\begin{array}{ccc} 1 & -1 & -1 \\ 0 & 0 & 0 \\ 0 & 0 & 0 \end{array}$$

D =

$$\begin{array}{ccc} -1 & 0 & 0 \\ 0 & -1 & 0 \\ 0 & 0 & -1 \end{array}$$

Diagonal matrix D confirms that the only eigenvalue is $\lambda = -1$ with multiplicity three. The three columns of V confirm the eigenspace $\mathbb{E}_{-1} = \text{span}\{(1,0,0)\}$. ☐

The matrices in Example 7.1.13 all have repeated eigenvalues. For these repeated eigenvalues the corresponding eigenspaces happen to be all one-dimensional. This contrasts with the case of symmetric matrices where the eigenspaces always have the same dimensionality as the multiplicity of the eigenvalue, as illustrated by Examples 4.1.14 and 4.1.20. Subsequent sections work towards Theorem 7.3.14 which establishes that for non-symmetric matrices an eigenspace has dimensionality between one and the multiplicity of the corresponding eigenvalue.

Example 7.1.14 By hand, find the eigenvalues and eigenspaces of the matrix

$$A = \begin{bmatrix} 0 & 3 & 0 & 0 & 0 \\ 1 & 0 & 3 & 0 & 0 \\ 0 & 1 & 0 & 3 & 0 \\ 0 & 0 & 1 & 0 & 3 \\ 0 & 0 & 0 & 1 & 0 \end{bmatrix}$$

(Example 7.1.15 confirms the answer using eig () in MATLAB/Octave.)

Solution: Adopt Procedure 7.1.11.

(a) The characteristic equation is

$\det(A - \lambda I)$

$$= \begin{vmatrix} -\lambda & 3 & 0 & 0 & 0 \\ 1 & -\lambda & 3 & 0 & 0 \\ 0 & 1 & -\lambda & 3 & 0 \\ 0 & 0 & 1 & -\lambda & 3 \\ 0 & 0 & 0 & 1 & -\lambda \end{vmatrix}$$

$$= (-\lambda) \begin{vmatrix} -\lambda & 3 & 0 & 0 \\ 1 & -\lambda & 3 & 0 \\ 0 & 1 & -\lambda & 3 \\ 0 & 0 & 1 & -\lambda \end{vmatrix} - 3 \begin{vmatrix} 1 & 3 & 0 & 0 \\ 0 & -\lambda & 3 & 0 \\ 0 & 1 & -\lambda & 3 \\ 0 & 0 & 1 & -\lambda \end{vmatrix} \quad \text{(by first row expansion (6.4))}$$

$$= (-\lambda)^2 \begin{vmatrix} -\lambda & 3 & 0 \\ 1 & -\lambda & 3 \\ 0 & 1 & -\lambda \end{vmatrix} - (-\lambda)3 \begin{vmatrix} 1 & 3 & 0 \\ 0 & -\lambda & 3 \\ 0 & 1 & -\lambda \end{vmatrix} - 3 \begin{vmatrix} -\lambda & 3 & 0 \\ 1 & -\lambda & 3 \\ 0 & 1 & -\lambda \end{vmatrix} \quad \begin{matrix} \text{(by first row and} \\ \text{first column,} \\ \text{respectively)} \end{matrix}$$

$$= (\lambda^2 - 3) \begin{vmatrix} -\lambda & 3 & 0 \\ 1 & -\lambda & 3 \\ 0 & 1 & -\lambda \end{vmatrix} + 3\lambda \begin{vmatrix} -\lambda & 3 \\ 1 & -\lambda \end{vmatrix} \quad \begin{matrix} \text{(by common factor, and} \\ \text{first column expansion)} \end{matrix}$$

$$= (\lambda^2 - 3)[(-\lambda)^3 + 0 + 0 - 0 + 3\lambda + 3\lambda] + 3\lambda(\lambda^2 - 3) \quad \text{(using (6.1))}$$

$$= (\lambda^2 - 3)(-\lambda^3 + 9\lambda) = -\lambda(\lambda^2 - 3)(\lambda^2 - 9) = 0.$$

The five eigenvalues are thus $\lambda = 0, \pm\sqrt{3}, \pm 3$, all of multiplicity one.

(b) Consider each eigenvalue in turn.

$\lambda = 0$ Solve

$$(A - 0I)x = \begin{bmatrix} 0 & 3 & 0 & 0 & 0 \\ 1 & 0 & 3 & 0 & 0 \\ 0 & 1 & 0 & 3 & 0 \\ 0 & 0 & 1 & 0 & 3 \\ 0 & 0 & 0 & 1 & 0 \end{bmatrix} x = 0.$$

The last row requires $x_4 = 0$. The fourth row requires $x_3 + 3x_5 = 0$, that is, $x_3 = -3x_5$. The third row requires $x_2 + 3x_4 = 0$, that is, $x_2 = -3x_4 = -3 \cdot 0 = 0$. The second row requires $x_1 + 3x_3 = 0$, that is, $x_1 = -3x_3 = 9x_5$. The first row requires $3x_2 = 0$, which is satisfied as $x_2 = 0$. Since all eigenvectors are of the form $(9x_5, 0, -3x_5, 0, x_5)$, the eigenspace $\mathbb{E}_0 = \text{span}\{(9, 0, -3, 0, 1)\}$.

$\lambda = \pm\sqrt{3}$ Being careful with the upper and lower signs, solve $(A \mp \sqrt{3}I)x = 0$, that is,

$$\begin{bmatrix} \mp\sqrt{3} & 3 & 0 & 0 & 0 \\ 1 & \mp\sqrt{3} & 3 & 0 & 0 \\ 0 & 1 & \mp\sqrt{3} & 3 & 0 \\ 0 & 0 & 1 & \mp\sqrt{3} & 3 \\ 0 & 0 & 0 & 1 & \mp\sqrt{3} \end{bmatrix} x = 0.$$

The last row requires $x_4 \mp \sqrt{3}x_5 = 0$, that is, $x_4 = \pm\sqrt{3}x_5$. The fourth row requires $x_3 \mp \sqrt{3}x_4 + 3x_5 = 0$, that is, $x_3 = \pm\sqrt{3}x_4 - 3x_5 = 3x_5 - 3x_5 = 0$. The third row requires $x_2 \mp \sqrt{3}x_3 + 3x_4 = 0$, that is, $x_2 = \pm\sqrt{3}x_3 - 3x_4 = \mp3\sqrt{3}x_5$. The second row requires $x_1 \mp \sqrt{3}x_2 + 3x_3 = 0$, that is, $x_1 = \pm\sqrt{3}x_2 - 3x_3 = -9x_5$. The first row requires $\mp\sqrt{3}x_1 + 3x_2 = 0$, which is satisfied as $\mp\sqrt{3}(-9x_5) + 3(\mp3\sqrt{3}x_5) = 0$. Since all eigenvectors are of the form $(-9x_5, \mp3\sqrt{3}x_5, 0, \pm\sqrt{3}x_5, x_5)$, the eigenspaces $\mathbb{E}_{\pm\sqrt{3}} = \mathrm{span}\{(-9, \mp3\sqrt{3}, 0, \pm\sqrt{3}, 1)\}$.

$\lambda = \pm 3$ Being careful with the upper and lower signs, solve $(A \mp 3I)x = 0$, that is,

$$\begin{bmatrix} \mp3 & 3 & 0 & 0 & 0 \\ 1 & \mp3 & 3 & 0 & 0 \\ 0 & 1 & \mp3 & 3 & 0 \\ 0 & 0 & 1 & \mp3 & 3 \\ 0 & 0 & 0 & 1 & \mp3 \end{bmatrix} x = 0.$$

The last row requires $x_4 \mp 3x_5 = 0$, that is, $x_4 = \pm3x_5$. The fourth row requires $x_3 \mp 3x_4 + 3x_5 = 0$, that is, $x_3 = \pm3x_4 - 3x_5 = 9x_5 - 3x_5 = 6x_5$. The third row requires $x_2 \mp 3x_3 + 3x_4 = 0$, that is, $x_2 = \pm3x_3 - 3x_4 = \pm9x_5$. The second row requires $x_1 \mp 3x_2 + 3x_3 = 0$, that is, $x_1 = \pm3x_2 - 3x_3 = 9x_5$. The first row requires $\mp3x_1 + 3x_2 = 0$, which is satisfied as $\mp3(9x_5) + 3(\pm9x_5) = 0$. Since all eigenvectors are of the form $(9x_5, \pm9x_5, 6x_5, \pm3x_5, x_5)$, the eigenspaces $\mathbb{E}_{\pm3} = \mathrm{span}\{(9, \pm9, 6, \pm3, 1)\}$.

$\square$

Example 7.1.15 Use MATLAB/Octave to confirm the eigenvalues and eigenvectors found for the matrix of Example 7.1.14.

Solution: In MATLAB/Octave execute

```
A= [0 3 0 0 0
 1 0 3 0 0
 0 1 0 3 0
 0 0 1 0 3
 0 0 0 1 0]
[V,D] =eig (A)
```

to obtain the report (2 d.p.)

```
V =
     0.62   -0.62    0.94   -0.85   -0.85
     0.62    0.62   -0.00    0.49   -0.49
     0.42   -0.42   -0.31   -0.00    0.00
     0.21    0.21   -0.00   -0.16    0.16
     0.07   -0.07    0.10    0.09    0.09
D =
     3.00       0       0       0       0
        0   -3.00       0       0       0
        0       0   -0.00       0       0
        0       0       0   -1.73       0
        0       0       0       0    1.73
```

The eigenvalues in D agree with the hand calculations of $\lambda = 0, \pm\sqrt{3}, \pm 3$. To confirm the hand calculation of eigenvectors in Example 7.1.14, here divide each column of V by its last element, V(5,:), via the auto-replication in V./V(5,:), which gives the more appealing matrix of eigenvectors (2 d.p.)

```
ans =
     9.00    9.00    9.00   -9.00   -9.00
     9.00   -9.00    0.00    5.20   -5.20
     6.00    6.00   -3.00    0.00    0.00
     3.00   -3.00    0.00   -1.73    1.73
     1.00    1.00    1.00    1.00    1.00
```

These also agree with the hand calculation.   ☐

7.1.2 Repeated eigenvalues are sensitive

Albeit hidden in Example 7.1.10, repeated eigenvalues are exquisitely sensitive to errors in either the matrix or the computation. If the matrix or the computation has an error ϵ, then expect a repeated eigenvalue of multiplicity m to appear as m eigenvalues all within about $\epsilon^{1/m}$ of each other. Consequently, when we find or compute m eigenvalues all within about $\epsilon^{1/m}$, then *suspect* them to be one eigenvalue of multiplicity m.

This optional subsection does not prove the sensitivity: it uses examples to introduce and illustrate.

Example 7.1.16 Explore the eigenvalues of the matrix $A = \begin{bmatrix} a & 1 \\ 0.0001 & a \end{bmatrix}$ for every parameter a.

Solution: By hand, the characteristic equation is

$$\det \begin{bmatrix} a - \lambda & 1 \\ 0.0001 & a - \lambda \end{bmatrix} = (a - \lambda)^2 - 0.0001 = 0.$$

Rearranging gives $(\lambda - a)^2 = 0.0001$. Taking square-roots, $\lambda - a = \pm 0.01$; that is, the two eigenvalues are $\lambda = a \pm 0.01$.

Alternatively, if we consider that the entry 0.0001 in the matrix is an error, and that the entry should really be zero, then the eigenvalues should really be one repeated eigenvalue $\lambda = a$ of multiplicity two. However, the 'error' 0.0001 splits the repeated eigenvalue into two by an amount $0.01 = \sqrt{0.0001} = \sqrt{\text{error}}$. □

Further, since computers work to a relative error of about 10^{-15}, then expect a repeated eigenvalue of multiplicity m to appear as m eigenvalues within about $10^{-15/m}$ of each other—even when there are no experimental errors in the matrix. Repeat some of the previous cases of Example 7.1.10, preceded by the MATLAB/Octave command format long, to see that repeated eigenvalues are sensitive to computational errors.

Example 7.1.17 Use MATLAB/Octave to compute eigenvalues of the following matrices and comment on the effect on repeated eigenvalues of errors in the matrix and/or the computation.

$$\text{(a) } B = \begin{bmatrix} 3 & 0 & -2 & 0 & 0 \\ -1 & 5 & 0 & -1 & 3 \\ -1 & 2 & 4 & 0 & 1 \\ 5 & -1 & 4 & 1 & -1 \\ 3 & 2 & 1 & -2 & 2 \end{bmatrix}$$

Solution: In MATLAB/Octave execute

```
eig([3    0 -2    0    0
    -1    5    0 -1    3
    -1    2    4    0    1
     5 -1    4    1 -1
     3    2    1 -2    2])
```

to get something like

```
ans =
    -3.5176e-08
    3.5176e-08
    6.4142
    5.0000
    3.5858
```

There are three eigenvalues of multiplicity one, namely $5 \pm \sqrt{2}$ and 5. The two values $\pm 3.5176\text{e-08}$ at first sight appear to be two eigenvalues, $\pm 3.5176 \cdot 10^{-8}$, each of multiplicity one. However, computers usually work to about 15 digits accuracy, that is, the typical error is about 10^{-15}, so an eigenvalue of multiplicity two is generally computed as two eigenvalues separated by about $\sqrt{10^{-15}} \approx 3 \cdot 10^{-8}$. Thus we suspect that the two values $\pm 3.5176\text{e-08}$ actually represent one eigenvalue $\lambda = 0$ with multiplicity two. □

(b) Suppose the above matrix B is obtained from some experiment where there are experimental errors in the entries with error roughly about 0.0001. Randomly perturb the entries in matrix B to see the effects of such errors on the eigenvalues (use `randn()`, Table 3.1).

Solution: In MATLAB/Octave execute

```
B= [3   0 -2   0   0
    -1   5   0 -1   3
    -1   2   4   0   1
     5 -1   4   1 -1
     3   2   1 -2   2]
eig(B+0.0001*randn(5))
```

to get something like

```
ans =
    -0.0226
     0.0225
     6.4145
     4.9999
     3.5860
```

The repeated eigenvalue $\lambda = 0$ splits into two eigenvalues, $\lambda = \pm 0.0226$, of magnitude roughly $0.01 = \sqrt{0.0001}$. The other eigenvalues are also perturbed by the errors but only by amounts of magnitude roughly 0.0001.

Depending upon the random numbers, other possible answers are like

```
ans =
    0.0001 + 0.0157i
    0.0001 - 0.0157i
    6.4146 + 0.0000i
    4.9993 + 0.0000i
    3.5860 + 0.0000i
```

where the repeated eigenvalue of zero splits to be a pair of non-real complex valued eigenvalues of roughly $\pm i\sqrt{0.0001} = \pm i0.01$.   □

(c) $C = \begin{bmatrix} -1 & 0 & 0 & 0 \\ -1 & 2 & -3 & 3 \\ 3 & 1 & -1 & 0 \\ 0 & 3 & -2 & 1 \end{bmatrix}$ perturbed by errors of magnitude 10^{-6}.

Solution: In MATLAB/Octave execute

```
C= [-1  0  0  0
    -1  2 -3  3
```

```
     3  1 -1  0
     0  3 -2  1]
eig(C+1e-6*randn(4))
```

to get something like

```
ans =
    4.0000 + 0.0000i
   -1.0156 + 0.0000i
   -0.9922 + 0.0139i
   -0.9922 - 0.0139i
```

The eigenvalue 4 of multiplicity one is not noticeably affected by the errors about 10^{-6}. However, the repeated eigenvalue of $\lambda = -1$ with multiplicity three is split into three close eigenvalues (two non-real complex valued) all differing by roughly 0.01 which is indeed $(10^{-6})^{1/3}$—the cube-root of the error 10^{-6}. ☐

Activity 7.1.18 In an experiment, measurements are made to three decimal place accuracy. Then in analysing the results, a 5×5 matrix is formed from the measurements, and its eigenvalues computed by MATLAB/Octave to be

$$-0.9851, \quad 0.1266, \quad 0.9954, \quad 1.0090, \quad 1.0850.$$

What should you suspect is the number of different (distinct) eigenvalues?

(a) four   (b) five   (c) two   (d) three

But symmetric matrices are OK   The eigenvalues of a symmetric matrix are not so sensitive. This lack of sensitivity is fortunate as many applications give rise to symmetric matrices (Chapter 4). For symmetric matrices, the eigenvalues and eigenvectors are reasonably robust to both computational perturbations and experimental errors.

Example 7.1.19 For perhaps the simplest example, consider the 2×2 symmetric matrix $A = \begin{bmatrix} a & 0 \\ 0 & a \end{bmatrix}$. Being diagonal, matrix A has eigenvalue $\lambda = a$ (multiplicity two). Now perturb the matrix by 'experimental' error to say $B = \begin{bmatrix} a & 10^{-4} \\ 10^{-4} & a \end{bmatrix}$. The characteristic equation of B is

$$\det(B - \lambda I) = (a - \lambda)^2 - 10^{-8} = 0.$$

Rearrange this equation to $(\lambda - a)^2 = 10^{-8}$. Taking square-roots gives $\lambda - a = \pm 10^{-4}$, that is, the eigenvalues of B are $\lambda = a \pm 10^{-4}$. Because a perturbation to the symmetric matrix of magnitude 10^{-4} only changes the eigenvalues by a similar amount, the eigenvalues are *not* sensitive.

Activity 7.1.20 What are the eigenvalues of matrix $\begin{bmatrix} a & 0.01 \\ -0.01 & a \end{bmatrix}$?

(a) $a \pm 0.01$ (b) $a \pm 0.1\,i$ (c) $a \pm 0.01\,i$ (d) $a \pm 0.1$

Example 7.1.21 Use MATLAB/Octave to compute the eigenvalues of the symmetric matrix

$$A = \begin{bmatrix} 1 & 1 & 0 & 2 \\ 1 & 0 & 2 & -1 \\ 0 & 2 & 1 & 4 \\ 2 & -1 & 4 & 1 \end{bmatrix}$$

and see that matrix A has an eigenvalue of multiplicity two. Explore the effects on the eigenvalues of errors in the matrix by perturbing the entries by random amounts roughly of magnitude 0.0001.

Solution: In MATLAB/Octave execute

```
A=[1 1 0 2
   1 0 2 -1
   0 2 1 4
   2 -1 4 1]
eig(A)
eig(A+0.0001*randn(4))
```

to firstly get, for the unperturbed matrix, the eigenvalues

```
ans =
  -4.6235
   1.0000
   1.0000
   5.6235
```

showing that the eigenvalue $\lambda = 1$ has multiplicity two. Whereas, secondly, the eigenvalues of the perturbed matrix depend upon the random numbers and so might be

```
ans =
  -4.6236
   5.6235
   0.9998
   0.9999
```

or perhaps

```
ans =
   -4.6236 +  0.0000i
    5.6234 +  0.0000i
    1.0001 +  0.0000i
    1.0001 -  0.0000i
```

In either case, for this symmetric matrix the perturbation by amounts roughly of magnitude 0.0001 only change the eigenvalues, whether repeated or not, by an amount of about the same magnitude. ◻

7.1.3 Application: discrete dynamics of populations

Age-structured populations are one case where matrix properties and methods are crucial. The approach of this section is also akin to much mathematical modelling of diseases and epidemics. This section aims to show how to derive and use a matrix-vector model for the change in time t of interesting properties y of a population. Specifically, this subsection derives and analyses the model $y(t+1) = Ay(t)$.

For a given species, and at every time t, let's define

- $y_1(t)$ to be the number of juveniles (including infants),
- $y_2(t)$ the number of adolescents, and
- $y_3(t)$ the number of adults.

Biologists usually only count the female population as females are the determining sex for reproduction.[4] How do these numbers of females evolve over time? from generation to generation? First we need to choose a basic time interval (the unit of time): it could be one year, one month, one day, or maybe six months. Whatever we choose as convenient, we then quantify the events that happen to the females in each time interval as shown schematically in the diagram below:

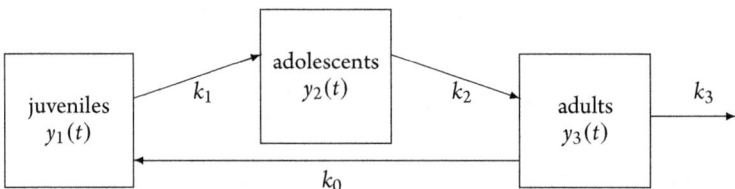

Over any one time interval, and only counting females:

- a fraction k_1 of the juveniles become adolescents;
- a fraction k_2 of the adolescents become adults;
- a fraction k_3 of the adults die;
- but adults also give birth to juveniles at rate k_0 per adult.

Model this scenario with a system of discrete dynamical equations, which are of the form that the numbers at the next time, $t+1$, depend upon the numbers at the time t:

---

[4] Although some bacteria/algae have seven sexes!

$$y_1(t+1) = \cdots,$$
$$y_2(t+1) = \cdots,$$
$$y_3(t+1) = \cdots.$$

Let's fill in the right-hand sides from the given information about the rate of particular events per time interval.

- A fraction k_1 of the juveniles $y_1(t)$ becoming adolescents also means a fraction $(1-k_1)$ of the juveniles remain juveniles, hence

$$y_1(t+1) = (1-k_1)y_1(t) + \cdots,$$
$$y_2(t+1) = +k_1 y_1(t) + \cdots,$$
$$y_3(t+1) = \cdots.$$

- A fraction k_2 of the adolescents $y_2(t)$ becoming adults also means a fraction $(1-k_2)$ of the adolescents remain adolescents, hence additionally

$$y_1(t+1) = (1-k_1)y_1(t) + \cdots,$$
$$y_2(t+1) = +k_1 y_1(t) + (1-k_2)y_2(t),$$
$$y_3(t+1) = +k_2 y_2(t) + \cdots.$$

- A fraction k_3 of the adults die means that a fraction $(1-k_3)$ of the adults remain adults, hence

$$y_1(t+1) = (1-k_1)y_1(t) + \cdots,$$
$$y_2(t+1) = +k_1 y_1(t) + (1-k_2)y_2(t),$$
$$y_3(t+1) = +k_2 y_2(t) + (1-k_3)y_3(t).$$

- But adults also give birth to juveniles at rate k_0 per adult so the number of juveniles increases by $k_0 y_3(t)$ from births:

$$y_1(t+1) = (1-k_1)y_1(t) + k_0 y_3(t),$$
$$y_2(t+1) = +k_1 y_1(t) + (1-k_2)y_2(t),$$
$$y_3(t+1) = +k_2 y_2(t) + (1-k_3)y_3(t).$$

This is our mathematical model of the age structure of the population.

Finally, write the mathematical model as the matrix-vector system

$$\begin{bmatrix} y_1(t+1) \\ y_2(t+1) \\ y_3(t+1) \end{bmatrix} = \begin{bmatrix} 1-k_1 & 0 & k_0 \\ k_1 & 1-k_2 & 0 \\ 0 & k_2 & 1-k_3 \end{bmatrix} \begin{bmatrix} y_1(t) \\ y_2(t) \\ y_3(t) \end{bmatrix},$$

that is, $y(t+1) = Ay(t)$. Such a model empowers predictions.

Example 7.1.22 *(orangutans)* From the following extract of the Wikipedia (2014) entry on orangutans (Cawthon Lang, 2005) derive a mathematical model for the age structure of the orangutans from one year to the next.

Gestation lasts for 9 months, with females giving birth to their first off-spring between the ages of 14 and 15 years. Female orangutans have [seven to] eight-year intervals between births, the longest interbirth intervals among the great apes.... Infant orangutans are completely dependent on their mothers for the first two years of their lives. The mother will carry the infant during travelling, as well as feed it and sleep with it in the same night nest. For the first four months, the infant is carried on its belly and never relieves physical contact. In the following months, the time an infant spends with its mother decreases. When an orangutan reaches the age of two, its climbing skills improve and it will travel through the canopy holding hands with other orangutans, a behaviour known as "buddy travel". Orangutans are juveniles from about two to five years of age and will start to temporarily move away from their mothers. Juveniles are usually weaned at about four years of age. Adolescent orangutans will socialize with their peers while still having contact with their mothers. Typically, orangutans live over 30 years in both the wild and captivity.

Suppose the initial population of orangutans in some area at year zero of a study is that of 30 adolescent females and 15 adult females. Use the mathematical model to predict the population for the next five years.

Solution: This solution illustrates the general steps.

Choose level: we choose a time unit of one year, and choose to model three age categories.[5]

Define:

- $y_1(t)$ is the number of juvenile females (including infant females) at time t (years), say age ≤ 5 years;
- $y_2(t)$ is the number of adolescent females, say $6 \leq$ age ≤ 14 years;
- $y_3(t)$ is the number of adult females, say age ≥ 15 years.

Quantify changes in a year: from the given information (albeit with numbers slightly modified here to make the results numerically simpler):

- Orangutans are juvenile for say 5 years, so in any one year a fraction $1/5$ of them grow to be adolescent and a fraction $4/5$ remain as juveniles. That is,

$$y_1(t+1) = \tfrac{4}{5}y_1(t) + \cdots, \quad y_2(t+1) = \tfrac{1}{5}y_1(t) + \cdots.$$

Say an adult female gives birth every 7–8 years, so on average she gives birth to a juvenile female every 15 years. Thus a fraction $1/15$ of adults give birth to a juvenile female in any

---

[5] A coarse model considers just the total number (of females) in a species; a fine model might count the number of each age in years (here 30 years); a 'micro' model might simulate each and every orangutan as an individual.

year. Consequently, we model juveniles as

$$y_1(t+1) = \tfrac{4}{5}y_1(t) + \tfrac{1}{15}y_3(t).$$

- Adolescent orangutans become breeding adults after another 9–10 years, so in any one year a fraction 1/10 of them grow to adults and 9/10ths remain adolescents. That is,

$$y_2(t+1) = \tfrac{1}{5}y_1(t) + \tfrac{9}{10}y_2(t), \quad y_3(t+1) = \tfrac{1}{10}y_2(t) + \cdots.$$

- Orangutans live to 30 years, about 15 years of adulthood so in any one year a fraction 14/15 of adult females live to the next year. Consequently,

$$y_3(t+1) = \tfrac{1}{10}y_2(t) + \tfrac{14}{15}y_3(t).$$

Our mathematical model of the age structure is then, in terms of the vector $y = (y_1, y_2, y_3)$,

$$y(t+1) = Ay(t) = \begin{bmatrix} \tfrac{4}{5} & 0 & \tfrac{1}{15} \\ \tfrac{1}{5} & \tfrac{9}{10} & 0 \\ 0 & \tfrac{1}{10} & \tfrac{14}{15} \end{bmatrix} y(t)$$

To predict the population we need to know the current population. The given information is that there are initially 30 adolescent females and 15 adult females. This information specifies that at the initial time, say time zero, the population vector $y(0) = (0, 30, 15)$ in the study area.

(a) Then the rule $y(t+1) = Ay(t)$ with time $t = 0$ gives

$$y(1) = Ay(0)$$

$$= \begin{bmatrix} \tfrac{4}{5} & 0 & \tfrac{1}{15} \\ \tfrac{1}{5} & \tfrac{9}{10} & 0 \\ 0 & \tfrac{1}{10} & \tfrac{14}{15} \end{bmatrix} \begin{bmatrix} 0 \\ 30 \\ 15 \end{bmatrix} = \begin{bmatrix} 1 \\ 27 \\ 17 \end{bmatrix}.$$

That is, during the first year we predict that there is one birth of a female juvenile, three adolescents matured to adults, and one adult died.

(b) Then the rule $y(t+1) = Ay(t)$ with time $t = 1$ year gives

$$y(2) = Ay(1)$$

$$= \begin{bmatrix} \tfrac{4}{5} & 0 & \tfrac{1}{15} \\ \tfrac{1}{5} & \tfrac{9}{10} & 0 \\ 0 & \tfrac{1}{10} & \tfrac{14}{15} \end{bmatrix} \begin{bmatrix} 1 \\ 27 \\ 17 \end{bmatrix} = \begin{bmatrix} \tfrac{29}{15} \\ \tfrac{49}{2} \\ \tfrac{557}{30} \end{bmatrix} = \begin{bmatrix} 1.93 \\ 24.50 \\ 18.57 \end{bmatrix} \text{ (2 d.p.)}.$$

In real life we cannot have such fractions of an orangutan. These predictions are averages, or expectations on average, and must be interpreted as such. Thus after two years, we expect on average nearly two juveniles, 24 or 25 adolescents, and 18 or 19 adults.

(c) Continuing on with the aid of MATLAB/Octave or calculator, the rule $y(t+1) = Ay(t)$ with time $t = 2$ years gives

$$y(3) = Ay(2)$$

$$= \begin{bmatrix} \frac{4}{5} & 0 & \frac{1}{15} \\ \frac{1}{5} & \frac{9}{10} & 0 \\ 0 & \frac{1}{10} & \frac{14}{15} \end{bmatrix} \begin{bmatrix} 1.93 \\ 24.50 \\ 18.57 \end{bmatrix} = \begin{bmatrix} 2.78 \\ 22.44 \\ 19.78 \end{bmatrix} \text{ (2 d.p.).}$$

In MATLAB/Octave do this calculation via

```
A=[4/5 0 1/15;1/5 9/10 0;0 1/10 14/15]
y0=[0;30;15]
y1=A*y0
y2=A*y1
y3=A*y2
y4=A*y3
y5=A*y4
```

(d) Consequently, the rule $y(t+1) = Ay(t)$ with time $t = 3$ years gives

$$y(4) = Ay(3)$$

$$= \begin{bmatrix} \frac{4}{5} & 0 & \frac{1}{15} \\ \frac{1}{5} & \frac{9}{10} & 0 \\ 0 & \frac{1}{10} & \frac{14}{15} \end{bmatrix} \begin{bmatrix} 2.78 \\ 22.44 \\ 19.78 \end{bmatrix} = \begin{bmatrix} 3.55 \\ 20.75 \\ 20.70 \end{bmatrix} \text{ (2 d.p.).}$$

(e) Lastly, the rule $y(t+1) = Ay(t)$ with time $t = 4$ years gives (to complete the plot on the right)

$$y(5) = Ay(4)$$

$$= \begin{bmatrix} \frac{4}{5} & 0 & \frac{1}{15} \\ \frac{1}{5} & \frac{9}{10} & 0 \\ 0 & \frac{1}{10} & \frac{14}{15} \end{bmatrix} \begin{bmatrix} 3.55 \\ 20.75 \\ 20.70 \end{bmatrix} = \begin{bmatrix} 4.22 \\ 19.38 \\ 21.40 \end{bmatrix} \text{ (2 d.p.).}$$

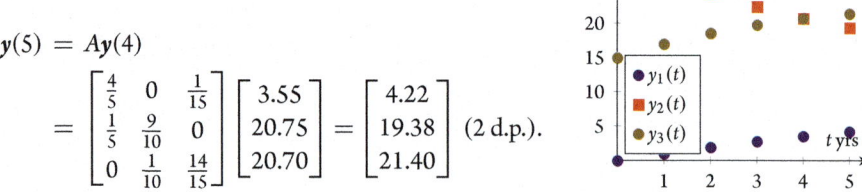

Thus after five years the mathematical model predicts about 4 juveniles, 19 adolescents, and 21 adults (on average).

The five-year population of 44 females (45 if you add all the fractions) is the same as the starting population. This nearly static total population is no accident, as we next see, and is one reason why orangutans are critically endangered. □

The mathematical model $y(t+1) = Ay(t)$ does predict/forecast future populations. However, to make predictions for many years and for general initial populations we prefer the formula solution given by the upcoming Theorem 7.1.25 and introduced in the next example.

Example 7.1.23 A vector $y(t) \in \mathbb{R}^2$ changes with time t according to the model

$$y(t+1) = Ay(t) = \begin{bmatrix} 1 & -1 \\ -4 & 1 \end{bmatrix} y(t).$$

First, what is $y(3)$ if the initial value $y(0) = (0,1)$? Second, find a general formula for $y(t)$ from every initial $y(0)$.

Solution: First, given $y(0) = (0,1)$ just compute (as drawn to the right)

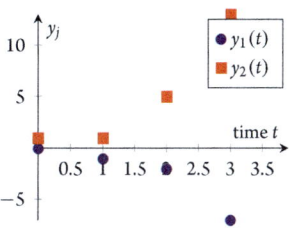

$$y(1) = Ay(0) = \begin{bmatrix} 1 & -1 \\ -4 & 1 \end{bmatrix} \begin{bmatrix} 0 \\ 1 \end{bmatrix} = \begin{bmatrix} -1 \\ 1 \end{bmatrix},$$

$$y(2) = Ay(1) = \begin{bmatrix} 1 & -1 \\ -4 & 1 \end{bmatrix} \begin{bmatrix} -1 \\ 1 \end{bmatrix} = \begin{bmatrix} -2 \\ 5 \end{bmatrix},$$

$$y(3) = Ay(2) = \begin{bmatrix} 1 & -1 \\ -4 & 1 \end{bmatrix} \begin{bmatrix} -2 \\ 5 \end{bmatrix} = \begin{bmatrix} -7 \\ 13 \end{bmatrix}.$$

Second, let's suppose there may be solutions in the form $y = \lambda^t v$ for some nonzero scalar λ and some vector $v \in \mathbb{R}^2$. Substitute into the model $y(t+1) = Ay(t)$ to find $\lambda^{t+1} v = A\lambda^t v$. Divide by (nonzero) λ^t to find that we require $Av = \lambda v$: this is an eigen-problem. That is, for every eigenvalue λ of matrix A, with corresponding eigenvector v, there is a solution $y = \lambda^t v$ of the model.

To find the eigenvalues here solve the characteristic equation

$$\det(A - \lambda I) = \begin{vmatrix} 1 - \lambda & -1 \\ -4 & 1 - \lambda \end{vmatrix} = (1 - \lambda)^2 - 4 = 0.$$

Rearrange to $(\lambda - 1)^2 = 4$ and take square-roots to give $\lambda - 1 = \pm 2$, that is, eigenvalues $\lambda = 1 \pm 2 = -1, 3$.

• For eigenvalue $\lambda_1 = -1$ a corresponding eigenvector has to satisfy

$$(A + I)v_1 = \begin{bmatrix} 2 & -1 \\ -4 & 2 \end{bmatrix} v_1 = 0.$$

That is, $v_1 \propto (1,2)$; let's take $v_1 = (1,2)$.

- For eigenvalue $\lambda_2 = 3$ a corresponding eigenvector has to satisfy

$$(A - 3I)v_2 = \begin{bmatrix} -2 & -1 \\ -4 & -2 \end{bmatrix} v_2 = \mathbf{0}.$$

That is, $v_2 \propto (1, -2)$; let's take $v_2 = (-1, 2)$.

Summarizing so far (with dashes denoting different formulas): $y'(t) = (-1)^t(1,2)$ and $y''(t) = 3^t(-1,2)$ are both solutions of the model $y(t+1) = Ay(t)$.

Further, the model $y(t+1) = Ay(t)$ is a *linear* equation in y, so every linear combination of solutions is also a solution. To see this, let's try $y(t) = c_1 y'(t) + c_2 y''(t)$. Substituting

$$
\begin{aligned}
y(t+1) &= c_1 y'(t+1) + c_2 y''(t+1) \\
&= c_1 Ay'(t) + c_2 Ay''(t) \\
&= A(c_1 y'(t) + c_2 y''(t)) \\
&= Ay(t),
\end{aligned}
$$

as required. So, $y(t) = c_1(-1)^t(1,2) + c_2 3^t(-1,2)$ are solutions of the model for every constant c_1 and c_2, and over all time t.

What values of constants c_1 and c_2 should be chosen for any given initial $y(0)$? Substitute time $t = 0$ into $y(t) = c_1(-1)^t(1,2) + c_2 3^t(-1,2)$ to require $y(0) = c_1(-1)^0(1,2) + c_2 3^0(-1,2)$. Since the zero powers $(-1)^0 = 3^0 = 1$, this requires $y(0) = c_1(1,2) + c_2(-1,2)$. Write as the matrix-vector system

$$\begin{bmatrix} 1 & -1 \\ 2 & 2 \end{bmatrix} \begin{bmatrix} c_1 \\ c_2 \end{bmatrix} = y(0) \iff \begin{bmatrix} c_1 \\ c_2 \end{bmatrix} = \begin{bmatrix} \frac{1}{2} & \frac{1}{4} \\ -\frac{1}{2} & \frac{1}{4} \end{bmatrix} y(0)$$

upon invoking the inverse (Theorem 3.2.7) of the matrix of eigenvectors, $P = \begin{bmatrix} v_1 & v_2 \end{bmatrix}$. That is, because the matrix of eigenvectors is invertible, we find constants c_1 and c_2 to suit any initial $y(0)$.

For example, with initial $y(0) = (0,1)$ the above formula gives $(c_1, c_2) = (\frac{1}{4}, \frac{1}{4})$ and so the corresponding formula solution is $y(t) = \frac{1}{4} \cdot (-1)^t(1,2) + \frac{1}{4} \cdot 3^t(-1,2)$. To check, evaluate at say $t = 3$ to find $y(3) = -\frac{1}{4}(1,2) + \frac{27}{4}(-1,2) = (-7,13)$, as in the direct calculation. In general, as here, as time t increases, the solution $y(t)$ grows like 3^t with a little oscillation from the $(-1)^t$ term. ☐

Activity 7.1.24 For Example 7.1.23, what is the particular solution when $y(0) = (1,1)$?

(a) $y = -\frac{1}{4} \cdot (-1)^t(1,2) + \frac{3}{4} \cdot 3^t(-1,2)$

(b) $y = \frac{3}{4} \cdot (-1)^t(1,2) - \frac{1}{4} \cdot 3^t(-1,2)$

(c) $y = 4 \cdot 3^t(-1,2)$

(d) $y = \frac{3}{4} \cdot (-1)^t(-1,2) - \frac{1}{4} \cdot 3^t(1,2)$

Now we establish that the same sort of general solution occurs for all such models.

Theorem 7.1.25 *Suppose the $n \times n$ square matrix A governs the dynamics of $y(t) \in \mathbb{R}^n$ according to $y(t+1) = Ay(t)$.*

(a) *Let λ_1, λ_2, ..., λ_m be eigenvalues of A and v_1, v_2, ..., v_m be corresponding eigenvectors, then a solution of $y(t+1) = Ay(t)$ for all time is the linear combination*

$$y(t) = c_1 \lambda_1^t v_1 + c_2 \lambda_2^t v_2 + \cdots + c_m \lambda_m^t v_m \qquad (7.2)$$

for every value of the constants c_1, c_2, ..., c_m.

(b) *Further, if the matrix of eigenvectors $P = \begin{bmatrix} v_1 & v_2 & \cdots & v_m \end{bmatrix}$ is invertible, then the general linear combination (7.2) is a **general solution** in that unique constants c_1, c_2, ..., c_m may be found for every given initial value $y(0)$.*

Proof.

7.1.25(a) Just premultiply (7.2) by matrix A to find that

$$Ay(t) = A(c_1 \lambda_1^t v_1 + c_2 \lambda_2^t v_2 + \cdots + c_m \lambda_m^t v_m)$$

(then using distributivity Theorem 3.1.24)

$$= c_1 \lambda_1^t A v_1 + c_2 \lambda_2^t A v_2 + \cdots + c_m \lambda_m^t A v_m$$

(then as eigenvectors $Av_j = \lambda_j v_j$)

$$= c_1 \lambda_1^t \lambda_1 v_1 + c_2 \lambda_2^t \lambda_2 v_2 + \cdots + c_m \lambda_m^t \lambda_m v_m$$

$$= c_1 \lambda_1^{t+1} v_1 + c_2 \lambda_2^{t+1} v_2 + \cdots + c_m \lambda_m^{t+1} v_m$$

which is the given formula (7.2) for $y(t+1)$. Hence (7.2) is a solution of $y(t+1) = Ay(t)$ for every constant c_1, c_2, ..., c_m.

7.1.25(b) For every given initial value $y(0)$, the solution (7.2) holds if we can find constants c_1, c_2, ..., c_m such that the solution (7.2) evaluates to $y(0)$ at time $t = 0$. Let's do this given the preconditions that the matrix $P = \begin{bmatrix} v_1 & v_2 & \cdots & v_m \end{bmatrix}$ is invertible. First, since matrix P is invertible, it must be square, and hence $m = n$ (that is, there must be n eigenvectors and n terms in (7.2)). Second, evaluating the solution (7.2) at $t = 0$ gives, since the zeroth power $\lambda_j^0 = 1$,

$$y(0) = c_1 v_1 + c_2 v_2 + \cdots + c_n v_n,$$

as an equation to be solved. Writing as a matrix-vector system this equation requires $Pc = y(0)$ for constant vector $c = (c_1, c_2, \ldots, c_n)$. The theorem's precondition is that matrix P is invertible. So $Pc = y(0)$ always has the unique solution $c = P^{-1}y(0)$ (Theorem 3.4.43) which determines the requisite constants. □

Activity 7.1.26 The matrix $A = \begin{bmatrix} 1 & 1 \\ a^2 & 1 \end{bmatrix}$ has eigenvectors $(1, a)$ and $(1, -a)$. For what value(s) of a does Theorem 7.1.25 *not* provide a general solution to $y(t+1) = Ay(t)$?

(a) $a = 1$ (b) $a = -1$ (c) $a = 0$ (d) $a = \pm 1$

Example 7.1.27 Consider the dynamics of $y(t+1) = Ay(t)$ for matrix $A = \begin{bmatrix} 1 & 3 \\ -1 & 1 \end{bmatrix}$. First, what is $y(3)$ when the initial value $y(0) = (1,0)$? Second, find a general solution.

Solution: First, just compute (as illustrated)

$$y(1) = Ay(0) = \begin{bmatrix} 1 & 3 \\ -1 & 1 \end{bmatrix}\begin{bmatrix} 1 \\ 0 \end{bmatrix} = \begin{bmatrix} 1 \\ -1 \end{bmatrix},$$

$$y(2) = Ay(1) = \begin{bmatrix} 1 & 3 \\ -1 & 1 \end{bmatrix}\begin{bmatrix} 1 \\ -1 \end{bmatrix} = \begin{bmatrix} -2 \\ -2 \end{bmatrix},$$

$$y(3) = Ay(2) = \begin{bmatrix} 1 & 3 \\ -1 & 1 \end{bmatrix}\begin{bmatrix} -2 \\ -2 \end{bmatrix} = \begin{bmatrix} -8 \\ 0 \end{bmatrix}.$$

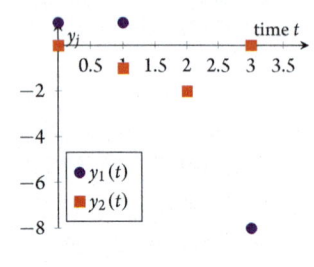

Interestingly, after three steps in time $y(3)$ is (-8) times the initial $y(0)$. This suggests that after six steps in time $y(6)$ is $(-8)^2 = 64$ times the initial $y(0)$, and so on. Perhaps the solution grows in magnitude roughly like 2^t but in some irregular manner—let's see via a general solution.

Second, find a general solution via the eigenvalues and eigenvectors of the matrix A. Its characteristic equation is

$$\det(A - \lambda I) = \begin{vmatrix} 1 - \lambda & 3 \\ -1 & 1 - \lambda \end{vmatrix} = (1 - \lambda)^2 + 3 = 0.$$

That is, $(\lambda - 1)^2 = -3$ which, upon taking square-roots, gives the complex conjugate pair of eigenvalues $\lambda = 1 \pm i\sqrt{3}$. Theorem 7.1.25 applies for complex eigenvalues and eigenvectors so we proceed.

- For eigenvalue $\lambda_1 = 1 + i\sqrt{3}$ the corresponding eigenvectors v_1 satisfy

$$\left(A - (1 + i\sqrt{3})I\right)v_1 = \begin{bmatrix} -i\sqrt{3} & 3 \\ -1 & -i\sqrt{3} \end{bmatrix} v_1 = 0.$$

Eigenvectors are proportional to $v_1 = (-i\sqrt{3}, 1)$.

Theorem 7.1.25 then establishes that a solution to $y(t+1) = Ay(t)$ is

$$y(t) = c_1(1 + i\sqrt{3})^t \begin{bmatrix} -i\sqrt{3} \\ 1 \end{bmatrix} + c_2(1 - i\sqrt{3})^t \begin{bmatrix} +i\sqrt{3} \\ 1 \end{bmatrix}.$$

This is a general solution since the matrix of the two eigenvectors (albeit nonreal) is invertible:

$$P = \begin{bmatrix} v_1 & v_2 \end{bmatrix} = \begin{bmatrix} -i\sqrt{3} & i\sqrt{3} \\ 1 & 1 \end{bmatrix} \quad \text{has } P^{-1} = \begin{bmatrix} \frac{i}{2\sqrt{3}} & \frac{1}{2} \\ \frac{-i}{2\sqrt{3}} & \frac{1}{2} \end{bmatrix}$$

as its inverse (Theorem 3.2.7).

For example, if $y(0) = (1,0)$, then the coefficient constants are $(c_1, c_2) = P^{-1}(1,0) = (i, -i)/(2\sqrt{3})$. Then the solution becomes

$$y(t) = \frac{i}{2\sqrt{3}}(1 + i\sqrt{3})^t \begin{bmatrix} -i\sqrt{3} \\ 1 \end{bmatrix} - \frac{i}{2\sqrt{3}}(1 - i\sqrt{3})^t \begin{bmatrix} +i\sqrt{3} \\ 1 \end{bmatrix}$$

$$= \tfrac{1}{2}(1 + i\sqrt{3})^t \begin{bmatrix} 1 \\ \frac{i}{\sqrt{3}} \end{bmatrix} + \tfrac{1}{2}(1 - i\sqrt{3})^t \begin{bmatrix} 1 \\ -\frac{i}{\sqrt{3}} \end{bmatrix}.$$

Through the magic of the complex conjugate form of the two terms in this expression, the nonreal parts cancel to always give a real result. For example, this complex formula predicts that at time step $t = 1$

$$y(1) = \tfrac{1}{2}(1 + i\sqrt{3}) \begin{bmatrix} 1 \\ \frac{i}{\sqrt{3}} \end{bmatrix} + \tfrac{1}{2}(1 - i\sqrt{3}) \begin{bmatrix} 1 \\ -\frac{i}{\sqrt{3}} \end{bmatrix}$$

$$= \frac{1}{2} \begin{bmatrix} 1 + i\sqrt{3} + 1 - i\sqrt{3} \\ \frac{i}{\sqrt{3}} - 1 - \frac{i}{\sqrt{3}} - 1 \end{bmatrix} = \begin{bmatrix} 1 \\ -1 \end{bmatrix},$$

as computed directly at the start of this example. □

One crucial qualitative aspect we need to know is whether components in the general solution (7.2) grow, decay, or stay the same magnitude as time increases. The growth or decay is determined by the eigenvalues; the reason is that λ_j^t is the only place that the time appears in the general formula (7.2).

- For example, in the general solution for Example 7.1.23, $y(t) = c_1(-1)^t(1,2) + c_2 3^t(-1,2)$, the 3^t factor grows in time since $3^1 = 3$, $3^2 = 9$, $3^3 = 27$, and so on. Whereas the $(-1)^t$ factor just oscillates in time since $(-1)^1 = -1$, $(-1)^2 = 1$, $(-1)^3 = -1$, and so on. Thus for long times, large t, we know that the term involving the factor 3^t, since it grows rapidly in t, soon dominates the solution.

- In Example 7.1.27, with complex conjugate eigenvalues, the situation is more complicated. Let's write every given complex eigenvalue in polar form $\lambda = r(\cos\theta + i\sin\theta)$ where magnitude $r = |\lambda|$ and angle θ is such that $\tan\theta = (\text{imag-part }\lambda)/(\text{real-part }\lambda)$. For example, $3 + 4i$ has magnitude $r = |3 + 4i| = \sqrt{3^2 + 4^2} = 5$ and angle $\theta = 53.15°$ since $\tan(53.15°) = 4/3$.

> Go to Mr. De Moivre;[6] he knows these things better than I do.   *Isaac Newton*

Question: how does this help us to understand the solution which has λ_j^t in it? Answer: De Moivre's theorem says that if $\lambda = r[\cos\theta + i\sin\theta]$, then $\lambda^t = r^t[\cos(\theta t) + i\sin(\theta t)]$. Since the magnitude $|\cos(\theta t) + i\sin(\theta t)| = \sqrt{\cos^2(\theta t) + \sin^2(\theta t)} = \sqrt{1} = 1$, the magnitude $|\lambda^t| = r^t$. For example, the magnitude $|(3 + 4i)^2| = |3 + 4i|^2 = 5^2 = 25$. We may

---

[6] Abraham De Moivre (1667–1754) was a mathematician born in Champagne, France. Due to religious persecution, De Moivre's family moved to London. During his stay in London he studied and became a proficient mathematician.

check this by directly computing $(3 + 4i)^2 = 3^2 + 2 \cdot 3 \cdot 4i + 4^2 i^2 = -7 + 24i$, and then $|-7 + 24i| = \sqrt{7^2 + 24^2} = \sqrt{625} = 25$.

In Example 7.1.27, the eigenvalue $\lambda_1 = 1 + i\sqrt{3}$ so its magnitude is $r_1 = |\lambda_1| = |1 + i\sqrt{3}| = \sqrt{1+3} = 2$. Hence the magnitude $|\lambda_1^t| = 2^t$ at every time step t. Similarly, the magnitude $|\lambda_2^t| = 2^t$ at every time step t. Consequently, the general solution

$$y(t) = c_1 \lambda_1^t \begin{bmatrix} -i\sqrt{3} \\ 1 \end{bmatrix} + c_2 \lambda_2^t \begin{bmatrix} +i\sqrt{3} \\ 1 \end{bmatrix}$$

grows in magnitude roughly like 2^t as both components grow like 2^t. It is a 'rough' growth because the components $\cos(\theta t)$ and $\sin(\theta t)$ cause 'oscillations' in time t. Nonetheless the overall growth like $|\lambda_1|^t = |\lambda_2|^t = 2^t$ is inexorable—and seen previously in the particular solution where we observe that $y(3)$ is eight times the magnitude of $y(0)$.

In general, for both real or complex eigenvalues λ, a term involving the factor λ^t, as time t increases,

- grows to infinity if $|\lambda| > 1$,
- decays to zero if $|\lambda| < 1$, and
- remains the same magnitude if $|\lambda| = 1$.

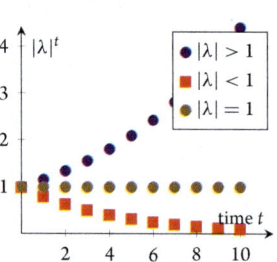

Activity 7.1.28 For which of the following values of λ, as time t increases, does λ^t grow in an oscillatory fashion?

(a) $\lambda = 1.5$ (b) $\lambda = -\frac{4}{5} + i\frac{4}{5}$ (c) $\lambda = \frac{3}{5} + i\frac{4}{5}$ (d) $\lambda = -0.8$

Example 7.1.29 (*orangutans over many years*) Extend the orangutan analysis of Example 7.1.22. Use Theorem 7.1.25 to predict the population over many years: from an initial population of 30 adolescent females and 15 adult females; and from a general initial population.

Solution: Example 7.1.22 derived that the age structure population $y = (y_1, y_2, y_3)$ satisfies $y(t + 1) = Ay(t)$ for matrix

$$A = \begin{bmatrix} \frac{4}{5} & 0 & \frac{1}{15} \\ \frac{1}{5} & \frac{9}{10} & 0 \\ 0 & \frac{1}{10} & \frac{14}{15} \end{bmatrix}.$$

Let's find the eigenvalues and eigenvectors of the matrix A using MATLAB/Octave via

```
A=[4/5 0 1/15;1/5 9/10 0;0 1/10 14/15]
[V,D]=eig(A)
```

to find

```
V =
  -0.3077+0.2952i  -0.3077-0.2952i   0.2673+0.0000i
   0.7385+0.0000i   0.7385+0.0000i   0.5345+0.0000i
  -0.4308-0.2952i  -0.4308+0.2952i   0.8018+0.0000i
D =
   0.8167+0.0799i   0.0000+0.0000i   0.0000+0.0000i
   0.0000+0.0000i   0.8167-0.0799i   0.0000+0.0000i
   0.0000+0.0000i   0.0000+0.0000i   1.0000+0.0000i
```

Evidently, from diag D, there is one real eigenvalue of $\lambda_3 = 1$ and two complex conjugate eigenvalues $\lambda_{1,2} = 0.8167 \pm i0.0799$. Corresponding eigenvectors are the columns v_j of V. Theorem 7.1.25 then assures us that a general solution for the orangutan population is

$$y(t) = c_1 \lambda_1^t v_1 + c_2 \lambda_2^t v_2 + c_3 \lambda_3^t v_3.$$

- For the initial population $y(0) = (0, 30, 15)$ we need to find constants $c = (c_1, c_2, c_3)$ such that $Vc = y(0)$. Solve this linear equation in MATLAB/Octave with Procedure 2.2.5:

```
y0=[0;30;15]
rcond(V)
c=V\y0
```

which gives the answer

```
ans =
     0.1963
c =
   10.1550+2.1175i
   10.1550-2.1175i
   28.0624+0.0000i
```

The rcond value of 0.1963 indicates that matrix V is invertible. Then the backslash operator computes the above coefficients c. Via the almost magical cancellation of complex conjugates, the real population of orangutans is for all times predicted to be (2 d.p.)

$$y(t) = (10.16 + 2.12\,i)(0.82 + 0.08\,i)^t \begin{bmatrix} -0.31 + 0.30\,i \\ 0.74 \\ -0.43 - 0.30\,i \end{bmatrix}$$

$$+ (10.16 - 2.12\,i)(0.82 - 0.08\,i)^t \begin{bmatrix} -0.31 - 0.30\,i \\ 0.74 \\ -0.43 + 0.30\,i \end{bmatrix}$$

$$+ 28.06 \begin{bmatrix} 0.27 \\ 0.53 \\ 0.80 \end{bmatrix}$$

since $\lambda_3^t = 1^t = 1$.

Since the magnitude $|\lambda_1| = |\lambda_2| = 0.82$ (2 d.p.), the first two terms in this expression decay to zero as time t increases. For example, $|\lambda_1^{12}| = |\lambda_2^{12}| = 0.09$. Hence the model predicts that over long times the population

$$\boldsymbol{y}(t) \approx 28.06 \begin{bmatrix} 0.27 \\ 0.53 \\ 0.80 \end{bmatrix} = \begin{bmatrix} 7.5 \\ 15.0 \\ 22.5 \end{bmatrix}$$

This prediction of a static population means that the orangutans are very sensitive to any small effects not included in the mathematical model.

- Such unfortunate sensitivity is typical for orangutans. It is not a quirk of the initial population. Recall that the general prediction for the orangutans is

$$\boldsymbol{y}(t) = c_1 \lambda_1^t \boldsymbol{v}_1 + c_2 \lambda_2^t \boldsymbol{v}_2 + c_3 \lambda_3^t \boldsymbol{v}_3.$$

The initial population determines the constants $\boldsymbol{c}$. However, the long-term population is always predicted to be static. The reason is that the magnitude of the eigenvalues $|\lambda_1| = |\lambda_2| = 0.82$ and so the first two terms in this general solution always decay to zero in time t. Further, the remaining third eigenvalue has magnitude $|\lambda_3| = |1| = 1$ and so the third term in the population prediction is always constant in time t. That is, over long times the population is always $\boldsymbol{y}(t) \approx c_3 \boldsymbol{v}_3$.

Such a static population means that the orangutans *are always* sensitive to other effects such as disease, or deforestation, or chance events, and so on. □

Example 7.1.30 *(servals grow)* The serval is a member of the cat family that lives in Africa. Given next is an extract from Wikipedia (2014) of a serval's Reproduction and Life History (Sunquist and Sunquist, 2002, pp. 142–151).

Kittens are born shortly before the peak breeding period of local rodent populations. A serval is able to give birth to multiple litters throughout the year, but commonly does so only if the earlier litters die shortly after birth. Gestation lasts from 66 to 77 days and commonly results in the birth of two kittens, although sometimes as few as one or as many as four have been recorded.

The kittens are born in dense vegetation or sheltered locations such as abandoned aardvark burrows. If such an ideal location is not available, a place beneath a shrub may be sufficient. The kittens weigh around 250 gm at birth, and are initially blind and helpless, with a coat of greyish woolly hair. They open their eyes at 9 to 13 days of age, and begin to take solid food after around a month. At around six months, they acquire their permanent canine teeth and begin to hunt for themselves; they leave their mother at about 12 months of age. They may reach sexual maturity from 12 to 25 months of age.

Life expectancy is about 10 years in the wild.

From the information in this extract, create a plausible, age structured, population model of servals: give reasons for estimates of the coefficients in the model. Choose three age categories of kittens, juveniles, and sexually mature adults. What does the model predict over long times?

Solution: Recall that we only model the number of *female* servals as females are the limiting breeders. Define

- $y_1(t)$ is the number of female kittens, less than 0.5 years old from when they "begin to hunt for themselves";

- $y_2(t)$ is the number of female juveniles, between 0.5 years and 1.5 years which is when they "reach sexual maturity" on average;

- $y_3(t)$ is the number of female breeding adults, older than 1.5 years, and dying at about the "life expectancy" of 10 years;

- since servals transition from one age category to another in multiples of six months (0.5 years), let the unit of time be six months, equivalently a half-year. Consequently, time $t+1$ is the time a half-year later than time t.

Modelling of the servals leads to the following equations.

- All kittens age to juveniles after 0.5 years, so none remain as kittens. Hence the model has

$$y_1(t+1) = 0 y_1(t) + \cdots, \quad y_2(t+1) = 1 y_1(t) + \cdots.$$

Kittens are commonly born once a year to each female, and the common litter size is two, so *on average* one female kitten is born per year per adult female, that is, on average $\frac{1}{2}$ female kitten is born per half-year per adult female: hence the kitten model is

$$y_1(t+1) = 0 y_1(t) + \tfrac{1}{2} y_3(t).$$

- Juveniles mature from the kittens, and age to an adult after about one year: that is, every half-year half of them remain juveniles, and half become adults, on average. So the model has

$$y_2(t+1) = y_1(t) + \tfrac{1}{2} y_2(t), \quad y_3(t+1) = \tfrac{1}{2} y_2(t) + \cdots.$$

- Adults mature from the juveniles, and die after about 8.5 years, which is about a rate $1/8.5$ per year: that is, a rate of $\frac{1}{17}$ per half-year leaving $\frac{16}{17}$ of them to live into the next half-year. So the adult model completes to

$$y_3(t+1) = \tfrac{1}{2} y_2(t) + \tfrac{16}{17} y_3(t).$$

Bring these equations together so that the age structure population $y = (y_1, y_2, y_3)$ satisfies $y(t+1) = A y(t)$ for matrix

$$A = \begin{bmatrix} 0 & 0 & \frac{1}{2} \\ 1 & \frac{1}{2} & 0 \\ 0 & \frac{1}{2} & \frac{16}{17} \end{bmatrix}.$$

Find the eigenvalues and eigenvectors of the matrix A using MATLAB/Octave via

```
A=[0 0 1/2;1 1/2 0;0 1/2 16/17]
[V,D]=eig(A)
```

to find

```
V =
 -0.3066+0.3439i  -0.3066-0.3439i   0.3352+0.0000i
  0.7838+0.0000i   0.7838+0.0000i   0.4633+0.0000i
 -0.3684-0.1942i  -0.3684+0.1942i   0.8203+0.0000i
D =
  0.1088+0.4387i   0.0000+0.0000i   0.0000+0.0000i
  0.0000+0.0000i   0.1088-0.4387i   0.0000+0.0000i
  0.0000+0.0000i   0.0000+0.0000i   1.2236+0.0000i
```

Evidently, from diag D, there is one real eigenvalue of $\lambda_3 = 1.2236$ and two complex conjugate eigenvalues $\lambda_{1,2} = 0.1088 \pm i 0.4387$. Corresponding eigenvectors are the columns v_j of V. Thus a general solution for the serval population is (Theorem 7.1.25)

$$y(t) = c_1 \lambda_1^t v_1 + c_2 \lambda_2^t v_2 + c_3 \lambda_3^t v_3.$$

In this general solution, the first two terms decay in time to zero. The reason is that the magnitudes

$$|\lambda_1| = |\lambda_2| = |0.1088 \pm i 0.4387| = \sqrt{0.1088^2 + 0.4387^2} = 0.4520,$$

and since this magnitude is less than one, then both λ_1^t and λ_2^t decay to zero with increasing time t. However, the third term increases in time as $\lambda_3 = 1.2236 > 1$. The model predicts the serval population increases by about 22% per half-year (about 50% per year). □

Predation, disease, and food shortages are just some processes not included in this model which act to limit the serval's population in ways not included in this model.

Mathematical modelling in application   Examples 7.1.29 and 7.1.30 introduce some of the real-life complexities of mathematical modelling. Bliss et al. (2016) discuss mathematical modelling in *Guidelines for Assessment and Instruction in Mathematics Modeling Education* and some of their comments are relevant here.

- "*Modelling (like real life) is open-ended and messy*" [p.23]: in our two examples here you have to extract the important factors from many unneeded details, and use them in the context of an imperfect model.

- Modellers "*must be making genuine choices*": in these problems, as in all modelling, there are choices that lead to different models—we have to operate and sensibly predict with such uncertainty.

- Lastly, they recommend to "*focus on the process, not the product*": depending upon your choices and interpretations you may develop alternative plausible models in these scenarios—it is the process of forming plausible models and interpreting the results that is important.

A crucial mathematical feature used in this section and its applications—so that we find a solution for all initial values—is that the matrix of eigenvectors is invertible. The upcoming Section 7.2 relates the invertibility of a matrix of eigenvectors to the new concept of 'linear independence' (Theorem 7.2.41).

7.1.4 Extension: SVDs connect to eigen-problems

Recall that Chapter 4 starts by illustrating the close connection between the SVD of a symmetric matrix and the eigenvalues and eigenvectors of that symmetric matrix. This subsection establishes that an SVD of a general matrix is closely connected to the eigenvalues and eigenvectors of a specific matrix of double the size. The connection depends upon determinants and solving linear systems and so, in principle, is an approach to computing an SVD distinct from the inductive maximization of Section 3.3.3.

Example 7.1.31 Compute the eigenvalues and eigenvectors of the (symmetric) matrix

$$B = \begin{bmatrix} 0 & 0 & 10 & 2 \\ 0 & 0 & 5 & 11 \\ 10 & 5 & 0 & 0 \\ 2 & 11 & 0 & 0 \end{bmatrix}.$$

Compare these with an SVD of matrix $A = \begin{bmatrix} 10 & 2 \\ 5 & 11 \end{bmatrix}$ from Example 3.3.2.

Solution: In MATLAB/Octave execute

```
B= [0 0 10 2
    0 0 5 11
    10 5 0 0
    2 11 0 0]
[V,D] =eig (B)
```

and obtain (2 d.p.)

```
V =
     0.42     0.57     0.57    -0.42
     0.57    -0.42    -0.42    -0.57
    -0.50    -0.50     0.50    -0.50
    -0.50     0.50    -0.50    -0.50
D =
    -14.14        0        0        0
        0    -7.07        0        0
        0        0     7.07        0
        0        0        0    14.14
```

The eigenvalues are the pairs ± 7.07 and ± 14.14, with corresponding eigenvector pairs $(0.57, -0.42, \pm 0.50, \mp 0.50)$ and $(\mp 0.42, \mp 0.57, -0.50, -0.50)$.

These eigenvalues/vectors occur in $\pm$ pairs because this matrix has the form

$$B = \begin{bmatrix} O_2 & A \\ A^T & O_2 \end{bmatrix}, \quad \text{here for matrix } A = \begin{bmatrix} 10 & 2 \\ 5 & 11 \end{bmatrix}.$$

Observe that not only are the eigenvectors orthogonal, because B is symmetric, but also, in each eigenvector, the two 'halves' are orthogonal:

- the components $(0.57, -0.42)$ from the first pair is orthogonal to $(\mp 0.42, \mp 0.57)$ from the second pair; and

- the components $(\pm 0.50, \mp 0.50)$ from the first pair is orthogonal to $(-0.50, -0.50)$ from the second pair.

The next Theorem 7.1.32 establishes how these properties relate to an SVD for the matrix A.

$\square$

Procedure 7.1.11 computes eigenvalues and eigenvectors by hand (albeit not practical for large matrices). The procedure is independent of the SVD. Let's now invoke this procedure to establish another method to find an SVD distinct from the inductive maximization of the proof in Section 3.3.3. The following Theorem 7.1.32 is a step towards an efficient numerical computation of an SVD (Trefethen and Bau, 1997, p.234).

Theorem 7.1.32 (SVD as an eigen-problem) *For every real $m \times n$ matrix A, the singular values of A are the non-negative eigenvalues of the $(m+n) \times (m+n)$ symmetric matrix $B = \begin{bmatrix} O_m & A \\ A^T & O_n \end{bmatrix}$. Each corresponding eigenvector $w \in \mathbb{R}^{m+n}$ of B gives corresponding singular vectors of A, namely $w = (u, v)$ for singular vectors $u \in \mathbb{R}^m$ and $v \in \mathbb{R}^n$.*

Proof. First prove that the SVD of A gives eigenvalues/vectors of B, and second prove the converse. For simplicity this proof addresses only the case $m = n$; the case $m \neq n$ is similar but the more intricate details are of little interest.

First, let $n \times n$ matrix $A = USV^T$ be an SVD (Theorem 3.3.6) for $n \times n$ orthogonal $U = \begin{bmatrix} u_1 & u_2 & \cdots & u_n \end{bmatrix}$, orthogonal $V = \begin{bmatrix} v_1 & v_2 & \cdots & v_n \end{bmatrix}$, and diagonal $S = \mathrm{diag}(\sigma_1, \sigma_2, \ldots, \sigma_n)$. Post-multiply the SVD by orthogonal V gives $AV = US$. Also, transpose the SVD to $A^T = (USV^T)^T = VS^T U^T = VSU^T$ and then post-multiply by orthogonal U to give $A^T U = VS$. Now consider each of the $\pm$ cases of

$$B \begin{bmatrix} U \\ \pm V \end{bmatrix} = \begin{bmatrix} O_n & A \\ A^T & O_n \end{bmatrix} \begin{bmatrix} U \\ \pm V \end{bmatrix} = \begin{bmatrix} \pm AV \\ A^T U \end{bmatrix} = \begin{bmatrix} \pm US \\ VS \end{bmatrix} = \begin{bmatrix} U \\ \pm V \end{bmatrix} (\pm S).$$

Set vector $w = (u_j, \pm v_j) : w \neq 0$ since u_j, v_j are unit vectors. Since S is diagonal, then the jth column of the above equation is $Bw = \pm \sigma_j w$ and hence $(u_j, \pm v_j)$ is an eigenvector of B corresponding to eigenvalue $\pm \sigma_j$, respectively, for each $j = 1, \ldots, n$ and each of the $\pm$ cases. This completes the first part of the proof.

Second, let $w \in \mathbb{R}^{2n}$ be an eigenvector of B corresponding to an eigenvalue λ (real as B is symmetric, Theorem 4.2.9) and scaled so that $|w| = \sqrt{2}$ (you will see why). Separate the components of w into two: $w = (u, v)$ for $u, v \in \mathbb{R}^n$. Then the fundamental eigen-problem $Bw = \lambda w$ (Definition 4.1.1) separates into

$$\begin{bmatrix} O & A \\ A^{\mathsf{T}} & O \end{bmatrix} \begin{bmatrix} u \\ v \end{bmatrix} = \lambda \begin{bmatrix} u \\ v \end{bmatrix} \iff Av = \lambda u \text{ and } A^{\mathsf{T}} u = \lambda v. \tag{7.3}$$

Further, the eigenvalues and eigenvectors of B come in pairs since the eigenvalue $\lambda' = -\lambda$ corresponds to eigenvector $w' = (u, -v)$: substitute into (7.3) to check, $A(-v) = -Av = -\lambda u = \lambda' u$ and $A^{\mathsf{T}} u = \lambda v = (-\lambda)(-v) = \lambda'(-v)$. If the eigenvalue $\lambda \neq 0$, then corresponding to the distinct eigenvalues $\pm \lambda$ the eigenvectors $(u, \pm v)$ of symmetric B are orthogonal (Theorem 4.2.11) and so $0 = (u, v) \cdot (u, -v) = u \cdot u - v \cdot v = |u|^2 - |v|^2$ and hence $|u| = |v| = 1$ as $|w|^2 = |u|^2 + |v|^2 = 2$ (see Example 7.1.31). Further, for any two distinct eigenvalues $|\lambda_i| \neq |\lambda_j| \neq 0$, the eigenvectors (u_i, v_i) and $(u_j, \pm v_j)$ are also orthogonal, hence $0 = (u_i, v_i) \cdot (u_j, \pm v_j) = u_i \cdot u_j \pm v_i \cdot v_j$. Taking the sum and the difference of the $\pm$ cases of this equation gives $u_i \cdot u_j = v_i \cdot v_j = 0$; that is, (Definition 3.2.38) $\{u_1, u_2, \ldots, u_n\}$ and $\{v_1, v_2, \ldots, v_n\}$ are each orthogonal sets. In the cases when matrix B has $2n$ distinct nonzero eigenvalues, choose (u_j, v_j) to be either eigenvector corresponding to positive eigenvalue λ_j, $j = 1, \ldots, n$, and a length $\sqrt{2}$. Upon setting $U = \begin{bmatrix} u_1 & u_2 & \cdots & u_n \end{bmatrix}$, $V = \begin{bmatrix} v_1 & v_2 & \cdots & v_n \end{bmatrix}$, and $S = \text{diag}(\lambda_1, \lambda_2, \ldots, \lambda_n)$, equation (7.3) gives the columns of $AV = US$ which since the columns of U and V are orthonormal gives the SVD $A = USV^{\mathsf{T}}$ for singular vectors $u_j, v_j \in \mathbb{R}^n$ and singular values $\lambda_j (> 0)$.

Extensions of this proof cater for the cases when zero is an eigenvalue and/or eigenvalues are repeated and/or the dimensions $m \neq n$. $\qquad \square$

7.1.5 Application: Exponential interpolation discovers dynamics

This optional subsection develops a method useful in many modern applied disciplines.

Many applications require identification of rates and frequencies from measured data (Pereyra and Scherer, 2010, e.g.): as a played musical note decays, what is its frequency? in the observed vibrations of a complicated bridge, what are its natural modes? in measurements of complicated bio-chemical reactions, what rates can be identified? All such tasks require fitting a sum of exponential functions to the data.

Example 7.1.33 This introductory example is the simplest case of fitting one exponential to two data points. Suppose we take two measurements of some process:

- at time $t_1 = 1$ we measure the value $f_1 = 5$, and
- at time $t_2 = 3$ we measure the value $f_2 = 10$.

Find an exponential fit to this data of the form $f(t) = ce^{rt}$ for some as yet unknown coefficients c and rate r.

Solution: The classic approach is to take logarithms of both sides:

$$f = ce^{rt} \iff (\log f) = (\log c) + rt,$$

and then determine the two constants $\log c$ and r by fitting a straight line through the two data points $(t, \log f) = (1, \log 5)$ and $(3, \log 10)$ respectively (recall that "$\log(x)$" denotes the natural logarithm of x, computed in MATLAB/Octave with $\log(x)$, see Table 3.2). This approach has the great virtue that the approach generalizes to fitting lots more noisy data (Section 3.5). But here we take a different approach—an approach that generalizes to fitting multiple exponentials.

The following basic steps correspond to complicated steps in the general procedure developed next for fitting multiple exponentials.

(a) We start with the question: is there a transformation that gives f_2 as a linear function of f_1? That is, can we write $f_2 = \lambda f_1$ for some constant λ? Answer: yes; since $f_1 = 5$ and $f_2 = 10$ we need $10 = \lambda 5$ with solution $\lambda = 2$.

(b) Since $f_2 = 2f_1$ we presume extrapolation to future values is reasonable via $f_3 = 2f_2 = 2 \cdot 2f_1 = 2^2 f_1$, and $f_4 = 2f_3 = 2 \cdot 2^2 f_1 = 2^3 f_1$, and so on. That is, $f_j = 2^{j-1} f_1$ for $j = 1, 2, 3, \ldots$.

(c) But we are given that $f_1 = 5$, so the exponential fit is $f_j = 5 \cdot 2^{j-1}$.

(d) Now these values occur at times $t_1 = 1, t_2 = 3$, and so on; that is, $t_j = 2j - 1$. Rearranging, for any time t the corresponding index $j = (t + 1)/2$. The corresponding f_j then gives f as a function of t, namely $f(t) = 5 \cdot 2^{(t+1)/2-1} = 5 \cdot 2^{(t-1)/2}$.

Equivalently, we write this $f(t)$ in terms of the exponential function. Recall the identity that $2^x = e^{x \log 2}$. Hence the fitted function $f(t) = 5 \cdot 2^{t/2-1/2} = (5/\sqrt{2}) e^{(t/2) \log 2}$. Since $5/\sqrt{2} = 3.5255$ and $\frac{1}{2} \log 2 = 0.3466$, the exponential fit is $f(t) = 3.5255 \, e^{0.3466 t}$. □

Activity 7.1.34 Plotted to the right are some points from a function $f(t)$. By inspection, decide which of the following exponentials best represents the data plotted?

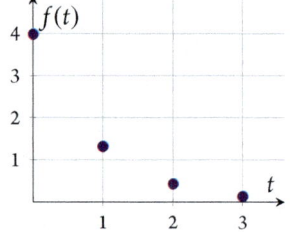

(a) $f \propto 1/3^t$

(b) $f \propto 1/2^t$

(c) $f \propto e^{-3t}$

(d) $f \propto e^{-t/2}$

Now let's develop the approach of Example 7.1.33 to the more complicated and interesting example of fitting the linear combination of two exponentials to four data points.

Example 7.1.35 Suppose in some chemical or biochemical experiment you measure the concentration of a key chemical at four times (as illustrated): at the start of the experiment, time $t_1 = 0$ you measure concentration $f_1 = 1$ (in some units); at time $t_2 = 1$ the measurement is $f_2 = 1$ (again); at $t_3 = 2$ the measurement is $f_3 = \frac{2}{3} = 0.6667$; and at $t_4 = 3$ the measurement is $f_4 = \frac{7}{18} = 0.3889$ (as plotted to the right). We generally expect chemical reactions to decay exponentially in time. So our task is to find a specific function of the form

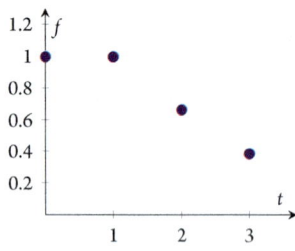

$$f(t) = c_1 e^{r_1 t} + c_2 e^{r_2 t},$$

that explains the data. Our aim is for this function to fit the four known data points, as plotted. The four unknown coefficients, c_1 and c_2 and rates r_1 and r_2, need to be determined from the four data points. That is, let's find these four unknowns from the data that

$$
\begin{aligned}
f(0) = 1 &\iff c_1 + c_2 = 1, \\
f(1) = 1 &\iff c_1 e^{r_1} + c_2 e^{r_2} = 1, \\
f(2) = \tfrac{2}{3} &\iff c_1 e^{r_1 2} + c_2 e^{r_2 2} = \tfrac{2}{3}, \\
f(3) = \tfrac{7}{18} &\iff c_1 e^{r_1 3} + c_2 e^{r_2 3} = \tfrac{7}{18}.
\end{aligned}
$$

These are four nonlinear equations for the unknown coefficients. However, some algebraic tricks empower us to use our beautiful linear algebra to find the solution. After solving these four nonlinear equations, we ultimately plot the function $f(t)$ that interpolates between the data as also shown to the right.

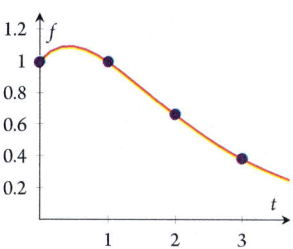

Solution: This solution has several twists and several new ideas that turns these four nonlinear equations into two linear algebra problems!

(a) First form the data into vectors of consecutive values:

$$
f_1 = \begin{bmatrix} f_1 \\ f_2 \end{bmatrix} = \begin{bmatrix} 1 \\ 1 \end{bmatrix}, \quad
f_2 = \begin{bmatrix} f_2 \\ f_3 \end{bmatrix} = \begin{bmatrix} 1 \\ \tfrac{2}{3} \end{bmatrix}, \quad
f_3 = \begin{bmatrix} f_3 \\ f_4 \end{bmatrix} = \begin{bmatrix} \tfrac{2}{3} \\ \tfrac{7}{18} \end{bmatrix},
$$

and, if only we had more data, notionally on to $f_4, f_5, \ldots$.

(b) Question: is there a linear transformation (Section 3.6) that gives both f_2 as a function of f_1, and f_3 as a function of f_2? That is, is there a 2×2 matrix K such that both $f_2 = Kf_1$ and $f_3 = Kf_2$?

Answer: if there is such a matrix K, then these two equations require both

$$
\begin{bmatrix} f_2 \\ f_3 \end{bmatrix} = K \begin{bmatrix} f_1 \\ f_2 \end{bmatrix} \iff \begin{bmatrix} 1 \\ \tfrac{2}{3} \end{bmatrix} = K \begin{bmatrix} 1 \\ 1 \end{bmatrix},
$$

$$
\begin{bmatrix} f_3 \\ f_4 \end{bmatrix} = K \begin{bmatrix} f_2 \\ f_3 \end{bmatrix} \iff \begin{bmatrix} \tfrac{2}{3} \\ \tfrac{7}{18} \end{bmatrix} = K \begin{bmatrix} 1 \\ \tfrac{2}{3} \end{bmatrix}.
$$

Put the two columns of these two equations side by side, we require the following combined equation to hold:

$$
\underbrace{\begin{bmatrix} f_2 & f_3 \\ f_3 & f_4 \end{bmatrix}}_{} = K \underbrace{\begin{bmatrix} f_1 & f_2 \\ f_2 & f_3 \end{bmatrix}}_{} \iff \underbrace{\begin{bmatrix} 1 & \tfrac{2}{3} \\ \tfrac{2}{3} & \tfrac{7}{18} \end{bmatrix}}_{A} = K \underbrace{\begin{bmatrix} 1 & 1 \\ 1 & \tfrac{2}{3} \end{bmatrix}}_{B}.
$$

This last equation has the form $A = KB$ for known matrices A and B (named A for 'After' values, and B for 'Before' values).

Post-multiplying $A = KB$ by the inverse of matrix B gives the matrix of the transformation as $K = AB^{-1}$. Here one readily calculates (Theorem 3.2.7)

$$B^{-1} = \begin{bmatrix} -2 & 3 \\ 3 & -3 \end{bmatrix}, \quad \text{so} \quad K = AB^{-1} = \begin{bmatrix} 0 & 1 \\ -\frac{1}{6} & \frac{5}{6} \end{bmatrix}.$$

(c) Since, for this matrix K, we know that $\boldsymbol{f}_2 = K\boldsymbol{f}_1, \boldsymbol{f}_3 = K\boldsymbol{f}_2$, we extrapolate into the future to predict $\boldsymbol{f}_4 = K\boldsymbol{f}_3, \boldsymbol{f}_5 = K\boldsymbol{f}_4$, and so on, where the vectors $\boldsymbol{f}_j = (f_j, f_{j+1})$. That is, our model through the four data points is summarized by $\boldsymbol{f}_{j+1} = K\boldsymbol{f}_j$ for $j = 1, 2, 3, \ldots$. This is the same class of dynamic models as explored by Section 7.1.3, albeit here not for populations. Consequently, Theorem 7.1.25 gives us a general algebraic solution of the model $\boldsymbol{f}_{j+1} = K\boldsymbol{f}_j$ in terms of the eigenvalues and eigenvectors of matrix K. Such an algebraic solution interpolates and extrapolates (with caution) the data.

Let's find the general solution here.

   i. The eigenvalues of matrix K satisfy the characteristic equation

$$\det(K - \lambda I) = \begin{vmatrix} -\lambda & 1 \\ -\frac{1}{6} & \frac{5}{6} - \lambda \end{vmatrix} = \lambda^2 - \frac{5}{6}\lambda + \frac{1}{6} = (\lambda - \frac{1}{2})(\lambda - \frac{1}{3}) = 0.$$

So the eigenvalues are $\lambda_1 = \frac{1}{2}$ and $\lambda_2 = \frac{1}{3}$.

   ii. For eigenvalue $\lambda_1 = \frac{1}{2}$ every corresponding eigenvector satisfies

$$(K - \tfrac{1}{2}I)\boldsymbol{v}_1 = \begin{bmatrix} -\frac{1}{2} & 1 \\ -\frac{1}{6} & \frac{1}{3} \end{bmatrix} \boldsymbol{v}_1 = \boldsymbol{0}.$$

That is, $\boldsymbol{v}_1 \propto (2, 1)$; let's take $\boldsymbol{v}_1 = (2, 1)$.

   iii. For eigenvalue $\lambda_2 = \frac{1}{3}$ every corresponding eigenvector satisfies

$$(K - \tfrac{1}{3}I)\boldsymbol{v}_2 = \begin{bmatrix} -\frac{1}{3} & 1 \\ -\frac{1}{6} & \frac{1}{2} \end{bmatrix} \boldsymbol{v}_2 = \boldsymbol{0}.$$

That is, $\boldsymbol{v}_2 \propto (3, 1)$; let's take $\boldsymbol{v}_2 = (3, 1)$.

By Theorem 7.1.25, a general solution of $\boldsymbol{f}_{j+1} = K\boldsymbol{f}_j$ is

$$\boldsymbol{f}_j = c_1 \boldsymbol{v}_1 \lambda_1^j + c_2 \boldsymbol{v}_2 \lambda_2^j = c_1 \begin{bmatrix} 2 \\ 1 \end{bmatrix} \frac{1}{2^j} + c_2 \begin{bmatrix} 3 \\ 1 \end{bmatrix} \frac{1}{3^j}.$$

(d) The data values also determine the constants c_1 and c_2. Recall that $f_1 = (1,1)$, so substituting $j = 1$ in the above general solution gives

$$\begin{bmatrix} 1 \\ 1 \end{bmatrix} = c_1 \begin{bmatrix} 1 \\ \frac{1}{2} \end{bmatrix} + c_2 \begin{bmatrix} 1 \\ \frac{1}{3} \end{bmatrix} = \begin{bmatrix} 1 & 1 \\ \frac{1}{2} & \frac{1}{3} \end{bmatrix} \begin{bmatrix} c_1 \\ c_2 \end{bmatrix}.$$

Solving this system of equations (for example, by subtracting twice the second row from the first, and subtracting three times the second row from the first) gives solution $c_1 = 4$ and $c_2 = -3$. Substitute these coefficients into the general solution to obtain the specific model

$$f_j = \begin{bmatrix} 8 \\ 4 \end{bmatrix} \frac{1}{2^j} + \begin{bmatrix} -9 \\ -3 \end{bmatrix} \frac{1}{3^j}.$$

(e) Recall that this example's quest is to fit a function $f(t)$ to the data. Here $f_j = (f_j, f_{j+1})$ and so the first component of the above specific model gives $f_j = 8(\frac{1}{2})^j - 9(\frac{1}{3})^j$. Then recall that $t_1 = 0$, $t_2 = 1$, $t_3 = 2$, and so on; that is, $t_j = j - 1$. Reverting this relation gives the index $j = t + 1$ and so in terms of time t the model is $f(t) = 8(\frac{1}{2})^{t+1} - 9(\frac{1}{3})^{t+1}$.

Consequently, the ultimate exponential fit to the data is (as previously plotted)

$$f(t) = 4(\tfrac{1}{2})^t - 3(\tfrac{1}{3})^t = 4e^{-0.6932\,t} - 3e^{-1.0986\,t}$$

since $\log \frac{1}{2} = -0.6932$ and $\log \frac{1}{3} = -1.0986$. □

Example 7.1.35 shows one way that fitting exponentials to data can be done with eigenvalues and eigenvectors. But one undesirable attribute of the example is the need to invert the matrix B to form matrix $K = AB^{-1}$. We avoid this inversion by generalizing eigen-problems as introduced by the following reworking of parts of Example 7.1.35.

Example 7.1.36 *(two short cuts)* Recall that Section 7.1.3 derived general solutions of dynamic equations such as $f_{j+1} = Kf_j$ by seeking solutions of the form $f_j = v\lambda^j$. For the previous Example 7.1.35 let's instead seek solutions of the form $f_j = Bw\lambda^j$. Substituting this form, the dynamic equation $f_{j+1} = Kf_j$ becomes $Bw\lambda^{j+1} = KBw\lambda^j$; then factoring λ^j, recognizing that $KB = A$, and swapping sides, this equation becomes $Aw = \lambda Bw$. This $Aw = \lambda Bw$ forms a generalized eigen-problem because it reduces to the standard eigen-problem in cases when the matrix $B = I$. Rework parts of Example 7.1.35 via this generalized eigen-problem.

Solution: Restart the analysis in Example 7.1.35(c): instead of the eigen-problem $Kv = \lambda v$, let's solve the generalized eigen-problem $Aw = \lambda Bw$. Here matrices

$$A = \begin{bmatrix} 1 & \frac{2}{3} \\ \frac{2}{3} & \frac{7}{18} \end{bmatrix}, \quad B = \begin{bmatrix} 1 & 1 \\ 1 & \frac{2}{3} \end{bmatrix}.$$

So the equation $Aw = \lambda Bw$ becomes

$$\begin{bmatrix} 1 & \frac{2}{3} \\ \frac{2}{3} & \frac{7}{18} \end{bmatrix} w = \lambda \begin{bmatrix} 1 & 1 \\ 1 & \frac{2}{3} \end{bmatrix} w = \begin{bmatrix} \lambda & \lambda \\ \lambda & \frac{2}{3}\lambda \end{bmatrix} w.$$

Subtract the matrix on the right-hand side from both sides to obtain

$$\begin{bmatrix} 1 & \frac{2}{3} \\ \frac{2}{3} & \frac{7}{18} \end{bmatrix} w - \begin{bmatrix} \lambda & \lambda \\ \lambda & \frac{2}{3}\lambda \end{bmatrix} w = \begin{bmatrix} 1-\lambda & \frac{2}{3}-\lambda \\ \frac{2}{3}-\lambda & \frac{7}{18}-\frac{2}{3}\lambda \end{bmatrix} w = 0$$

Being a homogeneous linear equation, this last equation only has nontrivial solutions w when the determinant is zero:

$$\det \begin{bmatrix} 1-\lambda & \frac{2}{3}-\lambda \\ \frac{2}{3}-\lambda & \frac{7}{18}-\frac{2}{3}\lambda \end{bmatrix} = (1-\lambda)(\tfrac{7}{18}-\tfrac{2}{3}\lambda) - (\tfrac{2}{3}-\lambda)^2$$

$$= \tfrac{2}{3}\lambda^2 - \tfrac{19}{18}\lambda + \tfrac{7}{18} - \lambda^2 + \tfrac{4}{3}\lambda - \tfrac{4}{9}$$

$$= -\tfrac{1}{3}\lambda^2 + \tfrac{5}{18}\lambda - \tfrac{1}{18}$$

$$= -\tfrac{1}{18}(6\lambda^2 - 5\lambda + 1)$$

$$= -\tfrac{1}{18}(2\lambda - 1)(3\lambda - 1).$$

This determinant is zero only for multipliers $\lambda_1 = \frac{1}{2}$ and $\lambda_2 = \frac{1}{3}$. These are precisely the same eigenvalues as in Example 7.1.35(c).

Another simplification is to avoid forming the full vector solution (Examples 7.1.35(d) and 7.1.35(e)): we only use just the first vector-component in the final answer for $f(t)$, so the full vector is redundant. Instead, given the multipliers $\lambda_1 = \frac{1}{2}$ and $\lambda_2 = \frac{1}{3}$, just seek the penultimate form $f_j = c_1(\frac{1}{2})^j + c_2(\frac{1}{3})^j$. Given the data values, evaluating this form of solution at $j = 1,2$ gives two equations:

$$j = 1, \quad c_1\tfrac{1}{2} + c_2\tfrac{1}{3} = f_1 = f(0) = 1;$$
$$j = 2, \quad c_1\tfrac{1}{4} + c_2\tfrac{1}{9} = f_2 = f(1) = 1.$$

Subtracting twice the second from the first, and subtracting three times the second from the first gives the coefficients $c_1 = 8$ and $c_2 = -9$. Hence the fit to the data is $f_j = 8(\frac{1}{2})^j - 9(\frac{1}{3})^j$ as in Example 7.1.35(e). Then, as before, index $j = t+1$ so the function fit becomes $f(t) = 4(\frac{1}{2})^t - 3(\frac{1}{3})^t$ to match the conclusion of the previous Example 7.1.35. $\square$

Generalized eigen-problem

As introduced by Example 7.1.36, let's generalize the Definition 4.1.1 of eigenvalues and eigenvectors. Such generalized eigen-problems also occur in the design analysis of complicated structures, such as buildings and bridges, where the second matrix B represents the various masses of the various elements making up a structure.

Definition 7.1.37 *Let A and B be $n \times n$ square matrices. The **generalized eigen-problem** is to find scalar eigenvalues λ and corresponding nonzero eigenvectors v such that $Av = \lambda Bv$.*

Example 7.1.38 Given $A = \begin{bmatrix} -2 & 2 \\ 3 & 1 \end{bmatrix}$ and $B = \begin{bmatrix} 3 & -3 \\ 2 & 0 \end{bmatrix}$, what eigenvalue corresponds to the eigenvector $v_1 = (1,1)$ of the *generalized* eigen-problem $Av = \lambda Bv$? Repeat this question for $v_2 = (-3, 13)$.

Solution: To test a proposed eigenvector in this generalized eigen-problem, multiply by the matrices and compare results.

- To test $v_1 = (1,1)$:

$$Av_1 = \begin{bmatrix} -2 & 2 \\ 3 & 1 \end{bmatrix}\begin{bmatrix} 1 \\ 1 \end{bmatrix} = \begin{bmatrix} 0 \\ 4 \end{bmatrix}; \qquad Bv_1 = \begin{bmatrix} 3 & -3 \\ 2 & 0 \end{bmatrix}\begin{bmatrix} 1 \\ 1 \end{bmatrix} = \begin{bmatrix} 0 \\ 2 \end{bmatrix}.$$

Since Av_1 is twice Bv_1 then v_1 is an eigenvector corresponding to eigenvalue $\lambda_1 = 2$.

- To test $v_2 = (-3, 13)$:

$$Av_2 = \begin{bmatrix} -2 & 2 \\ 3 & 1 \end{bmatrix}\begin{bmatrix} -3 \\ 13 \end{bmatrix} = \begin{bmatrix} 32 \\ 4 \end{bmatrix}; \qquad Bv_2 = \begin{bmatrix} 3 & -3 \\ 2 & 0 \end{bmatrix}\begin{bmatrix} -3 \\ 13 \end{bmatrix} = \begin{bmatrix} -48 \\ -6 \end{bmatrix}.$$

Since $32 = -\frac{2}{3}(-48)$ and $4 = -\frac{2}{3}(-6)$, then Av_2 is $-2/3$ times Bv_2, and so v_2 is an eigenvector corresponding to eigenvalue $\lambda_2 = -2/3$. ☐

Activity 7.1.39 Which of the following vectors is an eigenvector of the generalized eigen-problem

$$\begin{bmatrix} -2 & 0 \\ 1 & 0 \end{bmatrix} v = \lambda \begin{bmatrix} 3 & -1 \\ -1 & 1 \end{bmatrix} v?$$

(a) $(2, -1)$ (b) $(0, 0)$ (c) $(1, -1)$ (d) $(1, 1)$

The standard eigen-problem is the case when matrix $B = I$. Many of the properties for standard eigenvalues and eigenvectors also hold for generalized eigen-problems, although there are some differences. Perhaps the most important familiar property, albeit without proof, is that provided matrix B is invertible, then counted according to multiplicity there are n eigenvalues of a generalized eigen-problem in $n \times n$ matrices.

Example 7.1.40 Find all eigenvalues and corresponding eigenvectors of the generalized eigen-problem $A\mathbf{v} = \lambda B\mathbf{v}$ for matrices

$$A = \begin{bmatrix} -1 & 1 \\ -4 & 0 \end{bmatrix}, \quad B = \begin{bmatrix} -1 & 1 \\ -1 & 2 \end{bmatrix}$$

Solution: Now solving $A\mathbf{v} = \lambda B\mathbf{v}$ is the same as solving $(A - \lambda B)\mathbf{v} = \mathbf{0}$, which requires $\det(A - \lambda B) = 0$ (Theorem 6.1.29). Hence determine the eigenvalues by finding the zeros of the characteristic polynomial

$$\det(A - \lambda B) = \det \begin{bmatrix} -1 + \lambda & 1 - \lambda \\ -4 + \lambda & 0 - 2\lambda \end{bmatrix} = (-1 + \lambda)(-2\lambda) - (-4 + \lambda)(1 - \lambda)$$

$$= -2\lambda^2 + 2\lambda + \lambda^2 - 5\lambda + 4 = -\lambda^2 - 3\lambda + 4 = -(\lambda + 4)(\lambda - 1).$$

Hence the only eigenvalues are $\lambda_1 = -4$ and $\lambda_2 = 1$.

- For the case $\lambda_1 = -4$ the corresponding eigenvectors satisfy

$$(A + 4B)\mathbf{v} = \begin{bmatrix} -5 & 5 \\ -8 & 8 \end{bmatrix} \mathbf{v} = \mathbf{0}.$$

That is, the two components of $\mathbf{v}$ must be the same: corresponding eigenvectors are $\mathbf{v}_1 \propto (1, 1)$.

- For the case $\lambda_2 = 1$ the corresponding eigenvectors satisfy

$$(A - B)\mathbf{v} = \begin{bmatrix} 0 & 0 \\ -3 & -2 \end{bmatrix} \mathbf{v} = \mathbf{0}.$$

That is, $-3v_1 - 2v_2 = 0$ which rearranged is $v_1 = -\frac{2}{3}v_2$: corresponding eigenvectors are $\mathbf{v}_2 \propto (-\frac{2}{3}, 1)$.

□

Example 7.1.41 Find all eigenvalues and corresponding eigenvectors of the generalized eigen-problem $A\mathbf{v} = \lambda B\mathbf{v}$ for matrices

$$A = \begin{bmatrix} 2 & 2 \\ -1 & 0 \end{bmatrix}, \quad B = \begin{bmatrix} -1 & 1 \\ 1 & -1 \end{bmatrix}$$

Solution: Now solving $A\mathbf{v} = \lambda B\mathbf{v}$ is the same as solving $(A - \lambda B)\mathbf{v} = \mathbf{0}$, which requires $\det(A - \lambda B) = 0$ (Theorem 6.1.29). Hence determine the eigenvalues by finding the zeros of the characteristic polynomial

$$\det(A - \lambda B) = \det \begin{bmatrix} 2 + \lambda & 2 - \lambda \\ -1 - \lambda & 0 + \lambda \end{bmatrix} = (2 + \lambda)(\lambda) - (2 - \lambda)(-1 - \lambda)$$

$$= \lambda^2 + 2\lambda - 2 - \lambda^2 + \lambda + 2 = 3\lambda + 2.$$

Hence, from $3\lambda + 2 = 0$, the only eigenvalue is $\lambda = -2/3$. Here the matrix B is not invertible—its determinant is zero—and consequently in this example we do not get the typical full complement of two eigenvalues.

For the only eigenvalue $\lambda = -2/3$ the corresponding eigenvectors satisfy

$$(A + \tfrac{2}{3}B)v = \begin{bmatrix} \frac{4}{3} & \frac{8}{3} \\ -\frac{1}{3} & -\frac{2}{3} \end{bmatrix} v = 0.$$

That is, the two components of v must satisfy $\tfrac{1}{3}v_1 + \tfrac{2}{3}v_2 = 0$ which rearranged is $v_1 = -2v_2$: corresponding eigenvectors are $v \propto (-2, 1)$. $\qquad\square$

General fitting of exponentials

Suppose that some experiment or other observation has given us $2n$ data values $f_1, f_2, \ldots, f_{2n}$ at equi-spaced 'times' $t_1, t_2, \ldots, t_{2n}$, where the spacing $t_{j+1} - t_j = h$. The general aim is to fit the multi-exponential function (Cuyt, 2015, §2.6, e.g.)

$$f(t) = c_1 e^{r_1 t} + c_2 e^{r_2 t} + \cdots + c_n e^{r_n t}. \tag{7.4}$$

for some coefficients $c_1, c_2, \ldots, c_n$ and some rates $r_1, r_2, \ldots, r_n$ to be determined (possibly nonreal complex valued for oscillations). In general, finding the coefficients and rates is a delicate nonlinear task outside the remit of this book. However, as Examples 7.1.33 and 7.1.35 illustrate, in these circumstances we instead invoke our powerful linear algebra methods.

Because the data is sampled at equi-spaced times, h apart, then instead of seeking r_k we seek multipliers $\lambda_k = e^{r_k h}$. Then $r_k = (\log \lambda_k)/h$.

Procedure 7.1.42 (exponential interpolation) *Given measured data $f_1, f_2, \ldots, f_{2n}$ at $2n$ equi-spaced times $t_1, t_2, \ldots, t_{2n}$ where time $t_j = (j-1)h$ for time-spacing h (and starting from time $t_1 = 0$ without loss of applicability), we seek to fit the data with a sum of exponentials (7.4).*

1. *From the $2n$ data points, form two $n \times n$ (symmetric) Hankel matrices*

$$A = \begin{bmatrix} f_2 & f_3 & \cdots & f_{n+1} \\ f_3 & f_4 & \cdots & f_{n+2} \\ \vdots & \vdots & & \vdots \\ f_{n+1} & f_{n+2} & \cdots & f_{2n} \end{bmatrix}, \qquad B = \begin{bmatrix} f_1 & f_2 & \cdots & f_n \\ f_2 & f_3 & \cdots & f_{n+1} \\ \vdots & \vdots & & \vdots \\ f_n & f_{n+1} & \cdots & f_{2n-1} \end{bmatrix}.$$

   `A=hankel(f(2:n+1),f(n+1:2*n))` *and* `B=hankel(f(1:n),` `f(n:2*n-1))` *form these two matrices in Matlab/Octave (Table 7.1).*[7]

2. *Find the eigenvalues of the generalized eigen-problem $Av = \lambda Bv$:*
   - *by hand on small problems solve $\det(A - \lambda B) = 0$;*
   - *invoke* `lambda=eig(A,B)`, *and then* `r=log(lambda)/h` *in* MATLAB/*Octave.*

---

[7] This Hankel function is also invoked in exploring El Niño, Example 3.4.27, and walking gaits, Exercise 3.4.14. The reason is the same: the columns of matrices A and B appearing here are precisely the 'sliding windows' appearing in the two Singular Spectrum Analysis problems.

Table 7.1 As well as the MATLAB/Octave commands and operations listed in Tables 1.2, 2.3, 3.1 to 3.3, 3.7 and 5.1 this section invokes these functions.

- `hankel (x,y)` for two vectors of the same size, $x = (x_1, x_2, \ldots, x_n)$ and $y = (y_1, y_2, \ldots, yn)$, when $x_n = y_1$, forms the $n \times n$ matrix

$$\begin{bmatrix} x_1 & x_2 & x_3 & \cdots & x_n \\ x_2 & x_3 & \cdots & x_n & y_2 \\ x_3 & & \ddots & \ddots & \vdots \\ \vdots & \ddots & \ddots & & y_{n-1} \\ x_n & y_2 & \cdots & y_{n-1} & y_n \end{bmatrix}.$$

- `eig (A, B)` for $n \times n$ matrices A and B computes a vector in $\mathbb{R}^n$ of eigenvalues λ such that $\det(A - \lambda B) = 0$. Some of the computed eigenvalues in the vector may be $\pm\texttt{Inf}$ (if B is not invertible) which denotes that a corresponding eigenvalue does not exist.
The command `[V, D] =eig (A, B)` solves the generalized eigen-problem $Ax = \lambda Bx$ for eigenvalues λ returned in the diagonal of matrix D (non-existent eigenvalues are denoted by $\pm\texttt{Inf}$), and corresponding eigenvectors returned in the corresponding column of V.

This generalized eigen-problem typically determines n multipliers $\lambda_1, \lambda_2, \ldots, \lambda_n$, and thence the n rates $r_k = (\log \lambda_k)/h$.

3. *Determine the corresponding n coefficients $c_1, c_2, \ldots, c_n$ from any n point subset of the 2n data points. For example, the first n data points give the linear system*

$$\begin{bmatrix} 1 & 1 & \cdots & 1 \\ \lambda_1 & \lambda_2 & \cdots & \lambda_n \\ \lambda_1^2 & \lambda_2^2 & \cdots & \lambda_n^2 \\ \vdots & \vdots & & \vdots \\ \lambda_1^{n-1} & \lambda_2^{n-1} & \cdots & \lambda_n^{n-1} \end{bmatrix} \begin{bmatrix} c_1 \\ c_2 \\ \vdots \\ c_n \end{bmatrix} = \begin{bmatrix} f_1 \\ f_2 \\ f_3 \\ \vdots \\ f_n \end{bmatrix}.$$

In MATLAB/Octave *one may construct the matrix U appearing here with auto-replication* `U= (lambda.^(0:n-1)).'` *when* `lambda` *is a column vector of the eigenvalues. Since the eigenvalues λ may be nonreal complex valued we need the transpose* `".'"` *not the complex conjugate transpose* `"'"` *(Table 3.1).*

Proof. This proof establishes the generic case. Denote the columns of matrices A and B by vectors of consecutive data values $f_j = (f_j, f_{j+1}, \ldots, f_{j+n-1}) \in \mathbb{R}^n$ for $j = 1, 2, \ldots, n+1$ (Step 1 in Procedure 7.1.42).

- First seek a linear transformation, by some $n \times n$ matrix K, from one data vector to the next in the form $f_{j+1} = Kf_j$ for $j = 1, 2, \ldots, n$. Arrange these n equations $f_{j+1} = Kf_j$ into the one matrix equation

$$\begin{bmatrix} f_2 & f_3 & \cdots & f_4 \end{bmatrix} = \begin{bmatrix} Kf_1 & Kf_2 & \cdots & Kf_3 \end{bmatrix} = K \begin{bmatrix} f_1 & f_2 & \cdots & f_3 \end{bmatrix},$$

after factoring the matrix K. That is, matrix K satisfies $A = KB$ (which in principle we could typically solve to obtain $K = AB^{-1}$).

- Second, view $f_{j+1} = Kf_j$ as a dynamic system in j and modify the approach of Section 7.1.3. Seek solutions in the form $f_j = Bv\lambda^j$ for some nonzero vector v and multiplier λ. Then $f_{j+1} = Kf_j$ becomes $Bv\lambda^{j+1} = KBv\lambda^j$. This equation becomes the generalized eigen-problem $Av = \lambda Bv$ after factoring out λ^j (provided nonzero), and recalling that $A = KB$.

- Typically, Step 2 in Procedure 7.1.42, the eigen-problem $Av = \lambda Bv$ has n distinct eigenvalues λ_1, λ_2, $\ldots$, λ_n (possibly nonreal) with some corresponding eigenvectors v_1, v_2, $\ldots$, v_n. By the linearity of $f_{j+1} = Kf_j$, a general solution is then the linear combination $f_j = C_1 Bv_1 \lambda_1^j + C_2 Bv_2 \lambda_2^j + \cdots + C_n Bv_n \lambda_n^j$. We only need one component of this general solution, say the first, and so the data values $f_j = c_1 \lambda_1^{j-1} + c_2 \lambda_2^{j-1} + \cdots + c_n \lambda_n^{j-1}$ for some constants c_1, c_2, $\ldots$, c_n to be found ($c_i = C_i \lambda_i (Bv_i)_1$ in terms of C_i).

- Determine the constants c_1, c_2, $\ldots$, c_n by matching the general solution to n data values. Using the first n data values gives the n equations:

$$f_1 = c_1 + c_2 + \cdots + c_n,$$
$$f_2 = c_1 \lambda_1^1 + c_2 \lambda_2^1 + \cdots + c_n \lambda_n^1,$$
$$f_3 = c_1 \lambda_1^2 + c_2 \lambda_2^2 + \cdots + c_n \lambda_n^2,$$
$$\vdots$$
$$f_n = c_1 \lambda_1^{n-1} + c_2 \lambda_2^{n-1} + \cdots + c_n \lambda_n^{n-1}.$$

Form as a matrix-vector system for the unknown vector $(c_1, c_2, \ldots, c_n)$ to give the system solved by Step 3.

Lastly, rearrange $t_j = (j-1)h$ to $j-1 = t_j/h$, then $\lambda_k^{j-1} = e^{(\log \lambda_k) t_j/h} = e^{r_k t_j}$ for rates $r_k = (\log \lambda_k)/h$ (Step 7.1.42). So the solution $f_j = c_1 \lambda_1^{j-1} + c_2 \lambda_2^{j-1} + \cdots + c_n \lambda_n^{j-1}$ becomes the exponential interpolation $f(t) = c_1 e^{r_1 t} + c_2 e^{r_2 t} + \cdots + c_n e^{r_n t}$ as required. $\square$

Example 7.1.43 A damped piano string is struck and the sideways displacement of the string is measured at four times, 5 ms apart. The measurements (in mm) are $f_1 = 1.0000$, $f_2 = -0.3766$, $f_3 = -0.5352$ and $f_4 = 0.7114$ (as illustrated). Determine, by hand calculation, the frequency and damping of the string.[8]

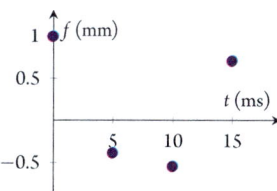

Recall Euler's formula that $e^{i\theta} = \cos \theta + i \sin \theta$ so the oscillations here are captured by non-real complex valued exponentials.

Solution: Follow Procedure 7.1.42 by hand.

---

[8] Some of you may know that identification of frequencies is most commonly done by what is called a Fourier transform. However, with a limited amount of good data, or for decaying oscillations, this approach should be better.

(a) Form the two Hankel matrices from the four data values f_1, f_2, f_3 and f_4:

$$A = \begin{bmatrix} f_2 & f_3 \\ f_3 & f_4 \end{bmatrix} = \begin{bmatrix} -0.3766 & -0.5352 \\ -0.5352 & 0.7114 \end{bmatrix}, \quad B = \begin{bmatrix} f_1 & f_2 \\ f_2 & f_3 \end{bmatrix} = \begin{bmatrix} 1.0000 & -0.3766 \\ -0.3766 & -0.5352 \end{bmatrix}.$$

(b) With some arithmetic, the determinant

$$\det(A - \lambda B)$$

$$= \det \begin{bmatrix} -0.3766 - \lambda & -0.5352 + 0.3766\lambda \\ -0.5352 + 0.3766\lambda & 0.7114 + 0.5352\lambda \end{bmatrix}$$

$$= (-0.3766 - \lambda)(0.7114 + 0.5352\lambda)$$

$$- (-0.5352 + 0.3766\lambda)^2$$

$$\vdots$$

$$= -0.6770\lambda^2 - 0.5098\lambda - 0.5544.$$

More arithmetic with the well-known formula for solving quadratic equations finds that this determinant is zero for complex conjugate eigenvalues $\lambda = -0.3765 \pm 0.8228\,i$. Such non-real complex eigenvalues are characteristic of oscillations. The complex logarithm of these complex values, divided by the time step of $0.005\,s = 5\,ms$, gives the two complex rates $r = -20.0 \pm 400.0\,i$ (to three significant digits).

(c) To find the corresponding coefficients, solve the complex linear equations

$$c_1 + c_2 = 1,$$

$$(-0.3765 + 0.8228\,i)c_1 + (-0.3765 - 0.8228\,i)c_2 = -0.3766.$$

By inspection the solution is $c_1 = c_2 = \frac{1}{2}$ (to three decimal places).

We conclude that the exponential fit is, as plotted to the right on two differing time scales,

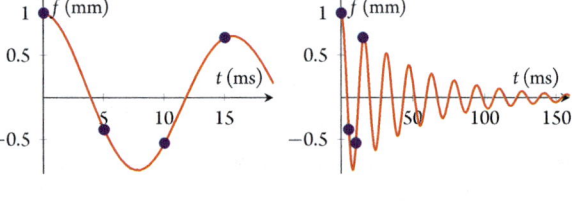

$$f(t) = \tfrac{1}{2}e^{-20t+400it} + \tfrac{1}{2}e^{-20t-400it}$$

$$= \left(\tfrac{1}{2}e^{400it} + \tfrac{1}{2}e^{-400it}\right)e^{-20t}$$

$$= \cos(400t)e^{-20t}.$$

Interpreting, the cosine factor indicates that the piano string oscillates at 400 radians per second which is $400/(2\pi) = 63.66$ cycles/sec. However, the piano string is damped by the factor e^{-20t} so that in just a fraction of a second the oscillations, and the sound, stop.   □

Example 7.1.44 For the data of the previous Example 7.1.43, determine the frequency and damping of the piano string using MATLAB/Octave.

Solution: Follow Procedure 7.1.42 using MATLAB/Octave (Table 7.1).

(a) Form the Hankel matrices from the data with commands

```
f=[1.0000
  -0.3766
  -0.5352
   0.7114]
A=hankel(f(2:3),f(3:4))
B=hankel(f(1:2),f(2:3))
```

(b) Compute the multipliers as the eigenvalues of the generalized problem, and then determine the rates with

```
lambda=eig(A,B)
r=log(lambda)/0.005
```

giving the results

```
lambda =
  -0.3765 + 0.8228i
  -0.3765 - 0.8228i
r =
  -19.99 + 399.99i
  -19.99 - 399.99i
```

(c) Compute the coefficients in the exponential fit with the following, as the value rcond=0.43 is good (Procedure 2.2.5),

```
U=[1 1; lambda(1) lambda(2)]
rcond(U)
c=U\f(1:2)
```

giving results

```
U =
   1.0000 + 0.0000i    1.0000 + 0.0000i
  -0.3765 + 0.8228i   -0.3765 - 0.8228i
ans =   0.4319
c =
   0.5000 + 0.0000i
   0.5000 - 0.0000i
```

This gives the same answer as by hand, namely

$$f(t) = \tfrac{1}{2}e^{-20t+400it} + \tfrac{1}{2}e^{-20t-400it}.$$

As before, Euler's formula transforms this to a real form.   □

Example 7.1.45 In a biochemical experiment every three seconds we measure the concentration of an output chemical as tabulated below (and illustrated to the right). Fit a sum of four exponentials to this data.

secs	concentration
0	0.0000
3	0.1000
6	0.2833
9	0.4639
12	0.6134
15	0.7277
18	0.8112
21	0.8705

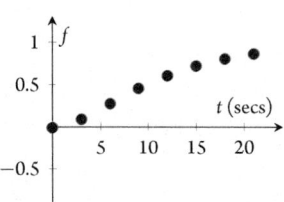

Solution: Follow Procedure 7.1.42 using MATLAB/Octave since the required matrices, 4×4, are too big for hand calculation (Table 7.1).

(a) Form the Hankel matrices from the data with commands

```
f=[0.0000
   0.1000
   0.2833
   0.4639
   0.6134
   0.7277
   0.8112
   0.8705]
A=hankel(f(2:5),f(5:8))
B=hankel(f(1:4),f(4:7))
```

(b) Compute the multipliers as the eigenvalues of the generalized eigen-problem, and then determine the rates with

```
lambda=eig(A,B)
r=log(lambda)/3
```

giving the results

```
lambda =
    0.9990
   -0.4735
    0.6736
    0.4922
r =
   -0.0003 +  0.0000i
   -0.2492 +  1.0472i
   -0.1317 +  0.0000i
   -0.2363 +  0.0000i
```

(c) Compute the coefficients in the exponential fit with the following (Table 5.1 introduces auto-replication of `.^`)

```
U=(lambda.^(0:3)).'
rcond(U)
c=U\f(1:4)
```

giving results

```
U  =
    1.0000    1.0000    1.0000    1.0000
    0.9990   -0.4735    0.6736    0.4922
    0.9981    0.2242    0.4538    0.2422
    0.9971   -0.1061    0.3057    0.1192
ans =   0.007656
c  =
    1.0117
   -0.0001
   -2.2756
    1.2641
```

However, the value `rcond=0.008` is poor (Procedure 2.2.5). This `rcond` suggests that the results have two less significant digits than the original data (Theorem 3.3.30). Since the original data is specified to four decimal places, the results are probably only accurate to two decimal places.

Consequently, this analysis fits the data with the exponential sum (as illustrated to the right)

$$f(t) \approx 1.01 \cdot 1^{t/3} + 0 \cdot (-0.47)^{t/3} - 2.28 \cdot 0.67^{t/3} + 1.26 \cdot 0.49^{t/3}$$
$$\approx 1.01 - 2.28 e^{-0.13t} + 1.26 e^{-0.24t}.$$

As with any data fitting, in practical applications be careful about the reliability of the results. Sound statistical analysis (taught in Statistics courses) needs to supplement Procedure 7.1.42 to inform us about expected errors and sensitivity. This problem of fitting exponentials to data is often sensitive to errors.

The techniques and theory of this subsection generalize to cater for noisy data and for complex system interactions with vast amounts of data. The generalization, *Dynamic Mode Decomposition* (Kutz et al., 2016, e.g.), applies across many areas such as fluid mechanics, video processing, epidemiology, neuroscience, and financial trading.

7.1.6 Exercises

Exercise 7.1.1 For each of the following lists of numbers, could the numbers be all the eigenvalues of a 4×4 matrix? Justify your answer.

(a) $-1.2, -0.6, 0.2, -1.4$

(b) $\pm 2, -3$

(c) $0, 3, \pm 5, 8$

(d) $0, 3 \pm 5i, 8$

Exercise 7.1.2 Find the trace of each of the following matrices.

(a) $\begin{bmatrix} 1 & 3 \\ -1 & -2 \end{bmatrix}$

(b) $\begin{bmatrix} -1 & -2 \\ 6 & 1 \end{bmatrix}$

(c) $\begin{bmatrix} 3.2 & -0.9 & -4.3 \\ 0.8 & -0.1 & 2.3 \\ -0.9 & 0.8 & -0.2 \end{bmatrix}$

(d) $\begin{bmatrix} -1.5 & 0.6 & 0.5 & 1.8 \\ 1.5 & -1.9 & -0.1 & 2.8 \\ -2.1 & -1.4 & -3.3 & -0.3 \\ -1.4 & -2.8 & 0.8 & -2.5 \end{bmatrix}$

Exercise 7.1.3 For each of the following matrices, determine the two highest-order terms and the constant term in the characteristic polynomial of the matrix.

(a) $\begin{bmatrix} -7 & 1 \\ -2 & 2 \end{bmatrix}$

(b) $\begin{bmatrix} 3 & -3 \\ 6 & 2 \end{bmatrix}$

(c) $\begin{bmatrix} 0 & 0 & 2 \\ 1 & 0 & 2 \\ 4 & 2 & 0 \end{bmatrix}$

(d) $\begin{bmatrix} 3 & -3 & 6 \\ -1 & 4 & 0 \\ 0 & 0 & -4 \end{bmatrix}$

(e) $\begin{bmatrix} -3 & 0 & 0 & 1 \\ 0 & 0 & 2 & 0 \\ 0 & 0 & 0 & 1 \\ -4 & -4 & 0 & 4 \end{bmatrix}$

(f) $\begin{bmatrix} 0 & 4 & 0 & 1 \\ 4 & -5 & 0 & -3 \\ 0 & -4 & 0 & 0 \\ 0 & 0 & -1 & 0 \end{bmatrix}$

Exercise 7.1.4 For each of the following characteristic polynomials, write down the size of the corresponding matrix, and the matrix's trace and determinant.

(a) $\lambda^2 + 5\lambda - 6$

(b) $\lambda^2 + 2\lambda - 10$

(c) $-\lambda^3 + 28\lambda - 199$

(d) $-\lambda^3 + 8\lambda^2 - 5\lambda$

(e) $\lambda^4 + 3\lambda^3 + 143\lambda - 56$

(f) $\lambda^4 + 5\lambda^2 - 41\lambda - 5$

Exercise 7.1.5 Some matrices have the following characteristic polynomials. In each case, write down the eigenvalues and their multiplicity.

(a) $(\lambda + 1)\lambda^4(\lambda - 2)^3$

(b) $(\lambda + 1)(\lambda + 2)^2(\lambda - 4)^3(\lambda - 1)^2$

(c) $(\lambda + 0.8)^4(\lambda - 2.6)^3(\lambda + 0.2)(\lambda - 0.7)^2$

Exercise 7.1.6 For each the following matrices, determine the characteristic polynomial by hand, and hence find all the eigenvalues of the matrix and their multiplicity. Show your working.

(a) $\begin{bmatrix} 0 & -3 \\ -1 & -2 \end{bmatrix}$

(b) $\begin{bmatrix} 0 & 5 \\ -2 & 2 \end{bmatrix}$

(c) $\begin{bmatrix} 4.5 & 16 \\ -1 & -3.5 \end{bmatrix}$

(d) $\begin{bmatrix} -1 & 1 & -1 \\ -6 & -6 & 2 \\ -5 & -3 & -3 \end{bmatrix}$

(e) $\begin{bmatrix} -2 & -5 & -1 \\ 0 & 3 & 1 \\ 0 & -6 & -2 \end{bmatrix}$

(f) $\begin{bmatrix} 9 & 3 & 0 \\ -12 & -3 & 0 \\ 2 & -4 & -2 \end{bmatrix}$

(g) $\begin{bmatrix} -10 & -10 & -16 \\ 4 & 4 & 6 \\ 3 & 3 & 5 \end{bmatrix}$

(h) $\begin{bmatrix} 1 & -15 & 7 \\ -1 & -1 & -1 \\ -5 & -15 & -1 \end{bmatrix}$

Exercise 7.1.7 For each the following matrices, use MATLAB/Octave to find all eigenvalues of the matrix and their multiplicity.

(a) $\begin{bmatrix} -2.7 & 0 & 1.6 \\ 6.3 & -1 & -27.8 \\ -0.1 & 0 & -1.9 \end{bmatrix}$

(b) $\begin{bmatrix} 4.9 & -8.1 & 5.4 \\ 8 & -11.2 & 8 \\ 3.9 & -3.9 & 3.4 \end{bmatrix}$

(c) $\begin{bmatrix} -6.7 & -0.6 & -6.6 & 3.6 \\ 3 & 0.1 & 3 & -2 \\ 2.8 & 0.6 & 2.7 & -1.6 \\ -6 & 0 & -6 & 3.1 \end{bmatrix}$

(d) $\begin{bmatrix} 11 & 17.9 & -33.4 & 46.4 \\ 1.2 & 0.9 & 2.8 & 2.2 \\ -12.8 & -21 & 37.2 & -54.8 \\ -12.3 & -19.7 & 33.7 & -51.3 \end{bmatrix}$

(e) $\begin{bmatrix} 309.4 & -29.7 & 451.3 & 337.3 & 305.9 \\ -217.6 & 20.3 & -313.5 & -236.1 & -215.9 \\ 3 & 0 & 0.7 & 3 & 3 \\ -232.6 & 24.1 & -336 & -254.9 & -230.9 \\ -83.6 & 5.6 & -119.8 & -89.2 & -81.8 \end{bmatrix}$

Exercise 7.1.8 For each of the following matrices, find by hand the eigenspace of the nominated eigenvalue. Confirm your answer with MATLAB/Octave. Show your working.

(a) $\begin{bmatrix} -12 & 10 \\ -15 & 13 \end{bmatrix}, \lambda = 3$

(b) $\begin{bmatrix} -1 & 9 \\ -1 & 5 \end{bmatrix}, \lambda = 2$

(c) $\begin{bmatrix} -12 & -82 & -17 \\ 3 & 18 & 3 \\ -6 & -26 & -1 \end{bmatrix}, \lambda = 0$

(d) $\begin{bmatrix} -4 & 0 & -4 \\ -2 & -4 & 0 \\ 8 & 4 & 6 \end{bmatrix}, \lambda = -2$ (e) $\begin{bmatrix} -3 & 2 & -6 \\ -4 & 3 & -6 \\ 2 & -1 & 4 \end{bmatrix}, \lambda = 1$

Exercise 7.1.9 For each of the following matrices, Use MATLAB/Octave to find their eigenvalues, with multiplicity, and to find eigenvectors corresponding to each eigenvalue (2 d.p.). (The next Section 7.2 discusses that for repeated eigenvalues we generally want to record the so-called 'linearly independent' eigenvectors.)

(a) $\begin{bmatrix} 14 & 3 & -6 \\ 14 & -2 & -2 \\ 31 & 4 & -11 \end{bmatrix}$

(c) $\begin{bmatrix} -144 & -374 & 316 & 18 \\ 21 & 45 & -42 & 0 \\ -49 & -138 & 112 & 9 \\ 134 & 336 & -286 & -13 \end{bmatrix}$

(b) $\begin{bmatrix} -1 & 0 & 0 \\ 5 & 5 & 6 \\ -5 & -6 & -7 \end{bmatrix}$

(d) $\begin{bmatrix} 75 & 7 & -13 & -51 & -129 \\ 120 & 12 & -24 & -84 & -208 \\ 62 & 6 & -12 & -42 & -106 \\ -48 & -5 & 9 & 32 & 83 \\ 62 & 6 & -11 & -42 & -107 \end{bmatrix}$

Exercise 7.1.10 Consider the following three matrices, which are perturbed versions of the matrix $\begin{bmatrix} 2 & 1 \\ 0 & 2 \end{bmatrix}$. Which perturbations show the high sensitivity of the repeated eigenvalue? Give reasons.

(a) $A = \begin{bmatrix} 2 & 1 \\ -0.0001 & 2 \end{bmatrix}$

(c) $C = \begin{bmatrix} 2.0001 & 1 \\ 0 & 2 \end{bmatrix}$

(b) $B = \begin{bmatrix} 2 & 1.0001 \\ 0 & 2 \end{bmatrix}$

Exercise 7.1.11 For each of the following matrices, use MATLAB/Octave to compute the eigenvalues, and also to compute the eigenvalues of matrices obtained by adding random perturbations that are roughly of magnitude 0.0001 (use randn). Give reasons for which eigenvalues appear sensitive and which appear to be not sensitive.

(a) $A = \begin{bmatrix} 0 & -3 \\ -1 & -2 \end{bmatrix}$

(d) $D = \begin{bmatrix} 2 & 1 & 1 \\ 1 & 2 & 1 \\ 1 & 1 & 2 \end{bmatrix}$

(b) $B = \begin{bmatrix} 0 & -3 \\ 3 & 6 \end{bmatrix}$

(e) $E = \begin{bmatrix} -1 & 1 & -1 \\ -6 & -6 & 2 \\ -5 & -3 & -3 \end{bmatrix}$

(c) $C = \begin{bmatrix} -10 & -10 & -16 \\ 4 & 4 & 6 \\ 3 & 3 & 5 \end{bmatrix}$

(f) $F = \begin{bmatrix} -6.7 & -0.6 & -6.6 & 3.6 \\ 3 & 0.1 & 3 & -2 \\ 2.8 & 0.6 & 2.7 & -1.6 \\ -6 & 0 & -6 & 3.1 \end{bmatrix}$

Exercise 7.1.12 Consider the evolving system $y(t+1) = Ay(t)$ for each of the following cases. Predict $y(1), y(2)$ and $y(3)$, for the given initial $y(0)$.

(a) $A = \begin{bmatrix} 3 & 0 \\ 3 & 2 \end{bmatrix}, y(0) = \begin{bmatrix} 6 \\ -1 \end{bmatrix}$

(c) $A = \begin{bmatrix} 11 & -14 & -4 \\ 7 & -10 & -2 \\ 4 & -4 & -2 \end{bmatrix}, y(0) = \begin{bmatrix} 0 \\ 2 \\ -2 \end{bmatrix}$

(b) $A = \begin{bmatrix} 0 & -1 \\ -4 & 2 \end{bmatrix}, y(0) = \begin{bmatrix} 3 \\ 0 \end{bmatrix}$

(d) $A = \begin{bmatrix} 9 & 7 & -5 \\ -16 & -8 & 8 \\ 10 & 10 & -6 \end{bmatrix}, y(0) = \begin{bmatrix} -1 \\ 1 \\ 1 \end{bmatrix}$

Exercise 7.1.13 For each of the matrices of the previous Exercise 7.1.12, find a general solution of $y(t+1) = Ay(t)$, if possible. Then use the corresponding given initial $y(0)$ to find a formula for the specific $y(t)$. Finally, check that the formula reproduces the values of $y(1), y(2)$, and $y(3)$ found in Exercise 7.1.12. Show your working.

Exercise 7.1.14 Reconsider the mathematical modelling of the serval, Example 7.1.30. Derive the matrix A in the model $y(t+1) = Ay(t)$ for the following cases:

(a) choose the unit of time to be three months $(1/4\ \text{year})$;
(b) choose the unit of time to be one month $(1/12\ \text{year})$.

Exercise 7.1.15 From the following partial description of the Tasmanian Devil, derive a mathematical model in the form $y(t+1) = Ay(t)$ for the age structure of the Tasmanian Devil. By finding eigenvalues, and a corresponding eigenvector for each, predict the long-term growth of the population, and predict the long-term relative numbers of Devils of various ages.

Devils are not monogamous, and their reproductive process is very robust and competitive. Males fight one another for the females, and then guard their partners to prevent female infidelity. Females can ovulate three times in as many weeks during the mating season, and 80% of two-year-old females are seen to be pregnant during the annual mating season. Females average four breeding seasons in their life and give birth to 20–30 live young after three weeks' gestation. The newborn are pink, lack fur, have indistinct facial features and weigh around 0.20 g (0.0071 oz) at birth. As there are only four nipples in the pouch, competition is fierce and few newborns survive. The young grow rapidly and are ejected from the pouch after around 100 days, weighing roughly 200 g (7.1 oz). The young become independent after around nine months, so the female spends most of her year in activities related to birth and rearing.

Wikipedia 2016, (Owen and Pemberton, 2005, pp. 64–6)

Exercise 7.1.16 From the following partial description of the elephant, derive a mathematical model in the form $y(t+1) = Ay(t)$ for the age structure of the elephant. By finding eigenvalues, and a corresponding eigenvector for each, predict the long-term growth of the population, and predict the long-term relative numbers of elephants of various ages.

Gestation in elephants typically lasts around two years with interbirth intervals usually lasting four to five years. Births tend to take place during the wet season. Calves are born 85 cm (33 in) tall and weigh around 120 kg (260 lb). Typically, only a single young is born, but twins sometimes occur. The relatively long pregnancy is maintained by five corpus luteums (as opposed to one in most mammals) and gives the foetus more time to develop, particularly the brain and trunk. As such, newborn elephants are precocial and quickly stand and walk to follow their mother and family herd. A new calf is usually the centre of attention for herd members. Adults and most of the other young will gather around the newborn, touching and caressing it with their trunks. For the first few days, the mother is intolerant of other herd members near her young. Alloparenting—where a calf is cared for by someone other than its mother— takes place in some family groups. Allomothers are typically two to twelve years old. When a predator is near, the family group gathers together with the calves in the centre.

For the first few days, the newborn is unsteady on its feet, and needs the support of its mother. It relies on touch, smell and hearing, as its eyesight is poor. It has little precise control over its trunk, which wiggles around and may cause it to trip. By its second week of life, the calf can walk more firmly and has more control over its trunk. After its first month, a calf can pick up, hold and put objects in its mouth, but cannot suck water through the trunk and must drink directly through the mouth. It is still dependent on its mother and keeps close to her.

For its first three months, a calf relies entirely on milk from its mother for nutrition after which it begins to forage for vegetation and can use its trunk to collect water. At the same time, improvements in lip and leg coordination occur. Calves continue to suckle at the same rate as before until their sixth month, after which they become more independent when feeding. By nine months, mouth, trunk and foot coordination is perfected. After a year, a calf's abilities to groom, drink, and feed itself are fully developed. It still needs its mother for nutrition and protection from predators for at least another year. Suckling bouts tend to last 2–4 min/hr for a calf younger than a year and it continues to suckle until it reaches three years of age or older. Suckling after two years may serve to maintain growth rate, body condition and reproductive ability. Play behaviour in calves differs between the sexes; females run or chase each other, while males play-fight. The former are sexually mature by the age of nine years while the latter become mature around 14–15 years. Adulthood starts at about 18 years of age in both sexes. Elephants have long lifespans, reaching 60–70 years of age. Wikipedia 2016 (Sukumar, 2003)

Exercise 7.1.17 From the following partial description of the dolphin (Indo-Pacific bottlenose dolphin), derive a mathematical model in the form $y(t+1) = Ay(t)$ for the age structure of the dolphin. (Assume only one calf is born at a time.) By finding eigenvalues, and a corresponding eigenvector for each, predict the long-term growth of the population, and predict the long-term relative numbers of dolphins of various ages.

Indo-Pacific bottlenose dolphins live in groups that can number in the hundreds, but groups of five to 15 dolphins are most common. In some parts of their range, they associate with the common bottlenose dolphin and other dolphin species, such as the humpback dolphin.

The peak mating and calving seasons are in the spring and summer, although mating and calving occur throughout the year in some regions. Gestation period is about 12 months. Calves are between 0.84 and 1.5 metres (2.8 and 4.9 ft) long, and weigh between 9 and 21 kilograms (20 and 46 lb). The calves are weaned between 1.5 and two years, but can remain with their mothers for up to five years. The interbirth interval for females is typically four to six years.

In some parts of its range, this dolphin is subject to predation by sharks; its life span is more than 40 years. Wikipedia 2016 (Reeves et al., 2002, pp. 362–5)

Exercise 7.1.18 You are given that a mathematical model of the age structure of some animal population is

$$y_1(t+1) = 0.5y_1(t) + y_3(t),$$
$$y_2(t+1) = 0.5y_1(t) + 0.7y_2(t),$$
$$y_3(t+1) = 0.3y_2(t) + 0.9y_3(t).$$

Invent an animal species, and a time scale, and create details of a plausible scenario for the breeding and life cycle of the species that could lead to this mathematical model. Write a coherent paragraph about the breeding and life cycle of the species with enough information that someone could deduce this mathematical model from your description. Be creative.

Exercise 7.1.19 For X denoting each of the following matrices, find by hand calculation the eigenvalues and eigenvectors of the larger matrix $\begin{bmatrix} O & X \\ X^T & O \end{bmatrix}$. Show your working. Relate these to an SVD of the particular matrix X.

(a) $A = \begin{bmatrix} 3 \\ 4 \end{bmatrix}$

(b) $B = \begin{bmatrix} -5 & 12 \end{bmatrix}$

(c) $C = \begin{bmatrix} 1 & 0 \\ 0 & -2 \end{bmatrix}$

(d) $D = \begin{bmatrix} 0 & 1 \\ -4 & 0 \end{bmatrix}$

Exercise 7.1.20 Find by hand calculation all eigenvalues and corresponding eigenvectors of the generalized eigen-problem $Av = \lambda Bv$ for the following pairs of matrices. Check your calculations with MATLAB/Octave.

(a) $A = \begin{bmatrix} 1 & 3 \\ 0 & 0 \end{bmatrix}, B = \begin{bmatrix} 0 & 3 \\ 4 & 4 \end{bmatrix}$

(b) $A = \begin{bmatrix} -1 & 0 \\ 5 & 1 \end{bmatrix}, B = \begin{bmatrix} 1 & 2 \\ -1 & -2 \end{bmatrix}$

(c) $A = \begin{bmatrix} 3 & -1 \\ -1 & -2 \end{bmatrix}, B = \begin{bmatrix} -3 & -3 \\ 1 & -1 \end{bmatrix}$

(d) $A = \begin{bmatrix} 1 & -1 \\ 1 & -4 \end{bmatrix}, B = \begin{bmatrix} 1 & -0 \\ 2 & -2 \end{bmatrix}$

(e) $A = \begin{bmatrix} 0 & 1 & -2 \\ -2 & -1 & -2 \\ 1 & 0 & 2 \end{bmatrix}, B = \begin{bmatrix} 1 & 0 & 1 \\ 0 & -2 & 1 \\ -1 & -3 & -1 \end{bmatrix}$

(f) $A = \begin{bmatrix} 0 & -2 & 1 \\ -4 & 2 & 3 \\ 2 & -2 & -1 \end{bmatrix}, B = \begin{bmatrix} 0 & 0 & -1 \\ 0 & 0 & -1 \\ -2 & 0 & 0 \end{bmatrix}$

(g) $A = \begin{bmatrix} -1 & 1 & -2 \\ -4 & 0 & 3 \\ 0 & -3 & 0 \end{bmatrix}, B = \begin{bmatrix} -1 & 1 & 1 \\ 5 & -2 & 1 \\ 2 & -2 & -2 \end{bmatrix}$

(h) $A = \begin{bmatrix} -1 & 2 & -1 \\ 0 & -2 & -1 \\ 3 & 2 & 1 \end{bmatrix}, B = \begin{bmatrix} -1 & 2 & -1 \\ 0 & 2 & 1 \\ 3 & 1 & -2 \end{bmatrix}$

Exercise 7.1.21 Use MATLAB/Octave to find and describe all eigenvalues and corresponding eigenvectors of the generalized eigen-problem $Av = \lambda Bv$ for the following pairs of matrices.

(a) $A = \begin{bmatrix} -2 & 1 & -2 & -2 \\ -2 & 2 & -1 & 0 \\ -1 & 1 & 2 & 1 \\ 1 & -2 & -1 & -1 \end{bmatrix}, B = \begin{bmatrix} -1 & 0 & -3 & 1 \\ -1 & 1 & -1 & 1 \\ -1 & 1 & 0 & 1 \\ 3 & 5 & 0 & -2 \end{bmatrix}$

(b) $A = \begin{bmatrix} 3 & 1 & -2 & -4 \\ -2 & -3 & -3 & 2 \\ 1 & 1 & 1 & 2 \\ 1 & 2 & 0 & 0 \end{bmatrix}, B = \begin{bmatrix} -2 & 1 & 1 & 1 \\ 1 & -1 & 1 & -1 \\ 2 & 0 & -1 & 0 \\ 0 & 0 & 1 & 0 \end{bmatrix}$

(c) $A = \begin{bmatrix} 0 & 2 & -1 & -1 \\ 0 & 0 & -1 & 0 \\ 3 & -3 & -1 & 1 \\ 2 & -3 & -2 & -3 \end{bmatrix}, B = \begin{bmatrix} 1 & 2 & 1 & 1 \\ 1 & 0 & 0 & -1 \\ 0 & 0 & 0 & 0 \\ 2 & -1 & -1 & 0 \end{bmatrix}$

(d) $A = \begin{bmatrix} 4 & 4 & 0 & 3 \\ 1 & -2 & 0 & 1 \\ 0 & 2 & 4 & 2 \\ 0 & 2 & -1 & 1 \end{bmatrix}, B = \begin{bmatrix} -1 & 2 & -1 & 1 \\ -2 & 2 & 1 & 1 \\ -1 & 0 & 1 & -2 \\ 1 & 1 & 1 & -1 \end{bmatrix}$

Exercise 7.1.22 Use the properties of determinants (Chapter 6), and that an nth degree polynomial has exactly n zeros (when counted according to multiplicity), to explain why the generalized eigen-problem $Av = \lambda Bv$, for real $n \times n$ matrices A and B, has n eigenvalues iff matrix B is invertible.

Exercise 7.1.23 Consider the generalized eigen-problem $Av = \lambda Bv$ and let $\lambda_1 \neq \lambda_2$ be distinct eigenvalues with corresponding eigenvectors v_1 and v_2, respectively. Invent two real *symmetric* matrices A and B and record your working which demonstrates that although v_1 and v_2 are not orthogonal, nonetheless $v_1^T B v_2 = 0$.

Exercise 7.1.24 Consider the generalized eigen-problem $Av = \lambda Bv$ for real *symmetric* $n \times n$ matrices A and B. Let $\lambda_1 \neq \lambda_2$ be distinct eigenvalues with corresponding eigenvectors v_1 and v_2, respectively. Prove that $v_1^T B v_2 = 0$ always holds.

Exercise 7.1.25 Prove that if both A and B are real *symmetric* $n \times n$ matrices, and the eigenvalues of B all have the same sign, then the eigenvalues of the generalized eigen-problem $Av = \lambda Bv$ are all real. Briefly explain why it is necessary to have the proviso that the eigenvalues of B are either all positive or all negative.

Exercise 7.1.26 In view of the preceding Exercise 7.1.25, invent real symmetric matrices A and B such that the generalized eigen-problem $Av = \lambda Bv$ has nonreal complex valued eigenvalues.

Exercise 7.1.27 Explain briefly how the properties established in the previous two Exercises 7.1.24 and 7.1.25 generalize important properties of the standard eigen-problem $A\boldsymbol{x} = \lambda\boldsymbol{x}$ for symmetric A.

Exercise 7.1.28 Consider the specified data values $\boldsymbol{f}$ at the specified times, and either by hand or using MATLAB/Octave fit a sum of exponentials (7.4), $f(t) = c_1 e^{r_1 t} + c_2 e^{r_2 t} + \cdots + c_n e^{r_n t}$. Plot the data and the curve you have fitted.

(a) For times $0, 1, 2, 3$ the data values are
$\boldsymbol{f} = (3, 2.75, 2.5625, 2.4219)$.

(b) For times $0, 1, 2, 3$ the data values are
$\boldsymbol{f} = (1.5833, 1.3333, 1.3056, 1.4954)$.

(c) For times $0, 2, 4, 6$ the data values are
$\boldsymbol{f} = (-1, 0.222222, 0.172840, 0.085048)$.

(d) For times $0, 0.5, 1, 1.5$ the data values
are $\boldsymbol{f} = (1.3, 0.75355, 0.45, 0.27678)$.

(e) For times $0, 1, 2, \ldots, 5$ the data values
are

$$\boldsymbol{f} = \begin{bmatrix} -1.00000 \\ 0.12500 \\ 0.71875 \\ 1.03906 \\ 1.21680 \\ 1.31885 \end{bmatrix}.$$

(f) For times $0, 1, 2, \ldots, 5$ the data values
are

$$\boldsymbol{f} = \begin{bmatrix} 3.250000 \\ 2.041667 \\ 1.225694 \\ 0.673900 \\ 0.300178 \\ 0.046636 \end{bmatrix}.$$

(g) For times $0, 2, 4, \ldots, 10$ the data values
are

$$\boldsymbol{f} = \begin{bmatrix} 0.16667 \\ 1.38889 \\ 2.72984 \\ 5.81259 \\ 12.88104 \\ 28.87004 \end{bmatrix}.$$

(h) For times $0, 0.5, 1, \ldots, 2.5$ the data
values are

$$\boldsymbol{f} = \begin{bmatrix} 0.75000 \\ 0.89846 \\ 1.08333 \\ 1.30323 \\ 1.55903 \\ 1.85343 \end{bmatrix}.$$

Exercise 7.1.29 Consider the specified data values $\boldsymbol{f}$ of some decaying oscillations at the specified times (in seconds). Use MATLAB/Octave to fit a sum of exponentials (7.4), $f(t) = c_1 e^{r_1 t} + c_2 e^{r_2 t} + \cdots + c_n e^{r_n t}$. Confirm that your fit reproduces the data values. What frequencies do you detect in the fitted constants?

(a) For times $0, 1, 2, \ldots, 7$ the data values are

$$f = \begin{bmatrix} -0.153161 \\ 0.484787 \\ -0.124780 \\ -0.283690 \\ 0.201896 \\ 0.174832 \\ -0.200418 \\ -0.092107 \end{bmatrix}.$$

(c) For times $0, 0.5, 1, \ldots, 3.5$ the data values are

$$f = \begin{bmatrix} 0.0060901 \\ 0.1753701 \\ 0.2385516 \\ 0.1787653 \\ 0.0404368 \\ -0.0998597 \\ -0.1742818 \\ -0.1552607 \end{bmatrix}.$$

(b) For times $0, 1, 2, \ldots, 7$ the data values are

$$f = \begin{bmatrix} 1.225432 \\ 0.499044 \\ 0.034164 \\ -0.093971 \\ -0.103784 \\ -0.028558 \\ 0.033825 \\ 0.026439 \end{bmatrix}.$$

(d) For times $0, 2, 4, \ldots, 14$ the data values are

$$f = \begin{bmatrix} 1.297867 \\ -0.800035 \\ 0.487305 \\ -0.285714 \\ 0.161499 \\ -0.090411 \\ 0.051244 \\ -0.028936 \end{bmatrix}.$$

Exercise 7.1.30 In astronomy, a Type Ia supernova explodes, peaking in luminosity in a few days, and then its luminosity declines over months. It is conjectured that the decline is powered by the radioactive decay from radioactive nickel to cobalt to iron. For the supernova SN1991T six measurements, starting on Julian day 2448366 and taken approximately 8 days apart, give the following luminosity values (in some units) (Pereyra and Scherer, 2010, p.146):

$$f = \begin{bmatrix} 0.6543 \\ 1.5783 \\ 1.6406 \\ 0.3678 \\ 0.05918 \\ 0.008227 \end{bmatrix}.$$

Detect the exponential decay in this supernova data by fitting a sum of exponentials $(7.4), f(t) = c_1 e^{r_1 t} + c_2 e^{r_2 t} + c_3 e^{r_3 t}$. Comment on the result.

Exercise 7.1.31 Recall that Example 3.4.27 introduced the phenomenon of El Niño which makes a large impact on the world's weather. El Niño is correlated significantly with the difference in atmospheric pressure between Darwin and Tahiti—the so-called Southern Oscillation Index (SOI). Figure 3.1 plots a ('smoothed') yearly average SOI each year for 50 years up to 1993. Here detect the cycles in the SOI by analysing the plotted data as a sum of exponentials $(7.4), f(t) = c_1 e^{r_1 t} + c_2 e^{r_2 t} + \cdots + c_n e^{r_n t}$.

(a) Enter the data into MATLAB/Octave:

```
year=(1944:1993)';
soi=[-0.03; 0.74; 6.37; -7.28; 0.44; -0.99; 1.32
6.42; -6.51; 0.07; -1.96; 1.72; 6.49; -5.61
-0.24; -2.90; 1.92; 6.54; -4.61; -0.47; -3.82
1.94; 6.56; -3.53; -0.59; -4.69; 1.76; 6.53
-2.38; -0.59; -5.48; 1.41; 6.41; -1.18; -0.45
-6.19; 0.89; 6.19; 0.03; -0.16; -6.78; 0.21; 5.84
1.23; 0.30; -7.22; -0.60; 5.33; 2.36; 0.91 ];
```

(b) Use Procedure 7.1.42 in MATLAB/Octave to compute the 25 complex rates r and 25 complex coefficients c that fit the data.

(c) Find the four relatively large coefficients, when compared to the relatively small coefficients of magnitude < 0.015.

(d) Explain why these four coefficients indicate that the SOI data appears to be dominantly composed of oscillations with periods about 5.1 years and 2.5 years.

Exercise 7.1.32 In a few sentences, answer/discuss each of the following.

(a) Given an $n \times n$ matrix, what leads to the characteristic polynomial being of nth degree?

(b) How does the trace of a matrix appear in its characteristic polynomial?

(c) What is the importance of the multiplicity of eigenvalues of a matrix?

(d) What is the relation between the characteristic polynomial of a matrix and its characteristic equation?

(e) Why is it beautifully useful to cater for complex valued eigenvalues and eigenvectors of real matrices arising in real problems?

(f) What is the evidence for repeated eigenvalues generally being sensitive to computational and experimental errors?

(g) What causes $y(t) = c_1 \lambda_1^t v_1 + c_2 \lambda_2^t v_2 + \cdots + c_m \lambda_m^t v_m$ to form a general solution to the evolving system $y(t+1) = Ay(t)$?

(h) How can the singular values of a matrix arise from an eigen-problem?

(i) Describe some scenarios that require fitting a sum of exponentials to data.

7.2 Linearly independent vectors may form a basis

In Chapter 4 on symmetric matrices, the eigenvectors from distinct eigenvalues are proved to be always orthogonal—because of matrix symmetry. For general matrices, the eigenvectors are not orthogonal—as introduced at the start of this Chapter 7. But the orthogonal property is extremely useful. Question: is there some analogue of orthogonality that is similarly useful for general matrices? Answer: yes. We now extend "orthogonal" to the more general concept of "linearly independent".

One reason that orthogonal vectors are useful is that they can form an orthonormal basis and hence act as the unit vectors of an orthogonal coordinate systems. Analogously, the concept of linearly independent vectors is closely connected to coordinate systems that are not orthogonal.

Subspace coordinate systems   In any given mathematical problem, an application wants two things from a general solution:

- firstly, the general solution must encompass every possibility (the solution must span the possibilities); and
- secondly, each possible solution should have a unique algebraic form in the general solution.

For an example of the need for a unique algebraic form, let's suppose we wanted to find solutions to the differential equation $d^2y/dt^2 - y = 0$. You might find $y = 3e^x + 2e^{-x}$, whereas I find $y = 5\cosh x + \sinh x$, and a friend finds $y = e^x + 4\cosh x$. It appears from these disparate algebraic forms that we all disagree. Should we all go and search for errors in the solution process? No. The reason is that all these solutions are the same. The apparent differences arise only because you choose exponentials to represent the solution, whereas I choose hyperbolic functions, and the friend a mixture; the solutions are the same, it is only the algebraic representation that appears different. In general, when we cannot immediately distinguish identical solutions, all algebraic manipulation becomes immensely more difficult due to algebraic ambiguity. To avoid such immense difficulties we need to introduce, in both calculus and linear algebra, the concept of linear independence.

Linear independence empowers us, often implicitly, to use a non-orthogonal coordinate system in a subspace. We replace orthogonal unit vectors by any suitable set of basis vectors. For example, in the plane any two vectors at an angle to each other suffice to be able to describe uniquely every vector (point) in the plane. As illustrated to the right, every point in the plane (end-point of a vector) is a unique linear combination of the two drawn basis vectors v_1 and v_2. Such a pair of basis vectors, termed a linearly independent pair, avoids the difficulties of algebraic ambiguity.

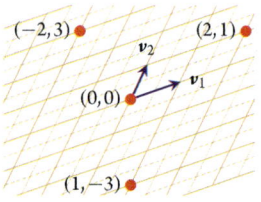

7.2.1 Linearly (in)dependent sets

This section defines "linearly dependent" and "linearly independent", and then relates the concept to homogeneous linear equations, orthogonality, and sets of eigenvectors.

Example 7.2.1 *(2D non-orthogonal coordinates)* Show that every vector in the plane $\mathbb{R}^2$ can be written uniquely as a linear combination of the two vectors $v_1 = (2,1)$ and $v_2 = (1,2)$ that are shown to the right.

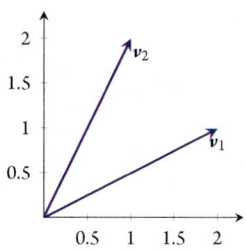

Solution: Let's start with some specific example vectors.

(a) The vector $(0,2)$ may be written as the linear combination $(0,2) = -\frac{2}{3}v_1 + \frac{4}{3}v_2$ as shown.

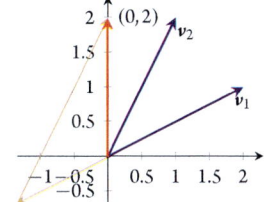

(b) The vector $(2,2)$ may be written as the linear combination $(2,2) = \frac{2}{3}v_1 + \frac{2}{3}v_2$ as shown.

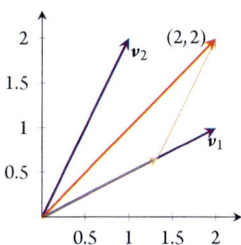

(c) The vector $(1,-1)$ may be written as the linear combination $(1,-1) = v_1 - v_2$ as shown.

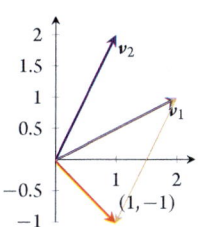

(d) The vector $(-3,-3)$ may be written as the linear combination $(-3,-3) = -v_1 - v_2$ as shown.

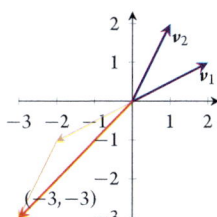

Now proceed to consider a general vector (x,y) and seek it as a linear combination of v_1 and v_2, namely $(x,y) = c_1 v_1 + c_2 v_2$. That is, let's write each and every point in the plane as a linear combination of v_1 and v_2 as illustrated to the right. Rewrite the equation in matrix-vector form as

$$\begin{bmatrix} v_1 & v_2 \end{bmatrix} \begin{bmatrix} c_1 \\ c_2 \end{bmatrix} = \begin{bmatrix} x \\ y \end{bmatrix}, \quad \text{that is,} \quad Vc = \begin{bmatrix} x \\ y \end{bmatrix} \text{ for } V = \begin{bmatrix} 2 & 1 \\ 1 & 2 \end{bmatrix}.$$

For any given (x,y), $Vc = (x,y)$ is a system of linear equations for the coefficients c. Theorem 3.4.43 asserts that the system has a unique solution c if and only if the matrix V is invertible. Here the unique solution is then that the vector of coefficients

$$c = V^{-1} \begin{bmatrix} x \\ y \end{bmatrix} = \begin{bmatrix} \frac{2}{3} & -\frac{1}{3} \\ -\frac{1}{3} & \frac{2}{3} \end{bmatrix} \begin{bmatrix} x \\ y \end{bmatrix}.$$

Equivalently, Theorem 3.4.43(c) asserts the system has a unique solution c—unique coefficients c—if and only if the homogeneous system $Vc = 0$ has only the zero solution $c = 0$. It is this last statement that leads to the upcoming Definition 7.2.4 of linearly independent vectors. ☐

Activity 7.2.2 The vector shown to the right is which of the following linear combinations of shown vectors v_1 and v_2?

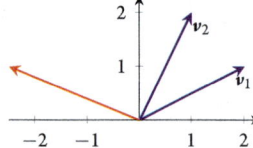

(a) $-2v_1 + 1.5v_2$ (c) $-1.5v_1 + v_2$

(b) $-v_1 + v_2$ (d) $-2.5v_1 + 2v_2$

Example 7.2.3 *(3D failure)* Show that vectors in $\mathbb{R}^3$ are not written uniquely as a linear combination of $v_1 = (-1,1,0)$, $v_2 = (1,-2,1)$, and $v_3 = (0,1,-1)$.

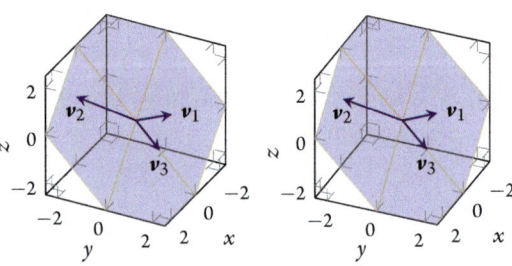

One reason for the failure is that these three vectors only span a plane, as shown to the right in stereo. The solution here looks at the different issue of unique representation.

Solution: As one example, consider the vector $(1,0,-1)$:

$$(1,0,-1) = -1v_1 + 0v_2 + 1v_3;$$
$$(1,0,-1) = 1v_1 + 2v_2 + 3v_3;$$
$$(1,0,-1) = -2v_1 - 1v_2 + 0v_3;$$
$$(1,0,-1) = (-1+t)v_1 + tv_2 + (1+t)v_3, \quad \text{for every } t.$$

This last combination shows that there are an infinite number of ways to write $(1,0,-1)$ as a linear combination of v_1, v_2, and v_3. Such an infinity of linear combinations means that v_1, v_2, and v_3 cannot form the basis for a useful 'coordinate system' because we cannot easily distinguish between the different combinations all describing the same vector. The reason for the infinity of combinations is that there is a nontrivial linear combination of v_1, v_2, and v_3 which is zero, namely here $v_1 + v_2 + v_3 = 0$. It is this last statement that leads to the Definition 7.2.4 of linearly dependent vectors.   □

Definition 7.2.4   *A set of vectors $\{v_1, v_2, \ldots, v_k\}$ is **linearly dependent** if there are scalars $c_1, c_2, \ldots, c_k$, at least one of which is nonzero, such that $c_1 v_1 + c_2 v_2 + \cdots + c_k v_k = 0$. A set of vectors that is not linearly dependent is called **linearly independent** (characterized by only the linear combination with $c_1 = c_2 = \cdots = c_k = 0$ gives the zero vector).*

When reading the terms "linearly in/dependent" be very careful: it is all too easy to misread the presence or absence of the crucial "in" syllable. The presence or absence of the "in" syllable makes all the difference between the property and its opposite.

Example 7.2.5   Are the following sets of vectors linearly dependent or linearly independent. Give reasons.

(a) $\{(-1,1,0), (1,-2,1), (0,1,-1)\}$

   Solution: The set is linearly dependent as the linear combination $(-1,1,0) + (1,-2,1) + (0,1,-1) = (0,0,0)$.   □

(b) $\{(2,1), (1,2)\}$

   Solution: The set is linearly independent because the linear combination equation $c_1(2,1) + c_2(1,2) = (0,0)$ is equivalent to the homogeneous matrix-vector system $\begin{bmatrix} 2 & 1 \\ 1 & 2 \end{bmatrix} c = 0$ which has *only* the zero solution $c = 0$.   □

(c) $\{(-2,4,1,-1,0)\}$

   Solution: This set of one vector in $\mathbb{R}^5$ is linearly independent as $c_1(-2,4,1,-1,0) = 0$ can only be satisfied with $c_1 = 0$. Indeed, any one nonzero vector v in $\mathbb{R}^n$ forms a linearly independent set, $\{v\}$, for the same reason.   □

(d) $\{(2,1), (0,0)\}$

Solution: The set is linearly dependent because the linear combination $0(2,1) + c_2(0,0) = (0,0)$ for every nonzero c_2. ☐

(e) $\{0, v_2, v_3, \ldots, v_k\}$

Solution: Every set that includes the zero vector is linearly dependent as $c_1 0 + 0v_2 + \cdots + 0v_k = 0$ for every nonzero c_1. ☐

(f) $\{e_1, e_2, e_3\}$, the set of standard unit vectors in $\mathbb{R}^3$.

Solution: This set is linearly independent as $c_1 e_1 + c_2 e_2 + c_3 e_3 = (c_1, c_2, c_3) = 0$ only when all three components are zero, $c_1 = c_2 = c_3 = 0$. ☐

(g) $\{(\frac{1}{3}, \frac{2}{3}, \frac{2}{3}), (\frac{2}{3}, \frac{1}{3}, -\frac{2}{3})\}$

Solution: This set is linearly independent. Seek some linear combination $c_1(\frac{1}{3}, \frac{2}{3}, \frac{2}{3}) + c_2(\frac{2}{3}, \frac{1}{3}, -\frac{2}{3}) = 0$. Take the dot product of both sides of this equation with $(1, 2, 2)$:

$$c_1 \begin{bmatrix} \frac{1}{3} \\ \frac{2}{3} \\ \frac{2}{3} \end{bmatrix} \cdot \begin{bmatrix} 1 \\ 2 \\ 2 \end{bmatrix} + c_2 \begin{bmatrix} \frac{2}{3} \\ \frac{1}{3} \\ -\frac{2}{3} \end{bmatrix} \cdot \begin{bmatrix} 1 \\ 2 \\ 2 \end{bmatrix} = 0 \cdot \begin{bmatrix} 1 \\ 2 \\ 2 \end{bmatrix}$$

$$\implies c_1 3 + c_2 0 = 0 \implies c_1 = 0.$$

Also, take the dot product with $(2, 1, -2)$:

$$c_1 \begin{bmatrix} \frac{1}{3} \\ \frac{2}{3} \\ \frac{2}{3} \end{bmatrix} \cdot \begin{bmatrix} 2 \\ 1 \\ -2 \end{bmatrix} + c_2 \begin{bmatrix} \frac{2}{3} \\ \frac{1}{3} \\ -\frac{2}{3} \end{bmatrix} \cdot \begin{bmatrix} 2 \\ 1 \\ -2 \end{bmatrix} = 0 \cdot \begin{bmatrix} 2 \\ 1 \\ -2 \end{bmatrix}$$

$$\implies c_1 0 + c_2 3 = 0 \implies c_2 = 0.$$

Hence $c_1 = c_2 = 0$ is the *only* possibility and so the two given vectors are linearly independent. ☐

These last two cases generalize to the next Theorem 7.2.8 about the linear independence of every orthonormal set of vectors.

Activity 7.2.6 Which of the following sets of vectors is linearly independent?

(a) $\{(-1, 1), (0, 1)\}$

(b) $\{(0, 1), (0, -1)\}$

(c) $\{(-1, 2), (-2, 4)\}$

(d) $\{(0, 0), (-2, 1)\}$

Example 7.2.7 *(calculus extension)* In calculus the notion of a function corresponds to the notion of a vector in our linear algebra. For the purposes of this example, consider 'vector' and 'function' to be synonymous, and that 'all components' and 'all x' are synonymous. Show that the set $\{e^x, e^{-x}, \cosh x, \sinh x\}$ is linearly dependent. What is a subset that is linearly independent?

Solution: The definition of the hyperbolic functions, namely that $\cosh x = (e^x + e^{-x})/2$ and $\sinh x = (e^x - e^{-x})/2$, immediately gives two nontrivial linear combinations that are zero for all x, namely $2\cosh x - e^x - e^{-x} = 0$ and $2\sinh x - e^x + e^{-x} = 0$ for all x. Either one of these implies that the set $\{e^x, e^{-x}, \cosh x, \sinh x\}$ is linearly dependent.

Because e^x and e^{-x} are not proportional to each other, there is no linear combination which is zero for all x, and hence the set $\{e^x, e^{-x}\}$ is linearly independent (as are any other pairs of the four functions). ☐

Theorem 7.2.8 *Every orthonormal set of vectors (Definition 3.2.38) is linearly independent.*

Proof. Let $\{v_1, v_2, \ldots, v_k\}$ be an orthonormal set of vectors in $\mathbb{R}^n$. Let's find all possible scalars $c_1, c_2, \ldots, c_k$ such that $c_1 v_1 + c_2 v_2 + \cdots + c_k v_k = \mathbf{0}$. Taking the dot product of this equation with v_1 requires $c_1 v_1 \cdot v_1 + c_2 v_2 \cdot v_1 + \cdots + c_k v_k \cdot v_1 = \mathbf{0} \cdot v_1$; by orthonormality this equation becomes $c_1 1 + c_2 0 + \cdots + c_k 0 = 0$; that is, $c_1 = 0$. Similarly, taking the dot product with v_2 requires $c_1 v_1 \cdot v_2 + c_2 v_2 \cdot v_2 + \cdots + c_k v_k \cdot v_2 = \mathbf{0} \cdot v_2$; by orthonormality this equation becomes $c_1 0 + c_2 1 + \cdots + c_k 0 = 0$; that is, $c_2 = 0$. And so on for every vector in the set, implying the coefficients $c_1 = c_2 = \cdots = c_k = 0$ is the only possibility. By Definition 7.2.4, the orthonormal set must be linearly independent. ☐

In contrast to orthonormal vectors which are always linearly independent, a set of two vectors proportional to each other is always linearly dependent as seen in the following examples. This linear dependence of proportional vectors then generalizes in the forthcoming Theorem 7.2.11.

Example 7.2.9 Show that the following sets are linearly dependent.

(a) $\{(1,2), (3,6)\}$

>*Solution:*   Since $(3,6) = 3(1,2)$ then the linear combination $1(3,6) - 3(1,2) = \mathbf{0}$ and the set is linearly dependent. ☐

(b) $\{(2.2, -2.1, 0, 1.5), (-8.8, 8.4, 0, -6)\}$

>*Solution:*   Since $(-8.8, 8.4, 0, -6) = -4(2.2, -2.1, 0, 1.5)$ then the linear combination
>$$(-8.8, 8.4, 0, -6) + 4(2.2, -2.1, 0, 1.5) = \mathbf{0},$$
>and so the set is linearly dependent. ☐

Activity 7.2.10 For what value of c is the set $\{(-3c, -2 + 2c), (1,2)\}$ linearly dependent?

(a) $c = 1$ (b) $c = 0$ (c) $c = \frac{1}{4}$ (d) $c = -\frac{1}{3}$

Theorem 7.2.11 *A set of vectors* $\{v_1 , v_2 , \ldots , v_m\}$ *is linearly dependent if and only if at least one of the vectors can be expressed as a linear combination of the other vectors. In particular, a set of two vectors* $\{v_1, v_2\}$ *is linearly dependent if and only if one of the vectors is a scalar multiple of the other.*

Proof. Exercise 7.2.4 establishes the particular case of a set of two vectors.

In the general case of m vectors, first establish that if one of the vectors can be expressed as a linear combination of the others, then the set is linearly dependent. Let's label the set of vectors so that it is vector v_1 which is a linear combination of the others; that is, $v_1 = c_2 v_2 + c_3 v_3 + \cdots + c_m v_m$. Rearranging, $(-1)v_1 + c_2 v_2 + c_3 v_3 + \cdots + c_m v_m = 0$; that is, there is a non-trivial (as at least $c_1 = -1 \neq 0$) linear combination of the set of vectors which is zero. Hence the set is linearly dependent.

Second, establish the converse. Given the set is linearly dependent, there exist coefficients, not all zero, such that $c_1 v_1 + c_2 v_2 + \cdots + c_m v_m = 0$. Suppose that we have labelled the vectors so that $c_1 \neq 0$. Then rearranging the equation gives $c_1 v_1 = -c_2 v_2 - c_3 v_3 - \cdots - c_m v_m$. Divide by the nonzero c_1 to deduce $v_1 = -(c_2/c_1)v_2 - (c_3/c_1)v_3 - \cdots - (c_m/c_1)v_m$; that is, v_1 is a linear combination of the other vectors. ☐

Example 7.2.12 Invoke Theorem 7.2.11 to deduce whether the following sets are linearly independent or linearly dependent.

(a) $\{(-1,1,0), (1,-2,1), (0,1,-1)\}$

    Solution: Since $(1,-2,1) = -(-1,1,0) - (0,1,-1)$ the set must be linearly dependent. ☐

(b) The set of two vectors shown to the right.

    Solution: Since they are not proportional to each other, we cannot write either as a scalar multiple of the other, and so the pair are linearly independent. ☐

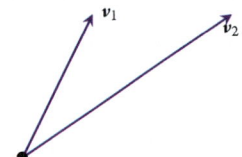

(c) The set of two vectors shown to the right.

    Solution: Since they appear proportional to each other, $v_2 \approx (-3)v_1$, so the pair appear linearly dependent. ☐

(d) $\{(1,3,0,-1), (1,0,-4,2), (-2,3,0,-3), (0,6,-4,-2)\}$

    Solution: Notice that the last vector is the sum of the first three, $(0,6,-4,-2) = (1,3,0,-1) + (1,0,-4,2) + (-2,3,0,-3)$, and so the set is linearly dependent. ☐

Recall that Theorem 4.2.11 established that for every two distinct eigenvalues of a symmetric matrix A, any corresponding two eigenvectors are orthogonal. Consequently, for a symmetric matrix A, a set of eigenvectors from distinct eigenvalues forms an orthogonal set. The following

Theorem 7.2.13 generalizes this property to non-symmetric matrices using the concept of linear independence. That the corresponding eigenvalues are all different is crucial.

Theorem 7.2.13 *For every $n \times n$ matrix A, let λ_1, λ_2, ..., λ_m be distinct eigenvalues of A with corresponding eigenvectors v_1, v_2, ..., v_m. Then the set $\{v_1 , v_2 , \ldots , v_m\}$ is linearly independent.*

Proof. Use contradiction. Assume the set $\{v_1 , v_2 , \ldots , v_m\}$ is linearly dependent. Choose $k < m$ such that the set $\{v_1 , v_2 , \ldots , v_k\}$ is linearly independent, whereas the set $\{v_1 , v_2 , \ldots , v_{k+1}\}$ is linearly dependent. Hence there exists non-trivial coefficients such that

$$c_1 v_1 + c_2 v_2 + \cdots + c_k v_k + c_{k+1} v_{k+1} = 0;$$

further, $c_{k+1} \neq 0$ as $\{v_1 , v_2 , \ldots , v_k\}$ is linearly independent. Pre-multiply the linear combination by matrix A:

$$c_1 A v_1 + c_2 A v_2 + \cdots + c_k A v_k + c_{k+1} A v_{k+1} = A 0$$
$$\implies c_1 \lambda_1 v_1 + c_2 \lambda_2 v_2 + \cdots + c_k \lambda_k v_k + c_{k+1} \lambda_{k+1} v_{k+1} = 0.$$

Now subtract $\lambda_{k+1} \times$ the original linear combination:

$$c_1 \lambda_1 v_1 + c_2 \lambda_2 v_2 + \cdots + c_k \lambda_k v_k + c_{k+1} \lambda_{k+1} v_{k+1}$$
$$- (c_1 \lambda_{k+1} v_1 + c_2 \lambda_{k+1} v_2 + \cdots + c_k \lambda_{k+1} v_k + c_{k+1} \lambda_{k+1} v_{k+1}) = 0$$
$$\implies c_1 (\lambda_1 - \lambda_{k+1}) v_1 + c_2 (\lambda_2 - \lambda_{k+1}) v_2 + \cdots$$
$$+ c_k (\lambda_k - \lambda_{k+1}) v_k + c_{k+1} \underbrace{(\lambda_{k+1} - \lambda_{k+1})}_{=0} v_{k+1} = 0$$
$$\implies c_1' v_1 + c_2' v_2 + \cdots + c_k' v_k = 0$$

for coefficients $c_j' = c_j (\lambda_j - \lambda_{k+1})$. Since all the eigenvalues are distinct, $\lambda_j - \lambda_{k+1} \neq 0$, and since the coefficients c_j are not all zero, hence c_j' are not all zero. Thus we have created a non-trivial linear combination of v_1, v_2, ..., v_k which is zero, and so the set $\{v_1 , v_2 , \ldots , v_k\}$ is linearly dependent. This contradiction of the choice of k proves that the assumption must be wrong. Hence the set $\{v_1 , v_2 , \ldots , v_m\}$ is linearly independent, as required.

Activity 7.2.14 The matrix $\begin{bmatrix} 2 & 1 \\ a^2 & 2 \end{bmatrix}$ has eigenvectors proportional to $(1, a)$, and proportional to $(1, -a)$. For what values of a does the matrix have two distinct eigenvalues?

(a) $a \neq 2$ (b) $a \neq 1$ (c) $a \neq -1$ (d) $a \neq 0$

Example 7.2.15 For each of the following matrices, show that the eigenvectors from distinct eigenvalues form linearly independent sets.

(a) Consider the matrix $B = \begin{bmatrix} -1 & 1 & -2 \\ -1 & 0 & -1 \\ 0 & -3 & 1 \end{bmatrix}$ from Example 7.1.13.

Solution: In MATLAB/Octave, executing

```
B= [-1  1 -2
    -1  0 -1
     0 -3  1]
[V,D]=eig(B)
```

gives eigenvectors and corresponding eigenvalues in

V =

-0.5774	0.7071	-0.7071
-0.5774	0.0000	0.0000
-0.5774	-0.7071	0.7071

D =

-2	0	0
0	1	0
0	0	1

Recognizing $0.7071 = 1/\sqrt{2}$, the last two eigenvectors, $(1/\sqrt{2}, 0, -1/\sqrt{2})$ and $(-1/\sqrt{2}, 0, 1/\sqrt{2})$, form a linearly dependent set because they are proportional to each other. This linear dependence does not confound Theorem 7.2.16 because the corresponding eigenvalues are the same, not distinct, namely $\lambda = 1$. The theorem only applies to eigenvectors of distinct eigenvalues.

Here the two distinct eigenvalues are $\lambda = -2$ and $\lambda = 1$. Recognizing $0.5774 = 1/\sqrt{3}$, two corresponding eigenvectors are $(-1/\sqrt{3}, -1/\sqrt{3}, -1/\sqrt{3})$ and $(1/\sqrt{2}, 0, -1/\sqrt{2})$. Because the zero component in the second corresponds to a nonzero component in the first, these cannot be proportional to each other, and so the pair form a linearly independent set. □

(b) Example 7.1.14 found the eigenvalues and eigenvectors of matrix

$$A = \begin{bmatrix} 0 & 3 & 0 & 0 & 0 \\ 1 & 0 & 3 & 0 & 0 \\ 0 & 1 & 0 & 3 & 0 \\ 0 & 0 & 1 & 0 & 3 \\ 0 & 0 & 0 & 1 & 0 \end{bmatrix}$$

In MATLAB/Octave execute

```
A= [0 3 0 0 0
    1 0 3 0 0
    0 1 0 3 0
    0 0 1 0 3
    0 0 0 1 0]
[V,D]=eig(A)
```

to obtain the report (2 d.p.)

```
V =
    0.62   -0.62    0.94   -0.85   -0.85
    0.62    0.62   -0.00    0.49   -0.49
    0.42   -0.42   -0.31   -0.00    0.00
    0.21    0.21   -0.00   -0.16    0.16
    0.07   -0.07    0.10    0.09    0.09
D =
    3.00       0       0       0       0
       0   -3.00       0       0       0
       0       0   -0.00       0       0
       0       0       0   -1.73       0
       0       0       0       0    1.73
```

The five eigenvalues are all distinct, so Theorem 7.2.16 asserts that a set of corresponding eigenvectors is linearly independent. The five columns of V, call them v_1, v_2, ..., v_5, are a set of corresponding eigenvectors.

To confirm their linear independence let's seek a linear combination being zero, that is, $c_1 v_1 + c_2 v_2 + \cdots + c_5 v_5 = 0$. Written as a matrix-vector system we seek $c = (c_1, c_2, \ldots, c_5)$ such that $Vc = 0$. Because the five singular values of square matrix V are all nonzero,[9] obtained from svd (V) as

```
ans =
    1.7703
    1.1268
    0.6542
    0.3625
    0.1922
```

consequently Theorem 3.4.43 asserts that $Vc = 0$ has only the zero solution. Hence, by Definition 7.2.4 the set of eigenvectors in the columns of V is linearly independent.

This last case of the preceding Example 7.2.15(b) connects the concept of linear in/dependence to the existence or otherwise of nonzero solutions to a homogeneous system of linear equations, $Vc = 0$. So does Example 7.2.5(b). The great utility of this connection is that we understand a lot about homogeneous systems of linear equations. The next Theorem 7.2.16 establishes this connection in general.

Theorem 7.2.16   Let v_1, v_2, ..., v_m be vectors in $\mathbb{R}^n$, and let the $n \times m$ matrix $V = \begin{bmatrix} v_1 & v_2 & \cdots & v_m \end{bmatrix}$. Then the set $\{v_1, v_2, \ldots, v_m\}$ is linearly dependent if and only if the homogeneous system $Vc = 0$ has a nonzero solution c.

Proof.   Now $\{v_1, v_2, \ldots, v_m\}$ is linearly dependent if and only if there are scalars, not all zero, such that the equation $c_1 v_1 + c_2 v_2 + \cdots + c_m v_m = 0$ holds (Definition 7.2.4). Let the vector

---

[9] One could alternatively compute the determinant det (V) $= 0.09090$ and because it is nonzero Theorem 7.2.41 asserts that the equation has only the solution $c = 0$.

$c = (c_1, c_2, \ldots, c_m)$, then this equation is equivalent to the statement $Vc = 0$. That is, if and only if $Vc = 0$ has a nonzero solution. ☐

Recall Theorem 1.3.25, that in $\mathbb{R}^n$ there can be no more that n vectors in an orthogonal set of vectors. The following theorem is the generalization: in $\mathbb{R}^n$ there can be no more than n vectors in a linearly independent set of vectors.

Activity 7.2.17 Which of the following sets of vectors are linearly dependent?

(a)

(b)

(c) None of these sets.

(d)

Theorem 7.2.18 *Every set of m vectors in $\mathbb{R}^n$ is linearly dependent when the number of vectors $m > n$.*

Proof. Form the m vectors $v_1, v_2, \ldots, v_m \in \mathbb{R}^n$ into the $n \times m$ matrix $V = \begin{bmatrix} v_1 & v_2 & \cdots & v_m \end{bmatrix}$. Consider the homogeneous system $Vc = 0$: as $m > n$, Theorem 2.2.30 (with the meaning of m and n swapped) asserts that $Vc = 0$ has infinitely many solutions. Thus $Vc = 0$ has nonzero solutions, so Theorem 7.2.16 implies that the set of eigenvectors $\{v_1, v_2, \ldots, v_m\}$ is linearly dependent. ☐

Example 7.2.19 Determine if the following sets of vectors are linearly dependent or independent. Give reasons.

(a) $\{(-1,-2), (-1,4), (0,5), (2,3)\}$

    Solution: As there are four vectors in $\mathbb{R}^2$ so Theorem 7.2.18 asserts the set is linearly dependent. ☐

(b) $\{(-6,-4,-1,-2), (2,0,1,-2), (2,-1,-1,1)\}$

    Solution: In MATLAB/Octave form the matrix with these vectors as columns

```
V= [-6    2    2
    -4    0   -1
    -1    1   -1
    -2   -2    1]
svd (V)
```

    and find that the three singular values are all nonzero (namely 7.7568, 2.7474, and 2.2988). Hence there are *no* free variables when solving $Vc = 0$ (Procedure 3.3.15), and consequently there is only the unique solution $c = 0$. By Theorem 7.2.16, the set of vectors is linearly independent. □

(c) $\{(-1,-2,2,-1), (1,3,1,-1), (-2,-4,4,-2)\}$

Solution: By inspection, the third vector is twice the first. Hence the linear combination $2(-1,-2,2,-1) + 0(1,3,1,-1) - (-2,-4,4,-2) = 0$ and so the set of vectors is linearly dependent. □

(d) $\{(3,3,-1,-1), (0,-3,-1,-7), (1,2,0,2)\}$

Solution: In MATLAB/Octave form the matrix with these vectors as columns

```
V= [3    0    1
    3   -3    2
   -1   -1    0
   -1   -7    2]
svd (V)
```

and find that the three singular values are 8.1393, 4.6638, and 0.0000. The zero singular value implies that there is a free variable when solving $Vc = 0$ (Procedure 3.3.15), and consequently that there are infinitely many nonzero c that solve $Vc = 0$. By Theorem 7.2.16, the set of vectors is linearly dependent. □

(e) $\{(10,3,3,1), (2,-3,0,-1), (1,-1,2,-1), (2,-1,-3,0), (-2,0,2,-1)\}$

Solution: There are five vectors in $\mathbb{R}^4$ so Theorem 7.2.18 asserts that the set is linearly dependent. □

(f) $\{(-0.4, -1.8, -0.2, 0.7, -0.2), (-1.1, 2.8, 2.7, -3.0, -2.6), (-2.3, -2.3, 4.1, 3.4, -1.6), (-2.6, -5.3, -3.3, -1.3, -4.1), (1.4, 5.2, -6.9, -0.7, 0.6)\}$

Solution: In MATLAB/Octave form the matrix V with these vectors as columns

```
V= [-0.4  -1.1  -2.3  -2.6  1.4
    -1.8   2.8  -2.3  -5.3   5.2
    -0.2   2.7   4.1  -3.3  -6.9
     0.7  -3.0   3.4  -1.3  -0.7
    -0.2  -2.6  -1.6  -4.1   0.6]
svd (V)
```

and find that the five singular values are 10.6978, 8.0250, 5.5920, 3.0277, and 0.0024. As the singular values are all nonzero, the homogeneous system $Vc = 0$ has the unique solution $c = 0$ (Procedure 3.3.15), and hence the set of five vectors are linearly independent.

However, the answer depends upon the context. In the strict mathematical context the vectors are unequivocally linearly independent. But in the context of practical problems, where errors in matrix entries are likely, there are 'shades of grey'. Here, one of the singular values is quite small, namely 0.0024. If the context informs us that the entries in the matrix had errors of say 0.01, then this singular value is effectively zero (Section 5.2). In the context of such errors, this set of five vectors would be linearly dependent. ☐

7.2.2 Form a basis for subspaces

Recall that Sections 2.3 and 3.4 defined subspaces and the span, namely that a subspace is a set of vectors closed under addition and scalar multiplication, and a span gives a subspace as all linear combinations of a set of vectors. Also, Definition 3.4.18 defined an "orthonormal basis" for a subspace to be a set of orthonormal vectors that span the subspace. This section generalizes the concept of an "orthonormal basis" by relaxing the requirement of orthonormality to result in the concept of a "basis".

Definition 7.2.20 *A **basis** for a subspace* $\mathbb{W}$ *of* $\mathbb{R}^n$ *is a set of vectors such that the set both spans* $\mathbb{W}$ *and is linearly independent.*

Example 7.2.21

(a) Recall that Example 7.2.5(b) and Example 7.2.1 showed that the two vectors $(2,1)$ and $(1,2)$ are linearly independent and span $\mathbb{R}^2$. Hence the set $\{(2,1)\,,\,(1,2)\}$ is a basis of $\mathbb{R}^2$.

(b) Recall that Example 7.2.5(a) showed that the set $\{(-1,1,0)\,,\,(1,-2,1)\,,\,(0,1,-1)\}$ is linearly dependent, so this set cannot be a basis.

   However, remove one vector from the set, such as the middle one, and consider the set $\{(-1,1,0)\,,\,(0,1,-1)\}$. As the two vectors are not proportional to each other, this set is linearly independent (Theorem 7.2.11). Also, the plane $x+y+z=0$ is a subspace, say $\mathbb{W}$. The plane $\mathbb{W}$ is characterized by $y=-x-z$. So every vector in $\mathbb{W}$ can be written as $(x,-x-z,z)=(x,-x,0)+(0,-z,z)=(-x)(-1,1,0)+(-z)(0,1,-1)$. That is, $\mathrm{span}\{(-1,1,0)\,,\,(0,1,-1)\}=\mathbb{W}$. Hence $\{(-1,1,0)\,,\,(0,1,-1)\}$ is a basis for the plane $\mathbb{W}$.

(c) Find a basis for the line given parametrically as $x=2.1t$, $y=1.3t$ and $z=-1.1t$ (shown to the right in stereo).

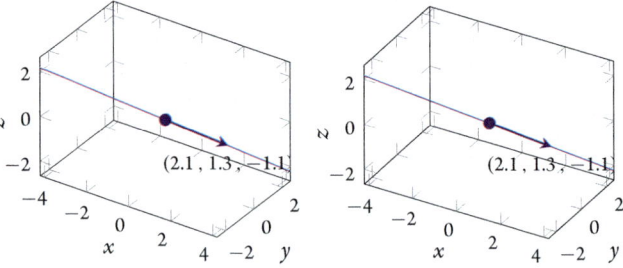

   Solution: The vectors in the line may be written as
$x=(x,y,z)=(2.1t,\,1.3t,\,-1.1t)=(2.1\,,\,1.3\,,\,-1.1)t$. Since the parameter t may vary over all values, vectors in the line form $\mathrm{span}\{(2.1\,,\,1.3\,,\,-1.1)\}$. Since $\{(2.1\,,\,1.3\,,\,-1.1)\}$ is a linearly independent set of vectors (Example 7.2.5(c)), it thus forms a basis for the vectors in the given line. ☐

(d) Find a basis for the line given parametrically as $x = 5.7t - 0.6$ and $y = 6.8t + 2.4$.

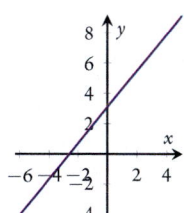

Solution: The vectors in the line may be written as $x = (5.7t - 0.6, 6.8t + 2.4)$. But this does not form a subspace as it does not include the zero vector $\mathbf{0}$ (as illustrated to the right). Since this line is not a subspace, it cannot have a basis. □

(e) Find a basis for the plane $3x - 2y + z = 0$.

Solution: Writing the equation of the plane as $z = -3x + 2y$ we

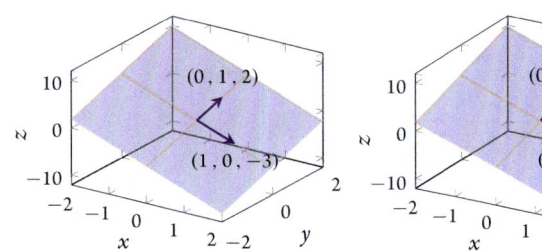

then write the plane parametrically (Section 1.3.4) as the vectors $x = (x, y, -3x + 2y) = (x, 0, -3x) + (0, y, 2y) = x(1, 0, -3) + y(0, 1, 2)$. Since x and y may vary over all values, the plane is the subspace span$\{(1, 0, -3), (0, 1, 2)\}$ (as illustrated above-right in stereo). Since $(1, 0, -3)$ and $(0, 1, 2)$ are not proportional to each other, they form a linearly independent set. Hence $\{(1, 0, -3), (0, 1, 2)\}$ is a basis for the plane. □

(f) Prove that every orthonormal basis of a subspace $\mathbb{W}$ is also a basis of $\mathbb{W}$.

Solution: Theorem 7.2.8 establishes that every orthonormal basis is linearly independent. By Definition 3.4.18, an orthonormal basis of $\mathbb{W}$ spans $\mathbb{W}$. Hence an orthonormal basis of $\mathbb{W}$ is also a basis of $\mathbb{W}$. □

Activity 7.2.22 Which of the following sets of vectors forms a basis for $\mathbb{R}^2$, but is not an orthonormal basis for $\mathbb{R}^2$?

(a)   (b)

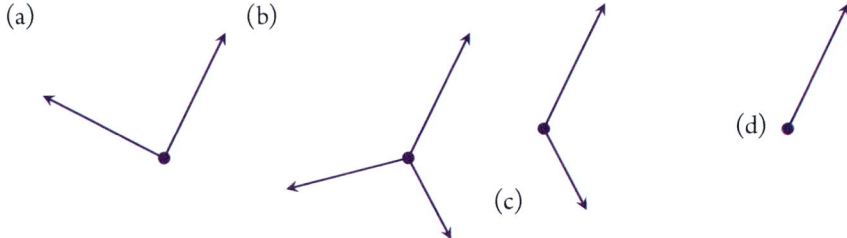

(d)

(c)

Recall that Theorem 3.4.28 establishes that an orthonormal basis of a given subspace always has the same number of vectors. The following theorem establishes that the same is true for general bases. The proof directly generalizes that for Theorem 3.4.28.

Theorem 7.2.23 *Every basis for a given subspace has the same number of vectors.*

Proof. Let $\mathcal{U} = \{u_1 , u_2 , \ldots , u_r\}$ and $\mathcal{V} = \{v_1 , v_2 , \ldots , v_s\}$ be any two bases for a subspace in $\mathbb{R}^n$. We prove the number of vectors $r = s$ by contradiction. In the first case, assume $r < s$. Since $\mathcal{U}$ is a basis for the subspace every vector in the set $\mathcal{V}$ can be written as a linear combination of vectors in $\mathcal{U}$ with some coefficients a_{ij}:

$$v_1 = u_1 a_{11} + u_2 a_{21} + \cdots + u_r a_{r1},$$
$$v_2 = u_1 a_{12} + u_2 a_{22} + \cdots + u_r a_{r2},$$
$$\vdots$$
$$v_s = u_1 a_{1s} + u_2 a_{2s} + \cdots + u_r a_{rs}.$$

Write each of these, such as the first one, in the form

$$v_1 = \begin{bmatrix} u_1 & u_2 & \cdots & u_r \end{bmatrix} \begin{bmatrix} a_{11} \\ a_{21} \\ \vdots \\ a_{r1} \end{bmatrix} = U a_1,$$

where $n \times r$ matrix $U = \begin{bmatrix} u_1 & u_2 & \cdots & u_r \end{bmatrix}$. Similarly for the other equations $v_2 = \cdots = U a_2$ through to $v_s = \cdots = U a_s$. Then the $n \times s$ matrix

$$V = \begin{bmatrix} v_1 & v_2 & \cdots & v_s \end{bmatrix} = \begin{bmatrix} U a_1 & U a_2 & \cdots & U a_s \end{bmatrix} = U \begin{bmatrix} a_1 & a_2 & \cdots & a_s \end{bmatrix} = UA$$

for the $r \times s$ matrix $A = \begin{bmatrix} a_1 & a_2 & \cdots & a_s \end{bmatrix}$. By assumption $r < s$ and so Theorem 2.2.30 assures us that the homogeneous system $Ax = 0$ has infinitely many solutions; choose any non-trivial solution $x \neq 0$. Consider

$$Vx = UAx \quad \text{(from above)}$$
$$= U0 \quad \text{(since } Ax = 0\text{)}$$
$$= 0.$$

The identity $Vx = 0$ implies that there is a nontrivial linear combination of the columns v_1, $v_2 , \ldots , v_s$ of V which gives zero, hence the set $\mathcal{V}$ is linearly dependent (Theorem 7.2.16). But this is a contradiction, so we cannot have $r < s$.

Second, a corresponding argument establishes we cannot have $s < r$. Hence $s = r$: all bases of a given subspace must have the same number of vectors. □

Example 7.2.24 Consider the plane $x+y+z=0$ in $\mathbb{R}^3$. Each of the following is a basis for the plane:

- $\{(-1,1,0), (1,-2,1)\}$;
- $\{(1,-2,1), (0,1,-1)\}$;
- $\{(0,1,-1), (-1,1,0)\}$.

The reasons are that all three vectors involved are in the plane, and that each pair is linearly independent (in each pair, one is not proportional to the other).

However, consider the set $\{(-1,1,0), (1,-2,1), (0,1,-1)\}$. Although each of the three vectors is in the plane $x+y+z=0$, this set is not a basis because the set is not linearly independent (Example 7.2.5(a)). Each individual vector, say $(-1,1,0)$, cannot form a basis for the plane because the span of one vector, such as $\text{span}\{(-1,1,0)\}$, is a line not the whole plane.

The orthonormal basis $\{(1,0,-1)/\sqrt{2}, (1,-2,1)/\sqrt{6}\}$ is another basis for the plane $x+y+z=0$; both vectors satisfy $x+y+z=0$ and are orthogonal and so linearly independent (Theorem 7.2.8). All these bases possess two vectors. □

That all bases for a given subspace, including orthonormal bases, have the same number of vectors (Theorem 7.2.23) leads to the following theorem about the dimensionality.

Theorem 7.2.25 *For every subspace W of $\mathbb{R}^n$, the* **dimension** *of W, denoted $\dim W$, is the number of vectors in any basis for W.*

Proof. Recall that Definition 3.4.30 defined $\dim W$ to be the number of vectors in any orthonormal basis for W. Theorem 7.2.8 certifies that all orthonormal bases are also bases (Definition 7.2.20), so Theorem 7.2.23 implies that every basis of W has $\dim W$ vectors. □

Activity 7.2.26 Which of the following sets forms a basis for a subspace of dimension two?

(a) $\{(1,-2,1), (1,0,-1)\}$

(b) $\{(1,1,-2), (2,2,-4)\}$

(c) $\{(-1,0,2), (0,0,1), (-1,2,0)\}$

(d) $\{(1,2)\}$

Procedure 7.2.27 (basis for a span) *Find a basis for the subspace $\mathbb{A} = \text{span}\{a_1, a_2, \ldots, a_n\}$ given that $\{a_1, a_2, \ldots, a_n\}$ is a set of n vectors in $\mathbb{R}^m$. Recall Procedure 3.4.23 underpins finding an orthonormal basis by the following.*

1. *Form $m \times n$ matrix $A := \begin{bmatrix} a_1 & a_2 & \cdots & a_n \end{bmatrix}$.*

2. *Factorize A into its SVD, $A = USV^{\mathsf{T}}$, and let $r = \text{rank} A$ be the number of nonzero singular values (or effectively nonzero when the matrix has experimental errors, Section 5.2).*

3. *The set $\{u_1, u_2, \ldots, u_r\}$ (where u_j denotes the columns of U) is a basis, specifically an orthonormal basis, for the r-dimensional subspace $\mathbb{A}$.*

Alternatively, if the rank $r = n$, then the set $\{a_1, a_2, \ldots, a_n\}$ is linearly independent and spans the subspace $\mathbb{A}$, and so is also a basis for the n-dimensional subspace $\mathbb{A}$.

Example 7.2.28 Apply Procedure 7.2.27 to find a basis for the following sets.

(a) Recall that Example 7.2.24 identified that every pair of vectors in the set $\{(-1,1,0),$ $(1,-2,1),$ $(0,1,-1)\}$ forms a basis for the plane that they span. Find another basis for the plane.

Solution: In MATLAB/Octave form the matrix with these vectors as columns:

```
A= [-1  1  0
     1 -2  1
     0  1 -1]
[U,S,V] =svd (A)
```

Then the SVD obtains (2 d.p.)

```
U =
  -0.41   -0.71    0.58
   0.82    0.00    0.58
  -0.41    0.71    0.58
S =
   3.00       0        0
      0    1.00        0
      0       0     0.00
V = ...
```

The two nonzero singular values determine rank $A = 2$ and hence the first two columns of U form an (orthonormal) basis for span$\{(-1,1,0),$ $(1,-2,1),$ $(0,1,-1)\}$. That is, $\{0.41(-1,2,-1),$ $0.71(-1,0,1)\}$ is a (orthonormal) basis for the two-dimensional plane. □

(b) Find a basis for the span of the three vectors

$$(-2,0,-4,1,1),\ (7,1,2,-1,-5),\ (-5,-1,2,3,-2).$$

Solution: In MATLAB/Octave it is often easiest to enter such vectors as rows, and then transpose with the dash operator to form the matrix with the vectors as columns:

```
A= [ -2  0 -4  1  1
      7  1  2 -1 -5
     -5 -1  2  3 -2]'
[U,S,V] =svd (A)
```

Then the SVD obtains (2 d.p.)

```
U =
   0.86   -0.32   -0.02    0.40    0.02
   0.12   -0.11   -0.06   -0.40    0.90
   0.22    0.65    0.72    0.07    0.13
```

$$\begin{array}{ccccc} -0.23 & 0.35 & -0.38 & 0.73 & 0.37 \\ -0.38 & -0.59 & 0.58 & 0.37 & 0.19 \end{array}$$

$S =$

$$\begin{array}{ccc} 10.07 & 0 & 0 \\ 0 & 5.87 & 0 \\ 0 & 0 & 3.01 \\ 0 & 0 & 0 \\ 0 & 0 & 0 \end{array}$$

$V = \ldots$

The three nonzero singular values determine rank $A = 3$ and so the original three vectors are linearly independent. Consequently, the original three vectors form a basis for their span.

If you prefer an orthonormal basis, then use the first three columns of U as an orthonormal basis. □

The procedure is different if the subspace of interest is defined by a system of equations instead of the span of some vectors.

Example 7.2.29 Find a basis for the solutions of the system in $\mathbb{R}^3$ of $3x + y = 0$ and $3x + 2y + 3z = 0$.

Solution: By hand manipulation, the first equation gives $y = -3x$; which when substituted into the second gives $3x - 6x + 3z = 0$, namely $z = x$. That is, all solutions are of the form $(x, -3x, x)$, namely span$\{(1, -3, 1)\}$. Thus a basis for the subspace of solutions is $\{(1, -3, 1)\}$.

Other possible bases are $\{(-1, 3, -1)\}$, $\{(2, -6, 2)\}$, and so on: there are an infinite number of possible answers. □

Example 7.2.30 Find a basis for the solutions of $-2x - y + 3z = 0$ in $\mathbb{R}^3$.

Solution: Rearrange the equation so that one variable is a function of the others, say $y = -2x + 3z$. Then the vector form of solutions are $(x, y, z) = (x, -2x + 3z, z) = (1, -2, 0)x + (0, 3, 1)z$ in terms of free variables x and z. Since $(1, -2, 0)$ and $(0, 3, 1)$ are not proportional to each other, they are linearly independent, and so a basis for the solutions is $\{(1, -2, 0), (0, 3, 1)\}$. (Infinitely many other bases are possible answers.) □

Activity 7.2.31 Which of the following is *not* a basis for the line $3x + 7y = 0$?

(a) $\{(-7, 3)\}$ (b) $\{(3, 7)\}$ (c) $\{(1, -\frac{3}{7})\}$ (d) $\{(-\frac{7}{3}, 1)\}$

Procedure 7.2.32 (basis from equations) *Suppose we seek a basis for a subspace W specified as the solutions of a system of equations.*

1. *Rewrite the system of equations as the homogeneous system $Ax = 0$. Then the subspace W is the nullspace of $m \times n$ matrix A.*

2. *Adapting Procedure 3.3.15 for the specific case of homogeneous systems, first find an SVD factorization $A = USV^{\mathsf{T}}$ and let $r = \operatorname{rank} A$ be the number of nonzero singular values (or effectively nonzero when the matrix has experimental errors, Section 5.2).*

3. *Then $y = (0,\ldots,0,y_{r+1},\ldots,y_n)$ is a general solution of $Sy = z = 0$. Consequently, all possible solutions $x = Vy$ are spanned by the last $n - r$ columns of V, which thus form an orthonormal basis for the subspace $\mathbb{W}$.*

Example 7.2.33 Find a basis for all solutions to each of the following systems of equations.

(a) $3x + y = 0$ and $3x + 2y + 3z = 0$ from Example 7.2.29.

Solution: Form matrix $A = \begin{bmatrix} 3 & 1 & 0 \\ 3 & 2 & 3 \end{bmatrix}$ and compute an SVD with $[U, S, V] = \mathrm{svd}\,(A)$ to obtain (2 d.p.)

```
U = ...
S =
    5.34        0        0
       0     1.86        0
V =
    0.77     0.56     0.30
    0.42    -0.09    -0.90
    0.48    -0.82     0.30
```

The two nonzero singular values determine $\operatorname{rank} A = 2$. Hence the solutions of the system are spanned by the last one column of V. That is, a basis for the solutions is $\{(0.3, -0.9, 0.3)\}$. ☐

(b) $7x = 6y + z + 3$ and $4x + 9y + 2z + 2 = 0$.

Solution: This system is not homogeneous (due to the constant terms, Definition 2.2.27), therefore $x = 0$ is not a solution. Consequently, the solutions of the system cannot form a subspace (Definition 3.4.3). Thus the concept of a basis does not apply (Definition 7.2.20). ☐

(c) $w + x = z$, $3w = x + y + 5z$, $4x + y + 2z = 0$.

Solution: Rearrange to the matrix-vector system $Ax = 0$ for vector $x = (w, x, y, z) \in \mathbb{R}^4$ and matrix

```
A = [1   1   0  -1
     3  -1  -1  -5
     0   4   1   2]
```

Enter into MATLAB/Octave as above and then find an SVD with [U,S,V]=svd(A) to obtain (2 d.p.)

```
U  =  ...
S  =
      6.77        0        0        0
         0     3.76        0        0
         0        0     0.00        0
V  =
     -0.40     0.45     0.09     0.80
      0.41     0.86    -0.19    -0.25
      0.20     0.10     0.97    -0.07
      0.80    -0.24    -0.10     0.54
```

There are two nonzero singular values, so rank $A = 2$. There are thus $4 - 2 = 2$ free variables in solving $Ax = 0$ leading to a 2D subspace with orthonormal basis of the last two columns of V. That is, an orthonormal basis for the subspace of all solutions is $\{(0.09, -0.19, 0.97, -0.10), (0.80, -0.25, -0.07, 0.54)\}$. □

Recall that this Section 7.2 started by discussing the need to have a unique representation of solutions to problems. If we do not have uniqueness, then the ambiguity in algebraic representation ruins basic algebra. The forthcoming theorem assures us that the linear independence of a basis ensures the unique representation that we need. In essence it says that every basis, whether orthogonal or not, can be used to form a coordinate system.

Example 7.2.34 *(a tale of two coordinate systems)* To the right are plotted three vectors and the origin. Take the view that these are fixed physically meaningful vectors; the issue of this example is how do we code such vectors in mathematics?

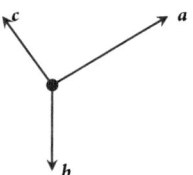

In the orthogonal standard coordinate system these three vectors and the origin have coordinates as plotted on the right by their end-points. Consequently, we write $a = (7,4)$, $b = (0,-5)$, and $c = (-3,4)$.

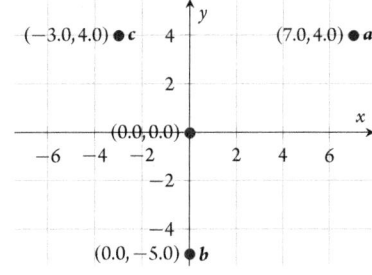

Now use the (red) basis $\mathcal{B} = \{v_1, v_2\}$ to form a non-orthogonal coordinate system (represented by the dotted grid). Then in this system the three vectors have coordinates $a = (2,1)$, $b = (1,-3)$, and $c = (-2,3)$.

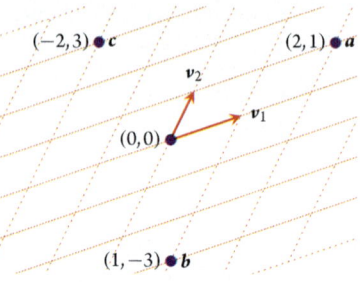

But surely we cannot say both $a = (7,4)$ and $a = (2,1)$; it appears nonsense. The reason for the different coordinate numbers representing the one vector a is that the underlying coordinate systems are different. For example, we can say both $a = 7e_1 + 4e_2$ and $a = 2v_1 + v_2$ without any contradiction: these two statements explicitly recognize the underlying standard unit vectors in the first expression and the underlying non-orthogonal basis vectors in the second.

Consequently, mathematicians, scientists, and engineers invented a more precise notation. We write $[a]_\mathcal{B} = (2,1)$ to represent that the coordinates of vector a in the basis $\mathcal{B}$ are $(2,1)$. Correspondingly, letting $\mathcal{E} = \{e_1, e_2\}$ denote the basis of the standard unit vectors, we write $[a]_\mathcal{E} = (7,4)$ to represent that the coordinates of vector a in the standard basis $\mathcal{E}$ are $(7,4)$. Similarly, $[b]_\mathcal{E} = (0,-5)$ and $[b]_\mathcal{B} = (1,-3)$; and $[c]_\mathcal{E} = (-3,4)$ and $[c]_\mathcal{B} = (-2,3)$.

The endemic practice of just writing $a = (7,4)$, $b = (0,-5)$ and $c = (-3,4)$ is rationalized in this more precise notation by the convention that if no basis is explicitly specified, then we assume the standard basis $\mathcal{E}$.[10]

Theorem 7.2.35 *For any given subspace* $\mathbb{W}$ *of* $\mathbb{R}^n$ *let* $\mathcal{B} = \{v_1, v_2, \ldots, v_k\}$ *be a basis for* $\mathbb{W}$. *Then there is exactly one way to write each and every vector* $w \in \mathbb{W}$ *as a linear combination of the basis vectors:* $w = c_1 v_1 + c_2 v_2 + \cdots + c_k v_k$. *The coefficients* $c_1, c_2, \ldots, c_k$ *are called the* **coordinates of** w **with respect to** $\mathcal{B}$, *and the column vector* $[w]_\mathcal{B} = (c_1, c_2, \ldots, c_k)$ *is called the* **coordinate vector of** w **with respect to** $\mathcal{B}$.

Proof. Consider any vector $w \in \mathbb{W}$. Since $\{v_1, v_2, \ldots, v_k\}$ is a basis for the subspace $\mathbb{W}$, w can be written as a linear combination of the basis vectors. Let two such linear combinations be

$$w = c_1 v_1 + c_2 v_2 + \cdots + c_k v_k,$$
$$w = d_1 v_1 + d_2 v_2 + \cdots + d_k v_k.$$

Subtract the second of these equations from the first, grouping common vectors:

$$0 = (c_1 - d_1) v_1 + (c_2 - d_2) v_2 + \cdots + (c_k - d_k) v_k.$$

Since $\{v_1, v_2, \ldots, v_k\}$ is linearly independent, this equation implies all the coefficients in parentheses are zero:

$$0 = (c_1 - d_1) = (c_2 - d_2) = \cdots = (c_k - d_k).$$

---

[10] Given that the values of the components in a vector change with the coordinate basis, some of you may wonder whether the same thing happens for matrices. The answer is yes: for a given linear transformation (Section 3.6), the values of the components in its matrix also depend upon the underlying coordinate basis.

That is, $c_1 = d_1$, $c_2 = d_2$, ..., $c_k = d_k$, and the two linear combinations are identical. Consequently, there is exactly one way to write a vector $w \in W$ as a linear combination of the basis vectors. □

Example 7.2.36

(a) Consider the diagram to the right of six labelled vectors. Estimate the coordinates of the four shown vectors a, b, c, and d in the shown basis $\mathcal{B} = \{v_1, v_2\}$.

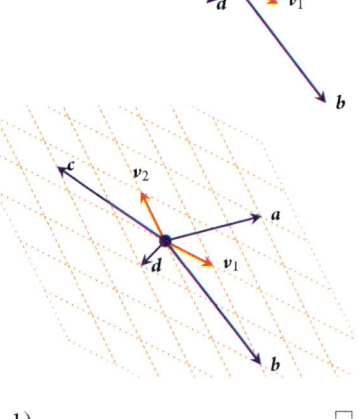

Solution: Draw in a grid corresponding to multiples of v_1 and v_2 in both directions, and parallel to v_1 and v_2, as shown to the right. Then from the grid, estimate that $a \approx 3v_1 + 2v_2$ hence the coordinates $[a]_\mathcal{B} \approx (3, 2)$. Similarly, $b \approx v_1 - 2v_2$ hence the coordinates $[b]_\mathcal{B} \approx (1, -2)$. Also, $c \approx -2v_1 + 0.5v_2$ hence the coordinates $[c]_\mathcal{B} \approx (-2, 0.5)$. And lastly, $d \approx -v_1 - v_2$ hence the coordinates $[d]_\mathcal{B} \approx (-1, -1)$. □

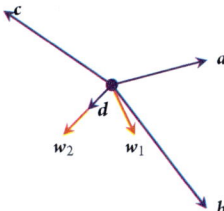

(b) Consider the same four vectors but with a pair of different basis vectors: let's see that although the vectors are the same, the coordinates in the different basis are different. Estimate the coordinates of the four shown vectors a, b, c, and d in the shown basis $\mathcal{W} = \{w_1, w_2\}$.

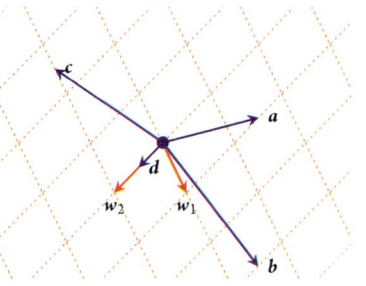

Solution: Draw in a grid corresponding to multiples of w_1 and w_2 in both directions, and parallel to w_1 and w_2, as shown to the right. Then from the grid, estimate that $a \approx w_1 - 1.5w_2$ hence the coordinates $[a]_\mathcal{W} \approx (1, -1.5)$. Similarly, $b \approx 3w_1 - 0.5w_2$ hence the coordinates $[b]_\mathcal{W} \approx (3, -0.5)$. Also, $c \approx -2.5w_1 + w_2$ hence the coordinates $[c]_\mathcal{W} \approx (-2.5, 1)$. And lastly, $d \approx 0.5w_2$ hence the coordinates $[d]_\mathcal{W} \approx (0, 0.5)$. □

Activity 7.2.37 For the vector x shown to the right, estimate the coordinates of x in the shown basis $B = \{b_1, b_2\}$.

(a) $[x]_B = (3, 2)$

(b) $[x]_B = (4, 1)$

(c) $[x]_B = (2, 3)$

(d) $[x]_B = (1, 4)$

Example 7.2.38 Let the basis $B = \{v_1, v_2, v_3\}$ for the three given vectors $v_1 = (-1, 1, -1)$, $v_2 = (1, -2, 0)$, and $v_3 = (0, 4, 5)$ (each of these is specified in the standard basis $\mathcal{E}$ of the standard unit vectors e_1, e_2, and e_3).

(a) What is the vector with coordinates $[a]_B = (3, -2, 1)$?

Solution: Since a coordinate system is not specified for the answer, we answer with the default of the standard basis $\mathcal{E}$. The vector $a = 3v_1 - 2v_2 + v_3$ which has standard coordinates $[a]_{\mathcal{E}} = 3(-1, 1, -1) - 2(1, -2, 0) + (0, 4, 5) = (-5, 11, 2)$. ☐

(b) What is the vector with coordinates $[b]_B = (-1, 1, 1)$?

Solution: The vector $b = -v_1 + v_2 + v_3$ which has standard coordinates (the default coordinates as the question does not specify) $[b]_{\mathcal{E}} = -(-1, 1, -1) + (1, -2, 0) + (0, 4, 5) = (2, 1, 6)$. ☐

(c) What are the coordinates in the basis B of the vector c where $[c]_{\mathcal{E}} = (-1, 3, 3)$ in the standard basis $\mathcal{E}$?

Solution: We seek coordinate values c_1, c_2, c_3 such that $c = c_1 v_1 + c_2 v_2 + c_3 v_3$. Expressed in the standard basis $\mathcal{E}$ this is the vector equation

$$\begin{bmatrix} -1 \\ 3 \\ 3 \end{bmatrix} = \begin{bmatrix} -1 \\ 1 \\ -1 \end{bmatrix} c_1 + \begin{bmatrix} 1 \\ -2 \\ 0 \end{bmatrix} c_2 + \begin{bmatrix} 0 \\ 4 \\ 5 \end{bmatrix} c_3.$$

A small system like this we solve by hand (Section 2.2.2): write in component form as

$$\begin{cases} -c_1 + c_2 & = -1 \\ c_1 - 2c_2 + 4c_3 = 3 \\ -c_1 \quad\quad + 5c_3 = 3 \end{cases}$$

(add 1st row to 2nd and take from 3rd)

$$\begin{cases} -c_1 + c_2 & = -1 \\ -c_2 + 4c_3 = 2 \\ -c_2 + 5c_3 = 4 \end{cases}$$

(subtract 2nd row from 3rd)

$$\begin{cases} -c_1 + c_2 & = -1 \\ -c_2 + 4c_3 = 2 \\ c_3 = 2 \end{cases}$$

Solving this triangular system gives $c_3 = 2$, $c_2 = 4c_3 - 2 = 6$, and $c_1 = c_2 + 1 = 7$. Thus the coordinates $[c]_B = (7, 6, 2)$ in the basis B. □

(d) What are the coordinates in the basis B of the vector d where $[d]_\mathcal{E} = (-3, 2, 0)$ in the standard basis $\mathcal{E}$?

Solution: We seek coordinate values d_1, d_2, d_3 such that $d = d_1 v_1 + d_2 v_2 + d_3 v_3$. Expressed in the standard basis $\mathcal{E}$ this equation is

$$\begin{bmatrix} -3 \\ 2 \\ 0 \end{bmatrix} = \begin{bmatrix} -1 \\ 1 \\ -1 \end{bmatrix} d_1 + \begin{bmatrix} 1 \\ -2 \\ 0 \end{bmatrix} d_2 + \begin{bmatrix} 0 \\ 4 \\ 5 \end{bmatrix} d_3.$$

To solve this system in MATLAB/Octave (Procedure 2.2.5), enter the matrix (easiest by transposing rows of v_1, v_2, and v_3) and the standard coordinates of d:

```
A= [-1  1  -1
     1 -2  0
     0  4  5]'
d= [-3 ; 2 ; 0]
```

Then compute the coordinates $[d]_B$ with dB=A\d to determine $[d]_B = (20, 17, 4)$.

But remember, before using A\ always check rcond(A), which here is the poor 0.0053 (Procedure 2.2.5). Interestingly, such a poor small value of rcond indicates that although the basis vectors in B are linearly independent, they are 'only just' linearly independent. A small change, or error, might make the basis vectors linearly dependent and thus B would be ruined as a basis for $\mathbb{R}^3$. The poor rcond indicates that B is a poor basis in practical application. □

Activity 7.2.39 What are the coordinates in the basis $B = \{(1,1), (1,-1)\}$ of the vector d where $[d]_\mathcal{E} = (2, -4)$ in the standard basis $\mathcal{E}$?

(a) $[d]_B = (-2, 6)$ (b) $[d]_B = (-1, 3)$ (c) $[d]_B = (1, 3)$ (d) $[d]_B = (3, -1)$

Example 7.2.40 You are given a basis $W = \{w_1, w_2, w_3\}$ for a 3D subspace W of $\mathbb{R}^5$, where the three basis vectors are $w_1 = (1, 3, -4, -3, 3)$, $w_2 = (-4, 1, -2, -4, 1)$, and $w_3 = (-1, 1, 0, 2, -3)$ (in the standard basis $\mathcal{E}$).

(a) What are the coordinates in the standard basis of the vector $a = 2w_1 + 3w_2 + w_3$?

Solution: In the standard basis

$$[a]_\mathcal{E} = 2 \begin{bmatrix} 1 \\ 3 \\ -4 \\ -3 \\ 3 \end{bmatrix} + 3 \begin{bmatrix} -4 \\ 1 \\ -2 \\ -4 \\ 1 \end{bmatrix} + \begin{bmatrix} -1 \\ 1 \\ 0 \\ 2 \\ -3 \end{bmatrix} = \begin{bmatrix} -11 \\ 10 \\ -14 \\ -16 \\ 6 \end{bmatrix}.$$ □

(b) What are the coordinates in the basis $\mathcal{W}$ of the vector $b = (-1, 2, -6, -11, 10)$ (in the standard coordinates $\mathcal{E}$).

Solution: We need to find coefficients c_1, c_2, c_3 such that $b = c_1 w_1 + c_2 w_2 + c_3 w_3$. This forms the set of linear equations

$$
\begin{bmatrix} -1 \\ 2 \\ -6 \\ -11 \\ 10 \end{bmatrix} = \begin{bmatrix} 1 \\ 3 \\ -4 \\ -3 \\ 3 \end{bmatrix} c_1 + \begin{bmatrix} -4 \\ 1 \\ -2 \\ -4 \\ 1 \end{bmatrix} c_2 + \begin{bmatrix} -1 \\ 1 \\ 0 \\ 2 \\ -3 \end{bmatrix} c_3.
$$

Form as the matrix-vector system

$$
\begin{bmatrix} 1 & -4 & -1 \\ 3 & 1 & 1 \\ -4 & -2 & 0 \\ -3 & -4 & 2 \\ 3 & 1 & -3 \end{bmatrix} \begin{bmatrix} c_1 \\ c_2 \\ c_3 \end{bmatrix} = \begin{bmatrix} -1 \\ 2 \\ -6 \\ -11 \\ 10 \end{bmatrix},
$$

and perhaps solve with MATLAB/Octave. Since there are more equations than unknowns, we should use an SVD in order to check that the system is consistent, namely, to check that $b \in \mathbb{W}$.

i. Code the matrix and the vector:

```
W= [1 -4 -1
    3 1 1
   -4 -2 0
   -3 -4 2
    3 1 -3]
b= [-1;2;-6;-11;10]
```

ii. Then obtain an SVD with $[U, S, V] = svd(W)$ (2 d.p.)

```
U =
    0.18   -0.88    0.04   -0.24   -0.37
   -0.32   -0.09    0.65    0.62   -0.28
    0.51    0.11   -0.46    0.56   -0.44
    0.64   -0.18    0.34    0.22    0.63
   -0.44   -0.42   -0.49    0.44    0.45
S =
    8.18       0       0
       0    4.52       0
```

```
              0        0     3.11
              0        0       0
              0        0       0
V =
          -0.74    -0.51    0.43
          -0.62     0.77   -0.15
           0.26     0.38    0.89
```

The three nonzero singular values establish that the three vectors in the basis W are indeed linearly independent (and since no singular value is small, then the vectors are robustly linearly independent).

iii. Find $z = U^T b$ with z=U' *b to get

```
z =
     -15.3413
      -2.2284
      -4.6562
      -0.0000
       0.0000
```

The last two values of z being zero confirm the system of equations is consistent, and so vector b is in the range of W, that is, b is in the subspace W.

iv. Find $y_j = z_j/\sigma_j$ with y=z(1:3)./diag(S) to get

```
y =
     -1.8761
     -0.4929
     -1.4958
```

v. Lastly, find the coefficients $[b]_W = Vy$ with bw=V*y to get

```
bw =
      1.00000
      1.00000
     -2.00000
```

That is, $[b]_W = (1, 1, -2)$.

That $[b]_W$ has three components and $[b]_\varepsilon$ has five components is not a contradiction. The difference in components occurs because the subspace W is 3D but lies in $\mathbb{R}^5$. Using the basis W implicitly builds in the information that the vector b is in a lower dimensional space, and so needs fewer components. □

Revisit unique solutions

Lastly, with all these extra concepts of determinants, eigenvalues, linear independence, and a basis, we now revisit the issue of when there is a unique solution to a set of linear equations.

Theorem 7.2.41 (*Unique Solutions: version 3*) *For every $n \times n$ square matrix A, and extending Theorems 3.3.27 and 3.4.43, the following statements are equivalent:*

(a) *A is invertible;*

(b) *$Ax = b$ has a **unique solution** for every $b \in \mathbb{R}^n$;*

(c) *$Ax = 0$ has only the zero solution;*

(d) *all n singular values of A are nonzero;*

(e) *the condition number of A is finite ($\mathtt{rcond} > 0$);*

(f) *$\operatorname{rank} A = n$;*

(g) *nullity $A = 0$;*

(h) *the column vectors of A span $\mathbb{R}^n$;*

(i) *the row vectors of A span $\mathbb{R}^n$.*

(j) *$\det A \neq 0$;*

(k) *0 is not an eigenvalue of A;*

(l) *the n column vectors of A are linearly independent;*

(m) *the n row vectors of A are linearly independent.*

Proof. Recall Theorems 3.3.27 and 3.4.43 establish the equivalence of 7.2.41(a)–7.2.41(i), and Theorem 6.1.29 proves the equivalence of 7.2.41(a) and 7.2.41(j). To establish that Property 7.2.41(k) is equivalent to 7.2.41(j), recall that Theorem 7.1.4 proves that $\det A$ equals the product of the eigenvalues of A. Hence $\det A$ is not zero if and only if all the eigenvalues are nonzero.

Lastly, Property 7.2.41(h) says the n column vectors span $\mathbb{R}^n$, so they must be a basis, and hence linearly independent. Conversely, if the n columns of A are linearly independent then they must span $\mathbb{R}^n$. Hence Property 7.2.41(l) is equivalent to 7.2.41(h). Similarly for Property 7.2.41(i) and the row vectors of 7.2.41(m). □

7.2.3 Exercises

Exercise 7.2.1 By inspection or basic arguments, decide whether the following sets of vectors are linearly dependent, or linearly independent. Give reasons.

(a) $\{(-2,3,3), (-1,2,-1)\}$

(b) $\{(0,2), (2,-2), (0,-1)\}$

(c) $\{(-3,0,-3), (3,2,-2)\}$

(d) $\{(0,2,2)\}$

(e) $\{(-2,-1,1)\,,\,(-2,-2,2)\,,\,(2,-1,-1)\,,\,(2,-2,-2)\}$

(f) $\{(2,4)\,,\,(1,2)\}$

(g) $\{(1,-2,3)\,,\,(1,1,2)\}$

Exercise 7.2.2 Compute an svd to decide whether the following sets of vectors are linearly dependent, or linearly independent. Give reasons.

(a) $\begin{bmatrix} 5 \\ 0 \\ -1 \end{bmatrix}, \begin{bmatrix} -2 \\ 5 \\ -1 \end{bmatrix}$

(b) $\begin{bmatrix} -2 \\ -1 \\ 1 \end{bmatrix}, \begin{bmatrix} 3 \\ -4 \\ -1 \end{bmatrix}, \begin{bmatrix} 5 \\ -3 \\ -2 \end{bmatrix}$

(c) $\begin{bmatrix} 2 \\ -4 \\ 1 \\ 6 \end{bmatrix}, \begin{bmatrix} -1 \\ 1 \\ 10 \\ -5 \end{bmatrix}, \begin{bmatrix} 1 \\ 6 \\ -2 \\ 4 \end{bmatrix}$

(d) $\begin{bmatrix} 1 \\ 0 \\ -5 \\ 2 \end{bmatrix}, \begin{bmatrix} -1 \\ 0 \\ -2 \\ 4 \end{bmatrix}, \begin{bmatrix} -1 \\ 3 \\ 1 \\ 1 \end{bmatrix}, \begin{bmatrix} 3 \\ 4 \\ -4 \\ -4 \end{bmatrix}$

Exercise 7.2.3 Prove that every orthogonal set of vectors is also a linearly independent set.

Exercise 7.2.4 Prove the particular case of Theorem 7.2.11, namely that a set of two vectors $\{v_1, v_2\}$ is linearly dependent if and only if one of the vectors is a scalar multiple of the other.

Exercise 7.2.5 Prove that every (non-empty) subset of a linearly independent set is also linearly independent. (Perhaps use contradiction.)

Exercise 7.2.6 Let $\{v_1, v_2, \ldots, v_m\}$ be a linearly independent set of vectors in $\mathbb{R}^n$. Given that a vector $u = c_1 v_1 + c_2 v_2 + \cdots + c_m v_m$ with coefficient $c_1 \neq 0$, prove that the set $\{v_2, v_3, \ldots, v_m, u\}$ is linearly independent.

Exercise 7.2.7 For each of the following systems of equations find by hand two different bases for their solution set (among the infinitely many bases that are possible). Show your working.

(a) $-x - 5y = 0$ and $y - 3z = 0$

(b) $6x + 4y + 2z = 0$ and $-2x - y - 2z = 0$

(c) $-2x + 3y - 6z = 0$

(d) $9x + 4y - 9z = 6$

Exercise 7.2.8 Use Procedure 7.2.32 to compute in MATLAB/Octave an orthonormal basis for all solutions to each of the following systems of equations.

(a) $2x - 6y - 9z = 0,\ 2x - 2z = 0,\ -x + z = 0$

(b) $3x - 3y + 8z = 0,\ -2x - 4y + 2z = 0,\ -4x + y - 7z = 0$

(c) $-2w + x + 2y - 6z = 0,\ -2w + 3x + 4y = 0,\ -2w + 2x + 3y - 3z = 0$

(d) $-w - 2x - 3y + 2z = 0,\ 2x + 2y - 2z = 0,\ -w - 3x - 4y + 3z = 0$

Exercise 7.2.9 Recall that Theorem 4.2.15 establishes that there are at most n eigenvalues of an $n \times n$ symmetric matrix. Adapt the proof of that theorem, using linear independence, to prove that there are at most n eigenvalues of an $n \times n$ non-symmetric matrix. (This is an alternative to the given proof of Theorem 7.1.1.)

Exercise 7.2.10 For each diagram, estimate roughly the components of each of the four vectors a, b, c, and d in the basis $\mathcal{P} = \{p_1, p_2\}$.

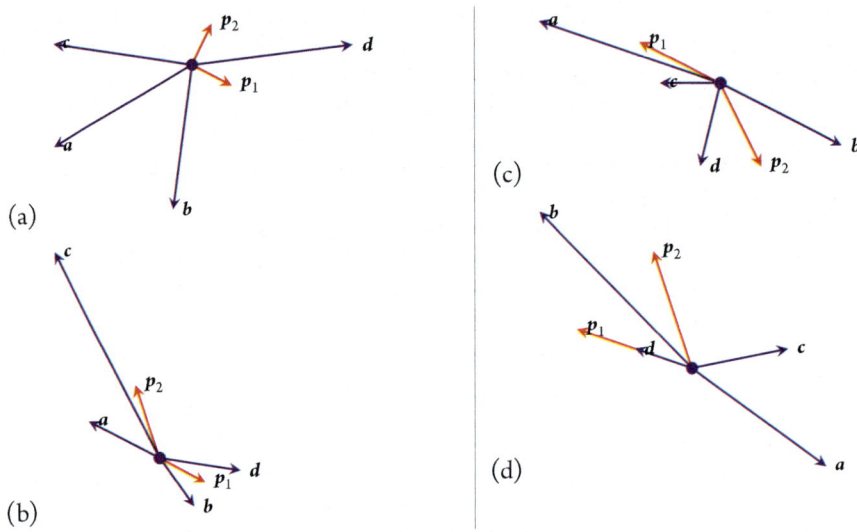

Exercise 7.2.11 Let the three given vectors $b_1 = (-1,1,-1)$, $b_2 = (1,-2,0)$, and $b_3 = (0,4,5)$ form a basis $\mathcal{B} = \{b_1, b_2, b_3\}$ (where these vectors are specified in the standard basis $\mathcal{E}$ of the standard unit vectors e_1, e_2, and e_3). For each of the following vectors with specified coordinates in basis $\mathcal{B}$, what is the vector when written in the standard basis?

(a) $[p]_\mathcal{B} = (1,-1,2)$

(b) $[q]_\mathcal{B} = (0,2,3)$

(c) $[r]_\mathcal{B} = (1,-3,-2)$

(d) $[s]_\mathcal{B} = (1,2,1)$

(e) $[t]_\mathcal{B} = (1/2,-1/2,1)$

(f) $[u]_\mathcal{B} = (-1/2,1/2,-1/2)$

Exercise 7.2.12 Repeat Exercise 7.2.11 but with the three basis vectors $b_1 = (6,2,1)$, $b_2 = (-2,-1,-2)$, and $b_3 = (-3,-1,5)$.

Exercise 7.2.13 Let the two given vectors $b_1 = (1,-2,2)$ and $b_2 = (1,-1,-1)$ form a basis $\mathcal{B} = \{b_1, b_2\}$ for the subspace $\mathbb{B}$ of $\mathbb{R}^3$ (specified in the standard basis $\mathcal{E}$ of the standard unit vectors e_1, e_2, and e_3). For each of the following vectors, specified in the standard basis $\mathcal{E}$, what is the vector when written in the basis $\mathcal{B}$ if it is possible?

(a) $[p]_\mathcal{E} = (0,1,3)$

(b) $[q]_\mathcal{E} = (-4,9,-11)$

(c) $[r]_\mathcal{E} = (-4,7,-5)$

(d) $[s]_\mathcal{E} = (0,2,-4)$

(e) $[t]_\mathcal{E} = (0,-2,5)$

(f) $[u]_\mathcal{E} = (-2/3,0,8/3)$

Exercise 7.2.14 Repeat Exercise 7.2.13 but with the two basis vectors $b_1 = (-2, 3, -1)$ and $b_2 = (0, -1, 3)$.

Exercise 7.2.15 Let the three vectors $b_1 = (-1, -2, 5, 3, -2)$, $b_2 = (-2, -2, 2, -1, -1)$, and $b_3 = (-4, 6, -4, 2, -1)$ form a basis $\mathcal{B} = \{b_1, b_2, b_3\}$ for the subspace $\mathbb{B}$ in $\mathbb{R}^5$ (each basis vector specified in the standard basis $\mathcal{E}$). For each of the following vectors, use MATLAB/Octave to find the requested coordinates, if possible.

(a) Find $[p]_\mathcal{E}$ when $[p]_\mathcal{B} = (2, 2, -1)$.
(b) Find $[q]_\mathcal{E}$ when $[q]_\mathcal{B} = (-5, 0, -2)$.
(c) Find $[r]_\mathcal{E}$ when $[r]_\mathcal{B} = (0, 3, 3)$.
(d) Find $[s]_\mathcal{E}$ when $[s]_\mathcal{B} = (-1, 5, 0)$.
(e) Find $[t]_\mathcal{B}$ when $[t]_\mathcal{E} = (-31, 26, -5, 19, -14)$.
(f) Find $[u]_\mathcal{B}$ when $[u]_\mathcal{E} = (-1, 6, 4, 14, -5)$.

Exercise 7.2.16 Repeat Exercise 7.2.15 but with basis vectors $b_1 = (-3, 8, -9, -1, 1)$, $b_2 = (10, -20, 14, -7, 2)$, and $b_3 = (-1, -2, 5, 3, -2)$ (specified in the standard basis $\mathcal{E}$).

Exercise 7.2.17 In a few sentences, answer/discuss each of the following.

(a) Why is it important to establish a mathematical framework in which solutions have a unique algebraic form?

(b) How is Definition 7.2.4 of linear independence linked to a unique algebraic form?

(c) How does linear independence compare to orthogonality of a set of vectors?

(d) For distinct eigenvalues of a matrix A, how does the linear independence of the corresponding eigenvectors arise?

(e) How can an SVD be useful in testing whether a set of vectors is linearly dependent, or not?

(f) How does a "basis" compare to an "orthonormal basis"?

(g) What problems might arise if, for a given subspace, there were bases for the subspace with different numbers of vectors?

(h) How does an SVD help us to find bases for subspaces?

(i) Why is new notation needed for vectors when we invoke the possibility of different bases?

7.3 Diagonalization identifies the transformation

Population modelling Recall that this Chapter 7 started by introducing the dynamics of two interacting species of animals. Recall that we let $y(t)$ and $z(t)$ be the number of female animals in each of the two species at time t (years). Modelling might deduce that the populations interact according to the rule that the population one year later is $y(t+1) = 2y(t) - 4z(t)$ and $z(t+1) = -y(t) + 2z(t)$. Then seeking solutions proportional to λ^t led to the eigen-problem

$$\begin{bmatrix} 2 & -4 \\ -1 & 2 \end{bmatrix} x = \lambda x.$$

This section introduces an alternate equivalent approach that provides a new view.

The alternate approach invokes non-orthogonal coordinates. Start by writing the population model as a system in terms of vector $y(t) = (y(t), z(t))$, namely

$$\begin{bmatrix} y(t+1) \\ z(t+1) \end{bmatrix} = \begin{bmatrix} 2y(t) - 4z(t) \\ -y(t) + 2z(t) \end{bmatrix}, \qquad \text{that is, } y(t+1) = \begin{bmatrix} 2 & -4 \\ -1 & 2 \end{bmatrix} y(t).$$

Now let's ask if there is a basis $\mathcal{P} = \{p_1, p_2\}$ for the yz-plane that simplifies this matrix-vector system. In such a basis every vector may be written as $y = Y_1 p_1 + Y_2 p_2$ for some scalar components Y_1 and Y_2. That is, $[y]_{\mathcal{P}} = (Y_1, Y_2)$. But to simplify writing we use the vector symbol $Y = (Y_1, Y_2)$ in place of $[y]_{\mathcal{P}}$. Write the relation $y = Y_1 p_1 + Y_2 p_2$ as the matrix-vector product $y = PY$ where matrix $P = [p_1 \;\; p_2]$ and vector $Y = (Y_1, Y_2)$. The populations y depends upon time t, and hence so does Y since $Y = [y]_{\mathcal{P}}$; that is, $y(t) = PY(t)$. Substitute this identity into the system of equations:

$$y(t+1) = PY(t+1) = \begin{bmatrix} 2 & -4 \\ -1 & 2 \end{bmatrix} PY(t).$$

Multiply both sides by P^{-1} (which exists by linear independence of the columns, Theorem 7.2.41(1)) to give

$$Y(t+1) = P^{-1} \underbrace{\begin{bmatrix} 2 & -4 \\ -1 & 2 \end{bmatrix} PY}_{P^{-1}AP}.$$

The question then becomes, for a given square matrix A, such as this, can we find a matrix P such that $P^{-1}AP$ is somehow simple? The answer is yes, in most cases: using eigenvalues and eigenvectors, the product $P^{-1}AP$ can usually be made into a simple diagonal matrix.

7.3.1 Eigenvectors achieve diagonalization

Recall that (Section 4.2.2) for a symmetric matrix A we could always factor $A = VDV^{\mathsf{T}} = VDV^{-1}$ for orthogonal matrix V and diagonal matrix D; thus a symmetric matrix is always orthogonally diagonalizable (Definition 4.2.17). For non-symmetric matrices, a diagonalization can mostly be done (although not always), the difference being that we need an invertible matrix, typically called P, instead of an orthogonal matrix V. Such a matrix A is generally termed 'diagonalizable' instead of the more specific 'orthogonally diagonalizable'.

Definition 7.3.1 *An $n \times n$ square matrix A is **diagonalizable** if there exists a diagonal matrix D and an invertible matrix P such that $A = PDP^{-1}$, equivalently $AP = PD$ or $P^{-1}AP = D$.*

Example 7.3.2

(a) Show that $A = \begin{bmatrix} 0 & 1 \\ 2 & -1 \end{bmatrix}$ is diagonalizable by matrix $P = \begin{bmatrix} 1 & -1 \\ 1 & 2 \end{bmatrix}$.

Solution: First, find the 2×2 inverse (Theorem 3.2.7)

$$P^{-1} = \frac{1}{3}\begin{bmatrix} 2 & 1 \\ -1 & 1 \end{bmatrix}.$$

Second, compute the product

$$P^{-1}AP = P^{-1}\begin{bmatrix} 1 & 2 \\ 1 & -4 \end{bmatrix} = \frac{1}{3}\begin{bmatrix} 3 & 0 \\ 0 & -6 \end{bmatrix} = \begin{bmatrix} 1 & 0 \\ 0 & -2 \end{bmatrix}.$$

Since this product is diagonal, $\mathrm{diag}(1,-2)$, the matrix A is diagonalizable. ☐

(b) Use a contradiction to prove that $B = \begin{bmatrix} 0 & 1 \\ 0 & 0 \end{bmatrix}$ is not diagonalizable.

Solution: Assume that B is diagonalizable by an invertible matrix $P = \begin{bmatrix} a & b \\ c & d \end{bmatrix}$. Being invertible, P has inverse $P^{-1} = \frac{1}{ad-bc}\begin{bmatrix} d & -b \\ -c & a \end{bmatrix}$ (Theorem 3.2.7). Then the product

$$P^{-1}BP = P^{-1}\begin{bmatrix} c & d \\ 0 & 0 \end{bmatrix} = \frac{1}{ad-bc}\begin{bmatrix} cd & d^2 \\ -c^2 & -cd \end{bmatrix}.$$

For the matrix $P^{-1}BP$ to be diagonal requires the off-diagonal elements to be zero: $d^2 = -c^2 = 0$. This requires both $c = d = 0$, but then the determinant $ad - bc = 0 - 0 = 0$ and so matrix P is not invertible (Theorem 3.2.7). This contradiction means that matrix B is not diagonalizable. ☐

(c) Is matrix $C = \begin{bmatrix} 1.2 & 3.2 & 2.3 \\ 2.2 & -0.5 & -2.2 \end{bmatrix}$ diagonalizable?

Solution: No, as it is not a square matrix. (Perhaps an SVD could answer the needs of whatever problem led to this question.) ☐

Example 7.3.3 Example 7.3.2(a) showed that matrix $P = \begin{bmatrix} 1 & -1 \\ 1 & 2 \end{bmatrix}$ diagonalizes matrix $A = \begin{bmatrix} 0 & 1 \\ 2 & -1 \end{bmatrix}$ to matrix $D = \mathrm{diag}(1,-2)$. As a prelude to the next Theorem 7.3.5, show that the columns of P are eigenvectors of A.

Solution: Invoke the original Definition 4.1.1 of an eigenvector for a matrix.

- The first column of P is $\boldsymbol{p}_1 = (1,1)$. Multiplying $A\boldsymbol{p}_1 = (0 + 1, 2 - 1) = (1,1) = 1\boldsymbol{p}_1$, so the first column vector $\boldsymbol{p}_1$ is an eigenvector of A corresponding to the eigenvalue 1. Correspondingly, this eigenvalue 1 is the first entry in the diagonal D.

- The second column of P is $p_2 = (-1,2)$. Multiplying $Ap_2 = (0+2,-2-2) = (2,-4) = -2p_2$, so the second column vector p_2 is an eigenvector of A corresponding to the eigenvalue -2. Correspondingly, this eigenvalue -2 is the second entry in the diagonal D.

□

Activity 7.3.4 Given that matrix $F = \begin{bmatrix} 5 & 8 \\ -4 & -7 \end{bmatrix}$ has eigenvectors $(-1,1)$ and $(2,-1)$ corresponding to respective eigenvalues -3 and 1, what matrix diagonalizes F to $D = \mathrm{diag}(-3,1)$?

(a) $\begin{bmatrix} 2 & -1 \\ 1 & -1 \end{bmatrix}$
(b) $\begin{bmatrix} 2 & -1 \\ -1 & 1 \end{bmatrix}$
(c) $\begin{bmatrix} -1 & 1 \\ 2 & -1 \end{bmatrix}$
(d) $\begin{bmatrix} -1 & 2 \\ 1 & -1 \end{bmatrix}$

Theorem 7.3.5 *For every $n \times n$ square matrix A, the matrix A is diagonalizable if and only if A has n linearly independent eigenvectors. If A is diagonalizable, with diagonal matrix $D = P^{-1}AP$, then the diagonal entries of D are eigenvalues, and the columns of P are corresponding eigenvectors.*

Proof. First, let matrix A be diagonalizable by invertible P and diagonal D. Write $P = \begin{bmatrix} p_1 & p_2 & \cdots & p_n \end{bmatrix}$ in terms of its columns, and let $D = \mathrm{diag}(\lambda_1, \lambda_2, \ldots, \lambda_n)$ in terms of its diagonal entries. Then $AP = PD$ becomes

$$A\begin{bmatrix} p_1 & p_2 & \cdots & p_n \end{bmatrix} = \begin{bmatrix} p_1 & p_2 & \cdots & p_n \end{bmatrix}\begin{bmatrix} \lambda_1 & 0 & \cdots & 0 \\ 0 & \lambda_2 & & 0 \\ \vdots & & \ddots & \vdots \\ 0 & 0 & \cdots & \lambda_n \end{bmatrix}.$$

Multiplying the matrix-column products on both sides gives

$$\begin{bmatrix} Ap_1 & Ap_2 & \cdots & Ap_n \end{bmatrix} = \begin{bmatrix} \lambda_1 p_1 & \lambda_2 p_2 & \cdots & \lambda_n p_n \end{bmatrix}.$$

Equating corresponding columns implies $Ap_1 = \lambda_1 p_1, Ap_2 = \lambda_2 p_2, \ldots, Ap_n = \lambda_n p_n$. As the matrix P is invertible, all its columns must be nonzero (Theorem 6.2.5(a)). Hence $p_1, p_2, \ldots, p_n$ are eigenvectors of matrix A corresponding to eigenvalues $\lambda_1, \lambda_2, \ldots, \lambda_n$. Since matrix P is invertible, Theorem 7.2.41(1) asserts that the n columns vectors $p_1, p_2, \ldots, p_n$ are linearly independent.

Second, suppose matrix A has n linearly independent eigenvectors $p_1, p_2, \ldots, p_n$ with corresponding eigenvalues $\lambda_1, \lambda_2, \ldots, \lambda_n$. Then follow the above argument backwards to deduce $AP = PD$ for invertible matrix $P = \begin{bmatrix} p_1 & p_2 & \cdots & p_n \end{bmatrix}$, and hence A is diagonalizable. Lastly, in these arguments, P is the matrix of eigenvectors and the diagonal of D are the corresponding eigenvalues, as required.

□

Example 7.3.6 Recall that Example 7.0.3 found the triangular matrix $A = \begin{bmatrix} -3 & 2 & 0 \\ 0 & -4 & 2 \\ 0 & 0 & 4 \end{bmatrix}$ has eigenvalues $-3, -4$ and 4 (from its diagonal) and corresponding eigenvectors proportional to $(1,0,0), (-2,1,0)$, and $(\frac{1}{14}, \frac{1}{4}, 1)$. Is matrix A diagonalizable?

Solution: These three eigenvectors are linearly independent as they correspond to distinct eigenvalues (Theorem 7.2.13). Hence the matrix is diagonalizable.

The previous statement answers the question. But, further, forming these eigenvectors into the columns of matrix

$$P = \begin{bmatrix} 1 & -2 & \frac{1}{14} \\ 0 & 1 & \frac{1}{4} \\ 0 & 0 & 1 \end{bmatrix},$$

we know that (Theorem 7.3.5) $P^{-1}AP = \mathrm{diag}(-3, -4, 4)$ where the eigenvalues appear in the same order as that of the eigenvectors in P.

One may check this by hand or with MATLAB/Octave. Enter the matrices with

```
A=[-3 2 0;0 -4 2;0 0 4]
P=[1 -2 1/14;0 1 1/4;0 0 1]
```

then compute $D = P^{-1}AP$ with D=P\A*P to find as required the following diagonal result

```
D =
   -3.0000    0.0000    0.0000
    0.0000   -4.0000    0.0000
    0.0000    0.0000    4.0000
```

□

Example 7.3.7   Recall the Sierpinski network of Example 4.1.20 (shown to the right). The following 9×9 matrix A encodes the network. Is the matrix diagonalizable?

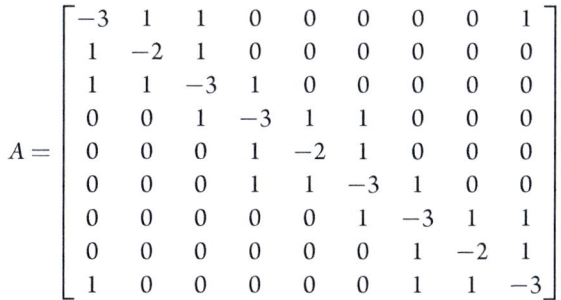

$$A = \begin{bmatrix} -3 & 1 & 1 & 0 & 0 & 0 & 0 & 0 & 1 \\ 1 & -2 & 1 & 0 & 0 & 0 & 0 & 0 & 0 \\ 1 & 1 & -3 & 1 & 0 & 0 & 0 & 0 & 0 \\ 0 & 0 & 1 & -3 & 1 & 1 & 0 & 0 & 0 \\ 0 & 0 & 0 & 1 & -2 & 1 & 0 & 0 & 0 \\ 0 & 0 & 0 & 1 & 1 & -3 & 1 & 0 & 0 \\ 0 & 0 & 0 & 0 & 0 & 1 & -3 & 1 & 1 \\ 0 & 0 & 0 & 0 & 0 & 0 & 1 & -2 & 1 \\ 1 & 0 & 0 & 0 & 0 & 0 & 1 & 1 & -3 \end{bmatrix}.$$

Solution: In that example we used MATLAB/Octave command [V,D] =eig(A) to compute a matrix of eigenvectors V and the corresponding diagonal matrix of eigenvalues D where (2 d.p.)

V =

-0.41	0.51	-0.16	-0.21	-0.45	0.18	-0.40	0.06	0.33
0.00	-0.13	0.28	0.63	0.13	-0.18	-0.58	-0.08	0.33
0.41	-0.20	-0.49	-0.42	0.32	0.01	-0.36	-0.17	0.33
-0.41	-0.11	0.52	-0.42	0.32	0.01	0.14	-0.37	0.33
-0.00	-0.18	-0.26	0.37	-0.22	0.51	0.36	-0.46	0.33
0.41	0.53	0.07	0.05	-0.10	-0.51	0.33	-0.23	0.33
-0.41	-0.39	-0.36	0.05	-0.10	-0.51	0.25	0.31	0.33
0.00	0.31	-0.03	0.16	0.55	0.34	0.22	0.55	0.33
0.41	-0.33	0.42	-0.21	-0.45	0.18	0.03	0.40	0.33

D =

-5.00	0	0	0	0	0	0	0	0
0	-4.30	0	0	0	0	0	0	0
0	0	-4.30	0	0	0	0	0	0
0	0	0	-3.00	0	0	0	0	0
0	0	0	0	-3.00	0	0	0	0
0	0	0	0	0	-3.00	0	0	0
0	0	0	0	0	0	-0.70	0	0
0	0	0	0	0	0	0	-0.70	0
0	0	0	0	0	0	0	0	-0.00

Since matrix A is symmetric, MATLAB/Octave computes for us an orthogonal matrix V with columns eigenvectors. Since V is orthogonal, its column vectors are orthonormal and hence its columns are linearly independent (Theorem 7.2.8). Since there exist nine linearly independent eigenvectors, the nine column vectors of V, the matrix A is diagonalizable. Further, the product $V^{-1}AV = D$ for the above diagonal matrix D of eigenvalues in the order of the eigenvectors in V. (Also, since V is orthogonal, $V^T AV = D$.) □

Example 7.3.8 Recall that Example 7.1.13 found eigenvalues and corresponding eigenspaces for various matrices. Revisit these cases and show that none of the matrices are diagonalizable.

(a) Matrix $A = \begin{bmatrix} 3 & 1 \\ 0 & 3 \end{bmatrix}$ had one eigenvalue $\lambda = 3$ with multiplicity two and corresponding eigenspace $\mathbb{E}_3 = \text{span}\{(1,0)\}$. This matrix is not diagonalizable as it has only one linearly independent eigenvector, such as $(1,0)$ or any nonzero multiple, and it needs two to be diagonalizable.

(b) Matrix $B = \begin{bmatrix} -1 & 1 & -2 \\ -1 & 0 & -1 \\ 0 & -3 & 1 \end{bmatrix}$ has eigenvalues $\lambda = -2$ (multiplicity one) and $\lambda = 1$ (multiplicity two). The corresponding eigenspaces are $\mathbb{E}_{-2} = \text{span}\{(1,1,1)\}$ and $\mathbb{E}_1 = \text{span}\{(-1,0,1)\}$. Thus the matrix has only two linearly independent eigenvectors, one from each eigenspace, and it needs three to be diagonalizable.

(c) Matrix $C = \begin{bmatrix} -1 & 0 & -2 \\ 0 & -3 & 2 \\ 0 & -2 & 1 \end{bmatrix}$ has only the eigenvalue $\lambda = -1$ with multiplicity three. The corresponding eigenspace $\mathbb{E}_{-1} = \text{span}\{(1,0,0)\}$. With only one linearly independent eigenvector, the matrix is not diagonalizable. □

Example 7.3.9 Use the results of Example 7.1.14 to show that the following matrix is diagonalizable:

$$A = \begin{bmatrix} 0 & 3 & 0 & 0 & 0 \\ 1 & 0 & 3 & 0 & 0 \\ 0 & 1 & 0 & 3 & 0 \\ 0 & 0 & 1 & 0 & 3 \\ 0 & 0 & 0 & 1 & 0 \end{bmatrix}$$

Solution: Example 7.1.14 derived that the five eigenvalues are $\lambda = 0, \pm\sqrt{3}, \pm 3$, all of multiplicity one. Further, the corresponding eigenspaces are

$$\mathbb{E}_0 = \text{span}\{(9,0,-3,0,1)\},$$
$$\mathbb{E}_{\pm\sqrt{3}} = \text{span}\{(-9,\mp 3\sqrt{3},0,\pm\sqrt{3},1)\},$$
$$\mathbb{E}_{\pm 3} = \text{span}\{(9,\pm 9,6,\pm 3,1)\}.$$

Here there are five linearly independent eigenvectors, one from each distinct eigenspace (Theorem 7.2.13). Since A is a 5×5 matrix it is thus diagonalizable. Further, Theorem 7.3.5 establishes that the matrix formed from the columns of the five eigenvectors is a possible matrix

$$P = \begin{bmatrix} 9 & -9 & -9 & 9 & 9 \\ 0 & -3\sqrt{3} & 3\sqrt{3} & 9 & -9 \\ -3 & 0 & 0 & 6 & 6 \\ 0 & \sqrt{3} & -\sqrt{3} & 3 & -3 \\ 1 & 1 & 1 & 1 & 1 \end{bmatrix}.$$

(One could also scale each column of P by a different arbitrary nonzero constant, and the diagonalization still holds.) Lastly, Theorem 7.3.5 establishes the diagonal matrix $P^{-1}AP = D = \text{diag}(0, \sqrt{3}, -\sqrt{3}, 3, -3)$ is that of the eigenvalues in the order corresponding to the eigenvectors in P. □

These examples illustrate a widely useful property. The 5×5 matrix in Example 7.3.9 has five distinct eigenvalues whose corresponding eigenvectors are necessarily linearly independent (Theorem 7.2.13) and so diagonalize the matrix (Theorem 7.3.5). The 3×3 matrix in Example 7.3.6 has three distinct eigenvalues whose corresponding eigenvectors are necessarily linearly independent (Theorem 7.2.13) and so diagonalize the matrix (Theorem 7.3.5). However, the matrices of Examples 7.3.7 and 7.3.8 have repeated eigenvalues—eigenvalues of multiplicity two or more—and these matrices may (Example 7.3.7) or may not (Example 7.3.8) be diagonalizable.

The following theorem confirms that matrices with as many *distinct* eigenvalues as the size of the matrix are always diagonalizable.

Theorem 7.3.10 *For every* $n \times n$ *square matrix* A, *if* A *has* n *distinct eigenvalues, then* A *is diagonalizable. Consequently, and allowing nonreal complex eigenvalues, a real non-diagonalizable matrix must be non-symmetric and must have at least one repeated eigenvalue (an eigenvalue with multiplicity two or more).*

Proof. First, let v_1, v_2, $\ldots$, v_n be eigenvectors corresponding to the n distinct eigenvalues of matrix A. (Recall that Theorem 7.1.1 establishes that there cannot be more than n eigenvalues.) As the corresponding eigenvalues are distinct, Theorem 7.2.13 establishes that these eigenvectors are linearly independent. Theorem 7.3.5 then establishes that the matrix A is diagonalizable.

Second, since an $n \times n$ matrix has n eigenvalues, when counted accordingly to multiplicity and allowing for nonreal complex valued eigenvalues (Procedure 7.1.11), a non-diagonalizable matrix must have at least one repeated eigenvalue. Further, by Theorem 4.2.19 a real symmetric matrix is always diagonalizable: hence a non-diagonalizable matrix must also be non-symmetric. □

Example 7.3.11 From the given information, are the matrices diagonalizable?

(a) The only eigenvalues of a 4×4 matrix are 1.8, -3, 0.4, and 3.2.

   Solution: Theorem 7.3.10 implies the matrix must be diagonalizable. □

(b) The only eigenvalues of a 5×5 matrix are 1.8, -3, 0.4, and 3.2.

   Solution: Here there are only four distinct eigenvalues of the 5×5 matrix. Theorem 7.3.10 does not apply as the precondition that there be five distinct eigenvalues is not met: the matrix may or may not be diagonalizable—it is unknowable on this information. □

(c) The only eigenvalues of a 3×3 matrix are 1.8, -3, 0.4, and 3.2.

   Solution: An error has been made in determining the eigenvalues because a 3×3 matrix has at most three distinct eigenvalues (Theorem 7.1.1). Because of the error, we cannot answer. □

Activity 7.3.12 A 3×3 matrix A depends upon a parameter a and has eigenvalues 6, $3 - 3a$, and $2 + a$. For which of the following values of parameter a may the matrix be *not* diagonalizable?

(a) $a = 4$ (b) $a = 2$ (c) $a = 1$ (d) $a = 3$

Extension: what are the two other values of a for which the matrix *may not* be diagonalizable?

Example 7.3.13 MATLAB/Octave computes the eigenvalues of matrix

$$A = \begin{bmatrix} -1 & 2 & -2 & 1 & -2 \\ -3 & -1 & -2 & 5 & 6 \\ 3 & 1 & 6 & -2 & -1 \\ 1 & 1 & 2 & 1 & -1 \\ 7 & 5 & -3 & 0 & 0 \end{bmatrix}$$

via eig(A) and reports them to be (2 d.p.)

```
ans =
  -3.45 + 3.50i
  -3.45 - 3.50i
   5.00
   5.00
   1.91
```

Is the matrix diagonalizable?

Solution: The matrix appears to have only four distinct eigenvalues (two of them non-real complex valued), and so on the given information Theorem 7.3.10 cannot determine whether the matrix is diagonalizable or not.

However, upon reporting the eigenvalues to four decimal places we find the two eigenvalues of 5.00 (2 d.p.) are more precisely two separate eigenvalues of 5.0000 and 4.9961. Hence this matrix has five distinct eigenvalues and so Theorem 7.3.10 implies that the matrix is diagonalizable.[11]   □

Recall that Definition 7.1.7 defined the multiplicity of every eigenvalue in terms of the characteristic polynomial. Also, the earlier Definition 4.1.15 identified that for every *symmetric* matrix the dimension of an eigenspace, $\dim \mathbb{E}_{\lambda_j}$, is equal to the multiplicity of the corresponding eigenvalue λ_j. However, for general non-symmetric matrices this equality between multiplicity and eigenspace dimension does not necessarily hold.

Theorem 7.3.14   *For every square matrix A, and for each eigenvalue λ_j of A, the corresponding eigenspace $\mathbb{E}_{\lambda_j}$ has dimension less than or equal to the multiplicity of λ_j; that is, $1 \le \dim \mathbb{E}_{\lambda_j} \le$ multiplicity of λ_j.*

Proof. Suppose λ_j is an eigenvalue of $n \times n$ matrix A and $\dim \mathbb{E}_{\lambda_j} = p < n$ (the case $p = n$ is proved by Exercise 7.3.7). Because its dimension is p, the eigenspace may be spanned

---

[11] Nonetheless, in an application where errors are significant then the matrix may be effectively non-diagonalizable. Such effective non-diagonalizability is indicated by poor conditioning of the matrix of eigenvectors which here has the poor rcond of 0.0004 (Procedure 2.2.5).

by p vectors. Then choose p orthonormal vectors v_1, v_2, ..., v_p to span $\mathbb{E}_{\lambda_j}$; these vectors v_1, v_2, ..., v_p are eigenvectors as they are in the eigenspace. Let

$$P = \begin{bmatrix} v_1 & \cdots & v_p & w_{p+1} & \cdots & w_n \end{bmatrix}$$

be any orthogonal matrix with v_1, v_2, ..., v_p as its first p columns. Equivalently, write as the partitioned matrix $P = \begin{bmatrix} V & W \end{bmatrix}$ for corresponding $n \times p$ and $n \times (n-p)$ matrices. Since P is orthogonal, its inverse $P^{-1} = P^{\mathsf{T}} = \begin{bmatrix} V^{\mathsf{T}} \\ W^{\mathsf{T}} \end{bmatrix}$. Since the columns of V are eigenvectors corresponding to eigenvalue λ_j, $AV = A\begin{bmatrix} v_1 & \cdots & v_p \end{bmatrix} = \begin{bmatrix} Av_1 & \cdots & Av_p \end{bmatrix} = \begin{bmatrix} \lambda_j v_1 & \cdots & \lambda_j v_p \end{bmatrix} = \lambda_j \begin{bmatrix} v_1 & \cdots & v_p \end{bmatrix} = \lambda_j V$.

Now consider

$$P^{-1}AP = \begin{bmatrix} V^{\mathsf{T}} \\ W^{\mathsf{T}} \end{bmatrix} A \begin{bmatrix} V & W \end{bmatrix} = \begin{bmatrix} V^{\mathsf{T}}AV & V^{\mathsf{T}}AW \\ W^{\mathsf{T}}AV & W^{\mathsf{T}}AW \end{bmatrix}$$

$$= \begin{bmatrix} \lambda_j V^{\mathsf{T}}V & V^{\mathsf{T}}AW \\ \lambda_j W^{\mathsf{T}}V & W^{\mathsf{T}}AW \end{bmatrix} = \begin{bmatrix} \lambda_j I_p & V^{\mathsf{T}}AW \\ O & W^{\mathsf{T}}AW \end{bmatrix}$$

where the last equality follows from the orthonormality of columns of $P = \begin{bmatrix} V & W \end{bmatrix}$. Then the characteristic polynomial of matrix A becomes

$$\det(A - \lambda I_n)$$
$$= \det(PP^{-1}APP^{-1} - \lambda PP^{-1}) \quad \text{(as } PP^{-1} = I_n)$$
$$= \det[P(P^{-1}AP - \lambda I_n)P^{-1}]$$
$$= \det P \det(P^{-1}AP - \lambda I_n) \det(P^{-1}) \quad \text{(product Theorem 6.1.16)}$$
$$= \det P \det(P^{-1}AP - \lambda I_n) \frac{1}{\det P} \quad \text{(inverse Theorem 6.1.29)}$$
$$= \det(P^{-1}AP - \lambda I_n)$$
$$= \det \begin{bmatrix} (\lambda_j - \lambda)I_p & V^{\mathsf{T}}AW \\ O & W^{\mathsf{T}}AW - \lambda I_{n-p} \end{bmatrix} \quad \text{(by above } P^{-1}AP)$$
$$= (\lambda_j - \lambda)^p \det(W^{\mathsf{T}}AW - \lambda I_{n-p})$$

by p successive first column expansions of the determinant (Theorem 6.2.24). Because of the factor $(\lambda_j - \lambda)^p$ in the characteristic polynomial of A, the eigenvalue λ_j must have multiplicity of at least $p = \dim \mathbb{E}_{\lambda_j}$—there may be more factors of $(\lambda_j - \lambda)$ hidden within $\det(W^{\mathsf{T}}AW - \lambda I_{n-p})$. $\qquad \square$

Example 7.3.15 Show that the matrix $A = \begin{bmatrix} 0 & 5 & 6 \\ -8 & 22 & 24 \\ 6 & -15 & -16 \end{bmatrix}$ has one eigenvalue of multiplicity three, and that the corresponding eigenspace has dimension two.

Solution: Find eigenvalues via the characteristic polynomial

$$\det(A - \lambda I) = \begin{vmatrix} -\lambda & 5 & 6 \\ -8 & 22 - \lambda & 24 \\ 6 & -15 & -16 - \lambda \end{vmatrix}$$

$$= -\lambda(22 - \lambda)(-16 - \lambda) + 5 \cdot 24 \cdot 6 + 6(-8)(-15)$$
$$- 6(22 - \lambda)6 + \lambda 24(-15) - 5(-8)(-16 - \lambda)$$

$$\vdots$$

$$= -\lambda^3 + 6\lambda^2 - 12\lambda + 8 = -(\lambda - 2)^3.$$

This characteristic polynomial is zero only for eigenvalue $\lambda = 2$ which is of multiplicity three.

The corresponding eigenspace comes from solving $(A - \lambda I)x = 0$ which here is

$$\begin{bmatrix} -2 & 5 & 6 \\ -8 & 20 & 24 \\ 6 & -15 & -18 \end{bmatrix} x = 0.$$

Observe that the second row is just four times the first row, and the third row is (-3) times the first row, hence all three equations in this system are equivalent to just the one from the first row, namely $-2x_1 + 5x_2 + 6x_3 = 0$. A general solution of this equation is $x_1 = \frac{5}{2}x_2 + 3x_3$. That is, all solutions are $x = (\frac{5}{2}x_2 + 3x_3, x_2, x_3) = x_2(\frac{5}{2}, 1, 0) + x_3(3, 0, 1)$. Hence all solutions form the two-dimensional eigenspace $\mathbb{E}_2 = \text{span}\{(\frac{5}{2}, 1, 0), (-3, 0, 1)\}$.   □

Example 7.3.16   Use MATLAB/Octave to find the eigenvalues and the dimension of the eigenspaces of the matrix

$$B = \begin{bmatrix} 344 & -1165 & -149 & -1031 & 1065 & -2816 \\ 90 & -306 & -38 & -272 & 280 & -742 \\ -45 & 140 & 12 & 117 & -115 & 302 \\ 135 & -470 & -70 & -421 & 445 & -1175 \\ -165 & 555 & 67 & 493 & -506 & 1338 \\ -105 & 360 & 48 & 322 & -335 & 886 \end{bmatrix}.$$

Solution: In MATLAB/Octave enter the matrix with

```
B= [344  -1165   -149  -1031   1065  -2816
     90   -306    -38   -272    280   -742
    -45    140     12    117   -115    302
    135   -470    -70   -421    445  -1175
   -165    555     67    493   -506   1338
   -105    360     48    322   -335    886]
```

Then [V,D] =eig (B) computes something like the following (2 d.p.)

V =

-0.19	0.19	-0.45	0.75	-0.20	0.15
-0.38	0.38	0.12	0.26	-0.03	0.00
-0.58	0.58	0.08	-0.00	-0.54	0.56
-0.00	-0.00	-0.83	0.28	-0.50	0.52
0.58	-0.58	-0.29	-0.45	-0.64	0.63
0.38	-0.38	0.08	-0.29	-0.04	0.03

D =

4.00	0	0	0	0	0
0	4.00	0	0	0	0
0	0	4.00	0	0	0
0	0	0	-1.00	0	0
0	0	0	0	-1.00	0
0	0	0	0	0	-1.00

Evidently, the matrix B has two eigenvalues, $\lambda = 4$ and $\lambda = -1$, both of multiplicity three. Although due to round-off error MATLAB/Octave reports these with errors of about 10^{-5} (Section 7.1.2)—possibly complex errors in which case ignore small complex parts. For each eigenvalue that MATLAB/Octave reports, the three corresponding columns of V contain corresponding eigenvectors. Each set of these three eigenvectors does span the corresponding eigenspace, but are not necessarily linearly independent.

- For eigenvalue $\lambda = 4$ the first two columns of V are clearly the negative of each other, and so are essentially the same eigenvector. The third column of V is clearly not proportional to the first two columns and so is linearly independent (Theorem 7.2.11). Thus MATLAB/Octave has computed only two linearly independent eigenvectors, either the first and third column, or the second and third column. Consequently, the dimension dim $\mathbb{E}_4 = 2$.

  One can confirm this dimensionality by computing the singular values of the first three columns of V with svd (V(:,1:3)) to find they are 1.4278, 0.9806, and 0.0000. The two nonzero singular values indicate that the dimension of the span is two (Procedure 3.4.23).

- For eigenvalue $\lambda = -1$ the last three columns of V look linearly independent and so we suspect the eigenspace dimension dim $\mathbb{E}_{-1} = 3$.

  To confirm the dimension of this eigenspace via Procedure 3.4.23, compute svd (V(:,4:6)) to find that the three singular values are 1.4136, 1.0001, and 0.0414. Since all three singular values are nonzero, dim $\mathbb{E}_{-1} = 3$.

  □

The MATLAB/Octave function eig () may produce for you a quite different matrix V of eigenvectors (possibly with complex parts). As discussed by Section 7.1.2, repeated eigenvalues are very sensitive, and this sensitivity means that small variations in the hidden MATLAB/Octave algorithm may produce quite large changes in the matrix V for repeated eigenvalues. However, each eigenspace spanned by the appropriate columns of V is robust.

7.3.2 Solve systems of differential equations

Population modelling   The population modelling seen so far (Section 7.1.3) expressed the changes of the population over discrete intervals in time via discrete time equations such as $y(t+1) = \cdots$ and $z(t+1) = \cdots$. One such example is to describe the population numbers year by year. The alternative is to model the changes in the population *continuously* in time. This alternative invokes and analyses differential equations. Such continuous time, differential equation models are common for exploring the interaction between different species, such as between humans and viruses. Such differential equations are also common throughout engineering and science.

Let's start with a continuous time version of the population modelling discussed at the start of this Chapter 7. Let two species interact continuously in time with populations $y(t)$ and $z(t)$ at time t (years). Suppose they interact according to differential equations $dy/dt = y - 4z$ and $dz/dt = -y + z$ (instead of the discrete time equations $y(t+1) = \cdots$ and $z(t+1) = \cdots$). Analogous to the start of this Section 7.3, we now ask the following question: is there a matrix transformation to new variables, the vector $Y(t)$, such that $(y,z) = PY$ for some as yet unknown matrix P, where the differential equations for Y are simple? Equivalently, is there a different basis $\mathcal{P} = \{p_1, p_2\}$ for the yz-plane in which the differential equations for $Y = [y]_{\mathcal{P}}$ are simple?

- First, form the differential equations into a matrix-vector system:

$$\begin{bmatrix} dy/dt \\ dz/dt \end{bmatrix} = \begin{bmatrix} y - 4z \\ -y + z \end{bmatrix} = \begin{bmatrix} 1 & -4 \\ -1 & 1 \end{bmatrix} \begin{bmatrix} y \\ z \end{bmatrix}.$$

So using vector $y = (y,z)$, this system is

$$\frac{dy}{dt} = Ay \quad \text{for matrix } A = \begin{bmatrix} 1 & -4 \\ -1 & 1 \end{bmatrix}.$$

- Second, let's see what happens when we transform to some, as yet unknown, new vector variable $Y(t)$ such that $y = PY$ for some constant invertible matrix P. Under such a transform: $\frac{dy}{dt} = \frac{d}{dt}PY = P\frac{dY}{dt}$; also $Ay = APY$. Hence substituting such an assumed transformation into the matrix-vector form of the differential equation leads to

$$P\frac{dY}{dt} = APY, \quad \text{that is} \quad \frac{dY}{dt} = \left(P^{-1}AP\right)Y.$$

To simplify this system for Y, we diagonalize the matrix on the right-hand side. The procedure is to choose the columns of P to be eigenvectors of the matrix A (Theorem 7.3.5).

- Third, find the eigenvectors of A by hand as it is a 2×2 matrix. Here the matrix $A = \begin{bmatrix} 1 & -4 \\ -1 & 1 \end{bmatrix}$ has characteristic polynomial $\det(A - \lambda I) = (1 - \lambda)^2 - 4$. This is zero for $(1 - \lambda)^2 = 4$, that is, $(1 - \lambda) = \pm 2$. Hence the eigenvalues $\lambda = 1 \pm 2 = 3, -1$.

  - For eigenvalue $\lambda_1 = 3$ the corresponding eigenvectors satisfy

$$(A - \lambda_1 I)p_1 = \begin{bmatrix} -2 & -4 \\ -1 & -2 \end{bmatrix} p_1 = 0,$$

with general solution $p_1 \propto (2, -1)$.

– For eigenvalue $\lambda_2 = -1$ the corresponding eigenvectors satisfy

$$(A - \lambda_2 I)p_2 = \begin{bmatrix} 2 & -4 \\ -1 & 2 \end{bmatrix} p_2 = 0,$$

with general solution $p_2 \propto (2, 1)$.
Thus setting transformation matrix

$$P = [p_1 \ \ p_2] = \begin{bmatrix} 2 & 2 \\ -1 & 1 \end{bmatrix} \implies \frac{dY}{dt} = \begin{bmatrix} 3 & 0 \\ 0 & -1 \end{bmatrix} Y$$

(every nonzero scalar multiple of the two columns of P also work).

- Fourth, having diagonalized the matrix, expand this diagonalized set of differential equations to write this system in terms of components $Y = (Y_1, Y_2)$:

$$\frac{dY_1}{dt} = 3Y_1 \quad \text{and} \quad \frac{dY_2}{dt} = -Y_2.$$

Each of these differential equations has well-known exponential solutions, respectively $Y_1 = c_1 e^{3t}$ and $Y_2 = c_2 e^{-t}$, for every value of the constants c_1 and c_2.

- Lastly, what does this mean for the original problem? Recall the relation

$$\begin{bmatrix} y \\ z \end{bmatrix} = y = PY = \begin{bmatrix} 2 & 2 \\ -1 & 1 \end{bmatrix} \begin{bmatrix} c_1 e^{3t} \\ c_2 e^{-t} \end{bmatrix} = \begin{bmatrix} 2c_1 e^{3t} + 2c_2 e^{-t} \\ -c_1 e^{3t} + c_2 e^{-t} \end{bmatrix}.$$

That is, a general solution of the original system of differential equations is $y(t) = 2c_1 e^{3t} + 2c_2 e^{-t}$ and $z(t) = -c_1 e^{3t} + c_2 e^{-t}$.

The diagonalization of the matrix empowers us to solve complicated systems of differential equations as a set of simple systems.

Such a general solution makes predictions. For example, suppose at time zero $(t = 0)$ the initial population (female) of y-animals is 22 and the population of z-animals is 9. From the above general solution we then know that at time $t = 0$

$$\begin{bmatrix} 22 \\ 9 \end{bmatrix} = \begin{bmatrix} y(0) \\ z(0) \end{bmatrix} = \begin{bmatrix} 2c_1 e^{3\cdot 0} + 2c_2 e^{-0} \\ -c_1 e^{3\cdot 0} + c_2 e^{-0} \end{bmatrix} = \begin{bmatrix} 2c_1 + 2c_2 \\ -c_1 + c_2 \end{bmatrix}$$

This equation determines the coefficients: $2c_1 + 2c_2 = 22$ and $-c_1 + c_2 = 9$. Adding the first to twice the second gives $4c_2 = 40$, that is, $c_2 = 10$. Then either equation determines $c_1 = 1$. Consequently, the particular solution from this initial population is

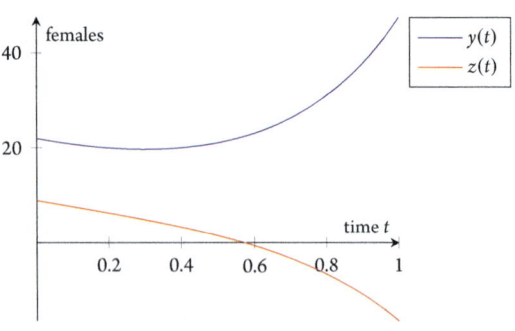

$$\begin{bmatrix} y \\ z \end{bmatrix} = \begin{bmatrix} 2 \cdot 1e^{3t} + 2 \cdot 10e^{-t} \\ -1e^{3t} + 10e^{-t} \end{bmatrix} = \begin{bmatrix} 2e^{3t} + 20e^{-t} \\ -e^{3t} + 10e^{-t} \end{bmatrix}.$$

The graph, above-right, of this solution shows that the population of y-animals grows in time, whereas the population of z-animals crashes and becomes extinct at about time 0.6 years.[12]

The forthcoming Theorem 7.3.18 confirms that the same approach solves general systems of differential equations: it is analogous to Theorem 7.1.25 for discrete dynamics.

Activity 7.3.17 A given population model is expressed as the differential equations $dx/dt = x + y - 3z$, $dy/dt = -2x + z$, and $dz/dt = -2x + y + 2z$. This may be written in matrix-vector form $dx/dt = Ax$ for vector $x(t) = (x, y, z)$ and which of the following matrices?

(a) $\begin{bmatrix} 1 & -2 & 2 \\ 0 & -2 & 1 \\ 1 & 1 & -3 \end{bmatrix}$ (b) $\begin{bmatrix} 1 & -3 & 1 \\ -2 & 1 & 0 \\ -2 & 2 & 1 \end{bmatrix}$ (c) $\begin{bmatrix} 1 & 1 & -3 \\ -2 & 0 & 1 \\ -2 & 1 & 2 \end{bmatrix}$ (d) $\begin{bmatrix} 1 & 1 & -3 \\ 0 & -2 & 1 \\ 1 & -2 & 2 \end{bmatrix}$

Theorem 7.3.18 *Let $n \times n$ square matrix A be diagonalizable by the invertible matrix $P = \begin{bmatrix} p_1 & p_2 & \cdots & p_n \end{bmatrix}$ whose columns are eigenvectors corresponding to eigenvalues $\lambda_1, \lambda_2, \ldots, \lambda_n$ of matrix A. Then a general solution $x(t)$ to the differential equation system $dx/dt = Ax$ is the linear combination*

$$x(t) = c_1 p_1 e^{\lambda_1 t} + c_2 p_2 e^{\lambda_2 t} + \cdots + c_n p_n e^{\lambda_n t} \tag{7.5}$$

for arbitrary constants $c_1, c_2, \ldots, c_n$.

Proof. First, instead of finding solutions for $x(t)$ directly, let's write the differential equations in terms of the alternate basis for $\mathbb{R}^n$, basis $\mathcal{P} = \{p_1, p_2, \ldots, p_n\}$ (as $p_1, p_2, \ldots, p_n$ are linearly independent). That is, solve for the coordinates $X(t) = [x(t)]_\mathcal{P}$ with respect to basis $\mathcal{P}$. From Theorem 7.2.35 recall that $X = [x]_\mathcal{P}$ means that $x = X_1 p_1 + X_2 p_2 + \cdots + X_n p_n = PX$. Substituting this into the differential equation $dx/dt = Ax$ requires $\frac{d}{dt}(PX) = A(PX)$ which is the same as $P\frac{dX}{dt} = APX$. Since matrix P is invertible, this equation is the same as $\frac{dX}{dt} = P^{-1}APX$. Because the columns of matrix P are eigenvectors, the product $P^{-1}AP$ is the diagonal matrix $D = \text{diag}(\lambda_1, \lambda_2, \ldots, \lambda_n)$, hence the system becomes $\frac{dX}{dt} = DX$. Because matrix D is diagonal, this is a much simpler system of differential equations. The n rows of the system $\frac{dX}{dt} = DX$ are

$$\frac{dX_1}{dt} = \lambda_1 X_1, \quad \frac{dX_2}{dt} = \lambda_2 X_2, \quad \ldots, \quad \frac{dX_n}{dt} = \lambda_n X_n.$$

Each of these has general solution

$$X_1 = c_1 e^{\lambda_1 t}, \quad X_2 = c_2 e^{\lambda_2 t}, \quad \ldots, \quad X_n = c_n e^{\lambda_n t},$$

[12] After time 0.6 years the differential equation model and its predictions becomes meaningless as there is no biological meaning to a negative number of animals z.

where c_1, c_2, ..., c_n are arbitrary constants. To rewrite this solution for the original coordinates x use

$$x = PX = \begin{bmatrix} p_1 & p_2 & \cdots & p_n \end{bmatrix} \begin{bmatrix} c_1 e^{\lambda_1 t} \\ c_2 e^{\lambda_2 t} \\ \vdots \\ c_n e^{\lambda_n t} \end{bmatrix} = c_1 p_1 e^{\lambda_1 t} + c_2 p_2 e^{\lambda_2 t} + \cdots + c_n p_n e^{\lambda_n t}$$

to derive the solution (7.5).

Second, being able to use the constants $c = (c_1, c_2, \ldots, c_n)$ to match every given initial condition shows that formula (7.5) is a general solution. Suppose the value of $x(0)$ is given. Recalling $e^0 = 1$, formula (7.5) evaluated at $t = 0$ requires

$$\begin{aligned} x(0) &= c_1 p_1 e^{\lambda_1 0} + c_2 p_2 e^{\lambda_2 0} + \cdots + c_n p_n e^{\lambda_n 0} \\ &= c_1 p_1 + c_2 p_2 + \cdots + c_n p_n = \begin{bmatrix} p_1 & p_2 & \cdots & p_n \end{bmatrix} c = Pc. \end{aligned}$$

Since matrix P is invertible, choose constants $c = P^{-1}x(0)$ for any given $x(0)$. □

Activity 7.3.19 Recall that the differential equations $dy/dt = y - 4z$ and $dz/dt = -y + z$ have a general solution $y(t) = 2c_1 e^{3t} + 2c_2 e^{-t}$ and $z(t) = -c_1 e^{3t} + c_2 e^{-t}$. What are the values of these constants given that $y(0) = 2$ and $z(0) = 3$?

(a) $c_1 = -2, c_2 = 0$ (b) $c_1 = c_2 = 1$ (c) $c_1 = -1, c_2 = 2$ (d) $c_1 = 0, c_2 = -1$

Example 7.3.20 Use matrix analysis to find (by hand) a general solution to the system of differential equations $\frac{du}{dt} = -2u + 2v$, $\frac{dv}{dt} = u - 2v + w$, and $\frac{dw}{dt} = 2v - 2w$.

Solution: Let vector $u = (u, v, w)$, and then form the differential equations into the matrix-vector system

$$\frac{du}{dt} = \begin{bmatrix} \frac{du}{dt} \\ \frac{dv}{dt} \\ \frac{dw}{dt} \end{bmatrix} = \begin{bmatrix} -2u + 2v \\ u - 2v + w \\ 2v - 2w \end{bmatrix} = \begin{bmatrix} -2 & 2 & 0 \\ 1 & -2 & 1 \\ 0 & 2 & -2 \end{bmatrix} \begin{bmatrix} u \\ v \\ w \end{bmatrix} = \underbrace{\begin{bmatrix} -2 & 2 & 0 \\ 1 & -2 & 1 \\ 0 & 2 & -2 \end{bmatrix}}_{A} u.$$

To use Theorem 7.3.18 we need eigenvalues and eigenvectors of the matrix A. Here the characteristic polynomial of A is, using (6.1),

$$\begin{aligned} \det(A - \lambda I) &= \det \begin{bmatrix} -2 - \lambda & 2 & 0 \\ 1 & -2 - \lambda & 1 \\ 0 & 2 & -2 - \lambda \end{bmatrix} \\ &= -(2 + \lambda)^3 + 0 + 0 - 0 + 2(2 + \lambda) + 2(2 + \lambda) \end{aligned}$$

$$= (2+\lambda)\left[-(2+\lambda)^2 + 4\right]$$
$$= (2+\lambda)\left[-\lambda^2 - 4\lambda\right] = -\lambda(\lambda+2)(\lambda+4).$$

This determinant is only zero for eigenvalues $\lambda = 0, -2, -4$.

- For eigenvalue $\lambda = 0$, corresponding eigenvectors p satisfy

$$(A - 0I)p = \begin{bmatrix} -2 & 2 & 0 \\ 1 & -2 & 1 \\ 0 & 2 & -2 \end{bmatrix} p = 0.$$

The last row of this equation requires $p_3 = p_2$, and the first row requires $p_1 = p_2$. Hence all solutions may be written as $p = (p_2, p_2, p_2)$. Choose any nonzero one, say $p = (1, 1, 1)$.

- For eigenvalue $\lambda = -2$, corresponding eigenvectors p satisfy

$$(A + 2I)p = \begin{bmatrix} 0 & 2 & 0 \\ 1 & 0 & 1 \\ 0 & 2 & 0 \end{bmatrix} p = 0.$$

The first and last rows of this equation require $p_2 = 0$, and the second row requires $p_3 = -p_1$. Hence all solutions may be written as $p = (p_1, 0, -p_1)$. Choose any nonzero one, say $p = (1, 0, -1)$.

- For eigenvalue $\lambda = -4$, corresponding eigenvectors p satisfy

$$(A + 4I)p = \begin{bmatrix} 2 & 2 & 0 \\ 1 & 2 & 1 \\ 0 & 2 & 2 \end{bmatrix} p = 0.$$

The last row of this equation requires $p_3 = -p_2$, and the first row requires $p_1 = -p_2$. Hence all solutions may be written as $p = (-p_2, p_2, -p_2)$. Choose any nonzero one, say $p = (-1, 1, -1)$.

With these three distinct eigenvalues, corresponding eigenvectors are linearly independent, and so Theorem 7.3.18 gives a general solution of the differential equations as

$$\begin{bmatrix} u \\ v \\ w \end{bmatrix} = c_1 \begin{bmatrix} 1 \\ 1 \\ 1 \end{bmatrix} e^{0t} + c_2 \begin{bmatrix} 1 \\ 0 \\ -1 \end{bmatrix} e^{-2t} + c_3 \begin{bmatrix} -1 \\ 1 \\ -1 \end{bmatrix} e^{-4t}.$$

That is, $u(t) = c_1 + c_2 e^{-2t} - c_3 e^{-4t}$, $v(t) = c_1 + c_3 e^{-4t}$, and $w(t) = c_1 - c_2 e^{-2t} - c_3 e^{-4t}$ for every values of the constants c_1, c_2, and c_3. □

Example 7.3.21 Use the general solution derived in Example 7.3.20 to predict the solution of the differential equations $\frac{du}{dt} = -2u + 2v$, $\frac{dv}{dt} = u - 2v + w$, and $\frac{dw}{dt} = 2v - 2w$ given the initial conditions that $u(0) = v(0) = 0$ and $w(0) = 4$.

Solution: Evaluating the general solution

$$
\begin{bmatrix} u \\ v \\ w \end{bmatrix} = c_1 \begin{bmatrix} 1 \\ 1 \\ 1 \end{bmatrix} + c_2 \begin{bmatrix} 1 \\ 0 \\ -1 \end{bmatrix} e^{-2t} + c_3 \begin{bmatrix} -1 \\ 1 \\ -1 \end{bmatrix} e^{-4t}
$$

at time $t = 0$ gives, using the initial conditions and $e^0 = 1$,

$$
\begin{bmatrix} 0 \\ 0 \\ 4 \end{bmatrix} = c_1 \begin{bmatrix} 1 \\ 1 \\ 1 \end{bmatrix} + c_2 \begin{bmatrix} 1 \\ 0 \\ -1 \end{bmatrix} + c_3 \begin{bmatrix} -1 \\ 1 \\ -1 \end{bmatrix} = \begin{bmatrix} c_1 + c_2 - c_3 \\ c_1 + c_3 \\ c_1 - c_2 - c_3 \end{bmatrix}.
$$

Solving by hand, the second row requires $c_3 = -c_1$, so the first row then requires $c_1 + c_2 + c_1 = 0$, that is, $c_2 = -2c_1$. Putting both of these into the third row requires $c_1 + 2c_1 + c_1 = 4$, that is, $c_1 = 1$. Then $c_2 = -2$ and $c_3 = -1$. Consequently, as drawn to the right, the particular solution is

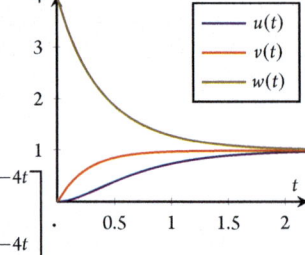

$$
\begin{bmatrix} u \\ v \\ w \end{bmatrix} = \begin{bmatrix} 1 \\ 1 \\ 1 \end{bmatrix} - 2 \begin{bmatrix} 1 \\ 0 \\ -1 \end{bmatrix} e^{-2t} - \begin{bmatrix} -1 \\ 1 \\ -1 \end{bmatrix} e^{-4t} = \begin{bmatrix} 1 - 2e^{-2t} + e^{-4t} \\ 1 - e^{-4t} \\ 1 + 2e^{-2t} + e^{-4t} \end{bmatrix}.
$$

$\square$

Example 7.3.22 Use MATLAB/Octave to find a general solution to the system of differential equations

$$
\begin{aligned}
dx_1/dt &= -\tfrac{1}{2}x_1 - \tfrac{1}{2}x_2 + x_3 + 2x_4, \\
dx_2/dt &= -\tfrac{1}{2}x_1 - \tfrac{1}{2}x_2 + 2x_3 + x_4, \\
dx_3/dt &= x_1 + 2x_2 - \tfrac{1}{2}x_3 - \tfrac{1}{2}x_4, \\
dx_4/dt &= 2x_1 + x_2 - \tfrac{1}{2}x_3 - \tfrac{1}{2}x_4.
\end{aligned}
$$

What is the particular solution that satisfies the initial conditions $x_1(0) = -5$, $x_2(0) = -1$ and $x_3(0) = x_4(0) = 0$? Record your commands and give reasons.

Solution: Write the system in matrix-vector form $\frac{dx}{dt} = Ax$ for vector

$$
x = \begin{bmatrix} x_1 \\ x_2 \\ x_3 \\ x_4 \end{bmatrix} \quad \text{and matrix } A = \begin{bmatrix} -\tfrac{1}{2} & -\tfrac{1}{2} & 1 & 2 \\ -\tfrac{1}{2} & -\tfrac{1}{2} & 2 & 1 \\ 1 & 2 & -\tfrac{1}{2} & -\tfrac{1}{2} \\ 2 & 1 & -\tfrac{1}{2} & -\tfrac{1}{2} \end{bmatrix}.
$$

Enter the matrix into MATLAB/Octave and then find its eigenvalues and eigenvectors as follows

```
A= [-1/2 -1/2 1 2
 -1/2 -1/2 2 1
 1 2 -1/2 -1/2
 2 1 -1/2 -1/2]
[V,D] =eig (A)
```

MATLAB/Octave tells us the eigenvectors and eigenvalues:

```
V =
   -0.5000    0.5000   -0.5000   -0.5000
   -0.5000   -0.5000    0.5000   -0.5000
    0.5000    0.5000    0.5000   -0.5000
    0.5000   -0.5000   -0.5000   -0.5000
D =
   -4.0000         0         0         0
         0   -1.0000         0         0
         0         0    1.0000         0
         0         0         0    2.0000
```

Then Theorem 7.3.18 gives that a general solution of the differential equations is

$$
x = c_1 \begin{bmatrix} -\frac{1}{2} \\ -\frac{1}{2} \\ \frac{1}{2} \\ \frac{1}{2} \end{bmatrix} e^{-4t} + c_2 \begin{bmatrix} \frac{1}{2} \\ -\frac{1}{2} \\ \frac{1}{2} \\ -\frac{1}{2} \end{bmatrix} e^{-t} + c_3 \begin{bmatrix} -\frac{1}{2} \\ \frac{1}{2} \\ \frac{1}{2} \\ \frac{1}{2} \end{bmatrix} e^{t} + c_4 \begin{bmatrix} -\frac{1}{2} \\ -\frac{1}{2} \\ -\frac{1}{2} \\ -\frac{1}{2} \end{bmatrix} e^{2t}.
$$

Given the specified initial conditions at $t = 0$, when all the above exponentials reduce to $e^0 = 1$, we just need to find the linear combination of the eigenvectors that equals the initial vector $x(0) = (-5, -1, 0, 0)$; that is, solve $Vc = x(0)$. In MATLAB/Octave, after checking that $\text{rcond}(V)$ is the good value 0.25, compute $c=V\backslash[-5;-1;0;0]$ to find the vector of coefficients is $c = (3, -2, 2, 3)$. Hence the particular solution is

$$
x = \begin{bmatrix} -\frac{3}{2} \\ -\frac{3}{2} \\ \frac{3}{2} \\ \frac{3}{2} \end{bmatrix} e^{-4t} + \begin{bmatrix} -1 \\ 1 \\ -1 \\ 1 \end{bmatrix} e^{-t} + \begin{bmatrix} -1 \\ 1 \\ 1 \\ 1 \end{bmatrix} e^{t} + \begin{bmatrix} -\frac{3}{2} \\ -\frac{3}{2} \\ -\frac{3}{2} \\ -\frac{3}{2} \end{bmatrix} e^{2t}.
$$

□

Example 7.3.23 Use matrix analysis to find (by hand) a general solution to the system of differential equations $\frac{dy}{dt} = z$ and $\frac{dz}{dt} = -4y$.

Solution: Let vector $y = (y, z)$, and then form the differential equations into the matrix-vector system

$$\frac{d\mathbf{y}}{dt} = \begin{bmatrix} \frac{dy}{dt} \\ \frac{dz}{dt} \end{bmatrix} = \begin{bmatrix} z \\ -4y \end{bmatrix} = \begin{bmatrix} 0 & 1 \\ -4 & 0 \end{bmatrix} \begin{bmatrix} y \\ z \end{bmatrix} = \underbrace{\begin{bmatrix} 0 & 1 \\ -4 & 0 \end{bmatrix}}_{A} \mathbf{y}.$$

To use Theorem 7.3.18 we need eigenvalues and eigenvectors of the matrix A. Here the characteristic polynomial of A is

$$\det(A - \lambda I) = \det \begin{bmatrix} -\lambda & 1 \\ -4 & -\lambda \end{bmatrix} = \lambda^2 + 4.$$

This determinant is only zero for $\lambda^2 = -4$, that is, $\lambda = \pm 2i$ (where $i = \sqrt{-1}$)—a pair of complex conjugate eigenvalues.

- For eigenvalue $\lambda = +2i$ the corresponding eigenvectors $\mathbf{p}$ satisfy

$$(A - \lambda I)\mathbf{p} = \begin{bmatrix} -2i & 1 \\ -4 & -2i \end{bmatrix} \mathbf{p} = \mathbf{0}.$$

The second row of this matrix is $-2i$ times the first row so we just need to satisfy the first row equation $\begin{bmatrix} -2i & 1 \end{bmatrix} \mathbf{p} = 0$. This equation is $-2ip_1 + p_2 = 0$, that is, $p_2 = 2ip_1$. Hence all eigenvectors are of the form $\mathbf{p} = (1, 2i)p_1$. Choose any one, say $\mathbf{p} = (1, 2i)$.

- Similarly, for eigenvalue $\lambda = -2i$ the corresponding eigenvectors $\mathbf{p}$ satisfy

$$(A - \lambda I)\mathbf{p} = \begin{bmatrix} 2i & 1 \\ -4 & 2i \end{bmatrix} \mathbf{p} = \mathbf{0}.$$

The second row of this matrix is $2i$ times the first row so we just need to satisfy the first row equation $\begin{bmatrix} 2i & 1 \end{bmatrix} \mathbf{p} = 0$. This equation is $2ip_1 + p_2 = 0$, that is, $p_2 = -2ip_1$. Hence all eigenvectors are of the form $\mathbf{p} = (1, -2i)p_1$. Choose any one, say $\mathbf{p} = (1, -2i)$.

With these two distinct eigenvalues, corresponding eigenvectors are linearly independent, and so Theorem 7.3.18 gives a general solution of the differential equations as

$$\begin{bmatrix} y \\ z \end{bmatrix} = c_1 \begin{bmatrix} 1 \\ 2i \end{bmatrix} e^{i2t} + c_2 \begin{bmatrix} 1 \\ -2i \end{bmatrix} e^{-i2t}.$$

That is, $y(t) = c_1 e^{i2t} + c_2 e^{-i2t}$ and $z(t) = 2ic_1 e^{i2t} - 2ic_2 e^{-i2t}$ for all constants c_1 and c_2. These formulas answer the exercise. The next examples show that because of the complex exponentials, this solution describes oscillations in time t. ☐

Example 7.3.24 Further consider Example 7.3.23. Suppose we additionally know that $y(0) = 3$ and $z(0) = 0$. Find the particular solution that satisfies these two initial conditions.

Solution: Use the derived general solution that $y(t) = c_1 e^{i2t} + c_2 e^{-i2t}$ and $z(t) = 2ic_1 e^{i2t} - 2ic_2 e^{-i2t}$. We find the constants c_1 and c_2 so that these satisfy the conditions $y(0) = 3$ and $z(0) = 0$. Substitute $t = 0$ into the general solution to require, using $e^0 = 1$,

$$y(0) = 3 = c_1 e^{i2 \cdot 0} + c_2 e^{-i2 \cdot 0} = c_1 + c_2,$$
$$z(0) = 0 = 2i c_1 e^{i2 \cdot 0} - 2i c_2 e^{-i2 \cdot 0} = 2i c_1 - 2i c_2,$$

The second of these equations requires $2i c_1 = 2i c_2$, that is, $c_1 = c_2$. The first, $c_1 + c_2 = 3$, then requires that $2c_1 = 3$, that is, $c_1 = 3/2$ and so $c_2 = 3/2$. Hence the particular solution is

$$y(t) = \frac{3}{2} e^{i2t} + \frac{3}{2} e^{-i2t} \quad \text{and} \quad z(t) = 3i e^{i2t} - 3i e^{-i2t}.$$

But recall Euler's formula that $e^{i\theta} = \cos\theta + i\sin\theta$ for every θ. Invoking Euler's formula the above particular solution simplifies:

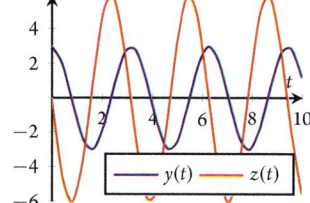

$$
\begin{aligned}
y(t) &= \frac{3}{2}[\cos 2t + i\sin 2t] + \frac{3}{2}[\cos(-2t) + i\sin(-2t)] \\
&= \frac{3}{2}\cos 2t + \frac{3}{2} i\sin 2t + \frac{3}{2}\cos 2t - \frac{3}{2} i\sin 2t \\
&= 3\cos 2t, \\
z(t) &= 3i[\cos 2t + i\sin 2t] - 3i[\cos(-2t) + i\sin(-2t)] \\
&= 3i\cos 2t - 3\sin 2t - 3i\cos 2t - 3\sin 2t \\
&= -6\sin 2t.
\end{aligned}
$$

Because $y(t)$ and $z(t)$ are just trigonometric functions of t, they oscillate in time t, as illustrated above-right. □

Example 7.3.25 In a real application the complex numbers of the general solution to Example 7.3.23 are usually inconvenient. Instead, we typically express the solution solely in terms of real quantities as just done in the previous Example 7.3.24. Use Euler's formula, that $e^{i\theta} = \cos\theta + i\sin\theta$ for every θ, to rewrite the general solution of Example 7.3.23 in terms of real functions.

Solution: Here use Euler's formula in the expression for

$$
\begin{aligned}
y(t) &= c_1 e^{i2t} + c_2 e^{-i2t} \\
&= c_1[\cos 2t + i\sin 2t] + c_2[\cos(-2t) + i\sin(-2t)] \\
&= c_1 \cos 2t + i c_1 \sin 2t + c_2 \cos 2t - i c_2 \sin 2t \\
&= (c_1 + c_2)\cos 2t + (i c_1 - i c_2)\sin 2t \\
&= C_1 \cos 2t + C_2 \sin 2t
\end{aligned}
$$

for constants $C_1 = c_1 + c_2$ and $C_2 = i(c_1 - c_2)$. Let's view the arbitrariness in c_1 and c_2 as having been 'transferred' to C_1 and C_2, then $y(t) = C_1 \cos 2t + C_2 \sin 2t$ is a general solution to the differential equations, and is expressed purely in real factors. In such a real form we explicitly see the oscillations in time t through the trigonometric functions $\cos 2t$ and $\sin 2t$.

The function $z(t) = 2ic_1e^{i2t} - 2ic_2e^{-i2t}$ has a corresponding form found by replacing c_1 and c_2 in terms of the same C_1 and C_2. From above, the constants $C_1 = c_1 + c_2$ and $C_2 = i(c_1 - c_2)$. Adding the first to $\pm i$ times the second determines $C_1 - iC_2 = 2c_1$ and $C_1 + iC_2 = 2c_2$, so that $c_1 = (C_1 - iC_2)/2$ and $c_2 = (C_1 + iC_2)/2$. Then the expression for

$$
\begin{aligned}
z(t) &= 2ic_1e^{i2t} - 2ic_2e^{-i2t} \\
&= 2i\frac{C_1 - iC_2}{2}[\cos 2t + i\sin 2t] - 2i\frac{C_1 + iC_2}{2}[\cos 2t - i\sin 2t] \\
&= (iC_1 + C_2)[\cos 2t + i\sin 2t] + (-iC_1 + C_2)[\cos 2t - i\sin 2t] \\
&= iC_1\cos 2t - C_1\sin 2t + C_2\cos 2t + iC_2\sin 2t \\
&\quad - iC_1\cos 2t - C_1\sin 2t + C_2\cos 2t - iC_2\sin 2t \\
&= -2C_1\sin 2t + 2C_2\cos 2t.
\end{aligned}
$$

That is, the corresponding general solution $z(t) = -2C_1\sin 2t + 2C_2\cos 2t$ is now also expressed in real factors. □

7.3.3 Exercises

Exercise 7.3.1 Which of the following matrices diagonalize the matrix $Z = \begin{bmatrix} 7 & 12 \\ -2 & -3 \end{bmatrix}$? Show your working.

(a) $P_a = \begin{bmatrix} -2 & 3 \\ 1 & -1 \end{bmatrix}$

(b) $P_b = \begin{bmatrix} 3 & -2 \\ -1 & 1 \end{bmatrix}$

(c) $P_c = \begin{bmatrix} 1 & -1 \\ -2 & 3 \end{bmatrix}$

(d) $P_d = \begin{bmatrix} 1 & 3 \\ 1 & 2 \end{bmatrix}$

(e) $P_e = \begin{bmatrix} 4 & 3 \\ -2 & -1 \end{bmatrix}$

(f) $P_f = \begin{bmatrix} -2 & 3 \\ 2 & -2 \end{bmatrix}$

(g) $P_g = \begin{bmatrix} -1 & 1 \\ 3 & -2 \end{bmatrix}$

(h) $P_h = \begin{bmatrix} 3 & 1 \\ 2 & 1 \end{bmatrix}$

Exercise 7.3.2 Redo Exercise 7.3.1 by finding which matrices $P_a, \ldots, P_h$ diagonalize each of the following matrices.

(a) $A = \begin{bmatrix} 5 & 12 \\ -2 & -5 \end{bmatrix}$

(b) $B = \begin{bmatrix} -3 & -12 \\ 2 & 7 \end{bmatrix}$

(c) $C = \begin{bmatrix} -1 & -1 \\ 6 & 4 \end{bmatrix}$

(d) $D = \begin{bmatrix} 4 & 6 \\ -1 & -1 \end{bmatrix}$

Exercise 7.3.3 Following Example 7.3.2(b), prove that for every scalar k the matrix $\begin{bmatrix} k & 1 \\ 0 & k \end{bmatrix}$ is not diagonalizable.

Exercise 7.3.4 In each of the following cases, you are given three linearly independent eigen-vectors and corresponding eigenvalues for some 3×3 matrix A. Write down three different matrices P that diagonalize the matrix A, and for each, write down the corresponding diagonal matrix $D = P^{-1}AP$.

(a) $\lambda_1 = -1, p_1 = (3,2,-1); \lambda_2 = 1, p_2 = (-4,-2,2); \lambda_3 = 3, p_3 = (-1,0,2).$

(b) $\lambda_1 = -1, p_1 = (2,1,2); \lambda_2 = -1, p_2 = (0,3,1); \lambda_3 = 2, p_3 = (4,-2,-2).$

(c) $\lambda_1 = 1, p_1 = (3,-7,2); \lambda_2 = -2, p_2 = (-4,-5,1); \lambda_3 = 4, p_3 = (1,2,3).$

Exercise 7.3.5 From the given information, are each of the matrices diagonalizable? Give reasons.

(a) The only eigenvalues of a 2×2 matrix are 2.2 and 0.1.

(b) The only eigenvalues of a 4×4 matrix are 2.2, 1.9, -1.8 and -1.

(c) The only eigenvalue of a 2×2 matrix is 0.7.

(d) The only eigenvalues of a 6×6 matrix are $-1.6, 0.3, 0.1$ and -2.3.

(e) The MATLAB/Octave function eig(A) returns the result

```
ans =
    3.0821 + 0.0000i
   -2.7996 + 0.0000i
   -0.7429 + 1.6123i
   -0.7429 - 1.6123i
```

(f) The MATLAB/Octave function eig(A) returns the result

```
ans =
   -1.0000
    1.0000
    2.0000
   -1.0000
```

Exercise 7.3.6 For each of the following 3×3 matrices, show with hand algebra that each matrix has one eigenvalue of multiplicity three, and then determine the dimension of the cor-responding eigenspace. Also compute the eigenvalue and eigenvectors with MATLAB/Octave: comment on any limitations in the computed 'eigenvalues' and 'eigenvectors'.

(a) $A = \begin{bmatrix} 2 & 0 & 0 \\ 0 & 2 & 0 \\ 0 & 0 & 2 \end{bmatrix}$

(b) $B = \begin{bmatrix} 2 & 1 & -2 \\ -8 & -4 & 8 \\ -2 & -1 & 2 \end{bmatrix}$

(c) $C = \begin{bmatrix} -2 & 1 & 0 \\ 0 & -2 & 0 \\ 0 & 1 & -2 \end{bmatrix}$

(d) $D = \begin{bmatrix} 7 & 1 & 22 \\ -6 & -1 & -20 \\ -1 & 0 & -3 \end{bmatrix}$

(e) $E = \begin{bmatrix} -3 & 2 & 7 \\ 2 & -3 & -9 \\ -1 & 1 & 3 \end{bmatrix}$

(f) $F = \begin{bmatrix} -1 & 0 & 0 \\ 0 & -1 & 0 \\ 0 & 0 & -1 \end{bmatrix}$

Exercise 7.3.7 For a given $n \times n$ square matrix A, suppose λ_1 is an eigenvalue with n corresponding linearly independent eigenvectors $p_1, p_2, \ldots, p_n$. Adapt parts of the proof of Theorem 7.3.14 to prove that the characteristic polynomial of matrix A is $\det(A - \lambda I) = (\lambda_1 - \lambda)^n$. Then deduce that λ_1 is the only eigenvalue of A, and is of multiplicity n.

Exercise 7.3.8 For each of the following matrices, use MATLAB/Octave to find the eigenvalues, their multiplicity, and the dimension of the eigenspaces. Give reasons (remember computational error).

(a) $A = \begin{bmatrix} -30.5 & 25.5 & 22 & -5 & -48.5 \\ -111 & 88 & 76 & -20 & -173 \\ 87.5 & -70.5 & -61 & 16 & 136.5 \\ 51 & -43 & -36 & 9 & 81 \\ -4.5 & 2.5 & 2 & -1 & -6.5 \end{bmatrix}$

(b) $B = \begin{bmatrix} -8 & 744 & -564 & 270 & 321 \\ 1 & -43 & 33 & -15 & -19 \\ 1 & 29 & -21 & 12 & 12 \\ 5 & -431 & 327 & -156 & -186 \\ -5 & 535 & -405 & 195 & 231 \end{bmatrix}$

(c) $C = \begin{bmatrix} 15.6 & 31.4 & -11.6 & -14.6 & -10.6 \\ -16.6 & -32.4 & 11.6 & 14.6 & 10.6 \\ 33.6 & 56.4 & -19.6 & -27.6 & -15.6 \\ 38.6 & 75.4 & -28.6 & -35.6 & -26.6 \\ -122 & -226 & 82 & 106 & 73 \end{bmatrix}$

(d) $D = \begin{bmatrix} -208 & 420 & -518 & 82 & 264 \\ 655 & -1336 & 1642 & -260 & -838 \\ 229 & -467 & 574 & -91 & -294 \\ -171 & 348 & -428 & 66 & 218 \\ -703 & 1431 & -1760 & 279 & 896 \end{bmatrix}$

Exercise 7.3.9 For each of the following systems of differential equations, find eigenvalues and eigenvectors to derive a general solution of the system of differential equations. Show your working.

(a) $\frac{dx}{dt} = x - 1.5y,\ \frac{dy}{dt} = 4x - 4y$

(b) $\frac{dx}{dt} = x,\ \frac{dy}{dt} = -12x + 5y$

(c) $\frac{dx}{dt} = 7x - 3y$

(d) $\frac{dx}{dt} = 0.2x + 1.2z,\ \frac{dy}{dt} = -x,$
$\frac{dz}{dt} = 1.8x + 0.8z$

(e) $\frac{du}{dt} = 4.5u + 7.5v + 7.5w,$
$\frac{dv}{dt} = 3u + 4v + 5w,$
$\frac{dw}{dt} = -7.5u - 11.5v - 12.5w$

(f) $\frac{dp}{dt} = -13p + 30q + 6r,$
$\frac{dq}{dt} = -32p + 69q + 14r,$
$\frac{dr}{dt} = 125p - 265q - 54r$

Exercise 7.3.10 In each of the following, a general solution to a differential equation is given. Find the particular solution that satisfies the specified initial conditions. Show your working.

(a) $(x,y) = c_1(0,1)e^{-t} + c_2(1,3)e^{2t}$ where $x(0) = 2$ and $y(0) = 1$

(b) $(x,y) = c_1(0,1)e^{-2t} + c_2(1,3)e^{t}$ where $x(0) = 0$ and $y(0) = 2$

(c) $(x,y,z) = c_1(0,0,1) + c_2(1,-1,1)e^{2t} + c_3(-2,3,1)e^{-2t}$ where $x(0) = 3$, $y(0) = -4$ and $z(0) = 0$

(d) $(x,y,z) = c_1(0,0,1)e^{-3t} + c_2(1,-1,1) + c_3(-2,3,1)e^{-t}$ where $x(0) = 3$, $y(0) = -4$ and $z(0) = 2$

Exercise 7.3.11 For each of the following systems of differential equations, use MAT-LAB/Octave to find eigenvalues and eigenvectors and hence derive a general solution of the system of differential equations. Record your working.

(a)
$$\frac{dx_1}{dt} = 15.8x_1 + 17.1x_2 - 119.7x_3 + 153.9x_4,$$
$$\frac{dx_2}{dt} = 1.4x_1 + 0.1x_2 + 12.9x_3 - 17.1x_4,$$
$$\frac{dx_3}{dt} = 6.2x_1 + 6.2x_2 - 43x_3 + 57.6x_4,$$
$$\frac{dx_4}{dt} = 3.4x_1 + 3.4x_2 - 25x_3 + 34x_4.$$

(b)
$$\frac{dx_1}{dt} = -12.2x_1 + 53.7x_2 + 50.1x_3 - 22.8x_4,$$
$$\frac{dx_2}{dt} = 0.6x_1 - 2.3x_2 - 2.7x_3 + 1.2x_4,$$
$$\frac{dx_3}{dt} = -20.4x_1 + 93.8x_2 + 90.2x_3 - 40.8x_4,$$
$$\frac{dx_4}{dt} = -38.2x_1 + 177.9x_2 + 170.7x_3 - 77.2x_4.$$

(c)
$$\frac{dx_1}{dt} = x_1 + 29.4x_2 - 3.2x_3 - 12.9x_4,$$
$$\frac{dx_2}{dt} = 1.4x_1 - 38.4x_2 + 5.6x_3 + 18.2x_4,$$
$$\frac{dx_3}{dt} = 2.3x_1 - 80.3x_2 + 12.3x_3 + 36.7x_4,$$
$$\frac{dx_4}{dt} = 2.4x_1 - 65x_2 + 9.2x_3 + 31.1x_4.$$

Exercise 7.3.12 In a few sentences, answer/discuss each of the following.

(a) What is the relation between diagonalization of a matrix and using non-standard basis vectors for a space?

(b) Why is diagonalization useful in population modelling? and systems of differential equations?

(c) Suppose you generate a matrix at random: why would you expect the matrix to be diagonalizable?

(d) How does it occur that the dimension of an eigenspace may be strictly less than the multiplicity of the corresponding eigenvalue?

(e) Why is the possibility of complex valued eigenvalues and eigenvectors important in an otherwise real problem?

7.4 Summary of general eigen-problems

- In the case of non-symmetric matrices, eigenvectors are usually not orthogonal, eigenvalues and eigenvectors are sometimes complex valued, and sometimes there are not as many eigenvectors as we expect.

- ★ The diagonal entries of a triangular matrix are the only eigenvalues of the matrix (Theorem 7.0.2). The corresponding eigenvectors of distinct eigenvalues are *generally* not orthogonal.

Find eigenvalues and eigenvectors of matrices

★ For every $n \times n$ square matrix A we call $\det(A - \lambda I)$ the **characteristic polynomial** of A (Theorem 7.1.1):
  - the characteristic polynomial of A is a polynomial of nth degree in λ;
  - there are at most n distinct eigenvalues of A.
• For every $n \times n$ matrix A (Theorem 7.1.4):
  - the product of the eigenvalues equals $\det A$ and equals the constant term in the characteristic polynomial;
  - the sum of the eigenvalues equals $(-1)^{n-1}$ times the coefficient of λ^{n-1} in the characteristic polynomial and equals the **trace** of the matrix, defined as the sum of the diagonal elements $a_{11} + a_{22} + \cdots + a_{nn}$.
• An eigenvalue λ_0 of a matrix A is said to have **multiplicity** m if the characteristic polynomial factorizes to $\det(A - \lambda I) = (\lambda - \lambda_0)^m g(\lambda)$ where $g(\lambda_0) \neq 0$ (Definition 7.1.7). Every eigenvalue of multiplicity $m \geq 2$ may also be called a **repeated eigenvalue**.
★ Procedure 7.1.11 finds by hand eigenvalues and eigenvectors of a (small) square matrix A:

  1. find all eigenvalues (possibly complex) by solving the **characteristic equation**, $\det(A - \lambda I) = 0$ —for an $n \times n$ matrix there are n eigenvalues when counted according to multiplicity and allowing complex eigenvalues;

  2. for each eigenvalue λ, solve the homogeneous linear equation $(A - \lambda I)x = 0$ to find the eigenspace $\mathbb{E}_\lambda$;

  3. write each eigenspace as the span of a few chosen eigenvectors.

In MATLAB/Octave, for a given square matrix A, execute [V,D]=eig(A), then the diagonal entries of D, diag(D), are the eigenvalues of A. Corresponding to the eigenvalue D(j,j) is an eigenvector $v_j = $ V(:,j), the jth column of V.
★ If a non-symmetric matrix or computation has an error ϵ, then expect a repeated eigenvalue of multiplicity m to appear as m eigenvalues all within about $\epsilon^{1/m}$ of each other. Thus when we find or compute m eigenvalues all within about $\epsilon^{1/m}$, then suspect them to actually be one eigenvalue of multiplicity m.
★★ In modelling populations, one often seeks the number of animals of various ages as a function of time. Define $y_j(t)$ to be the number of females in age category j at time t, and form into the vector $y(t) = (y_1, y_2, \ldots, y_n)$. Then encoding expected births, ageing, and deaths into mathematics leads to the matrix-vector population model that $y(t + 1) = Ay(t)$. This model empowers predictions.
★ Suppose the $n \times n$ square matrix A governs the dynamics of $y(t) \in \mathbb{R}^n$ according to $y(t + 1) = Ay(t)$ (Theorem 7.1.25).
  - Let $\lambda_1, \lambda_2, \ldots, \lambda_m$ be eigenvalues of A and $v_1, v_2, \ldots, v_m$ be corresponding eigenvectors, then a solution of $y(t + 1) = Ay(t)$ is the linear combination

$$y(t) = c_1 \lambda_1^t v_1 + c_2 \lambda_2^t v_2 + \cdots + c_m \lambda_m^t v_m$$

  for all constants $c_1, c_2, \ldots, c_m$.
  - Further, if the number of eigenvectors $m = n$ (the size of A), and the matrix of eigenvectors $P = \begin{bmatrix} v_1 & v_2 & \cdots & v_n \end{bmatrix}$ is invertible, then the general linear combination

is a **general solution**, in that unique constants c_1, c_2, ..., c_n may be found for every given initial value $y(0)$.

- In applications, to population models for example, and for both real and complex eigenvalues λ, the jth term in the solution $y(t) = c_1 \lambda_1^t v_1 + c_2 \lambda_2^t v_2 + \cdots + c_m \lambda_m^t v_m$ will, as time t increases,
  - grow to infinity if $|\lambda_j| > 1$,
  - decay to zero if $|\lambda_j| < 1$, and
  - remain the same magnitude if $|\lambda_j| = 1$.

- For every real $m \times n$ matrix A, the singular values of A are the non-negative eigenvalues of the $(m+n) \times (m+n)$ symmetric matrix $B = \begin{bmatrix} O_m & A \\ A^T & O_n \end{bmatrix}$ (Theorem 7.1.32). Writing an eigenvector $w \in \mathbb{R}^{m+n}$ of B as $w = (u, v)$ gives corresponding singular vectors of A, $u \in \mathbb{R}^m$ and $v \in \mathbb{R}^n$.

- After measuring musical notes, vibrations of complicated buildings, or bio-chemical reactions, we often need to fit exponential functions to the data.

- Procedure 7.1.42 fits the exponential interpolation $c_1 e^{r_1 t} + c_2 e^{r_2 t} + \cdots + c_n e^{r_n t}$ to the measured data $f_1, f_2, \ldots, f_{2n}$ at $2n$ equi-spaced times $t_1, t_2, \ldots, t_{2n}$ where time $t_j = jh$ for time-spacing h.

  1. From the $2n$ data points, form two $n \times n$ (symmetric) Hankel matrices

  $$A = \begin{bmatrix} f_2 & f_3 & \cdots & f_{n+1} \\ f_3 & f_4 & \cdots & f_{n+2} \\ \vdots & \vdots & & \vdots \\ f_{n+1} & f_{n+2} & \cdots & f_{2n} \end{bmatrix}, \qquad B = \begin{bmatrix} f_1 & f_2 & \cdots & f_n \\ f_2 & f_3 & \cdots & f_{n+1} \\ \vdots & \vdots & & \vdots \\ f_n & f_{n+1} & \cdots & f_{2n-1} \end{bmatrix}.$$

  Use A=hankel(f(2:n+1),f(n+1:2*n)) and B=hankel(f(1:n), f(n:2*n-1)) in MATLAB/Octave.

  2. Find the eigenvalues of the so-called **generalized eigen-problem** $Av = \lambda Bv$:
     - by hand on small problems solve $\det(A - \lambda B) = 0$;
     - in MATLAB/Octave invoke lambda=eig(A,B), and then r=log(lambda)/h.
     This eigen-problem typically determines n multipliers λ_1, λ_2, ..., λ_n, and thence the n rates $r_k = (\log \lambda_k)/h$.

  3. Determine the corresponding n coefficients c_1, c_2, ..., c_n from any n point subset of the $2n$ data points. For example, the first n data points give the linear system

  $$\begin{bmatrix} 1 & 1 & \cdots & 1 \\ \lambda_1 & \lambda_2 & \cdots & \lambda_n \\ \lambda_1^2 & \lambda_2^2 & \cdots & \lambda_n^2 \\ \vdots & \vdots & & \vdots \\ \lambda_1^{n-1} & \lambda_2^{n-1} & \cdots & \lambda_n^{n-1} \end{bmatrix} \begin{bmatrix} c_1 \\ c_2 \\ \vdots \\ c_n \end{bmatrix} = \begin{bmatrix} f_1 \\ f_2 \\ f_3 \\ \vdots \\ f_n \end{bmatrix}$$

  Construct this matrix with U=(lambda.^(0:n-1)).' in MATLAB/Octave (see Table 5.1).

Linearly independent vectors may form a basis

* ★ A set of vectors $\{v_1 , v_2 , \ldots , v_k\}$ is **linearly dependent** if there are scalars $c_1 , c_2 , \ldots , c_k$, at least one of which is nonzero, such that $c_1 v_1 + c_2 v_2 + \cdots + c_k v_k = 0$ (Definition 7.2.4). A set of vectors that is not linearly dependent is called **linearly independent**.
* ★ Every orthonormal set of vectors is linearly independent (Theorem 7.2.8).
* A set of vectors $\{v_1 , v_2 , \ldots , v_m\}$ is linearly dependent if and only if at least one of the vectors can be expressed as a linear combination of the other vectors (Theorem 7.2.11). In particular, a set of two vectors $\{v_1, v_2\}$ is linearly dependent if and only if one of the vectors is a scalar multiple of the other.
* ★★ For every $n \times n$ matrix A, let $\lambda_1 , \lambda_2 , \ldots , \lambda_m$ be distinct eigenvalues of A with corresponding eigenvectors $v_1 , v_2 , \ldots , v_m$. Then the set $\{v_1 , v_2 , \ldots , v_m\}$ is linearly independent (Theorem 7.2.13).
* ★ Let $v_1 , v_2 , \ldots , v_m$ be vectors in $\mathbb{R}^n$, and let the $n \times m$ matrix $V = \begin{bmatrix} v_1 & v_2 & \cdots & v_m \end{bmatrix}$. Then the set $\{v_1 , v_2 , \ldots , v_m\}$ is linearly dependent if and only if the homogeneous system $Vc = 0$ has a nonzero solution c (Theorem 7.2.16).
* Every set of m vectors in $\mathbb{R}^n$ is linearly dependent when the number of vectors $m > n$ (Theorem 7.2.18).
* ★ A **basis** for a subspace $\mathbb{W}$ of $\mathbb{R}^n$ is a set of vectors such that the set both spans $\mathbb{W}$ and is linearly independent (Definition 7.2.20).
* Every basis for a given subspace has the same number of vectors (Theorem 7.2.23).
* For every subspace $\mathbb{W}$ of $\mathbb{R}^n$, the **dimension** of $\mathbb{W}$ is the number of vectors in any basis for $\mathbb{W}$ (Theorem 7.2.25).
* ★ Procedure 7.2.27 finds a basis for the subspace $\mathbb{A} = \mathrm{span}\{a_1 , a_2 , \ldots , a_n\}$ for every given set of n vectors in $\mathbb{R}^m$, $\{a_1 , a_2 , \ldots , a_n\}$.

  1. Form $m \times n$ matrix $A := \begin{bmatrix} a_1 & a_2 & \cdots & a_n \end{bmatrix}$.
  2. Factorize A into a SVD, $A = USV^\mathsf{T}$, and let $r = \mathrm{rank}\, A$ be the number of nonzero singular values (or effectively nonzero when the matrix has experimental errors).
  3. The first r columns of U form a basis, specifically an orthonormal basis, for the r-dimensional subspace $\mathbb{A}$.

  Alternatively, if the rank $r = n$, then the set $\{a_1 , a_2 , \ldots , a_n\}$ is linearly independent and span the subspace $\mathbb{A}$, and so is also a basis for the n-dimensional subspace $\mathbb{A}$.
* ★ Procedure 7.2.32 finds a basis for a subspace $\mathbb{W}$ specified as the solutions of a system of equations.

  1. Rewrite the system of equations as the homogeneous system $Ax = 0$ so that the subspace $\mathbb{W}$ is the nullspace of $m \times n$ matrix A.
  2. Find an SVD factorization $A = USV^\mathsf{T}$ and let $r = \mathrm{rank}\, A$ be the number of nonzero singular values (or effectively nonzero when the matrix has experimental errors).
  3. The last $n - r$ columns of V form an orthonormal basis for the subspace $\mathbb{W}$.
* For any subspace $\mathbb{W}$ of $\mathbb{R}^n$ let $\mathcal{B} = \{v_1 , v_2 , \ldots , v_k\}$ be a basis for $\mathbb{W}$. Then there is exactly one way to write each and every vector $w \in \mathbb{W}$ as a linear combination of the basis vectors: $w = c_1 v_1 + c_2 v_2 + \cdots + c_k v_k$. The coefficients $c_1 , c_2 , \ldots , c_k$ are called the **coordinates**

of w with respect to $\mathcal{B}$, and the column vector $[w]_{\mathcal{B}} = (c_1, c_2, \ldots, c_k)$ is called the **coordinate vector of w with respect to** $\mathcal{B}$.

★ For every $n \times n$ square matrix A, and extending Theorems 3.3.27 and 3.4.43, the following statements are equivalent (Theorem 7.2.41):

- A is invertible;
- $Ax = b$ has a unique solution for every $b \in \mathbb{R}^n$;
- $Ax = 0$ has only the zero solution;
- all n singular values of A are nonzero;
- the condition number of A is finite $(\texttt{rcond} > 0)$;
- $\operatorname{rank} A = n$;
- $\operatorname{nullity} A = 0$;
- the column vectors of A span $\mathbb{R}^n$;
- the row vectors of A span $\mathbb{R}^n$.
- $\det A \neq 0$;
- 0 is not an eigenvalue of A;
- the n column vectors of A are linearly independent;
- the n row vectors of A are linearly independent.

Diagonalization identifies the transformation

- An $n \times n$ square matrix A is **diagonalizable** if there exists a diagonal matrix D and an invertible matrix P such that $A = PDP^{-1}$, equivalently $AP = PD$ or $P^{-1}AP = D$ (Definition 7.3.1).
- ★★ For every $n \times n$ square matrix A, the matrix A is diagonalizable if and only if A has n linearly independent eigenvectors (Theorem 7.3.5). If A is diagonalizable, with diagonal matrix $D = P^{-1}AP$, then the diagonal entries of D are eigenvalues, and the columns of P are corresponding eigenvectors.
- ★ For every $n \times n$ square matrix A, if A has n distinct eigenvalues, then A is diagonalizable (Theorem 7.3.10). Consequently, and allowing complex eigenvalues, a real non-diagonalizable matrix must be non-symmetric and must have at least one repeated eigenvalue.
- ★ For every square matrix A, and for each eigenvalue λ_j of A, the corresponding eigenspace $\mathbb{E}_{\lambda_j}$ has dimension less than or equal to the multiplicity of λ_j; that is, $1 \leq \dim \mathbb{E}_{\lambda_j} \leq$ multiplicity of λ_j (Theorem 7.3.14).
- Mathematical models of interaction populations of animals, plants, and diseases are often written as differential equations in continuous time. Letting $y(t)$ be the vector of numbers of each species at time t, the basic model is a linear system of differential equations $dy/dt = Ay$.

  Using Newton's second law, that mass $\times$ acceleration $=$ force, many mechanical systems may be modelled by differential equations also in the form of the linear system $dy/dt = Ay$.
- ★★ Let $n \times n$ square matrix A be diagonalizable by the invertible matrix $P = \begin{bmatrix} p_1 & p_2 & \cdots & p_n \end{bmatrix}$ whose columns are eigenvectors corresponding to eigenvalues $\lambda_1, \lambda_2, \ldots, \lambda_n$. Then a general solution $x(t)$ to the differential equation system $dx/dt = Ax$ is the linear combination

$$x(t) = c_1 p_1 e^{\lambda_1 t} + c_2 p_2 e^{\lambda_2 t} + \cdots + c_n p_n e^{\lambda_n t}$$

for arbitrary constants $c_1, c_2, \ldots, c_n$ (Theorem 7.3.18).

Answers to selected activities

7.0.4d, 7.1.2a, 7.1.5c, 7.1.8c, 7.1.12c, 7.1.18d, 7.1.20c, 7.1.24b, 7.1.26c, 7.1.28b, 7.1.34a, 7.1.39c, 7.2.2a, 7.2.6a, 7.2.10c, 7.2.14d, 7.2.17a, 7.2.22c, 7.2.26a, 7.2.31b, 7.2.37a, 7.2.39b, 7.3.4d, 7.3.12a, 7.3.17c, 7.3.19c,

Answers to selected exercises

7.0.1b complex

7.0.1d complex

7.0.2b $\mathbb{E}_2 = \text{span}\{(0 , 1)\}$.

7.0.2d $\mathbb{E}_{-2} = \text{span}\{(0 , 0 , 1)\}$,
   $\mathbb{E}_{-4} = \text{span}\{(0 , 2 , -5)\}$,
   $\mathbb{E}_0 = \text{span}\{(8 , -6 , -11)\}$.

7.0.2f $\mathbb{E}_0 = \text{span}\{(1 , 0 , 0 , 0)\}$,
   $\mathbb{E}_7 = \text{span}\{(-2 , 7 , 0 , 0)\}$,
   $\mathbb{E}_3 = \text{span}\{(-22 , 3 , 12 , 0)\}$.

7.1.1b yes

7.1.1d yes

7.1.2b 0

7.1.2d -9.2

7.1.3b $\lambda^2 - 5\lambda + 24$

7.1.3d $-\lambda^3 + 3\lambda^2 + \cdots - 36$

7.1.3f $\lambda^4 + 5\lambda^3 + \cdots - 16$

7.1.4b $2 \times 2, -2, -10$

7.1.4d $3 \times 3, 8, 0$

7.1.4f $4 \times 4, 0, -5$

7.1.5b -1, one; -2, two; 1, two; 4, three

7.1.6a 1 once, -3 once

7.1.6c 1/2 twice

7.1.6e 0 once, -2 once, 1 once

7.1.6g -1 once, 0 twice

7.1.7a -1 once, -2.3 twice

7.1.7c -0.9 once, -0.1 once, 0.1 twice

7.1.7e 1.8 twice, -2.3 once, -3.8 twice

7.1.8b $\mathbb{E}_2 = \text{span}\{(3 , 1)\}$

7.1.8d $\mathbb{E}_{-2} = \text{span}\{(-2 , 2 , 1)\}$

7.1.9a $\lambda = -3$, once, $(0 , -0.89 , -0.45)$;
   $\lambda = 2$, twice, $(0.27 , 0.53 , 0.80)$.

7.1.9c $\lambda = 5$, once,
   $(-0.82 , 0 , -0.41 , 0.41)$; $\lambda = 3$, once,
   $(0.55 , 0.28 , 0.55 , 0.55)$; $\lambda = -4$,
   twice, $(0.67 , 0 , 0.33 , -0.67)$.

7.1.10a sensitive

7.1.10c not sensitive

7.1.11b $\lambda = 3$ sensitive

7.1.11d neither sensitive, symmetric

7.1.11f -0.9 and -0.1 not sensitive, 0.1 sensitive

7.1.12b $(0 , -12)$, $(12 , -24)$ and $(24 , -96)$

7.1.12d $(-7 , 16 , -6)$, $(79 , -64 , 126)$ and $(-367 , 256 , -606)$

7.1.14b $\begin{bmatrix} \frac{5}{6} & 0 & \frac{1}{12} \\ \frac{1}{6} & \frac{11}{12} & 0 \\ 0 & \frac{1}{12} & \frac{101}{102} \end{bmatrix}$

7.1.19b $\lambda = 0, \pm 13$, $(0 , \frac{12}{13} , \frac{5}{13})$ and $(\pm 1 , -\frac{5}{13} , \frac{12}{13})$

7.1.19d $\lambda = \pm 1$, ± 4, $(1 , 0 , 0 , \pm 1)$ and $(0 , 1 , \mp 1 , 0)$

7.1.20b eigenvalue $\frac{1}{7}$; eigenvectors proportional to $(1 , -4)$

7.1.20d eigenvalues $1 \pm i/\sqrt{2}$, eigenvectors proportional to $(1 , \mp i/\sqrt{2})$, respectively

7.1.20f eigenvalues $0 , -2$; eigenvectors proportional to $(1 , \frac{1}{2} , 1)$, $(0 , -\frac{1}{2} , 1)$, respectively

7.1.20h eigenvalues $1 , \frac{12}{17} , -1$; eigenvectors proportional to $(1 , 0 , 0)$, $(4 , 1 , -2)$, $(1 , -7 , -15)$, respectively

7.1.21b eigenvalues -5.26, 1.53, -0.98 (2 d.p.)

7.1.21d eigenvalues 2.74, -2.20, $-0.50 \pm 0.62\,\mathrm{i}$ (2 d.p.)

7.1.28a $f = (3/4)^t + 2$

7.1.28c $f = -2(1/3)^t + (2/3)^t$

7.1.28e $f = -2(1/2)^t - \frac{1}{2}(3/4)^t + \frac{3}{2}$

7.1.28g $f = \frac{1}{2}(3/2)^t - (1/3)^t + \frac{2}{3}(3/4)^t$

7.1.29a frequencies 1.20 and 1.70 radians/sec (2 d.p.)

7.1.29c frequencies 1.94 and 1.46 radians/sec (2 d.p.)

7.1.30 $f =$ $1459e^{-0.36t} - 2044e^{-0.33t} + 585e^{-0.27t}$ where t is days post-2448366; but rcond is very poor!

7.2.1b lin. dep.

7.2.1d lin. indep.

7.2.1f lin. dep.

7.2.2a lin. indep.

7.2.2c lin. indep.

7.2.7a Two possibilities are $\{(-1, 1/5, 1/15)\}$ and $\{(15, -3, -1)\}$.

7.2.7c Two possibilities are $\{(1/2, 1, 1/3), (-1, 1/3, 1/2)\}$ and $\{(3, -8, -5), (12, 10, 1)\}$.

7.2.8a $\{(0.55, -0.64, 0.55)\}$ (2 d.p.)

7.2.8c $\{(0.76, -0.27, 0.58, -0.10), (-0.51, -0.78, 0.33, 0.15)\}$ (2 d.p.)

7.2.10a $(-2, -3), (1, -3), (-3, -1), (3, 2)$

7.2.10c $(2.5, 0.5), (-1.5, 0), (1, 0.5), (1, 1.5)$

7.2.11a $[p]_{\mathcal{E}} = (-2, 11, 9)$

7.2.11c $[r]_{\mathcal{E}} = (-4, -1, -11)$

7.2.11e $[t]_{\mathcal{E}} = (-1, 11/2, 9/2)$

7.2.13a $[p]_{\mathcal{B}} = (-1, 1)$

7.2.13c $[r]_{\mathcal{B}} = (-3, -1)$

7.2.13e not in $\mathcal{B}$

7.2.15a $[p]_{\mathcal{E}} = (-2, -14, 18, 2, -5)$

7.2.15c $[r]_{\mathcal{E}} = (-18, 12, -6, 3, -6)$

7.2.15e $[t]_{\mathcal{B}} = (3, 2, 6)$

7.3.1b yes

7.3.1d no

7.3.1f no

7.3.1h no

7.3.5b yes

7.3.5d unknown

7.3.5f unknown

7.3.6b $\lambda = 0$, two; errors 10^{-8} and two eigenvectors effectively the same

7.3.6d $\lambda = 1$, one; errors 10^{-5}, all three eigenvectors effectively the same

7.3.6f $\lambda = -1$, three; all good

7.3.8b $\lambda = 0$ twice, dim $\mathbb{E}_0 = 2$; $\lambda = 1$ thrice, dim $\mathbb{E}_1 = 2$

7.3.8d $\lambda = -1 \pm \mathrm{i}$ once each, dim $\mathbb{E}_{-1\pm\mathrm{i}} = 1$; $\lambda = -2$ thrice, dim $\mathbb{E}_{-2} = 2$

7.3.9b $(x, y) = c_1(0, 1)e^{5t} + c_2(1, 3)e^t$

7.3.9d $(x, y, z) = c_1(0, 1, 0) + c_2(1, 1, -1)e^{-t} + c_3(-2, 1, 3)e^{2t}$

7.3.9f $(p, q, r) = c_1(2, 2, -5)e^{2t} + c_2(3, 1, 2)e^t + c_3(0, -1, 5)e^{-t}$

7.3.10b $(x, y) = (0, 2)e^{-2t}$

7.3.10d $(x, y, z) = (0, 0, 2)e^{-3t} + (1, -1, 1) + (2, -3, -1)e^{-t}$

7.3.11b (2 d.p.) $x = c_1(0.63, 0.63, -0.42, 0.21)e^{0.5t} + c_2(-0.23, -0.23, -0.23, -0.92)e^{0.4t} + c_3(0.64, 0, 0.43, 0.64)e^{-1.6t} + c_4(-0.89, 0, 0, 0.45)e^{0.8t}$

References

Alpers, B., Demlova, M., Fant, C.-H., Gustafsson, T., Lawson, D., Mustoe, L., Olsson-Lehtonen, B., Robinson, C. and Velichova, D. (2013), A framework for mathematics curricula in engineering education, Technical report, European Society for Engineering Education (SEFI). http://sefi.htw-aalen.de/curriculum.htm

Anton, H. and Rorres, C. (1991), *Elementary linear algebra. Applications version*, 6th edn, Wiley.

Arnold, V. I. (2014), *Mathematical understanding of nature*, Amer. Math. Soc.

Arrow, K. J. (1950), 'A difficulty in the concept of social welfare', *Journal of Political Economy* **58**(4), 328–346.

Berry, M. W., Dumais, S. T. and O'Brien, G. W. (1995), 'Using linear algebra for intelligent information retrieval', *SIAM Review* **37**(4), 573–595. http://epubs.siam.org/doi/abs/10.1137/1037127

Bliss, K., Fowler, K., Galluzzo, B., Garfunkel, S., Giordano, F., Godbold, L., Gould, H., Levy, R., Libertini, J., Long, M., Malkevitch, J., Montgomery, M., Pollak, H., Teague, D., van der Kooij, H. and Zbiek, R. (2016), GAIMME—Guidelines for Assessment and Instruction in Mathematics Modeling Education, Technical report, SIAM and COMAP. http://www.siam.org/reports/gaimme.php?_ga=1

Bressoud, D. M., Friedlander, E. M. and Levermore, C. D. (2014), 'Meeting the challenges of improved post-secondary education in the mathematical sciences', *Notices of the AMS* **61**(5), 502–3.

Cawthon Lang, K. A. (2005), 'Primate factsheets: Orangutan (pongo) taxonomy, morphology, & ecology', http://pin.primate.wisc.edu/factsheets/entry/orangutan.

Chartier, T. (2015), *When life is linear: from computer graphics to bracketology*, Math Assoc Amer. http://www.maa.org/press/books/when-life-is-linear-from-computer-graphics-to-bracketology

Cowen, C. C. (1997), On the centrality of linear algebra in the curriculum, Technical report, Mathematical Association of America. http://www.maa.org/centrality-of-linear-algebra

Cuyt, A. (2015), Approximation theory, in N. J. Higham, M. R. Dennis, P. Glendinning, P. A. Martin, F. Santosa and J. Tanner, eds, 'Princeton Companion to Applied Mathematics', Princeton, chapter IV.9, pp. 248–262.

Davis, B. and Uhl, J. (1999), *Matrices, Geometry and Mathematica*, Wolfram Research.

Donoho, D. L. and Stodden, V. (2015), Reproducible research in the mathematical sciences, in N. J. Higham, M. R. Dennis, P. Glendinning, P. A. Martin, F. Santosa and J. Tanner, eds, 'Princeton Companion to Applied Mathematics', Princeton, chapter VIII.5, pp. 916–925.

Driscoll, T. A. and Maki, K. L. (2007), 'Searching for rare growth factors using multicanonical monte carlo methods', *SIAM Review* **49**(4), 673–692. http://link.aip.org/link/?SIR/49/673/1

Dua, D. and Graff, C. (2019), 'UCI machine learning repository'. http://archive.ics.uci.edu/ml

Fara, P. (2009), *Science: a four thousand year history*, OUP.

Ganesan, K., Zhai, C. and Han, J. (2010), A graph based approach to abstractive summarization of highly redundant opinions, in 'Proceedings of the 23rd International Conference on Computational Linguistics (COLING 2010)', Beijing, pp. 340–348.

Gorodetski, V. I., Popyack, L. J. and Samoilov, V. (2001), SVD-based approach to transparent embedding data into digital images, in V. I. Gorodetski, V. A. Skormin and L. J. Popyack, eds, 'Information assurance in computer netwroks: methods, models, and architectures for network security', Vol. 2052 of *Lecture Notes in Computer Science*, Springer, pp. 263–274.

Halpern, D. F. and Hakel, M. D. (2003), 'Applying the science of learning to the university and beyond: Teaching for long-term retention and transfer', *Change: The Magazine of Higher Learning* **35**(4), 36–41.

Hannah, J. (1996), 'A geometric approach to determinants', *The American Mathematical Monthly* **103**(5), 401–409. http://www.jstor.org/stable/2974931

Higham, N. J. (1996), *Accuracy and stability of numerical algorithms*, SIAM.

Higham, N. J. (2015), Numerical linear algebra and matrix analysis, *in* N. J. Higham, M. R. Dennis, P. Glendinning, P. A. Martin, F. Santosa and J. Tanner, eds, 'Princeton Companion to Applied Mathematics', Princeton, chapter IV.10, pp. 263–281.

Holt, J. (2013), *Linear algebra with applications*, Freeman.

Horton, N., Chance, B., Cohen, S., Grimshaw, S., Hardin, J., Hesterberg, T., Hoerl, R., Malone, C., Nichols, R. and Nolan, D. (2014), Curriculum guidelines for undergraduate programs in statistical science, Technical report, American Statistical Association. http://www.amstat.org/education/curriculumguidelines.cfm

Hughes-Hallett, D., Gleason, A. M. and McCallum, et al., W. G. (2013), *Calculus: single and multivariable*, 6th edn, Wiley.

Kappeler, P., Rasoloarison, R., Razafimanantsoa, L., Walter, L. and Roos, C. (2005), 'Morphology, behaviour and molecular evolution of giant mouse lemurs (mirza spp.) gray 1870, with description of a new species', *Primate Report* **71**, 3–26.

Kleiber, M. (1947), 'Body size and metabolic rate', *Physiological Reviews* **27**, 511–541.

Kress, R. (2015), Integral equations, *in* N. J. Higham, M. R. Dennis, P. Glendinning, P. A. Martin, F. Santosa and J. Tanner, eds, 'Princeton Companion to Applied Mathematics', Princeton, chapter IV.4, pp. 200–208.

Kutz, J. N., Brunton, S. L., Brunton, B. W. and Proctor, J. L. (2016), *Dynamic Mode Decomposition: Data-driven Modeling of Complex Systems*, number 149 *in* 'Other titles in applied mathematics', SIAM, Philadelphia.

Larson, R. (2013), *Elementary linear algebra*, 7th edn, Brooks/Cole Cengage Learning.

Lay, D. C. (2012), *Linear Algebra and its Applications*, 4th edn, Addison-Wesley.

Mandelbrot, B. B. (1982), *The fractal geometry of nature*, W. H. Freeman.

Mittermeier, R., Nash, S., Louis, E., Richardson, M., Schwitzer, C., Langrand, O., Rylands, A., Hawkim, F., Rajaobelina, S., Ratsimbazafy, J., Rasoloarison, R., Roos, C., Kapplelar, P. and Mackinnon, J. (2010), *Lemurs of Madagascar*, 3rd edition edn, Conservation International.

Moody, D. L. (2009), 'The "physics" of notations: Towards a scientific basis for constructing visual notations in software engineering', *IEEE Trans. Soft. Engrg.* **35**(5), 756–778.

Nakos, G. and Joyner, D. (1998), *Linear algebra with applications*, Brooks/Cole.

Owen, D. and Pemberton, D. (2005), *Tasmanian Devil: A unique and threatened animal*, Allen & Unwin; Crows Nest, New South Wales.

Pashler, H., Bain, P. M., Bottge, B. A., Graesser, A., Koedinger, K., McDaniel, M. and Metcalfe, J. (2007), Organizing instruction and study to improve student learning: IES practice guide, Technical report, National Center for Education Research, Institute of Education Sciences, U.S. Dept of Education. http://files.eric.ed.gov/fulltext/ED498555.pdf

Pereyra, V. and Scherer, G. (2010), *Exponential Data Fitting and its Applications*, Bentham Science. http://ebooks.benthamscience.com/book/9781608050482/

Poole, D. (2015), *Linear algebra: A modern introduction*, 4th edn, Cengage Learning.

Quarteroni, A. and Saleri, F. (2006), *Scientific Computing with MATLAB and Octave*, Vol. 2 of *Texts in Computational Science and Engineering*, 2nd edn, Springer.

Reeves, R., Stewart, B., Clapham, P. and Powell, J. (2002), *National Audubon Society Guide to Marine Mammals of the World*, Alfred A. Knopf, New York.

Roulstone, I. and Norbury, J. (2013), *Invisible in the storm: the role of mathematics in understanding weather*, Princeton.

Schonefeld, S. (1995), 'Eigenpictures: Picturing the eigenvector problem', *The College Mathematics Journal* **26**(4), 316–319. http://www.jstor.org/stable/2687037

Schumacher, C. S., Siegel, M. J. and Zorn, P. (2015), 2015 CUPM curriculum guide to majors in the mathematical sciences, Technical report, The Mathematical Association of America. http://www.maa.org/programs/faculty-and-departments/curriculum-department-guidelines-recommendations/cupm

Stewart, G. W. (1993), 'On the Early History of the Singular Value Decomposition', *SIAM Review* **35**(4), 551–566.

Sukumar, R. (2003), *The Living Elephants: Evolutionary Ecology, Behaviour, and Conservation*, Oxford University Press, USA.

Sunquist, M. and Sunquist, F. (2002), *Wild Cats of the World*, University of Chicago Press.

Trefethen, L. N. and Bau, III, D. (1997), *Numerical linear algebra*, SIAM.

Turner, P. R., Crowley, J. M., Humpherys, J., Levy, R., Socha, K. and Wasserstein, R. (2015), Modeling across the curriculum II. report on the second SIAM-NSF workshop, Alexandria, VA, Technical report, [http://www.siam.org/reports/ModelingAcross Curr_2014.pdf].

Uhlig, F. (2002), A new unified, balanced, and conceptual approach to teaching linear algebra, Technical report, Department of Mathematics, Auburn University, http://www.auburn.edu/ uhligfd/TLA/download/tlateach.pdf.

Westfall, R. S. (2012), 'Cardano, girolamo', The Galileo Project https://web.archive.org/web/20120728110911/ http://galileo.rice.edu/Catalog/NewFiles/cardano.html.

Will, T. (2004), Introduction to the singular value decomposition, Technical report, [http://www.uwlax.edu/faculty/will/svd].

Index

The manufacturer's authorised representative in the EU for product
safety is Oxford University Press España S.A. of El Parque Empresarial
San Fernando de Henares, Avenida de Castilla, 2 - 28830 Madrid
(www.oup.es/en or product.safety@oup.com). OUP España S.A. also acts
as importer into Spain of products made by the manufacturer.
Printed and bound by CPI Group (UK) Ltd, Croydon, CR0 4YY

22/04/2026

02094922-0002